The Ecology of Plant Litter Decomposition in Stream Ecosystems

Christopher M. Swan · Luz Boyero ·
Cristina Canhoto
Editors

The Ecology of Plant Litter Decomposition in Stream Ecosystems

Springer

Editors
Christopher M. Swan
Geography and Environmental Sciences
University of Maryland Baltimore County
Baltimore, MD, USA

Luz Boyero
Faculty of Science and Technology
University of the Basque Country
Leioa, Bizkaia, Spain

Cristina Canhoto
Department of Life Sciences
University of Coimbra
Centre for Functional Ecology
Coimbra, Portugal

ISBN 978-3-030-72853-3 ISBN 978-3-030-72854-0 (eBook)
https://doi.org/10.1007/978-3-030-72854-0

Cover image: Stream flowing through birch forest in Canencia, Guadarrama Mountains, Spain
Photo credit: Tomás Calle

This Springer imprint is published by the registered company Springer Nature Switzerland AG
The registered company address is: Gewerbestrasse 11, 6330 Cham, Switzerland

Foreword

To what extent is knowledge about the plant litter decomposition process in streams important to the scientific community? Whoever reads this book probably knows that primary production in streams flowing through forest areas is limited, as a consequence of the scarcity of light under the tree canopy and the low concentration of dissolved nutrients. In contrast, unidirectional flows promoted by gravity move great amounts of dead plant organic matter from the terrestrial environment to the stream (vertical and lateral inputs) and within the stream (downstream transport). This plant litter represents the energetic basis to foodwebs in many streams, which are widely recognised as heterotrophic systems. The retentive capacity of a stream together with a series of physical, chemical and biological factors, determine the efficiency with which plant litter is used by detritivores and decomposers, its consequences on biodiversity, productivity and nutrient dynamics, and its interaction with global change drivers such as pollution and climate change. In consequence, it is not surprising that the plant litter decomposition process has attracted the attention of stream ecologists for decades.

The study of plant litter decomposition in stream ecosystems greatly developed since the last quarter of the past century, after the publication of several conceptual works that highlighted the relevance of terrestrial organic matter to stream functioning. Among these, "The stream and its valley" (Hynes, 1975) and "The river continuum concept" (Vannote et al., 1980) are worth mentioning. Later on, the impressive growth of research teams and scientific publications on this topic culminated in the organization of the first specialised scientific meeting in 1997: the Litter Breakdown in Rivers and Streams meeting (later renamed as Plant Litter Processing in Freshwaters, PLPF), which was held in Bilbao (Spain). The first edition was followed by others in Lunz (Austria, 1999), Szentendre (Hungary, 2002), Toulouse (France, 2005), Coimbra (Portugal, 2008), Cracow (Poland, 2011), London (United Kingdom, 2014) and, again, Bilbao (2017), coinciding with the 20th anniversary of the first meeting. The 9th edition should have already been celebrated in Braga (Portugal), but the Covid-19 pandemic has temporarily prevented it. These events have regularly and succesully joined together scientists from all over the world, demonstrating that the study of plant litter decomposition in streams continues to attract great attention from researchers, and underscoring the crucial role that this process has for our planet's biogeochemistry.

The PLPF meetings have been fruitful not only in sharing scientific findings, but also in promoting collaboration among scientists from around all the world, all working on plant litter decomposition but from different perspectives and scales, from molecular to global. This book is proof of such outcome: while not being the proceedings of a PLPF meeting, its idea came to light in the last meeting held in Bilbao. As a result, a selected group of experts has produced this state-of-the-art compendium, which reviews the basic knowledge and the progress being made in the last few decades about the process of plant litter decomposition in stream ecosystems, in the face of a changing world. Now it's time for you to enjoy it, learn from it and put it into practice!

Bilbao, Spain Jesús Pozo

References

Hynes, H. B. N. (1975). The stream and its valley. *Verhandlungen des Internationalen Verein Limnologie, 19*, 1–15.

Vannote, R. L., Minshall, G. W., Cummins, K. W., Sedell, J. R., & Cushing, C. E. (1980). The river continuum concept. *Canadian Journal of Fisheries and Aquatic Sciences, 37*, 130–137.

Contents

Part I
General Overview on Plant Litter Decomposition in Streams

Chapter 1
The Ecology of Plant Litter Decomposition in Stream Ecosystems: An Overview

Christopher M. Swan, Luz Boyero, and Cristina Canhoto

Abstract The decomposition of plant litter of terrestrial origin is a key process for the functioning of many stream ecosystems with notable relevance for global biogeochemical cycles. The process has received much attention in the literature but, since the iconic paper of Kaushik and Hynes (1971), we lack a comprehensive review of its patterns of variation and drivers. This book provides an updated compendium of the ecology of plant litter decomposition in streams, through 22 chapters grouped in four sections that focus on (I) the analysis of the patterns and drivers of decomposition, the importance of (II) biodiversity and (III) multiple aspects of global change, and (IV) the methodological approaches used to study litter decomposition and its applications. By summarizing decades of fruitful research, we hope this will be a reference textbook for ecologists and students that will stimulate further research and promote collaboration among researchers in this field.

Ninety percent of the global terrestrial plant biomass production ends up entering the detrital pool, supporting detritus-based food webs in both terrestrial and aquatic ecosystems (Gessner et al., 2010). Understanding how this organic matter is processed within ecosystems is of utmost importance, not only because this process underlies fundamental ecosystem services (Mancinelli & Mulder, 2015), but also

C. M. Swan (✉)
Department of Geography and Environmental Systems,
University of Maryland, Baltimore County, Baltimore, MD 21250, USA
e-mail: chris.swan@umbc.edu

L. Boyero
Department of Plant Biology and Ecology, University of the Basque
Country (UPV/EHU), Leioa, Spain

Faculty of Science and Technology, University of the Basque Country, Leioa, Bizkaia, Spain
e-mail: luz.boyero@ehu.eus

C. Canhoto
Department of Life Sciences, Centre for Functional Ecology, University
of Coimbra, Calçada Martim de Freitas, 3000-456 Coimbra, Portugal
e-mail: ccanhoto@ci.uc.pt

C. M. Swan et al. (eds.), *The Ecology of Plant Litter Decomposition in Stream Ecosystems*, https://doi.org/10.1007/978-3-030-72854-0_1

because it determines the amount of carbon (C) that is mineralized or sequestered, and hence potential feedbacks with the global climate (Heimann & Reichstein, 2008).

Stream ecosystems are estimated to receive, process, and transport nearly 1.9 Pg of terrestrially-derived C per year globally (Cole et al., 2007). Many streams rely on this allochthonous source of C for the maintenance of their food webs and various functions, given that primary production is often severely limited, due to the combination of high riparian shading and low nutrient availability (Vannote et al., 1980). In consequence, 'the fate of dead leaves that fall into streams' stands out as a major topic in stream ecology, that served as title for the seminal contribution of Kaushik and Hynes (1971), and was recently revisited in Jane Marks' review paper (Marks, 2019).

This book intends to provide a comprehensive, contemporary compendium of the patterns (Part I), factors (Parts II and III) and approaches (Part IV) that govern the process of plant litter decomposition in streams, adding up to the several reviews available in the literature (e.g., Barlöcher & Sridhar, 2014; Boulton & Boon, 1991; Gessner et al., 1999; Graça et al., 2015; Marks, 2019; Royer & Minshall, 2003; Tank et al., 2010; Webster & Benfield, 1986), but going into much more detail, as allowed by a book format. The last few decades have seen an explosion of research addressing key questions about the ecological interactions at play in this process. A plethora of basic and applied ecological questions have been tested using plant litter decomposition as a study system, solely guided by a comprehensive book focused on litter decomposition methodological approaches, which has been recently re-edited (Bärlocher et al., 2020). The need for a complementary, more theoretical and updated approach was thought to be needed and is the rationale for this book.

The book integrates the many authors that shared and/or received the inputs of a common road of high-quality investigations presented and discussed in cozy, highly interactive and scientifically fruitful meetings—"Plant Litter Processing in Freshwaters" (PLPF). These triennially held meetings joined stream ecologists from all over the world and clearly contributed to stimulate stream ecological research, litter decomposition in particular, constituting grounds for the 22 chapters presented herein.

The expression "(plant/leaf) litter decomposition" is transversally used in this book by all authors. The majority of the book covers stream ecosystems, with a main focus on permanent headwater streams flowing through forested areas, because these are the streams that most rely on allochthonous plant litter as their main energy source (Vannote et al., 1980). However, some chapters deal with other types of streams (e.g., intermittent rivers and ephemeral streams), and some describe decomposition in terrestrial ecosystems for a comparative view.

This book is divided into four Parts that focus on different aspects of the plant litter decomposition process, namely its patterns and drivers (Part I), the role of biodiversity on litter decomposition outcomes (Part II) and on the consequences of the multiple facets and dimensions of global change on the process dynamics (Part III); the last Part considers the importance of methodological approaches and applications of this vital process, calling the attention to more mechanistic and future prospects of the presently used litter decomposition assessments and as a central tenet for human wellbeing (Part IV). We hope that this compendium of chapters, which overall review the knowledge on plant litter decomposition in streams that has resulted from decades of research, can serve as a reference textbook for ecologists and students. Furthermore, we hope that this book can stimulate further research, new lines of inquiry on this topic and promote collaboration among researchers in this field.

References

Bärlocher, F., Gessner, M. O., & Graça, M. A. S. (2020). *Methods to Study Litter Decomposition: A Practical Guide* (2nd ed.). Springer.

Barlöcher, F., & Sridhar, K. R. (2014). Association of animals and fungi in leaf decomposition. In E. B. G. Jones, K. D. Hyde, & K. L. Pang (Eds.), *Freshwater Fungi and Fungus-like Organisms.* (pp. 405–433). De Gruyter.

Boulton, A. J., & Boon, P. I. (1991). A review of methodology used to measure leaf litter decomposition in lotic environments: Time to turn over an old leaf? *Australian Journal of Marine and Freshwater Research, 42*, 1–43.

Cole, J. J., Prairie, Y. T., Caraco, N. F., McDowell, W. H., Tranvik, L. J., Striegl, R. G., Duarte, C. M., Kortelainen, P., Downing, J. A., Middelburg, J. J., & Melack, J. (2007). Plumbing the global carbon cycle: Integrating inland waters into the terrestrial carbon budget. *Ecosystems, 10*, 172–185.

Gessner, M. O., Chauvet, E., & Dobson, M. (1999). A perspective on leaf litter breakdown in streams. *Oikos, 85*, 377–384.

Gessner, M. O., Swan, C. M., Dang, C. K., McKie, B. G., Bardgett, R. D., Wall, D. H., & Hättenschwiler, S. (2010). Diversity meets decomposition. *Trends in Ecology and Evolution, 25*, 372–380.

Graça, M. A. S., Ferreira, V., Canhoto, C., Encalada, A. C., Gerrero-Bolaño, F., Wantzen, K. M., & Boyero, L. (2015). A conceptual model of litter breakdown in low order streams. *International Review of Hydrobiology, 100*, 1–12.

Heimann, M., & Reichstein, M. (2008). Terrestrial ecosystem carbon dynamics and climate feedbacks. *Nature, 451*, 289–292.

Kaushik, N. K., & Hynes, H. B. N. (1971). The fate of dead leaves that fall into streams. *Archiv für Hydrobiologie, 68*, 465–515.

Mancinelli, G., & Mulder, C. (2015). Detrital dynamics and cascading effects on supporting ecosystem services. *Advances in Ecological Research, 53*, 97–160.

Marks, J. C. (2019). Revisiting the fates of dead leaves that fall into streams. *Annual Review of Ecology, Evolution, and Systematics, 50*, 547–568.

Royer, T. V., & Minshall, G. W. (2003). Controls on leaf processing in streams from spatial-scaling and hierarchical perspectives. *Journal of the North American Benthological Society, 22*, 352–358.

Tank, J. L., Rosi-Marshall, E. J., Griffiths, N. A., Entrekin, S. A., & Stephen, M. L. (2010). A review of allochthonous organic matter dynamics and metabolism in streams. *Journal of the North American Benthological Society, 29*, 118–146.

Vannote, R. L., Minshall, G. W., Cummins, K. W., Sedell, J. R., & Cushing, C. E. (1980). The river continuum concept. *Canadian Journal of Fisheries and Aquatic Sciences, 37*, 130–137.

Webster, J. R., & Benfield, E. F. (1986). Vascular plant breakdown in freshwater ecosystems. *Annual Review of Ecology and Systematics, 17*, 567–594.

Chapter 2
Multi-Scale Biophysical Factors Driving Litter Dynamics in Streams

Alan M. Tonin, José F. Gonçalves Júnior, Richard G. Pearson, Manuel A. S. Graça, Javier Pérez, and Luz Boyero

Abstract Terrestrial litter that decomposes in streams is critical to carbon and nutrient fluxes and aquatic food web dynamics. Litter dynamics is influenced by biogeochemical, morphological, environmental and climatic factors, making it challenging to understand how these factors relate to each other and to litter decomposition across different spatial scales. Here, we present a hierarchical framework that accommodates the links among a wide variety of local and regional factors (e.g., litter quality, water chemistry, flow) in relation to climate, geology, biogeography and phylogeny. These factors ultimately influence the agents or processes (e.g., microbes,

A. M. Tonin (✉) · J. F. Gonçalves Júnior
Limnology-Aquaripária Lab, Department of Ecology, IB, University of Brasília (UnB), Brasília, Brazil

J. F. Gonçalves Júnior
e-mail: jfjunior@unb.br

R. G. Pearson
Centre for Tropical Water and Aquatic Ecosystem Research (TropWATER), James Cook University, Townsville, QLD, Australia

College of Marine and Environmental Sciences, James Cook University, Townsville, QLD, Australia

R. G. Pearson
e-mail: richard.pearson@jcu.edu.au

M. A. S. Graça
Department of Life Sciences and Marine and Environmental Sciences Centre (MARE), University of Coimbra, Coimbra, Portugal
e-mail: mgraca@ci.uc.pt

J. Pérez · L. Boyero
Department of Plant Biology and Ecology, University of the Basque Country (UPV/EHU), Leioa, Spain
e-mail: javier.perezv@ehu.es

L. Boyero
e-mail: luz.boyero@ehu.eus

L. Boyero
IKERBASQUE, Bilbao, Spain

C. M. Swan et al. (eds.), *The Ecology of Plant Litter Decomposition in Stream Ecosystems*, https://doi.org/10.1007/978-3-030-72854-0_2

detritivores, physical fragmentation, retention capacity) that govern litter inputs, storage and decomposition in streams. This framework highlights the dependence of litter dynamics on spatial scale and cautions against extrapolations across scales without quantifying the influence of biophysical variables on the different agents and processes. The framework can be used as a basis for experimental and observational studies of those interactions to develop broader mechanistic understanding of litter dynamics.

2.1 Streams as Hotspots of Organic Matter Processing

Fresh waters (i.e., wetlands, estuaries, lakes, rivers and streams) comprise only 0.01% of the Earth's water and approximately 0.8% of its surface area (Gleick, 1996). Yet this small fraction of global water supports a disproportionately high diversity of plants and animals (Dudgeon et al., 2006) and significantly contributes to the carbon cycle, at both regional and global scales (Cole et al., 2007; Hotchkiss et al., 2015; Raymond et al., 2013). Within fresh waters, 1st to 3rd order streams comprise over 75% of river network length (Raymond et al., 2013) and are closely linked to the terrestrial landscapes, so they are hotspots of organic matter processing because of the large amounts of terrestrial organic matter they receive, their typically low in-stream primary production, their high retentive capacity and their efficient decomposer communities (Battin et al., 2008).

Forest streams are typically net heterotrophic ecosystems, where overall stream respiration surpasses primary production (limited by riparian shading) and secondary production is fuelled by inputs of terrestrial organic carbon (Hall et al., 2000; Neres-Lima et al., 2017). Riparian vegetation, which supplies large amounts of plant litter streams and riparian soils (Tonin et al., 2017), also reduces light penetration and primary productivity in the stream, and may reduce water temperature extremes during the hottest and coldest periods of the year. The decomposition of this plant litter is the basis of key stream ecosystem processes, namely nutrient and carbon cycling and secondary production (Marks, 2019; Wardle et al., 2004). However, there are major gaps in our knowledge of organic matter dynamics in streams (e.g., timing and magnitude of inputs and biophysical factors acting on decomposition), especially in understudied areas of the world such as the tropics. For instance, little is known about the connection of litter fluxes with the otherwise well-known decomposition process, with only a few studies evaluating the inputs and outputs of litter (e.g., Fisher & Likens, 1972, 1973; Pozo et al., 1997; Webster & Meyer, 1997), despite their relevance for global carbon and nutrient cycling.

2.2 Dynamics of Litter Inputs and Storage in Streams

Terrestrial organic matter that enters streams is generally classified as coarse particulate organic matter, CPOM (>1 mm); fine particulate organic matter, FPOM

(0.45 μm–1 mm); and dissolved organic matter, DOM (<0.45 μm) (Webster et al., 1999). This material reaches streams via direct fall or lateral transport from the riparian zone. Most material is senescent, although large quantities of green leaves may reach streams in the tropics after storms (Wootton et al., 2019). In this chapter we focus on leaf litter because it is the dominant plant litter flux in streams (>60% of total fluxes of streams from different biomes, Abelho, 2001; Tonin et al., 2017), it is renewed annually, it has more rapid decomposition than woody material (Neres-Lima et al., 2017; Wallace et al., 1997) and it is the focus of most studies of decomposition in streams.

Litter falling from trees as a result of abscission or storm damage (hereafter 'litterfall') may reach the stream directly, but most litter falls to the ground, given its much larger area than that of streams. A proportion of this litter eventually is transported to streams by wind, water, gravity or animals, constituting lateral inputs to the stream. Although neglected in some litter studies, lateral inputs may represent a high proportion of total litter inputs to the stream (e.g., up to 55% in several tropical streams; Tonin et al., 2017), but not all streams (e.g., < 7%; Benson & Pearson, 1993). Lateral inputs and litterfall may differ in nutritional quality for stream consumers depending on the residence time on riparian soils, where litter undergoes physical and biological degradation (e.g., García-Palacios et al., 2016). Litter may also enter a stream reach by transport within the stream, here termed 'upstream input'.

After litter enters a stream, it can be retained by in-stream structures (e.g., rocks, roots, logs), increasing its residence time and enhancing decomposition by leaching and consumption by microbes and invertebrates. Litter may be redistributed or removed from streams by floods, although a considerable proportion may remain in pools, backwaters and debris dams. It may also be buried in sediments (e.g., in the hyporheic zone—the interface between surface stream and groundwater; Boulton et al., 1998). Here we refer to retained material as 'storage'. Litter accumulates in the streambed when inputs (i.e., litterfall, lateral inputs and upstream inputs) are higher than outputs (i.e., downstream transport and litter decomposition). The most complex of these litter fluxes is loss of litter mass by decomposition, resulting from chemical, physical and biological agents and their interactions.

2.3 Mechanisms of Litter Fluxes in Streams: Local and Regional Scales

Here, we present a framework describing the connections among litter inputs, storage and decomposition. We use a hierarchical spatial perspective to outline the links between biophysical factors that influence litter dynamics at different scales, like other frameworks of decomposition (Graça et al., 2015; Royer & Minshall, 2003), but differing in the inclusion of key related processes, components and the complexity of interactions across scales (Fig. 2.1). The framework is structured across three spatial

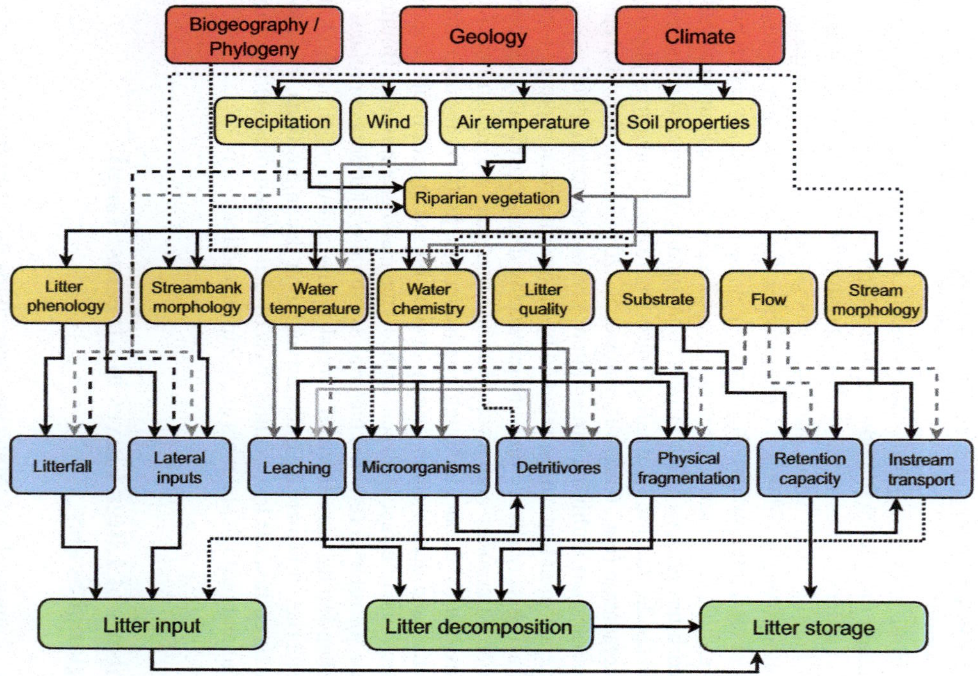

Fig. 2.1 Conceptual framework outlining the linkages of multiple biophysical factors driving litter dynamics (i.e., litter inputs, decomposition and storage; green boxes) at different spatial scales: regional (red boxes), local (orange boxes) and fine (blue boxes). Arrows of different shades or patterns (solid, dashed or dotted) are used to facilitate visual representation. This framework can apply to any chosen timeframe. Given the complexity of our conceptual framework, some linkages between factors may have not been depicted.

scales: regional, which accommodates variation in geology, climate and biogeography; local, including, for example, stream morphology, water quality and riparian vegetation identity; and fine, the scale at which many biophysical agents, such as litter fall, leaching, fragmentation consumption influence litter storage and decomposition. Most interactions within scales are not shown for simplicity (e.g., flow and stream morphology). The different components of the framework are described and expanded in the following sections. The framework is not explicitly linked to a particular timeframe, and could be applied at any time scale, from instantaneous to monthly, annual or multiannual.

2.3.1 *Litter Inputs*

Litter inputs are influenced by a variety of factors and their interactions, including litter phenology and production, stream and streambank morphology, precipitation, wind, retention capacity of streams and water flow. Firstly, litter production determines the amount of litterfall and depends on forest physiognomy, plant species diversity and vegetation composition and phenology, which are shaped by climatic factors, geomorphology, soils, and plant phylogeny and biogeography. Litter fall is caused by natural abscission (to conserve water or photosynthetic efficiency) and storms, which may release leaves, fruits, flowers and wood, especially in tropical regions (Benson & Pearson, 1993; Covich, 1988). Greater litterfall is expected in productive forests, which occur mostly on fertile soils and in warmer and wetter environments, such as rainforests (which also harbor higher plant diversity), than in less productive forests on infertile soils or in environments limited by water or temperature, such as cold dry forests, although the extended lifespan of leaves of evergreen tropical trees may reduce the inputs.

Secondly, streambank morphology (e.g., heterogeneity and slope) regulates lateral litter transport to the stream through its retention capacity, in relation to topography, hydrology (Leopold et al., 1992), and interactions with riparian vegetation. The heterogeneity of the streambank is determined by the presence of obstacles that hamper litter transport to the stream, such as depressions in the ground, living and dead trunks and buttresses, exposed roots, rocks and saplings and other small plants. Bank slope influences litter transport as steeper slopes facilitate litter movement by combinations of gravity, wind and enhanced surface runoff, especially during heavy rainfall. Thirdly, stream morphology and water flow affect litter inputs from upstream, through transport and retention mechanisms.

2.3.2 *Litter Storage*

The amount of litter stored on the streambed is regulated by interactions among three main factors: litter inputs, retention capacity of streams and litter decomposition.

Firstly, litter inputs would increase litter storage linearly if a stream retained all inputs. This might occur during periods of very low flow, but is otherwise unlikely because of high stream heterogeneity and the variable influence of flow and retention capacity.

Secondly, retention capacity determines litter storage as it reduces instream transport. The retention capacity of a stream is a function of stream morphology (e.g., width, depth, slope, sinuosity), water flow, substrate (including the type, size and quantity of retentive structures such as rocks and fallen tree trunks or branches, which may remain in the stream for years; Díez et al., 2000; Wallace et al., 1995), streambank structures (including roots and living trunks), and their interactions. Stream morphology is determined by geomorphological and hydrological processes, as well as riparian vegetation, which can provide flow resistance and increase bank strength by means of roots and woody debris (Hupp et al., 2016). Volume and velocity of water flow are influenced by precipitation and cross-sectional area and slope of the stream, which can determine whether retention structures are over-topped, and whether the stream has power to dislodge material from retention structures. In general, the retentive capacity of the streambed increases with substrate size—for example, boulders and cobbles are more efficient in retaining litter than gravel and sand (Jones, 1997). Large wood may greatly increase retention by redirecting flow and physically retaining litter directly or in pools.

Thirdly, litter decomposition decreases litter storage through the transformation of coarse litter into fine and dissolved material (Gessner et al., 1999), which is more easily transported by water flow, buried in the sediments (Webster et al., 1999) or incorporated into microbial and animal biomass. The components affecting litter decomposition are explored below. In summary, high retention is expected in small, shallow, sinuous and low-gradient streams; in low-flow conditions; and in streams with large substrates, high litter inputs and low decomposition rates. Retention is determined by the interactions among these variables.

2.3.3 Litter Decomposition

The processes that contribute to litter decomposition and affect its rate are leaching, physical fragmentation, microbial processing and detritivore consumption (Gessner et al., 1999; Marks, 2019).

2.3.3.1 Leaching

Leaching is the dissolution of water-soluble compounds from litter and may account for up to 40% of mass loss within a week of immersion, with the greatest loss typically occurring in the first 24–48 h (Gomes, 2015; Taylor & Bärlocher, 1996). However, greatest mass loss due to leaching may occur during the first 10 days of immersion in litter from some tropical plants (Gomes, 2015). Leaching is influenced by litter quality, and water chemistry, temperature, velocity and turbulence. Firstly, litter quality includes nutrient content (especially nitrogen and phosphorus), carbon recalcitrance (due to high content of lignin, cellulose and hemicellulose) and content of secondary metabolites (e.g., repellent or toxic substances such as phenols used in protection or competition by living plants). The degree of leaching may be determined by the concentration of soluble components, especially nutrients, low-mass carbon molecules and some secondary compounds, and their resistance to dissolution. For example, thick litter cuticles and superficial waxes may slow the dissolution of soluble compounds (Kuiters & Sarink, 1986; Schreeg et al., 2013). Litter quality is determined by the species composition of riparian vegetation and their physiological and morphological traits, which are influenced by climate, landscape, biogeography and phylogeny (Boyero et al., 2017; Siefert et al., 2015). Importantly, apparently similar plant communities may differ in litter quality as a result of soil characteristics (via different nutrient resorption efficiencies; Vergutz et al., 2012) or species interactions (e.g., competition for nutrients; Casper & Jackson, 1997).

Secondly, water chemistry affects leaching through variation in pH, hardness and mineral concentrations (Essington, 2005). Water hardness and mineral concentrations may affect different chemical compounds in different ways (e.g., polyphenols bind to hard-water minerals; Gebely, 2016). Water chemistry relates to lithology (i.e., characteristics of parent rock), soil properties and riparian vegetation (through regulation of dissolved inorganic and organic molecules). Thirdly, water temperature influences the solubility of molecules in the water (Chergui & Pattee, 1988). Water temperature is driven primarily by climate (through solar radiation), but riparian canopy density is also important in affecting stream shading. Finally, water flow enhances dissolution of water-soluble compounds. In summary, greater leaching is expected in litter with higher amounts of water-soluble compounds and lower protection from dissolution (e.g., absence of superficial waxes and low carbon recalcitrance), and in alkaline, warmer, faster and more turbulent waters. However, leaching is the least studied component of decomposition but compelling evidence of its importance is limited.

2.3.3.2 Microbial Decomposition

Fungi and bacteria are important contributors to decomposition and may have complementary roles (e.g., fungi facilitate the penetration of bacteria into leaf tissue). Fungi represent the largest proportion of microbial biomass (Findlay & Arsuffi, 1989; Hieber & Gessner, 2002). The contribution of microbes to decomposition is determined by biogeography and phylogeny, water temperature, water chemistry and litter quality. Firstly, biogeography and phylogeny may be responsible for community composition of fungi and bacteria (and, thus, different efficiencies in degrading leaf litter carbon), although there is controversy about the relative roles of the legacies of historical events (e.g., dispersal limitation and past environmental conditions) versus contemporary environmental conditions in shaping their communities (Martiny et al., 2006; O'Malley, 2007). Secondly, water temperature influences microorganisms by regulating the distribution of species (Dang et al., 2009), their metabolic and sporulation rates and their biomass (Ferreira & Chauvet, 2011). Thirdly, microbes increase their activity and biomass with increased availability of dissolved nutrients (through the maximization of carbon intake; Suberkropp & Chauvet, 1995), pH and alkalinity (by increasing the activity of different enzyme types associated with leaf softening and maceration; Chamier, 1987; Jenkins & Suberkropp, 1995). Microorganisms also perform better in softer litter, which is more susceptible to enzymatic degradation, and in litter that is rich in macro and micronutrients (such as calcium and magnesium) and not chemically defended (Schneider et al., 2012). Additionally, the selective feeding activity of detritivores on litter may affect microbial diversity and biomass by consuming some fungal species and rejecting others (e.g., Arsuffi & Suberkropp, 1989; Barlocher, 2005).

2.3.3.3 Fragmentation by Detritivorous Invertebrates

Detritivorous invertebrates are key organisms in litter decomposition, often responsible for a large proportion of total litter decomposition (e.g., 51–64% of litter mass loss; Hieber & Gessner, 2002), although this proportion is generally lower in tropical streams (Boyero et al., 2015; Gonçalves et al., 2007; Tonin, Hepp, et al., 2018). They consume litter directly and produce large amounts of FPOM via maceration and defaecation (Graça, 2001), which is consumed by other invertebrates (Cummins & Klug, 1979). Decomposition is affected by detritivore biogeography and phylogeny, water temperature, water chemistry, litter quality, water flow and streambed substrate. Large-scale drivers such as biogeography determine detritivore species distributions. For example, some taxa are more abundant and diverse in particular biogeographic realms (Boyero et al., 2011), including high abundance and diversity of caddisflies in the Australian realm; beetles in the Neotropics; and stoneflies and amphipods in the Palearctic. Detritivore contributions to decomposition tend to increase with their density, biomass and diversity (e.g., Jonsson & Malmqvist, 2000; Tonello et al., 2016; Tonin et al., 2014). Although detritivore density and diversity are typically higher in colder climates (Boyero et al., 2011), high biomass of efficient detritivores may compensate for lower diversity in some tropical streams (Cheshire et al., 2005; Tonin et al., 2014). Detritivore community composition may also have an effect on decomposition, mostly through the presence of efficient consumers (e.g., some caddisflies, stoneflies or amphipods; Tonin et al., 2014). Sometimes overlooked in studies of decomposition by detritivores are the larger crustaceans, which may be particularly abundant in tropical streams but not always clearly accounted for in field and laboratory experiments (Cogo & Santos, 2013). These animals may be of high biomass compared with insects and important consumers of litter in many systems (e.g., Coughlan et al., 2010; Crowl et al., 2001; Moulton et al., 2010).

Water chemistry may also shape detritivore communities. For example, some caddisfly and amphipod species are sensitive to acid waters (e.g., Dangles et al., 2004; Herrmann et al., 1993), whereas stoneflies are tolerant of them (e.g., Dangles & Guérold, 1999). Litter quality influences the consumption rates of detritivores and their C:N:P ratios, growth and survival (e.g., Graça et al., 2001; Tonin et al., 2017). They usually prefer litter that is soft and nutrient-rich (including macro and micronutrients), with low concentrations of secondary metabolites. Water flow and streambed substrate may also regulate detritivore distribution within streams, as different species occur in different substrate types (rocky *vs.* leaf litter substrates; Cheshire et al., 2005) and detritivores usually aggregate in areas of high litter accumulation, frequently in pools or backwaters, or behind retention structures (Heino et al., 2004). Therefore, litter storage and its spatial distribution often influence detritivore contribution to decomposition (e.g., Tonin, Hepp, et al., 2018). Finally, detritivores typically benefit from microbial colonization of litter (i.e., microbial conditioning; Casotti et al., 2019), because microbes increase the nutritional quality of litter and convert indigestible material into more labile compounds (Bärlocher, 1985), although fresh green leaves are consumed in some situations (Wootton et al., 2019).

2.3.3.4 Physical Fragmentation

Physical fragmentation is an important component of litter decomposition, but it can be difficult to separate its effects from those of other components, particularly from detritivore-mediated decomposition (Rader et al., 1994). Physical fragmentation depends on litter toughness, flow and flow-substrate interactions (Fonseca et al., 2013; Hoover et al., 2006). Litter toughness is increased by high concentrations of lignin, cellulose and hemicellulose. Flow affects physical fragmentation by abrading the litter surface with suspended particulate material, especially small particles, which are more likely to be transported by the current (Ferreira et al., 2006; Heard et al., 1999). Flow also causes turbulence, which may increase shear stress and enhance litter fragmentation, but there is limited information on this process.

2.4 Future Research Needs

The influence of scale is a major issue in ecology, both in terms of scientific understanding and subsequent management application (Levin, 1992). Frequently, local field studies and laboratory experiments are the only ones able to investigate patterns and processes in detail, so many decomposition studies have been undertaken at single stream sites. The major disadvantage is that conclusions cannot be confidently scaled up to generate broad conclusions. Similar studies undertaken at many sites help building a picture from which we can generalise, although the use of different methods can sometimes hinder comparisons. In response to such issues, in the last decade there has been a move to undertake parallel decomposition studies at multiple sites across the planet using standardised methods (Boyero & Pearson, 2017).

For development of general paradigms, understanding variability across scales is important (Tonin et al., 2019). Variability may be due to the physical environment operating at several scales, illustrated in the conceptual framework of Graça et al. (2015), but also to many biotic variables operating at fine scales, such as intraspecific competition (Boyero & Pearson, 2006), resource partitioning (Tonin, Pozo, et al., 2018), presence of large crustaceans (Coughlan et al., 2010), predation (Boyero et al., 2008), litter diversity (Bastian et al., 2008), decomposer diversity (Bastian et al., 2008), life cycle (Nolen & Pearson, 1992) and size or developmental stage (Nolen & Pearson, 1993). A hierarchical approach is required, preferably in several regions that may be regarded as representative globally. A limited number of detailed but biogeographically restricted studies of this nature have been undertaken in examination of diversity (e.g., Heino et al., 2018) and decomposition processes (Rezende et al., 2014; Tiegs et al., 2009; Tonin, Hepp, et al., 2018). Such hierarchical studies might provide a biogeographical, geomorphological and geological basis for developing a classification of streams relating to food webs, in a more globally inclusive scheme than, for example, the river continuum concept (Vannote et al., 1980), recognising that it is difficult to apply a single conceptual framework to all lotic systems.

Tank et al. (2010), in an extensive review, nominated a number of areas where more information was required on the dynamics of organic material in streams. A decade later, these gaps largely remain, particularly when considering issues at different scales. They include understanding of the dynamics and importance of dissolved organic matter, materials and energy budgets, inputs from floodplains and retention/transport generally. Those that we have highlighted above and others directly relevant to the decomposition process include: improved models of litter budgets (inputs, transport, storage) across scales; rates of organic matter decomposition across scales and biomes (mostly tropical ones), despite major advances in the last 10 years; multiple biophysical influences on decomposition; types of litter other than leaves (flowers, fruits and especially wood) across scales; fate of fine particulate organic matter (mineralization and incorporation in food webs); influence of variation within species and within individual trees in leaf chemistry on processing rates at different scales; effects of anthropogenic changes on litter inputs, decomposition and storage, which is an expanding field but has no systematic approach at the global scale (includes vegetation clearing, exotic species, species invasion, agriculture, urbanisation and changes in temperature and hydrology with climate); development of spatially explicit models at large scales (e.g., ecoregions, biomes), which provide the opportunity to formulate new hypotheses; and development of general models/frameworks of litter dynamics across scales, as elucidated in this paper. We look forward to future global experiments, meta-analyses and syntheses towards developing a more comprehensive framework that will enhance our understanding of the variable importance of organic matter and its sources in stream ecosystems, providing the capacity to predict effects of environmental change and inform improved management.

References

Abelho, M. (2001). From litterfall to breakdown in streams: A review. *Scientific World Journal, 1*, 656–680.

Arsuffi, T. L., & Suberkropp, K. (1989). Selective feeding by shredders on leaf-colonizing stream fungi: Comparison of macroinvertebrate taxa. *Oecologia, 79*, 30–37.

Bärlocher, F. (1985). The role of fungi in the nutrition of stream invertebrates. *Botanical Journal of the Linnean Society, 91*, 83–94.

Bärlocher, F. (2005). Freshwater fungal communities. In J. Dighton & J. F. White (Eds.), *The fungal community: Its organization and role in the ecosystem* (p. 39). CRC Press.

Bastian, M., Pearson, R. G., & Boyero, L. (2008). Effects of diversity loss on ecosystem function across trophic levels and ecosystems: A test in a detritus-based tropical food web. *Austral Ecology, 33*, 301–306.

Battin, T. J., Kaplan, L. A., Findlay, S., Hopkinson, C. S., Marti, E., Packman, A. I., Newbold, J. D., & Sabater, F. (2008). Biophysical controls on organic carbon fluxes in fluvial networks. *Nature Geoscience, 1*, 95–100.

Benson, L. J., & Pearson, R. G. (1993). Litter inputs to a tropical Australian rainforest stream. *Austral Ecology, 18*, 377–383.

Boulton, A. J., Findlay, S., Marmonier, P., Stanley, E. H., & Valett, M. H. (1998). The functional significance of the hyporheic zone in streams and rivers. *Annual Review of Ecology and Systematics, 29*, 59–81.

Boyero, L., Graça, M. A. S., Tonin, A. M., Pérez, J., Swafford, J. A., Ferreira, V., Landeira-Dabarca, A., Alexandrou, A. M., Gessner, M. O., McKie, B. G., Albariño, R. J., Barmuta, L. A., Callisto, M., Chará, J., Chauvet, E., Colón-Gaud, C., Dudgeon, D., Encalada, A. C., Figueroa, R., … Pearson, R. G. (2017). Riparian plant litter quality increases with latitude. *Scientific Reports, 7*, 10562.

Boyero, L., & Pearson, R. G. (2006). Intraspecific interference in a tropical stream shredder guild. *Marine and Freshwater Research, 57*, 201–206.

Boyero, L., & Pearson, R. G. (2017). Global-scale coordinated networks as a tool for exploring the functioning of stream ecosystems. *Limnetica, 36*, 557–565.

Boyero, L., Pearson, R. G., Dudgeon, D., Graça, M. A. S., Gessner, M. O., Albariño, R., Ferreira, V., Yule, C. M., Boulton, A. J., Arunachalam, M., Callisto, M., Chauvet, E., Ramírez, A., Chará, J., Moretti, M. S., Gonçalves, J. F. J., Helson, J. E., Chará-Serna, A., Encalada, A. C., … Pringle, C. M. (2011). Global distribution of a key trophic guild contrasts with common latitudinal diversity patterns. *Ecology, 92*, 1839–1848.

Boyero, L., Pearson, R. G., Gessner, M. O., Barmuta, L. A., Ferreira, V., Graça, M. A. S., Dudgeon, D., Boulton, A. J., Callisto, M., Chauvet, E., Helson, J. E., Bruder, A., Albariño, R. J., Yule, C. M., Arunachalam, M., Davies, J. N., Figueroa, R., Flecker, A. S., Ramírez, A., … West, D. C. (2011). A global experiment suggests climate warming will not accelerate litter decomposition in streams but might reduce carbon sequestration. *Ecology Letters, 14*, 289–294.

Boyero, L., Pearson, R. G., Gessner, M. O., Dudgeon, D., Ramírez, A., Yule, C. M., Callisto, M., Pringle, C. M., Encalada, A. C., Arunachalam, M., Mathooko, J., Helson, J. E., Rincón, J., Bruder, A., Cornejo, A., Flecker, A. S., Mathuriau, C., M'Erimba, C., Gonçalves Jr, J. F., … Jinggut, T. (2015). Leaf-litter breakdown in tropical streams: is variability the norm? *Freshwater Science, 34*, 759–769.

Boyero, L., Rincón, P. A., & Pearson, R. G. (2008). Effects of a predatory fish on a tropical detritus-based food web. *Ecological Research, 23*, 649–655.

Casotti, C. G., Kiffer, W. P., Costa, L. C., Barbosa, P., & Moretti, M. S. (2019). The longer the conditioning, the better the quality? The effects of leaf conditioning time on aquatic hyphomycetes and performance of shredders in a tropical stream. *Aquatic Ecology, 53*, 163–178.

Casper, B. B., & Jackson, R. B. (1997). Plant competition underground. *Annual Review of Ecology and Systematics, 28*, 545–570.

Chamier, A.-C. (1987). Effect of pH on microbial degradation of leaf litter in seven streams of the English Lake District. *Oecologia, 71*, 491–500.
Chergui, H., & Pattee, E. (1988). Effect of water current on the decomposition of dead leaves and needles. *Verhandlungen Des Internationalen Verein Limnologie, 23*, 1294–1298.
Cheshire, K., Boyero, L., & Pearson, R. G. (2005). Food webs in tropical Australian streams: Shredders are not scarce. *Freshwater Biology, 50*, 748–769.
Cogo, G. B., & Santos, S. (2013). The role of aeglids in shredding organic matter in Neotropical streams. *Journal of Crustacean Biology, 33*, 519–526.
Cole, J. J., Prairie, Y. T., Caraco, N. F., McDowell, W. H., Tranvik, L. J., Striegl, R. G., Duarte, C. M., Kortelainen, P., Downing, J. A., & Middelburg, J. J. (2007). Plumbing the global carbon cycle: Integrating inland waters into the terrestrial carbon budget. *Ecosystems, 10*, 172–185.
Coughlan, J. F., Pearson, R. G., & Boyero, L. (2010). Crayfish process leaf litter in tropical streams even when shredding insects are common. *Marine and Freshwater Research, 61*, 541–548.
Covich, A. P. (1988). Geographical and historical comparisons of neotropical streams: Biotic diversity and detrital processing in highly variable habitats. *Journal of the North American Benthological Society, 7*, 361–386.
Crowl, T. A., McDowell, W. H., Covich, A. P., & Johnson, S. L. (2001). Freshwater shrimp effects on detrital processing and nutrients in a tropical headwater stream. *Ecology, 82*, 775–783.
Cummins, K. W., & Klug, M. J. (1979). Feeding ecology of stream invertebrates. *Annual Review of Ecology and Systematics, 10*, 147–172.
Dang, C. K., Schindler, M., Chauvet, E., & Gessner, M. O. (2009). Temperature oscillation coupled with fungal community shifts can modulate warming effects on litter decomposition. *Ecology, 90*, 122–131.
Dangles, O., Gessner, M. O., Guerold, F., & Chauvet, E. (2004). Impacts of stream acidification on litter breakdown: Implications for assessing ecosystem functioning. *Journal of Applied Ecology, 41*, 365–378.
Dangles, O., & Guérold, F. (1999). Impact of headwater stream acidification on the trophic structure of macroinvertebrate communities. *International Review of Hydrobiology, 84*, 287–297.
Díez, J., Larrañaga, S., Elosegi, A., & Pozo, J. (2000). Effect of removal of wood on streambed stability and retention of organic matter. *Journal of the North American Benthological Society, 19*, 621–632.
Dudgeon, D., Arthington, A. H., Gessner, M. O., Kawabata, Z., Knowler, D. J., Leveque, C., Naiman, R. J., Prieur-Richard, A. H., Soto, D., Stiassny, M. L., & Sullivan, C. A. (2006). Freshwater biodiversity: Importance, threats, status and conservation challenges. *Biological Reviews of the Cambridge Philosophical Society, 81*, 163–182.
Essington, M. E. (2005). *Soil and water chemistry: An integrative approach*. CRC Press.
Ferreira, V., & Chauvet, E. (2011). Synergistic effects of water temperature and dissolved nutrients on litter decomposition and associated fungi. *Global Change Biology, 17*, 551–564.
Ferreira, V., Graça, M. A. S., de Lima, J. L. M. P., & Gomes, R. (2006). Role of physical fragmentation and invertebrate activity in the breakdown rate of leaves. *Archiv für Hydrobiologie, 165*, 493–513.
Findlay, S. E. G., & Arsuffi, T. L. (1989). Microbial growth and detritus transformations during decomposition of leaf litter in a stream. *Freshwater Biology, 21*, 261–269.
Fisher, S. G., & Likens, G. E. (1972). Stream ecosystem: Organic energy budget. *BioScience, 22*, 33–35.
Fisher, S. G., & Likens, G. E. (1973). Energy flow in Bear Brook, New Hampshire: An integrative approach to stream ecosystem metabolism. *Ecological Monographs, 43*, 421–439.
Fonseca, A. L. S., Bianchini, I. J., Pimenta, C. M. M., Soares, C. B. P., & Mangiavacchi, N. (2013). The flow velocity as driving force for decomposition of leaves and twigs. *Hydrobiologia, 703*, 59–67.
García-Palacios, P., Shaw, E. A., Wall, D. H., & Hättenschwiler, S. (2016). Temporal dynamics of biotic and abiotic drivers of litter decomposition. *Ecology Letters, 19*, 554–563.
Gebely, T. (2016). *Tea: A user's guide*. Eggs and Toast Media, LCC.

Gessner, M. O., Chauvet, E., & Dobson, M. (1999). A perspective on leaf litter breakdown in streams. *Oikos, 85*, 377–384.

Gleick, P. H. (1996). Water resources. *Encyclopedia of Climate and Weather, 2*, 817–823.

Gomes, P. P. (2015). *Influência da química do detrito foliar e da água sobre a comunidade de hifomicetos aquáticos*. PhD Thesis, University of Brasília.

Gonçalves, J. F. J., Graça, M. A. S., & Callisto, M. (2007). Litter decomposition in a Cerrado savannah stream is retarded by leaf toughness, low dissolved nutrients and a low density of shredders. *Freshwater Biology, 52*, 1440–1451.

Graça, M. A. S. (2001). The role of invertebrates on leaf litter decomposition in streams—A review. *International Review of Hydrobiology, 86*, 383–393.

Graça, M. A. S., Cressa, C., Gessner, M. O., Feio, M. J., Callies, K. A., & Barrios, C. (2001). Food quality, feeding preferences, survival and growth of shredders from temperate and tropical streams. *Freshwater Biology, 46*, 947–957.

Graça, M. A. S., Ferreira, V., Canhoto, C., Encalada, A. C., Guerrero-Bolaño, F., Wantzen, K. M., & Boyero, L. (2015). A conceptual model of litter breakdown in low order streams. *International Review of Hydrobiology, 100*, 1–12.

Hall, R. O., Wallace, J. B., & Eggert, S. L. (2000). Organic matter flow in stream food webs with reduced detrital resource base. *Ecology, 81*, 3445–3463.

Heard, S. B., Schultz, G. A., Ogden, C. B., & Griesel, T. C. (1999). Mechanical abrasion and organic matter processing in an Iowa stream. *Hydrobiologia, 400*, 179–186.

Heino, J., Louhi, P., & Muotka, T. (2004). Identifying the scales of variability in stream macroinvertebrate abundance functional composition and assemblage structure. *Freshwater Biology, 49*, 1230–1239.

Heino, J., Melo, A. S., Jyrkänkallio-Mikkola, J., Petsch, D. K., Saito, V. S., Tolonen, K. T., Bini, L. M., Landeiro, V. L., Silva, T. S. F., & Pajunen, V. (2018). Subtropical streams harbour higher genus richness and lower abundance of insects compared to boreal streams, but scale matters. *Journal of Biogeography, 45*, 1983–1993.

Herrmann, J., Degerman, E., Gerhardt, A., Johansson, C., Lingdell, P. E., & Muniz, I. P. (1993). Acid-stress effects on stream biology. *Ambio, 22*, 298–307.

Hieber, M., & Gessner, M. O. (2002). Contribution of stream detrivores, fungi, and bacteria to leaf breakdown based on biomass estimates. *Ecology, 83*, 1026–1038.

Hoover, T. M., Richardson, J. S., & Yonemitsu, N. (2006). Flow-substrate interactions create and mediate leaf litter resource patches in streams. *Freshwater Biology, 51*, 435–447.

Hotchkiss, E. R., Hall, R. O., Jr., Sponseller, R. A., Butman, D., Klaminder, J., Laudon, H., Rosvall, M., & Karlsson, J. (2015). Sources of and processes controlling CO_2 emissions change with the size of streams and rivers. *Nature Geoscience, 8*, 696–699.

Hupp, C. R., Dufour, S., & Bornette, G. (2016). Vegetation as a tool in the interpretation of fluvial geomorphic processes and landforms. In G. M. Kondolf & H. Piégay (Eds.), *Tools in fluvial geomorphology* (pp. 210–226). Wiley.

Jenkins, C. C., & Suberkropp, K. (1995). The influence of water chemistry on the enzymatic degradation of leaves in streams. *Freshwater Biology, 33*, 245–253.

Jones, J. B. (1997). Benthic organic matter storage in streams: Influence of detrital import and export, retention mechanisms, and climate. *Journal of the North American Benthological Society, 16*, 109–119.

Jonsson, M., & Malmqvist, B. (2000). Ecosystem process rate increases with animal species richness: Evidence from leaf-eating, aquatic insects. *Oikos, 89*, 519–523.

Kuiters, A. T., & Sarink, H. M. (1986). Leaching of phenolic compounds from leaf and needle litter of several deciduous and coniferous trees. *Soil Biology and Biochemistry, 18*, 475–480.

Leopold, L. B., Wolman, M. G., & Miller, J. P. (1992). *Fluvial processes in geomorphology*. Courier Corporation.

Levin, S. A. (1992). The problem of pattern and scale in ecology: The Robert H. MacArthur Award Lecture. *Ecology, 73*, 1943–1967.

Marks, J. C. (2019). Revisiting the fates of dead leaves that fall into streams. *Annual Review of Ecology, Evolution, and Systematics, 50*, 547–568.

Martiny, J. B. H., Bohannan, B. J. M., Brown, J. H., Colwell, R. K., Fuhrman, J. A., Green, J. L., Horner-Devine, M. C., Kane, M., Krumins, J. A., Kuske, C. R., Morin, P. J., Naeem, S., Ovreas, L., Reysenbach, A.-L., Smith, V. H., & Staley, J. T. (2006). Microbial biogeography: Putting microorganisms on the map. *Nature Reviews Microbiology, 4*, 102–112.

Moulton, T., Magalhães-Fraga, S. P., Brito, E., & Barbosa, F. (2010). Macroconsumers are more important than specialist macroinvertebrate shredders in leaf processing in urban forest streams of Rio de Janeiro, Brazil. *Hydrobiologia, 638*, 55–66.

Neres-Lima, V., Machado-Silva, F., Baptista, D. F., Oliveira, R. B. S., Andrade, P. M., Oliveira, A. F., Sasada-Sato, C. Y., Silva-Junior, E. F., Feijó-Lima, R., Angelini, R., Camargo, P. B., & Moulton, T. P. (2017). Allochthonous and autochthonous carbon flows in food webs of tropical forest streams. *Freshwater Biology, 62*, 1012–1023.

Nolen, J., & Pearson, R. (1992). Life history studies of *Anisocentropus kirramus* Neboiss (Trichoptera: Calamoceratidae) in a tropical Australian rainforest stream. *Aquatic Insects, 14*, 213–221.

Nolen, J., & Pearson, R. (1993). Factors affecting litter processing by *Anisocentropus kirramus* (Trichoptera: Calamoceratidae) from an Australian tropical rainforest stream. *Freshwater Biology, 29*, 469–479.

O'Malley, M. A. (2007). The nineteenth century roots of 'everything is everywhere'. *Nature Reviews Microbiology, 5*, 647.

Pozo, J., González, E., Díez, J., & Elosegi, A. (1997). Leaf-litter budgets in two contrasting forested streams. *Limnetica, 13*, 77–84.

Rader, R. B., McArthur, J. V., & Aho, J. M. (1994). Relative importance of mechanisms determining decomposition in a Southeastern blackwater stream. *The American Midland Naturalist, 132*, 19–31.

Raymond, P. A., Hartmann, J., Lauerwald, R., Sobek, S., McDonald, C., Hoover, M., Butman, D., Striegl, R., Mayorga, E., & Humborg, C. (2013). Global carbon dioxide emissions from inland waters. *Nature, 503*, 355–359.

Rezende, R. S., Petrucio, M. M., & Gonçalves, J. F., Jr. (2014). The effects of spatial scale on breakdown of leaves in a tropical watershed. *PLoS ONE, 9*, e97072.

Royer, T. V., & Minshall, G. W. (2003). Control on Leaf processing in streams from spatial-scaling and hierarchical perspectives. *Journal of the North American Benthological Society, 22*, 352–358.

Schneider, T., Keiblinger, K. M., Schmid, E., Sterflinger-Gleixner, K., Ellersdorfer, G., Roschitzki, B., Richter, A., Eberl, L., Zechmeister-Boltenstern, S., & Riedel, K. (2012). Who is who in litter decomposition? Metaproteomics reveals major microbial players and their biogeochemical functions. *The ISME journal, 6*, 1749–1762.

Schreeg, L. A., Mack, M. C., & Turner, B. L. (2013). Nutrient-specific solubility patterns of leaf litter across 41 lowland tropical woody species. *Ecology, 94*, 94–105.

Siefert, A., Violle, C., Chalmandrier, L., Albert, C. H., Taudiere, A., Fajardo, A., Aarssen, L. W., Baraloto, C., Carlucci, M. B., Cianciaruso, M. V., de L. Dantas, V., de Bello, F., Duarte, L. D. S., Fonseca, C. R., Freschet, G. T., Gaucherand, S., Gross, N., Hikosaka, K., Jackson, B., … Wardle, D. A. (2015). A global meta-analysis of the relative extent of intraspecific trait variation in plant communities. *Ecology Letters, 18*, 1406–1419.

Suberkropp, K., & Chauvet, E. (1995). Regulation of leaf breakdown by fungi in streams: Influences of water chemistry. *Ecology, 76*, 1433–1445.

Tank, J. L., Rosi-Marshall, E. J., Griffiths, N. A., Entrekin, S. A., & Stephen, M. L. (2010). A review of allochthonous organic matter dynamics and metabolism in streams. *Journal of the North American Benthological Society, 29*, 118–146.

Taylor, B. R., & Bärlocher, F. (1996). Variable effects of air-drying on leaching losses from tree leaf litter. *Hydrobiologia, 325*, 173–182.

Tiegs, S. D., Akinwole, P. O., & Gessner, M. O. (2009). Litter decomposition across multiple spatial scales in stream networks. *Oecologia, 161*, 343–351.

Tonello, G., Naziloski, L. A., Tonin, A. M., Restello, R. M., & Hepp, L. U. (2016). Effect of *Phylloicus* on leaf breakdown in a subtropical stream. *Limnetica, 35*, 243–252.
Tonin, A. M., Boyero, L., Bambi, P., Pearson, R. G., Correa-Araneda, F., & Gonçalves Jr., J.F. (2019). High within-stream replication is needed to predict litter fluxes in wet–dry tropical streams. *Freshwater Biology, 65*, 688–697.
Tonin, A. M., Boyero, L., Monroy, S., Basaguren, A., Pérez, J., Pearson, R. G., Cardinale, B. J., Gonçalves, J. F., & Pozo, J. (2017). Stream nitrogen concentration, but not plant N-fixing capacity, modulates litter diversity effects on decomposition. *Functional Ecology, 31*, 1471–1481.
Tonin, A. M., Gonçalves Jr., J. F., Bambi, P., Couceiro, S. R. M., Feitoza, L. A. M., Fontana, L. E., Hamada, N., Hepp, L. U., Lezan-Kowalczuk, V. G., Leite, G. F. M., Lemes-Silva, A. L., Lisboa, L. K., Loureiro, R. C., Martins, R. T., Medeiros, A. O., Morais, P. B., Moretto, Y., Oliveria, P. C. A., Pereira, E. B., … Boyero, L. (2017). Plant litter dynamics in the forest-stream interface: Precipitation is a major control across tropical biomes. *Scientific Reports, 7*, 10799.
Tonin, A. M., Hepp, L. U., & Gonçalves Jr., J. F. (2018). Spatial variability of plant litter decomposition in stream networks: From litter bags to watersheds. *Ecosystens, 21*, 567–581.
Tonin, A. M., Hepp, L. U., Restello, R. M., & Gonçalves, J. F., Jr. (2014). Understanding of colonization and breakdown of leaves by invertebrates in a tropical stream is enhanced by using biomass as well as count data. *Hydrobiologia, 740*, 79–88.
Tonin, A. M., Pozo, J., Monroy, S., Basaguren, A., Pérez, J., Gonçalves, J. F., Jr., Pearson, R., Cardinale, B. J., & Boyero, L. (2018). Interactions between large and small detritivores influence how biodiversity impacts litter decomposition. *Journal of Animal Ecology, 87*, 1465–1474.
Vannote, R. L., Minshall, G. W., Cummins, K. W., Sedell, J. R., & Cushing, C. E. (1980). The river continuum concept. *Canadian Journal of Fisheries and Aquatic Sciences, 37*, 130–137.
Vergutz, L., Manzoni, S., Porporato, A., Novais, R. F., & Jackson, R. B. (2012). Global resorption efficiencies and concentrations of carbon and nutrients in leaves of terrestrial plants. *Ecological Monographs, 82*, 205–220.
Wallace, J. B., Eggert, S. L., Meyer, J. L., & Webster, J. R. (1997). Multiple trophic levels of a forest stream linked to terrestrial litter inputs. *Science, 277*, 102–104.
Wallace, J. B., Webster, J. R., & Meyer, J. L. (1995). Influence of log additions on physical and biotic characteristics of a mountain stream. *Canadian Journal of Fisheries and Aquatic Sciences, 52*, 2120–2137.
Wardle, D. A., Bardgett, R. D., Klironomos, J. N., Setälä, H., van der Putten, W. H., & Wall, D. H. (2004). Ecological linkages between aboveground and belowground biota. *Science, 304*, 1629–1633.
Webster, J. R., Benfield, E. F., Ehrman, T. P., Schaeffer, M. A., Tank, J. L., Hutchens, J. J., & D'Angelo, D. J. (1999). What happens to allochthonous material that falls into streams? A synthesis of new and published information from Coweeta. *Freshwater Biology, 41*, 687–705.
Webster, J. R., & Meyer, J. L. (1997). Stream organic matter budgets: An introduction. *Journal of the North American Benthological Society, 16*, 3–13.
Wootton, A., Pearson, R. G., & Boyero, L. (2019). Patterns of flow, leaf litter and shredder abundance in a tropical stream. *Hydrobiologia, 826*, 353–365.

Chapter 3
Stoichiometry of Plant Litter Decomposition in Stream Ecosystems

Michael Danger, Julio Arce-Funck, Miriam Beck, Clément Crenier, Vincent Felten, Ziming Wang, and Florence Maunoury-Danger

Abstract Ecological stoichiometry is an approach of ecology aimed at understanding the causes and consequences of elemental imbalances in nature (mainly focusing carbon, nitrogen, and phosphorus), from molecular to ecosystem levels. Like most detritus in nature, plant litters are nutrient depleted, and both microbial decomposers and detritivores are exposed to large stoichiometric constraints. These nutritional constraints will ultimately control litter decomposition rates, nutrient mineralization, and affect, in turn, decomposers' community structures. To consider such stoichiometric constraints could greatly help understanding the functioning of detritus-based ecosystems. In this chapter, focused on leaf litter, diverse examples of stoichiometric constraints at play at the detritus-decomposer interface in aquatic ecosystems are presented. The different steps involving stoichiometric processes and ultimately conducting to litter decomposition, from the production of leaf litter in the riparian zones of freshwater ecosystems to its incorporation in aquatic food webs and its potential recycling in freshwater ecosystems are discussed. Stoichiometric constraints arising between detritus and decomposers in freshwater ecosystems are then placed into the context of current global change. Finally, we highlight

M. Danger (✉) · J. Arce-Funck · M. Beck · C. Crenier · V. Felten · Z. Wang ·
F. Maunoury-Danger
Université de Lorraine, CNRS, LIEC, 57000 Metz, France
e-mail: michael.danger@univ-lorraine.fr

LTSER France, Zone Atelier du Bassin de La Moselle, Vandoeuvre-lès-Nancy, France

M. Beck
e-mail: miriam.beck@univ-lorraine.fr

V. Felten
e-mail: vincent.felten@univ-lorraine.fr

Z. Wang
e-mail: ziming.wang@univ-lorraine.fr

F. Maunoury-Danger
e-mail: florence.maunoury-danger@univ-lorraine.fr

M. Danger
Institut Universitaire de France, Paris, France

C. M. Swan et al. (eds.), *The Ecology of Plant Litter Decomposition in Stream Ecosystems*, https://doi.org/10.1007/978-3-030-72854-0_3

the limits of the stoichiometric approach as well as some complementary approaches and perspectives of work are proposed.

3.1 Ecological Stoichiometry: Conceptual Bases in Detritus-Based Ecosystems

Ecological stoichiometry is an approach of ecology aimed at understanding the causes and consequences of imbalances between several chemical elements during organisms' interactions and ecological processes, from molecular to ecosystem levels (Elser et al., 1996; Sterner & Elser, 2002). It relies on the observation that all living organisms are composed of the same chemical elements (e.g., carbon (C), nitrogen (N), phosphorus (P), iron (Fe), silica (Si), potassium (K), etc.) in more or less important and variable amounts. Ecological stoichiometry approaches then consider the elemental compositions (often expressed as molecular elemental ratios, for example C:N or N:P ratios) of basal resources and living organisms involved in ecological interactions. Any mismatch between consumer requirements and elements available in their resources constitutes a stoichiometric constraint that can alter consumers' life history traits (survival, growth, reproduction), nutrient transfer efficiency and ecosystem productivity, but also nutrient recycling (Cebrian et al., 2009; Sterner & Elser, 2002; Vanni, 2002). It can also be efficiently applied to community ecology, since stoichiometric constraints alter competitive interaction outcomes between species and participate to shape community structures (Moe et al., 2005; Tilman, 1982). Taking explicitly into account the couplings between biogeochemical cycles and organisms via chemical elements, ecological stoichiometry permits to relate several sub-disciplines of ecology, such as ecophysiology, population and community ecology, and ecosystem ecology.

Although formalized quite recently (Sterner & Elser, 2002), ecological stoichiometry finds its origin in much older approaches. As early as 1925, Lotka in his book entitled *Elements of Physical Biology*, evoked the importance of considering the elemental composition of living organisms in the study of their interactions and their impacts on their environment. Results from Redfield (1958) then largely questioned the couplings between elements in organic matter and their importance for biogeochemical cycles. In 1986, Reiners first proposed a mechanistic view of the connections between the elemental composition of organisms, environmental constraints, and ecosystem processes. This approach has then been successfully tested into diverse ecosystems, from lakes and rivers to marine and terrestrial ecosystems (Sterner & Elser, 2002). Despite the apparent generality of ecological stoichiometry concepts, most studies have been restricted to the plant—herbivore interface. Fewer studies investigated the importance and the impacts of stoichiometric constraints for upper trophic levels (predators) and for detritus consumers. While stoichiometric constraints are expected to be reduced at higher trophic levels due to the higher elemental quality of resources (animal prey), those arising at the detritus—detritivore interface are expected to be extremely high (Danger, 2020; Evans-White & Halvorson, 2017; Martinson et al., 2008). Detritus includes all types of dead animals, dead microorganisms, and dead plant tissues (e.g., dead leaves, dead wood, macrophytes, and dead algae), but also *faeces* and dissolved organic matter excreted or exuded (e.g., exopolysaccharides, dissolved organic matter, root or leaf exudates; Moore et al., 2004). Due to the central importance of plant production on Earth, a large majority of detritus found in ecosystems have a plant origin (Moore et al., 2004). For numerous vascular plants, plant tissues or substances released as detritus are particularly nutrient-poor since most plants have long been selected for reabsorbing and retaining the most limiting nutrients before detritus release (Killingbeck, 1996; Noodén et al., 1997). In particular, litter of terrestrial plants generally contains extremely low N and P concentrations (Cross et al., 2005). Stoichiometric constraints are thus expected to be particularly exacerbated at the detritus—detritivore interface (Martinson et al., 2008).

Because of their generally concave profiles, aquatic ecosystems are more subject to inputs of allochthonous organic matter than terrestrial ecosystems (Leroux & Loreau, 2008; Polis & Strong, 1996). In freshwater ecosystems, a dominant proportion of this organic matter is composed of plant detritus from terrestrial origin, mainly under the form of large particles (Coarse Particulate Organic Matter, CPOM), such as leaf litter or dead wood, but also as dissolved organic material originating, for example, from root exudates or leaf litter leaching (Meyer et al., 1998). The detritus have long been acknowledged as important or even preponderant resources for freshwater ecosystem functioning (Lindeman, 1942; Wetzel, 1995). Headwater streams draining forest catchments, for example, rely for a large part on leaf litter inputs from adjacent terrestrial ecosystems (Fisher & Likens, 1973; Wallace et al., 1999). Despite their generally refractory nature and their reduced nutrient contents, detritus largely sustain the development of aquatic food webs (Wetzel, 1995). Studying the stoichiometric constraints at the detritus—decomposer interface and their ecological consequences in freshwater ecosystems is thus of critical importance, and has been the topic of an increasing number of studies in the past two decades.

Rather than proposing an exhaustive review of all stoichiometric studies carried out on detritus decomposition in aquatic ecosystems, this chapter will gather diverse examples of the stoichiometric constraints occurring at the detritus—decomposer interface in aquatic ecosystems. Detritus will be restricted to leaf litter for simplification, but stoichiometric constraints will also generally apply for other detritus types, more or less pronounced depending on the stoichiometric imbalance between detritus and their consumer (microbial and metazoan detritivores) requirements. This chapter will thus be constructed by discussing the different steps involving stoichiometric processes and ultimately conducting to litter decomposition, from the production of leaf litter in the riparian zones of freshwater ecosystems to its incorporation in aquatic food webs and its potential mass and nutrient recycling in freshwater ecosystems (Fig. 3.1). Finally, stoichiometric constraints arising between detritus and decomposers in freshwater ecosystems will then be placed in the context of current global change, and some perspectives of work will be proposed.

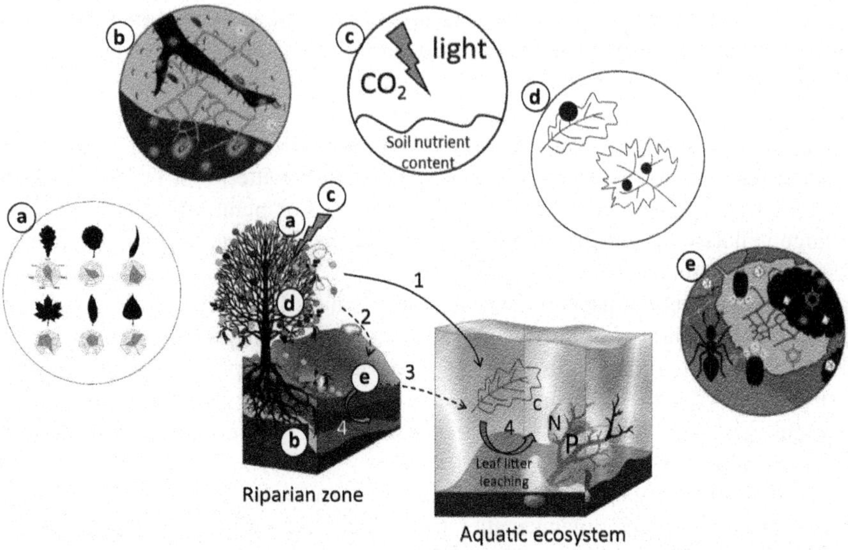

Fig. 3.1 Non exhaustive presentation of factors susceptible to affect leaf litter stoichiometry, from its production in the riparian zone to leaf litter deposition in aquatic ecosystems: (**a**) riparian plant identity, leaf litter traits including stoichiometric traits largely differing between species, (**b**) microbial processes occuring in the vicinity of riparian plant roots, such as nitrogen fixation, (**c**) Environmental factors, such as light availability, CO_2 concentrations, or nutrient levels in soils, (**d**) predation or parasitism occuring on green leaves, (**e**) biological processes occuring at the soil surface (see text for details and references). Leaf litter can either enter aquatic ecosystem directly (arrow 1) or indirectly (arrows 2 and 3), after a variable period of decomposition on soil surface. Leaf litter leaching (arrows 4) will occur quickly in aquatic medium, and might be delayed on soil depending on weather conditions. This leaching is generally higher for P than for N and C, altering leaf litter stoichiometry

3.2 From the Riparian Zone to Freshwaters: The Stoichiometry of Leaf Litter

Litter inputs often represent the main energy and nutrient sources for many forested aquatic ecosystems (Wallace et al., 1999). These inputs largely vary between biomes and seasons, ranging from large seasonal inputs in temperate, broad-leaved forests, to more annually distributed inputs in evergreen or coniferous forests. Also, the relative abundance of broad-leaved trees, evergreen or coniferous trees can largely impact instream leaf litter decomposition, especially due to its impacts on litter inputs seasonality and the overall litter chemistry (Ferreira et al., 2016). To date, most stoichiometric studies have concerned the litter decomposition of deciduous broadleaf species in temperate ecosystems, and most examples cited hereafter will concern this litter category. In any case, in comparison with green leaves, leaf litter is generally considered as a low stoichiometric quality resource, i.e., containing low nutrient contents and exhibiting high C:nutrient ratios (Cross et al., 2005; Sariyildiz &

Anderson, 2005). These high C:nutrient ratios are mainly attributed to plant nutrient remobilization at abscission (Chapin & Kedrowski, 1983; Killingbeck, 1996; Nooden et al., 1997). Nevertheless, other parameters have been evidenced as playing a role in leaf litter elemental composition.

First of all, leaf litter stoichiometry can vary much between species (Hladyz et al., 2009; Melillo et al., 1982; Ostrofsky, 1997, Figs. 3.1a and 3.2). As an illustration, litters coming from N-fixing species are known to present, on average, higher N contents than litters coming from non-fixing ones (Hladyz et al., 2009; Fig. 3.1b). While alder (*Alnus glutinosa*, L.) leaves can reach C:N molar ratios as low as 14, other deciduous species like ash (*Fraxinus excelsior*, L.) or beech (*Fagus sylvatica*, L.) generally exhibit C:N ratios higher than 40 (Hladyz et al., 2009). Such high N contents in leaf litter of N-fixing plants can be explained by the absence of plant

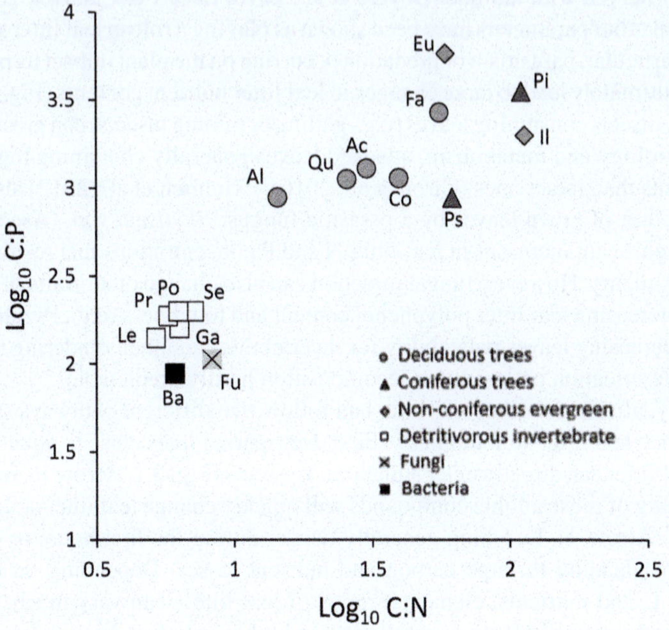

Fig. 3.2 Illustration of the elemental imbalance between diverse leaf litters and their potential consumers, including some common detritivorous invertebrate taxa, bacteria, and fungi. Leaf litter data include data from deciduous trees (Al: *Alnus glutinosa*, Fa: *Fagus sylvatica*, Qu: *Quercus robur*, Ac: *Acer pseudoplatanus*, Co: *Corylus avellana*, data from Hladyz et al., 2009), conifers (Ps: *Pinus sylvestris*, Pi: *Picea* sp., data from Enríquez et al., 1993), and non-coniferous evergreen trees (Eu: *Eucalyptus globulus*, data from Kiffer et al., 2018; Il: *Ilex aquifolium*, data from Hladyz et al., 2009). Detritivorous invertebrate data correspond to whole body tissue analyses carried out on some of the dominant taxa in north-eastern France headwater streams (Ga: *Gammarus fossarum*, Se: *Sericostoma personatum*, Pr: *Protonemura* sp., Po: *Potamophylax* sp., Le: *Leuctra* sp., data from Beck et al., unpublished data). Bacteria and fungi data correspond to microbial cultures of natural isolates (Mouginot et al., 2014). Note that bacteria and fungi elemental ratios correspond to mean values and that these values can be highly variable (see Danger et al., 2016)

limitation by this element and thus an absence or a reduction of N remobilization of this element prior to abscission. Second, beyond interspecific differences, litter stoichiometry can also widely differ between individuals of a single species. Several environmental parameters have been shown to affect plant tissue stoichiometry, such as climate, CO_2 concentration, light availability, or nutrient availability in soils (Ågren, 2008; Biasi et al., 2017; Graça & Poquet, 2014; Norby et al., 2001; Fig. 3.1c). Litter stoichiometry then more or less closely relates green leaves stoichiometry with a general reduction of nutrient contents (McGroddy et al., 2004), the intensity of nutrient decrease depending largely on plant resorption activity (Killingbeck, 1996). From a global viewpoint, litter stoichiometry seems to depend on latitude, with litters being more P depleted in the tropics than in higher latitudes (Boyero et al., 2017). Such large scale patterns, generally explained by variations of soil P availability with latitude, could at least partly drive differences in leaf litter decomposition already observed with latitude (Boyero et al., 2016; Lecerf & Chauvet, 2008).

Several other parameters have been shown as playing a role in leaf litter stoichiometry. In particular, parasitism or predation occurring on the plant impact its physiology and can ultimately lead to large changes in leaf litter nutrient contents (Fig. 3.1d). For example, insects consuming leaves (e.g., galling or mining insects) can greatly change leaf physiology and metabolism, attacked leaves generally containing higher levels of nutrients than intact ones (Giron et al., 2016). Grimmett et al. (2012) also showed that infection of green leaves by a parasitic fungus, *Rhytisma spp.* (Ascomycetes), lead to significant increases in leaf litter N and P concentrations that result in higher leaf litter quality. However, fungal infection can also change other traits of leaf litter, such as increasing leaf litter polyphenol content and toughness (e.g., Pazianoto et al., 2019), decreasing leaves palatability for invertebrates and then rendering the impact of parasite infection on leaf litter decomposition hardly predictable.

Finally, after abscission, leaf litter can follow two different pathways susceptible to influence leaf litter stoichiometry. First, for riparian trees close to aquatic ecosystems, leaf litter can directly enter aquatic ecosystems (Fig. 3.1, Arrow 1). In that case, the leaching of most soluble compounds will quickly change leaf litter stoichiometry (Fig. 3.1, Arrow 4). Leaching generally occurs during the first hours or days after immersion, leading to large carbon and nutrient losses. Depending on respective losses of C and nutrients, elemental ratios of leaf litters can vary much. However, since N and especially P are generally lost in higher amounts through leaching than C (Maunoury-Danger et al., 2018; Schreeg et al., 2013), leaching most often results in increases of leaf litters C:N and C:P ratios (Danger, Arce Funck, et al., 2013). Second, leaf litter can fall on the soil and remain for a variable time in terrestrial ecosystems before being transferred to aquatic ecosystems (Fig. 3.1e, Arrows 2 and 3). These lateral inputs can be as high as direct leaf litter inputs in small forested streams (Wallace et al., 1999). Depending on the duration and the context of leaf litter exposure to soil conditions (e.g., moisture, presence of microbial and invertebrate decomposers), the consequences on leaf litter stoichiometry should largely vary, influencing in turn the degradability of leaf litter leachates and leaf litter decomposition in aquatic ecosystems (del Campo et al., 2020). Note also that even if a large part of carbon and nutrients leached from leaf litter stay in the surrounding terrestrial ecosystem, a non-negligible amount of these compounds might ultimately arrive in aquatic ecosystems, and ultimately be used by microbial decomposers (Meyer et al., 1998).

3.3 Stoichiometry of Litter Microbial Decomposition in Freshwaters

When entering the aquatic ecosystem, concomitantly with leaf litter leaching, leaf litter will be promptly colonized by microbial decomposers (Fig. 3.3a, b).

Microbial decomposers are composed of both aquatic fungi (e.g., aquatic hyphomycetes) and bacteria. Aquatic fungi are generally considered as the first and most important leaf litter decomposers in rivers (in terms of both biomass and activity, e.g., Baldy et al., 1995; Hieber & Gessner, 2002) while bacteria become increasingly abundant at later stages of litter decomposition. Microbial activity ultimately leads to

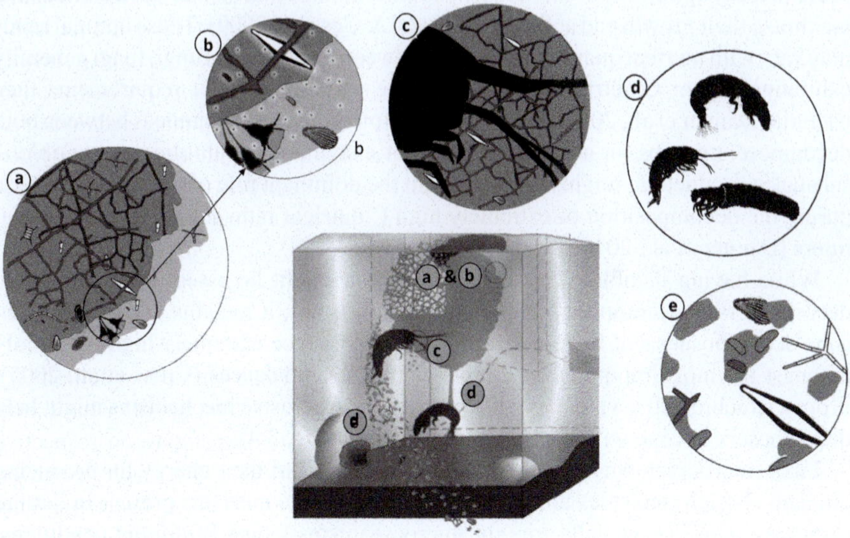

Fig. 3.3 Microbial- and detritivore-driven processes affected by or affecting leaf litter stoichiometry. (**a**) Leaf litter stoichiometry can select for particular taxa of decomposers and microbial activity progressively changes leaf litter elemental composition, (**b**) in addition to fungi and bacteria, other organisms can change leaf litter quality for detritivores, such as protozoans or microalgae developing on leaf litter surface, (**c**) detritivores can selectively feed on patches of higher stoichiometric quality and change the elemental composition of remaining material, (**d**) leaf litter stoichiometry will affect not only detritivores growth but also the taxonomic composition of detritivores communities, altering in turn leaf litter decomposition, (**e**) microbial and detritivores activity release fine particles with variable stoichiometry (fungal conidia, leaf litter debris, detritivore *faeces*, etc.)—see text for details and references

direct litter mass loss but is also particularly important for ensuring leaf litter conditioning, i.e., the enhancement of leaf palatability for detritivores (Gessner et al., 1999). Microbially-conditioned leaf litters generally become softer, more nutritious, and more palatable to detritivores (Graça, 2001). Despite the central role of microbial decomposers in leaf litter conditioning and decomposition in aquatic ecosystems, the stoichiometric determinants of their decomposing activity received attention only recently.

Most studies dealing with aquatic fungi stoichiometry have shown that at the strain level, fungi are largely non-homeostatic, i.e., they can store large amounts of nutrients in their biomass, resulting in widely variable mycelium stoichiometry (Danger & Chauvet, 2013; Danger et al., 2016). This elemental plasticity is especially important for P, fungal biomass C:N ratios being far much more constrained than C:P ones (Gulis et al., 2017). Results are more contrasted for bacteria, the degree of homeostasis varies between bacterial strains (Chrzanowski & Kyle, 1996; Godwin & Cotner, 2014). However, when considering microbial communities instead of individual strains, stoichiometric flexibility of microbial biomass seems to be the rule (Danger et al., 2008). Despite this flexibility, microbial decomposers communities could present optimal nutrient ratios, i.e., ratios of nutrients in their environment that maximize their growth and activity (Güsewell & Gessner, 2009). These optimal ratios may vary with nutrient quantity, and differ between bacteria and fungi, fungi generally exhibiting higher C:nutrient ratios and more variable nutrient requirements than bacteria (Danger et al., 2016, Fig. 3.2). Such stoichiometric differences between both decomposers groups, in addition to different enzymatic capabilities and antibiotic substance production, might help to explain the dominant role of fungi over bacteria during the decomposition of extremely high C:nutrient ratio substrates such as leaf litters (Danger et al., 2016).

While having flexible elemental composition might be essential for microbial decomposers to decompose high C:nutrient substrates, it remains that microorganisms decomposing leaf litters and dead wood must face extremely high—certainly amongst the most important on Earth—elemental imbalances (Cross et al., 2005). From a stoichiometric viewpoint, three main, non-exclusive mechanisms might help decomposers to cope with these large imbalances (Fig. 3.4).

First, microorganisms can invest an important part of their energy for producing efficient (N-rich) enzymes aimed at recovering the rare nutrients present in detrital resources. This is especially true for microorganisms facing P-limiting conditions, where available N and energy can be invested in the production of large amounts of phosphatase (Clivot et al., 2013). Second, microorganisms can strongly reduce their Carbon-Use Efficiency (CUE, i.e., the ratio of microbial growth to total microbial carbon assimilation; see Sinsabaugh et al., 2013) and eliminate a part of C in excess through higher carbon release via respiratory processes (Manzoni et al., 2012). Finally, microorganisms can largely use dissolved inorganic nutrients from their environment to balance their stoichiometric requirements (Cheever et al., 2012; Howarth & Fisher, 1976). This process, referred as microbial nutrient immobilization, will largely participate to commonly observed patterns of nutrient enrichment of leaf litters during decomposition (Howarth & Fisher, 1976). Inorganic nitrogen

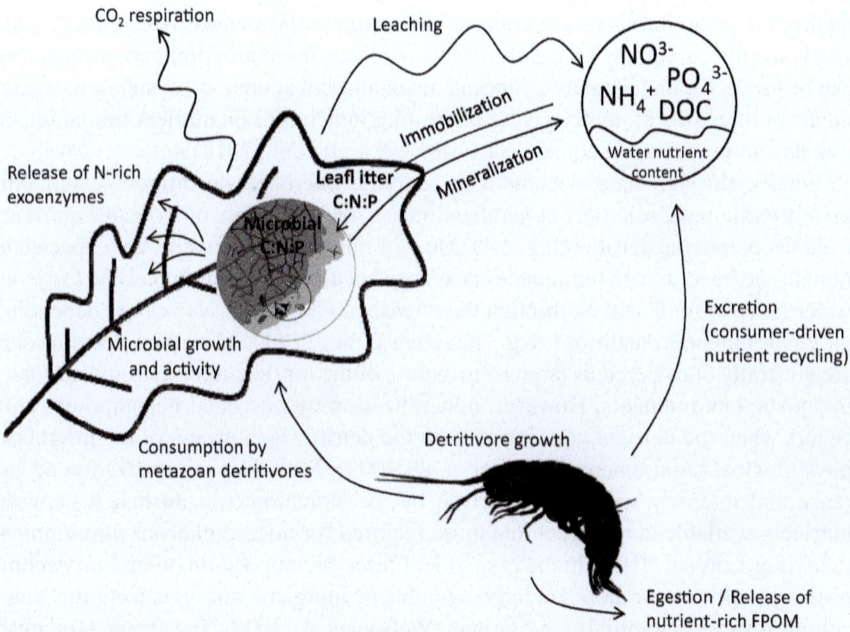

Fig. 3.4 Roles of microbial decomposers and metazoan detritivores on leaf litter stoichiometry. Leaching first leads to the release of soluble carbon and nutrients in water. Microbial decomposers being non-homeostatic, they first immobilize nutrients from the water column to balance their stoichiometric requirements before mineralizing leaf litter (i.e., producing a net release of mineral nutrients). During their growth, exoenzymes released by decomposers can lead to large losses of N-rich compounds. High respiration rates and CO_2 production can help microorganisms to balance their stoichiometric requirements. In anoxic conditions, some C and N might also be lost as CH_4 or N_2/N_2O. Finally, metazoan detritivores consume leaf litter more or less conditioned by microorganisms. Detritivores can be highly selective, feeding on nutrient-rich patches on leaf litter, permitting to balance their nutrient demand. A part of these nutrients will be retained in detritivores biomass, and consumers' metabolism, physiology, and stoichiometry will ultimately control the amount and stoichiometry of nutrients released under the form of dissolved nutrients or *faeces*. Excreted nutrients produced by detritivores can represent a large part of the total leaf litter remineralization process (a process commonly called the 'consumer-driven nutrient recycling'). The repackaging of leaf litter nutrients in fecal pellets can delay the remineralization of nutrients by detritivores, and the balance between excretion *vs.* egestion processes might play an important but understudied role in aquatic ecosystems functioning (see text for details and references)

immobilization from water has been suggested to participate to the increase of fungal biomass while P would be controlled less homeostatically and stored in microbial biomass, leading to large variations in the P content of leaf litter decomposers (microbial C:N remaining quite constant when compared to C:P; Gulis et al., 2017; Manning et al., 2015). Large microbial immobilization capabilities permit to understand the general stimulatory effect of nutrient availability on litter decomposition (Ferreira et al., 2015; Gulis & Suberkropp, 2003). This immobilization also leads to reductions in litter C:nutrient ratios, which will then represent resources of higher stoichiometric quality for detritivorous invertebrates (Cross et al., 2003; Danger, Arce Funck, et al., 2013; see the paragraph 3.4 below). Also, note that these immobilization capacities can be useful for understanding nutrient availability in aquatic ecosystems, measurements of litter stoichiometry giving interesting indications on nutrient limitations or imbalances occurring in aquatic ecosystems (Farell et al., 2018).

Finally, through their decomposing activity, microbial decomposers will ultimately participate to detritus mineralization, i.e., the net release of inorganic nutrients from decomposing detritus (Fig. 3.4). Most of mineralized elements are expected to remain and be re-used in the aquatic ecosystem but a part might ultimately be lost from water (at least for C and N, through the release of CO_2, CH_4, N_2 or N_2O depending on environmental conditions, e.g., Beaulieu et al., 2011). Microbial decomposers are generally considered as large contributors of the nutrient release from leaf litters in aquatic environments. However, mineralization by microbial decomposers only occurs when the nutrient concentration in the detritus is in excess of the microbes' physiological requirements (Webster et al., 2009). Microbial mineralization occurrence and intensity will thus be driven by the stoichiometric mismatch between nutrients available in resources and those required for microorganisms development (Chérif & Loreau, 2013). In the case of leaf litter decomposition, microbial decomposers should first immobilize large amounts of inorganic nutrients from the water column during their initial active growth (Webster et al., 2009). The demand for nutrients should then peak and decline with microbial biomass when most degradable parts of the litter have been colonized. Net nutrient release from decomposing leaf litter should thus occur at a point during decomposition when the stoichiometric demand for microbial growth has been fulfilled (Cheever et al., 2012; Webster et al., 2009). Nutrient availability and stoichiometry in both detrital resources and water are thus supposed to be important drivers of leaf litter decomposition, also controlling the shift from net immobilization to net mineralization (Güsewell & Gessner, 2009). While such simple stoichiometric models are attractive, they still lack large experimental assessments. Reality might appear as more complex in particular since microbial decomposers are not homeostatic, and since microbial communities can be highly diversified, including numerous taxa (even primary producers such as benthic algae) with diverse biological and ecological traits. In particular, predictions of simple stoichiometric models might be largely complexified by community dynamics arising during leaf litter decomposition, species replacements commonly occurring and reducing the stoichiometric imbalance between resources and microbial requirements (Danger et al., 2008; Kaiser et al., 2014).

3.4 Stoichiometry of Metazoan Detritivores

In many aquatic ecosystems, most litter resources are considered as becoming available for metazoan detritivores consumption when conditioned by microbial decomposers (Gessner et al., 1999; Fig. 3.3c). This microbial conditioning is generally accompanied by an increase in litter stoichiometric quality (see the above Sect. 3.3).

In contrast with microbial decomposers, metazoans are generally assumed to be quite homeostatic in comparison with their detrital resources (Cross et al., 2005; Evans-White et al., 2005). This means that they are able to maintain a quite stable elemental composition, even when the nutrient concentrations in their food are variable. This has been confirmed for some detritivorous invertebrate species, both in laboratory experiments (Danger, Arce Funck, et al., 2013) and in the field (Evans-White et al., 2005). However, several studies questioned this principle, and some observations revealed deviations from a strict homeostasis (Cross et al., 2003; Small & Pringle, 2010). Other influencing factors on stoichiometry have indeed been evidenced such as sex, body size or ontogeny (Halvorson et al., 2015). In all cases, phylogenetic signals remain important in determining invertebrate stoichiometry (Evans-White et al., 2005; Gonzalez et al., 2018), and invertebrate taxa elemental compositions are on average far less variable than those of their resources.

To date, elemental imbalance between resource stoichiometry and consumer requirements has generally been simply evaluated as the difference between consumers' body elemental ratios and those from resources (e.g., Cross et al., 2005, Fig. 3.2). Yet, by doing this, the metabolic and biochemical costs necessary for a consumer to process and assimilate the diet are totally omitted (Danger, 2020; Frost et al., 2006). In particular, carbon loss due to the energetic costs of nutrient assimilation is neglected. Stoichiometric requirements of a metazoan (also called Threshold Elemental Ratios, TER) can be either evaluated from mathematical models incorporating carbon and nutrient assimilation efficiencies, elemental ingestion rates and mass-specific respiration rates (for details on TERs calculation, see Frost et al., 2006), or be measured experimentally using controlled resources (Khattak et al., 2018; Ruiz et al., 2020). Nevertheless, even if true consumers' stoichiometric requirements differ from consumers' biomass elemental composition, these two parameters generally remain quite proportional (Frost et al., 2006). In addition to the difficulty of evaluating consumers' stoichiometric requirements, it is often hard to evaluate the stoichiometric quality of detritivores' resources. These organisms are indeed far more selective than what is generally thought, most detritivorous invertebrates select for example the highest quality patches on leaf litter surface (Lauridsen et al., 2014). Measuring the elemental composition of bulk leaf litter might thus sometimes give erroneous information on actual elemental imbalances undergone by detritivores.

Since leaf litters generally have extremely high C:nutrient ratios, especially the least microbially-conditioned ones, stoichiometric constraints are expected to be extremely important for detritivores (Martinson et al., 2008). Using leaf litters controlled for their elemental contents (using the large P-immobilization capacities of microbial decomposers), Danger, Arce Funck, et al. (2013) showed that lower detrital C:P ratios strongly increased the survival and growth of an invertebrate detritivore (the freshwater crustacean, *Gammarus fossarum*). Lower stoichiometric constraints were also associated with significant improvements of invertebrates' locomotor activity, energetic status, and reproduction (Arce-Funck et al., 2018; Rollin et al., 2018). Such findings, already found for herbivorous species, are consistent with the Growth Rate Hypothesis (GRH), initially proposed by Elser et al. (1996). High growth rates are associated with high rates of protein synthesis, which require high amounts of P-rich rRNA, and rRNA often represents a major part of the entire cellular P content (Elser et al., 2003). The resulting demand for P of fast-growing organisms are generally displayed in the elemental composition of the body tissue of consumers. Therefore, organisms with high growth rates (i.e., a high demand for P) generally exhibit low C:P and N:P ratios, while organisms that grow at slower rates which require less phosphorus have higher C:P and N:P body stoichiometry (e.g., Main et al., 1997).

At the community level, large elemental imbalances between detritivores' requirements and their detrital resources and their negative impacts on consumers' life history traits should have important consequences (Fig. 3.3d). In particular, it might be expected that taxa exhibiting the largest elemental imbalances will be replaced by taxa that are better adapted to the stoichiometric conditions. For example, one could expect that the least homeostatic taxa (i.e., the taxa exhibiting the highest elemental composition plasticity) would be less impacted by changes in nutrient availabilities. Cross et al. (2003) showed for example that some taxa were more flexible in terms of elemental stoichiometry than others, representing a potential mechanism reducing the intensity of stoichiometric constraints undergone by these taxa. One could also expect that ecosystems with the highest P concentrations should select for taxa that have the highest P contents and the lowest body C:P ratios. In the past two decades, a few experimental approaches were developed to test this hypothesis, using either *in situ* gradients of nutrients or ecosystem-level artificial nutrient enrichments. Due to nutrient immobilization by microbial decomposers (see the above Sect. 3.3), higher nutrient levels or nutrient enrichments in water are expected to decrease leaf litter C:P and C:N ratios and thereby increase its quality for detritivorous invertebrates (e.g., Cross et al., 2003). Several studies succeeded in showing an impact of resources C:P ratios on detritivores community structure and the selection of P-rich taxa (Dang et al., 2009; Evans-White et al., 2009; Prater et al., 2015; Singer & Battin, 2007) while other studies failed (e.g., Demi et al., 2019). In most studies, shifts in community composition were accompanied by shifts in other macroinvertebrate traits. For example, Cross et al. (2006) and Prater et al. (2015) showed that in nutrient rich conditions, small and fast-growing taxa were favored and dominated the community, while under nutrient-poor conditions, taxa were larger and expressed slower growth rates. These shifts follow the assumptions of the GRH, as explained above. Up to now, it is unclear how detritivores' body stoichiometry is related to other invertebrates' functional traits. Some studies already reported differences in body stoichiometry between different invertebrates' feeding-groups (Cross et al., 2003; Evans-White et al., 2005). Developing trait-based approaches on detritivorous invertebrates incorporating stoichiometric traits would certainly help to disentangle the importance of stoichiometric traits relative to other functional traits (Meunier et al., 2017).

3.5 Stoichiometry for Linking Organisms Requirements to Freshwater Ecosystems Functioning

How and to what extent stoichiometric changes in detritus, decomposers and detritivores affect the overall ecosystem functioning represent an important and partly unsolved scientific question (Woodward, 2009). The conceptual framework of ecological stoichiometry could help predict some ecological consequences of stoichiometric imbalances at the ecosystem scale (Fig. 3.4). Stoichiometric models could in particular help to understand and to predict the role of leaf litters microbial decomposers in nutrient mineralization (e.g., Daufresne & Loreau, 2001; Manzoni et al., 2010), i.e., the net release of inorganic nutrients that could then be rendered available in ecosystems (see the above Sect. 3.3 for more details). Similarly a release of stoichiometric constraints for detritivorous invertebrate, potentially associated with higher detritivores' feeding rates (Cornut et al., 2015; Flores et al., 2014), will ultimately lead to faster leaf litter mass loss and to increased detritivores' production (Cross et al., 2006; Demi et al., 2019; Greenwood et al., 2007). Such increases in prey production could in turn lead to significant increases in predators' production and deeply modify food web structures, as already shown after experimental stream nutrient enrichments (Cross et al., 2006; Demi et al., 2018; Greenwood et al., 2007). However, this response seems to partly depend on prey species: when increases in detritivores production are associated with large shifts in detritivores biological traits (e.g., changes in prey body size; Davis et al., 2010), detritivores might ultimately be outside the range of suitable/ optimal prey for prevailing predators. In this case, response of higher trophic levels might be delayed or decoupled from the stoichiometry-induced increased energy flow through the food web.

Another important input of ecological stoichiometry for understanding the impact of leaf litter stoichiometry on ecosystem functioning concerns the so-called Consumer-Driven Nutrient Recycling (Atkinson et al., 2017; Elser & Urabe, 1999). As discussed earlier, since microbial decomposers first immobilize nutrients from the water column to fulfill their stoichiometric requirements during leaf litter decomposition, nutrient mineralization often occurs in the latest stages of microbial decomposition. In numerous ecosystems, leaf litters are consumed by detritivores when microbial decomposers biomass peak in the decomposing material, i.e., before or at the beginning of net nutrient release by microorganisms. Thus, detritivorous organisms, through their nutrient excretion, often play a major role in nutrient release (Fuller et al., 2015; Halvorson et al., 2015). Several studies indicated that the stoichiometry of nutrients excreted was related to the elemental composition of the consumers (e.g., Vanni, 2002). However, if stoichiometric imbalance between detritivores and detrital resources plays a role in the nature and quantity of nutrient released, consumers metabolism and body mass also represent central parameters controlling consumer-driven nutrient recycling (Alves et al., 2010; Vanni & McIntyre, 2016).

Leaf litter consumers not only release inorganic nutrient in their environment but also produce some fine detrital particles, store nutrients in their biomass—depending on their elemental composition—and transform detrital organic matter through their digestive activity (Halvorson et al., 2015; Figs. 3.3e, 3.4). Therefore, in addition to nutrient excretion, these processes also play key roles in ecosystems functioning. First, nutrients bound in leaf litter consumers biomass can represent important nutrient sinks that can then be consumed by predators. Also, in the case of detritivorous insects, a part of the nutrients initially stored in their biomass can be exported in adjacent terrestrial ecosystems through their emergence from the stream (e.g., Grimm, 1988). Such nutrient transfers might be important for terrestrial ecosystems functioning, as already pointed out for energy fluxes (Nakano & Murakami, 2001). Finally, during their feeding activity, detritivores eating leaf litters will release large amounts of fine organic matter particles (up to 80% of detritivore mediated decomposition; López-Rojo et al., 2018), either under the form of non-ingested fragmented parts of leaf litter, or as non-assimilated organic matter into *faeces* (also called egesta). Nutrients released as fecal particles instead of excreted, dissolved nutrients, might in particular represent a major part of the nutrients released by detritivores in their environment (Grimm, 1988; Halvorson et al., 2017). This "nutrient repackaging" induced by leaf litter consumers could lead to the production of large, stable sinks of nutrients in aquatic ecosystems (Halvorson et al., 2017). The ratio of egested to excreted nutrients, which largely depends on detritivores' physiology and stoichiometric requirements, represent an understudied but essential parameter explaining detritivores impacts on ecosystems functioning.

3.6 Conclusions and Main Perspectives of Research

Still in development, the conceptual bases of ecological stoichiometry offers an interesting frame for the study of litter decomposition. From the understanding of microbial processes to the construction of ecosystem nutrient budgets, several studies have proven the great interest of this approach. Litters being most often particularly nutrient-depleted, stoichiometric imbalances are expected to play central roles in numerous processes related to litter decomposition. The above, non-exhaustive review of studies dealing with stoichiometric controls of leaf litter decomposition and their consequences at the individual, community, and ecosystem levels clearly underlines the great scientific potential of such approaches. It should also not hide that not all in litter decomposition can be predicted by this quite simple approach, stoichiometric approaches having their own limits. Below are listed some important perspectives of research that might deserve more attention in the future, as well as some complementary approaches that might help to refine or to broaden ecological stoichiometry results.

3.6.1 Complementary Nutritional Constraints for Litter Decomposition

As described above, ecological stoichiometry is mainly a bottom-up approach trying to relate nutritional constraints to their consequences at different levels of biological organization, from individuals to communities and ecosystems. Nevertheless, other nutritional limitations might occur in aquatic ecosystems, these ones potentially interacting or interfering with stoichiometric constraints. Among the long list of nutritional constraints that have been discussed for litter decomposition, early studies dealt with leaf litter toughness and microbial conditioning (Barlocher & Kendrich, 1975; Kaushik & Hynes, 1971). Thick and hard tissues of freshly fallen leaf litter have been proposed as being hardly consumable by most detritivores. Through their enzymatic activity, microbial decomposers strongly change litter biochemical composition and decompose a part of refractory carbon compounds (Suberkropp et al., 1976), leading to softer leaf litter tissues and higher litter consumption rates by detritivores (Kaushik & Hynes, 1971). Such changes are generally accompanied by large changes in leaf litter palatability for detritivores, some aquatic fungi making leaf litter more attractive—or sometimes more repellent—to detritivorous invertebrates (Suberkropp et al., 1983). More recent works also showed that leaf litter conditioning leads to important shifts in litter lipidic profiles, increasing for example the abundance of some sterols and some long chain polyunsaturated fatty acids (Arce-Funck et al., 2015; Torres-Ruiz & Wehr, 2010). All these changes in leaf litter quality co-varying with stoichiometric indices, it is sometimes hard to disentangle the effects of stoichiometry and those from other nutritional constraints. One way to limit the occurrence of such confounding factors is to experimentally control leaf litter stoichiometry by acting on nutrient immobilization by leaf litter decomposers during short-term nutrient pulses (Danger, Arce Funck, et al., 2013). Another potentially important research avenue would be to investigate the co-limitations probably occurring during leaf litter decomposition, by using gradient approaches (as already suggested for herbivores: Danger, 2020; Sperfeld et al., 2016). Finally, it must also be noted that contrary to detritivores that are quite well characterized in terms of nutritional requirements, far less is known for microbial decomposers. Some studies evidenced that microbial decomposers might have "optimal N:P ratios" maximizing their decomposing activity (Güsewell & Gessner, 2009). One could expect that microorganisms activity is also potentially co-limited by other micronutrients (García-Palacios et al., 2016; Powers & Salute, 2011) or biochemical compounds lacking in their environment, as suggested for example by the observed positive effects of phytosterol additions on aquatic fungi growth in laboratory cultures (Gessner & Chauvet, 1993). However, such investigations remain rare for microorganisms, with most approaches being generally restricted to carbon degradability and N and/or P availability.

3.6.2 Stoichiometric Interactions with Other Organisms

Leaf litter decomposition studies in aquatic ecosystems have been generally restricted to the evaluation of the respective roles of microbial decomposers and metazoan detritivores (Gessner et al., 1999). However, detritus-based food webs might be far more complex than simply considering these two groups of organisms (Fig. 3.3b). Among the far less studied organisms, protozoans have been shown as playing non-negligible roles in the fate of leaf litter in aquatic ecosystems, most often stimulating microbial activity and leaf litter decomposition (Ribblett et al., 2005; Risse-Buhl et al., 2012). Microalgae developing on leaf litter surface, through their release of labile carbon exudates, have also been shown to affect leaf litter decomposition. Depending on environmental conditions, such as light intensity or nutrient levels, the present algae might stimulate (Danger, Cornut, et al., 2013; Pope et al., 2020) or reduce microbial activity (Halvorson, Barry, et al., 2019, see Halvorson, Francoeur, et al., 2019 for a meta-analysis). These observations might be the results of complex priming effects, i.e., release of energetic limitations and/or shifts in C-sources utilization following C release by living primary producers that may also be partly controlled by stoichiometric constraints (Guenet et al., 2010). In addition to bringing essential fatty acids for detritivores (Crenier et al., 2017), microalgae development might thus greatly influence leaf litter decomposition. Brown and green food webs thus largely interact and ultimately influence stoichiometric processes and litter decomposition (Evans-White & Halvorson, 2017). The presence of top-predators might also directly, through their top-down control (Jabiol et al., 2014), or indirectly, through changes in detritivores behavior and elemental requirements (Hawlena et al., 2012), alter the stoichiometric balances at play during litter decomposition. Finally, a last illustration of the importance of taking into account the whole complexity of aquatic food webs for leaf litter decomposition concerns the observed stimulatory effect of flat worms on leaf litter decomposition (Majdi et al., 2014). Flatworms were indeed shown to accelerate litter decomposition through positive effects on microbial decomposers. Developing stoichiometrically explicit models of aquatic food webs, including their whole complexity, might thus represent an important research perspective for a deeper understanding of litter decomposition. This would also certainly represent an important pre-requisite for understanding ecological services rendered by detritus-based headwater streams to adjacent ecosystems and downstream habitats (e.g., Piccolo & Wipfli, 2002).

3.6.3 Stoichiometry of Litter Decomposition in a Changing World

In addition to its fundamental interest, the study of the different stoichiometric aspects of leaf litter decomposition appears as especially important when considering the emergency of predicting ecosystem responses to ongoing global changes. In particular, human-induced alterations of nutrient biogeochemical cycles tend to totally imbalance global nutrient stoichiometry (Elser et al., 2009; Penuelas et al., 2013). Such global changes in the relative abilities of nutrients in ecosystems, in association with other parameters of global changes, might change ecosystem processes in very complex ways (Cross et al., 2015). In particular, current temperature increase in ecosystems might change consumers' stoichiometric requirements concomitantly with changes in nutrient availabilities, rendering predictions far more complex than when considering parameters of global changes independently. For example, recent theoretical developments showed that ectotherm metazoans see their optimal C:P ratio decrease then increase with temperature due to partly decoupled stimulations of animals metabolism and growth (Ruiz et al., 2020). Predictions for microbial decomposers are much more complex since results seems to depend on the degree of control exerted on microbial decomposers (e.g., totally controlled micrococosms, Ferreira & Chauvet, 2011a *vs.* semi-controlled mescocosms conditions, Gossiaux et al., 2020). Understanding and predicting nutrient requirements of microbial decomposers seems thus to represent an important perspective of work. In addition, increases in CO_2 concentrations might reduce leaf litter stoichiometric quality but also chemical quality for leaf litter decomposers (Tuchman et al., 2002), even if temperature increase might sometimes play a greater role on litter decomposition than CO_2-induced litter quality changes (Ferreira & Chauvet, 2011b). Pollutants can also reduce litter decomposition through their impacts on microbial decomposers (Arce-Funck, Clivot, et al., 2013), metazoan detritivores (life history traits, feeding activity, physiology: Arce-Funck, Clivot, et al., 2013; Arce-Funck, Danger, et al., 2013; Arce-Funck et al., 2018; Felten et al., 2008) and food quality (Arce-Funck, Danger, et al., 2013), selecting the traits of the tolerant species (including stoichiometric traits). Ecological stoichiometry offers the conceptual basis for understanding the consequences of the impacts of pollutants (leaves decomposition, organisms, community) on food webs, biogeochemical cycles (stocks and fluxes), and ecosystem functioning (Danger & Maunoury-Danger, 2013).

3.6.4 More Conceptualization to Disentangle Stoichiometric Controls and Other Mechanisms at Play

As a general conclusion of this chapter, it must be reminded that litter decomposition in aquatic ecosystems is definitely a multifactorial process, and that stoichiometric constraints represent one factor, among others, which can control litter decomposition. As already underlined for ecological processes occurring at the herbivores-primary producers interface, our predictions of litter decomposition might greatly gain from merging the different conceptual models that are currently available. For example, Sperfeld et al. (2017) proposed to relate the nutritional geometry and the ecological stoichiometry frameworks, since both approaches appear as largely complementary. The first approach generally considers the differently sized categories of food components such as macronutrients (proteins, carbohydrates and lipids) and the consequences of their imbalance on consumers' behaviors or fitness (Simpson & Raubenheimer, 2012). The second approach, ecological stoichiometry, more directly relates nutritional constraints to ecosystem functioning. Bridging these frameworks together could permit making ecological stoichiometry more accurate by considering the true nature of molecules behind their C, N, and P content, while nutritional geometry studies could more closely relate to ecosystem processes. Another promising approach would consist in merging the metabolic theory of ecology and ecological stoichiometry. This might indeed help understanding the carbon metabolism and nutrient requirements of both microbial decomposers and metazoan detritivores, permitting in turn to predict community structures and ecosystem processes (Ott et al., 2014; Ruiz et al., 2020). This might be particularly valuable when considering changes in organisms' sizes or responses to temperature increases. It has also been recently proposed to extend the niche concept to stoichiometric niches (González et al., 2017). Such an approach permits to easily visualize elemental composition and variability of different taxa, helping for example to anticipate some phylogenetic differences between consumer requirements. Finally, the development of a trait-based approach of ecological stoichiometry (sensu Meunier et al., 2017) would certainly be useful for predicting microbial decomposers and detritivores community structures in response to litter stoichiometry. This could also allow to more closely relate stoichiometric traits of taxa with other functional traits, and ultimately broaden our understanding of functional consequences of species shifts in ecosystems. This list is obviously non-exhaustive, but should definitely help to deepen our understanding of litter decomposition and permit to predict its underlying causes and its consequences for aquatic ecosystems.

References

Ågren, G. I. (2008). Stoichiometry and nutrition of plant growth in natural communities. *Annual Review of Ecology Evolution, and Systematics, 39*, 153–170. https://doi.org/10.1146/annurev.ecolsys.39.110707.173515.

Alves, J. M., Caliman, A., Guariento, R. D., Figueiredo-Barros, M. P., Carneiro, L. S., Farjalla, V. F., Bozelli, R. L., & Esteves, F. A. (2010). Stoichiometry of benthic invertebrate nutrient recycling: Interspecific variation and the role of body mass. *Aquatic Ecology, 44*(2), 421–430. https://doi.org/10.1007/s10452-009-9302-3.

Arce-Funck, J., Bec, A., Felten, V., Perrière, F., & Danger, M. (2015). Biochemical composition of aquatic hyphomycetes: A potential source of polyunsaturated fatty acids in detritus based streams ? *Fungal Ecology, 13*, 205–210. https://doi.org/10.1016/j.funeco.2014.09.004.

Arce-Funck, J., Clivot, H., Felten, V., Rousselle, P., Guérold, F., & Danger, M. (2013). Phosphorus availability modulates the toxic effect of silver on aquatic fungi and leaf litter decomposition. *Aquatic Toxicology, 144–145*, 199–207. https://doi.org/10.1016/j.aquatox.2013.10.001.

Arce-Funck, J., Crenier, C., Danger, M., Billoir, E., Usseglio-Polatera, P., & Felten, V. (2018). High stoichiometric food quality increases moulting organism vulnerability to pollutant impacts:

An experimental test with *Gammarus fossarum* (Crustacea: Amphipoda). *Science of the Total Environment, 645*, 1484–1495. https://doi.org/10.1016/j.scitotenv.2018.07.227.

Arce-Funck, J., Danger, M., Gismondi, E., Cossu-Leguille, C., Guerold, F., & Felten, V. (2013). Behavioural and physiological responses of *Gammarus fossarum* (Crustacea Amphipoda) exposed to silver. *Aquatic Toxicology, 142–143*, 73–84. https://doi.org/10.1016/j.aquatox.2013.07.012.

Atkinson, C. L., Capps, K. A., Rugenski, A. T., & Vanni, M. J. (2017). Consumer-driven nutrient dynamics in freshwater ecosystems: From individuals to ecosystems. *Biological Reviews, 92*(4), 2003–2023. https://doi.org/10.1111/brv.12318.

Baldy, V., Gessner, M. O., & Chauvet, E. (1995). Bacteria, fungi and the breakdown of leaf litter in a large river. *Oikos, 74*(1), 93–102. https://doi.org/10.2307/3545678.

Bärlocher, F., & Kendrich, B. (1975). Assimilation efficiency of *Gammarus pseudolimnaeus* (Amphipoda) feeding on fungal mycelium or autumn-shed leaves. *Oikos, 26*, 55–59. https://doi.org/10.2307/3543277.

Biasi, C., Graça, M. A., Santos, S., & Ferreira, V. (2017). Nutrient enrichment in water more than in leaves affects aquatic microbial litter processing. *Oecologia, 184*(2), 555–568. https://doi.org/10.1007/s00442-017-3869-5.

Beaulieu, J. J., Tank, J. L., Hamilton, S. K., Wollheim, W. M., Hall, R. O., Mulholland, P. J., Peterson, B. J., Ashkenas, L. R., Cooper, L. W., Dahm, C. N., Dodds, W. K., Grimm, N. B., Johnson, S. L., McDowell, W. H., Poole, G. C., Valett, H. M., Arango, C. P., Bernot, M. J., Burgin, A. J., ... Thomas, S. M. (2011). Nitrous oxide emission from denitrification in stream and river networks. *Proceedings of the National Academy of Sciences, 108*(1), 214–219. https://doi.org/10.1073/pnas.1011464108.

Boyero, L., Graça, M. A., Tonin, A. M., Pérez, J., Swafford, A. J., Ferreira, V., ... & Pearson, R. G. (2017). Riparian plant litter quality increases with latitude. *Scientific Reports, 7*(1), 1–10.

Boyero, L., Pearson, R. G., Hui, C., Gessner, M. O., Pérez, J., Alexandrou, M. A., Graça, M. A. S., Cardinale, B. J., Albariño, R. J., Arunachalam, M., Barmuta, L. A., Boulton, A. J., Bruder, A., Callisto, M., Chauvet, E., Death, R. G., Dudgeon, D., Encalada, A. C., Ferreira, V., … Yule, C. M. (2016). Biotic and abiotic variables influencing plant litter breakdown in streams: A global study. *Proceedings of the Royal Society B: Biological Sciences, 283*(1829), 20152664. https://doi.org/10.1098/rspb.2015.2664.

Cebrian, J., Shurin, J. B., Borer, E. T., Cardinale, B. J., Ngai, J. T., Smith, M. D., & Fagan, W. F. (2009). Producer nutritional quality controls ecosystem trophic structure. *PLoS ONE, 4*(3), e4929. https://doi.org/10.1371/journal.pone.0004929.

Chapin, F. S., III., & Kedrowski, R. A. (1983). Seasonal changes in nitrogen and phosphorus fractions and autumn retranslocation in evergreen and deciduous taiga trees. *Ecology, 64*, 376–391. https://doi.org/10.2307/1937083.

Cheever, B. M., Kratzer, E. B., & Webster, J. R. (2012). Immobilization and mineralization of N and P by heterotrophic microbes during leaf decomposition. *Freshwater Science, 31*, 133–147. https://doi.org/10.1899/11-060.1.

Cherif, M., & Loreau, M. (2013). Plant–herbivore–decomposer stoichiometric mismatches and nutrient cycling in ecosystems. *Proceedings of the Royal Society B: Biological Sciences, 280*(1754), 20122453. https://doi.org/10.1098/rspb.2012.2453.

Chrzanowski, T. H., & Kyle, M. (1996). Ratios of carbon, nitrogen and phosphorus in *Pseudomonas fluorescens* as a model for bacterial element ratios and nutrient regeneration. *Aquatic Microbial Ecology, 10*, 115–122. https://doi.org/10.3354/ame010115.

Clivot, H., Danger, M., Pagnout, C., Wagner, P., Rousselle, P., Poupin, P., & Guérold, F. (2013). Impaired leaf litter processing in acidified streams. *Microbial Ecology, 65*(1), 1–11. https://doi.org/10.1007/s00248-012-0107-x.

Cornut, J., Ferreira, V., Gonçalves, A. L., Chauvet, E., & Canhoto, C. (2015). Fungal alteration of the elemental composition of leaf litter affects shredder feeding activity. *Freshwater Biology, 60*(9), 1755–1771. https://doi.org/10.1111/fwb.12606.

Crenier, C., Arce-Funck, J., Bec, A., Perrière, F., Leflaive, J., Guérold, F., Felten, V., & Danger, M. (2017). Minor food sources can be major drivers of secondary production in detritus-based ecosystems. *Freshwater Biology, 62*(7), 1155–1167. https://doi.org/10.1111/fwb.12933.

Cross, W. F., Benstead, J. P., Frost, P. C., & Thomas, S. A. (2005). Ecological stoichiometry in freshwater benthic systems: Recent progress and perspectives. *Freshwater Biology, 50*(11), 1895–1912. https://doi.org/10.1111/j.1365-2427.2005.01458.x.

Cross, W. F., Benstead, J. P., Rosemond, A. D., & Wallace, J. B. (2003). Consumer-resource stoichiometry in detritus-based streams. *Ecology Letters, 6*(8), 721–732. https://doi.org/10.1046/j.1461-0248.2003.00481.x.

Cross, W. F., Hood, J. M., Benstead, J. P., Huryn, A. D., & Nelson, D. (2015). Interactions between temperature and nutrients across levels of ecological organization. *Global Change Biology, 21*, 1025–1040. https://doi.org/10.1111/gcb.12809.

Cross, W. F., Wallace, J. B., Rosemond, A. D., & Eggert, S. L. (2006). Whole-system nutrient enrichment increases secondary production in a detritus-based ecosystem. *Ecology, 87*(6), 1556–1565. https://doi.org/10.1890/0012-9658(2006)87[1556:WNEISP]2.0.CO;2.

Dang, C. K., Harrison, S., Sturt, M. M., Giller, P. S., & Jansen, M. A. K. (2009). Is the elemental composition of stream invertebrates a determinant of tolerance to organic pollution? *Journal of the North American Benthological Society, 28*, 778–784. https://doi.org/10.1899/08-163.1.

Danger, M. (2020). Ecological stoichiometry in detritus-based ecosystems with a special focus on forested headwater streams. *Limnetica, 39*, 339–352. https://doi.org/10.23818/limn.39.22.

Danger, M., Arce Funck, J., Devin, S., Heberlé, J., & Felten, V. (2013). Phosphorus content in detritus controls life history traits of a detritivore. *Functional Ecology, 27*, 807–815. https://doi.org/10.1111/1365-2435.12079.

Danger, M., & Chauvet, E. (2013). Elemental composition and degree of homeostasis of fungi: Are aquatic hyphomycetes more like metazoans, bacteria, or plants? *Fungal Ecology, 6*, 453–457. https://doi.org/10.1016/j.funeco.2013.05.007.

Danger, M., Cornut, J., Chauvet, E., Chavez, P., Elger, A., & Lecerf, A. (2013). Benthic algae stimulate leaf litter decomposition in detritus-based headwater streams: A case of aquatic priming effect? *Ecology, 94*(7), 1604–1613. https://doi.org/10.1890/12-0606.1.

Danger, M., Daufresne, T., Lucas, F., Pissart, S., & Lacroix, G. (2008). Does Liebig's law of the minimum scale up from species to communities? *Oikos, 117*, 1741–1751. https://doi.org/10.1111/j.1600-0706.2008.16793.x.

Danger, M., Gessner, M. O., & Bärlocher, F. (2016). Ecological stoichiometry of aquatic fungi: Current knowledge and perspectives. *Fungal Ecology, 19*, 100–111. https://doi.org/10.1016/j.funeco.2015.09.004.

Danger, M., & Maunoury-Danger, F. (2013). Ecological Stoichiometry. In J.-F. Férard & C. Blaise (Eds.), *Encyclopedia of Aquatic Ecotoxicology* (pp. 317–326). Springer. https://doi.org/10.1007/978-94-007-5704-2.

Daufresne, T., & Loreau, M. (2001). Ecological stoichiometry, primary producer-decomposer interactions, and ecosystem persistence. *Ecology, 82*, 3069–3082. https://doi.org/10.1890/0012-9658(2001)082[3069:ESPPDI]2.0.CO;2.

Davis, J. M., Rosemond, A. D., Eggert, S. L., Cross, W. F., & Wallace, J. B. (2010). Long-term nutrient enrichment decouples predator and prey production. *Proceedings of the National Academy of Sciences, 107*, 121–126. https://doi.org/10.1073/pnas.0908497107.

del Campo, R., Martí, E., Bastias, E., Barberá, G. G., Sánchez-Montoya, M. M., & Gomez, R. (2020). Floodplain Preconditioning of Leaf Litter Modulates the Subsidy of Terrestrial C and Nutrients in Fluvial Ecosystems. *Ecosystems*. https://doi.org/10.1007/s10021-020-00508-5.

Demi, L. M., Benstead, J. P., Rosemond, A. D., & Maerz, J. C. (2018). Litter P content drives consumer production in detritus-based streams spanning an experimental N: P gradient. *Ecology, 99*(2), 347–359. https://doi.org/10.1002/ecy.2118.

Demi, L. M., Benstead, J. P., Rosemond, A. D., & Maerz, J. C. (2019). Experimental N and P additions alter stream macroinvertebrate community composition *via* taxon-level responses to shifts in detrital resource stoichiometry. *Functional Ecology, 33*(5), 855–867. https://doi.org/10.1111/1365-2435.13289.

Elser, J. J., Acharya, K., Kyle, M., Cotner, J., Makino, W., Markow, T., Watts, T., Hobbie, S., Fagan, W., Schade, J., Hood, J., & Sterner, R. W. (2003). Growth rate-stoichiometry couplings in diverse biota. *Ecology Letters, 6*, 936–943. https://doi.org/10.1046/j.1461-0248.2003.00518.x.

Elser, J. J., Andersen, T., Baron, J. S., Bergström, A. K., Jansson, M., Kyle, M., Nydick, K. R., Steger, L., & Hessen, D. O. (2009). Shifts in lake N: P stoichiometry and nutrient limitation driven by atmospheric nitrogen deposition. *Science, 326*(5954), 835–837. https://doi.org/10.1126/science.1176199.

Elser, J. J., Dobberfuhl, D., MacKay, N. A., & Schampel, J. H. (1996). Organism size, life history, and N: P stoichiometry: Towards a unified view of cellular and ecosystem processes. *BioScience, 46*, 674–684. https://doi.org/10.2307/1312897.

Elser, J. J., & Urabe, J. (1999). The stoichiometry of consumer-driven nutrient recycling: Theory, observations, and consequences. *Ecology, 80*(3), 735–751. https://doi.org/10.1890/0012-9658(1999)080[0735:TSOCDN]2.0.CO;2.

Enríquez, S. C. M. D., Duarte, C. M., & Sand-Jensen, K. A. J. (1993). Patterns in decomposition rates among photosynthetic organisms: The importance of detritus C: N: P content. *Oecologia, 94*(4), 457–471. https://doi.org/10.1007/BF00566960.

Evans-White, M. A., Dodds, W. K., Huggins, D. G., & Baker, D. S. (2009). Thresholds in macroinvertebrate biodiversity and stoichiometry across water-quality gradients in Central Plains (USA) streams. *Journal of the North American Benthological Society, 28*, 855–868. https://doi.org/10.1899/08-113.1.

Evans-White, M. A., & Halvorson, H. M. (2017). Comparing the ecological stoichiometry in green and brown food webs–a review and meta-analysis of freshwater food webs. *Frontiers in Microbiology, 8*, 1184. https://doi.org/10.3389/fmicb.2017.01184.

Evans-White, M. A., Stelzer, R. S., & Lamberti, G. A. (2005). Taxonomic and regional patterns in benthic macroinvertebrate elemental composition in streams. *Freshwater Biology, 50*, 1786–1799. https://doi.org/10.1111/j.1365-2427.2005.01455.x.

Farrell, K. J., Rosemond, A. D., Kominoski, J. S., Bonjour, S. M., Rüegg, J., Koenig, L. E., Baker, C. L., Trentman, M. T., Harms, T. K., & McDowell, W. H. (2018). Variation in detrital resource stoichiometry signals differential carbon to nutrient limitation for stream consumers across biomes. *Ecosystems, 21*(8), 1676–1691. https://doi.org/10.1007/s10021-018-0247-z.

Felten, V., Charmantier, G., Mons, R., Geffard, A., Rousselle, P., Coquery, M., Garric, J., & Geffard, O. (2008). Physiological and behavioural responses of *Gammarus pulex* (Crustacea Amphipoda) exposed to cadmium. *Aquatic Toxicology, 83*, 413–425. https://doi.org/10.1016/j.aquatox.2007.12.002.

Ferreira, V., & Chauvet, E. (2011a). Synergistic effects of water temperature and dissolved nutrients on litter decomposition and associated fungi. *Global Change Biology, 17*(1), 551–564. https://doi.org/10.1111/j.1365-2486.2010.02185.x.

Ferreira, V., & Chauvet, E. (2011b). Future increase in temperature more than decrease in litter quality can affect microbial litter decomposition in streams. *Oecologia, 167*(1), 279–291. https://doi.org/10.1007/s00442-011-1976-2.

Ferreira, V., Castagneyrol, B., Koricheva, J., Gulis, V., Chauvet, E., & Graça, M. A. (2015). A meta-analysis of the effects of nutrient enrichment on litter decomposition in streams. *Biological Reviews, 90*(3), 669–688. https://doi.org/10.1111/brv.12125.

Ferreira, V., Koricheva, J., Pozo, J., & Graça, M. A. (2016). A meta-analysis on the effects of changes in the composition of native forests on litter decomposition in streams. *Forest Ecology and Management, 364,* 27–38.

Fisher, S. G., & Likens, G. E. (1973). Energy flow in Bear Brook, New Hampshire: An integrative approach to stream ecosystem metabolism. *Ecological Monographs, 43*, 421–439. https://doi.org/10.2307/1942301.

Flores, L., Larrañaga, A., & Elosegi, A. (2014). Compensatory feeding of a stream detritivore alleviates the effects of poor food quality when enough food is supplied. *Freshwater Science, 33*(1), 134–141. https://doi.org/10.1086/674578.

Frost, P. C., Benstead, J. P., Cross, W. F., Hillebrand, H., Larson, J. H., Xenopoulos, M. A., & Yoshida, T. (2006). Threshold elemental ratios of carbon and phosphorus in aquatic consumers. *Ecology Letters, 9*(7), 774–779. https://doi.org/10.1111/j.1461-0248.2006.00919.x.

Fuller, C. L., Evans-White, M. A., & Entrekin, S. A. (2015). Growth and stoichiometry of a common aquatic detritivore respond to changes in resource stoichiometry. *Oecologia, 177*(3), 837–848. https://doi.org/10.1007/s00442-014-3154-9.

García-Palacios, P., McKie, B. G., Handa, I. T., Frainer, A., & Hättenschwiler, S. (2016). The importance of litter traits and decomposers for litter decomposition: A comparison of aquatic and terrestrial ecosystems within and across biomes. *Functional Ecology, 30*, 819–829. https://doi.org/10.1111/1365-2435.12589.

Gessner, M. O., & Chauvet, E. (1993). Ergosterol-to-biomass conversion factors for aquatic hyphomycetes. *Applied and Environmental Microbiology, 59*(2), 502–507. https://doi.org/10.1128/AEM.59.2.502-507.1993.

Gessner, M. O., Chauvet, E., & Dobson, M. (1999). A perspective on leaf litter breakdown in streams. *Oikos*, 377–384. https://doi.org/10.2307/3546505.

Giron, D., Huguet, E., Stone, G. N., & Body, M. (2016). Insect-induced effects on plants and possible effectors used by galling and leaf-mining insects to manipulate their host-plant. *Journal of Insect Physiology, 84*, 70–89. https://doi.org/10.1016/j.jinsphys.2015.12.009.

Godwin, C. M., & Cotner, J. B. (2014). Carbon: Phosphorus homeostasis of aquatic bacterial assemblages is mediated by shifts in assemblage composition. *Aquatic Microbial Ecology, 73*(3), 245–258. https://doi.org/10.3354/ame01719.

González, A. L., Céréghino, R., Dézerald, O., Farjalla, V. F., Leroy, C., Richardson, B. A., Richardson, M. J., Romero, G. Q., & Srivastava, D. S. (2018). Ecological mechanisms and phylogeny shape invertebrate stoichiometry: A test using detritus-based communities across Central and South America. *Functional Ecology, 32*(10), 2448–2463. https://doi.org/10.1111/1365-2435.13197.

González, A. L., Dézerald, O., Marquet, P. A., Romero, G. Q., & Srivastava, D. S. (2017). The multidimensional stoichiometric niche. *Frontiers in Ecology and Evolution, 5*, 110. https://doi.org/10.3389/fevo.2017.00110.

Gossiaux, A., Rollin, M., Guérold, F., Felten, V., Laviale, M., Bachelet, Q., Poupin, P., Chauvet, E., Bec, A., & Danger, M. (2020). Temperature and nutrient effects on the relative importance of brown and green pathways for stream ecosystem functioning: A mesocosm approach. *Freshwater Biology,*, 1–17. https://doi.org/10.1111/fwb.13474.

Graça, M. A. (2001). The role of invertebrates on leaf litter decomposition in streams–a review. *International Review of Hydrobiology, 86*(4–5), 383–393. https://doi.org/10.1002/1522-2632(200107)86:4/5%3c383::AID-IROH383%3e3.0.CO;2-D.

Graça, M. A. S., & Poquet, J. M. (2014). Do climate and soil influence phenotypic variability in leaf litter, microbial decomposition and shredder consumption? *Oecologia, 174*(3), 1021–1032. https://doi.org/10.1007/s00442-013-2825-2.

Greenwood, J. L., Rosemond, A. D., Wallace, J. B., Cross, W. F., & Weyers, H. S. (2007). Nutrients stimulate leaf breakdown rates and detritivore biomass: Bottom-up effects *via* heterotrophic pathways. *Oecologia, 151*(4), 637–649. https://doi.org/10.1007/s00442-006-0609-7.

Grimm, N. B. (1988). Role of macroinvertebrates in nitrogen dynamics of a desert stream. *Ecology, 69*(6), 1884–1893. https://doi.org/10.2307/1941165.

Grimmett, I. J., Smith, K. A., & Bärlocher, F. (2012). Tar-spot infection delays fungal colonization and decomposition of maple leaves. *Freshwater Science, 31*(4), 1088–1095. https://doi.org/10.1899/12-034.1.

Guenet, B., Danger, M., Abbadie, L., & Lacroix, G. (2010). Priming Effect: Bridging the gap between terrestrial and aquatic ecology. *Ecology, 9*(10), 2850–2861. https://doi.org/10.1890/09-1968.1.

Gulis, V., Kuehn, K. A., Schoettle, L. N., Leach, D., Benstead, J. P., & Rosemond, A. D. (2017). Changes in nutrient stoichiometry, elemental homeostasis and growth rate of aquatic litter-associated fungi in response to inorganic nutrient supply. *The ISME Journal, 11*(12), 2729. https://doi.org/10.1038/ismej.2017.123.

Gulis, V., & Suberkropp, K. (2003). Leaf litter decomposition and microbial activity in nutrient-enriched and unaltered reaches of a headwater stream. *Freshwater Biology, 48*(1), 123–134. https://doi.org/10.1046/j.1365-2427.2003.00985.x.

Güsewell, S., & Gessner, M. O. (2009). N: P ratios influence litter decomposition and colonization by fungi and bacteria in microcosms. *Functional Ecology, 23*, 211–219. https://doi.org/10.1111/j.1365-2435.2008.01478.x.

Halvorson, H. M., Barry, J. R., Lodato, M. B., Findlay, R. H., Francoeur, S. N., & Kuehn, K. A. (2019). Periphytic algae decouple fungal activity from leaf litter decomposition *via* negative priming. *Functional Ecology, 33*(1), 188–201. https://doi.org/10.1111/1365-2435.13235.

Halvorson, H. M., Francoeur, S. N., Findlay, R. H., & Kuehn, K. A. (2019). Algal-mediated priming effects on the ecological stoichiometry of leaf litter decomposition: A meta-analysis. *Frontiers in Earth Science, 7*, 76. https://doi.org/10.3389/feart.2019.00076.

Halvorson, H. M., Fuller, C., Entrekin, S. A., & Evans-White, M. A. (2015). Dietary influences on production, stoichiometry and decomposition of particulate wastes from shredders. *Freshwater Biology, 60*(3), 466–478. https://doi.org/10.1111/fwb.12462.

Halvorson, H. M., Hall, D. J., & Evans-White, M. A. (2017). Long-term stoichiometry and fates highlight animal egestion as nutrient repackaging, not recycling, in aquatic ecosystems. *Functional Ecology, 31*(9), 1802–1812. https://doi.org/10.1111/1365-2435.12875.

Hawlena, D., Strickland, M. S., Bradford, M. A., & Schmitz, O. J. (2012). Fear of predation slows plant-litter decomposition. *Science, 336*(6087), 1434–1438. https://doi.org/10.1126/science.1220097.

Hieber, M., & Gessner, M. O. (2002). Contribution of stream detrivores, fungi, and bacteria to leaf breakdown based on biomass estimates. *Ecology, 83*(4), 1026–1038. https://doi.org/10.1890/0012-9658(2002)083[1026:COSDFA]2.0.CO;2.

Hladyz, S., Gessner, M. O., Giller, P. S., Pozo, J., & Woodward, G. (2009). Resource quality and stoichiometric constraints on stream ecosystem functioning. *Freshwater Biology, 54*(5), 957–970. https://doi.org/10.1111/j.1365-2427.2008.02138.x.

Howarth, R. W., & Fisher, S. G. (1976). Carbon, nitrogen, and phosphorus dynamics during leaf decay in nutrient-enriched stream microecosystems. *Freshwater Biology, 6*(3), 221–228. https://doi.org/10.1111/j.1365-2427.1976.tb01608.x.

Jabiol, J., Cornut, J., Danger, M., Jouffroy, M., Elger, A., & Chauvet, A. (2014). Litter identity mediates predator impacts on the functioning of an aquatic detritus-based food web. *Oecologia, 176*, 225–235. https://doi.org/10.1007/s00442-014-2990-y.

Kaiser, C., Franklin, O., Dieckmann, U., & Richter, A. (2014). Microbial community dynamics alleviate stoichiometric constraints during litter decay. *Ecology Letters, 17*(6), 680–690. https://doi.org/10.1111/ele.12269.

Kaushik, N. K., & Hynes, H. B. N. (1971). The fate of the dead leaves that fall into streams. *Archiv für Hydrobiologie, 68*, 465–515. https://doi.org/10.1146/annurev-ecolsys-110218-024755.

Khattak, H. K., Prater, C., Wagner, N. D., & Frost, P. C. (2018). The threshold elemental ratio of carbon and phosphorus of *Daphnia magna* and its connection to animal growth. *Scientific Reports, 8*(1), 1–8. https://doi.org/10.1038/s41598-018-27758-7.

Kiffer, W. P., Jr., Mendes, F., Casotti, C. G., Costa, L. C., & Moretti, M. S. (2018). Exotic *Eucalyptus* leaves are preferred over tougher native species but affect the growth and survival of shredders in an Atlantic Forest stream (Brazil). *PLoS ONE, 13*(1), e0190743.

Killingbeck, K. T. (1996). Nutrients in senesced leaves: Keys to the search for potential resorption and resorption proficiency. *Ecology, 77*(6), 1716–1727. https://doi.org/10.2307/2265777.

Lauridsen, R. B., Edwards, F. K., Cross, W. F., Woodward, G., Hildrew, A. G., & Jones, J. I. (2014). Consequences of inferring diet from feeding guilds when estimating and interpreting consumer resource stoichiometry. *Freshwater Biology, 59*, 1497–1508. https://doi.org/10.1111/fwb.12361.

Lecerf, A., & Chauvet, E. (2008). Intraspecific variability in leaf traits strongly affects alder leaf decomposition in a stream. *Basic and Applied Ecology, 9*, 598–605. https://doi.org/10.1016/j.baae.2007.11.003.

Leroux, S. J., & Loreau, M. (2008). Subsidy hypothesis and strength of trophic cascades across ecosystems. *Ecology Letters, 11*(11), 1147–1156. https://doi.org/10.1111/j.1461-0248.2008.01235.x.

Lindeman, R. L. (1942). The trophic-dynamic aspect of ecology. *Ecology, 23*, 399–418. https://doi.org/10.1016/S0092-8240(05)80045-X.

López-Rojo, N., Martínez, A., Pérez, J., Basaguren, A., Pozo, J., & Boyero, L. (2018). Leaf traits drive plant diversity effects on litter decomposition and FPOM production in streams. *PLoS ONE, 13*(5), e0198243. https://doi.org/10.1371/journal.pone.0198243.

Lotka, A. J. (1925). *Elements of Physical Biology*. Williams and Wilkins.

Main, T., Dobberfuhl, D. R., & Elser, J. J. (1997). N: P stoichiometry and ontogeny in crustacean zooplankton: A test of the growth rate hypothesis. *Limnology and Oceanography, 42*, 1474–1478. https://doi.org/10.4319/lo.1997.42.6.1474.

Majdi, N., Boiché, A., Traunspurger, W., & Lecerf, A. (2014). Predator effects on a detritus-based food web are primarily mediated by non-trophic interactions. *Journal of Animal Ecology, 83*(4), 953–962. https://doi.org/10.1111/1365-2656.12189.

Manning, D. W., Rosemond, A. D., Kominoski, J. S., Gulis, V., Benstead, J. P., & Maerz, J. C. (2015). Detrital stoichiometry as a critical nexus for the effects of streamwater nutrients on leaf litter breakdown rates. *Ecology, 96*(8), 2214–2224. https://doi.org/10.1890/14-1582.1.

Manzoni, S., Taylor, P., Richter, A., Porporato, A., & Ågren, G. I. (2012). Environmental and stoichiometric controls on microbial carbon-use efficiency in soils. *New Phytologist, 196*(1), 79–91. https://doi.org/10.1111/j.1469-8137.2012.04225.x.

Manzoni, S., Trofymow, J. A., Jackson, R. B., & Porporato, A. (2010). Stoichiometric controls on carbon, nitrogen, and phosphorus dynamics in decomposing litter. *Ecological Monographs, 80*(1), 89–106. https://doi.org/10.1890/09-0179.1.

Martinson, H. M., Schneider, K., Gilbert, J., Hines, J. E., Hamback, P. A., & Fagan, W. F. (2008). Detritivory: Stoichiometry of a neglected trophic level. *Ecological Research, 23*, 487–491. https://doi.org/10.1007/s11284-008-0471-7.

Maunoury-Danger, F., Felten, V., Bojic, C., Fraysse, F., Ponce, M. C., Dedourge-Geffard, O., Geffard, A., Guérold, F., & Danger, M. (2018). Metal release from contaminated leaf litter and leachate toxicity for the freshwater crustacean *Gammarus fossarum*. *Environmental Science and Pollution Research, 25*(12), 11281–11294. https://doi.org/10.1007/s11356-017-9452-0.

McGroddy, M. E., Daufresne, T., & Hedin, L. O. (2004). Scaling of C: N: P stoichiometry in forests worldwide: Implications of terrestrial Redfield-type ratios. *Ecology, 85*(9), 2390–2401. https://doi.org/10.1890/03-0351.

Melillo, J. M., Aber, J. D., & Muratore, J. F. (1982). Nitrogen and lignin control of hardwood leaf litter decomposition dynamics. *Ecology, 63*(3), 621–626. https://doi.org/10.2307/1936780.

Meunier, C. L., Boersma, M., El-Sabaawi, R., Halvorson, H. M., Herstoff, E. M., Van de Waal, D. B., Vogt, R. J., & Litchman, E. (2017). From elements to function: Toward unifying ecological stoichiometry and trait-based ecology. *Frontiers in Environmental Science, 5*, 18. https://doi.org/10.3389/fenvs.2017.00018.

Meyer, J. L., Wallace, J. B., & Eggert, S. L. (1998). Leaf litter as a source of dissolved organic carbon in streams. *Ecosystems, 1*(3), 240–249. https://doi.org/10.1007/s100219900019.

Moe, S. J., Stelzer, R. S., Forman, M. R., Harpole, W. S., Daufresne, T., & Yoshida, T. (2005). Recent advances in ecological stoichiometry: Insights for population and community ecology. *Oikos, 109*, 29–39. https://doi.org/10.1111/j.0030-1299.2005.14056.x.

Moore, J. C., Berlow, E. L., Coleman, D. C., de Ruiter, P. C., Dong, Q., Hastings, A., Johnson, N. C., McCann, K. S., Melville, K., Morin, P. J., Nadelhoffer, K., Rosemond, A. D., Post, D. M., Sabo, J. L., Scow, K. M., Vanni, M. J., & Wall, D. H. (2004). Detritus, trophic dynamics and biodiversity. *Ecology Letters, 7*(7), 584–600. https://doi.org/10.1111/j.1461-0248.2004.00606.x.

Mouginot, C., Kawamura, R., Matulich, K. L., Berlemont, R., Allison, S. D., Amend, A. S., & Martiny, A. C. (2014). Elemental stoichiometry of Fungi and Bacteria strains from grassland leaf litter. *Soil Biology and Biochemistry, 76*, 278–285. https://doi.org/10.1016/j.soilbio.2014.05.011.

Nakano, S., & Murakami, M. (2001). Reciprocal subsidies: Dynamic interdependence between terrestrial and aquatic food webs. *Proceedings of the National Academy of Sciences, 98*(1), 166–170. https://doi.org/10.1073/pnas.98.1.166.

Noodén, L. D., Guiamét, J. J., & John, I. (1997). Senescence mechanisms. *Physiologia Plantarum, 101*(4), 746–753. https://doi.org/10.1111/j.1399-3054.1997.tb01059.x.

Norby, R. J., Cotrufo, M. F., Ineson, P., O'Neill, E. G., & Canadell, J. G. (2001). Elevated CO_2, litter chemistry, and decomposition: A synthesis. *Oecologia, 127*(2), 153–165. https://doi.org/10.1007/s004420000615.

Ostrofsky, M. L. (1997). Relationship between chemical characteristics of autumn-shed leaves and aquatic processing rates. *Journal of the North American Benthological Society, 16*(4), 750–759. https://doi.org/10.2307/1468168.

Ott, D., Digel, C., Rall, B. C., Maraun, M., Scheu, S., & Brose, U. (2014). Unifying elemental stoichiometry and metabolic theory in predicting species abundances. *Ecology Letters, 17*(10), 1247–1256. https://doi.org/10.1111/ele.12330.

Pazianoto, L. H., Solla, A., & Ferreira, V. (2019). Leaf litter decomposition of sweet chestnut is affected more by oomycte infection of trees than by water temperature. *Fungal Ecology, 41*, 269–278. https://doi.org/10.1016/j.funeco.2019.07.005.

Penuelas, J., Poulter, B., Sardans, J., Ciais, P., Van Der Velde, M., Bopp, L., Boucher, O., Godderis, Y., Hinsinger, P., Llusia, J., Nardin, E., Vicca, S., Obersteiner, M., & Janssens, I. A. (2013). Human-induced nitrogen–phosphorus imbalances alter natural and managed ecosystems across the globe. *Nature Communications, 4*(1), 1–10. https://doi.org/10.1038/ncomms3934.

Piccolo, J. J., & Wipfli, M. S. (2002). Does red alder (*Alnus rubra*) in upland riparian forests elevate macroinvertebrate and detritus export from headwater streams to downstream habitats in southeastern Alaska? *Canadian Journal of Fisheries and Aquatic Sciences, 59*, 503–513. https://doi.org/10.1139/f02-019.

Polis, G. A., & Strong, D. R. (1996). Food web complexity and community dynamics. *The American Naturalist, 147*, 813–846. https://doi.org/10.1086/285880.

Pope, C. A., Halvorson, H. M., Findlay, R. H., Francoeur, S. N., & Kuehn, K. A. (2020). Light and temperature mediate algal stimulation of heterotrophic activity on decomposing leaf litter. *Freshwater Biology*,, 1–13. https://doi.org/10.1111/fwb.13465.

Powers, J. S., & Salute, S. (2011). Macro-and micronutrient effects on decomposition of leaf litter from two tropical tree species: Inferences from a short-term laboratory incubation. *Plant and Soil, 346*(1–2), 245–257. https://doi.org/10.1007/s11104-011-0815-x.

Prater, C., Norman, E. J., & Evans-White, M. A. (2015). Relationships among nutrient enrichment, detritus quality and quantity, and large-bodied shredding insect community structure. *Hydrobiologia, 753*(1), 219–232. https://doi.org/10.1007/s10750-015-2208-2.

Redfield, A. C. (1958). The biological control of chemical factors in the environment. *American Scientist, 46*(3), 205–221.

Reiners, W. A. (1986). Complementary models for ecosystems. *The American Naturalist, 127*, 59–73. https://doi.org/10.1086/284467.

Ribblett, S. G., Palmer, M. A., & Coats, D. W. (2005). The importance of bacterivorous protists in the decomposition of stream leaf litter. *Freshwater Biology, 50*, 516–526. https://doi.org/10.1111/j.1365-2427.2005.01338.x.

Risse-Buhl, U., Karsubke, M., Schlief, J., Baschien, C., Weitere, M., & Mutz, M. (2012). Aquatic protists modulate the microbial activity associated with mineral surfaces and leaf litter. *Aquatic Microbial Ecology, 66*(2), 133–147. https://doi.org/10.3354/ame01564.

Rollin, M., Coulaud, R., Danger, M., Sohm, B., Flayac, J., Chaumot, A., Geffard, O., & Felten, V. (2018). Additive effect of calcium depletion and low resource quality on *Gammarus fossarum* (Crustacea, Amphipoda) life history traits. *Environmental Science and Pollution Research, 25*(12), 11264–11280. https://doi.org/10.1007/s11356-017-9390-x.

Ruiz, T., Koussoroplis, A. M., Danger, M., Aguer, J. P., Morel-Desrosiers, N., & Bec, A. (2020). U-shaped response unifies views on temperature dependency of stoichiometric requirements. *Ecology Letters, 23*(5), 860–869. https://doi.org/10.1111/ele.13493.

Sariyildiz, T., & Anderson, J. M. (2005). Variation in the chemical composition of green leaves and leaf litters from three deciduous tree species growing on different soil types. *Forest Ecology and Management, 210*(1–3), 303–319. https://doi.org/10.1016/j.foreco.2005.02.043.

Schreeg, L. A., Mack, M. C., & Turner, B. L. (2013). Nutrient-specific solubility patterns of leaf litter across 41 lowland tropical woody species. *Ecology, 94*(1), 94–105. https://doi.org/10.2307/23435672.

Simpson, S. J., & Raubenheimer, D. (2012). The Nature of Nutrition: A Unifying Framework From Animal Adaptation to Human Obesity. *Princeton University Press*. https://doi.org/10.1002/ajhb.22363.

Singer, G. A., & Battin, T. J. (2007). Anthropogenic subsidies alter stream consumer-resource stoichiometry, biodiversity, and food chains. *Ecological Applications, 17*, 376–389. https://doi.org/10.1890/06-0229.

Sinsabaugh, R. L., Manzoni, S., Moorhead, D. L., & Richter, A. (2013). Carbon use efficiency of microbial communities: Stoichiometry, methodology and modelling. *Ecology Letters, 16*(7), 930–939. https://doi.org/10.1111/ele.12113.

Small, G. E., & Pringle, C. M. (2010). Deviation from strict homeostasis across multiple trophic levels in an invertebrate consumer assemblage exposed to high chronic phosphorus enrichment in a Neotropical stream. *Oecologia, 162*(3), 581–590. https://doi.org/10.1007/s00442-009-1489-4.

Sperfeld, E., Raubenheimer, D., & Wacker, A. (2016). Bridging factorial and gradient concepts of resource co-limitation: Towards a general framework applied to consumers. *Ecology Letters, 19*(2), 201–215. https://doi.org/10.1111/ele.12554.

Sperfeld, E., Wagner, N. D., Halvorson, H. M., Malishev, M., & Raubenheimer, D. (2017). Bridging ecological stoichiometry and nutritional geometry with homeostasis concepts and integrative models of organism nutrition. *Functional Ecology, 31*(2), 286–296. https://doi.org/10.1111/1365-2435.12707.

Sterner, R. W., & Elser, J. J. (2002). Ecological stoichiometry: The biology of elements from molecules to the biosphere. *Princeton University Press*. https://doi.org/10.1515/9781400885695.

Suberkropp, K., Arsuffi, T. L., & Anderson, J. P. (1983). Comparison of degradative ability, enzymatic activity, and palatability of aquatic hyphomycetes grown on leaf litter. *Applied and Environmental Microbiology, 46*(1), 237–244. https://doi.org/10.1128/aem.46.1.237-244.1983.

Suberkropp, K., Godshalk, G. L., & Klug, M. J. (1976). Changes in the chemical composition of leaves during processing in a woodland stream. *Ecology, 57*(4), 720–727. https://doi.org/10.2307/1936185.

Tilman, D. (1982). *Resource competition and community structure*. Princeton University Press, 360pp. https://doi.org/10.4319/lo.1983.28.5.1043.

Torres-Ruiz, M., & Wehr, J. D. (2010). Changes in the nutritional quality of decaying leaf litter in a stream based on fatty acid content. *Hydrobiologia, 651*(1), 265–278. https://doi.org/10.1007/s10750-010-0305-9.

Tuchman, N. C., Wetzel, R. G., Rier, S. T., Wahtera, K. A., & Teeri, J. A. (2002). Elevated atmospheric CO_2 lowers leaf litter nutritional quality for stream ecosystem food webs. *Global Change Biology, 8*(2), 163–170. https://doi.org/10.1046/j.1365-2486.2002.00460.x.

Vanni, M. J. (2002). Nutrient cycling by animals in freshwater ecosystems. *Annual Review of Ecology and Systematics, 33*(1), 341–370. https://doi.org/10.1146/annurev.ecolsys.33.010802.150519.

Vanni, M. J., & McIntyre, P. B. (2016). Predicting nutrient excretion of aquatic animals with metabolic ecology and ecological stoichiometry: A global synthesis. *Ecology, 97*(12), 3460–3471. https://doi.org/10.1002/ecy.1582.

Wallace, J. B., Eggert, S. L., Meyer, J. L., & Webster, J. R. (1999). Effects of resource limitation on a detrital-based ecosystem. *Ecological Monographs, 69*(4), 409–442. https://doi.org/10.1890/0012-9615(1999)069[0409:EORLOA]2.0.CO;2.

Webster, J. R., Newbold, J. D., Thomas, S. A., Valett, H. M., & Mulholland, P. J. (2009). Nutrient uptake and mineralization during leaf decay in streams—A model simulation. *International Review of Hydrobiology, 94*(4), 372–390. https://doi.org/10.1002/iroh.200811158.

Wetzel, R. G. (1995). Death, detritus, and energy flow in aquatic ecosystems. *Freshwater Biology, 33*(1), 83–89. https://doi.org/10.1111/j.1365-2427.1995.tb00388.x.

Woodward, G. (2009). Biodiversity, ecosystem functioning and food webs in fresh waters: Assembling the jigsaw puzzle. *Freshwater Biology, 54*, 2171–2187. https://doi.org/10.1111/j.1365-2427.2008.02081.x.

Chapter 4
Global Patterns of Plant Litter Decomposition in Streams

Luz Boyero, Mark O. Gessner, Richard G. Pearson, Eric Chauvet, Javier Pérez, Scott D. Tiegs, Alan M. Tonin, Francisco Correa-Araneda, Naiara López-Rojo, and Manuel A. S. Graça

Abstract Understanding ecological patterns and processes at the global scale is becoming increasingly important in view of the rapid pace of environmental change and consequent impacts on ecosystems. This chapter reviews current knowledge about how plant litter decomposition—a key stream ecosystem process—and its major biotic and abiotic drivers vary globally along geographic gradients. The

L. Boyero (✉) · J. Pérez · N. López-Rojo
Department of Plant Biology and Ecology, University of the Basque Country (UPV/EHU), Leioa, Spain
e-mail: luz.boyero@ehu.eus

J. Pérez
e-mail: javier.perezv@ehu.eus

N. López-Rojo
e-mail: naiara.lopez@ehu.eus

L. Boyero
IKERBASQUE, Bilbao, Spain

M. O. Gessner
Department of Experimental Limnology, Leibniz Institute of Freshwater Ecology and Inland Fisheries (IGB), Berlin, Germany

Department of Ecology, Berlin Institute of Technology (TU Berlin), Berlin, Germany

M. O. Gessner
e-mail: gessner@igb-berlin.de

R. G. Pearson
School of Science and Engineering and Centre for Tropical Water and Aquatic Ecosystem Research (TropWater), James Cook University, Townsville, QLD, Australia
e-mail: richard.pearson@jcu.edu.au

E. Chauvet
Laboratoire Écologie Fonctionnelle et Environnement, Université de Toulouse-CNRS, Toulouse, France
e-mail: eric.chauvet@univ-tlse3.fr

S. D. Tiegs
Department of Biological Sciences, Oakland University, Rochester, MI, USA
e-mail: tiegs@oakland.edu

C. M. Swan et al. (eds.), *The Ecology of Plant Litter Decomposition in Stream Ecosystems*, https://doi.org/10.1007/978-3-030-72854-0_4

evidence available suggests that thermal regime is the main factor influencing microbial decomposition rate, which tends to increase with temperature. The presence or absence of litter-consuming detritivores is a major determinant of overall decomposition rate, the scarcity of these invertebrates being a common pattern in tropical streams. However, relating detritivore-mediated decomposition rates to environmental factors is complex, because of the interplay between detritivore abundance, body size distribution, diversity and community composition, as well as plant litter traits and diversity, all of which are influenced by climate, geology and biogeography. Meta-analyses and a growing number of coordinated large-scale studies have greatly enhanced our understanding of geographical variation of litter decomposition in streams, and have enabled first projections of how climate warming and a range of other aspects of global environmental change will affect the process.

4.1 Introduction

The extent of the total stream network globally is estimated at 88 million km (Downing, 2012), a length equivalent to 200 times the distance from Earth to the moon. Many of these streams receive plant litter from riparian vegetation, which is deposited in the stream channel, partly transported downstream and partly used and decomposed by microbial decomposers and litter-consuming detritivores. Indeed, terrestrial plant litter represents the single most important source of carbon (C) and nutrients to many streams (Wallace et al., 1997). This is particularly true for streams surrounded by dense riparian vegetation, which not only provides abundant litter but also limits instream primary production (Vannote et al., 1980). Once entrained in stream channels, litter is broken down as a result of several processes occurring simultaneously (Gessner et al., 1999), including physicochemical processes (i.e., leaching of soluble litter constituents and fragmentation by shear stress and abrasion) and biological processes (i.e., decomposition mediated by microorganisms and litter-consuming detritivores). As a consequence, C compounds and nutrients are released and may be incorporated into stream food webs (Marks, 2019; Webster & Benfield, 1986), lost downstream, or emitted to the atmosphere as carbon dioxide (CO_2), methane (CH_4), or other gases after further biogeochemical transformations

A. M. Tonin
Limnology-Aquaripária Lab, University of Brasília (UnB), Brasília, Brazil

F. Correa-Araneda
Unidad de Cambio Climático y Medio Ambiente, Instituto Iberoamericano de Desarrollo Sostenible (IIDS), Universidad Autónoma de Chile, Temuco, Chile
e-mail: francisco.correa@uautonoma.cl

M. A. S. Graça
Department of Life Sciences and Marine and Environmental Sciences Centre (MARE), University of Coimbra, Coimbra, Portugal
e-mail: mgraca@ci.uc.pt

(e.g., molecular nitrogen, N_2; nitrous oxide, N_2O) (Battin et al., 2009; Mulholland et al., 2008). Uncovering the global patterns of these processes is important in order to quantify the role of stream ecosystems in global C and nutrient budgets, and to understand the hierarchy of factors governing rates of decomposition.

4.2 Assessing Global Patterns to Inform About Global Change

An understanding of large-scale variation in the patterns of litter decomposition in streams can provide insights into major ecosystem consequences of global environmental change. Clearly, global change is affecting ecosystems worldwide at unprecedented rates, with strong impacts on biodiversity and ecosystem functioning and potential impairment of ecosystem services to humans (IPCC, 2014; Vitousek et al., 1997). Corvalan et al. (2005) advocated that special attention be given to fresh waters as the most threatened ecosystems on Earth, with any change in their functioning likely altering provisioning, regulating and supporting ecosystem services. Among the large-scale changes imposed on ecosystems, those related to climate have been most prominently examined, and many are well documented and of serious concern. The periodically published reports of the Intergovernmental Panel on Climate Change (IPCC) have provided increasingly compelling evidence that anthropogenic emissions of greenhouse gases to the atmosphere cause global warming and alter precipitation patterns in many parts of the globe, and that these trends are projected to continue in the future (IPCC, 2018).

The general global warming trend, although not uniform in magnitude across biomes and regions, is unique in Earth's recent history (Neukom et al., 2019). Its consequences on ecosystems are not easy to study, but substituting space (across broad temperature gradients) for time has been widely used as a proxy approach to project changes in the structure of biological communities and fluxes of matter and energy in response to climate change (Pennings & Bertness, 1999). Limitations of this approach include influences of latitudinal changes in other environmental factors, whether related to temperature variation or other causes, including the physical setting of landscapes, water chemistry and biodiversity patterns (Willig et al., 2003). However, these limitations may not apply when temperature changes cause range shifts of species, as has been reported for plants, fishes and plankton (Deutsch et al., 2008; Mohseni et al., 2003; Pitelka & Group, 1997). Additionally, evolutionary responses to altered temperature regimes by species are to be expected (e.g., De Meester et al., 2018), although little pertinent information is currently available for stream organisms (Haase et al., 2019). Such range shifts and adaptations of species need to be incorporated into models designed to assess impacts of global environmental change on ecosystem structure and processes. For the process of litter decomposition in streams, this applies especially to riparian vegetation, microbial decomposers and litter-consuming detritivores.

4.3 Approaches to Determining Global Patterns

A variety of complementary approaches have been used to determine patterns and the underlying drivers of decomposition at large scales, ranging from regional to continental and global. This has included direct comparisons of litter decomposition between two or three regions that differ climatically (e.g., Ferreira et al., 2012; Gonçalves et al., 2006) or geographically (e.g., Camacho et al., 2009), as well as comprehensive literature reviews (e.g., Marks, 2019; Tank et al., 2010; Webster & Benfield, 1986). Two approaches have been particularly valuable in providing insights into large-scale geographical patterns in litter decomposition. First, meta-analyses and data syntheses have proved powerful in detecting and quantifying influences of particular environmental drivers, including factors affected by global change (Ferreira et al., 2015; Follstad Shah et al., 2017; Zhang et al., 2019). Second, coordinated distributed experiments at global or continental scales have been increasingly initiated over the last decade (Boyero et al., 2016; Boyero, Pearson, Gessner, et al., 2011; Handa et al., 2014; Tiegs et al., 2019; Woodward et al., 2012) and have provided invaluable comparative information. This includes the recognition that decomposition rates of different litter types vary with latitude, as do the roles of microorganisms and detritivores, despite high spatial variability (Fig. 4.1).

A key advantage of coordinated distributed experiments over literature-based meta-analyses for quantifying effects and pinpointing mechanisms is the opportunity to employ consistent protocols, and even identical materials (Boyero & Pearson, 2017), at a wide variety of locations (Fraser et al., 2013). This methodological standardization eliminates important sources of variation and hence enhances statistical power of hypothesis tests to facilitate detection of effects and generalities (Pennings & Bertness, 1999). The first coordinated study indicating temperature effects on litter decomposition in streams along a latitudinal gradient was conducted by Irons et al. (1994), although the experiments were restricted to three sites across North and Central America, with the possibility that any observed pattern was confounded by factors other than temperature. However, more comprehensive studies with sites distributed across all continents except Antarctica (e.g., Boyero, Pearson, Gessner, et al., 2011) have reached similar conclusions (see below).

Ideally, global assessments are made through large coordinated networks of researchers (e.g., GLoBE, 1000IRES, CELLDEX networks; Boyero & Pearson, 2017). This approach allows framing questions as hypotheses derived from previous observations at multiple sites, which can then be tested by employing identical methods. The approach has its own difficulties, however, including disparate geographic representation and effective coordination of teams and a lack of effective funding mechanisms for international networks. Nevertheless, multiple standardised global studies of litter decomposition in streams have been completed in recent years (Fig. 4.2), starting with publications based on a modest number (~25) of globally distributed sampling sites (Boyero et al., 2016; Boyero, Pearson, Dudgeon, et al., 2011). These studies have identified consistent global patterns in decomposition (Fig. 4.1) and provided insights into the influence of climatic and other environmental

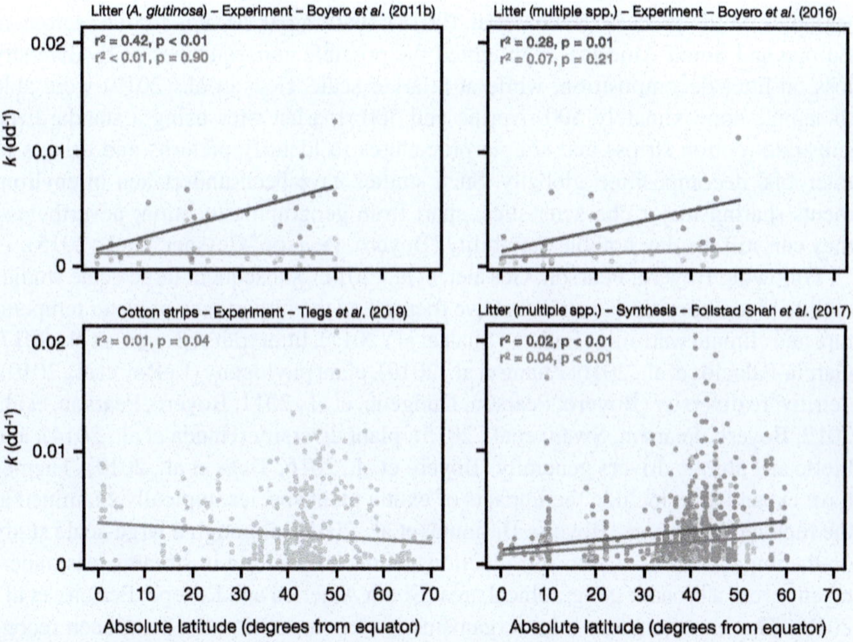

Fig. 4.1 Latitudinal variation in temperature-corrected decomposition rates (*k* per degree day) reported in several global studies. Light grey lines represent microbial decomposition of litter or cotton strips and dark grey lines represent total litter decomposition mediated by both microorganisms and detritivores; r^2 and p-values of linear regressions are shown

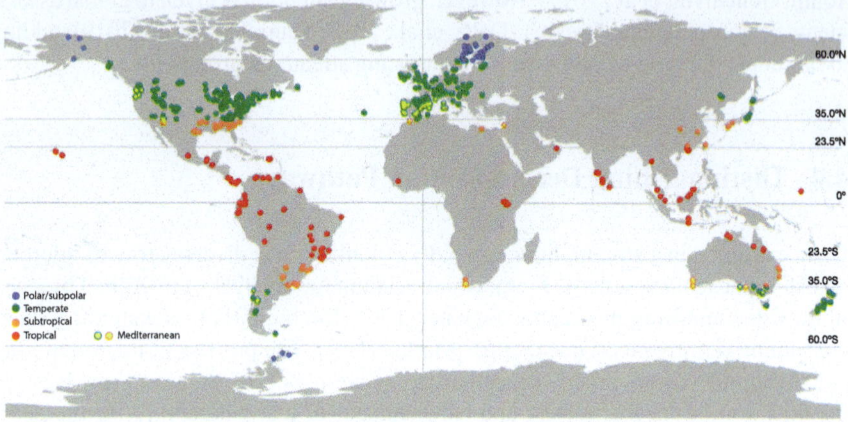

Fig. 4.2 Map showing locations of regions included in global decomposition studies (extracted from Boyero et al., 2021; Follstad Shah et al., 2017; LeRoy et al., 2019; Tiegs et al., 2019; Zhang et al., 2019; blue, green, orange and red colours represent latitudinal bands (respectively: polar/subpolar, >60°; temperate, 35–60°; subtropical, 23.5–35°; and tropical, 0–23.5°) and yellow represents Mediterranean regions within temperate or subtropical areas

variables on the process. Handa et al. (2014), working at 20 sites in five regions in Europe and South America, highlighted the possible consequences of biodiversity loss on litter decomposition, while at a larger scale Tiegs et al. (2019) were able to sample approximately 500 riverine and 500 riparian sites using a standardized substrate (cotton strips) and simple procedures to identify patterns and drivers of microbial decomposition globally. Such studies have been undertaken in environments sharing major characteristics, apart from geographic location; nevertheless, they can still display notable variability (Boyero, Pearson, Gessner, et al., 2015).

Following Boyero, Pearson, Gessner, et al. (2011), subsequent large-scale studies on litter decomposition in streams have focused on the importance of water temperature and climate warming (Follstad Shah et al., 2017), litter traits (Boyero et al., 2017; García-Palacios et al., 2016; Zhang et al., 2019), plant phylogeny (LeRoy et al., 2019), detritivore diversity (Boyero, Pearson, Dudgeon, et al., 2011; Boyero, Pearson, et al., 2012; Boyero, Pearson, Swan, et al., 2015), plant diversity (Handa et al., 2014), and biotic and abiotic drivers generally (Boyero et al., 2016; Tiegs et al., 2019). Themes have expanded to include the impacts of exotic plant species, typically examined at the regional scale (e.g., Boyero, Barmuta, et al., 2012), although a large-scale study on the impact of *Eucalyptus* introductions on decomposition in streams was undertaken across a broader geographical area (South America and Europe; Ferreira et al., 2018). The importance of microorganisms on decomposition has long been recognised both regionally and globally (e.g., Boyero, Pearson,Gessner, et al., 2011), and knowledge has expanded with the global investigation by Seena et al. (2019) into the diversity of fungi colonizing litter. Most commonly, studies have focused on perennial headwater forest streams (Fig. 4.3), removing much of the environmental variation that is characteristic of stream ecosystems (e.g., natural non-forest streams; Tonin, Goncalves, et al., 2017). However, growing attention is given to global assessments of non-perennial streams (Datry et al., 2018; Shumilova et al., 2019), adding to our understanding of processes in contrasting stream environments.

4.4 Distinguishing Decomposition Pathways

Once entrained in a stream, litter is rapidly colonized by a diverse array of microorganisms, particularly aquatic fungi, which account for 60–99% of microbial biomass in litter decomposing in streams (Gessner, 1997; Kuehn, 2016), as well as by litter-consuming detritivores (Graça, 2001; Marks, 2019). The process of microbial litter colonization is often called 'conditioning' (Cummins, 1974), which involves the modification of the leaf matrix and enhancement of leaf palatability for the detritivores (Gessner et al., 1999). Aquatic hyphomycetes are effective in both roles through, first, the production of extracellular enzymes that rapidly degrade complex litter constituents and, second, the production of microbial biomass containing essential nutrients for detritivores that are in short supply in the litter (Bärlocher & Kendrick, 1975; Marks, 2019).

Fig. 4.3 Examples of streams from around the world where decomposition studies have been conducted. (**a**) Ecuador (0°); (**b**) French Guiana, France (4 °N); (**c**) Queensland, Australia (19 °S); (**d**) Chile (37 °S); (**e**) Spain (37 °N); (**f**) Switzerland (46 °N); (**g**) Germany (54 °N); (**h**) Alaska, USA (56 °N); (**i**) Iceland (64 °N). Photos by J. Schreckinger (**a**, with permission), M. Schindler (**b**, with permission), L. Boyero (**c, e**), F. Correa-Araneda (**d**), M. O. Gessner (**f**), D. S. Finn (**g**, with permission) and S. D. Tiegs (**h, i**)

There is evidence that the fate of litter entering streams can differ depending on the relative importance of microbial and detritivore-mediated decomposition. Decomposition driven by fungi in streams, best understood through the study of aquatic hyphomycetes, involves the mineralization of a large fraction of the litter into CO_2 (Baldy et al., 2007; Suberkropp, 1991) and presumably other inorganic compounds. Notable amounts of dissolved (Meyer et al., 1998) and fine-particulate (Suberkropp & Klug, 1980) organic C and nutrients are also released (Gessner et al., 1999), especially as a result of effective litter maceration by fungal pectinases (Suberkropp & Klug, 1980). Similarly, detritivores can be extremely important agents driving litter decomposition (Wallace & Webster, 1996). By ingesting, digesting and egesting the organic matter, they transform part of the litter into biomass, not unlike the conversion of litter by microbes (Suberkropp et al., 2010), while simultaneously releasing CO_2 and mineral nutrients (Iversen, 1979). Moreover, the litter consumed by detritivores is fragmented during gut passage, resulting in the release of strikingly large amounts of fine-particulate organic matter (FPOM), which is a staple food source for many other animals (Wallace & Webster, 1996).

The relative proportions of these transformation products are likely to vary with litter recalcitrance and environmental context (e.g., temperature, nutrient availability, geochemistry, water pollution; Marks, 2019). Outcomes will also depend on the extent to which microbial decomposers and detritivores interact. If, for example, the microbial conditioning of litter is insufficient to narrow the typically large carbon-to-nitrogen (C:N) and carbon-to-phosphorus (C:P) ratios of the plant tissue (Graça, Ferreira, et al., 2015), detritivores will defecate a large portion of the ingested material as FPOM (Manzoni et al., 2010). Or, when detritivores rapidly crop the microbial biomass developing on decomposing litter, the contribution of microbes to decomposition will be small, even when conditions for microbial growth are favourable (Robinson et al., 1998). Thus, streams can experience very different rates of decomposition, resulting in spatial variation that will influence the global patterns of the process. Even when overall decomposition rates are similar, marked differences in decomposition pathways and resource-use efficiency could still occur, depending on whether microbial or detritivore-mediated decomposition dominate and which decomposition pathways prevail (Boyero, Pearson, Gessner, et al., 2011).

4.5 Global Patterns and Drivers of Microbial Decomposition

Microbial decomposition often constitutes a substantial fraction of the total amount of litter decomposed in streams (Hieber & Gessner, 2002; Webster & Benfield, 1986). While the contribution of microorganisms to litter decomposition is not homogenously distributed across the planet, it shows predictable large-scale patterns across biomes and latitudinal gradients. Graça, Ferreira, et al., (2015) reviewed these patterns using the hierarchical conceptual framework proposed by Royer and

Minshall (2003), and identified multiple factors, from local to global scales, influencing decomposition rates and the relative role of microbial decomposers and detritivores.

Among these factors, temperature stands out as a main driver of microbial decomposition and can explain much of this variability through its influence on metabolic rate (Brown et al., 2004). Thus, low-latitude streams (i.e., those in tropical and subtropical areas at latitudes between approximately 0 and 35°) generally show the highest microbial decomposition rates (Boyero, Pearson, Gessner, et al., 2011; Irons et al., 1994). However, variation within the tropics and subtropics is high (Boyero, Pearson, Gessner, et al., 2015; Tiegs et al., 2019), partly due to large variation in temperature across altitudinal gradients (Encalada et al., 2019). Furthermore, the non-limiting role of temperature for decomposition in many tropical streams raises the relative importance of other environmental factors such as nutrient availability or pH (Tiegs et al., 2019). At high latitudes, in contrast, low temperatures impose constraints on microbial activity, so microbial decomposition is slow and more consistent across cooler areas of the planet (Tiegs et al., 2019). The critical role of temperature is corroborated by results of studies quantifying decomposition rates per degree day (rather than per day), an approach that removes the effect of temperature and facilitates examining the influence of other factors (Gessner & Peeters, 2020). Thus, a global study showed no latitudinal gradient in microbially mediated litter decomposition rates expressed per degree day (Boyero, Pearson, Dudgeon, et al., 2011; but see Follstad Shah et al., 2017), and the latitudinal gradient found in cotton-strip decomposition became much less pronounced when examined on a per-degree-day basis (Tiegs et al., 2019).

A second factor varying at large scales and strongly influencing microbial decomposition is litter quality, involving tissue concentrations of nutrients and secondary compounds and leaf physical structure (e.g., toughness). Tropical streams often receive litter inputs of lower quality than their high-latitude counterparts (Boyero et al., 2017; Wantzen,Wagner, Suetfeld, et al., 2002). This includes lower concentrations of nutrients, especially P, which is probably related to the commonly low nutrient content of tropical soils (Alvarez-Clare & Mack, 2011) resulting from their old age (i.e., longer weathering) and high rates of nutrient leaching (Reich & Oleksyn, 2004), despite an efficient P resorption by tropical plants (Yuan & Chen, 2009). Moreover, litter with lower nutrient content is also often richer in refractory structural and potentially inhibitory secondary compounds, reflecting defenses against herbivory (Boyero et al., 2017; LeRoy et al., 2006). Litter quality tends to be low in semi-arid and arid zones of the planet, which occur mostly at latitudes between 30° and 40° in the northern and southern hemispheres (i.e., Mediterranean regions and deserts), where leaves are often leathery and have thick cuticles (Gallardo & Merino, 1993). However, many riparian plants could be exceptions to this rule as they are close to a source of water and, potentially, nutrients (Smith et al., 1998). Lastly, while broadleaf litter quality increases with latitude (Boyero et al., 2017), boreal forests are dominated by conifers characterized by highly recalcitrant litter, and beyond 60° of latitude (i.e., the tundra) few trees are found, so litter inputs to streams are scarce and mostly composed of grasses and bryophytes (Peterson et al., 1986).

Nutrients dissolved in stream water represent another important environmental factor governing differences in spatial patterns of microbial decomposition in streams (Suberkropp & Chauvet, 1995). Aquatic hyphomycetes can readily retrieve nutrients from water (Suberkropp, 1998), indicating that decomposition of litter can be enhanced in nutrient-rich waters, particularly when litter nutrient concentrations are low (Jabiol et al., 2019; Tonin, Boyero, et al., 2017). To date, however, data are insufficient to assess the repercussion globally, because water chemistry is influenced by an array of factors, including not only geology and soil quality, but also the degree of human intervention. Nevertheless, evidence from a comprehensive meta-analysis (Ferreira et al., 2015) and a coordinated experiment at the continental scale (Woodward et al., 2012) suggests that global effects are likely to be important.

The effectiveness of microbial decomposition is also affected by seasonality, which not only determines the timing of litter inputs but also temperature and streamflow regimes (including the occurrence of floods and droughts), which vary globally. Hydrological variability may be high in the tropics, either day to day near the equator (Yule & Pearson, 1996), or seasonally in other tropical areas, and floods reduce litter availability (Graça, Ferreira, et al., 2015). Seasonality is also critically important at mid latitudes ($\approx$35–50°), where most research on decomposition has been conducted to date. Studies at mid latitudes have generally taken place during and shortly after peak litter fall in the autumn and early winter as the most relevant season for decomposition to occur naturally. At higher latitudes, seasonal fluctuations are also high, but litter inputs are more variable, ranging from distinctly seasonal to irregular throughout the year (Benfield, 1997), although much less information on seasonal influences on litter decomposition in streams is available from these latitudes.

Microbial decomposition can be related to the biomass and productivity of aquatic hyphomycetes (Gessner & Chauvet, 1994), as well as their diversity and sporulation rate (Pérez et al., 2012). Contrasting with the general trend of increased biological diversity towards the tropics (Willig et al., 2003), aquatic hyphomycete diversity peaks at mid latitudes (Duarte et al., 2016; Jabiol et al., 2013; Seena et al., 2019), and is often low in streams outside the temperate zones (Graça, Hyde, et al., 2015).

4.6 Global Patterns and Drivers of Detritivore-Mediated Decomposition

Shredders, defined as invertebrates with mouthparts capable of cutting and chewing litter (Cummins et al., 1989), are a key functional feeding group in forested headwater streams (Vannote et al., 1980; Wallace et al., 1997). Here we extend the group of invertebrates exploiting leaf litter in streams, referring more generally to litter-consuming detritivores, which include taxa that may be classified in functional feeding groups other than shredders but may nonetheless feed on litter and thus contribute to decomposition (Boyero et al., 2020).

Many local studies conducted in temperate climates have reported high abundances and taxonomic richness of litter-consuming detritivores, and large contributions to total decomposition (20–60% of total litter mass loss; Andrushchenko et al., 2016; Bruder et al., 2014; Cuffney et al., 1990; Graça, 2001; Hieber & Gessner, 2002; Pascoal et al., 2005; Woodward et al., 2012). In contrast, studies from tropical areas have reported a wide variety of results: while most have emphasized a paucity of litter-consuming detritivores and low detritivore-mediated decomposition rates (<10% of total litter mass loss; Bruder et al., 2014; Dobson et al., 2002; Gonçalves et al., 2007; Irons et al., 1994; Pettit et al., 2011; Rueda-Delgado et al., 2006; Tenkiano & Chauvet, 2018; Zúñiga-Céspedes et al., 2018), others have found numbers of litter-consuming detritivores similar to those of streams in temperate regions, with important contributions to decomposition (Cheshire et al., 2005; Encalada et al., 2010; Fugère et al., 2018; Graça & Cressa, 2010; Jinggut & Yule, 2015; Masese et al., 2014; Tonin, Hepp, et al., 2018; Yule et al., 2009).

The large variability in the abundance and richness of these detritivores across tropical streams reported from local investigations has been further supported by global studies that have used identical methods across multiple sites (Boyero, Pearson, Dudgeon, et al., 2011; Boyero, Pearson, et al., 2012; Boyero, Pearson, Gessner, et al., 2015). Nevertheless, these studies have typically reported litter-consuming detritivore abundance and richness more than two-fold higher in temperate than tropical streams. Similarly, despite high variability in detritivore-mediated decomposition rates, global studies have demonstrated an overall increase with latitude (Boyero et al., 2016; Boyero, Pearson, Dudgeon, et al., 2011; Follstad Shah et al., 2017), confirming earlier results based on studies conducted at a restricted number of sites (Irons et al., 1994).

Several explanations have been invoked for the latitudinal gradient in detritivore numbers. First, many typical litter-consuming detritivores are evolutionarily adapted to cool waters, and may be scarce in the tropics because of limited tolerance to elevated temperature (Boyero, Pearson, Dudgeon, et al., 2011). The key role of temperature is supported by the observation that, along altitudinal gradients in some tropical areas, detritivore abundance and richness are greater at cooler sites at high altitudes (Yule et al., 2009). Moreover, cool water at high latitudes reduces microbial activity, possibly allowing litter to persist in streams for extended periods, especially if temperatures below freezing point prevent flushing by high streamflow, thus providing more reliable resources for litter-consuming detritivores to flourish (Dobson et al., 2002; Irons et al., 1994). However, low temperatures may not be advantageous if microbial conditioning is essential for detritivore feeding (Graça, 2001).

Second, seasonality may affect detritivore-mediated decomposition through effects on litter availability much more than microbial decomposition, because life cycles of invertebrates are longer and more complex. Flood disturbances can deplete streams from litter and result in unfavourable conditions to detritivores relying on this resource (Coughlan et al., 2010; Graça, Hyde, et al., 2015; Wantzen & Wagner, 2006; Yule, 1996). However, there are indications that stream detritivore communities can quickly recover after floods (Wootton et al., 2018), and floods are not necessarily more frequent in the tropics than at higher latitudes (Winterbourn et al., 1981).

Third, litter quality tends to decrease towards the equator, as reported by several local studies (Coley & Barone, 1996; Dobson et al., 2002; Stout, 1989; Wantzen, Wagner, et al., 2002; Marquis et al., 2012) and a global study examining 151 riparian tree species from 24 regions across a latitudinal gradient (Boyero et al., 2017). Tropical plants are often better protected against consumers: they are tougher and contain more toxic compounds, most of which remain active after senescence, and they tend to be more depleted in P relative to C (i.e., their C:P ratios are high) than their temperate counterparts (Boyero et al., 2017). However, reciprocal incubation experiments have produced contradictory evidence, with high-quality temperate litter decomposing at rates lower than or equal to tropical litter in tropical streams (Bruder et al., 2014; Ferreira et al., 2012).

A fourth explanation is related to the Bergman and temperature-size rules (Foster et al., 2011; Horne et al., 2015). Many invertebrates are smaller at lower latitudes, and smaller invertebrates may be less efficient at consuming litter because their mouthparts are not sufficiently robust to shred tough leaf tissue effectively. However, large crustaceans (e.g., Brachyura, Parastacidae, Palaemonidae, Atyidae) are common in tropical streams and can readily consume such decomposing litter (Coughlan et al., 2010; Dobson et al., 2002; Wantzen & Wagner, 2006). Other large litter-consuming detritivores at low latitudes include some snails and semi-aquatic cockroaches (Yule et al., 2009; M. Moretti, pers. comm.) and probably tadpoles (Schmidt et al., 2017). However, these animals are often too large to enter litterbags, and thus may have been overlooked in decomposition studies. Finally, many tropical taxa are likely to have been incorrectly assigned to functional feeding groups, given the lack of knowledge on the feeding ecology of these invertebrates and reliance on information from related taxa in temperate regions. For example, mayflies usually consume FPOM, or scrape off biofilms, but some genera such as *Atalophlebia* (Leptophlebiidae) in Australia (Cheshire et al., 2005), or *Acanthiops* (Baetidae) in Africa (Dobson et al., 2002) have been identified as litter consumers. Collectively, the above factors may thus have resulted in a general underestimation of litter-consuming detritivores in tropical streams.

4.7 Conclusion and Perspectives

Globally coordinated studies have contributed substantially to assessing the importance of litter decomposition as a pivotal ecosystem process in streams worldwide. Is, then, the process of global importance? The answer to this question is not yet clear. Despite its impressive length, the global stream network represents only a small area of the planet (approximately 0.6% of the non-glaciated land surface; Allen & Pavelsky, 2018). However, given the tight linkages to their catchments through direct or indirect inputs of terrestrial organic matter, and because of much higher decomposition rates than in terrestrial environments (Handa et al., 2014), streams could have a significantly greater influence on global C and nutrient cycles than their surface area implies. This would suggest that comprehensive global-scale analyses of litter decomposition are needed to elaborate robust estimates of the contribution of stream ecosystems to global biogeochemical cycles. The relative contribution of chemical and biological processes in inland waters to global atmospheric CO_2 is largely unknown at present (Hotchkiss et al., 2015; Raymond et al., 2013). However, von Schiller et al. (2019) presented a first estimate of CO_2 release from decomposing litter in the global network of intermittent streams, and Boyero, Pearson, Gessner, et al. (2011) concluded that C sequestration resulting from incomplete decomposition in streams before deposition in lake and ocean sediments increases with latitude.

Many major research gaps remain beyond the current lack of reliable global estimates, despite growing attention given to litter decomposition across broad geographical areas. In particular, wide areas of the planet remain greatly understudied, especially in Asia and Africa but also in Central and South America (Fig. 4.2). These gaps in geographic coverage are most evident in the tropics, and result in large uncertainties, not least because many of these regions are characterized by a particularly high diversity of habitats and species. Even within better-known regions, investigations into the decomposition of litter in streams have typically been limited to a few dozens of sites, data on which are insufficient as a basis for reliable broad-scale assessments.

Global studies may help to identify gaps that may be pursued at local or regional studies and in laboratory experiments. For example, large numbers of local studies have contributed to understanding the influence of temperature (Ferreira & Chauvet, 2011; Martínez et al., 2014), dissolved nutrients (Connolly & Pearson, 2013; Gulis et al., 2006; Rosemond et al., 2015), land-use change (Wild et al., 2019), and biotic factors such as biodiversity, phylogeny and plant traits (López-Rojo et al., 2018, 2019; Tonin, Boyero, et al., 2017) on decomposition, but few studies have addressed the effects of intra- and interspecific interactions or body size of detritivores (Boyero & Pearson, 2006, Tonin, Pozo, et al., 2018). The importance of nutrients other than N and P such as calcium and magnesium has been highlighted but has received little comprehensive study (García-Palacios et al., 2016). With regard to nutrients, there is extensive literature on their influence on decomposition in streams (e.g., Ferreira et al., 2015; Woodward et al., 2012), but there are no global comparisons of the influence of decomposition on nutrient dynamics in streams. Regarding climate change, some studies have directly addressed the effects of warming (e.g., Boyero, Pearson, Gessner, et al., 2011; Follstad Shah et al., 2017) and decreased flow (Sabater et al., 2018) on decomposition, suggesting that responses vary across biomes. Still to be considered are the effects of climate change on the diversity of litter and microbial decomposers and litter consuming detritivores. Finally, knowledge of decomposition responses to hydrological regime shifts, which are inconsistent amongst regions of the world, may be improved through climate modelling at regional scales.

Clearly, large-scale assessments must address substantial variability within regions resulting from variation in climate, geology, vegetation and human influences that shape regional and larger-scale patterns in decomposition rates, pathways and controlling factors. This variety of stream environments needs to be comprehensively documented and explicitly accounted for in analyses of relationships between environmental drivers and decomposition. For example, how do slow-flowing lowland streams compare with the swiftly flowing upland streams where litter decomposition has been studied most commonly? Or, do deforestation, nutrient enrichment, changes in hydrological regimes or channel morphology, biodiversity loss or species range shifts, warming, salinization, and other anthropogenic perturbations affect the process similarly in different biomes? Global surveys, meta-analyses or specifically targeted studies could address such questions. Particularly useful could be coordinated manipulative experiments designed to assess the importance of individual drivers and their combinations. Such experiments are logistically challenging at the global scale, difficult to fund, and lacking to date. They could prove extremely powerful, however, to achieve a quantitative understanding of the hierarchy of drivers (Graça, Hyde, et al., 2015) controlling litter decomposition in streams across the globe.

References

Allen, G. H., & Pavelsky, T. M. (2018). Global extent of rivers and streams. *Science, 361*, 585–588.

Alvarez-Clare, S., & Mack, M. C. (2011). Influence of precipitation on soil and foliar nutrients across nine Costa Rican forests. *Biotropica, 43*, 433–441.

Andrushchenko, I. V., Taylor, B. R., Toxopeus, J., & Wilson, E. (2016). Congregations of the leaf-shredding insect *Lepidostoma togatum* mediate exceptionally rapid mass loss from leaf litter in Nova Scotia rivers. *Hydrobiologia, 788*, 245–265.

Baldy, V., Gobert, V., Guerold, F., Chauvet, E., Lambrigot, J., & Charcosset, J.-Y. (2007). Leaf litter breakdown budgets in streams of various trophic statuses: Effects of dissolved organic nutrients in microorganisms and invertebrates. *Freshwater Biology, 52*, 1322–1335.

Bärlocher, F., & Kendrick, B. (1975). Leaf-conditioning by microorganisms. *Oecologia, 20*, 359–362.

Battin, T. J., Luyssaert, S., Kaplan, L. A., Aufdenkampe, A. K., Richter, A., & Tranvik, L. J. (2009). The boundless carbon cycle. *Nature Geoscience, 2*, 598–600.

Benfield, E. F. (1997). Comparison of litterfall input to streams. *Journal of the North American Benthological Society, 16*, 104–108.

Boyero, L., Barmuta, L. A., Ratnarajah, L., Schmidt, K., & Pearson, R. G. (2012). Effects of exotic riparian vegetation on leaf breakdown by shredders: A tropical–temperate comparison. *Freshwater Science, 31*, 296–303.

Boyero, L., Graça, M. A. S., Tonin, A. M., Pérez, J., Swafford, J. A., Ferreira, V., Landeira-Dabarca, A., Alexandrou, M. A., Gessner, M. O., McKie, B. G., Albarino, R. J., Barmuta, L. A., Callisto, M., Chara, J., Chauvet, E., Colon-Gaud, C., Dudgeon, D., Encalada, A.C., Figueroa, R., … Pearson, R. G. (2017). Riparian plant litter quality increases with latitude. *Scientific Reports, 7*, 10562.

Boyero, L., & Pearson, R. G. (2006). Intraspecific interference in a tropical stream shredder guild. *Marine and Freshwater Research, 57*, 201–206.

Boyero, L., & Pearson, R. G. (2017). Global-scale coordinated networks as a tool for exploring the functioning of stream ecosystems. *Limnetica, 36*, 557–565.

Boyero, L., Pearson, R. G., Albariño, R., Callisto, M., Correa-Araneda, F., Encalada, A. C., Masese, F., Moretti, M., Ramírez, A., Sparkman, A., Swan, C. M., Yule, C. M., & Graça, M. A. S. (2020). Identifying stream invertebrates as plant litter consumers. In F. Barlochër, M. O. Gessner, & M. A. S. Graça (Eds.),*Methods to Study Plant Litter Decomposition. A Practical Guide*. 2nd Edition. Springer, pp. 445–464.

Boyero, L., Pearson, R. G., Dudgeon, D., Ferreira, V., Graça, M. A. S., Gessner, M. O., Boulton, A. J., Chauvet, E., Yule, C. M., Albariño, R. J., Ramírez, A., Helson, J. E., Callisto, M., Arunachalam, M., Chará, J., Figueroa, R., Mathooko, J. M., Gonçalves, J. F., Moretti, M., … Barmuta, L. A. (2012). Global patterns of stream detritivore distribution: implications for biodiversity loss in changing climates. *Global Ecology and Biogeography, 21*, 134–141.

Boyero, L., Pearson, R. G., Dudgeon, D., Graça, M. A. S., Gessner, M. O., Albariño, R. J., Ferreira, V., Yule, C. M., Boulton, A. J., Arunachalam, M., Callisto, M., Chauvet, E., Ramírez, A., Chará, J., Moretti, M. S., Gonçalves, J. F., Helson, J. E., Chará-Serna, A. M., Encalada, A. C., … Pringle, C. M. (2011). Global distribution of a key trophic guild contrasts with common latitudinal diversity patterns. *Ecology, 92*, 1839–1848.

Boyero, L., Pearson, R. G., Gessner, M. O., Barmuta, L. A., Ferreira, V., Graça, M. A. S., Dudgeon, D., Boulton, A. J., Callisto, M., Chauvet, E., Helson, J. E., Bruder, A., Albariño, R. J., Yule, C. M., Arunachalam, M., Davies, J. N., Figueroa, R., Flecker, A. S., Ramírez, A., … West, D. C. (2011). A global experiment suggests climate warming will not accelerate litter decomposition in streams but may reduce carbon sequestration. *Ecology Letters, 14*, 289–294.

Boyero, L., Pearson, R. G., Gessner, M. O., Dudgeon, D., Ramírez, A., Yule, C. M., Callisto, M., Pringle, C. M., Encalada, A. C., Arunachalam, M., Mathooko, J., Helson, J. E., Rincón, J., Bruder, A., Cornejo, A., Flecker, A. S., Mathuriau, C., M'Erimba, C., Gonçalves, J. F., … Jinggut, T. (2015). Leaf-litter breakdown in tropical streams: is variability the norm? *Freshwater Science, 34*, 759–769.

Boyero, L., Pearson, R. G., Hui, C., Gessner, M. O., Pérez, J., Alexandrou, M. A., Graça, M. A. S., Cardinale, B. J., Albariño, R., Arunachalam, M., Barmuta, L. A., Boulton, A. J., Bruder, A., Callisto, M., Chauvet, E., Death, R. G., Dudgeon, D., Encalada, A. C., Ferreira, V., … Yule, C. M. (2016). Biotic and abiotic variables influencing plant litter breakdown in streams: a global study. *Proceedings of the royal society B: Biological sciences, 283*, 20152664.

Boyero, L., Pearson, R. G., Swan, C. M., Hui, C., Albariño, R. J., Arunachalam, M., Callisto, M., Chará, J., Chará-Serna, A. M., Chauvet, E., Cornejo, A., Dudgeon, D., Encalada, A. C., Ferreira, V., Gessner, M. O., Gonçalves, J. F., Graça, M. A. S., Helson, J. E., Mathooko, J. M., … Yule, C. M. (2015). Latitudinal gradient of nestedness and its potential drivers in stream detritivores. *Ecography,* 38.

Boyero, L., Pérez, J., López-Rojo, N., Tonin, A. M., Correa-Araneda, F., Pearson, R. G., Bosch, J., Albariño, R. J., Anbalagan, S., Barmuta, L.A., Beesley, L., Burdon, F. J., Caliman, A., Callisto, M., Campbell, I. C., Cardinale, B. J., Casas, J. J., Chará-Serna, A. M., Ciapała, S., Chauvet, E., … Yule, C. M. (2021). Latitude dictates plant diversity effects on instream decomposition. *Science Advances, 7*(13), eabe7860.

Brown, J. H., Gillooly, J. F., Allen, A. P., Savage, V. M., & West, G. B. (2004). Toward a metabolic theory of ecology. *Ecology, 85*, 1771–1789.

Bruder, A., Schindler, M. H., Moretti, M. S., & Gessner, M. O. (2014). Litter decomposition in a temperate and a tropical stream: The effects of species mixing, litter quality and shredders. *Freshwater biology, 59*, 438–449.

Camacho, R., Boyero, L., Cornejo, A., Ibáñez, A., & Pearson, R. G. (2009). Local variation in shredder distribution can explain their oversight in tropical streams. *Biotropica, 41*, 625–632.

Cheshire, K., Boyero, L., & Pearson, R. G. (2005). Food webs in tropical Australian streams: Shredders are not scarce. *Freshwater biology, 50*, 748–769.

Coley, P. D., & Barone, J. A. (1996). Herbivory and plant defenses in tropical forests. *Annual review of ecology and systematics, 27*, 305–335.

Connolly, N. M., & Pearson, R. G. (2013). Nutrient enrichment of a heterotrophic stream alters leaf litter nutritional quality and shredder physiological condition via the microbial pathway. *Hydrobiologia, 718*, 85–92.

Corvalan, C., Hales, S., & McMichael, A. (2005). *Ecosystems and human well-being: Health synthesis: A report of the Millennium Ecosystem Assessment*. World Health Organization.

Coughlan, J. F., Pearson, R. G., & Boyero, L. (2010). Crayfish process leaf litter in tropical streams even when shredding insects are common. *Marine and freshwater research, 61*, 541–548.

Cuffney, T. F., Wallace, J. B., & Lugthart, G. J. (1990). Experimental evidence quantifying the role of benthic invertebrates in organic matter dynamics of headwater streams. *Freshater biology, 23*, 281–299.

Cummins, K. W. (1974). Structure and function of stream ecosystems. *BioScience, 24*, 631–641.

Cummins, K. W., Wilzbach, M. A., Gates, D. M., Perry, J. B., & Taliaferro, W. B. (1989). Shredders and riparian vegetation. *BioScience, 39*, 24–30.

Datry, T., Foulquier, A., Corti, R., von Schiller, D., Tockner, K., Mendoza-Lera, C., Clément, J. C., Gessner, M. O., Moleón, M., Stubbington, R., Gücker, B., Albariño, R., Allen, D. C., Altermatt, F., Arce, M. I., Arnon, S., Banas, D., Banegas-Medina, A., Beller, E., ... Zoppini, A. (2018). A global analysis of terrestrial plant litter dynamics in non-perennial waterways. *Nature geoscience, 11*, 497–503.

De Meester, L., Sotks, R., & Brans, K. I. (2018). Genetic adaptation as a biological buffer against climate change: Potential and limitations. *Integrative zoology, 13*, 372–391.

Deutsch, C. A., Tewksbury, J. J., Huey, R. B., Sheldon, K. S., Ghalambor, C. K., Haak, D. C., & Martin, P. R. (2008). Impacts of climate warming on terrestrial ectotherms across latitude. *Proceedings of the national academy of sciences of the united states of america, 105*, 6668–6672.

Dobson, M., Magana, A., Mathooko, J. M., & Ndegwa, F. K. (2002). Detritivores in Kenyan highland streams: More evidence for the paucity of shredders in the tropics? *Freshwater biology, 47*, 909–919.

Downing, J. (2012). Global abundance and size distribution of streams and rivers. *Inland waters, 2*, 229–236.

Duarte, S., Barlochër, F., Pascoal, C., & Cássio, F. (2016). Biogeography of aquatic hyphomycetes: Current knowledge and future perspectives. *Fungal ecology, 19*, 169–181.

Encalada, A., Calles, J., Ferreira, V., Canhoto, C., & Graça, M. A. S. (2010). Riparian land use and the relationship between the benthos and litter decomposition in tropical montane forest streams. *Freshwater biology, 55*, 1719–1733.

Encalada, A. C., Flecker, A. S., Poff, N. L., Suárez, E., Herrera-R, G. A., Ríos-Touma, B., Jumani, S., Larson, E. I., & Anderson, E. P. (2019). A global perspective on tropical montane rivers. *Science, 365*, 1124–1129.

Ferreira, V., Boyero, L., Calvo, C., Correa, F., Figueroa, R., Gonçalves, J. F., Goyenola, G., Graça, M. A. S., Hepp, L. U., Kariuki, S., López-Rodríguez, A., Mazzeo, N., M'Erimba, C., Monroy, S., Peil, A., Pozo, J., Rezende, R., & Teixeira-de-Mello, F. (2018). A global assessment of the effects of eucalyptus plantations on stream ecosystem functioning. *Ecosystems, 22*, 629–642.

Ferreira, V., Castagneyrol, B., Koricheva, J., Gulis, V., Chauvet, E., & Graça, M. A. (2015). A meta-analysis of the effects of nutrient enrichment on litter decomposition in streams. *Biological reviews of the cambridge philosophical society, 90*, 669–688.

Ferreira, V., & Chauvet, E. (2011). Synergistic effects of water temperature and dissolved nutrients on litter decomposition and associated fungi. *Global change biology, 17*, 551–564.

Ferreira, V., Encalada, A. C., & Graça, M. A. S. (2012). Effects of litter diversity on decomposition and biological colonization of submerged litter in temperate and tropical streams. *Freshwater science, 31*, 945–962.

Follstad Shah, J. J., Kominoski, J. S., Ardon, M., Dodds, W. K., Gessner, M. O., Griffiths, N. A., Hawkins, C. P., Johnson, S. L., Lecerf, A., LeRoy, C. J., Manning, D. W. P., Rosemond, A. D., Sinsabaugh, R. L., Swan, C. M., Webster, J. R., & Zeglin, L. H. (2017). Global synthesis of the temperature sensitivity of leaf litter breakdown in streams and rivers. *Global Change Biology, 23*, 3064–3075.

Foster, J., Hirst, A. G., & Woodward, G. (2011). Growth and development rates have different thermal responses. *The American Naturalist, 178*, 668–778.

Fraser, L. H., Henry, H. A. L., Carlyle, C. N., White, S. R., Beierkuhnlein, C., Cahill, J. F., Casper, B. B., Cleland, E., Collins, S. L., Dukes, J. S., Knapp, A. K., Lind, E., Long, R., Luo, Y., Reich, P. B., Smith, M. D., Sternberg, M., & Turkington, R. (2013). Coordinated distributed experiments: An emerging tool for testing global hypotheses in ecology and environmental science. *Frontiers in Ecology and the Environment, 11*, 147–155.

Fugère, V., Jacobsen, D., Finestone, E. H., & Chapman, L. J. (2018). Ecosystem structure and function of afrotropical streams with contrasting land use. *Freshwater Biology, 63*, 1498–1513.

Gallardo, A., & Merino, J. (1993). Leaf decomposition in two Mediterranean ecosystems of sourthwest Spain: Influence of substrate quality. *Ecology, 74*, 152–161.

García-Palacios, P., McKie, B. G., Handa, I. T., Frainer, A., & Hättenschwiler, S. (2016). The importance of litter traits and decomposers for litter decomposition: A comparison of aquatic and terrestrial ecosystems within and across biomes. *Functional Ecology, 30*, 819–829.

Gessner, M. O. (1997). Fungal biomass, production and sporulation associated with particulate organic matter in streams. *Limnetica, 13*, 33–44.

Gessner, M. O., & Chauvet, E. (1994). Importance of stream microfungi in controlling breakdown rates of leaf litter. *Ecology, 75*, 1807–1817.

Gessner, M. O., Chauvet, E., & Dobson, M. (1999). A perspective on leaf litter breakdown in streams. *Oikos, 85*, 377–384.

Gessner, M. O., & Peeters, F. (2020). Determining temperature-normalized decomposition rates. In F. Barlochër, M. O. Gessner, & M. A. S. Graça (Eds.), *Methods to study litter decomposition: A practical guide.* (pp. 545–552). Springer International Publishing.

Gonçalves, J. F., Graça, M. A. F., & Callisto, M. (2006). Leaf-litter breakdown in 3 streams in temperate, Mediterranean, and tropical Cerrado climates. *Journal of the North American Benthological Society, 24*, 344–355.

Gonçalves, J. F., Graça, M. A. S., & Callisto, M. (2007). Litter decomposition in a Cerrado savannah stream is retarded by leaf toughness, low dissolved nutrients and a low density of shredders. *Freshwater Biology, 52*, 1440–1451.

Graça, M. A. S. (2001). The role of invertebrates on leaf litter decomposition in streams—A review. *International Review of Hydrobiology, 86*, 383–393.

Graça, M. A. S., & Cressa, C. (2010). Leaf quality of some tropical and temperate tree species as food resource for stream shredders. *International Review of Hydrobiology, 95*, 97–41.

Graça, M. A. S., Ferreira, V., Canhoto, C., Encalada, A. C., Gerrero-Bolaño, F., Wantzen, K. M., & Boyero, L. (2015). A conceptual model of litter breakdown in low order streams. *International Review of Hydrobiology, 100*, 1–12.

Graça, M. A. S., Hyde, K., & Chauvet, E. (2015). Aquatic hyphomycetes and litter decomposition in tropical—Subtropical low order streams. *Fungal Ecology, 19*, 182–189.

Gulis, V., Ferreira, V., & Graça, M. A. S. (2006). Stimulation of leaf litter decomposition and associated fungi and invertebrates by moderate eutrophication: Implications for stream assessment. *Freshwater Biology, 51*, 1655–1669.

Haase, P., Pilotto, F., Li, F., Sundermann, A., Lorenz, A. W., Tonkin, J. D., & Stoll, S. (2019). Moderate warming over the past 25years has already reorganized stream invertebrate communities. *Science of the Total Environment, 658*, 1531–1538.

Handa, I. T., Aerts, R., Berendse, F., Berg, M. P., Bruder, A., Butenschoen, O., Chauvet, E., Gessner, M. O., Jabiol, J., Makkonen, M., McKie, B. G., Malmqvist, B., Peeters, E. T., Scheu, S., Schmid, B., van Ruijven, J., Vos, V. C., & Hattenschwiler, S. (2014). Consequences of biodiversity loss for litter decomposition across biomes. *Nature, 509*, 218–221.

Hieber, M., & Gessner, M. O. (2002). Contribution of stream detrivores, fungi, and bacteria to leaf breakdown based on biomass estimates. *Ecology, 83*, 1026–1038.

Horne, C. R., Hirst, A. G., & Atkinson, D. (2015). Temperature-size responses match latitudinal clines in arthropodes, revealing critical differences between aquatic and terrestrial species. *Ecology Letters, 18*, 327–335.

Hotchkiss, E. R., Hall, R. O., Jr., Sponseller, R. A., Butman, D., Klaminder, J., Laudon, H., Rosvall, M., & Karlsson, J. (2015). Sources of and processes controlling CO_2 emissions change with the size of streams and rivers. *Nature Geoscience, 8*, 696–699.

IPCC. (2014). *Climate Change 2014: Synthesis Report. Contribution of Working Groups I, II and III to the Fifth Assessment Report of the Intergovernmental Panel on Climate Change.* IPCC, Geneva, Switzerland.

IPCC. (2018). *Global warming of 1.5°C. An IPCC Special Report on the impacts of global warming of 1.5°C above pre-industrial levels and related global greenhouse gas emission pathways, in the context of strengthening the global response to the threat of climate change, sustainable development, and efforts to eradicate poverty* (V. Masson-Delmotte, P. Zhai, H. O. Pörtner, D. Roberts, J. Skea, P. R. Shukla, A. Pirani, W. Moufouma-Okia, C. Péan, R. Pidcock, S. Connors, J. B. R. Matthews, Y. Chen, X. Zhou, M. I. Gomis, E. Lonnoy, T. Maycock, M. Tignor, & T. Waterfield, Eds.). Geneva, Switzerland.

Irons, J. G., Oswood, M. W., Stout, R. J., & Pringle, C. M. (1994). Latitudinal patterns in leaf litter breakdown: Is temperature really important? *Freshwater biology, 32*, 401–411.

Iversen, T. M. (1979). Laboratory energetics of larvae of Sericostoma personatum (Trichoptera). *Holarctic Ecology, 2*, 1–5.

Jabiol, J., Bruder, A., Gessner, M. O., Makkonen, M., McKie, B. G., Peeters, E. T. H. M., Vos, V. C. A., & Chauvet, E. (2013). Diversity patterns of leaf-associated aquatic hyphomycetes along a broad latitudinal gradient. *Fungal Ecology, 6*, 439–448.

Jabiol, J., Lecerf, A., Lamothe, S., Gessner, M. O., & Chauvet, E. (2019). Litter quality modulates effects of dissolved nitrogen on leaf decomposition by stream microbial communities. *Microbial Ecology, 77*, 959–966.

Jinggut, T., & Yule, C. M. (2015). Leaf-litter breakdown in streams of East Malaysia (Borneo) along an altitudinal gradient: Initial nitrogen content of litter limits shredder feeding. *Freshwater Science, 34*, 691–701.

Kuehn, K. A. (2016). Lentic and lotic habitats as templets for fungal communities: Traits, adaptations, and their significance to litter decomposition within freshwater ecosystems. *Fungal Ecology, 19*, 135–154.

LeRoy, C. J., Hipp, A. L., Lueders, K., Follstad Shah, J. J., Kominoski, J. S., Ardón, M., Dodds, W. K., Gessner, M. O., Griffiths, N. A., Lecerf, A., Manning, D. W. P., Sinsabaugh, R. L., & Webster, J. R. (2019). Plant phylogenetic history explains in-stream decomposition at a global scale. *Journal of Ecology, 108*, 17–35.

LeRoy, C. J., Whitham, T. G., Keim, P., & Marks, J. C. (2006). Plant genes link forests and streams. *Ecology, 87*, 255–261.

López-Rojo, N., Martínez, A., Pérez, J., Basaguren, A., Pozo, J., & Boyero, L. (2018). Leaf traits drive plant diversity effects on litter decomposition and FPOM production in streams. *PLoS ONE, 13*, e0198243.

López-Rojo, N., Pozo, J., Pérez, J., Basaguren, A., Martínez, A., Tonin, A. M., Correa-Araneda, F., & Boyero, L. (2019). Plant diversity loss affects stream ecosystem multifunctionality. *Ecology, 100*, e02847.

Manzoni, S., Trofymow, J. A., Jackson, R. B., & Porporato, A. (2010). Stoichiometric controls on carbon, nitrogen, and phosphorus dynamics in decomposing litter. *Ecological Monographs, 80*, 89–106.

Marks, J. C. (2019). Revisiting the fates of dead leaves that fall into streams. *Annual Review of Ecology, Evolution, and Systematics, 50*, 547–568.

Marquis, R. J., Ricklefs, R. E., & Abdala-Roberts, L. (2012). Testing the low latitude/high defense hypothesis for broad-leaved tree species. *Oecologia, 169*, 811–820.

Martínez, A., Larrañaga, A., Pérez, J., Descals, E., & Pozo, J. (2014). Temperature affects leaf litter decomposition in low-order forest streams: Field and microcosm approaches. *FEMS Microbiology Ecology, 87*, 257–267.

Masese, F. O., Kitaka, N., Kipkemboi, J., Gettel, G. M., Irvine, K., & McClain, M. E. (2014). Macroinvertebrate functional feeding groups in Kenyan highland streams: Evidence for a diverse shredder guild. *Freshwater Science, 33*, 435–450.

Meyer, J. L., Wallace, J. B., & Eggert, S. L. (1998). Leaf litter as a source of dissolved organic carbon in streams. *Ecosystems, 1*, 240–249.

Mohseni, O., Stefan, H. G., & Eaton, J. G. (2003). Global warming and potential changes in fish habitat in U.S. streams. *Climatic Change, 59*, 389–409.

Mulholland, P. J., Helton, A. M., Poole, G. C., Hall, R. O., Hamilton, S. K., Peterson, B. J., Tank, J. L., Ashkenas, L. R., Cooper, L. W., Dahm, C. N., Dodds, W. K., Findlay, S. E., Gregory, S. V., Grimm, N. B., Johnson, S. L., McDowell, W. H., Meyer, J. L., Valett, H. M., Webster, J. R., … Thomas, S. M. (2008). Stream denitrification across biomes and its response to anthropogenic nitrate loading. *Nature, 452*, 202–205.

Neukom, R., Steiger, N., Gomez-Navarro, J. J., Wang, J., & Werner, J. P. (2019). No evidence for globally coherent warm and cold periods over the preindustrial Common Era. *Nature, 571*, 550–554.

Pascoal, C., Cassio, F., Marcotegui, A., Sanz, B., & Gomes, P. (2005). Role of fungi, bacteria, and invertebrates in leaf litter breakdown in a polluted river. *Journal of the North American Benthological Society, 24*, 784–797.

Pennings, S. C., & Bertness, M. D. (1999). Using latitudinal variation to examine effects of climate on coastal salt marsh pattern and process. *Current Topics in Wetland Biogeochemistry, 3*, 100–111.

Pérez, J., Descals, E., & Pozo, J. (2012). Aquatic hyphomycete communities associated with decomposing alder leaf litter in reference headwater streams of the Basque Country (northern Spain). *Microbial Ecology, 64*, 279–290.

Peterson, B. J., Hobbie, J. E., & Corliss, T. L. (1986). Carblon flow in a tundra stream ecosystem. *Canadian Journal of Fisheries and Aquatic Sciences, 43*, 1259–1270.

Pettit, N. E., Davies, T., Fellman, J. B., Grierson, P. F., Warfe, D. M., & Davies, P. M. (2011). Leaf litter chemistry, decomposition and assimilation by macroinvertebrates in two tropical streams. *Hydrobiologia, 680*, 63–77.

Pitelka, L. F., & Group, P. M. W. (1997). Plant migration and climate change. *American Scientist, 86*, 464–473.

Raymond, P. A., Hartmann, J., Lauerwald, R., Sobek, S., McDonald, C., Hoover, M., Butman, D., Striegl, R., Mayorga, E., Humborg, C., Kortelainen, P., Durr, H., Meybeck, M., Ciais, P., & Guth, P. (2013). Global carbon dioxide emissions from inland waters. *Nature, 503*, 355–359.

Reich, P. B., & Oleksyn, J. (2004). Global patterns of plant leaf N and P in relation to temperature and latitude. *Proceedings of the National Academy of Sciences United State of America, 101*, 11001–11006.

Robinson, C. T., Gessner, M. O., & Ward, J. V. (1998). Leaf breakdown and associated macroinvertebrates in alpine glacial streams. *Freshwater Biology, 40*, 215–228.

Rosemond, A. D., Benstead, J. P., Bumpers, P. M., Gullis, V., Kominoski, J. S., Manning, D. W. P., Suberkropp, K., & Wallace, J. B. (2015). Experimental nutrient additions accelerate terrestrial carbon loss from stream ecosystems. *Science, 347*, 1142–1145.

Royer, T. V., & Minshall, G. W. (2003). Controls on leaf processing in streams from spatial-scaling and hierarchical perspectives. *Journal of the North American Benthological Society, 22*, 352–358.

Rueda-Delgado, G., Wantzen, K. M., & Tolosa, M. B. (2006). Leaf-litter decomposition in an Amazonian floodplain stream: Effects of seasonal hydrological changes. *Journal of the North American Benthological Society, 25*, 233–249.

Sabater, S., Bregoli, F., Acuna, V., Barcelo, D., Elosegi, A., Ginebreda, A., Marce, R., Munoz, I., Sabater-Liesa, L., & Ferreira, V. (2018). Effects of human-driven water stress on river ecosystems: A meta-analysis. *Scientific Reports, 8*, 11462.

Schiller, D., Datry, T., Corti, R., Foulquier, A., Tockner, K., Marcé, R., García-Baquero, G., Odriozola, I., Obrador, B., Elosegi, A., Mendoza-Lera, C., Gessner, M. O., Stubbington, R., Albariño, R., Allen, D. C., Altermatt, F., Arce, M. I., Arnon, S., Banas, D., … Zoppini, A. (2019). Sediment respiration pulses in intermittent rivers and ephemeral streams. *Global Biogeochemical Cycles, 33*(10), 1251–1263.

Schmidt, K., Blanchette, M., Pearson, R. G., Alford, R. A., & Davis, A. M. (2017). Trophic roles of tadpoles in tropical Australian streams. *Freshwater Biology, 62*, 1929–1941.

Seena, S., Bärlocher, F., Sobral, O., Gessner, M. O., Dudgeon, D., McKie, B. G., Chauvet, E., Boyero, L., Ferreira, V., Frainer, A., Bruder, A., Matthaei, C. D., Fenoglio, S., Sridhar, K. R., Albarino, R. J., Douglas, M. M., Encalada, A. C., Garcia, E., Ghate, S. D., … Graça, M. A. S.

(2019). Biodiversity of leaf litter fungi in streams along a latitudinal gradient. *Science of the Total Environment, 661*, 306–315.

Shumilova, O., Zak, D., Datry, T., von Schiller, D., Corti, R., Foulquier, A., Obrador, B., Tockner, K., Allan, D.C., Altermatt, F., Arce, M. I., Arnon, S., Banas, D., Banegas-Medina, A., Beller, E., Blanchette, M.L., Blanco-Libreros, J.F., Blessing, J., Boechat, I.G., … Zarfl, C. (2019). Simulating rewetting events in intermittent rivers and ephemeral streams: A global analysis of leached nutrients and organic matter. *Global Change Biology, 25*(5), 1591–1611.

Smith, S. D., Devitt, D. A., Sala, A., Cleverly, J. R., & Busch, D. E. (1998). Water relations of riparian plants from warm desert regions. *Wetlands, 18*, 687–696.

Stout, R. J. (1989). Effects of condensed tannins on leaf processing in mid-latitude and tropical streams: A theoretical approach. *Canadian Journal of Fisheries and Aquatic Sciences, 46*, 1097–1106.

Suberkropp, K. (1991). Relationships between growth and sporulation of aquatic hyphomycetes on decomposing leaf litter. *Mycological Research, 95*, 843–850.

Suberkropp, K. (1998). Effect of dissolved nutrients on two aquatic hyphomycetes growing on leaf litter. *Mycological Research, 102*, 998–1002.

Suberkropp, K., & Chauvet, E. (1995). Regulation of leaf breakdown by fungi in streams: Influences of water chemistry. *Ecology, 76*, 1433–1445.

Suberkropp, K., Gulis, V., Rosemond, A. D., & Benstead, J. P. (2010). Ecosystem and physiological scales of microbial responses to nutrients in a detritus-based stream: Results of a 5-year continuous enrichment. *Limnology and Oceanography, 55*, 149–160.

Suberkropp, K., & Klug, M. J. (1980). The maceration of deciduous leaf litter by aquatic hyphomycetes. *Canadian Journal of Botany, 58*, 1025–1031.

Tank, J. L., Rosi-Marshall, E. J., Griffiths, N. A., Entrekin, S. A., & Stephen, M. L. (2010). A review of allochthonous organic matter dynamics and metabolism in streams. *Journal of the North American Benthological Society, 29*, 118–146.

Tenkiano, N. S. D., & Chauvet, E. (2018). Leaf litter decomposition in Guinean savannah streams. *Inland Waters, 8*, 413–421.

Tiegs, S. D., Costello, D. M., Isken, M. W., Woodward, G., McIntyre, P. B., Gessner, M. O., Chauvet, E., Griffiths, N. A., Flecker, A. S., Acuña, V., Albariño, R., Allen, D. C., Alonso, C., Andino, P., Arango, C., Aroviita, J., Barbosa, M. V. M., Barmuta, L. A., Baxter, C. V., … Zwart, J.A. (2019). Global patterns and drivers of ecosystem functioning in rivers and riparian zones. *Science Advances*, 5, eaav0486.

Tonin, A. M., Boyero, L., Monroy, S., Basaguren, A., Pérez, J., Pearson, R. G., Cardinale, B. J., Gonçalves, J. F. J., & Pozo, J. (2017). Stream nitrogen concentration, but not plant N-fixing capacity, modulates litter diversity effects on decomposition. *Functional Ecology, 31*, 1471–1481.

Tonin, A. M., Gonçalves, J. F., Jr., Bambi, P., Couceiro, S. R. M., Feitoza, L. A. M., Fontana, L. E., Hamada, N., Hepp, L. U., Lezan-Kowalczuk, V. G., Leite, G. F. M., Lemes-Silva, A. L., Lisboa, L. K., Loureiro, R. C., Martins, R. T., Medeiros, A. O., Morais, P. B., Moretto, Y., Oliveria, P. C. A., Pereira, E. B., … Boyero, L. (2017). Plant litter dynamics in the forest-stream interface: precipitation is a major control across tropical biomes. *Scientific Reports, 7*, 10799.

Tonin, A. M., Hepp, L. U., & Gonçalves, J. F. (2018). Spatial variability of plant litter decomposition in stream networks: From litter bags to watersheds. *Ecosystems, 21*, 567–581.

Tonin, A. M., Pozo, J., Monroy, S., Basaguren, A., Pérez, J., Gonçalves, J. F., Jr., Pearson, R., Cardinale, B. J., & Boyero, L. (2018). Interactions between large and small detritivores influence how biodiversity impacts litter decomposition. *Journal of Animal Ecology, 87*, 1465–1474.

Vannote, R. L., Minshall, G. W., Cummins, K. W., Sedell, J. R., & Cushing, C. E. (1980). The River Continuum Concept. *Canadian Journal of Fisheries and Aquatic Sciences, 37*, 130–137.

Vitousek, P. M., Mooney, H. A., Lubchenco, J., & Melillo, J. M. (1997). Human Domination of Earth's Ecosystems. *Science, 277*, 494–499.

Wallace, J., Eggert, S., Meyer, J., & Webster, J. (1997). Multiple trophic levels of a forest stream linked to terrestrial litter inputs. *Science, 277*, 102–104.

Wallace, J. B., & Webster, J. R. (1996). The role of macroinvertebrates in stream ecosystem function. *Annual Review of Entomology, 41*, 115–139.
Wantzen, K. M., & Wagner, R. (2006). Detritus processing by invertebrate shredders: A neotropical–temperate comparison. *Journal of the North American Benthological Society, 25*, 216–232.
Wantzen, K. M., Wagner, R., & Junk, W. J. (2002). How do plant-herbivore interactions of tree influence coarse detritus processing by shredders in aquatic ecossystems of different latitudes? *Verhandlungen der Internationalen Vereinigung für Limnolologie, 28*, 815–821.
Wantzen, K. M., Wagner, R., Suetfeld, R., & Junk, W. J. (2002). How do plant-herbivore interactions of trees influence coarse detritus processing by shredders in aquatic ecosystems of different latitudes? *Verhandlungen der Internationalen Vereinigung für Limnolologie, 28*, 815–821.
Webster, J. R., & Benfield, E. F. (1986). Vascular plant breakdown in freshwater ecosystems. *Annual Review of Ecology and Systematics, 17*, 567–594.
Wild, R., Gücker, B., & Brauns, M. (2019). Agricultural land use alters temporal dynamics and the composition of organic matter in temperate headwater streams. *Freshwater Biology, 38*, 566–581.
Willig, M. R., Kaufman, D. M., & Stevens, R. D. (2003). Latitudinal gradients of biodiversity: Pattern, process, scale, and synthesis. *Annual Review of Ecology, Evolution, and Systematics, 34*, 273–309.
Winterbourn, M. J., Rounick, J. S., & Cowie, B. (1981). Are New Zealand stream ecosystems really different? *New Zealand Journal of Marine and Freshwater Research, 15*, 321–328.
Woodward, G., Gessner, M. O., Giller, P. S., Gullis, V., Hladyz, H., Lecerf, A., Malmqvist, B., McKie, B. G., Tiegs, S. D., Cariss, H., Dobson, M., Elosegi, A., Ferreira, V., Graça, M. A. S., Fleituch, T., Lacoursière, J. O., Nistorescu, M., Pozo, J., Risnoveanu, G., Schindler, M., … Chauvet, E. (2012). Continental-scale effects of nutrient pollution on stream ecosystem functioning. *Science, 336*, 1438–1440.
Wootton, A., Pearson, R. G., & Boyero, L. (2018). Patterns of flow, leaf litter and shredder abundance in a tropical stream. *Hydrobiologia, 826*, 353–365.
Yuan, Z., & Chen, H. Y. H. (2009). Global-scale patterns of nutrient resorption associated with latitude, temperature and precipitation. *Global Ecology and Biogeography, 18*, 11–18.
Yule, C. M. (1996). Trophic relationships and food webs of the benthic invertebrate fauna of two aseasonal tropical streams on Bougainville Island, Papua New Guinea. *Journal of Tropical Ecology, 12*, 517–534.
Yule, C. M., Leong, M. Y., Liew, K. C., Ratnarajah, L., Schmidt, K., Wong, H. M., Pearson, R. G., & Boyero, L. (2009). Shredders in Malaysia: Abundance and richness are higher in cool upland tropical streams. *Journal of the North American Benthological Society, 28*, 404–415.
Yule, C. M., & Pearson, R. G. (1996). Aseasonality of benthic invertebrates in a tropical stream on Bougainville Island, Papua New Guinea. *Archiv für Hydrobiologie, 137*, 95–117.
Zhang, M., Cheng, X., Geng, Q., Shi, Z., Luo, Y., Xu, X., & Morellato, P. (2019). Leaf litter traits predominantly control litter decomposition in streams worldwide. *Global Ecology and Biogeography, 28*, 1469–1486.
Zúñiga-Cépsedes, B., Zúñiga, M. C., & Chará, J. (2018). The effect of macroinvertebrate exclusion on leaf breakdown rates in two upland Colombian streams. *International Journal of Tropical Biology, 66*, 457–467.

Chapter 5
Plant Litter Decomposition in Intermittent Rivers and Ephemeral Streams

Rubén del Campo, Arnaud Foulquier, Gabriel Singer, and Thibault Datry

Abstract Intermittent rivers and ephemeral streams (hereafter IRES) are waterways that temporarily cease to flow and/or dry up. They represent half the length of the global river network and are expanding in time and space in response to global change. The hydrological regimes of IRES are characterized by alternating flowing, non-flowing and dry phases, which translate to varying importance of in-stream litter accumulation, processing and downstream transport. Decomposition agents, processes and rates dramatically change among these hydrological phases, leading to decomposition dynamics that differ markedly from perennial rivers and streams. As a result, IRES have a specific "biogeochemical heartbeat" characterized by high temporal and spatial variability of leaf decomposition, and so they can be idealized as pulsed bioreactors. The ecological effects of flow cessation and drying are sometimes visible far beyond rewetting, generating "legacy effects" that become apparent even during later flowing phases. Rewetting events can represent "hot moments" of litter decomposition due to the intense biological and physical activities, generating pulses of transport and decomposition. Upscaling the abundant reach-scale knowledge to larger river-network scales is probably one of the most challenging but timely paths for future research.

R. del Campo (✉) · G. Singer
Department of Ecology, University of Innsbruck, Technikerstrasse 25, 6020 Innsbruck, Austria
e-mail: ruben.del-campo@uibk.ac.at

G. Singer
e-mail: gabriel.singer@uibk.ac.at

A. Foulquier
University Grenoble Alpes, University Savoie Mont Blanc, CNRS, LECA, 38000 Grenoble, France
e-mail: arnaud.foulquier@univ-grenoble-alpes.fr

T. Datry
INRAE, UR RIVERLY, Centre de Lyon-Villeurbanne, 5 rue de la Doua CS70077, 69626 Villeurbanne cedex, France
e-mail: thibault.datry@inrae.fr

C. M. Swan et al. (eds.), *The Ecology of Plant Litter Decomposition in Stream Ecosystems*, https://doi.org/10.1007/978-3-030-72854-0_5

5.1 What Are Intermittent Rivers and Ephemeral Streams?

5.1.1 Habitat Mosaic and Hydrological Phases

Intermittent rivers and ephemeral streams (IRES) include all flowing waters that temporarily cease to flow in surface and/or dry at some point along their course (Datry, Bonada, et al., 2017, Fig. 5.1). With the cessation of surface flow, shallow water habitats dry and thereby generate a chain of isolated pools, potentially still connected by hyporheic flow. We refer here to this phase as the non-flowing period. Finally, the drying results in the complete disappearance of surface water in the channel (though hyporheic flow may still exist) and leads to the dry phase. In response to river discharge and groundwater levels (Datry et al., 2016; Jaeger et al., 2014; Stanley et al., 1997), the extent of and connectivity between flowing, non-flowing and dry habitats (Fig. 5.1) vary continuously across a river network, forming a dynamically shifting mosaic of terrestrial and aquatic habitats. These dynamics translate to hydrological phases, which are each characterised by different biota, ecological processes and ecosystem services (Datry et al., 2014). The phase shifts can represent hot moments (sensu McClain et al., 2003) for some biological processes, e.g., the increase of aquatic food resources available for terrestrial invertebrates when surface water flow disappears (McIntosh et al., 2017), or the transport of terrestrial organisms (Corti & Datry, 2012) and huge amounts of organic matter (Datry et al., 2018) to downstream ecosystems upon rewetting.

5.1.2 Abundance and Distribution

Classifying flow regimes is complex. The scientific literature shows many efforts (e.g., Gallart et al., 2012; Uys & O'Keeffe, 1997; Williams, 2006) to assign names to classes of intermittent rivers based on drying duration, frequency and predictability.

Fig. 5.1 Alternating flowing (**a**), non-flowing (**b**), dry (**c**), and rewetting phases (**d**) in an intermittent river in France (Calavon River). Photo credits: Bertrand Launay

However, a global consensus remains elusive, as many intermittently flowing waterways dry for widely different periods in different years, leading to variation of a single waterway among different categories (Datry, Bonada, et al., 2017). For the sake of simplicity, we refer to 'intermittent rivers and ephemeral streams', IRES, to refer to all flowing waters that cease to flow or dry completely at some point along their course. Arguably the world's most widespread type of flowing water (Datry et al., 2018; Larned et al., 2010), IRES range from small ephemeral streams that flow for a few days after heavy rain to large intermittent rivers that recede to isolated pools but might not dry completely. We acknowledge that some rivers can be ephemeral and streams intermittent and that many local names such as *winterbournes*, *wadis*, *arroyos* and *ramblas* (e.g., Steward et al., 2012) can be used to describe IRES. This diversity of names highlights the diversity and cultural importance of IRES to people living in their catchments (Fig. 5.2).

IRES occur on all continents, including Antarctica (Larned et al., 2010; Steward et al., 2012). Some global estimations calculate that IRES comprise 15% of the global

Fig. 5.2 Different types of IRES from across the world during non-flowing conditions: (**a**) unnamed karstic stream, West Coast, South Island, New Zealand, (**b**) Río Seco, Chaco, Bolivia, (**c**) Asse River, Provence, France, (**d**) unnamed gravel-bed stream, West Coast, South Island, New Zealand, (**e**) unnamed stream, Altiplano, Bolivia, (**f**) Chaki Mayu, Amazonia, Bolivia, (**g**) Clauge, Jura, France, (**h**) Calavon River, Provence, France, and (**i**) Hozgarganta River, Andalucía, Spain. Photo credits: Thibault Datry (**a**–**f**), Bertrand Launay (**g**, **h**), and Núria Bonada (**i**). Figure extracted from Datry, Corti, et al. (2017). Permission for reuse requested from Elsevier

river-network area (Raymond et al., 2013), representing >50% of the global river-network length (Lehner et al. personal communication). Indeed, every river network may include IRES virtually, since headwaters are usually at least partly intermittent (Fritz et al., 2013; Grill et al., 2019; Lowe & Likens, 2005). IRES are ubiquitous in most climatic regions (Bonada & Resh, 2013; Leigh et al., 2016; Sabater & Tockner, 2010; Vander Vorste et al., 2020) but are particularly conspicuous in arid landscapes, which constitute already more than a third of the Earth's land surface (Millenial Ecosystem Assessment, 2005; Tooth, 2000; Fig. 5.2). For example, more than the 70% of rivers in Australia are considered intermittent (Sheldon et al., 2010), 66 to 94% of river lengths in Southwest USA have an intermittent or ephemeral flow regime (Levick et al., 2008), and up to 35% of the French river network is prone to drying (Snelder et al., 2013).

5.1.3 Drivers of Flow Intermittence and Trends

Different processes generate natural flow intermittence in streams and rivers, acting individually or in combination: transmission loss (infiltration of surface water into porous streambeds), evapotranspiration, downward shifts in groundwater tables, hill-slope runoff recession, and freezing (Larned et al., 2010). Furthermore, there is growing evidence that the total number and length of IRES is increasing due to anthropogenic causes. Anthropogenic flow intermittence can be due to: alteration of land-use patterns, flow regulation and diversion, surface or groundwater extraction, and reduced precipitation and increased evaporation resulting from climate change (Datry, Bonada, et al., 2017; Palmer et al., 2008; Steward et al., 2012).

There is little doubt that IRES are going to be the dominant type of waterway in the near future (Datry et al., 2018) due to the predicted increase in flow intermittence worldwide as a consequence of natural and anthropogenic disturbances associated to climate and global change. Currently, many IRES are already experiencing longer and more frequent flow cessation and dry periods. For example, using a hydrological modelling coupled to climate change scenarios, Cipriani et al. (2014) estimated annual flow intermittence to increase on average by 5–10% by 2050 in the intermittent Albarine River in France. Water abstraction and impoundment have caused many formerly perennial rivers to become intermittent in the last 50 years, including large rivers such as the Nile, Indus, Yellow, Amu and Syr Darya, Rio Grande and Colorado (Meybeck, 2003; Postel, 2000). In the near future, the extent of IRES in fluvial networks will increase particularly in regions where severe climatic drying and/or water appropriation occurs (Cipriani et al., 2014; Döll & Schmied, 2012; Larned et al., 2010). Transitions of flow regime from perennial to intermittent are projected until the 2050s globally, for instance in Mediterranean regions, Southern and West Africa, Brazil, the Caribbean, California or north-eastern Australia (Döll & Schmied, 2012). On the contrary, some IRES might shift to a perennial flow regime due to warmer winters in some parts of Alaska, Canada or Siberia (Döll & Schmied, 2012). Similarly, the influence of the discharge of agricultural fields, industrial or municipal effluents, the controlled releases from dams and weirs, or the transfer of river water among basins may also turn some IRES into perennial waterways (Datry et al., 2014; Hassan & Egozi, 2001; Steward et al., 2012).

5.2 Rates, Agents and Processes of Leaf Litter Decomposition in IRES Habitats

5.2.1 Leaf Litter Decomposition in Flowing Water Conditions

Processes and agents involved in leaf litter decomposition in IRES during flowing phases are similar to perennial water courses. Leaf litter decomposition is driven by a combination of abiotic (leaching of soluble compounds, physical abrasion) and biotic processes (microbial- and macroinvertebrate-driven decomposition) (Fig. 5.3). However, preceding dry phases can exert a strong negative effect on the aquatic decomposition rates of leaf litter during the flowing phase of IRES. Various studies have reported that the increase of the dry-phase length and drying frequency may result in a drastic decrease of leaf litter decomposition rates in IRES compared to perennial streams (Table 5.1) (Datry et al., 2011; Maamri et al., 1997, 2001; Martínez et al., 2015; Monroy et al., 2016; Schlief & Mutz, 2011; but see Pinna et al., 2016). Lower rates of decomposition have been attributed to lower shredder abundances and biodiversity in IRES compared to perennial systems (Datry et al., 2011; Mariluan et al., 2015; Martínez et al., 2015; Monroy et al., 2016; Pinna et al., 2016; Schlief & Mutz, 2011). In contrast, the rates of microbial decomposer activity seem to be less affected by flow intermittence, suggesting a higher resistance and resilience of microbial communities to drying events than of macroinvertebrates (Datry et al., 2011; Mariluan et al., 2015; Pinna et al., 2016). High functional redundancy, shorter generation times or higher phenotypic plasticity may be reasons (Bonada et al., 2017; Zeglin, 2015), yet there may also be higher chances for adaptation for microbial communities to conditions in IRES (Gionchetta et al., 2019; Timoner et al., 2012, 2014). It is expected, for instance, that IRES should be dominated by fungal species with traits of higher desiccation resistance (see Shearer et al., 2007). So far, the most common strategies observed in decomposer communities in IRES to resist flow intermittence are the use of humid refugia (like leaf litter packs or hyporheos) during drying, or species with terrestrial and aquatic life cycle stages that pass to a dormant state during the dry phase (Romaní et al., 2017). Despite that, the few studies comparing microbial decomposer communities in intermittent and perennial rivers so far have found only very limited differences in diversity, richness and composition of species (Febria et al., 2015; Foulquier et al., 2015; Maamri et al., 2001). Although prokaryotic biofilm communities have been well studied in IRES during last years (see Romaní et al., 2017; Sabater et al., 2016), the number of studies comparing the diversity of fungal and bacterial decomposer communities in intermittent and perennial streams is still limited (see Romaní et al., 2017), so further research to

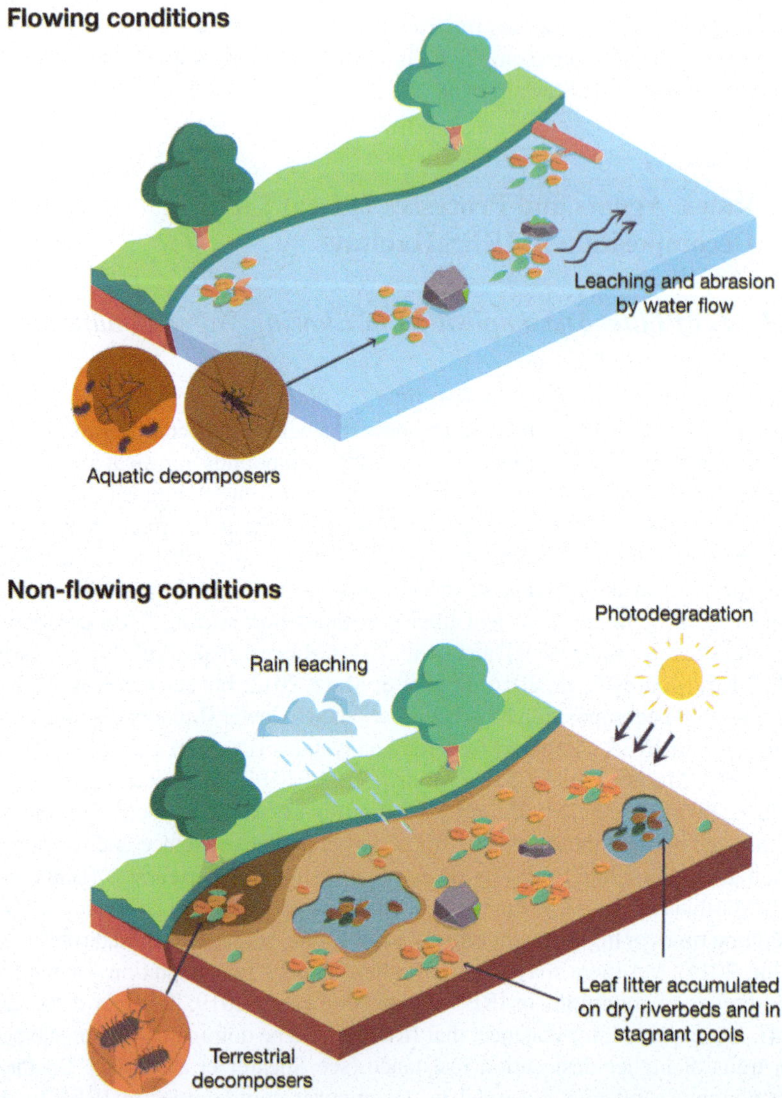

Fig. 5.3 Decomposition agents and factors driving leaf-litter decomposition in flowing water conditions and in the terrestrial-aquatic habitat mosaic appearing during drying. During flowing conditions, leaf litter is mainly processed by aquatic decomposers in a similar way to perennial streams, physical processes such as leaching and mechanical abrasion and fragmentation during transport with flowing water play additional roles. With drying, the recession of flow eventually shuts down downstream transport and leaf litter can accumulate in habitats with various environmental conditions (isolated pools, wet-shaded areas, irradiated and dry riverbed spots). In this dynamic terrestrial-aquatic habitat mosaic leaves can partially decompose under the influence of multiple biotic and abiotic factors, which can translate into a conditioning of the leaf material that affects its later decomposition after rewetting. See Table 5.2 for more information about changes in leaf-litter chemistry and biodegradability during the non-flowing period. Diagram made by Patricia Tudela Rosique

Table 5.1 Comparison of leaf litter decomposition rates ("k") between the flowing phases of IRES and perennial streams, and among terrestrial and aquatic habitats typically found in IRES across the different hydrological phases. Mean k values were extracted either from original values ("O") stated in the text or tables, or estimated ("E") from figures shown in the studies. The average and standard deviation ("SD") of the k value for each habitat and flow regime was computed using only k values from studies using coarse litterbags ("C") and day^{-1} ("d^{-1}") as unit of measurement. *Studies using only fine ("F") litterbags or degree-days^{-1} ("dd^{-1}") are shown but were not used to compute average values. **mean k values in Burrows et al. (2017) combine fine and coarse litterbags ("F + C"). Study cases in perennial rivers correspond to studies where decomposition in IRES and perennials streams is compared. To reach a similar number of studies in IRES and perennial rivers, we complemented the table with some studies carried out in perennial, low order streams with natural conditions

Flow regime	Habitat-specific average k (mean ± SD and range)	Study	Leaf litter species	Study-specific k (mean)	Observations
Intermittent	**Dry riverbed** 0.0073 ± 0.0113 (0.0005–0.036)	Abril et al. (2016)	*Populus nigra*	0.0111	C/E/d^{-1}
		Boulton (1991)	*Eucalyptus viminalis*	0.0042	C/E/d^{-1}
		Burrows et al. (2017)**	*Eucalyptus tereticornis* and *E. camaldulensis*	0.0360	F + C/O/d^{-1}
		Corti et al. (2011)	*Populus* sp.	0.0060	C/E/d^{-1}
		Maamri et al. (1997)	*Salix pedicellata*	0.0011	C/E/d^{-1}
			Nerium oleander	0.0011	C/E/d^{-1}
		Maamri et al. (2001)	*S. pedicellata*	0.0013	C/E/d^{-1}
		Mariluan et al. (2015)	*Nothofagus pumilio*	0.0044	C/E/d^{-1}
		Riedl et al. (2013)	*Populus trichocarpa*	0.0005	C/O/d^{-1}
	Hyporheic 0.0580 ± 0.0040 (0.048–0.071)	Burrows et al. (2017)**	*E. tereticornis* and *E camaldulensis*	0.0580	F + C/O/d^{-1}
	Isolated pool 0.0253 ± 0.0263 (0.0067–0.044)	Abril et al. (2016)	*P. nigra*	0.0440	C/O/d^{-1}
		Schlief and Mutz (2011)	*Alnus glutinosa*	0.0067	C/O/d^{-1}
	Pools connected to flow 0.0436 ± 0.0147 (0.0333–0.054)	Burrows et al. (2017)	*E. tereticornis* and *E camaldulensis*	0.0540	F + C/O/d^{-1}

(continued)

Table 5.1 (continued)

Flow regime	Habitat-specific average k (mean ± SD and range)	Study	Leaf litter species	Study-specific k (mean)	Observations
		Corti et al. (2011)	*Populus* sp.	0.0333	C/E/d^{-1}
	Running water 0.0210 ± 0.0181 (0.0048–0.0604)	Datry et al. (2011)*	*A. glutinosa*	0.0011	C/O/dd^{-1}
		Foulquier et al. (2015)*	*A. glutinosa*	0.0050	F/E/d^{-1}
		Monroy et al. (2016)*	*Quercus robur*	0.0003	C/O/dd^{-1}
			A. glutinosa	0.0014	C/O/dd^{-1}
		Boulton (1991)	*E. viminalis*	0.0101	C/E/d^{-1}
		Burrows et al. (2017)**	*E. tereticornis* and *E camaldulensis*	0.0490	F + C/O/d^{-1}
		Maamri et al. (1997)	*S. pedicellata*	0.0064	C/E/d^{-1}
			N. oleander	0.0048	C/E/d^{-1}
		Maamri et al. (2001)	*S. pedicellata*	0.0057	C/E/d^{-1}
		Mariluam et al. (2015)	*N. pumilio*	0.0062	C/O/d^{-1}
		Pinna et al. (2016)	*Phragmites australis*	0.0260	C/E/d^{-1}
		Riedl et al. (2013)	*P. trichocarpa*	0.0170	C/O/d^{-1}
		Schlief and Mutz (2011)	*A. glutinosa*	0.0167	C/O/d^{-1}
		Tenkiano and Chauvet (2018)	*Pterocarpus santalinoides*	0.0289	C/O/d^{-1}
			Alchornea cordifolia	0.0604	C/O/d^{-1}

(continued)

Table 5.1 (continued)

Flow regime	Habitat-specific average k (mean ± SD and range)	Study	Leaf litter species	Study-specific k (mean)	Observations
Perennial	**Running water** 0.0258 ± 0.0187 (0.0017–0.061)	Datry et al. (2011)*	*A. glutinosa*	0.0073	C/O/dd^{-1}
		Foulquier et al. (2015)*	*A. glutinosa*	0.0084	F/E/d^{-1}
		Monroy et al. (2016)*	*Q. robur*	0.0007	C/O/dd^{-1}
			A. glutinosa	0.0032	C/O/dd^{-1}
		Abril et al. (2016)	*P. nigra*	0.0531	C/O/d^{-1}
		Bunn et al. (1988)	*Eucalyptus marginata*	0.0017	C/O/d^{-1}
		Burrows et al. (2017)**	*E. camaldulensis*	0.0610	F + C/O/d^{-1}
		Cornut et al. (2010)	*A. glutinosa*	0.0292	C/O/d^{-1}
		Griffith and Tiegs (2016)	*Quercus alba*	0.0124	C/O/d^{-1}
			Liriodendron tulipifera	0.0259	C/O/d^{-1}
			Acer rubrum	0.0359	C/O/d^{-1}
		Gulis and Suberkropp (2003)	*Rhododendron maximum*	0.0018	C/O/d^{-1}
			A. rubrum	0.0048	C/O/d^{-1}
		Langhans et al. (2008)	*P. nigra*	0.0233	C/O/d^{-1}
		Mariluan et al. (2015)	*N. pumilio*	0.0067	C/O/d^{-1}
		Pinna et al. (2016)	*P. australis*	0.0266	C/E/d^{-1}

(continued)

Table 5.1 (continued)

Flow regime	Habitat-specific average k (mean ± SD and range)	Study	Leaf litter species	Study-specific k (mean)	Observations
		Quinn et al. (2000)	*Knightia excelsa*	0.0167	C/O/d^{-1}
			Populus canadensis	0.0374	C/O/d^{-1}
			Aristotelia serrata	0.0547	C/O/d^{-1}
		Risse-Buhl et al. (2017)	*A. glutinosa*	0.0343	C/E/d^{-1}
		Solagaistua et al. (2016)	*A. glutinosa*	0.0125	C/O/d^{-1}

identify and discern particularities of microbial decomposer communities in IRES is required.

5.2.2 *Leaf Litter Decomposition in the Terrestrial-Aquatic Habitat Mosaic During Drying*

The cessation of surface flow promotes the fragmentation of surface water and the formation of isolated pools along the stream channel. Water physicochemistry in isolated pools can be highly heterogeneous (see von Schiller et al., 2017), which can result in different implications for the processing of leaf litter in these habitats. In stagnant pools exposed to solar radiation, warmer water, lack of water renewal and the accumulation of riparian leaf litter enable high respiration rates and promote hypoxic conditions (Canhoto et al., 2013; von Schiller et al., 2011). The leaching of leaf litter also leads to higher nutrient concentrations, water acidification and an accumulation of potentially toxic compounds (Canhoto et al., 2013; von Schiller et al., 2011). These adverse environmental conditions can curb decomposition of leaf litter due to reduced microbial and detritivore activities (Table 5.1) (Canhoto et al., 2013; Corti et al., 2011; Schlief & Mutz, 2009). Some remnant pools, however, can remain sufficiently well connected to hyporheic flow and may thereby maintain as a favourable refuge for microbial decomposers and shredders (Bogan et al., 2019) during the non-flowing phase. These can then drive leaf litter decomposition even in the absence of flowing conditions (Abril et al., 2016; Corti et al., 2011; Langhans et al., 2008).

Besides remnant pools, dry and exposed stream habitats are an important habitat with ongoing drying. Eventually, the complete disappearance of surface water in the channel makes dry streambeds the dominant habitat in the dry phase of IRES. The consequent emersion of leaf litter considerably reduces its decomposition due to the drastic reduction of microbial decomposer and detritivore activities (Table 5.1) (e.g., Boulton, 1991; Bruder et al., 2011; Corti et al., 2011; Duarte et al., 2017; Foulquier et al., 2015; Maamri et al., 1997, 2001; Mora-Gómez et al., 2018). The disappearance of water means great physiological stress to aquatic organisms and results in marked changes in the composition and activity of microbial decomposers and detritivores (Duarte et al., 2017; Foulquier et al., 2015; Larned et al., 2007; Mora-Gómez et al., 2018). Aquatic shredders undergo a severe decline with the absence of water (eg. Abril et al., 2016; Corti et al., 2011; Datry, 2012; Martínez et al., 2015; Schlief & Mutz, 2011), but, conversely, these conditions allow colonization by terrestrial invertebrates (Corti & Datry, 2016; Corti et al., 2013; Sánchez-Montoya et al., 2016), which might then participate in the decomposition of accumulated leaf litter during the dry phase of IRES, although the evidence for this is still very scarce (see Bastow et al., 2002; Rosado et al., 2014). Reductions in water availability reduce microbial biomass and constrain microbial activities through a reduction of substrate and nutrient diffusion combined with osmotic stress when water potential declines

(Amalfitano et al., 2008; Duarte et al., 2017; Manzoni et al., 2012; Mora-Gómez et al., 2018; Schimel et al., 2007). Relatively high rates of decomposition may be maintained in relatively moist habitats such as shaded sediments close to riparian vegetation or subsurface zones (Abril et al., 2016; Arias-Real et al., 2019; Burrows et al., 2017; Gionchetta et al., 2019; Solagaistua et al., 2016).

While the absence of water reduces the leaching rates and physical abrasion associated to the flowing phases, leaf litter can be more exposed to other abiotic factors such as solar radiation, high temperatures or precipitation. In fact, during the dry phase, abiotic factors can have a greater relative importance on leaf litter than biotic ones (Steward et al., 2012). Local environmental conditions of the stream reach (for instance, shading by riparian vegetation canopy) can modulate the intensity of these abiotic factors on leaf litter and thus its processing during the dry phase (del Campo et al., 2019). On dry streambeds with low riparian cover, solar radiation can become the main factor affecting leaf litter (Abril et al., 2016; del Campo et al., 2019). It drives photodegradation through photolysis reactions of aromatic and condensed aromatic compounds (mainly lignin and phenols otherwise considered recalcitrant) that have a higher capacity to absorb solar radiation (Austin & Ballaré, 2010). This translates to photochemical mineralization of organic matter (Austin & Vivanco, 2006; Brandt et al., 2009; Rutledge et al., 2010) as well as chemical changes of the residual material with implications for later decomposition (see Table 5.2). Occasional rainfall events that rewet riverbeds without leading to the resumption of flow can promote certain mass loss of leaf litter by leaching (del Campo et al., 2019; Mora-Gómez et al., 2019), while at the same time they can trigger brief pulses of organic matter decomposition (Muñoz et al., 2018; Timoner et al., 2014) with peaks of CO_2-production (Datry et al., 2018; Gómez-Genner et al., 2016; Marcé et al., 2019). For arid lands similar phenomena are known as the "Birch effect", i.e., the abrupt increase of organic matter mineralization and associated CO_2 emissions following the rewetting of previously dry soils (Birch, 1958; Wilson & Baldwin, 2008).

5.3 Dynamics of Leaf Litter Decomposition in IRES

5.3.1 IRES Act Locally as Punctuated Biogeochemical Reactors

In IRES, processing of leaf litter and other detrital organic matter is highly dynamic, reflecting the hydrological dynamics of IRES, which, as coupled aquatic-terrestrial ecosystems, function as 'pulsed biogeochemical reactors' (Fig. 5.4). These are conceptualized as processing, storing and transporting organic matter in response to temporal flow fluctuations (Datry et al., 2014; Jacobson & Jacobson, 2013; Larned et al., 2010).

Leaf litter is processed mostly during flowing phases when both aquatic microbial decomposers and shredders reach their maximum densities and activities in IRES

Table 5.2 Examples of typical preconditioning situations that can affect the chemical composition and biodegradability of leaf litter accumulated in various riverbed habitats during drying. How leaf litter is conditioned in IRES during the non-flowing phase depends on locally heterogeneous environmental conditions. Due to the great spatiotemporal heterogeneity of these rivers, various of these habitats may even be found along the same stream reach, which can lead to a chemical diversification of leaf litter at small spatial scale. In the table, question marks indicate that the effect of abrasion and fragmentation of leaf litter on its biodegradation by aquatic decomposer communities have not been tested yet

Riverbed habitats	Decomposition agents or factors involved	Effect on leaf litter	Effect on leaf litter biodegradability
Open, exposed dry riverbeds	Solar radiation and heat	× Lignin loss by photodegradation[1,2,3]	+
		× Increase of organic matter solubility enhancing leaching loss[4,5,6] (combined action of sunlight and heat)	–
		× Polymerization and accumulation of phenolic compounds[6] (combined action of sunlight and heat)	–
	Rain	× Leaching of nutrients and more soluble C compounds[3,6]	–
	Other abiotic factors (wind, sediment burial…)	× Abrasion[7]	?
Humid riverbed sediments	Terrestrial invertebrates	× Fragmentation[8,9]	?
	Microbial activity	× Consumption of labile C resources and accumulation of recalcitrant compounds[5,10,11]	–
		× Increase of the nutrient content (mainly N, but also P) by microbial immobilization[3,11,12]	+
Stagnant pools	Anoxic, acid water	× Leaching of nutrients and more soluble C compounds[4,13]	–
		× Accumulation of phenolic compounds and cellulose due to acidic conditions[4]	–

[1]Austin et al. (2016), [2]Pu et al. (2014), [3]del Campo and Gómez (2016), [4]Dieter et al. (2013), [5]Abril et al. (2016), [6]del Campo et al. (2019), [7]Austin (2011), [8]Bastow et al. (2002), [9]Rosado et al. (2014), [10]Zheng et al. (2018), [11]Mora-Gómez et al. (2019), [12]Abelho and Descals (2019), [13]Dieter et al. (2011)

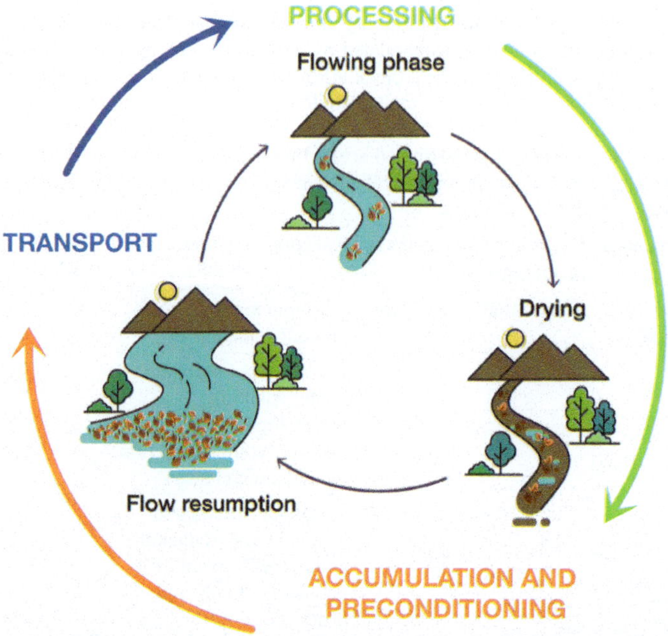

Fig. 5.4 Temporal changes in organic-matter dynamics associated with the hydrological phases of intermittent rivers. Due to their dynamic hydrology, IRES are considered to function as punctuated biogeochemical reactors. The continuous succession of drying/rewetting events promote a pulsed processing of organic matter in these streams, mainly during flowing water conditions. After the cessation of surface flow, leaf litter and other organic-matter substrates (woody litter, dead macrophytes) are retained and preconditioned under diverse terrestrial-aquatic habitat conditions until flow resumes and organic matter is transported downstream, where a new decomposition cycle can start again. Diagram made by Patricia Tudela Rosique

(Datry et al., 2014; Foulquier et al., 2015). As drying starts and flow first diminishes and then stops, leaf litter and other organic substrates (woody debris, algal mats, fine particulate organic matter) experience slower decomposition and start to accumulate on the riverbed (Datry et al., 2018; Dewson et al., 2007; Stanley et al., 1997). Particularly, in forested IRES with deciduous vegetation, massive amounts of riparian leaf litter can accumulate on dry riverbeds due to early abscission periods caused by the increase of water stress during summer (Acuña et al., 2007; Datry et al., 2018; Sanpera-Calbet et al., 2016).

The fragmentation of surface flow during drying prompts the emergence of a shifting mosaic of terrestrial and aquatic habitats (Datry et al., 2014; Stanley et al., 1997), where leaf litter can accumulate and remain immobilized, yet exposed to diverse environmental conditions (Figs. 5.3 and 5.4). These include highly irradiated dry riverbed areas, isolated pools with high temperature, cold pools connected to hyporheic flow, wet and shaded remnant sediments, or even areas subjected to recurrent wet-dry cycles. Most leaf litter accumulates under environmental conditions where it undergoes modest microbial decomposition and various physical and chemical conditions can lead to chemical alteration of leaf litter (e.g., dry riverbeds or isolated pools). This type of organic matter processing may be more aptly conceptualized as "preconditioning" as its most important consequence is the modulation of later decomposition rates in aquatic conditions after flow resumption (see Table 5.2 for some examples) (del Campo et al., 2019; Dieter et al., 2011, 2013; Mora-Gómez et al., 2019).

Once the dry phase finishes, the resumption of surface flow promotes the reactivation of aquatic decomposition but also the downstream transport of leaf litter (Fig. 5.4). The relative dominance of decomposition versus transport depends on the type of rewetting. The resumption of flow can be very variable in magnitude, type and timing, depending on the climate, hydrology and position in the catchment (Corti & Datry, 2012; Larned et al., 2010; von Schiller et al., 2017). The re-establishment of flow can arise from gradual increases of discharge following the rise of groundwater levels or be very pronounced and abrupt (e.g., flash flooding), if triggered by intense runoff events following storms. Flash floods can cause spectacular downstream transport of massive amounts of leaf litter, which form subsidies for downstream located ecosystems rather than entering local decomposer food chains. This is facilitated by more extreme flows, that appear during flash floods, driving physical abrasion (Corti & Datry, 2012) and constraining microbial activity due to shear stress (Zoppini et al., 2010). In contrast, gradual rewetting favours the onset of microbial decomposition over transport. The slow increase of water and nutrient availability promotes the release of physiological constraints imposed on microbial communities during the dry phase and thus a rapid stimulation of microbial decomposition, similarly to the Birch effect (Bruder et al., 2011; Maamri et al., 2001; Schlief & Mutz, 2011). Indeed, a recent study involving IRES worldwide, has shown that microbial communities of the litter itself can become activated within a few minutes upon rewetting and produce a peak of respiration within 24 h (Datry et al., 2018). According to estimations by Datry et al. (2018), these CO_2 respiration pulses associated to leaf litter rewetting can amount to 10% of the daily CO_2 emissions coming from perennial rivers and streams. Consequently, rewetting events are 'hot moments' (McClain et al., 2003) in river metabolism (von Schiller et al., 2019) due to the transport of large amounts of particulate organic matter from dry riverbeds and the release of huge loads of dissolved organic carbon and nutrients to the water column by leaching (Corti & Datry, 2012; Shumilova et al., 2019). Furthermore, increased microbial respiration on leaf material and dissolved organic matter might even lead to negative implications on stream functioning through the development of hypoxic conditions during blackwater events (flood events characterized by extremely high dissolved organic carbon values in the water column) (Hladyz et al., 2011). Finally, an important aspect of rewetting events is that the dominance of transport along the riverine continuum suggests the necessity to conceptualize leaf litter processing in IRES at the larger spatial scale of a river network instead of just at the reach scale.

5.3.2 *Leaf Litter Decomposition Across River Networks: IRES as Dynamic Metaecosystems*

Ecologists increasingly recognize river networks as metaecosystems, i.e., sets of connected local "component" ecosystems that exchange organisms (species) and resources (nutrients and organic matter) at a larger 'regional' landscape scale (Battin et al., 2008; Loreau et al., 2003). The application of the metaecosystem concept to the real world is still a challenging topic 17 years after its conceptual birth by Loreau et al. (2003) and would certainly benefit from a more intense exchange between theory and empirical research (Gounand et al., 2018). River networks are especially challenging in this respect due to various reasons: Most importantly, the spatial configuration of rivers assumes the form of a hierarchical dendritic network (Rodriguez-Iturbe & Rinaldo, 1997), which differs greatly from the "islands-in-a-matrix" configuration assumable for other metaecosystems. The river network´s topology also imposes constraints on connectivity as some material exchange or organism dispersal is bound to the river corridor, the water phase or even the flow direction (Benda et al., 2004). Unidirectional flow of water imposes asymmetry for exchange processes—for some more so (e.g., transport of solutes and microbes), for some less so (e.g., migration of fish or insects with a flying stage). The dynamic nature of IRES as a spatio-temporal mosaic of flowing, non-flowing, and dry habitats makes them strong candidates to further develop the metaecosystem perspective of river networks (Datry, Corti, et al., 2017). Indeed, this dynamic habitat mosaic creates a set of local ecosystems that variably contribute to the transport, accumulation, and processing (including preconditioning) of organic matter within individual reaches but also across whole river networks (Datry et al., 2014; Larned et al., 2010). The transport of organic matter is inhibited during non-flowing phases at several spatial scales, between dynamically appearing lentic and semiaquatic habitats but also from tributaries to the main stem. Even if isolated from the rest of the network, remaining aquatic habitats are not metabolically inactive during the dry phase. The resumption of water flow then reconnects dry and isolated aquatic habitats to the river network and leaf litter is transported and eventually stored in recipient downstream aquatic ecosystems, where it can (eventually again) be subjected to aquatic decomposition. The successful application of the metaecosystem concept to river networks comprising IRES requires careful consideration of the dynamics of connectivity across several spatial scales as well as the dynamics within 'local' ecosystems.

Ultimately, at the river-network scale, downstream organic matter fluxes are controlled by how local flow regimes merge to create river network-wide flow patterns, which in turn, shape the quantity and biodegradability of organic matter that is locally available in individual reaches of the network. The flow regime of an IRES modulates its capacity to transport, retain and process organic matter (Jacobson & Jacobson, 2013; Larned et al., 2010; Stanley et al., 1997). For instance, the long dry phases of ephemeral streams and the return of flow in the form of flash flood events promote the transport and mobilization of large amounts of organic matter over its processing, which—when measured as bulk mass loss—is minimal during the dry phase (see Box 5.1). Consequently, the spatial distribution of flow intermittence across the network can exert a great influence on organic matter fluxes at such large spatial scale. For instance, the concentration of IRES in the upper or the lower part of the fluvial networks can drastically affect the quantity and quality of organic matter at the outlet of the network. When IRES are present primarily in the lower part of the catchment, they receive lower amounts of organic matter that is already pre-processed (i.e., aged organic matter of low biodegradability) from upstream perennial systems. In contrast, when IRES are mainly distributed in the upper part of the network, downstream ecosystems can receive great amounts of non-processed particulate organic matter (i.e., fresh organic matter of high biodegradability). In fact, the quantity and quality of organic matter leaving the entire system at its most downstream point relative to input in the form of terrestrial subsidies may serve as an integrative result of a river network's organic matter-processing capacity.

As explained above, the aquatic decomposition of leaf litter after flow resumption is influenced by the environmental conditions during the previous drying period. Considering the great environmental heterogeneity of the terrestrial-aquatic habitat mosaic of IRES during the drying, the accumulation and preconditioning of leaf litter under such diversity of conditions might promote its chemical diversification. This could have important implications for later decomposition of mixed leaf litter that emerges by pooling components with contrasting chemical composition from various habitats. Decomposition of such mixed leaf litter in downstream recipient ecosystems may be influenced by non-additive effects of leaf litter diversity (see Gessner et al., 2010; López-Rojo et al., 2018; Stoler et al., 2016). Positive effects may exist in the form of priming of more recalcitrant, heavily pre-processed leaf material by more labile, younger or elsewise specifically preconditioned leaves. Fungi are especially capable of exploiting chemically diversified material (Gessner et al., 2010) if offered in close proximity as, for example, a leaf pack immobilized on a flow-obstructing structure or a leaf aggregation in a pool or reservoir. On the other hand, the action of selective invertebrates may be hindered by less-attractive resources and create a negative effect of leaf litter diversity.

Besides the movement of leaf litter across the river network, the dynamics of its potential decomposers also need to be considered in a spatially explicit manner. Metacommunity theory (Holyoak et al., 2005; Leibold et al., 2004) posits that local community composition and diversity result from various processes ranging across spatial scales, including classic deterministic niche definitions as well as neutral processes related to dispersal (i.e., immigration and extinction dynamics) (Hubbell, 2001; MacArthur & Wilson, 2001). Importantly, in a fluvial network, dispersal is constrained by the dendritic topology and can be uni-, bi- or multidirectional depending on the mobility traits of particular taxa (Brown & Swan, 2010; Crabot et al., 2019; Grant et al., 2007; Rodriguez-Iturbe et al., 2009). This results in distinct spatial patterns of biodiversity (e.g., Carrara et al., 2012), which notably differ between insects (Finn et al., 2011) and heterotroph bacteria (Besemer et al., 2013), two main groups of consumers driving leaf litter decomposition. For fungi specifically, such river network-wide patterns of biodiversity have not been investigated so far. And more importantly, even though effects of intermittent flow on freshwater communities are well-known at the local scale, virtually nothing is known on how drying alters the spatio-temporal organisation of biodiversity at the river network scale by dynamically disrupting hydrological connectivity (Datry et al., 2016). Understanding how biodiversity is affected by changes in river network-wide connectivity is vital, however, as leaf decomposition depends on the occurrence and functional diversity of decomposers. While in perennial reaches decomposers may suffer a shortage of resources when upstream tributaries are disconnected, in intermittent reaches aquatic or terrestrial decomposers need to first colonize their respective habitat before decomposing eventually delivered leaf material. Considering chemical diversity of leaf material as discussed above, the spatial patterns of such resource diversity may not align with those of biodiversity, potentially creating situations of inefficient leaf decomposition in local ecosystems caused by a mismatch between resource and consumer traits. Such mismatch may also just mean non-existent consumer traits for a too diverse resource pool, which could be considered as (functional) species undersaturation (Mateo et al., 2017). Conversely, this also suggests potential for positive consumer biodiversity effects on leaf decomposition by enabling complementary resource use (Tylianakis et al., 2008).

In conclusion, due to the potential of flow intermittence to alter organic matter fluxes as well as biodiversity in river networks, the perspective of IRES as metaecosystems should be developed in future models as well as empirical research. Without this, the effect of flow intermittence on regional C fluxes at the network scale will not be reliably estimated.

Box 5.1: Differences in organic matter dynamics between intermittent rivers and ephemeral streams

The decomposition and dynamics of organic matter in IRES are strongly controlled by the flow regime. Particularly, differences in the duration of the flowing and non-flowing periods determine the main factors driving organic matter decomposition in the different types of IRES. For instance, in intermittent rivers with a short dry phase (weeks to a couple months) and very long flowing periods, the decomposition of organic matter (mainly leaf litter) is mainly carried out by aquatic decomposer communities. On the other hand, in ephemeral streams where surface water flow only lasts for a few days after rain events, organic matter (mainly woody debris, especially in arid streams) accumulates on dry riverbeds for very long periods (months to years). Thus, in ephemeral streams, organic matter decomposition is principally carried out by abiotic processes such as photodegradation, and only by short pulses of microbial activity following rewetting events (Jacobson et al., 1999).

These differences in flow regime and decomposition agents in turn control the location, the dynamics and the lifespan of the organic matter as a trophic resource in IRES networks (Jacobson & Jacobson, 2013; Jacobson et al., 1999). In intermittent rivers, leaf litter may still be actively decomposed under wet or even flowing conditions, so any eventual and limited mobility of leaf litter along the network translates to a short-term resource pulse of carbon/energy for downstream systems. In contrast, the predominant terrestrial conditions of ephemeral rivers promote very slow processing rates of accumulated organic matter due to the limitation of biotic decomposition under conditions of water limitation. As a consequence, in these streams, organic matter is mainly relocated from upstream to downstream sites by recurrent spates and can represent a long-term C resource along the network (Fig. 5.5).

Fig. 5.5 On the left, leaf litter accumulated on a dry riverbed of a forested intermittent stream. On the right, leaf and woody debris accumulated on an ephemeral stream after a flash flood event. (**a**) Hagenbach stream in Göstling an der Ybbs, Austria; (**b**) River Chícamo in Murcia, Spain. Photo credits: Rubén del Campo and Rosa Gómez

5.4 Roadmap for Research and Applications

Although the number of ecological studies on IRES has grown exponentially during the last decades (Leigh et al., 2016), there are still many knowledge gaps concerning organic matter dynamics and decomposition in these ecosystems. So far, most research in IRES has focused on the effect of the cessation of surface flow on decomposer communities and decomposition rates. But beyond changes in surface water flow, the highly dynamic hydrology of IRES promotes other changes in environmental conditions, which may strongly affect the processing of organic matter, yet have remained unstudied so far. In particular, three main aspects of the processing of organic matter in IRES still remain unclear and require further research: (a) completing the analysis of the decomposition of leaf litter under the whole spectrum of heterogeneous environmental conditions of IRES; (b) a further understanding of the legacy effect of non-flowing phases on the processing of organic matter after flow resumption; and (c) a mechanistic understanding of the effect of flow intermittence on cycling of organic matter at the fluvial network scale, including as driven by functionally relevant biodiversity patterns.

These three topics are hierarchically connected and need to be further developed to reach a proper understanding of organic matter processing in fluvial networks comprising IRES. First, we need to further investigate understudied habitats of IRES, for instance isolated pools or hyporheic zones (but see Arias-Real et al., 2019; Burrows et al., 2017). Hyporheic environments, specifically under conditions of subsurface flow, may act as a refuge for heterotrophic communities and thus maintain the decomposition of buried leaf litter (Arias-Real et al., 2019; Burrows et al., 2017; Gionchetta et al., 2019; Solagaistua et al., 2016). Also, a deeper study of the diversity of microbial decomposer communities inhabiting terrestrial and aquatic habitats of IRES is necessary for a complete understanding of organic matter processing in these ecosystems.

The second gap consists of the unknown effect of non-flowing phases on the subsequent decomposition of leaf litter after flow resumption. Leaf litter accumulated during non-flowing states can undergo considerable changes in its chemical composition and biodegradability—for later aquatic decomposition further downstream these may best be conceptualized as 'preconditioning' forming the basis for legacy effects at much larger spatial scale (Table 5.2). So far, most of the studies carried out in this regard have been purely experimental (mostly in laboratories), so our understanding of how preconditioning affects subsequent decomposition after rewetting in natural circumstances is still limited. Existing evidence suggests that contrasting environmental conditions during non-flow periods could lead to a chemical diversification of leaf litter chemistry, which in turn, could cause non-additive effects on the decomposition of diversified leaf litter packs in downstream systems after flow resumption (see Gessner et al., 2010; López-Rojo et al., 2018; Stoler et al., 2016). Future studies should address the potential role of non-flowing phases or IRES as promoters of chemical diversity by studying and characterizing in situ the chemical composition of organic matter accumulated across terrestrial and aquatic habitats during non-flowing phases.

Finally, we still lack studies upscaling or modelling the effect of flow intermittence on organic matter cycling at the larger spatial scale of fluvial networks. To equip these models with a strong empirical basis, we need to invest into collecting data on (1) leaf litter dynamics (local mass loss as well as reach scale and regional mass balances) across whole river networks with IRES, and (2) decomposition rates of leaf litter across hydrological phases of IRES including the implications of phase transitions. The combination of such datasets could allow to build models simulating organic matter fluxes in river networks with different levels of flow intermittence and predict future changes associated with climate change.

IRES are a prominent element within landscapes globally and will become more abundant in the near future due to on-going climate and global change. To improve our understanding of how flow intermittence affects ecosystem processes, such as leaf litter decomposition, is pivotal to generate a complete knowledge of the role of fluvial networks on C cycling at large scales. Such knowledge will in turn be used for a successful adaptation of management strategies to protect and conserve river networks, their biodiversity, functional integrity and the ecosystem services they provide.

References

Abelho, M. & Descals, E. (2019). Litter movement pathways across terrestrial–aquatic ecosystem boundaries affect litter colonization and decomposition in streams. *Functional Ecology, 33*, 1785–1797.

Abril, M., Muñoz, I., & Menéndez, M. (2016). Heterogeneity in leaf litter decomposition in a temporary Mediterranean stream during flow fragmentation. *Science of the Total Environment, 553*, 330–339.

Acuña, V., Giorgi, A., Muñoz, I., Sabater, F., & Sabater, S. (2007). Meteorological and riparian influences on organic matter dynamics in a forested Mediterranean stream. *Journal of the North American Benthological Society, 26*, 54–69.

Amalfitano, S., Fazi, S., Zoppini, A., Barra Caracciolo, A., Grenni, P., & Puddu, A. (2008). responses of benthic bacteria to experimental drying in sediments from Mediterranean temporary rivers. *Microbial Ecology, 55*, 270–279.

Arias-Real, R., Muñoz, I., Gutierrez-Cánovas, C., Granados, V., Lopez-Laseras, P., & Menéndez, M. (2019). Subsurface zones in intermittent streams are hotspots of microbial decomposition during the non flow period. *Science of The Total Environment*, 135485.

Austin, A. T. (2011). Has water limited our imagination for aridland biogeochemistry? *Trends in Ecology & Evolution, 26*, 229–235.

Austin, A. T., & Ballaré, C. L. (2010). Dual role of lignin in plant litter decomposition in terrestrial ecosystems. *PNAS, 107*, 4618–4622.

Austin, A. T., Méndez, M. S., & Ballaré, C. L. (2016). Photodegradation alleviates the lignin bottleneck for carbon turnover in terrestrial ecosystems. *Proceedings of the National Academy of Sciences, 113*, 4392–4397.

Austin, A. T., & Vivanco, L. (2006). Plant litter decomposition in a semi-arid ecosystem controlled by photodegradation. *Nature, 442*, 555.

Battin, T. J., Kaplan, L. A., Findlay, S., Hopkinson, C. S., Marti, E., Packman, A. I., Newbold, J. D., & Sabater, F. (2008). Biophysical controls on organic carbon fluxes in fluvial networks. *Nature Geoscience, 1*, 95–100.

Bastow, J. L., Sabo, J. L., Finlay, J. C., & Power, M. E. (2002). A basal aquatic-terrestrial trophic link in rivers: algal subsidies via shore-dwelling grasshoppers. *Oecologia, 131*, 261–268.

Benda, L., Poff, N. L., Miller, D., Dunne, T., Reeves, G., Pess, G., & Pollock, M. (2004). The network dynamics hypothesis: How channel networks structure riverine habitats. *BioScience, 54*, 413–427.

Besemer, K., Singer, G., Quince, C., Bertuzzo, E., Sloan, W., & Battin, T. J. (2013). Headwaters are critical reservoirs of microbial diversity for fluvial networks. *Proceedings of the Royal Society B*, 280.

Birch, H. F. (1958). The effect of soil drying on humus decomposition and nitrogen availability. *Plant and Soil, 10*, 9–31.

Bogan, M. T., Leidy, R. A., Neuhaus, L., Hernández, C. J., & Carlson, S. M. (2019). Biodiversity value of remnant pools in an intermittent stream during the great California drought. *Aquatic Conservation: Marine and Freshwater Ecosystems, 29*, 976–989.

Bonada, N., Carlson, S. M., Datry, T., Finn, D. S., Leigh, C., Lytle, D. A., Monaghan, M. T., & Tedesco, P. A. (2017). Genetic, evolutionary, and biogeographical processes in intermittent rivers and ephemeral streams. In T. Datry, N. Bonada, & A. Boulton (Eds.), *Intermittent rivers and ephemeral streams* (pp. 405–431). Academic Press.

Bonada, N., & Resh, V. H. (2013). Mediterranean-climate streams and rivers: Geographically separated but ecologically comparable freshwater systems. *Hydrobiologia, 719*, 1–29.

Boulton, A. J. (1991). Eucalypt leaf decomposition in an intermittent stream in south-eastern Australia. *Hydrobiologia, 211*, 123–136.

Brandt, L. A., Bohnet, C., & King, J. Y. (2009). Photochemically induced carbon dioxide production as a mechanism for carbon loss from plant litter in arid ecosystems. *Journal of Geophysical Research: Biogeosciences, 114*, G02004.

Brown, B. L., & Swan, C. M. (2010). Dendritic network structure constrains metacommunity properties in riverine ecosystems. *Journal of Animal Ecology, 79*, 571–580.

Bruder, A., Chauvet, E., & Gessner, M. O. (2011). Litter diversity, fungal decomposers and litter decomposition under simulated stream intermittence. *Functional Ecology, 25*, 1269–1277.

Bunn, S. E. (1988). Processing of leaf litter in a northern jarrah forest stream Western Australia: I. Seasonal differences. *Hydrobiologia, 162*, 201–210.

Burrows, R. M., Rutlidge, H., Bond, N. R., Eberhard, S. M., Auhl, A., Andersen, M. S., Valdez, D. G., & Kennard, M. J. (2017). High rates of organic carbon processing in the hyporheic zone of intermittent streams. *Scientific Reports, 7*, 13198.

Canhoto, C., Calapez, R., Gonçalves, A. L., & Moreira-Santos, M. (2013). Effects of *Eucalyptus* leachates and oxygen on leaf-litter processing by fungi and stream invertebrates. *Freshwater Science, 32*, 411–424.

Carrara, F., Altermatt, F., Rodriguez-Iturbe, I., & Rinaldo, A. (2012). Dendritic connectivity controls biodiversity patterns in experimental metacommunities. *Proceedings of the National Academy of Sciences of the United States of America, 109*, 5761–5766.

Cipriani, T., Tilmant, F., Branger, F., Sauquet, E., & Datry, T. (2014). Impact of climate change on aquatic ecosystems along the Asse River Network. *FRIEND-Water 2014, Hydrology in a Changing World: Environmental and Human Dimensions*, 463–468.

Cornut, J., Elger, A., Lambrigot, D., Marmonier, P., & Chauvet, E. (2010). Early stages of leaf decomposition are mediated by aquatic fungi in the hyporheic zone of woodland streams. *Freshwater Biology, 55*, 2541–2556.

Corti, R., & Datry, T. (2012). Invertebrates and sestonic matter in an advancing wetted front travelling down a dry river bed (Albarine, France)—Portail documentaire. *Freshwater Science, 31*, 1187–1201.

Corti, R., & Datry, T. (2016). Terrestrial and aquatic invertebrates in the riverbed of an intermittent river: Parallels and contrasts in community organisation. *Freshwater Biology, 61*, 1308–1320.

Corti, R., Datry, T., Drummond, L., & Larned, S. T. (2011). Natural variation in immersion and emersion affects breakdown and invertebrate colonization of leaf litter in a temporary river. *Aquatic Sciences, 73*, 537–550.

Corti, R., Larned, S. T., & Datry, T. (2013). A comparison of pitfall-trap and quadrat methods for sampling ground-dwelling invertebrates in dry riverbeds. *Hydrobiologia, 717*, 13–26.

Crabot, J., Heino, J., Launay, B., & Datry, T. (2019). Drying determines the temporal dynamics of stream invertebrate structural and functional beta diversity. *Ecography, 43*, 620–635.

Datry, T. (2012). Benthic and hyporheic invertebrate assemblages along a flow intermittence gradient: Effects of duration of dry events. *Freshwater Biology, 57*, 563–574.

Datry, T., Bonada, N., & Boulton, A. J. (2017). General Introduction. In T. Datry, N. Bonada, & A. Boulton (Eds.), *Intermittent rivers and ephemeral streams* (pp. 1–20). Academic Press.

Datry, T., Bonada, N., & Heino, J. (2016). Towards understanding the organisation of metacommunities in highly dynamic ecological systems. *Oikos, 125*, 149–159.

Datry, T., Corti, R., Claret, C., & Philippe, M. (2011). Flow intermittence controls leaf litter breakdown in a French temporary alluvial river: The "drying memory." *Aquatic Sciences, 73*, 471–483.

Datry, T., Corti, R., Heino, J., Hugueny, B., Rolls, R. J., & Ruhí, A. (2017). Habitat fragmentation and metapopulation, metacommunity, and metaecosystem dynamics in intermittent rivers and ephemeral streams. In T. Datry, N. Bonada, & A. Boulton (Eds.), *Intermittent rivers and ephemeral streams* (pp. 377–403). Academic Press.

Datry, T., Foulquier, A., Corti, R., von, D., Schiller, Tockner, K., Mendoza-Lera, C., J., Clement, C., Gessner, M.O., Moleon, M., Stubbington, R., Gucker, B., Albarino, R., D., Allen, C., Altermatt, F., Arce, M.I., Arnon, S., Banas, D., Banegas-Medina, A., … Zoppini, A. (2018). A global analysis of terrestrial plant litter dynamics in non-perennial waterways. *Nature Geoscience, 11*, 497–503.

Datry, T., Larned, S. T., & Tockner, K. (2014). Intermittent rivers: A challenge for freshwater ecology. *BioScience, 64*, 229–235.

del Campo, R., & Gómez, R. (2016). Exposure of wood in floodplains affects its chemical quality and its subsequent breakdown in streams. *Science of the Total Environment, 543*, 652–661.

del Campo, R., Gómez, R., & Singer, G. (2019). Dry phase conditions prime wet-phase dissolved organic matter dynamics in intermittent rivers. *Limnology and Oceanography, 64*, 1966–1979.

Dewson, Z. S., James, A. B. W., & Death, R. G. (2007). Stream ecosystem functioning under reduced flow conditions. *Ecological Applications, 17*, 1797–1808.

Dieter, D., Frindte, K., Krüger, A., & Wurzbacher, C. (2013). Preconditioning of leaves by solar radiation and anoxia affects microbial colonisation and rate of leaf mass loss in an intermittent stream. *Freshwater Biology, 58*, 1981–1931.

Dieter, D., Schiller, D. v., García-Roger, E. M., Sánchez-Montoya, M. M., Gómez, R., Mora-Gómez, J., Sangiorgio, F., Gelbrecht, J., & Tockner, K. (2011). Preconditioning effects of intermittent stream flow on leaf litter decomposition. *Aquatic Sciences, 73*, 599–609.

Duarte, S., Mora-Gómez, J., Romaní Anna, M., Cássio, F., & Pascoal, C. (2017). Responses of microbial decomposers to drought in streams may depend on the environmental context. *Environmental Microbiology Reports, 9*, 756–765.

Döll, P., & Schmied, H. M. (2012). How is the impact of climate change on river flow regimes related to the impact on mean annual runoff? A global-scale analysis. *Environmental Research Letters, 7*, 14–37.

Febria, C. M., Hosen, J. D., Crump, B. C., Palmer, M. A., & Williams, D. D. (2015). Microbial responses to changes in flow status in temporary headwater streams: A cross-system comparison. *Frontiers in Microbiology, 6*, 522.

Finn, D. S., Bonada, N., Murria, C., & Hughes, J. M. (2011). Small but mighty: Headwaters are vital to stream network biodiversity at two levels of organization. *Journal of the North Americal Benthological Society, 30*, 963–980.

Foulquier, A., Artigas, J., Pesce, S., & Datry, T. (2015). Drying responses of microbial litter decomposition and associated fungal and bacterial communities are not affected by emersion frequency. *Freshwater Science, 34*, 1233–1244.

Fritz, K. M., Hagenbuch, E., D'Amico, E., Reif, M., Wigington, P. J., Jr., Leibowitz, S. G., Comeleo, R. L., Ebersole, J. L., & Nadeau, T.-L. (2013). Comparing the Extent and Permanence of Headwater Streams From Two Field Surveys to Values From Hydrographic Databases and Maps. *JAWRA Journal of the American Water Resources Association, 49*, 867–882.

Gallart, F., Prat, N., García-Roger, E. M., Latron, J., Rieradevall, M., Llorens, P., Barberá, G. G., Brito, D., De Girolamo, A. M., Lo Porto, A., Buffagni, A., Erba, S., Neves, R., Nikolaidis, N. P., Perrin, J. L., Querner, E. P., Quiñonero, J. M., Tournoud, M. G., Tzoraki, O., … Froebrich, J. (2012). A novel approach to analysing the regimes of temporary streams in relation to their controls on the composition and structure of aquatic biota. *Hydrology and Earth System Sciences, 16*, 3165–3182.

Gessner, M. O., Swan, C. M., Dang, C. K., McKie, B. G., Bardgett, R. D., Wall, D. H., & Hättenschwiler, S. (2010). Diversity meets decomposition. *Trends in Ecology & Evolution, 25*, 372–380.

Gionchetta, G., Romaní, A. M., Oliva, F., & Artigas, J. (2019). Distinct responses from bacterial, archaeal and fungal streambed communities to severe hydrological disturbances. *Scientific Reports, 9*, 13506.

Gómez-Gener, L., Obrador, B., Marcé, R., Acuña, V., Catalán, N., Casas-Ruiz, J. P., Sabater, S., Muñoz, I., & von Schiller, D. (2016). When water vanishes: Magnitude and regulation of carbon dioxide emissions from dry temporary streams. *Ecosystems, 19*, 710–723.

Gounand, I., Harvey, E., Little, C. H., & Altermatt, F. (2018). Meta-Ecosystems 2.0: Rooting the Theory into the Field. *Trends in Ecology and Evolution, 33*, 36–46.

Grant, E. H. C., Lowe, W. H., & Fagan, W. F. (2007). Living in the branches: Population dynamics and ecological processes in dendritic networks. *Ecology Letters, 10*, 165–175.

Griffiths, N. A., & Tiegs, S. D. (2016). Organic-matter decomposition along a temperature gradient in a forested headwater stream. *Freshwater Science, 35*, 518–533.

Grill, G., Lehner, B., Thieme, M., Geenen, B., Tickner, D., Antonelli, F., Babu, S., Borrelli, P., Cheng, L., Crochetiere, H., & Macedo, H. E. (2019). Mapping the world's free-flowing rivers. *Nature, 569*, 215–221.

Gulis, V., & Suberkropp, K. (2003). Leaf litter decomposition and microbial activity in nutrient-enriched and unaltered reaches of a headwater stream. *Freshwater Biology, 48*, 123–134.

Hassan, M. A., & Egozi, R. (2001). Impact of wastewater discharge on the channel morphology of ephemeral streams. *Earth Surface Processes and Landforms, 26*, 1285–1302.

Hladyz, S., Watkins, S. C., Whitworth, K. L., & Baldwin, D. S. (2011). Flows and hypoxic blackwater events in managed ephemeral river channels. *Journal of Hydrology, 401*, 117–125.

Holyoak, M., Leibold, M. A., Mouquet, N. M., Holt, R. D., & Hoopes, M. F. (2005). Metacommunities: A framework for large-scale community ecology. In M. Holyoak, M. A. Leibold, & R. D. Holt (Eds.), *Metacommunities*. (pp. 1–31). The University of Chicago Press.

Hubbell, S. P. (2001). *The unified neutral theory of biodiversity and biogeography*. (p. 375). Princeton University Press.

Jacobson, P. J., & Jacobson, K. M. (2013). Hydrologic controls of physical and ecological processes in Namib Desert ephemeral rivers: Implications for conservation and management. *Journal of Arid Environments, 93*, 80–93.

Jacobson, P. J., Jacobson, K. M., Angermeier, P. L., & Don, S. C. (1999). Transport, retention and ecological significance of woody debris within a large ephemeral river. *Journal of the North American Benthological Society, 18*, 429–444.

Jaeger, K. L., Olden, J. D., & Pelland, N. A. (2014). Climate change poised to threaten hydrologic connectivity and endemic fishes in dryland streams. *Proceedings of the National Academy of Sciences, 111*, 13894–13899.

Langhans, S. D., Tiegs, S. D., Gessner, M. O., & Tockner, K. (2008). Leaf-decomposition heterogeneity across a riverine floodplain mosaic. *Aquatic Sciences, 70*, 337–346.

Larned, S. T., Datry, T., Arscott, D. B., & Tockner, K. (2010). Emerging concepts in temporary-river ecology. *Freshwater Biology, 55*, 717–738.

Larned, S. T., Datry, T., & Robinson, C. T. (2007). Invertebrate and microbial responses to inundation in an ephemeral river reach in New Zealand: Effects of preceding dry periods. *Aquatic Sciences, 69*, 554–567.

Leibold, M. A., Holyoak, M., Mouquet, N., Amarasekare, P., Chase, J. M., Hoopes, M. F., Holt, R. D., Shurin, J. B., Law, R., Tilman, D., Loreau, M., & Gonzalez, A. (2004). The metacommunity concept: A framework for multi-scale community ecology. *Ecology Letters, 7*, 601–613.

Leigh, C., Boulton, A. J., Courtwright, J. L., Fritz, K., May, C. L., Walker, R. H., & Datry, T. (2016). Ecological research and management of intermittent rivers: An historical review and future directions. *Freshwater Biology, 61*, 1181–1199.

Levick, L. R., Goodrich, D. C., Hernández, M., Fonseca, J., Semmens, D. J., Stromberg, J. C., Tluczek, M., Leidy, R. A., Scianni, M., & Guertin, D. P. (2008). *The ecological and hydrological significance of ephemeral and intermittent streams in the arid and semi-arid American Southwest*. US Environmental Protection Agency, Office of Research and Development.

Lowe, W. H., & Likens, G. E. (2005). Moving Headwater Streams to the Head of the Class. *BioScience, 55*, 196–197.

López-Rojo, N., Martínez, A., Pérez, J., Basaguren, A., Pozo, J., & Boyero, L. (2018). Leaf traits drive plant diversity effects on litter decomposition and FPOM production in streams. *PLoS ONE, 13*, e0198243.

Loreau, M., Mouquet, N., & Holt, R. D. (2003). Meta-ecosystems: A theoretical framework for a spatial ecosystem ecology. *Ecology Letters, 6*, 673–679.

MacArthur, R. H., & Wilson, E. O. (2001). *The Theory of Island Biogeography*. (p. 203). Princeton University Press.

Maamri, A., Bärlocher, F., Pattee, E., & Chergui, H. (2001). Fungal and bacterial colonisation of *Salix pedicellata* leaves decaying in permanent and intermittent streams in eastern Morocco. *International Review of Hydrobiology, 86*, 337–348.

Maamri, A., Chergui, H., & Pattee, E. (1997). Leaf litter processing in a temporary northeastern. *Archive für Hydrobiologie, 140*, 513–531.

Manzoni, S., Schimel, J. P., & Porporato, A. (2012). Responses of soil microbial communities to water stress: Results from a meta-analysis. *Ecology, 93*, 930–938.

Marcé, R., Obrador, B., Gómez-Gener, L., Catalán, N., Koschorreck, M., Arce, M. I., Singer, G., & von Schiller, D. (2019). Emissions from dry inland waters are a blind spot in the global carbon cycle. *Earth-Science Reviews, 188*, 240–248.

Mariluan, G., Díaz Villanueva, V., & Albariño, R. (2015). Leaf litter breakdown and benthic invertebrate colonization affected by seasonal drought in headwater lotic systems of Andean Patagonia. *Hydrobiologia, 760*, 171–187.

Martínez, A., Pérez, J., Molinero, J., Sagarduy, M., & Pozo, J. (2015). Effects of flow scarcity on leaf-litter processing under oceanic climate conditions in calcareous streams. *Science of the Total Environment, 503–504*, 251–257.

Mateo, R. G., Mokany, K., & Guisan, A. (2017). Biodiversity models: What if unsaturation is the rule? *Trends in Ecology & Evolution, 32*, 556–566.

McClain, M. E., Boyer, E. W., Dent, C. L., Gergel, S. E., Grimm, N. B., Groffman, P. M., Hart, S. C., Harvey, J. W., Johnston, C. A., Mayorga, E., McDowell, W. H., & Pinay, G. (2003). Biogeochemical hot spots and hot moments at the interface of terrestrial and aquatic ecosystems. *Ecosystems, 6*, 301–312.

McIntosh, A. R., Leigh, C., Boersma, K. S., McHugh, P. A., Febria, C., & García-Berthou, E. (2017). Food webs and trophic interactions in intermittent rivers and ephemeral streams. In T. Datry, N. Bonada & A. Boulton (Eds.), *Intermittent rivers and ephemeral streams* (pp. 323–347). Academic Press.

Meybeck, M. (2003). Global analysis of river systems: From Earth system controls to Anthropocene syndromes. Philosophical Transactions of the Royal Society of London. *Series B, Biological Sciences, 358*, 1935–1955.

Millenial Ecosystem Assessment. (2005). *Ecosystems and human well-being: Synthesis*. Island, Washington, DC.

Monroy, S., Menéndez, M., Basaguren, A., Pérez, J., Elosegi, A., & Pozo, J. (2016). Drought and detritivores determine leaf litter decomposition in calcareous streams of the Ebro catchment (Spain). *Science of the Total Environment, 573*, 1450–1459.

Mora-Gómez, J., Boix, D., Duarte, S., Cássio, F., Pascoal, C., Elosegi, A. & Romaní, A. M. (2019). Legacy of summer drought on autumnal leaf litter processing in a temporary mediterranean stream. *Ecosystems, 23*, 989–1003.

Mora-Gómez, J., Duarte, S., Cássio, F., Pascoal, C., & Romaní, A. M. (2018). Microbial decomposition is highly sensitive to leaf litter emersion in a permanent temperate stream. *Science of the Total Environment, 621*, 486–496.

Muñoz, I., Abril, M., Casas-Ruiz Joan, P., Casellas, M., Gómez-Gener, L., Marcé, R., Menéndez, M., Obrador, B., Sabater, S., Schiller, D., & Acuña, V. (2018). Does the severity of non-flow periods influence ecosystem structure and function of temporary streams? A mesocosm study. *Freshwater Biology, 63*, 613–625.

Palmer, M. A., Reidy Liermann, C. A., Nilsson, C., Flörke, M., Alcamo, J., Lake, P. S., & Bond, N. (2008). Climate change and the world's river basins: Anticipating management options. *Frontiers in Ecology and the Environment, 6*, 81–89.

Pinna, M., Marini, G., Cristiano, G., Mazzotta, L., Vignini, P., Cicolani, B., & Di Sabatino, A. (2016). Influence of aperiodic summer droughts on leaf litter breakdown and macroinvertebrate assemblages: Testing the drying memory in a Central Apennines River (Aterno River, Italy). *Hydrobiologia, 782*, 111–126.

Postel, S. L. (2000). Entering an era of water scarcity: The challenges ahead. *Ecological Applications, 10*, 941–948.

Pu, G., Du, J., Ma, X., Lv, Y., Jia, Y., Jia, X., & Tian, X. (2014). Contribution of ambient atmospheric exposure to *Typha angustifolia* litter decomposition in aquatic environment. *Ecological Engineering, 67*, 144–149.

Quinn, J. M., Burrell, G. P., & Parkyn, S. M. (2000). Influences of leaf toughness and nitrogen content on in-stream processing and nutrient uptake by litter in a Waikato, New Zealand, pasture stream and streamside channels. *New Zealand Journal of Marine and Freshwater Research, 34*, 253–271.

Raymond, P. A., Hartmann, J., Lauerwald, R., Sobek, S., McDonald, C., Hoover, M., Butman, D., Striegl, R., Mayorga, E., Humborg, C., Kortelainen, P., Durr, H., Meybeck, M., Ciais, P., & Guth, P. (2013). Global carbon dioxide emissions from inland waters. *Nature, 503*, 355–359.

Riedl, H. L., Marczak, L. B., McLenaghan, N. A., & Hoover, T. M. (2013). The role of stranding and inundation on leaf litter decomposition in headwater streams. *Riparian Ecology and Conservation, 1*, 3–10.

Risse-Buhl, U., Mendoza-Lera, C., Norf, H., Pérez, J., Pozo, J., & Schlief, J. (2017). Contrasting habitats but comparable microbial decomposition in the benthic and hyporheic zone. *Science of the Total Environment, 605–606*, 683–691.

Rodriguez-Iturbe, I., & Rinaldo, A. (1997). *Fractal river basins - chance and self-organization.* (p. 547). Cambridge University Press.

Rodriguez-Iturbe, I., Muneepeerakul, R., Bertuzzo, E., Levin, S. A., & Rinaldo, A. (2009). River networks as ecological corridors: A complex systems perspective for integrating hydrologic, geomorphologic, and ecologic dynamics. *Water Resources Research, 45*, W01413.

Romaní, A. M., Chauvet, E., Febria, C., Mora-Gómez, J., Risse-Buhl, U., Timoner, X., Weitere, M., & Zeglin, L. (2017). The biota of intermittent rivers and ephemeral streams: prokaryotes, fungi, and protozoans. In T. Datry, N. Bonada, & A. Boulton (Eds.), *Intermittent rivers and ephemeral streams* (pp. 161–188). Academic Press.

Rosado, J., Morais, M., & Tockner, K. (2014). Mass dispersal of terrestrial organisms during first flush events in a temporary stream. *River Research and Applications, 31*, 912–917.

Rutledge, S., Campbell David, I., Baldocchi, D., & Schipper Louis, A. (2010). Photodegradation leads to increased carbon dioxide losses from terrestrial organic matter. *Global Change Biology, 16*, 3065–3074.

Sabater, S., Timoner, X., Borrego, C., & Acuña, V. (2016). Stream biofilm responses to flow intermittence: From cells to ecosystems. *Frontiers in Environmental Science, 4*, 14.

Sabater, S., & Tockner, K. (2010). Effects of Hydrologic Alterations on the Ecological Quality of River Ecosystems. In S. Sabater & D. Barceló (Eds.), *Water Scarcity in the Mediterranean: Perspectives Under Global Change.* (pp. 15–39). Springer.

Sánchez-Montoya, M. D. M., Schiller, D., Ruhí, A., Pechar, G. S., Proia, L., Miñano, J., & Tockner, K. (2016). Responses of ground-dwelling arthropods to surface flow drying in channels and adjacent habitats along Mediterranean streams. *Ecohydrology, 9*, 1376–1387.

Sanpera-Calbet, I., Acuña, V., Butturini, A., Marcé, R., & Muñoz, I. (2016). El Niño southern oscillation and seasonal drought drive riparian input dynamics in a Mediterranean stream. *Limnology and Oceanography, 61*, 214–226.

Schimel, J., Balser, T. C., & Wallenstein, M. (2007). Microbial stress-response physiology and its implications for ecosystem function. *Ecology, 88*, 1386–1394.

Schlief, J., & Mutz, M. (2009). Effect of sudden flow reduction on the decomposition of alder leaves (*Alnus glutinosa* [L.] Gaertn.) in a temperate lowland stream: A mesocosm study. *Hydrobiologia, 624*, 205–217.

Schlief, J., & Mutz, M. (2011). Leaf decay processes during and after a supra-seasonal hydrological drought in a temperate lowland stream. *International Review of Hydrobiology, 96*, 633–655.

Shearer, C. A., Descals, E., Kohlmeyer, B., Kohlmeyer, J., Marvanová, L., Padgett, D., Porter, D., Raja, H. A., Schmit, J. P., Thorton, H. A., & Voglymayr, H. (2007). Fungal biodiversity in aquatic habitats. *Biodiversity and Conservation, 16*, 49–67.

Sheldon, F., Bunn, S. E., Hughes, J. M., Arthington, A. H., Balcombe, S. R., & Fellows, C. S. (2010). Ecological roles and threats to aquatic refugia in arid landscapes: Dryland river waterholes. *Marine and Freshwater Research, 61*, 885–895.

Shumilova, O., Zak, D., Datry, T., von Schiller, D., Corti, R., Foulquier, A., Obrador, B., Tockner, K., Allan, D. C., Altermatt, F., Arce, M. I., Arnon, S., Banas, D., Banegas-Medina, A., Beller,

E., Blanchette, M. L., Blanco-Libreros, J. F., Blessing, J., Boëchat, I. G., … Zarfl, C. (2019). Simulating rewetting events in intermittent rivers and ephemeral streams: A global analysis of leached nutrients and organic matter. *Global Change Biology, 25*, 1591–1611.

Snelder, T. H., Datry, T., Lamouroux, N., Larned, S. T., Sauquet, E., Pella, H., & Catalogne, C. (2013). Regionalization of patterns of flow intermittence from gauging station records. *Hydrology and Earth System Sciences, 17*, 2685–2699.

Solagaistua, L., Arroita, M., Aristi, I., Larrañaga, A., & Elosegi, A. (2016). Changes in discharge affect more surface than subsurface breakdown of organic matter in a mountain stream. *Marine and Freshwater Research, 67*, 1826–1834.

Stanley, E. H., Fisher, S. G., & Grimm, N. B. (1997). Ecosystem Expansion and Contraction in StreamsDesert streams vary in both space and time and fluctuate dramatically in size. *BioScience, 47*, 427–435.

Steward, A. L., Schiller, D. V., Tockner, K., Marshall, J. C., & Bunn, S. E. (2012). When the river runs dry: Human and ecological values of dry riverbeds. *Frontiers in Ecology and the Environment, 10*, 202–209.

Stoler, A. B., Burke, D. J., & Relyea, R. A. (2016). Litter chemistry and chemical diversity drive ecosystem processes in forest ponds. *Ecology, 97*, 1783–1795.

Tenkiano, N. S. D., & Chauvet, E. (2018). Leaf litter decomposition in Guinean savannah streams. *Inland Waters, 8*, 413–421.

Timoner, X., Acuña, V., Frampton, L., Pollard, P., Sabater, S., & Bunn, S. E. (2014). Biofilm functional responses to the rehydration of a dry intermittent stream. *Hydrobiologia, 727*, 185–195.

Timoner, X., AcuÑA, V., Von Schiller, D., & Sabater, S. (2012). Functional responses of stream biofilms to flow cessation, desiccation and rewetting. *Freshwater Biology, 57*, 1565–1578.

Tooth, S. (2000). Process, form and change in dryland rivers: A review of recent research. *Earth-Science Reviews, 51*, 67–107.

Tylianakis, J. M., Rand, T. A., Kahmen, A., Klein, A.-M., Buchmann, N., Perner, J., & Tscharntke, T. (2008). Resource Heterogeneity Moderates the Biodiversity-Function Relationship in Real World Ecosystems. *PLoS Biology, 6*, 947–956.

Uys, M. C., & O'Keeffe, J. H. (1997). Simple Words and Fuzzy Zones: Early Directions for Temporary River Research in South Africa. *Environmental Management, 21*, 517–531.

Vander Vorste, R., Sarremejane, R., & Datry, T. (2020). Intermittent rivers and ephemeral streams: A unique biome with important contributions to biodiversity and ecosystem services. In M. I. Goldstein & D. A. DellaSala (Eds.), *Encyclopedia of the World's Biomes* (pp. 419–429). Elsevier.

von Schiller, D., Acuña, V., Graeber, D., Martí, E., Ribot, M., Sabater, S., Timoner, X., & Tockner, K. (2011). Contraction, fragmentation and expansion dynamics determine nutrient availability in a Mediterranean forest stream. *Aquatic Sciences, 73*, 485.

von Schiller, D., Bernal, S., Dahm, C. N., & Martí, E. (2017). Nutrient and organic matter dynamics in intermittent rivers and ephemeral streams. In T. Datry, N. Bonada, & A. Boulton (Eds.), *Intermittent rivers and ephemeral streams* (pp. 135–160). Academic Press.

von Schiller, D., Datry, T., Corti, R., Foulquier, A., Tockner, K., Marcé, R., García-Baquero, G., Odriozola, I., Obrador, B., Elosegi, A., Mendoza-Lera, C., Gessner, M. O., Stubbington, R., Albariño, R., Allen, D. C., Altermatt, F., Arce, M. I., Arnon, S., Banas, D., … Zoppini, A. (2019). Sediment respiration pulses in intermittent rivers and ephemeral streams. *Global Biogeochemical Cycles, 33*, 1251–1263.

Williams, D. D. (2006). *The biology of temporary waters*. Oxford University Press.

Wilson, J. S., & Baldwin, D. S. (2008). Exploring the 'Birch effect' in reservoir sediments: Influence of inundation history on aerobic nutrient release. *Chemistry and Ecology, 24*, 379–386.

Zeglin, L. H. (2015). Stream microbial diversity in response to environmental changes: Review and synthesis of existing research. *Frontiers in Microbiology, 6*, 454.

Zheng, J., Wang, Y., Chen, C., Jiang, Q., Han, S., & Xu, Z. (2018). Effects of inundation and stranding on leaf litter decomposition and chemical transformation. *Aquatic Sciences, 80*, 9.

Zoppini, A., Amalfitano, S., Fazi, S., & Puddu, A. (2010). Dynamics of a benthic microbial community in a riverine environment subject to hydrological fluctuations (Mulargia River, Italy). *Hydrobiologia, 657*, 37–51.

Chapter 6
Plant Litter Decomposition in Terrestrial Ecosystems Compared to Streams

Pablo García-Palacios, I. Tanya Handa, and Stephan Hättenschwiler

Abstract The decomposition of dead organic matter is critical for carbon and nutrient cycles across ecosystems from the bottom of oceans to mountain tops. Despite similarities in the driving abiotic and biotic factors, and interconnected flows of organic matter between streams and their surrounding riparian zones, litter decomposition has often been studied separately in aquatic and terrestrial ecosystems, with some notable exceptions. This collaborative research across systems has identified some common patterns, which is necessary to move towards a broader litter decomposition theory and to develop a global decomposition model. Here we compare terrestrial litter decomposition with that occurring in aquatic systems, by providing an overview of the terrestrial literature and highlighting commonalities and differences with decomposition in streams. Specifically, we look at (i) the influence of climate, decomposer communities (microbes and invertebrates) and leaf litter quality (chemical and morphological), (ii) the consequences of changing diversity (in decomposer and litter communities), and (iii) the effects of global change (climate warming, nitrogen deposition and biotic invasions) on litter decomposition. Lastly, we identify recent approaches developed in terrestrial ecosystems that may help to increase our understanding of the abiotic and biotic drivers, diversity effects and global change effects on litter decomposition in aquatic ecosystems.

P. García-Palacios (✉)
Instituto de Ciencias Agrarias, Consejo Superior de Investigaciones Científicas, Serrano 115 Bis, 28006 Madrid, Spain
e-mail: pablo.garcia@ica.csic.es

I. T. Handa
Département des Sciences Biologiques, Université du Québec à Montréal, C.P. 8888, Succ. Centre-Ville, Montréal, QC H3C 3P8, Canada

S. Hättenschwiler
CEFE, Université de Montpellier, CNRS, EPHE, IRD, Université Paul-Valéry Montpellier, Montpellier, France

C. M. Swan et al. (eds.), *The Ecology of Plant Litter Decomposition in Stream Ecosystems*, https://doi.org/10.1007/978-3-030-72854-0_6

6.1 Introduction

The decomposition of plant litter is one of the most important ecosystem processes in the biosphere, as it is critical for carbon and nutrient cycles across systems, including oceans, freshwater and terrestrial ecosystems. The litter produced by plants in terrestrial ecosystems may eventually reach inland surface waters such as streams and rivers, either by falling directly into the water bodies or by transport from the soil surface through runoff or wind. As a result, streams and their surrounding riparian zones are interconnected by flows of organic matter that represent a substantial contribution to the global carbon cycle (Battin et al., 2009). Common abiotic and biotic factors such as environmental conditions (climate, nutrient availability), litter quality (chemical and morphological traits) and local decomposer communities (microbes and detritivores) drive leaf litter decomposition in aquatic and terrestrial ecosystems. Consequently, current global anthropogenic changes such as biodiversity loss, warming, nitrogen enrichment and biotic invasions may impact litter decomposition in a broadly similar fashion. Unfortunately, only a handful of studies have addressed the decomposition of plant litter along the aquatic-terrestrial litter continuum. This has prevented the establishment of commonalities across ecosystems, which is important to formulate a theory on plant litter decomposition that is valid across systems.

Despite similarities in the driving abiotic and biotic factors, and interconnected flows of organic matter, litter decomposition has often been studied separately in aquatic and terrestrial ecosystems with some notable exceptions. For instance, the pioneering work of Merritt and Lawson (1992), Wagener et al. (1998), and Hutchens and Wallace (2002) encouraged communication between aquatic and terrestrial ecologists towards a more comprehensive understanding of litter decomposition. The European consortium "BioCycle", inspired by the review of Gessner et al. (2010) on diversity effects across systems, embraced this challenge and set up a collaborative research project with ecologists working in both aquatic and terrestrial ecosystems. They investigated litter decomposition in forest floors and nearby streams across five major biomes, from the tropics to the subarctic (Handa et al., 2014). Furthermore, follow-ups of this and other projects demonstrated that biotic and abiotic drivers play similar roles in aquatic and terrestrial decomposition across contrasting spatial (García-Palacios et al., 2016) and temporal (Yue et al., 2018) scales. All this and other recent research (Abelho & Descals, 2019) represent novel contributions towards the development of a global decomposition model, and although much work remains to be done, this chapter offers a first synthesis.

Here we review three major areas in litter decomposition research in terrestrial ecosystems. First, we address the main biotic and abiotic drivers of litter decomposition. Then, we assess how diversity in litter and decomposer communities affect litter decomposition rates. Finally, we review the response of litter decomposition to three global change drivers (climate warming, nitrogen enrichment and biotic invasions). Along each section, we provide insights on how previous findings in terrestrial ecosystems compare results from aquatic ecosystems. Our intention is not to review these topics in streams, which are covered in detail in other chapters of this book, nor to review terrestrial decomposition extensively, but to focus on the main commonalities and differences. We also identify recent approaches developed in terrestrial ecosystems that may help to improve our understanding of the abiotic and biotic drivers, diversity effects and global change effects on litter decomposition in aquatic ecosystems.

6.2 Main Biotic and Abiotic Drivers of Litter Decomposition in Terrestrial Ecosystems Compared to Streams

6.2.1 The Role of Litter Quality and Climatic Conditions

The triangular relationship among environmental parameters, the quality of decomposing litter, and decomposer organisms as the major control factors over decomposition is well established for terrestrial ecosystems (Berg & Laskowski, 2005; Swift et al., 1979; Fig. 6.1). It does not differ fundamentally compared to aquatic environments in general and to stream ecosystems in particular (Wagener et al., 1998). How terrestrial ecologists weighed the relative importance of these three groups of control factors, however, changed over the years and is still evolving. The classical view, that persisted over many decades, was that climatic variables control decomposition at large spatial scales among biomes and that within a specific climatic zone, litter quality was the most important driving factor (Berg et al., 1993; Meentemeyer, 1984; Fig. 6.1). This conceptual model was also referred to the Hierarchical Model of Decomposition (Lavelle et al., 1993). The spatial hierarchy of climate and litter quality control was then considered implicitly as a stratification of importance with climate being the primary control, and litter quality, the secondary control. This view was questioned in a broad synthesis of a large number of decomposition studies (Cornwell et al., 2008) showing that variability in decomposition due to differences in litter quality was at least twice as high as that observed across broad climatic gradients. In other words, the same litter type decomposing in highly contrasting climatic conditions across continental gradients may vary less in mass loss compared to contrasting litter types decomposing in the same location under identical climatic conditions. These observations required revisiting the hierarchical model of decomposition (Fig. 6.1).

The importance of litter quality effects compared to climate control also depends on how litter quality varies within and across climatic zones. Surprisingly, these differences are not that well documented presently, because studies that exhaustively report litter quality for all plant species of a given community along with their abundances, and thus their relative contribution to the overall litter pool, are rare. Community ecologists collecting good abundance data, commonly infer litter quality from live plant traits, which can at best be a proxy for the actual litter quality. Leaf

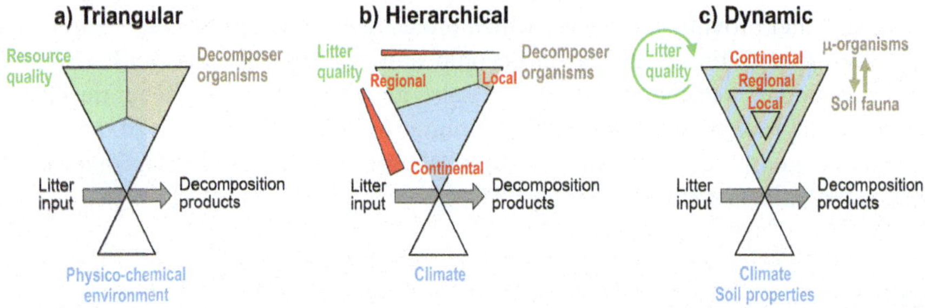

Fig. 6.1 The evolution of the conceptual model of litter decomposition over the past fifty years. The graphical representation is inspired by the original drawing by Swift et al. (1979) who used the regulation symbol (⧖) to superimpose the classical concept of the triangular control of the transformation of fresh litter input into decomposition products (**a**). The three colors referring to the three groups of factors have equal proportions because the Triangular Model did not initially propose an explicit hierarchy between the physico-chemical environment (edaphic and climatic factors), resource (litter) quality, and decomposer organisms. Over the following three decades (1980s through 2000s), studies at regional and continental scales reinforced the perception that climatic factors (essentially temperature and humidity) dominate litter quality control, and that decomposer organisms are merely reflecting climate and litter quality regulation without independent control, leading to the Hierarchical Model (**b**). The factor ranking incidentally implied a spatial stratification from local to continental scales with increasing dominance of climate control at increasing scales. The Hierarchical Model was revisited following studies that reported stronger litter quality than climate control (Cornwell et al., 2008 as a key study), which is not shown here. Instead, we propose the Dynamic Decomposition Model (**c**) emphasizing the dynamic switch among the three control factors during the decomposition process (e.g., García-Palacios et al., 2016) regardless of the spatial scale considered. In other words, all three factors can be dominant at some point during decomposition at the local as well as at the regional or continental scales. The Dynamic Model also emphasizes that litter quality changes strongly over time as do the specific traits that control decomposition. In addition, it appears inadequate to keep decomposer organisms in one single box. The distinction between at least microbial communities and soil fauna, which interact with each other, may significantly improve the understanding of decomposition and its controls

litter quality varies substantially from live foliage traits resulting from the important chemical and structural changes during leaf senescence, for example, when nutrients are resorbed (Aerts & Chapin, 2000) or when secondary metabolites undergo dynamic changes (Paaso et al., 2017). Another difficulty is that the standard set of live foliage traits commonly measured, and thus available for numerous plant species (e.g., carbon and nitrogen concentration, specific leaf area, leaf dry matter content), are not always relevant for litter decomposition. Less commonly measured litter quality traits such as condensed tannins, concentrations of Mg, Ca, Mn, or non-structural carbohydrates often predict decomposition better, depending on the type of ecosystem and spatial scale considered (Guerrero-Ramírez et al., 2016; Hättenschwiler & Jørgensen, 2010; Makkonen et al., 2012). On the other hand, soil ecologists or biogeochemists collecting good litter quality data typically focus on a few contrasting litter types without referring to their relative abundance in the local plant community, or alternatively, on community-level litter fall without distinguishing individual species. This latter approach allows the evaluation of ecosystem-scale processes and provides a good assessment of community-level decomposition. On the contrary, it permits only a limited understanding of the driving mechanisms, especially if non-additive effects on decomposition of mixtures of litter from a range of different plant species occur, which are not predictable from community weighed-mean litter traits. We will come back to this important issue later when we discuss the role of biodiversity in litter decomposition.

Overall, litter quality control of decomposition is similar in aquatic compared to terrestrial ecosystems, especially in lower order forest streams with low autochthonous primary production and sharing the same litter inputs as neighboring terrestrial ecosystems. In one of the rare field experiments comparing terrestrial to aquatic decomposition across a broad latitudinal gradient, García-Palacios et al. (2016) identified the same set of litter traits explaining litter mass loss in forest streams and adjacent forest floors. There were some biome-specific differences, but overall, Mg and Ca concentrations were the most important litter traits associated to variation in litter decomposition across biomes in both types of ecosystems. Another, more local study in an alpine environment found that the relative role of environmental factors and litter quality were consistent in aquatic and terrestrial decomposition even at different decomposition stages (Yue et al., 2018). Collectively, these studies suggest that decomposition dynamics can be predicted from the same variables, irrespective of whether litter decomposes in a stream or on the soil surface. It is important to note, however, that these comparative studies used spatially coupled pairs of low-order streams and terrestrial sites sharing the same local environmental conditions. It is less likely that a similar coherence in the relative importance of control factors persists at larger spatial integration, with for example, varying local nitrogen deposition, heterogeneous bedrock, and thus, distinct nutrient limitation, or predominant oxygen control in higher order and more slowly-flowing streams.

Compared to biotic control factors, i.e., litter quality and decomposer organisms, environmental control during the decomposition process seems more straightforward to quantify. Technically this is certainly true, but practically, temperature, as well as humidity, which in contrast to most aquatic ecosystems (with the exception of intermittent streams) is an additional important environmental factor regulating decomposer activity in terrestrial ecosystems, may not have been characterized sufficiently well in the past. This is because microclimatic variability was mostly neglected, assuming that data from the nearest climate station can represent the climatic conditions for a particular study site reasonably well. This may not be the case as some studies suggest (Bradford et al., 2014, 2016; Joly et al., 2017). Indeed, temperature and humidity may vary strongly at very small spatial scales of only a few meters or even at the individual litterbag scale (Bradford et al., 2016), depending on microtopography, exposition, plant presence or animal activities, leading to variations that can be as large as among climate stations at regional scales (Bradford et al., 2014). The unaccounted variability in microclimate blurred the understanding of the relative importance of climate control in terrestrial decomposition and the spatial scale at which it operates. This is probably less problematic for aquatic decomposition. In these systems, water temperature and oxygen concentration, two important environmental factors controlling decomposer activity during aquatic decomposition, are standard measurements at the microsite scale because these data are not readily available from climate stations, and oxygen can vary strongly at small spatial scales. Therefore, the established relative contribution of environmental control over decomposition appears more robust in aquatic than terrestrial ecosystems.

6.2.2 *The Role of Decomposer Organisms*

The approaches to study temperature and litter quality control over decomposition are comparable in terrestrial and aquatic ecosystems, and share broadly the same methodological and conceptual development. However, terrestrial and stream ecologists considered the role of decomposer organisms quite differently in the past (Wagener et al., 1998). Detritivorous animals and their contribution to decomposition initially received much more attention amongst stream ecologists (Graça, 2001; Wagener et al., 1998) than amongst terrestrial ecologists, who commonly excluded detritivores by the use of litterbag mesh sizes smaller than the body sizes of most detritivores, especially soil macrofauna (e.g., earthworms, millipedes, isopods, etc.). In fact, terrestrial decomposition has been traditionally considered mostly as a microbial-driven process (Wagener et al., 1998). Microbial decomposers were seen mostly like an engine with its performance depending on available fuel (litter quality) and suitable conditions (temperature and humidity), but not on the characteristics of the 'engine' itself. Indeed microbial diversity and physiology was largely neglected, because the tools to measure it did not suffice, and because of the long held paradigm that the same microorganisms are everywhere. This classical view changed considerably over the last 15 years with the development of molecular tools and extensive biogeographical assessments of the occurrence and diversity of soil microbes. Recent studies showed that soil microbial communities differ substantially in time (over the course of litter decomposition; Herzog et al., 2019) and space (Baldrian, 2017). Spatial variation of a rather basic microbial parameter, such as biomass, can be an important determinant of decomposition even at regional scales alongside with climatic factors (Bradford et al., 2017), further questioning the dominant role of climate control in the classical Hierarchical Decomposition Model (Fig. 6.1). Additionally, soil fauna are increasingly considered in litter decomposition studies, showing that they are major players beyond microbial communities also in terrestrial decomposition (García-Palacios et al., 2013; Hättenschwiler & Gasser, 2005; Wall et al., 2008). Numerous studies manipulating the presence, abundance, or diversity of soil fauna (e.g., Coulis et al., 2015; Handa et al., 2014; Heemsbergen et al., 2004) clearly showed that they need to be taken into account as a control factor on their own.

Collectively, the recent findings of how microorganisms and soil fauna contribute to decomposition call for a revised role of decomposers in the control of terrestrial litter decomposition and its integration in conceptual models (Fig. 6.1). Accounting for decomposer organisms in decomposition studies is not an easy task, because of the tremendous number of different taxa, organized in highly complex food webs, which depend on plant-derived litter as the main source of energy and matter. The complexity of terrestrial decomposer communities in terms of diversity and the number of trophic and non-trophic links appears to be higher, compared to that of streams, but perhaps not of other aquatic ecosystems such as marine benthos. Because of this complexity, but also because the composition and diversity of decomposer communities can vary substantially among different plant communities, it is presently difficult to incorporate soil organisms in predictive mechanistic decomposition models.

6.2.3 Temporal Dynamics of Biotic and Abiotic Drivers of Litter Decomposition

The relative impact of environmental factors, litter quality, and decomposer communities may vary over time with proceeding decomposition. For example, García-Palacios et al. (2016) showed that control by litter microbial and nematode communities dominated during early decomposition stages, while soil moisture and legacy effects of initial litter quality increased in importance during later stages of decomposition. Such temporal shifts in control mechanisms are important to consider for a better understanding of how environmental conditions, litter quality, and decomposer communities affect decomposition interactively in a revised triangular relationship (Fig. 6.1), yet studies addressing such temporal dynamics explicitly are still rare. Changing control through time appear to be even less studied for stream ecosystems (but see Yue et al., 2018), perhaps also because litter decomposition proceeds generally quicker compared to terrestrial ecosystems. The different time scales at which freshly fallen leaf litter disappears in terrestrial and aquatic ecosystems is one of the major differences affecting the interplay of different control factors and predictive modeling of decomposition in terrestrial and aquatic systems. In certain biomes and for certain litter types it may take more than ten years in terrestrial ecosystems (Parton et al., 2007) compared to only a few weeks in stream ecosystems (Gessner et al., 2010) until the leaf litter visually disappears. These large differences actually hide the fact that smaller leaf particles resulting from detritivore activity ("fragmentation" or "comminution") are easily washed downstream with flowing water in streams, while they remain longer in place under terrestrial conditions. This means that part of the decomposition process is 'delocalized' in streams and decomposition in the strict sense of mineralization of organic matter may in the end not differ as much between terrestrial and aquatic ecosystems. This is difficult to measure correctly in either system because dissolved organic compounds and particulate organic matter move down the soil profile and down the streams.

6.3 Diversity and Litter Decomposition in Terrestrial Ecosystems Compared to Streams

6.3.1 *Leaf Litter Diversity*

The rapid rate of biodiversity loss worldwide has prompted research efforts in recent decades directed towards understanding if and how biodiversity influences ecosystem functioning (Eisenhauer et al., 2019; Hooper et al., 2012). Most often, diversity metrics (e.g., species richness, functional diversity) are manipulated as the independent variable, and ecosystem functions are measured to understand whether the whole is greater than the sum of its parts (net diversity effect) and if so, whether mechanisms could be identified driving such effects (Eisenhauer et al., 2019). While primary productivity is the ecosystem function having received the most attention to date and for which strong positive net diversity effects have been observed across ecosystems (Tilman et al., 2014), litter decomposition has been studied too, recognized as an ecosystem function of key importance for both terrestrial and aquatic ecosystems. Manipulative biodiversity experiments on litter decomposition have been treated in meta-analyses (Cardinale et al., 2011; Gartner & Cardon, 2004; Lecerf & Kominoski, 2010) and through large-scale collaborative experimental studies (García-Palacios et al., 2016; Handa et al., 2014). Evidence to date in terrestrial ecosystems suggests that overall net diversity effects are weak (if present at all), in comparison to plant productivity responses (Cardinale et al., 2011; Handa et al., 2014), but there is increasing evidence (Lefcheck et al., 2015; Wagg et al., 2014) suggesting that positive biodiversity effects increase when considering the ability of ecosystems to maintain multiple functions simultaneously (i.e., multifunctionality). The number of studies addressing the biodiversity-ecosystem multifunctionality relationship in streams is scarce, but novel evidence suggest an important positive linkage similar to that found in terrestrial ecosystems (López-Rojo et al., 2019).

Net diversity effects, when observed, can represent either selection or complementarity effects of species mixtures within assembled biodiversity experiments (Loreau & Hector, 2001). Selection effects are an extension of the mass ratio hypothesis (Grime, 1998), which suggests that the local dominance of a species with particular traits present within a community can be determinant to the ecosystem function under study. Complementarity effects, on the other hand, point to species differences or species interactions within a community that result in synergistic or antagonistic effects on the measured ecosystem function (Loreau & Hector, 2001). Such complementarity effects reflect niche partitioning of species favouring specialized resource use or interactions such as facilitation or inhibition that translate into diversity effects. In their collaborative study across a latitudinal gradient, Handa et al. (2014) showed that complementarity rather than selection mechanisms drove diversity effects in both terrestrial and aquatic litter mixtures. For instance, litter mixtures combining a N-rich litter and a non-recalcitrant rapidly decomposing litter led to a positive net diversity effect in this cross-system study, likely due to translocation of nitrogen through fungal hyphae from N-rich to N-poor litter thus facilitating microbial decomposition. In a recent aquatic microcosm study with a riparian forest litter mixture, López-Rojo et al. (2019) also found that complementarity effects were more prevalent in explaining observed net diversity effects, but selection effects also played a role.

Quantifying selection and complementarity effects in litter mixtures requires the measurement of litter mass loss at the species level, which is time-consuming and potentially challenging if the stage of litter decay hinders species identification. The use of community-level metrics and plant traits can partially solve this issue by establishing indirect associations with selection and complementarity effects. In this line, selection effects have been associated with the community-weighted mean of functional traits that drive the decomposition process, while complementarity effects are typically associated with the functional dissimilarity of these traits within a community (García-Palacios et al., 2017). With litter quality as one of the key drivers of decomposition, it is not surprising that idiosyncratic responses are frequently observed, leading to the conclusion 'community composition matters'. However, most litter decomposition studies addressing the role of diversity in both terrestrial and aquatic ecosystems have been performed without explicitly considering underlying mechanisms by separating individual species from litter mixtures. While these studies comparing monocultures *vs.* increasing levels of litter diversity are useful for assessing if the whole is (or is not) greater than the sum of its parts, these assessments have limited our understanding of selection and complementarity effects in both soils and streams.

6.3.2 Multi-trophic Diversity

Evidently, when considering decomposition dynamics, diversity exists not only at the level of the plant communities that provide the majority of organic matter input, but also at the level of the decomposers themselves (microbial and faunal communities). Decomposers interact in complex networks on the basis of decomposing litter substrates and connected through multiple trophic and non-trophic interactions (Wagg et al., 2019). While there is compelling evidence that multi-trophic biodiversity in general (Lefcheck et al., 2015), and soil biodiversity in particular (Wagg et al., 2014), enhances multifunctionality across ecosystems, much remains to be discovered about how varying decomposer diversity influences litter decomposition as an ecosystem function.

Contrary to the community assembly approach when studying the effect of the diversity of plants on litter decomposition, addressing the effects of varying decomposer diversity in both terrestrial and aquatic systems has, to date, more frequently relied on exclusion experiments (Handa et al., 2014). This constraint has been particularly true for manipulating microbial diversity given the many inactive or dormant microbes sampled in the environment and a high percentage of unculturable microbes (Baldrian, 2017; Jansson & Hofmockel, 2018). In a recent terrestrial grassland study,

Wagg et al. (2019) sieved soils through varying mesh sizes to simplify microbial network complexity and demonstrated that reduced fungal-bacterial diversity slowed down litter decomposition. An alternate approach to simplify microbial diversity is through the dilution-to-extinction approach where, for example, Maron et al. (2018) added soil inoculum from grassland microbial communities to sterile soil in microcosms and demonstrated that decreasing diversity altered litter C cycling by favouring the decomposition of non-recalcitrant carbon over recalcitrant carbon. While the functional redundancy principle likely ensures functional stability below a particular threshold of species loss (Miki et al., 2014), a tremendous diversity of microbes contributes to litter decomposition and interact in a cross-kingdom functional succession of communities (Herzog et al., 2019).

When considering decomposer community complexity that includes meso- (up to 1-mm body size) and macrofauna (up to 5-mm body size) in addition to microbes, community exclusion litter microcosm experiments have been useful to demonstrate that increasing decomposer community complexity accelerates litter carbon and nitrogen cycling in both forests and freshwater stream ecosystems (Handa et al., 2014). However, other studies have used community assembly approaches to manipulate macrodetritivores in microcosms to assess litter decomposition. For example, in a study combining dissimilar soil detrivores such as a millipede and snail, De Oliveira et al. (2010) demonstrated both synergistic and antagonistic interactions depending on the stage of litter decomposition of Mediterranean forest litter. Other terrestrial studies have hinted at niche partitioning mechanisms reflected through functional dissimilarity (Coulis et al., 2015; Fontana et al., 2019; Heemsbergen et al., 2004) or strong litter preferences by detritivores that may account for diversity effects (Rouifed et al., 2010; Vos et al., 2011). Similarly, the literature addressing the effects of detritivores on litter decomposition in streams using exclusion procedures is also prominent (Gessner et al., 2010). More detailed analysis using community-level properties are less frequent, but Frainer et al. (2014) found that the functional diversity of detritivore communities had contrasting effects on litter decomposition in boreal streams across seasons and habitats.

In conclusion, the effects of leaf litter diversity on litter decomposition in terrestrial ecosystems and streams are reasonably well identified, but we still have a limited understanding of the biological mechanisms accounting for how plant diversity influences organic matter decomposition in both systems. With regard to the diversity of decomposer communities, emerging molecular tools and trait-based approaches are helping us move beyond exclusion experiments towards a more functional understanding of decomposer diversity in soil and stream detrital food webs.

6.4 Global Change and Litter Decomposition in Terrestrial Ecosystems Compared to Streams

The effects of global anthropogenic change on ecosystem processes has been a hot topic in the biogeochemical and ecological literature over the past 20 years, and litter decomposition is not an exception. From the multiple drivers of global change promoting planetary-scale shifts in the Earth system, here we focus on climate warming, nitrogen enrichment and biotic invasions. The effects of these three global change drivers on terrestrial litter decomposition have been synthesized in systematic reviews and quantitative meta-analyses, which are necessary to assess general patterns across different ecosystem types. Furthermore, warming, nitrogen enrichment and biotic invasions also influence the biotic and abiotic drivers of litter decomposition in streams, allowing us to look for commonalities in aquatic *vs.* terrestrial systems.

The interpretation of global change effects on terrestrial litter decomposition is not straightforward, as these effects are the result of both direct and indirect mechanisms. Specifically, climate warming and nitrogen enrichment influence litter decomposition though direct temperature effects on soil biological activity and direct nitrogen effects on soil chemistry, respectively (Wardle, 2004). However, warming, nitrogen enrichment and biotic invasions also play an indirect role mediated by shifts in plant and soil communities (Manning et al., 2006; Castro-Díez et al., 2014). Importantly, the magnitude of such indirect effects may be larger than that of direct effects mediated by changes in abiotic factors, strongly modifying the net effects of global change on litter decomposition. In short, both direct and indirect mechanisms must be considered.

6.4.1 Climate Warming

The fate of the soil carbon pool is a pressing issue under ongoing climate warming, as even subtle losses may represent a substantial contribution to the buildup of the atmospheric CO_2 pool, promoting a positive land carbon-climate feedback. Litter decomposition is a pathway of soil carbon loss to the atmosphere, and accordingly the number of studies addressing warming effects on litter decomposition increased in recent years. Different methods were used to experimentally simulate climate warming in terrestrial ecosystems, with open-top chambers, infrared lamps and soil heating cables among the most common (Fig. 6.2). Two recent global meta-analyses showed a slight positive effect of experimental warming on terrestrial litter decomposition (Lu et al., 2013; Yue et al., 2015). Across 34 different articles, Lu et al. (2013) reported a 6.8% increase, and from a slightly lower sample of 22 articles, Yue et al. (2015) reported a 4.4% increase. In streams, a recent quantitative literature review found a significant positive effect of warming on litter decomposition (Amani et al., 2019). However, as it is usually the case across different field studies,

Fig. 6.2 Different methods to experimentally simulate climate warming in terrestrial ecosystems. **a** Open-top chambers passively increasing temperature using hexagonal methacrylate sheets in Bogong High Plains, Australia (*source* https://jscamacresearch.wordpress.com/, photo credit Henrik Wahren). **b** Soil heating cables in Hubbard Brook Long-Term Ecological Research, USA (*source* https://hubbardbrook.org, photo credit: Rebecca Sanders-DeMott). **c** Infrared heaters in El Yunque National Forest, Puerto Rico (*source* https://blogs.agu.org/, photo credit: Stephanie Roe)

the overall positive warming effects on litter decomposition were strongly influenced by methodological differences among studies. For instance, the magnitude and direction of warming effects depended on the ecosystem type, study length, temperature increase, and method used to simulate elevated temperatures. When only direct effects of elevated temperatures on litter decomposition are considered, litter mass loss is typically higher in warming than in control plots, indicating higher decomposer activity under more favorable microsite conditions.

Addressing the combined direct and indirect warming effects on litter decomposition is far more complicated, and only a handful of studies have explicitly done so. For instance, Cornelissen et al. (2007) collected litter from 33 experimental warming studies in cold ecosystems and studied litter decomposition in two thermally contrasted sites in Sweden. This elegant experimental design allowed the authors to disentangle direct climate effects from indirect litter quality changes resulting from warming-induced changes in plant community composition. In fact, the site of litter incubation explained 50% of variation in litter mass loss, which was 42% higher in the warmer than colder site. Interestingly, warming promoted a shift in plant community composition from grass- and sedge-dominated communities to shrub-dominated communities. This plant community shift explained 30% of variation in litter mass loss. Increasing shrub dominance with warming resulted in an average decrease of litter mass loss of 40%. This result suggests that although climate warming seems to alleviate the temperature limitation over litter decomposition in cold ecosystems, the likely shrub expansion with warming in these high-latitude ecosystems may counteract such direct warming effects. In addition to changes in plant community composition, climate warming may also change decomposer communities, representing an additional indirect effect on litter decomposition (David & Handa, 2010). In fact, Boyero et al. (2011) found a stimulation of microbial decomposition but inhibition of detritivore-mediated decomposition with temperature in streams across a latitudinal gradient in six continents.

6.4.2 Nitrogen Enrichment

Soil nitrogen (N) enrichment through atmospheric N deposition is one of the major global change drivers affecting ecosystem functioning (Galloway et al., 2008). The effects of N enrichment are easier to simulate experimentally than those of warming, as simple inorganic N fertilization can efficiently mimic current N deposition rates. Consequently, a plethora of experimental field studies has addressed the effects of N enrichment on terrestrial litter decomposition. For instance, Knorr et al. (2005) and Zhang et al. (2018) synthesized the effects of N enrichment on litter decomposition using data from 24 and 55 articles, respectively. On average N enrichment had no significant effect on litter decomposition. Importantly, the relationship between the effect size of N enrichment and N application rate was negative, with stimulated litter decomposition at low levels of N application but suppressed litter decomposition at high levels. The same increasingly negative effects with higher N enrichment from atmospheric deposition and agricultural run-off has been demonstrated for streams with meta-analyses (Ferreira et al., 2015) and large-spatial scale observational studies (Woodward et al., 2012). High rates of N input to streams may have toxic effects for invertebrates counteracting the stimulating effect on microbial decomposition.

Nitrogen enrichment can influence litter decomposition rates via direct changes in fundamental controls of litter C mineralization and N release/immobilization patterns such as soil nutrient stoichiometry and pH, which commonly decreases with N enrichment. However, N enrichment also affects litter decomposition indirectly by altering the diversity and composition of plant and soil communities. The response of plant communities to N enrichment usually includes decreased species richness and compositional shifts towards resource acquisitive species (Isbell et al., 2013), with cascading effects on litter quality. Soil microbial communities can also mediate N enrichment effects on litter decomposition via reduced soil microbial biomass (Treseder, 2008) and oxidase enzymatic activities involved in the degradation of recalcitrant C compounds (Jian et al., 2016). Nevertheless, the relative contribution of such indirect effects on net decomposition rates vary across studies, with plant compositional shifts counteracting the direct effects of N enrichment on soil chemistry (Pichon et al., 2019) or playing a minor role compared with direct effects (Manning et al., 2006).

6.4.3 Biotic Invasions

The effects of biotic invasions on terrestrial litter decomposition rely mostly on indirect impacts via changes in the plant and soil community, as this global change driver does not commonly directly modify the environmental conditions of the invaded site. Here we focus on the indirect effects via the plant community, as studies addressing how non-native soil organisms alter litter decomposition rates are less numerous, which limits our ability to elaborate general conclusions.

Several reviews and meta-analysis have synthesized the effects of non-native plant invasions on litter decomposition, with contrasting results. The vote-counting approach of Ehrenfeld (2003) and quantitative reviews of Liao et al. (2008) and Castro-Díez et al. (2014) showed higher decomposition of non-native than native leaf litter. Specifically, Liao et al. (2008) found a 117% increase in non-native litter decomposition rates, which correlated well with the higher litter N and specific leaf area in non-native compared to native litter and green leaves. These results have contributed to the generalization that invasive plants decompose more quickly, leading to more rapid cycling of nutrients and C release in invaded ecosystems. However, neither the meta-analysis of Vilà et al. (2011) nor the multi-species study (78 deciduous forest species) of Jo et al. (2016) found significant differences in litter decomposition rates between non-native and native species. As pointed out by Zuukswert and Prescott (2016), the general perception that non-native litter decomposes faster may arise from a bias in current studies focusing primarily on plant species that are known to have a major influence on ecosystem functioning, such as N fixing species like *Myrica faya* (Liao et al., 2008).

Most litter decomposition studies addressing the impacts of biotic invasions in streams have focused on exotic trees. Recently, Ferreira et al. (2016) synthesized the available literature addressing the effects of tree plantations and tree invasions in natural ecosystems. They found that litter decomposition was 26% lower in streams flowing through *Eucalyptus globulus* plantations than in streams flowing through native forests. However, the effects were not significant when non-native tree species other than *Eucalyptus globulus* were included in the comparison.

In contrast with the two previous sections on decomposition drivers and diversity effects, we are not aware of any study jointly addressing the influence of global change in terrestrial and aquatic ecosystems. However, as warming, N enrichment and biotic invasions also influence the biotic and abiotic drivers of litter decomposition in streams, we may be able to infer consistent patterns across ecosystems by comparing the results found in meta-analyses and large-spatial scale studies. Although climate warming significantly increases litter decomposition in soils and streams, warming-induced effects can change substantially after accounting for the indirect changes via the plant and decomposer communities. The effects of N enrichment on litter decomposition seem to vary as a function of the amount of N addition in both soil and stream studies. This pattern may be the result of indirect effects via changes in plant and decomposer communities, although the number of studies is still quite limited for robust conclusions. With regard to biotic invasions, the effects of exotic plants on soil and stream litter decomposition seem to be species-specific.

6.5 Suggested Approaches for Future Studies

6.5.1 Future Studies Looking at Biotic and Abiotic Drivers

It appears that the role of climate was overestimated in the past, because climatic variables were not measured at a fine enough spatial resolution, and variables related to decomposer communities were largely ignored, especially in terrestrial studies (Bradford et al., 2016). Future studies would need to measure environmental factors (climate variables, but also soil or water characteristics) at appropriate spatial and temporal scales and should include a sufficiently large array of relevant litter quality traits and a reasonably detailed characterization of decomposer communities for a better understanding of how the different factors interact in controlling decomposition.

The understanding of the fate of decomposition products is another area of research that has received insufficient attention to date. The vast majority of decomposition studies have used and still use the litterbag approach or variants of it and assess decomposition as litter mass loss from these bags. Much of this lost litter material may actually not decompose (in the sense of mineralization) during the study, but transported out of the bags in the form of particulate or dissolved organic matter. Ignoring the fate of this organic matter is likely leading to erroneous assumptions in how carbon and nutrients are cycled through the ecosystem. It will be important to address the fate of "lost litter mass" more explicitly in future studies. This may be even more critical for stream ecosystems where flowing water may transport a large amount of particulate and dissolved litter material over relatively large distances. It is methodologically challenging even with the powerful approach of using isotopically labelled litter material to follow the decomposition products, in particular in stream ecosystems where they may not easily be recovered.

We believe that from the three groups of control factors, decomposer organisms require stronger attention in future work determining how the different drivers interactively affect decomposition, in particular in terrestrial ecosystems. It is now clear that decomposers do not simply "translate" the effects of litter quality and environmental conditions, but are an important driver on their own (Bradford et al., 2017; García-Palacios et al., 2016). Interactions between microbial communities and detritivores are particularly poorly understood. The recent finding that the transformation of fresh leaf litter material into fecal pellets by an abundant millipede species can change carbon and nitrogen release during further microbial decomposition as well as the relevant traits predicting decomposition (Joly et al., 2018) suggests that such interactions can have strong impacts on decomposition. However, data are extremely limiting, especially under field conditions.

The large majority of decomposition studies focused on leaf litter, neglecting other plant tissues such as wood and roots that quantitatively contribute at least as much to the overall litter produced by plants. Root decomposition studies are strikingly few in the literature compared to leaf litter studies, although there are recent advances in terrestrial ecosystems (Guerrero-Ramírez et al., 2016; Herzog et al., 2019; Jo et al., 2016; Sun et al., 2018). There are some obvious reasons for this because roots are more difficult to access and to collect, roots from different plant species are often difficult to distinguish, and roots do not seem to readily fall into streams. A recent study following leaf litter and first-order roots (these are the few millimeter long root tips, which turnover two to three times per year) from several woody species over six years reported substantially slower decomposition of first-order roots that was controlled by completely different traits compared to leaf litter (Sun et al., 2018). Thus, at the scale of the ecosystem, we may currently understand only part of the decomposition process reasonably well, calling for more root studies in the future. Even in aquatic systems, especially in low-order stream, root decomposition may be considerable as roots from riparian vegetation can grow into the water body, but appear to have received very little attention.

6.5.2 *Future Studies Looking at Diversity Effects*

One major challenge to improve our understanding of how both resource and consumer diversity influences litter decomposition dynamics will be to integrate our multi-trophic understanding of food web interactions into models so as to better predict process rates across ecosystems. While some synthesis efforts have suggested that top down effects of consumer diversity are stronger than bottom up effects of resource detrital diversity (Srivastava et al., 2009), others have pointed towards strong bottom up drivers, particularly nutrient limitation and stiochiometric constraints in freshwater food webs (Frainer et al., 2016). Given the challenges of resolving food webs and the desire to develop tools that allow us to extrapolate network structure towards predicting litter decomposition, one potential approach is trait matching of consumer feeding traits to resource palatability traits (Brousseau et al., 2018). Recent work assessing trait covariation of detrivores and their resources has pointed to over one third spatial covariation of consumer and resource traits for collembolans and leaf litter (Raymond-Léonard et al., 2019), as well as for litter-dwelling detritivorous and predatory macroarthropods with their respective resources (Brousseau et al., 2019). These results suggest certain predictive bottom-up structuring forces. Linking trait-matching models that can successfully predict litter-feeding interactions (Brousseau et al., 2018) with estimates of process rates like decomposition represents a promising future research direction. Additionally, refining our understanding of microbial networks and the function of microbes will be a high priority with metagenomic tools that indicate the functional potential of communities all the way to metaphenomics, which combine this information with that of available resources (Jansson & Hofmockel, 2018). Soil viruses have been largely understudied but as demonstrated through microbial loops in aquatic systems can dramatically change ecosystem process rates (Kuzyakov & Mason-Jones, 2018). Future studies will clearly benefit from embracing all this complexity and seeking predictive tools that integrate our understanding of these multi-trophic perspectives to predict litter decomposition in the context of a changing planet.

6.5.3 Future Studies Looking at Global Change Effects

In the global change section of this chapter, we highlighted the importance of disentangling the direct effects of climate warming and nitrogen enrichment on litter decomposition from the indirect effects via changes in plant community composition. Here we propose two methods inspired in previous litter decomposition studies performed in terrestrial ecosystems, one experimental and one analytical, which may also help to tease apart these two important mechanisms in litter decomposition studies performed in streams.

The first method is a mechanistic factorial experiment that has been used to test the effects of warming (Allison et al., 2013; Cornelissen et al., 2007) and nitrogen enrichment (Manning et al., 2006; Pichon et al., 2019) on terrestrial litter decomposition (Fig. 6.3). The experimental design consists in the following factors: 'plot environment' (direct global change effect) and 'litter origin' (indirect effect via changes in plant community). First, a global change field experiment with two levels (warming or nitrogen enrichment *vs.* control) is conducted for a period long enough to allow for plant community compositional shifts (i.e., 3–5 years in grasslands). This design is replicated in a few experimental blocks. Then, naturally-senesced leaf litter is collected from the global change and control plots, and used to establish a litter reciprocal transplant. A similar approach may be followed in aquatic ecosystems, using either experimental treatments in a single stream or multiple streams along

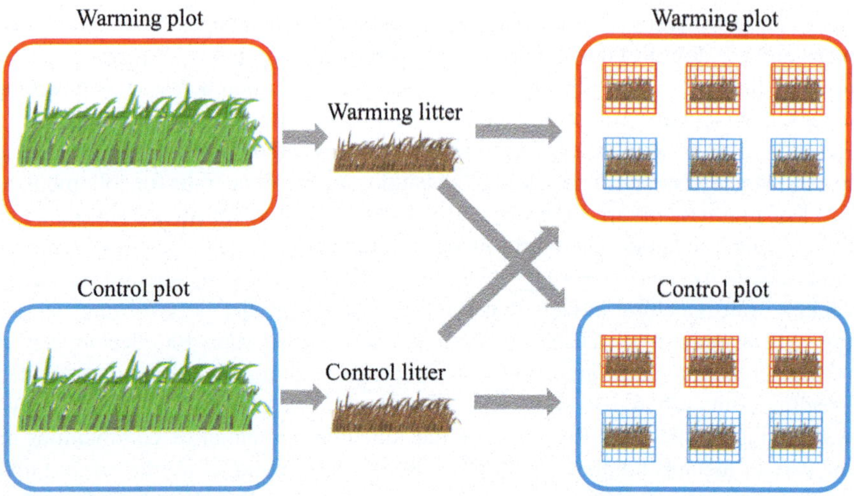

Fig. 6.3 Experimental design to disentangle the direct effects of global change (climate warming) on litter decomposition from those indirectly mediated by changes in plant community composition. First, leaf litter is harvest from warming and control plots. Then, a reciprocal transplant litterbag study is conducted to test the effects of 'plot environment' (warming *vs.* control plot) and 'litter origin' (warming *vs.* control litter). Several litterbags may be included in each block if sequential sampling is required (in this example we envisioned three harvests)

contrasting environmental conditions. Additionally, Allison et al. (2013) considered a third treatment ('microbe origin') assessing indirect effect via changes in the decomposer community. To do that, the authors gamma-sterilized all litterbags and litter material, and reinoculated them with a microbial inoculum extracted from the global change and control litter treatments. However, the validity of this approach is more uncertain due to litter microbial colonization during incubation in the field.

The second method is an analytical procedure to disentangle the direct and indirect effects of global change drivers in large-spatial scale observations. This approach assumes that the environmental gradient selected mainly encapsulates the variation in the variable of interest (i.e., temperature and nitrogen enrichment). Ideally, such environmental information should be recorded at each site in sufficient detail, knowing that environmental factors should be evaluated at an appropriate spatial resolution (Bradford et al., 2014). Then, naturally-senesced leaf litter of the dominant plant species is collected at each site, and used to fill site-specific litterbags containing litter species in the same proportions found in the native litter layer. Leaf litter traits such as nutrient concentration, morphology or stoichiometry may be used to quantify variation in plant community composition across sites. At each site, soil samples are also collected to measure physicochemical parameters as well as the abundance and diversity of microbial and animal decomposers. Finally, all the abiotic and biotic drivers of litter decomposition are linked in a conceptual path diagram similar to that shown in Fig. 6.4a. This conceptual model is then analyzed using path analyses and tested against field data. García-Palacios et al. (2017) followed a similar approach

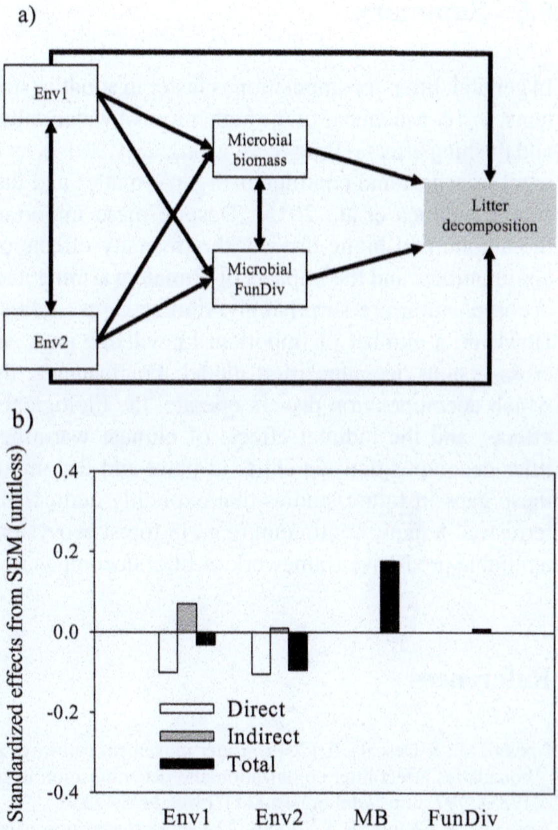

Fig. 6.4 A priori conceptual path diagram (**a**) depicting pathways by which environmental conditions (Env1 and Env2 are the two first axes of a PCA with climatic variables and soil physicochemistry), soil microbial biomass (MB) and soil microbial functional diversity (FunDiv; assessed with community-level physiological profiles) influence litter decomposition (% of litter C loss). Standardized direct, indirect and total effects (**b**) derived from the structural equation model. Redrawn from Journal of Ecology (García-Palacios et al., 2017)

when addressing the direct and indirect (mediated by the soil microbial community) effects of climate and soil parameters on litter decomposition across 10 sites located along a regional climatic gradient in southern France. Litterbags were harvested after a year of field incubation, and the total, direct and indirect effects of environment (climate and soil conditions) on litter carbon loss were calculated. The total effects of Env1 (accounting for MAP and MAT differences among sites) were small, as the indirect effects via changes in the soil microbial community partially offset the direct effects (Fig. 6.4b). This result supports the pattern found in Cornelissen et al. (2007) when assessing warming-induced effects on litter decomposition via shrub expansion in cold ecosystems. Studies at large spatial scales conducted in streams (e.g., Boyero et al., 2011) are also well-suited to perform a similar analysis.

6.6 Summary

In general, litter decomposition is faster in aquatic (stream) than in terrestrial ecosystems, as a consequence of the higher loss of water-soluble compounds due to leaching and flushing effects (Berg & McClaugherty, 2014), as well as due to abrasion by sediment transport and continuous organic matter and nutrient supplies from upstream sources (Graça et al., 2015). Despite these importance differences, it seems that the abiotic and biotic drivers, the diversity effects of plant litter and decomposer communities, and the impacts of climate warming and nitrogen enrichment on litter decomposition are surprisingly similar across terrestrial and aquatic ecosystems. However, a number of important knowledge gaps still limit the development of a cross-system decomposition model. For instance, the spatiotemporal variation at which decomposition drivers operate, the biological mechanisms behind diversity effects, and the indirect effects of climate warming and nitrogen enrichment on litter decomposition via shifts in plant and decomposer communities. Addressing these gaps in future studies that explicitly address litter decomposition across the terrestrial-aquatic continuum (e.g., in forest ecosystems) will contribute to generate a reliable predictive framework of litter decomposition at biome and global scale.

References

Abelho, M., & Descals, E. (2019). Litter movement pathways across terrestrial–aquatic ecosystem boundaries affect litter colonization and decomposition in streams. *Functional Ecology, 33*(9), 1785–1797. https://doi.org/10.1111/1365-2435.13356.

Aerts, R., & Chapin, F. S. (2000). The mineral nutrition of wild plants revisited: Re-evaluation of processes and patterns. *Advances in Ecological Research, 30,* 1–67. https://doi.org/10.1016/S0065-2504(08)60016-1.

Allison, S. D., Lu, Y., Weihe, C., Goulden, M. L., Martiny, A. C., Treseder, K. K., & Martiny, J. B. H. (2013). Microbial abundance and composition influence litter decomposition response to environmental change. *Ecology, 94*(3), 714–725. https://doi.org/10.1890/12-1243.1.

Amani, M., Graça, M. A. S., & Ferreira, V. (2019). Effects of elevated atmospheric CO_2 concentration and temperature on litter decomposition in streams: A meta-analysis. *International Review of Hydrobiology, 104*(1–2), 14–25. https://doi.org/10.1002/iroh.201801965.

Baldrian, P. (2017). Forest microbiome: Diversity, complexity and dynamics. *FEMS Microbiology Reviews, 41,* 109–130. https://doi.org/10.1093/femsre/fuw040.

Battin, T. J., Luyssaert, S., Kaplan, L. A., Aufdenkampe, A. K., Richter, A., & Tranvik, L. J. (2009). The boundless carbon cycle. *Nature Geoscience, 2*(9), 598–600. https://doi.org/10.1038/ngeo618.

Berg, B., Berg, M. P., Bottner, P., Box, E., Breymeyer, A., De Anta, R. C., Couteaux, M., Escudero, A., Gallardo, A., Kratz, W., Madeira, M., Mälkönen, E., Mcclaugherty, C., Meentemeyer, V., & Desanto, A. V. (1993). Litter mass-loss rates in pine forests of Europe and Eastern United-States—Some relationships with climate and litter quality. *Biogeochemistry, 20*(3), 127–159. https://doi.org/10.1007/BF00000785.

Berg, B., & Laskowski, R. (2005). Nitrogen dynamics in decomposing litter. *Advances in Ecological Research, 38*(05), 157–183. https://doi.org/10.1016/S0065-2504(05)38005-6.

Berg, B., & McClaugherty, C. (2014). *Plant litter: Decomposition, humus formation, carbon sequestration* (3rd ed.). Springer-Verlag. https://doi.org/10.1007/978-3-642-38821-7.

Boyero, L., Pearson, R. G., Gessner, M. O., Barmuta, L. A., Ferreira, V., Graça, M. A. S., Dudgeon, D., Boulton, A. J., Callisto, M., Chauvet, E., Helson, J. E., Bruder, A., Albariño, R. J., Yule, C. M., Arunachalam, M., Davies, J. N., Figueroa, R., Flecker, A. S., Ramírez, A., … West, D. C. (2011). A global experiment suggests climate warming will not accelerate litter decomposition in streams but might reduce carbon sequestration. *Ecology Letters*, *14*(3), 289–294. https://doi.org/10.1111/j.1461-0248.2010.01578.x.

Bradford, M. A., Berg, B., Maynard, D. S., Wieder, W. R., & Wood, S. A. (2016). Understanding the dominant controls on litter decomposition. *Journal of Ecology, 104*(1), 229–238. https://doi.org/10.1111/1365-2745.12507.

Bradford, M. A., Veen, G. F., Bonis, A., Bradford, E. M., Classen, A. T., Cornelissen, J. H. C., Crowther, T. W., De Long, J. R., Freschet, G. T., Kardol, P., Manrubia-Freixa, M., Maynard, D. S., Newman, G. S., Logtestijn, R. S. P., Viketoft, M., Wardle, D. A., Wieder, W. R., Wood, S. A., & van der Putten, W. H. (2017). A test of the hierarchical model of litter decomposition. *Nature Ecology & Evolution, 1*, 1836–1845. https://doi.org/10.1038/s41559-017-0367-4.

Bradford, M. A., Warren, R. J., Baldrian, P., Crowther, T. W., Maynard, D. S., Oldfield, E. E., Wieder, W. R., Wood, S. A., & King, J. R. (2014). Climate fails to predict wood decomposition at regional scales. *Nature Climate Change*, *4*(7), 625–630. https://doi.org/10.1038/nclimate2251.

Brousseau, P.-M., Gravel, D., & Handa, I. T. (2018). Trait-matching and phylogeny as predictors of predator–prey interactions involving ground beetles. *Functional Ecology, 32*(1), 192–202. https://doi.org/10.1111/1365-2435.12943.

Brousseau, P.-M., Gravel, D., & Handa, I. T. (2019). Traits of litter-dwelling forest arthropod predators and detritivores covary spatially with traits of their resources. *Ecology, 100*(10), https://doi.org/10.1002/ecy.2815.

Cardinale, B. J., Matulich, K. L., Hooper, D. U., Byrnes, J. E., Duffy, J. E., Gamfeldt, L., Balvanera, P., O'Connor, M. I., & Gonzalez, A. (2011). The functional role of producer diversity in ecosystems. *American Journal of Botany*, *98*(3), 572–592. https://doi.org/10.3732/ajb.1000364.

Castro-Díez, P., Godoy, O., Alonso, A., Gallardo, A., & Saldaña, A. (2014). What explains variation in the impacts of exotic plant invasions on the nitrogen cycle? A meta-analysis. *Ecology Letters, 17*(1), 1–12. https://doi.org/10.1111/ele.12197.

Cornelissen, J. H. C., Van Bodegom, P. M., Aerts, R., Callaghan, T. V., Van Logtestijn, R. S. P., Alatalo, J., Stuart Chapin F., Gerdol, R., Gudmundsson, J., Gwynn-Jones, D., Hartley, A. E., Hik, D. S., Hofgaard, A., Jónsdóttir, I. S., Karlsson, S., Klein, J. A., Laundre, J., Magnusson, B., Michelsen, A., … Team, M. O. L. (2007). Global negative vegetation feedback to climate warming responses of leaf litter decomposition rates in cold biomes. *Ecology Letters*, *10*(7), 619–627. https://doi.org/10.1111/j.1461-0248.2007.01051.x.

Cornwell, W. K., Cornelissen, J. H. C., Amatangelo, K., Dorrepaal, E., Eviner, V. T., Godoy, O., Hobbie, S. E., Hoorens, B., Kurokawa, H., Pérez-Harguindeguy, N., Quested, H. M., Santiago, L. S., Wardle, D. A., Wright, I. J., Aerts, R., Allison, S. D., Van Bodegom, P., Brovkin, V., Chatain, A., … Westoby, M. (2008). Plant species traits are the predominant control on litter decomposition rates within biomes worldwide. *Ecology Letters*, *11*(10), 1065–1071. https://doi.org/10.1111/j.1461-0248.2008.01219.x.

Coulis, M., Fromin, N., David, J. F., Gavinet, J., Clet, A., Devidal, S., Roy, J., & Hättenschwiler, S. (2015). Functional dissimilarity across trophic levels as a driver of soil processes in a Mediterranean decomposer system exposed to two moisture levels. *Oikos*, *124*(10), 1304–1316. https://doi.org/10.1111/oik.01917.

David, J. F., & Handa, I. T. (2010). The ecology of saprophagous macroarthropods (millipedes, woodlice) in the context of global change. *Biological Reviews, 85*(4), 881–895. https://doi.org/10.1111/j.1469-185X.2010.00138.x.

De Oliveira, T., Hättenschwiler, S., & Handa, I. T. (2010). Snail and millipede complementarity in decomposing Mediterranean forest leaf litter mixtures. *Functional Ecology, 24*(4), 937–946. https://doi.org/10.1111/j.1365-2435.2010.01694.x.

Eisenhauer, N., Schielzeth, H., Barnes, A. D., Barry, K. E., Bonn, A., Brose, U., Bruelheide, H., Buchmann, N., Buscot, F., Ebeling, A., Ferlian, O., Freschet, G. T., Giling, D. P., Hättenschwiler,

S., Hillebrand, H., Hines, J., Isbell, F., Koller-France, E., & Jochum, M. (2019). A multitrophic perspective on biodiversity–ecosystem functioning research. In N. Eisenhauer, D. A. Bohan, & A. J. Dumbrell (Eds.), *Advances in Ecological Research, 61*, 1–54. https://doi.org/10.1016/bs.aecr.2019.06.001.

Ehrenfeld, J. G. (2003). Effects of exotic plant invasions on soil nutrient cycling processes. *Ecosystems, 6*(6), 503–523. https://doi.org/10.1007/s10021-002-0151-3.

Ferreira, V., Castagneyrol, B., Koricheva, J., Gulis, V., Chauvet, E., & Graça, M. A. S. (2015). A meta-analysis of the effects of nutrient enrichment on litter decomposition in streams. *Biological Reviews, 90*(3), 669–688. https://doi.org/10.1111/brv.12125.

Ferreira, V., Koricheva, J., Pozo, J., & Graça, M. A. S. (2016). A meta-analysis on the effects of changes in the composition of native forests on litter decomposition in streams. *Forest Ecology and Management, 364*(15), 27–38. https://doi.org/10.1016/j.foreco.2016.01.002.

Fontana, S., Berg, M. P., & Moretti, M. (2019). Intraspecific niche partitioning in macrodetritivores enhances mixed leaf litter decomposition. *Functional Ecology, 33*(12), 2391–2401. https://doi.org/10.1111/1365-2435.13448.

Frainer, A., Jabiol, J., Gessner, M. O., Bruder, A., Chauvet, E., & McKie, B. G. (2016). Stoichiometric imbalances between detritus and detritivores are related to shifts in ecosystem functioning. *Oikos, 125*(6), 861–871. https://doi.org/10.1111/oik.02687.

Frainer, A., McKie, B. G., & Malmqvist, B. (2014). When does diversity matter? Species functional diversity and ecosystem functioning across habitats and seasons in a field experiment. *Journal of Animal Ecology, 83*(2), 460–469. https://doi.org/10.1111/1365-2656.12142.

Galloway, J., Towsend, A., Erisman, J.W., Bekunda, M., Zucong, C. Freney, J. R., … Sutton, M. A. (2008). Transformation of the nitrogen cycle: recent trends, questions, and potential solutions. *Science, 320*(5878), 889–892. https://doi.org/10.1126/science.1136674.

García-Palacios, P., Maestre, F. T., Kattge, J., & Wall, D. H. (2013). Climate and litter quality differently modulate the effects of soil fauna on litter decomposition across biomes. *Ecology Letters, 16*(8), 1045–1053. https://doi.org/10.1111/ele.12137.

García-Palacios, P., McKie, B. G., Handa, I. T., Frainer, A., & Hättenschwiler, S. (2016a). The importance of litter traits and decomposers for litter decomposition: A comparison of aquatic and terrestrial ecosystems within and across biomes. *Functional Ecology, 30*(5), 819–829. https://doi.org/10.1111/1365-2435.12589.

García-Palacios, P., Shaw, E. A., Wall, D. H., & Hättenschwiler, S. (2016b). Temporal dynamics of biotic and abiotic drivers of litter decomposition. *Ecology Letters, 19*(5), 554–563. https://doi.org/10.1111/ele.12590.

García-Palacios, P., Shaw, E. A., Wall, D. H., & Hättenschwiler, S. (2017). Contrasting mass-ratio vs. niche complementarity effects on litter C and N loss during decomposition along a regional climatic gradient. *Journal of Ecology, 105*(4), 968–978. https://doi.org/10.1111/1365-2745.12730..

Gartner, T. B., & Cardon, Z. G. (2004). Decomposition dynamics in mixed-species leaf litter. *Oikos, 104*(2), 230–246. https://doi.org/10.1111/j.0030-1299.2004.12738.x.

Gessner, M. O., Swan, C. M., Dang, C. K., McKie, B. G., Bardgett, R. D., Wall, D. H., & Hättenschwiler, S. (2010). Diversity meets decomposition. *Trends in Ecology & Evolution, 25*(6), 372–380. https://doi.org/10.1016/j.tree.2010.01.010.

Graça, M. A. S. (2001). The role of invertebrates on leaf litter decomposition in streams—A review. *International Review of Hydrobiology, 86*(4–5), 383–393. https://doi.org/10.1002/1522-2632(200107)86:4/5%3c383:AID-IROH383%3e3.0.CO;2-D.

Graça, M. A., Ferreira, V., Canhoto, C., Encalada, A. C., Guerrero-Bolaño, F., Wantzen, K. M., & Boyero, L. (2015). A conceptual model of litter breakdown in low order streams. *International Review of Hydrobiology, 100*(1), 1–12. https://doi.org/10.1002/iroh.201401757.

Grime, J. P. (1998). Benefits of plant diversity to ecosystems: Immediate, filter and founder effects. *Journal of Ecology, 86*(6), 902–910. https://doi.org/10.1046/j.1365-2745.1998.00306.x.

Guerrero-Ramírez, N. R., Craven, D., Messier, C., Potvin, C., Turner, B. L., & Handa, I. T. (2016). Root quality and decomposition environment, but not tree species richness, drive root decomposition in tropical forests. *Plant and Soil, 404,* 125–139. https://doi.org/10.1007/s11104-016-2828-y.

Handa, I. T., Aerts, R., Berendse, F., Berg, M. P., Bruder, A., Butenschoen, O., Chauvet, E., Gessner, M. O., Jabiol, J., Makkonen, M., McKie, B. G., Malmqvist, B., Peeters, E. T. H. M., Scheu, S., Schmid, B., van Ruijven, J., Vos, V. C. A., & Hättenschwiler, S. (2014). Consequences of biodiversity loss for litter decomposition across biomes. *Nature, 509*(7499), 218–221. https://doi.org/10.1038/nature13247.

Hättenschwiler, S., & Gasser, P. (2005). Soil animals alter plant litter diversity effects. *Proceedings of the National Academy of Sciences of the United States of America, 102*(5), 1519–1524. https://doi.org/10.1073/pnas.0404977102.

Hättenschwiler, S., & Jørgensen, H. B. (2010). Carbon quality rather than stoichiometry controls litter decomposition in a tropical rain forest. *Journal of Ecology, 98*(4), 754–763. https://doi.org/10.1111/j.1365-2745.2010.01671.x.

Heemsbergen, D. A., Berg, M. P., Loreau, M., van hal, J. R., Faber, J. H., & Verhoef, H. A. (2004). Biodiversity effects on soil processes explained by interspecific functional dissimilarity. *Science, 306*(5698), 1019–1020. https://doi.org/10.1126/science.1101865.

Herzog, C., Hartmann, M., Frey, B., Stierli, B., Rumpel, C., Buchmann, N., & Brunner, I. (2019). Microbial succession on decomposing root litter in a drought-prone Scots pine forest. *ISME Journal, 13*(9), 2346–2362. https://doi.org/10.1038/s41396-019-0436-6.

Hooper, D., Adair, E., Cardinale, B., Byrnes, J. E. K., Hungate, B. A., Matulich, K. L., Gonzalez, A., Duffy, J. E., Gamfeldt, L., & O'Connor, M. I. (2012). A global synthesis reveals biodiversity loss as a major driver of ecosystem change. *Nature, 486*, 105–108. https://doi.org/10.1038/nature11118.

Hutchens, J. J., & Wallace, J. B. (2002). Ecosystem linkages between southern Appalachian headwater streams and their banks: Leaf litter breakdown and invertebrate assemblages. *Ecosystems, 5*(1), 80–91. https://doi.org/10.1007/s10021-001-0057-5.

Isbell, F., Reich, P. B., Tilman, D., Hobbie, S. E., Polasky, S., & Binder, S. (2013). Nutrient enrichment, biodiversity loss, and consequent declines in ecosystem productivity. *Proceedings of the National Academy of Sciences of the United States of America, 110*(29), 11911–11916. https://doi.org/10.1073/pnas.1310880110.

Jansson, J. K., & Hofmockel, K. S. (2018). The soil microbiome—From metagenomics to metaphenomics. *Current Opinion in Microbiology, 43,* 162–168. https://doi.org/10.1016/j.mib.2018.01.013.

Jian, S., Li, J., Chen, J., Wang, G., Mayes, M. A., Dzantor, K. E., Hui, D., & Luo, Y. (2016). Soil extracellular enzyme activities, soil carbon and nitrogen storage under nitrogen fertilization: A meta-analysis. *Soil Biology and Biochemistry, 101*(October), 32–43. https://doi.org/10.1016/j.soilbio.2016.07.003.

Jo, I., Fridley, J. D., & Frank, D. A. (2016). More of the same? In situ leaf and root decomposition rates do not vary between 80 native and nonnative deciduous forest species. *New Phytologist, 209*(1), 115–122. https://doi.org/10.1111/nph.13619.

Joly, F., Kurupas, K. L., & Throop, H. L. (2017). Pulse frequency and soil-litter mixing alter the control of cumulative precipitation over litter decomposition. *Ecology, 98*(9), 2255–2260. https://doi.org/10.1002/ecy.1931.

Joly, F.-X., Coq, S., Coulis, M., Nahmani, J., & Hättenschwiler, S. (2018). Litter conversion into detritivore faeces reshuffles the quality control over C and N dynamics during decomposition. *Functional Ecology, 32*(11), 2605–2614. https://doi.org/10.1111/1365-2435.13178.

Knorr, M., Frey, S. D., & Curtis, P. S. (2005). Nitrogen additions and litter decomposition: A meta-analysis. *Ecology, 86*(12), 3252–3257. https://doi.org/10.1890/05-0150.

Kuzyakov, Y., & Mason-Jones, K. (2018). Viruses in soil: Nano-scale undead drivers of microbial life, biogeochemical turnover and ecosystem functions. *Soil Biology & Biochemistry, 127*(December), 305–317. https://doi.org/10.1016/j.soilbio.2018.09.032.

Lavelle, P., Blanchart, E., Martin, A., Martin, S., & Spain, A. (1993). A hierarchical model for decomposition in terrestrial ecosystems: Application to soils of the humid tropics. *Biotropica, 25*(2), 130–150. https://doi.org/10.2307/2389178.

Lecerf, A., & Kominoski, J. (2010). Functional leaf traits and biodiversity effects on litter decomposition in a stream: Comment. *Ecology, 91*(6), 1867–1869. https://doi.org/10.1890/0012-9658-91.6.1867.

Lefcheck, J., Byrnes, J., Isbell, F., Gamfeldt, L., Griffin, J. N., Eisenhauer, N., Hensel, M. J., Hector, A., Cardinale, B. J., & Duffy, J. E. (2015). Biodiversity enhances ecosystem multifunctionality across trophic levels and habitats. *Nature Communications, 6*, 6936. https://doi.org/10.1038/ncomms7936.

Liao, C., Peng, R., Luo, Y., Zhou, X., Wu, X., Fang, C., Chen, J., & Li, B. (2008). Altered ecosystem carbon and nitrogen cycles by plant invasion: A meta-analysis. *New Phytologist*, *177*(3), 706–714. https://doi.org/10.1111/j.1469-8137.2007.02290.x.

López-Rojo, N., Pozo, J., Pérez, J., Basaguren, A., Martínez, A., Tonin, A. M., Correa-Araneda, F., & Boyero, L. (2019). Plant diversity loss affects stream ecosystem multifunctionality. *Ecology, 100*(12). https://doi.org/10.1002/ecy.2847..

Loreau, M., & Hector, A. (2001). Partitioning selection and complementarity in biodiversity experiments. *Nature, 412,* 72–76. https://doi.org/10.1038/35083573.

Lu, M., Zhou, X., Yang, Q., Li, H., Luo, Y., Fang, C., Chen, J., Yang, X., & Li, B. (2013). Responses of ecosystem carbon cycle to experimental warming: A meta-analysis. *Ecology*, *94*(3), 726–738. https://doi.org/10.1890/12-0279.1.

Makkonen, M., Berg, M. P., Handa, I. T., Hättenschwiler, S., van Ruijven, J., van Bodegom, P. M., & Aerts, R. (2012). Highly consistent effects of plant litter identity and functional traits on decomposition across a latitudinal gradient. *Ecology Letters, 15*(9), 1033–1041. https://doi.org/10.1111/j.1461-0248.2012.01826.x.

Manning, P., Newington, J. E., Robson, H. R., Saunders, M., Eggers, T., Bradford, M. A., Bardgett, R. D., Bonkowski, M., Ellis, R. J., Gange, A. C., Grayston, S. J., Kandeler, E., Marhan, S., Reid, E., Tscherko, D., Godfray, H. C. J., & Rees, M. (2006). Decoupling the direct and indirect effects of nitrogen deposition on ecosystem function. *Ecology Letters*, *9*(9), 1015–1024. https://doi.org/10.1111/j.1461-0248.2006.00959.x.

Maron, P.-A., Sarr, A., Kaisermann, A., Lévêque, J., Mathieu, O., Guigue, J., Karimi, B., Bernard, L., Dequiedt, S., Terrat, S., Chabbi, A., & Ranjard, L. (2018). High microbial diversity promotes soil ecosystem functioning. *Applied and Environmental Microbiology, 84*, e02738-17. https://doi.org/10.1128/AEM.02738-17.

Meentemeyer, V. (1984). The geography of organic decomposition rates. *Annals of the Association of American Geographers, 74*(4), 551–560. https://doi.org/10.1111/j.1467-8306.1984.tb01473.x.

Merritt, R. W., & Lawson, D. L. (1992). The role of leaf litter macroinvertebrates in stream-floodplain dynamics. *Hydrobiologia, 248*(1), 65–77. https://doi.org/10.1007/BF00008886.

Miki, T., Yokokawa, T., & Matsui, K. (2014). Biodiversity and multifunctionality in a microbial community: A novel theoretical approach to quantify functional redundancy. *Proceedings of the Royal Society of London B, 281*(1776), 20132498. https://doi.org/10.1098/rspb.2013.2498.

Paaso, U., Keski-Saari, S., Keinänen, M., Karvinen, H., Silfver, T., Rousi, M., & Mikola, J. (2017). Intrapopulation genotypic variation of foliar secondary chemistry during leaf senescence and litter decomposition in silver birch (*Betula pendula*). *Frontiers in Plant Science, 8,* 1074. https://doi.org/10.3389/fpls.2017.01074.

Parton, W., Silver, W. L., Burke, I. C., Grassens, L., Harmon, M. E., Currie, W. S., King, J. Y., Adair, E. C., Brandt, L. A., Hart, S. C., & Fasth, B. (2007). Global-scale similarities in nitrogen release patterns during long-term decomposition. *Science, 315*(5810), 361–364. https://doi.org/10.1126/science.1134853.

Pichon, N. A., Cappelli, S. Soliveres, S., Hölzel, N., Klaus, V. H., Kleinebecker, T., & Allan, E. (2019). Decomposition disentangled: A test of the multiple mechanisms by which nitrogen enrichment alters litter decomposition. *bioRiv*. https://doi.org/10.1101/671545.

Raymond-Léonard, L. J., Gravel, D., & Handa, I. T. (2019). A novel set of traits to describe collembola mouthparts: Taking a bite out of the broad chewing mandible classification. *Soil Biology & Biochemistry, 138,* 2019107608. https://doi.org/10.1016/j.soilbio.2019.107608.

Rouifed, S., Handa, I. T., David, J., & Hättenschwiler, S. (2010). The importance of biotic factors in predicting global change effects on decomposition of temperate forest leaf litter. *Oecologia, 163*(1), 247–256. https://doi.org/10.1007/s00442-009-1528-1.

Srivastava, D. S., Cardinale, B. J., Downing, A. L., Duffy, J. E., Jouseau, C., Sankaran, M., & Wright, J. P. (2009). Diversity has stronger top-down than bottom-up effects on decomposition. *Ecology, 90*(4), 1073–1083. https://doi.org/10.1890/08-0439.1.

Sun, T., Hobbie, S. E., Berg, B., Zhang, H., Wang, Q., Wang, Z., & Hättenschwiler, S. (2018). Contrasting dynamics and trait controls in first-order root compared with leaf litter decomposition. *Proceedings of the National Academy of Sciences of the United States of America, 115*(41), 10392–10397. https://doi.org/10.1073/pnas.171659115.

Swift, M. J., Heal, O. W., & Anderson, J. M. (1979). *Decomposition in Terrestrial Ecosystems.* University of California Press.

Tilman, D., Isbell, F., & Cowles, J. M. (2014). Biodiversity and ecosystem function. *Annual Review of Ecology Evolution and Systematics, 45,* 471–493. https://doi.org/10.1146/annurev-ecolsys-120213-091917.

Treseder, K. K. (2008). Nitrogen additions and microbial biomass: A meta-analysis of ecosystem studies. *Ecology Letters, 11*(10), 1111–1120. https://doi.org/10.1111/j.1461-0248.2008.01230.x.

Vilà, M., Espinar, J. L., Hejda, M., Hulme, P. E., Jarošík, V., Maron, J. L., Pergl, J., Schaffner, U., Sun, Y., & Pyšek, P. (2011). Ecological impacts of invasive alien plants: A meta-analysis of their effects on species, communities and ecosystems. *Ecology Letters, 14*(7), 702–708. https://doi.org/10.1111/j.1461-0248.2011.01628.x.

Vos, V. C. A., van Ruijven, J., Berg, M. P., Peeters, E. T. H. M., & Berendse, F. (2011). Macro-detritivore identity drives leaf litter diversity effects. *Oikos, 120*(7), 1092–1098. https://doi.org/10.1111/j.1600-0706.2010.18650.x.

Wagener, S. M., Oswood, M. W., & Schimel, J. P. (1998). Rivers and soils: Parallels in carbon and nutrient processing. *BioScience, 48*(2), 104–108. https://doi.org/10.2307/1313135.

Wagg, C., Bender, S. F., Widmer, F., & van der Heijden, M. G. A. (2014). Soil biodiversity and soil community composition determine ecosystem multifunctionality. *Proceedings of the National Academy of Sciences of the United States of America, 111*(14), 5266–5270. https://doi.org/10.1073/pnas.132005411.

Wagg, C., Schlaeppi, K., Banerjee, S., Kuramae, E. E., & van der Heijden, M. G. A. (2019). Fungal-bacterial diversity and microbiome complexity predict ecosystem functioning. *Nature Communications, 10,* 4841. https://doi.org/10.1038/s41467-019-12798-y.

Wall, D. H., Bradford, M. A., St. John, M. G., Trofymow, J. A., Behan-Pelletier, V., Bignell, D. E., Dangerfield, J. M., Parton, W. J., Rusek, J., Voigt, W., Wolters, V., Gardel, H. Z., Ayuke, F. O., Bashford, R., Beljakova, O. I., Bohlen, P. J., Brauman, A., Flemming, S., Henschel, J. R., … Zou, X. (2008). Global decomposition experiment shows soil animal impacts on decomposition are climate-dependent. *Global Change Biology, 14*(11), 2661–2677. https://doi.org/10.1111/j.1365-2486.2008.01672.x.

Wardle, D.A., Bardgett, R.D., Klironomos, J.N., Setälä, H., van der Putten, W.H. & Wall, D.H. (2004). Ecological linkages between aboveground and belowground biota. Science, 304, 1629–1633. https://doi.org/10.1126/science.1094875.

Woodward, G., Gessner, M. O., Giller, P. S., Gulis, V., Hladyz, S., Lecerf, A., Malmqvist, B., McKie, B. G., Tiegs, S. D., Cariss, H., Dobson, M., Elosegi, A., Ferreira, V., Graça, M. A. S., Fleituch, T., Lacoursière, J. O., Nistorescu, M., Pozo, J., Risnoveanu, G., … Chauvet, E. (2012). Continental-scale effects of nutrient pollution on stream ecosystem functioning. *Science, 336*(6087), 1438–1440. https://doi.org/10.1126/science.1219534.

Yue, K., Peng, C., Yang, W., Peng, Y., Fang, J., & Wu, F. (2015). Study type and plant litter identity modulating the response of litter decomposition to warming, elevated CO_2, and elevated O_3: A meta-analysis. *Journal of Geophysical Research: Biogeosciences, 120*(3), 441–451. https://doi.org/10.1002/2014JG002885.

Yue, K., García-Palacios, P., Parsons, S. A., Yang, W., Peng, Y., Tan, B., Huang, C., & Wu, F. (2018). Assessing the temporal dynamics of aquatic and terrestrial litter decomposition in an alpine forest. *Functional Ecology*, *32*(10), 2464–2475. https://doi.org/10.1111/1365-2435.13143.

Zhang, T., Luo, Y., Chen, H. Y. H., & Ruan, H. (2018). Responses of litter decomposition and nutrient release to N addition: A meta-analysis of terrestrial ecosystems. *Applied Soil Ecology, 128*(July), 35–42. https://doi.org/10.1016/j.apsoil.2018.04.004.

Zuukswert, J., & Prescott, C. E. (2016). Invasive plant species and litter decomposition: Time to challenge assumptions. *New Phytologist, 209*(1), 5–7. https://doi.org/10.1111/nph.13741.

Part II
Biodiversity and Plant Litter Decomposition

Chapter 7
Biodiversity and Plant Litter Decomposition in Streams

Christopher M. Swan

Abstract The main factors influencing litter decomposition in streams are substrate quality, metazoan feeding, microbial activity and environmental context. However, the biodiversity of both resources (litter) and consumers (mostly detritivorous invertebrates) can also influence decomposition, with consequences for stream ecosystem functioning. With regard to leaf litter diversity, in general, decomposition rates increase with litter species richness, but this relationship shifts in response to the environmental context, e.g., nutrient availability in the water column, water flow, and differential shredder feeding rates. Increasing shredder diversity tends to result in faster decomposition rates, due to facilitative and complementarity effects related to intra- *vs.* interspecific interactions. Multitrophic diversity is studied the least, and justifiably so given the complexities of proper experimental needs. However, available evidence suggests that loss of taxa at multiple trophic levels results in altered rates decomposition compared to those expected from intact food webs.

7.1 Introduction

With up to 90% of global terrestrial plant production entering the dead organic matter pool, decomposition of organic carbon (C) in stream sediments stands out as central for ecosystem function in these ecosystems (Cebrian, 1999). This important ecosystem process is the means by which forested stream food webs are supported (Wallace et al., 1997). Streams can receive substantial inputs of terrestrially-derived leaf litter (hereafter litter), which can support a significant portion of total system secondary production (Wallace et al., 1997). For example, a litter exclusion study carried out in the eastern U.S. over many years not only led to the loss of invertebrate consumers, but also to the loss of higher trophic levels (Wallace et al., 1997). In a tropical stream, Rosemond et al. (2001) showed that multiple trophic levels interacted

C. M. Swan (✉)
Department of Geography and Environmental Systems, University of Maryland, Baltimore County, 216 Interdisciplinary Life Sciences Building, 1000 Hilltop Circle, Baltimore, MD 21250, USA
e-mail: chris.swan@umbc.edu

C. M. Swan et al. (eds.), *The Ecology of Plant Litter Decomposition in Stream Ecosystems*, https://doi.org/10.1007/978-3-030-72854-0_7

with nutrient loading to drive rates of litter decomposition. Decomposition of litter in rivers is the sum of a variety of biotic and abiotic factors (Webster & Benfield, 1986). Decomposition describes the rate at which energy from litter inputs from adjacent forests is liberated to higher trophic levels. This results in large and elaborate food webs.

7.2 What Limits Rates of Decomposition?

Breakdown of leaf litter from large to small particles and mineralization to carbon dioxide (CO_2) is driven by a number of factors. This complex process involves regional climate (which determines flow and temperature), foliar chemistry and consumers, both microbial and metazoan (Webster & Benfield, 1986). Decomposition rate varies with temperature, with higher rates associated with higher temperatures (Webster & Benfield, 1986). Areas depleted of oxygen can result in a slower decomposition rate (Webster & Benfield, 1986). This is more common in bogs and swamps where decomposition can cease overall (Webster & Benfield, 1986). Precipitation influences near-bed flow conditions where abrasion by mobile sediments can accelerate decay rates (Paul & Meyer, 2001; Swan et al., 2008; Webster & Benfield, 1986). Litter of different species is differentially susceptible to these physical factors.

The foliar chemistry of organic material is important to the breakdown process. Structural chemistries, such as lignin, hemicellulose, and cellulose, generally slow breakdown rates (Ostrofsky, 1993, 1997; Webster & Benfield, 1986). Secondary compounds, such as tannins, and structural compounds produced by plants to deter herbivory, remain in the leaf litter after senescence (Ostrofsky, 1993). However, alkaloids generally do not persist in litter after it has entered the stream due to their increased solubility compared to tannins, lignins, cellulose, etc. (Webster & Benfield, 1986). In contrast, tannins and phenolics remain an active deterrence to digestion of litter material by both macro- and microconsumers (Cummins & Klug, 1979). Leaf nutrient content, particularly nitrogen (N) and phosphorus (P) levels, increase breakdowns rates of litter (Driebe & Whitham, 2000; Lummer et al., 2012; Webster & Benfield, 1986). Ratios of C:nitrogen (N) and lignin:N are good predictors of breakdown rates (Melillo et al., 1982). Interspecific variation in structural, secondary and nutrient chemistries is of importance for how biodiversity influences the breakdown process (Lecerf et al., 2011).

The dominant detritivores in streams are bacteria, fungi, and invertebrates often referred to as "shredders" (Webster & Benfield, 1986). Shredders lack the enzymatic capability to digest lignin, so initial litter decomposers of organic matter in streams are the bacteria and fungi (Webster & Benfield, 1986). Their exoenzyme activity results in softening of the litter, making microbes themselves and the softened litter palatable to shredders (Webster & Benfield, 1986). Microbial immobilization of dissolved nutrients by these microbes can increase breakdown via enhanced growth and therefore their degradative ability, rendering the microbial-litter matrix more attractive to shredders (Rosemond et al., 2010). In oligotrophic streams, the contributions by invertebrates to decomposition is estimated as high as 40%, with microbial mineralization at 10% (Hieber & Gessner 2002). However, such estimates vary with feeding efficiency of the invertebrates, temperature and other abiotic conditions that can constrain the presence, abundance, and performance of detritivorous consumers (Webster & Benfield 1986).

7.3 Litter Diversity Effects on Decomposition

Litter decomposition and subsequent nutrient cycling is used as a measure of ecosystem functioning. Slow processing of the detritus and nutrient retention stabilizes energy transferred to higher trophic levels (Wallace et al., 1997). Decomposition is a suitable metric for evaluating ecosystem function, and litter diversity is known to be important to food web structure, as well as patterns of energy and nutrient dynamics (Jabiol et al., 2013), in addition to the influence it has on decomposition in streams (Gessner et al., 2010; Lecerf et al., 2007; Swan & Palmer 2004). Although litter diversity effects on decomposition reveal a variety of outcomes (i.e., either no effect, inhibitory or stimulatory), mixing leaf species that are functionally distinct leads to mass loss to occur at rates that are different from what is expected of individual species in isolation (Swan et al., 2009) (Fig. 7.1).

A large number of studies have tested whether rates of decomposition decrease when species are lost from litter (Gessner et al., 2010). The focus of these studies varies by changing the diversity of litter. Intraspecific variation in litter chemical composition is known to drive diversity effects on decomposition (Lecerf & Chauvet 2008). However, interspecific differences among plant species are typically greater, with some species being rich in nutrients, whereas others are nutrient-poor and/or contain high concentrations of lignin that resists degradation. Furthermore, secondary compounds can be detrimental to both microbial and detritivorous consumers by inhibiting digestion. Changes in the species composition and diversity of litter supplied to streams therefore entail profound changes in the patterns and rates of litter utilization and decomposition by both micro- and macroconsumers (Gessner et al., 2010). At the consumer level, microbes and shredders derive different resources from different types of litter via complementary resource use, when chemically divergent litter species are available (Swan & Palmer 2006c). In that way, optimizing nutrient acquisition from litter mixtures potentially alters overall decomposition rates compared to individual constituent litter species (Swan & Palmer, 2006b).

In several cases, an increase in litter richness has been linked to both increases and decreases in decomposition rate, which has been explained through a selection effect (Swan et al., 2009). For example, in the presence of a refractory species, such as an oak or sycamore, mixtures as a whole decomposed slower than expected. In other cases, however, oak species that typically decompose slowly revealed faster rates of litter decomposition when mixed with more labile species. When shedders

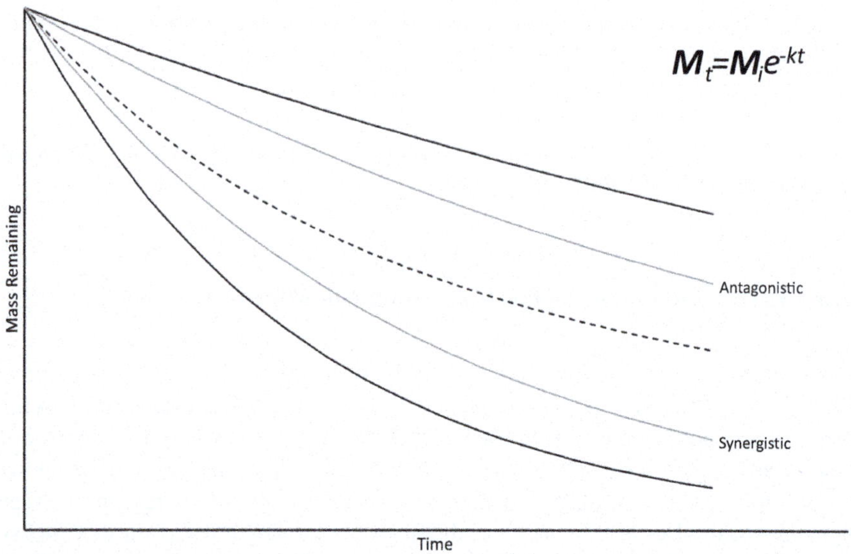

Fig. 7.1 Assuming exponential decay of leaf litter, and two species of leaves (dark lines), the expected decay of the mixture is represented by the dashed line. If the mixture decays faster than expected, or litter species decay synergistically, decay is represented by the lower gray line. In contrast, if the litter species decay slower than expected, or antagonistically, the result is the upper gray line

are present, their preferences can keep mixtures decomposing slower than expected (Swan, 2011). This is due to consumer avoidance of the refractory species (Swan & Palmer 2006a). This leads to an overall slowing of decomposition when species richness is high (Swan & Palmer, 2006a).

Another mechanism by which high species richness of litter can result in slower decomposition is simply due to the relative toughness of refractory leaves compared to labile leaves. For example, in an urban stream with high flow, the tougher species ended up protecting or armoring the more labile species (Swan et al., 2008). This is despite the fact that consumers still preferred the labile species in the mixture. In this case, flow and abrasion were more important in regulating decomposition of the mixture than were consumers (Paul & Meyer, 2001; Swan et al., 2008).

We only know of one study that considered evenness of litter mixtures on decomposition. The majority of litter mixing studies hold the mass of each species in the mixture constant. This is unrealistic since leaf litter species occur in different proportions in the stream and shifts seasonally (Swan & Palmer, 2004). Swan et al. (2009) tracked the relative inputs of litter to a small, headwater stream in Maryland, USA, and created experimental litter mixtures that reflected the proportional biomass of input to the stream. They compared these uneven mixtures to even mixtures together in a standard litter decomposition study. They found strong species identity effects, but effects of species richness were significantly higher only in mixtures with uneven, realistic combinations of litter species. This suggests that dominant species play a large role in driving litter diversity effects on decomposition rates.

7.4 Consumer Effects on Mixed Litter Decomposition

Shredders can influence both the rate of decomposition as well as the stability (e.g., the opposite of variability in mass loss) in mass loss (Swan, 2011). This happens in a number of ways. As stated, litter mixing translates into resource heterogeneity, which can present consumers with complementary resources, which is hypothesized to increase their feeding rate. This can be extended to resource attraction. Labile species in a mixture might attract a higher number of shredders, increasing decomposition of the entire mixture, including even the refractory species. Conversely, a more refractory species might deter shredder colonization, thereby reducing overall decomposition rate. In that way, species number and composition of the leaf litter can drive feeding and colonization dynamics (Ball et al., 2009). One different phenomenon can occur when mixed litter offers shredders choices that may either increase or decrease litter consumption at the mixture scale. Swan and Palmer (2006a) showed that a leaf shredding isopod, *Caecidotea communis*, avoidance of refractory litter in the presence of more labile species can actually reduce decomposition at the mixture scale. The degree of resource choice likely varies among shredder taxa, potentially keeping this effect from being pervasive (Gessner et al., 2010).

Given substantial variation in resource quality among senesced litter species, decomposition rates are known to change with litter species loss, as both microbial and invertebrate consumers respond to loss of litter diversity. However, the implications for such species loss on the stability of decomposition have been less studied. In a field experiment, Swan (2011) manipulated litter diversity as single- and mixed-species treatments in a full-factorial design with the presence/absence of the leaf-shredding consumer, the caddisfly *Pycnopsyche gentilis*. It was hypothesized that in the absence of the consumer, loss of litter species would result in higher variability (i.e., lower stability) in decomposition rates, owing to the portfolio effect commonly observed in plant communities. However, compensatory feeding by the consumer should offset the effect of leaf litter species loss. In addition, there was higher variation in litter processing among single-species litter treatments compared to among diverse mixtures. When *P. gentilis* had access, variation among single-species litter treatments was significantly reduced (i.e., stability increased), and was statistically indistinguishable from high diversity litter treatments. In small streams, how loss of streamside forest species influences stability of instream organic matter processing can be independent of important detritivorous consumers.

7.5 Nutrient Transfer, Immobilization and Litter Species Mixtures

Bacteria and fungi move nutrient from high- to low-quality litter species in mixtures, thus alleviating nutrient limitation for consumers colonizing nutrient-poor litters (Rosemond et al., 2001, 2010). However, fast and diffuse breakdown of litter may export nutrients downstream rather than immediately to adjacent litter within leaf packs (Gessner et al., 2010). Nutrient transfer between litter species of different nutrient contents may therefore be dependent on environmental factors, such as flow, as well as microbial processing rates. The result is higher nutrient content of the litter-microbial matrix on slow-decomposing species, and subsequently higher decomposition rate compared with the single species alone.

Bacteria and fungi colonize litter creating a litter-microbial matrix (Webster & Benfield, 1986). These organisms can derive their resources from the litter substrate or the water column (Rosemond et al., 2010; Webster & Benfield, 1986). As many of the mechanisms by which there are emergent effects of litter mixing rely on the degree of interspecific variation in litter quality, environmental context certainly plays a role. For example, under eutrophic conditions where nutrients are high in the water column, these microbes can immobilize nutrients creating a situation where the litter-microbial matrix is of higher resource quality to macroconsumers than would be if nutrients were low (Rosemond et al., 2010). In one of the few studies to explore this, Rosemond et al. (2010) exposed mixtures of *Liriodendron tulipifera*, *Acer rubrum* and *Rhododendron maximum* in all possible single, double, and triple combinations to a stream experimentally enriched in N and P, and compared any emergent effect of litter mixing to mixtures exposed to ambient, low nutrient conditions in an adjacent stream. As predicted, while they found a non-additive (antagonistic) effect of litter mixing on decomposition rates under reference conditions, this was not the case when nutrient levels were elevated. Differential responses among single-species litters to nutrient enrichment contributed to this result. Nutrient enrichment lowered the C:N ratio and had the greatest effect on the lowest-quality litter species (*R. maximum*) and the least effect on the highest-quality litter species (*L. tulipifera*), resulting in lower interspecific variation in the C:N ratio. Detritivore abundance was correlated with litter C:N ratio in the reference stream, potentially contributing to variation in decomposition rates. In the nutrient-enriched stream, on the other hand, detritivore abundance was higher for all litter species and was unrelated to C:N ratio. Thus, non-additive effects of litter mixing were suppressed by elevated stream water nutrients, which increased nutrient content of litter, reduced variation in C:N ratio among litter species, and increased detritivore abundance.

The importance of inhibitory compounds varies among species (Ostrofsky, 1993, 1997; Webster & Benfield, 1986), and the distribution of these compounds in mixed litter assemblages may explain how leaf litter diversity drives changes in breakdown rates. Intraspecific differences in condensed tannins (e.g., carbon based secondary compounds) can influence litter decomposition in aquatic ecosystems (LeRoy et al., 2007), whereas N-based secondary compounds (e.g., alkaloids) are lost from litter during instream decomposition and therefore less influential (Ardón & Pringle 2008). However, persistence, and even increases in relative concentrations of secondary compounds (condensed tannins, lignin) may not explain the patterns of decomposition observed among single- and mixed-species litter (Ball et al., 2009).

7.6 Structural Heterogeneity in Litter Mixtures

Besides contributing to variability in terms of chemical composition, physical complexity within litter mixtures translates into diverse microhabitats compared to monocultures (Hector et al., 2000). The importance of invertebrates to increasing litter breakdown is well documented (Hieber & Gessner, 2002; Wallace & Webster 1996) and, furthermore, how litter diversity effects breakdown rates can be facilitated by invertebrate activity (Swan & Palmer, 2006c). In addition, it has been shown that trophic interactions beyond that of consumer-resource are important to the decomposition process (Jabiol et al., 2013). Top-down responses to bottom-up leaf litter diversity, such as interactions between invertebrates and the microbial community, might be a mechanism that explains the litter-diversity effect on breakdown rate. In addition, structural heterogeneity in leaf litter assemblages confer higher leaf surface area, which could elicit higher abundances of decomposer populations that could then translate into faster litter processing rates (Hansen & Coleman 1998).

7.7 Litter Mixing Effects on Shredders

But what are the effects of litter diversity on consumer communities? Kominoski et al. (2009) examined effects of litter quality and species mixing on microbial community diversity and associated litter processing in a forested headwater stream. Single- and mixed-species litter from dominant tree species (*Liriodendron tulipifera, Acer rubrum, Quercus prinus, Rhododendron maximum*) were incubated for 50 days in a southern Appalachian headwater stream. Although mass loss of individual species was generally unaffected by mixing, microbial respiration was greater on *A. rubrum* and *Q. prinus* litter incubated with *R. maximum* compared to either species alone. High resource heterogeneity, which here was manipulated experimentally to achieve litter mixing of low- and higher-quality litter species, resulted in shifts in microbial community diversity on individual litter species. The results suggested that changes in tree species composition in riparian forests, and subsequent changes in litter resource heterogeneity, could alter stream microbial community diversity and function. Fernandes et al. (2015) in a microcosm experiment with leaves of alder, oak and eucalyptus that had been previously colonized by microbes in a mixed forest stream looked to test how loss of litter diversity and time affected leaf consumption by invertebrate shredders, the elemental composition of shredder tissues, and the quality of fine particulate organic matter (FPOM) egested by the shredders. They found that the number and identity of litter species affected consumption and FPOM production by shredders, which both increased with higher litter species diversity. C and N composition of invertebrate tissues changed with the litter identity. FPOM quality, taken as C:N ratio, was positively related to litter quality. Litter consumption by the animals decreased linearly with the increase in C:N ratio differences between litter and invertebrate tissues. Hence, their results revealed that changes in litter diversity affect the activity of shredders, via leaf consumption and production of FPOM, and the quality of food resources, via elemental composition of the shredders. There is therefore the potential for significant effects of litter mixing on consumers, both microbial and invertebrate, in stream ecosystems.

7.8 Decomposer Diversity Effects on Decomposition

7.8.1 Shredder Diversity

The diversity of shredders can have significant effects on litter decomposition. Many studies have shown that diversity of these organisms can lead to shifts in decomposition rate beyond what would be expected than when exploring decomposition by a single species detritivore (Gessner et al., 2010; Jonsson & Malmqvist, 2000). Such non-additive effects can occur when there is strong competition, facilitation, or niche complementarity (Gessner et al., 2010). The mode of feeding, size, feeding rate, and phenology are all among the suite of traits hypothesized to influence the nature of emergent biodiversity effects of shredders on litter decomposition (Gessner et al., 2010).

Most studies looking at how species richness of detritivores is related to litter decomposition were carried out in the laboratory. Jonsson et al. (2002) looked at a suite of shredder species in the same guild (Plecoptera). They found significant differences in feeding rate between species when alone, and significant increase in loss rate as richness increased. The authors proposed that facilitation and release from intra-specific interference were the two most likely mechanisms at work. In a follow-up study, they tested this "interference hypothesis" by manipulating both species richness and density of three Plecopteran taxa (Jonsson & Malmqvist, 2000). They showed that one species showed no effect, one a positive, marginally insignificant, effect, and a third species showed a significant, positive effect of decreasing density. Thus, their interference hypothesis partly explained the results from their previous study. In the same study, the researchers tested their "facilitation" hypothesis by sequentially introducing and removing two species. If this hypothesis were true, facilitation would be expressed in higher process rates than when replacing with individuals of the same species. They found that decomposition rate did increase when one species was introduced after the other species, while the opposite sequence did not show any increase, confirming of their previous results. They concluded that both intra-specific interference and inter-specific facilitation may explain the positive effect of species richness observed in their system, and that species loss may be expected to have negative consequences on ecosystem functioning if any species is lost.

In tropical streams, Boyero et al. (2007) uncovered interesting patterns of interspecific interactions. They manipulated richness, evenness and identity of four shredder species—three caddisflies and one mayfly—in microcosms and followed rates of litter decomposition. Species richness, evenness and species identity all affected leaf decomposition rates. Decomposition tended to increase with higher richness, but only for the three caddisflies, likely through a release of intraspecific interference. Species identity was more important than richness, suggesting that some species in this shredder guild are not redundant. Consequently, losing a particular species could have important consequences for litter decomposition, compared to losing species richness per se.

One interesting example showing how increasing diversity can lead to a decrease in decomposition rate was shown by Creed et al. (2009). They show that if much litter decomposition is performed by one species (i.e., a functionally dominant species) and this species is also a competitive dominant that excludes other taxa from a habitat, then it is possible to obtain a negative relationship between richness and litter decomposition. Results of their study of litter decomposition in a small stream suggested that the caddisfly *Pycnopsyche gentilis*, a common Trichopteran in the Piedmont region of the US, detritivorous insect in North American headwater streams, was both a functional and competitive dominant. A second experiment compared the effect of *P. gentilis* on litter decomposition to that of other detritivore taxa by enclosing them with leaf litter packs in a section of headwater stream in which they were uncommon, or a transplant study. Final leaf pack mass was significantly lower in the *P. gentilis* enclosure treatment. Litter exposed to a greater diversity of detritivores displayed little reduction in mass. Taken together, the results of these studies demonstrated that *P. gentilis* was a functionally dominant detritivore. A third experiment manipulated the density of *P. gentilis* and showed that this taxon was also a competitively dominant species. Leaf packs and large *P. gentilis* were placed in enclosures that were permeable to the majority of other detritivores but not *P. gentilis*. Litter mass lost increased with increasing *P. gentilis* density. However, leaf packs exposed to *P. gentilis* contained fewer detritivore taxa, which suggested that *P. gentilis* was also a competitive dominant. There was a negative relationship between three measures of diversity and litter decomposition in the *P. gentilis* density experiment. Thus, experiments conducted in natural communities that incorporate important species interactions, such as competition, may produce diversity-decomposition relationships other than the positive ones that are commonly reported.

Large-scale surveys of streams with differing levels of shredder species richness largely support the experimental results. Jonsson et al. (2001) looked at 23 boreal streams of varying sizes and found that, in addition to important environmental factors, shredder species richness was most strongly correlated with litter decomposition. Dangles et al. (2011) found similar results in 24 high altitude Neotropical streams; even after accounting for environmental factors like water flow, shredder richness and abundance had significant impacts on decomposition rates. This result was supported by a complementary field experiment manipulating shredder richness. The result was transgressive overyielding, i.e., that diverse communities decomposed litter at rates greater than any of the species in isolation, suggesting that complementary resource use or facilitation among species was the underlying mechanism.

While many of the tests of species loss have focused on hypotheses about niche complementarity and facilitation, subsequent conclusions of how disturbance—local or regional—can deplete species, with consequences for litter decomposition rates. There has been some work in this regard. Jonsson et al. (2002), in field microcosms, removed shredder species in the sequences in which they are predicted to disappear, in response to two common types of anthropogenic disturbances: acidification and organic pollution, and followed litter decomposition rates. Species identity significantly affected decomposition rates, while species richness as a whole was non-significant. Mckie and Malmqvist (2009) investigated effects of biodiversity and two abiotic perturbations on three related indices of ecosystem functioning: litter decomposition, detritivore processing efficiency and detritivore growth. In naturally acidic, nutrient poor streams, these variables were measured under two disturbance regimes: liming and nutrient enrichment, raising pH and nutrient levels, respectively. In contrast to expectations, leaf decomposition actually increased under the two disturbance regimes, likely due to a reduction in interspecific trait complementarity which were alleviated under the stress of the disturbance regimes. Moreover, species richness was associated with a decrease in decomposition, while shredder density increased decomposition rates. Lastly, Dangles and Malmqvist (2004) surveyed streams with different levels of species richness, identity and evenness. They found that shredder community composition influenced the diversity–decomposition relationship, with decomposition being much higher for a given species richness at sites with high species dominance than at sites where dominance was low. This was exacerbated by the identity of the dominant species. The dominance effects on decomposition varied seasonally and the number of species required for maintaining decomposition increased with increasing evenness. These results suggest that under a disturbance regime, loss of particular shredder taxa, or of the dominant shredder species, could be detrimental to litter loss rates if these taxa are differentially susceptible to disturbance. Overall, these field studies highlight the importance of environmental context for biodiversity–ecosystem functioning relationships.

7.8.2 Microbial Diversity

Fungi are the main group of microbial decomposers on litter. In fact, their exoenzyme activity softens leaves, making them more palatable to shredders (Webster & Benfield, 1986). There has been some work on how species loss of fungi can influence litter decomposition rates. Duarte et al. (2008) conducted a microcosm experiment with monocultures and all possible combinations of four aquatic hyphomycete species, *Articulospora tetracladia, Flagellospora curta, Geniculospora grandis* and *Heliscus submerses*. They focused on litter decomposition, fungal production and reproductive effort. They found that both species richness and identity significantly affected fungal biomass and conidial production, but only species identity had a significant effect on decomposition. In mixed cultures, all measures of fungal functions were greater than expected from the weighted performances of participating species in monoculture. The three examined aspects of fungal activity tended to increase with species richness, suggesting a complementary effect was taking place.

Alternatively, Dang et al. (2005) found, in a stream microcosm, that rates of litter decomposition and associated fungal spore production were unaffected by changes in decomposer diversity under either ambient or harsh environmental conditions. However, they did find that the magnitude in variability of process rates among communities increased when species numbers were reduced. The authors attributed this to the portfolio effect with the uneven species distribution typical of natural communities tending to weaken that effect. Curbing species extinctions to maintain ecosystem functioning can thus be important even in situations where process rates are unaffected over the short term.

7.9 Vertical Diversity

Biodiversity effects on litter decomposition have mostly been tested separately at different trophic levels - litter, microbial, or detritivore. But in a more realistic food-web context, this represents a focus on horizontal, within-trophic level diversity, which needs to be studied also in the context of vertical, across-trophic level diversity. There is no general prediction for how biodiversity at different trophic levels will influence decomposition. For example, if competitive hierarchies among detritivore taxa are driving biodiversity effects of the consumers on decomposition, the introduction of a predator might shift the hierarchy, perhaps removing the most efficient species. The result could be a change in how biodiversity is related to decomposition rates (see above). Alternatively, should a predator feed on the least efficient consumer, diversity of consumers might enhance decomposition. Furthermore, top predators may also have non-lethal effects on shredders (e.g., fish odor, see below), so that a diverse community is less affected by such indirect predator effects, and therefore even more important in the presence of a predator. This becomes all the more complicated by relative abundance of efficient *vs.* inefficient consumers. Taken together, the way food web actors interact to disrupt complementary versus facilitative effects will, in part, help explain how horizontal and vertical diversity reach down to control decomposition rates.

Exploration of the within- *vs.* between-trophic effects of biodiversity on litter decomposition rely on classic theory on top-down *vs.* bottom up effects. For example, in laboratory microcosms Costantini and Rossi (2010) showed a general increase in decomposition rate with increasing biodiversity, and that this was controlled by within- and between-trophic level interactions, supporting the hypothesis of both bottom-up and top-down effects of diversity on this process. Mixing litter species stimulated growth of a diverse fungal assemblage, but interactions between fungal species slowed down decomposition. However, diversity of shredders reduced fungal mass and accelerated leaf decomposition. Possible explanations posited by the investigators of the positive relationship between detritivore diversity and decomposition are a reduction in fungal dominance and a differentiation in the use of different resource patches promoted by higher fungal diversity.

Streams harbor suites of taxa with different feeding strategies (Cummins & Klug 1979). These strategies can depend on one another. For example, particle feeders—either by suspension or gathering from surfaces—depend on the delivery of particles from various sources. These can include feces, or frass, from shredders (Cummins & Klug 1979). Jonsson and Malmqvist (2005) manipulated the presence, absence and species richness of shredding stoneflies in laboratory microcosms containing litter to test whether the shredder species richness affected the performance of suspension-feeding black fly larvae. The presence of shredders alone increased the production of particles and contributed to higher black fly growth than in the absence of shredders. In one of the experiments they carried out, as species richness increased, so did black fly growth. This was in addition to species composition effects. Their results show that species richness and composition in one functional feeding group of consumers may affect another down the processing chain, most likely via effects on both quantity and size distribution of products derived from the process.

Probably the most elaborate study to date exploring horizontal and vertical diversity simultaneously is by Jabiol et al. (2013). They rightly hypothesized that simultaneous losses of species at different trophic levels may also result in interactive effects, with potentially complex outcomes for ecosystem functioning. They manipulated fungal decomposer diversity (0, 1, 5 species), detritivore diversity (0, 1, 3 species), and the presence of predatory fish scent. The results suggest that trophic complexity is key to eliciting diversity effects on litter decomposition. Specifically, although fungi and detritivores tended to promote decomposition individually, rates were highest in the most complete community where all trophic levels were represented at the highest possible species richness. In part, the effects were trait-mediated, reflected in the contrasting foraging responses of the detritivore species to predator scent. Their results highlight the importance of interactive effects of simultaneous species loss within multiple trophic levels on ecosystem functioning.

References

Ardón, M., & Pringle, C. M. (2008). Do secondary compounds inhibit microbial- and insect-mediated leaf breakdown in a tropical rainforest stream, Costa Rica? *Oecologia.*

Ball, B. A., Bradford, M. A., & Hunter, M. D. (2009). Nitrogen and phosphorus release from mixed litter layers is lower than predicted from single species decay. *Ecosystems, 12,* 87–100.

Boyero, L., Pearson, R. G., & Bastian, M. (2007). How biological diversity influences ecosystem function: A test with a tropical stream detritivore guild. *Ecological Research, 22,* 551–558.

Cebrian, J. (1999). Patterns in the fate of production in plant communities. *The American Naturalist, 154,* 449–468.

Costantini, M. L., & Rossi, L. (2010). Species diversity and decomposition in laboratory aquatic systems: The role of species interactions. *Freshwater Biology, 55,* 2281–2295.

Creed, R. P., Cherry, R. P., Pflaum, J. R., & Wood, C. J. (2009). Dominant species can produce a negative relationship between species diversity and ecosystem function. *Oikos, 118,* 723–732.

Cummins, K., & Klug, M. (1979). Feeding ecology of stream invertebrates. *Annual Review of Ecology and Systematics, 10,* 147–172.

Dang, C. K., Chauvet, E., & Gessner, M. O. (2005). Magnitude and variability of process rates in fungal diversity-litter decomposition relationships. *Ecology Letters, 8,* 1129–1137.

Dangles, O., Crespo-Pérez, V., Andino, P., Espinosa, R., Calvez, R., & Jacobsen, D. (2011). Predicting richness effects on ecosystem function in natural communities: Insights from high-elevation streams. *Ecology, 92,* 733–743.

Dangles, O., & Malmqvist, B. (2004). Species richness-decomposition relationships depend on species dominance. *Ecology Letters, 7,* 395–402.

Driebe, E. M., & Whitham, T. G. (2000). Cottonwood hybridization affects tannin and nitrogen content of leaf litter and alters decomposition. *Oecologia, 123,* 99–107.

Duarte, S., Pascoal, C., & Cássio, F. (2008). High diversity of fungi may mitigate the impact of pollution on plant litter decomposition in streams. *Microbial Ecology, 56,* 688–695.

Fernandes, I., Duarte, S., C ssio, F., & udia Pascoal, C. (2015). Plant litter diversity affects invertebrate shredder activity and the quality of fine particulate organic matter in streams. *Marine and Freshwater Research.*

Gessner, M. O., Swan, C. M., Dang, C. K., McKie, B. G., Bardgett, R. D., Wall, D. H., & Hättenschwiler, S. (2010). Diversity meets decomposition. *Trends in Ecology & Evolution, 25,* 372–380.

Hansen, R. A., & Coleman, D. C. (1998). Litter complexity and composition are determinants of the diversity and species composition of oribatid mites (Acari: Oribatida) in litterbags. *Applied Soil Ecology.*

Hector, A., Beale, A. J., Minns, A., Otway, S. J., & Lawton, J. H. (2000). Consequences of the reduction of plant diversity for litter decomposition: Effects through litter quality and microenvironment. *Oikos.*

Hieber, M., & Gessner, M. (2002). Contribution of stream detrivores, fungi, and bacteria to leaf breakdown based on biomass estimates. *Ecology, 83,* 1026–1038.

Jabiol, J., Mckie, B. G., Bruder, A., Bernadet, C., Gessner, M. O., & Chauvet, E. (2013). Trophic complexity enhances ecosystem functioning in an aquatic detritus-based model system. *Journal of Animal Ecology, 82,* 1042–1051.

Jonsson, M., Dangles, O., Malmqvist, B., & Guérold, F. (2002). Simulating species loss following perturbation: Assessing the effects on process rates. *Proceedings Biological Sciences/The Royal Society, 269,* 1047–1052.

Jonsson, M., & Malmqvist, B. (2000). Ecosystem process rate increases with animal species richness: Evidence from leaf-eating, aquatic insects. *Oikos, 89,* 519–523.

Jonsson, M., & Malmqvist, B. (2005). Species richness and composition effects in a detrital processing chain. *Journal of the North American Benthological Society, 24,* 798–806.

Jonsson, M., Malmqvist, B., & Hoffsten, P.-O. (2001). Leaf litter breakdown rates in boreal streams: does shredder species richness matter? *Freshwater Biology, 46,* 161–171.

Kominoski, J. S., Hoellein, T. J., Kelly, J. J., & Pringle, C. M. (2009). Does mixing litter of different qualities alter stream microbial diversity and functioning on individual litter species? *Oikos, 118,* 457–463. https://doi.org/10.1111/j.1600-0706.2008.17222.x.

Lecerf, A., & Chauvet, E. (2008). Intraspecific variability in leaf traits strongly affects alder leaf decomposition in a stream. *Basic and Applied Ecology, 9,* 598–605.

Lecerf, A., Marie, G., Kominoski, J. S., LeRoy, C. J., Bernadet, C., & Swan, C. M. (2011). Incubation time, functional litter diversity, and habitat characteristics predict litter-mixing effects on decomposition. *Ecology, 92,* 160–169.

Lecerf, A., Risnoveanu, G., Popescu, C., Gessner, M. O., & Chauvet, E. (2007). Decomposition of diverse litter mixtures in streams. *Ecology, 88,* 219–227.

LeRoy, C. J., Whitham, T. G., Wooley, S. C., & Marks, J. C. (2007). Within-species variation in foliar chemistry influences leaf-litter decomposition in a Utah river. *Journal of the North American Benthological Society*.
Lummer, D., Scheu, S., & Butenschoen, O. (2012). Connecting litter quality, microbial community and nitrogen transfer mechanisms in decomposing litter mixtures. *Oikos, 121*, 1649–1655.
Mckie, B. G., & Malmqvist, B. (2009). Assessing ecosystem functioning in streams affected by forest management: Increased leaf decomposition occurs without changes to the composition of benthic assemblages. *Freshwater Biology, 54*, 2086–2100.
Melillo, J. M., Aber, J. D., & Muratore, J. F. (1982). Nitrogen and lignin control of hardwood leaf litter decomposition dynamics. *Ecology, 63*, 621–626.
Ostrofsky, M. L. (1993). Effect of tannins on leaf processing and conditioning rates in aquatic ecosystems: An empirical approach. *Canadian Journal of Fisheries and Aquatic Sciences, 50*, 1176–1180.
Ostrofsky, M. L. (1997). Relationship between chemical characteristics of autumn-shed leaves and aquatic processing rates. *Journal of the North American Benthological Society, 16*, 750–759.
Paul, M. J., & Meyer, J. L. (2001). Streams in the urban landscape. *Annual Review of Ecology and Systematics, 32*, 333–365.
Rosemond, A., Pringle, C., & Ramirez, A. (2001). A test of top-down and bottom-up control in a detritus-based food web. *Ecology, 82*, 2279–2293.
Rosemond, A. D., Swan, C. M., Kominoski, J. S., & Dye, S. E. (2010). Non-additive effects of litter mixing are suppressed in a nutrient-enriched stream. *Oikos, 119*, 326–336.
Swan, C. M. (2011). Consumer presence and resource diversity independently induce stability of ecosystem function in a Piedmont stream. *Ecosphere, 2*, art136.
Swan, C. M., Gluth, M. A., & Horne, C. L. (2009). Leaf litter species evenness influences nonadditive breakdown in a headwater stream. *Ecology, 90*, 1650–1658.
Swan, C., Healey, B., & Richardson, D. (2008). The role of native riparian tree species in decomposition of invasive tree of heaven (Ailanthus altissima) leaf litter in an urban stream. *Ecoscience, 15*, 27–35.
Swan, C. M., & Palmer, M. A. (2004). Leaf diversity alters litter breakdown in a Piedmont stream. *Journal of the North American Benthological Society, 23*, 15–28.
Swan, C. M., & Palmer, M. A. (2006a). Preferential feeding by an aquatic consumer mediates non-additive decomposition of speciose leaf litter. *Oecologia, 149*, 107–114.
Swan, C. M., & Palmer, M. A. (2006b). Composition of speciose leaf litter alters stream detritivore growth, feeding activity and leaf breakdown. *Oecologia, 147*, 469–478.
Wallace, J. B., Eggert, S. L., Meyer, J. L., & Webster, J. R. (1997). Multiple trophic levels of a forest stream linked to terrestrial litter inputs. *Science, 277*, 102–104.
Wallace, J. B., & Webster, J. R. (1996). The role of macroinvertebrates in stream ecosystem function. *Annual Review of Entomology, 41*.
Webster, J., & Benfield, E. (1986). Vascular plant breakdown in freshwater ecosystems. *Annual Review of Ecology and Systematics, 17*, 567–594.

Chapter 8
The Role of Key Plant Species on Litter Decomposition in Streams: Alder as Experimental Model

Javier Pérez, Ana Basaguren, Naiara López-Rojo, Alan M. Tonin, Francisco Correa-Araneda, and Luz Boyero

Abstract A key or keystone species is defined as a species with disproportionately large effects on the ecosystem relative to its abundance. In freshwater ecology it is often used with a bottom-up perspective, to refer to riparian plant species whose litter resources are of particular importance for invertebrate communities and ecosystem processes. This includes fast-decomposing species that represent an important litter supply in terms of nutrients (e.g., alder) and slow-decomposing species that last for long in the stream and are able to sustain communities in periods were preferred resources have disappeared (e.g., oak). This chapter will focus on the major role that litter of the genus *Alnus* (i.e., alder) plays in the decomposition process, a crucial component of stream ecosystem functioning. Alder litter often determines overall decomposition rates and how these are affected by factors such as plant diversity as well as rates of nutrient cycling or secondary production. We take advantage of the wide use of alder litter in multiple studies conducted at different spatial scales (from

J. Pérez (✉) · A. Basaguren · N. López-Rojo · L. Boyero
Department of Ecology and Plant Biology, Faculty of Science and Technology, University of the Basque Country (UPV/EHU), 48940 Leioa, Spain
e-mail: javier.perezv@ehu.eus

A. Basaguren
e-mail: ana.basaguren@ehu.eus

N. López-Rojo
e-mail: naiara.lopez@ehu.eus

L. Boyero
e-mail: luz.boyero@ehu.eus

A. M. Tonin
Department of Ecology, IB, Universidade de Brasília, Asa Norte, Brasília, Distrito Federal CEP 70910-900, Brazil

F. Correa-Araneda
Unidad de Cambio Climático y Medio Ambiente, Instituto de Estudios Del Hábitat, Facultad de Arquitectura y Construcción, Universidad Autónoma de Chile, Temuco, Chile
e-mail: francisco.correa@uautonoma.cl

L. Boyero
IKERBASQUE, Basque Foundation for Science, Bilbao, Spain

C. M. Swan et al. (eds.), *The Ecology of Plant Litter Decomposition in Stream Ecosystems*, https://doi.org/10.1007/978-3-030-72854-0_8

local to global) and with different approaches (from laboratory to field studies) to illustrate how the presence and abundance of a key riparian plant species can drive stream ecosystem functioning.

8.1 The Key Species Concept

The concept of keystone species was coined by Robert T. Paine, a zoologist and ecologist at the University of Washington (USA), to describe the role of a predator species for shaping community structure in a rocky intertidal zone. In his words, "*...the species composition and physical appearance were greatly modified by the activities of a single native species high in the food web. These individual populations are the keystone of the community's structure, and the integrity of the community and its unaltered persistence through time... are determined by their activities and abundances*" (Paine, 1969). That was the first time this author used the keystone term, although he had previously described the important role of the starfish *Pisaster ochraceus* as predator of mussels in an intertidal community (Paine, 1966).

The keystone species term has been broadly applied since its first use, but its meaning has often varied (Mills et al., 1993). Key and keystone species have been generally used with similar meaning (but see Higdon, 2002), and we use them here as synonymous. While the original definition referred to top predators with large influence in the maintenance of community diversity and organization, in conservation ecology it has been typically used as synonym of the umbrella species concept, which implies the protection of overall communities or ecosystems through the management of a single species (e.g., Rohlf, 1991). Moreover, Mills et al. (1993), in a review about this concept, considered five categories of key species depending on the community compartment mainly affected by their loss: (1) predators (with effects on their prey or their competitors), (2) prey (with effects on their predators or other prey), (3) plants (with effects on animal consumers), (4) links (e.g., pollinators or seed dispersers, with effects on plants) and (5) modifiers (e.g., beavers, with effects on the habitat). In recent years, as a consequence of the growing rate of biodiversity loss (Barnosky et al., 2011; Loh & Wackernagel, 2004), the probability of extinction of key species has increased. Given the main ecological role of these species, their loss generally has important repercussions for ecosystem functioning, with the term keystone process species *sensu* Folke et al. (1996) also being used to refer to species that can drive critical ecosystem processes. Again, this term usually refers to players more than resources, with a top-down perspective (Davic, 2003).

However, in freshwater ecology the key species concept is often used with a bottom-up perspective, referring to riparian plant species whose litter is of particular importance for stream communities and ecosystem processes such as decomposition. Several studies have pointed out the key importance of certain riparian plant species in stream ecosystems. For example, Piccolo and Wipfli (2002) and Wipfli and Musslewhite (2004) highlighted the role of red alder (*Alnus rubra*) in upland stream reaches in Alaska to increase invertebrate and detritus subsidies to downstream sites and support higher salmonid biomass through trophic linkages. Swan and Palmer (2004) studied litter diversity effects on decomposition and concluded that the presence of a single key species, the American sycamore (*Platanus occidentalis*), caused non-additive effects in the decomposition of litter mixtures. These authors reported in a subsequent study that the loss of individual riparian species could influence consumer production and material processing, and suggested that the presence of this single slow-decomposing species (*P. occidentalis*) could inhibit detritivore growth and thus control consumer secondary production (Swan & Palmer, 2006). Also with a bottom-up perspective, França et al. (2009) studied litter inputs to a Brazilian stream in order to identify key riparian species for energy fluxes and nutrient cycling in tropical headwaters.

The review by Woodward (2009) about freshwater biodiversity, ecosystem functioning and food webs discussed the importance of litter variability in terms of ecological stoichiometry; thus, the presence of more recalcitrant litter types (e.g., oak) could enable consumers to persist once higher-quality resources (e.g., alder) have been depleted (see also Haapala et al., 2001). Kominoski et al. (2011) studied the effects of riparian species composition on stream ecosystem functioning, which was driven by high-quality litter inputs. The revision by Swan and Kominoski (2012) on plant biodiversity and decomposition highlighted several bottom-up effects on aquatic food webs. Boyero et al. (2014) showed that the identity of species that are lost from communities and of those that remain is a key driver of decomposition. The conceptual model of litter decomposition in low-order streams by Graça et al. (2015) suggested that riparian vegetation dominated by nitrogen (N) fixers (such as alder) accelerates overall decomposition in streams. Tonin et al. (2017a) assessed plant diversity effects on decomposition and identified different underling mechanisms (following Loreau & Hector, 2001), one of them being positive selection (i.e., positive diversity effects arising when the presence of a particular species with high decomposition rate dominates the decomposition rate of the mixture).

As shown in the above examples, the effects of key plant species on stream ecosystem functioning have been commonly explored in the literature. In these studies, the key species concept refers mainly to two types of species: (1) fast-decomposing species that represent an important litter supply in terms of nutrients (e.g., alder); and (2) slow-decomposing species that last for long in the stream and are able to sustain communities in periods when preferred resources have disappeared (e.g., oak). The latter species are not riparian species, at least in temperate areas, but they are abundant in the surrounding forest; thus, they might be dominant but not keystone species (*sensu* Higdon, 2002). The presence of alder litter often determines overall decomposition rates, or associated processes such as nutrient cycling or secondary production, and how these are affected by factors such as plant diversity or detritivore community density. A large number of plant litter decomposition experiments have used alder litter for two main reasons: (1) alder species are often dominant riparian species (mostly in Europe but also present in other areas; Fig. 8.1); and (2) they provide ecosystems with fast decomposing litter, hence being particularly suitable for short-term experiments lasting a few weeks. In this chapter, we mainly

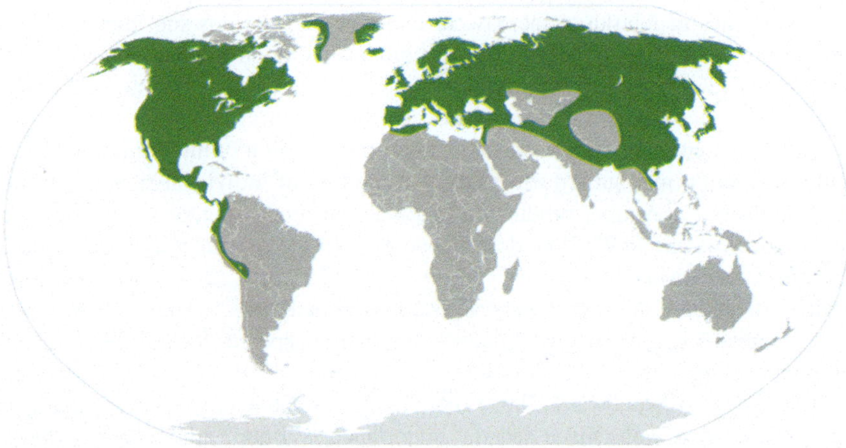

Fig. 8.1 Natural distribution of the alder species (genus *Alnus*) (*Image credit* Ninjatacoshell—commons.wikimedia.org/w/index.php?curid=27474274—based on Silvester [1977])

focus on the major role that alder litter plays in the process of litter decomposition in temperate streams.

8.2 Alder Litter in Field Experiments

Alder is the common name of a genus of riparian plants (*Alnus*) belonging to the family Betulaceae. The genus comprises about 35 species, some of which are dominant riparian species in Eurasia (mainly black alder, *Alnus glutinosa*, but also grey alder, *A. incana*). Moreover, the genus is also present in the American continent, including grey alder but with some endemic species such as Andean alder (*A. acuminata*), red alder (*A. rubra*) or Arizona alder (*A. oblongifolia*); and in Asia, including Japanese alder (*A. japonica*) and Manchurian alder (*A. hirsuta*). The N fixation capacity is the most relevant functional trait of alder trees, through the symbiotic relationship with *Frankia alni*, a filamentous actinomycete, N-fixing bacterium (Franche et al., 2008). Thus, they are scarcely limited by soil N content. These bacteria, found in tree root nodules, absorb atmospheric N_2, making it available to the tree in exchange of sugars. Therefore, alder (through roots and litter) improves the fertility of the soil where it grows, boosting ecological succession in poor soils. In aquatic ecosystems it constitutes an allochthonous resource of key importance, reaching to the stream in the form of litter with relatively low carbon (C):N and C:P ratios (Hladyz et al., 2009). Hence, hereafter we will mainly focus on alder species as examples of fast-decomposing, nutrient-rich litter that is of key importance in their distribution areas.

Taking advantage of the wide use of alder in multiple field studies using the litter bag technique (Bärlocher, 2005) at different spatial scales (from local to global; Table 8.1), we illustrate how the presence and abundance of this key riparian tree (as an example of fast-decaying species) can drive overall stream ecosystem functioning.

8.2.1 Alder and Stream Litter Processing Capacity

Alder litter is commonly used in studies dealing with land use (which implies the replacement or removal of riparian and/or surrounding vegetation), as it usually reflects the processing capacity of the stream detritivore community better than lower-quality litter. Thus, in Northern Spain, decomposition of black alder (*A. glutinosa*) was 20% lower in streams flowing through exotic pine plantations (*Pinus radiata*) than in streams flowing through mixed deciduous forests, possibly in relation to the lower detritivore densities in the former; in contrast, decomposition of pine needles was lower than that of alder and did not differ between stream types (Martínez et al., 2013). Similarly, litter of Andean alder (*A. acuminata*) decomposed faster than that of another N-fixing local tree species (cutlass guaba, *Inga spectabilis*) in forest and pasture streams of Ecuador; the difference was greater for coarse-mesh than for fine-mesh bags in forest streams (1.9- *vs.* 1.3-fold,) and in pasture than forest streams (1.8- *vs.* 1.4-fold), and alder litter harboured almost twice more species of microbial decomposers and showed higher sporulation rates (Encalada et al., 2010).

The above pattern, however, is not general, as other studies have shown higher sensitivity of poorer-quality litter to land use changes. For example, black alder decomposition did not differ between streams running through mixed deciduous forests and beech forests in France (Lecerf et al., 2005), in contrast to common oak litter (*Quercus robur*) decomposition, which was higher in mixed forest streams in coarse-mesh bags; still, alder decomposed faster than oak, in both coarse-mesh and fine-mesh bags. In a regional scale study, black alder also decomposed much faster than common oak in coarse-mesh and fine-mesh bags in 11 Mediterranean streams in Northeastern Spain with varying environmental conditions in terms of hydrology, water quality and invertebrate assemblages; alder decomposition was related to detritivore density and biomass, but that was not the case for oak (Monroy et al., 2016).

8.2.2 Dissolved Nutrients and Alder Decomposition

Alder litter has lower nutrient stoichiometric imbalance for stream consumers than most other litter types (Hladyz et al., 2009), so the availability of dissolved nutrients or the stream trophic status (*sensu* Dodds, 2007) generally has less effect on alder decomposition than on that of nutrient-poorer species. In consequence,

Table 8.1 Examples of field decomposition studies (streams and rivers) using leaf litter of alder and other species (also other substrates: wood and sticks). A brief statement about alder-derived results is given

References	Plant species	Alder-related result
Lecerf et al. (2005)	*Alnus glutinosa*	Decomposition rate of alder litter was around twice as fast as that of oak litter
	Quercus robur	
Ferreira et al. (2006)	*Alnus glutinosa*	Nitrate enrichment did not stimulate alder litter decomposition rates (as did for oak litter)
	Ochroma pyramidale (wood)	
	Quercus robur	
Gulis et al. (2006)	*Alnus glutinosa*	Low levels of eutrophication stimulated decomposition less for alder than for oak litter
	Quercus robur	
Abelho (2008)	*Alnus glutinosa*	Alder litter was colonized earlier and then macroinvertebrates shifted towards harder litter
	Castanea sativa	
	Quercus ilex	
Arroita et al. (2012)	*Alnus glutinosa*	Shedders were more abundant in alder bags and almost absent from poplar sticks. Poplar and alder litter belong to the fast processing group
	Fagus sylvatica	
	Eucalyptus camaldulensis	
	Platanus × *hispanica*	
	Populus nigra (leaves & sticks)	
	Quercus robur	
Ferreira et al. (2012)	*Alnus glutinosa*	Alder seemed to drive the decomposition of litter packs
	Castanea sativa	
	Inga punctata	
	Quercus robur	
	Triplaris dugandii	
	Zygia cataractae	
Martínez et al. (2013)	*Alnus glutinosa*	Pine plantations reduced litter processing capacity of streams
	Pinus radiata	
Pérez et al. (2014)	*Alnus glutinosa*	Transplantation to a nutrient-richer site enhanced eucalypt litter processing (but not alder one)
	Eucalyptus globulus	
Boyero et al. (2016)	*Alnus glutinosa*	Temperature had a large positive influence on microbial decomposition of alder litter
	+70 local species	
Monroy et al. (2016)	*Alnus glutinosa*	Alder litter decomposition was positively correlated to the density and biomass of shredders; that of oak litter was not
	Quercus robur	

using alder in comparison with poorer-quality litter allows examining the magnitude of such effects. In a Portuguese stream experimentally subjected to different levels of N enrichment across a stream reach, decomposition of black alder was not affected by enrichment, while decomposition of common oak (which was lower than that of alder) was enhanced by N enrichment (Ferreira et al., 2006). Similarly, in another study, decomposition of black alder did not consistently differ between 5 pairs of reference/eutrophic Portuguese streams, in contrast to the consistently faster decomposition of common oak in eutrophic streams (Gulis et al., 2006); the authors suggested that this could be due to a higher susceptibility of alder litter to mechanical fragmentation and erratic detritivore colonisation and feeding, and to the lower nutrient content of oak litter, which may trigger a more rapid microbial response to eutrophication. Similar results were observed in a litter-bag transplantation experiment in Northern Spain (Pérez et al., 2014), where black alder decomposition was less affected by the stream trophic status than that of blue gum eucalypt (*Eucalyptus globulus*); while eucalypt decomposed faster in a midstream than a headwater reach (the former being more affected by agroforestry), alder decomposition was similar between reaches, even when litter had been transplanted from the other reach. Nonetheless, other studies have shown that dissolved nutrient availability can influence variables related to alder decomposition such as the conditioning level reached [i.e., the microbially-mediated quality change; e.g., Pozo et al. (2011)].

8.2.3 Alder: The Top of the Class

Alder litter is often a preferred resource of decomposers, being usually the one that is decomposed fastest in studies comparing different types of litter (Casas et al., 2013). For example, in a Portuguese stream, black alder decomposed more than 3-times faster ($k = 0.033\ d^{-1}$) than sweet chestnut (*Castanea sativa*) or common oak ($k = 0.011\ d^{-1}$ in both cases) in coarse-mesh bags in a 2-month experiment; this study showed that detritivores colonized alder earlier (with a peak at day 28) than the other species (day 65), suggesting that they exploited their preferred resource first and then shifted to less palatable resources (Abelho, 2008). Similarly, in a pristine stream in Northern Spain, black alder and black poplar (*Populus nigra*) decomposed at a similar rate ($k = 0.012\ d^{-1}$ for both) and were faster than 4 other species [common oak, beech (*Fagus sylvatica*), London plane (*Platanus × hispanica*) and river red gum eucalypt (*Eucalyptus camaldulensis*): $k = 0.002$ to $0.008\ d^{-1}$] in coarse-mesh bags; furthermore, detritivore densities and fungal sporulation rates were highest in alder (Arroita et al., 2012). However, in the same study, the decomposition rate of alder in a nearby polluted stream was reduced by 48%, a greater reduction than that for most other species (5% for poplar, 24% for oak, 26% for beech and 36% for eucalypt), suggesting that deleterious effects of pollutants on detritivores were not compensated by nutrient subsidies for alder litter as they did for nutrient-poor litter types (Arroita et al., 2012).

8.2.4 Alder Is Always Welcome

Given the known palatability of alder litter, this substrate is a good candidate when aiming to select a common substrate for decomposition studies conducted at large spatial scales. For example, black alder decomposition rates were similar between a temperate (Portugal) and a tropical (Ecuador) stream, but faster in the temperate stream when data were adjusted by water temperature (Ferreira et al., 2012); it decomposed faster in coarse-mesh than in fine-mesh bags, and it showed higher fungal biomass, taxonomic richness and sporulation rates than other species, and higher densities of detritivores than common oak, in the temperate stream. In the same study, litter of other species incubated with alder in 3-species mixtures decomposed faster than in monoculture, so the presence of alder seemed to enhance decomposition; in contrast, the presence of oak in mixtures had the opposite effect (Ferreira et al., 2012).

In a large-scale study across Europe, decomposition of black alder and common oak in coarse-mesh bags (and less so in fine-mesh bags) showed a humped-shaped response (i.e., a subsidy stress response) to increasing nutrient concentrations; rates were low at the extremes, and low to high at intermediate concentrations, where nutrients were not limiting and other stressors were not too high, so other drivers became important (Woodward et al., 2012). In another large-scale study including multiple tropical sites around the world, black alder litter decomposed faster than most native litter mixtures, with few exceptions; differences between alder and mixtures occurred for both coarse-mesh and fine-mesh bags, and variation across sites was larger for alder than for mixtures, possibly reflecting differences in detritivore and microbial assemblages (Boyero et al., 2015).

In other global-scale studies, decomposition of black alder was mainly driven by temperature, mostly through its direct influence on microbial decomposition rate (Boyero et al., 2011, 2016). Alder decomposition was higher in more alkaline waters in warmer streams (possibly through effects on microorganisms and detritivores that are sensitive to low pH), but higher in more acidic waters in cooler streams (where detritivores typical of acidic waters dominate).

8.3 Alder Litter in Laboratory Experiments

Here we mainly refer to litter decomposition experiments conducted in microcosms, which are enclosed, simplified and relatively small ecological systems containing a reduced assemblage of living organisms (Canhoto et al., 2005). While artificial stream facilities (which are in between field and microcosms approaches) are also used sometimes (e.g., Zubrod et al., 2017), the low number of studies so far precludes any generalization. Microcosms can be of very different sizes, from little flasks or test tubes to big tanks; however, for litter decomposition studies, microcosms often consist of glass jars filled with stream water, sediment and litter as the main resource for detritivores and microorganisms, and provided with a natural photoperiod and aeration to mimic streams conditions (for further details see chapter 20 in this volume). The present section draws some generalizations derived from the use of alder litter in microcosm experiments using a selection of ten stream microcosm studies (Table 8.2).

8.3.1 Alder Is a Good Resource for Consumers

As occurred in field studies, alder litter usually is the highest-quality resource in microcosm experiments, due to its low C:N ratio. This implies that it is usually the resource preferred by detritivores, resulting in higher consumption, and rendering higher detritivore growth rates. Most microcosm experiments have measured consumption rates (and other associated variables) of litter types of contrasting quality, which are generally offered separately to consumers (monocultures). Thus, in most cases [but see Solagaistua et al. (2019)], alder has shown the highest consumption rate. This is despite the fact that different detritivorous consumers have been used, with caddisflies (e.g., *Sericostoma pyrenaicum, S. vittatum, Limnephilus atlanticus*) and amphipods (e.g., *Gammarus pullex, G. fossarum, Echinogammarus berilloni*) being the most common ones; an exception is Fidalgo et al. (2013), who used the red swamp crayfish *Procambarus clarkii.* All these microcosm experiments have allowed the study of variables associated with the decomposition process that are very difficult or impossible to measure in the field. This includes, for example, detritivore growth and survival rates (Balibrea et al., 2017; Larrañaga et al., 2014a), or nutrient assimilation efficiency (Santonja et al., 2018), all of which are usually higher when invertebrates feed on alder. These physiological variables appear to be accurate short-term ecological indicators of different stressors (Landeira-Dabarca et al., 2019; Pérez et al., 2021).

8.3.2 Alder Is a Key Driver of Litter Diversity Effects on Decomposition

Even if some relevant studies have focused on detritivore consumption preferences (e.g., Balibrea et al., 2017; Foucreau et al., 2013; Graça & Cressa, 2010), the majority of experiments conducted in the last couple of decades have examined the consequences of losing litter diversity on decomposition. These studies have shown that litter diversity loss significantly alters the rates of decomposition and other related processes, such as nutrient cycling and the production of fine particulate organic matter (Fernandes et al., 2015; López-Rojo et al., 2018, 2019). Some diversity effects might be driven by a small supply of alder (Larrañaga et al., 2020), stimulating the consumption of low quality litter. Furthermore, the simultaneous assessment of

Table 8.2 Examples of laboratory decomposition studies (stream microcosms mainly) using alder and other litter species. A brief statement of alder-derived results is given

References	Plant species	Alder related result
Villanueva et al. (2011)	*Alnus glutinosa*	Larvae lost phosphorus at higher temperature when fed oak or eucalypt litter, but not when fed alder litter
	Quercus robur	
	Eucalyptus globulus	
Jabiol and Chauvet (2012)	*Alnus glutinosa*	Mixing litter increased walnut consumption but not that of alder or oak litter
	Betula pendula	
	Juglans regia	
	Quercus robur	
Foucreau et al. (2013)	*Alnus gutinosa*	Consumption rates on alder litter were significantly higher
	Carpinus betulus	
	Quercus robur	
Larrañaga et al. (2014a)	*Alnus glutinosa*	Detritivores feeding on alder litter had higher growth rate and recovered the mass lost in reproduction in short time
	Eucalyptus globulus	
	Quercus robur	
Campos et al. (2014)	*Alnus glutinosa*	Cadmium exposure only affected the mass loss of alder litter, reducing detritivore feeding
	Eucalyptus globulus	
Arce-Funck et al. (2016)	*Acer pseudoplatanus*	Consumption and energetic storage, but not consumer tolerance to silver, was higher on alder litter
	Alnus glutinosa	
Tonin et al. (2017a)	*Alnus glutinosa*	Alder litter decomposed faster than other species. Detritivores fed preferentially on alder litter
	Populus nigra	
	Robinia pseudoacacia	
	Salix atrocinerea	
Santonja et al. (2018)	*Alnus glutinosa*	Consumption was higher in alder than in oak litter. Assimilation was twofold higher
	Quercus robur	
López-Rojo et al. (2018)	*Alnus glutinosa*	Alder litter monocultures showed the highest consumption and growth rate. Plant diversity effect increased in presence of alder litter
	Corylus avellana	
	Ilex aquifolium	
	Quercus robur	
Solagaistua et al. (2019)	*Alnus glutinosa*	Alder litter was the second preferred resource; consumption of European ash litter was higher, following a quality index
	Corylus avellana	
	Fagus sylvatica	
	Fraxinus excelsior	
	Quercus robur	

microcosms containing monocultures and mixtures has enabled to understand the mechanisms underlying such diversity effects. Nutrient-poor species such as oak can benefit from the presence of nutrient-rich species such as alder, which suggests possible mechanisms. For example, alder leaching could enhance microbial colonization and detritivore-mediated consumption of poor resources; thus, the presence of alder can increase the magnitude of diversity effects on decomposition and associated processes (López-Rojo et al., 2018).

Results of microcosm studies are not always consistent or easily comparable. This could be related to different methodological procedures, such as the use of different substrates, even within the same species, which might differ in nutrient contents (e.g., litter from different areas or years, Lecerf & Chauvet, 2008), or consumers with different conditions (e.g., life stage, physiological status, acclimation period). For example, some studies found alder consumption to be variable (sometimes lower than expected) when offered together with lower-quality litter in mixture (Jabiol & Chauvet, 2012; Little & Altermatt, 2018). Solagaistua et al. (2019) suggested that special care should be taken when comparing microcosm experiments with different incubation times.

8.3.3 Alder Can Inform About Early Effects of Environmental Change

As alder litter is a fast-decomposing litter, it can be used to detect early ecological consequences of environmental change such as increases in water temperature, dissolved nutrients or pollutants, using experiments of relatively short duration. For example, Villanueva et al. (2011) examined the metabolic rates of *Sericostoma vittatum* larvae feeding on black alder, blue gum eucalypt and common oak at two different temperatures, concluding that litter type modulated the effects of temperature; while larvae fed poor-quality litter lost P and reduced N assimilation efficiency at higher temperature, this did not occur for larvae fed alder. Also, Tonin et al. (2017a) showed that microbial activity was enhanced by dissolved N availability on N-poor poplar, but not on N-rich alder. Similarly, studies assessing effects of other pollutants on decomposition have been usually conducted using a single litter type, often black alder (e.g., Zubrod et al., 2017). Campos et al. (2014) examined the decomposition of black alder and eucalyptus that were pre-incubated in water with cadmium; alder, but not eucalyptus, was decomposed more slowly at higher cadmium concentrations. In contrast, Arce-Funck et al. (2016) conducted and experiment where *G. fossarum* previously fed black alder or sycamore maple (*Acer pseudoplatanus*) were exposed to different water silver concentrations; although consumption rate and energetic storage was higher for individuals fed alder, this did not increase their tolerance to dissolved silver, so alder litter was unable to compensate for the negative effects of this metal. Therefore, taking these studies into account, some toxicological effects could be litter type-dependent (Arce-Funck et al., 2018; Cornejo et al., 2020).

8.4 Comparisons Between Alder and Poor-Quality Litter

Some of the above sections have mentioned the utility of analysing patterns of decomposition of high-quality, fast-decomposing alder in comparison with those of poor-quality, slow-decomposing species (mainly oak). Here we explore the usefulness of calculating ratios between alder and oak decomposition rates (hereafter A/O) for different purposes. The use of decomposition ratios as indices of stream ecosystem functioning was suggested by Gessner and Chauvet (2002), who proposed the use of ratios between impact and reference sites as indicators of impairment, or between coarse-mesh and fine-mesh bags as indicators of detritivore activity. They also suggested that ratios of fast-decomposing and recalcitrant litter could be useful; for example, they could inform about different mechanisms contributing to decomposition, or about the different time scales required for impact assessment. Still, the use of such ratios is scarce, even if the experimental comparison of litter types of contrasting quality is a common practice. An exception is Larrañaga et al. (2014b), who explored black alder (A) and eucalyptus (E) decomposition rates and associated detritivore densities and presented both of them as A/E ratios.

While ratios cannot replace original data on decomposition rates of different species, they provide complementary information. For example, they can reflect the influence of different factors or the relative importance of different decomposition agents (e.g., alder may better reflect the processing capacity of detritivores, and oak the degree of microbial conditioning). Here, taking advantage of the abundance of decomposition field studies using black alder and common oak, especially deriving from the RivFunction European project (Chauvet et al., 2016), we present A/O ratios from coarse-mesh and fine-mesh bags (representing total and microbial decomposition, respectively) and explore the potential of such ratios to provide useful information about stream ecosystems. The following comments should be considered only as tentative explanations of the patterns evidenced by A/O ratios, supported by the empirical observations of studies revised here.

Firstly, we plotted mean values of A/O ratios of decomposition rates for the different regions (Fig. 8.2) included in Woodward et al. (2012), who explored effects of nutrient enrichment in 100 streams greatly differing in dissolved nutrient availability (the 1st workpackage of RivFunction). In most cases, alder decomposed considerably faster than oak. Only sites in Northern Sweden (with very low temperature and dissolved nutrient availability) presented A/O ratios below 2 in coarse-mesh and fine-mesh bags, meaning that decomposition of alder and oak was very similar; these low ratios (and rates) suggested biotic limitation, usually observed in such environmental conditions. In general, A/O ratios from both coarse-mesh and fine-mesh bags showed values between 2 and 6. Several regions of Europe, all of them within the upper range of water temperature values (Switzerland, Spain, France and Portugal), showed marked differences between coarse-mesh and fine-mesh A/O ratios, the latter being lower; this suggests a higher preference of detritivores for alder, while microbial decomposers appeared to be less determined by litter quality, as usually observed. The A/O ratios within each of these regions were rather similar,

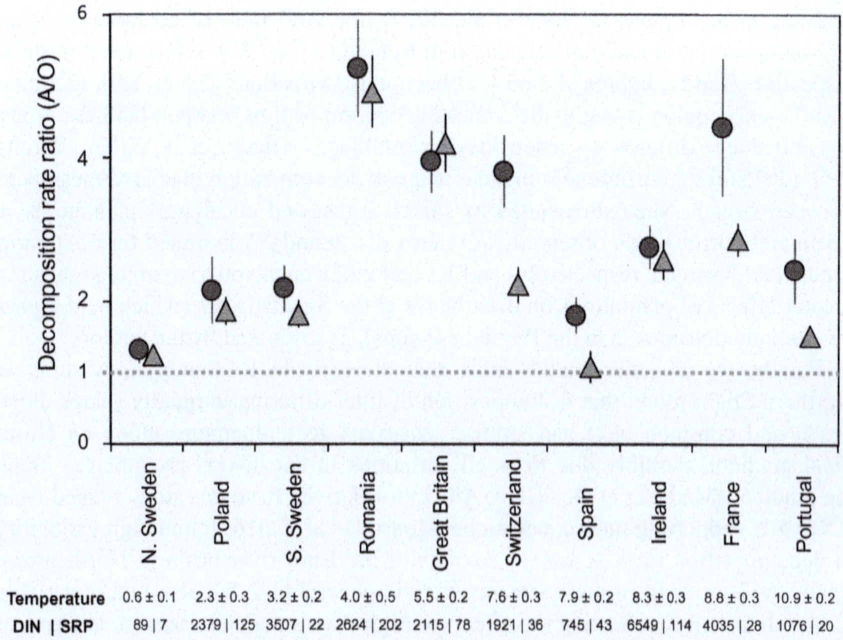

Fig. 8.2 Mean A/O ratios (±SE, n = 10 sites per region) of exponential decomposition rates in coarse-mesh (dark circles) and fine-mesh bags (light triangles), with regions ordered by mean water temperature (ºC). Mean dissolved nutrient availability (μg L^{-1}) in terms of dissolved inorganic nitrogen (DIN) and soluble reactive phosphorus (SRP). The grey dot line means no difference between alder and oak decomposition rate (i.e., A/O = 1) (Data credit Woodward et al. [2012])

with only French sites showing high intra-regional variability in coarse-mesh ratios. Regional similarity in A/O ratios could be related to ecogeographic features (e.g., climate, hydrology, biological communities, etc.) determining the processing capacities of the different regions, and intra-regional variability of the ratio might point to high environmental variability among sites.

Secondly, we discuss some coarse-mesh A/O ratios from studies derived from the 2nd workpackage of RivFunction, which was focused on the effects of riparian forest modification. Lecerf et al. (2005) compared effects of forest type (beech *vs.* mixed) on decomposition across sites in France. These authors indicated that there were significant differences in decomposition rates of coarse-mesh bags only when considering the interaction between forest type and litter type. Using data from this study, we obtained A/O ratios of ca. 4.0 in mixed forest sites and 1.8 in beech forest sites, indicating the important influence of riparian vegetation in supporting detritivore assemblages. Riipinen et al. (2010) examined alder and oak decomposition at 30 sites in 3 European regions; the A/O ratios in coarse-mesh bags were ca. 6.0 in England, but much lower in Ireland and Poland (2.5–3.0 and 1.0–1.5, respectively), regardless of the surrounding vegetation type. Similar results were found in a study comparing litter decomposition in streams surrounded by deciduous forest or pasture in 3 European regions (Hladyz et al., 2010): the A/O ratio of coarse-mesh bags was independent of riparian vegetation in Ireland (3.8 *vs.* 3.9 in forest and pasture, respectively) and Romania (4.3 *vs.* 4.5), but not in Switzerland (2.5 *vs.* 1.0). The latter was the only region showing differences in decomposition between land use types, possibly due to differences in detritivore assemblages (Hladyz et al., 2010). Ferreira et al. (2015) found differences in alder and oak decomposition in coarse-mesh bags between streams sites surrounded by mixed forest and eucalyptus plantations in Spain and Portugal; we observed A/O ratios of 2.9 and 2.5 in mixed forest streams (Spain and Portugal, respectively) and 1.1 and 2.0 in eucalyptus streams, suggesting greater effects of plantations on detritivores at the Spanish sites (which held higher invertebrate densities than the Portuguese sites), as discussed by the authors.

Finally, we consider studies other than those from RivFunction. A study in Northern Spain found that decomposition of litter differing in quality (black alder, beach and common oak) had similar sensitivity to temperature along an altitudinal gradient, possibly due to small variations in detritivore assemblages along the gradient (Martínez et al., 2016); A/O ratios derived from this study ranged from 4.5 to 4.7, supporting their conclusions. Monroy et al. (2016) found high variability in decomposition rates across 11 streams of the Ebro river basin in Northeastern Spain; A/O ratios from these streams were in general high but also highly variable (2.8–8.0), possibly reflecting the effects of high environmental variation in terms of hydrology, water quality and invertebrate assemblages.

8.5 Conclusions

We have reviewed knowledge about key riparian species, in particular alder, which presence and/or abundance are often main drivers of stream ecosystem functioning. This information may help with the selection of litter types to be used in decomposition experiments, depending on the study design and hypotheses to be tested in each case. Thus, the use of alder litter may be particularly useful when short-term patterns are of interest, while slow decomposers such as oak (or any other local recalcitrant species) may be more informative in the long term. Moreover, alder litter, which is a highly preferred resource, could act as 'bait' for detritivores in a reach depleted of good-quality litter (e.g., Martínez et al., 2015); this could be useful in some cases, but may also lead us to biased conclusions. Alder litter also has characteristics, other than its fast decomposition, which make it useful in experiments; this includes high nutrient concentrations derived from the N-fixing capacity of the tree, which makes this species particularly suitable for studies exploring the role of litter quality on decomposition, or the role of functional trait diversity (as alder presents special traits such as the N-fixing capacity).

The same could apply to other fast-decomposing species in areas where alder is not present. For example, in tropical regions (e.g., França et al., 2009), a similar role may be played by other (N-fixing or non-fixing) species, and most likely by a variety of species, as tropical riparian forests are more diverse (Boyero et al., 2017; Tonin et al., 2017b). Lianas (woody vines), which are structural parasites of shade-tolerant trees with fast-growing capacity, might be a good example of key plant species in tropical moist forest (van der Heijden et al., 2013) and also for litter decomposition in streams. Some fast-decaying *Ficus* species might play a similar key role in the tropical streams, as its decomposition rate is similar to that of *Alnus* species (López-Rojo et al., 2020). Overall, key plant species (alone or in combination with other species) deserve special consideration in studies assessing litter decomposition and related processes that are fundamental components of stream ecosystem functioning.

References

Abelho, M. (2008). Effects of leaf litter species on macroinvertebrate colonization during decomposition in a Portuguese stream. *International Review of Hydrobiology, 93,* 358–371.

Arce-Funck, J., Crenier, C., Danger, M., Billoir, E., Usseglio-Polatera, P., & Felten, V. (2018). High stoichiometric food quality increases moulting organism vulnerability to pollutant impacts: An experimental test with *Gammarus fossarum* (Crustacea: Amphipoda). *Science of the Total Environment, 645,* 1484–1495.

Arce-Funck, J., Crenier, C., Danger, M., Cossu-Leguille, C., Guérold, F., & Felten, V. (2016). Stoichiometric constraints modulate impacts of silver contamination on stream detritivores: An experimental test with *Gammarus fossarum. Freshwater Biology, 61,* 2075–2089.

Arroita, M., Aristi, I., Flores, L., Larrañaga, A., Díez, J., Mora, J., Romani, A. M., & Elosegi, A. (2012). The use of wooden sticks to assess stream ecosystem functioning: Comparison with leaf breakdown rates. *Science of the Total Environment, 440,* 115–122.

Balibrea, A., Ferreira, V., Gonçalves, V., & Raposeiro, P. M. (2017). Consumption, growth and survival of the endemic stream shredder *Limnephilus atlanticus* (Trichoptera, Limnephilidae) fed with distinct leaf species. *Limnologica, 64,* 31–37.

Bärlocher, F. (2005). Leaf mass loss estimated by litter bag technique. In M. A. S. Graça, F. Bärlocher, & M. O. Gessner (Eds.), *Methods to study litter decomposition: A practical guide* (pp. 37–42). Springer.

Barnosky, A. D., Matzke, N., Tomiya, S., Wogan, G. O., Swartz, B., Quental, T. B., Marshall, C., McGuire, J. L., Lindsey, E. L., Maguire, K. C., Mersey, B., & Ferrer, E. A. (2011). Has the Earth's sixth mass extinction already arrived? *Nature, 471,* 51–57.

Boyero, L., Cardinale, B. J., Bastian, M., & Pearson, R. G. (2014). Biotic vs. abiotic control of decomposition: A comparison of the effects of simulated extinctions and changes in temperature. *PLoS One, 9*(1), e87426.

Boyero, L., Graça, M. A. S., Tonin, A. M., Pérez, J., Swafford, A. J., Ferreira, V., Landeira-Dabarca, A., Alexandrou, M. A., Gessner, M. O., McKie, B. G., Albariño, R. J., Barmuta, L. A., Callisto, M., Chara, J., Chauvet, E., Colon-Gaud, C., Dudgeon, D., Encalada, A. C., Figueroa, R., … Pearson, R. G. (2017). Riparian plant litter quality increases with latitude. *Scientific Reports,* 7, 10562.

Boyero, L., Pearson, R. G., Gessner, M. O., Barmuta, L. A., Ferreira, V., Graça, M. A. S., Dudgeon, D., Boulton, A. J., Callisto, M., Chauvet, E., Helson, J. E., Bruder, A., Albariño, R. J., Yule, C. M., Arunachalam, M., Davies, J. N., Figueroa, R., Flecker, A. S., Ramírez, A., … West, D. C. (2011). A global experiment suggests climate warming will not accelerate litter decomposition in streams but may reduce carbon sequestration. *Ecology Letters, 14,* 289–294.

Boyero, L., Pearson, R. G., Gessner, M. O., Dudgeon, D., Ramírez, A., Yule, C. M., Callisto, M., Pringle, C. M., Encalada, A. C., Arunachalam, M., Mathooko, J., Helson, J. E., Rincón, J., Bruder, A., Cornejo, A., Flecker, A. S., Mathuriau, C., M'Erimba, C., Gonçalves, J. F., … Jinggut, T.

(2015). Leaf-litter breakdown in tropical streams: Is variability the norm? *Freshwater Science, 34*, 759–769.

Boyero, L., Pearson, R. G., Hui, C., Gessner, M. O., Pérez, J., Alexandrou, M. A., Graça, M. A., Cardinale, B. J., Albariño, R. J., & Arunachalam, M. (2016). Biotic and abiotic variables influencing plant litter breakdown in streams: A global study. *Proceedings of the Royal Society B: Biological Sciences, 283*, 20152664.

Campos, D., Alves, A., Lemos, M. F., Correia, A., Soares, A. M., & Pestana, J. L. (2014). Effects of cadmium and resource quality on freshwater detritus processing chains: A microcosm approach with two insect species. *Ecotoxicology, 23*, 830–839.

Canhoto, C., Graça, M. A., & Bärlocher, F. (2005). Feeding preferences. In: *Methods to study litter decomposition* (pp. 297–302). Springer.

Casas, J. J., Larrañaga, A., Menéndez, M., Pozo, J., Basaguren, A., Martínez, A., Pérez, J., González, J. M., Mollá, S., & Casado, C. (2013). Leaf litter decomposition of native and introduced tree species of contrasting quality in headwater streams: How does the regional setting matter? *Science of the Total Environment, 458*, 197–208.

Chauvet, E., Ferreira, V., Giller, P. S., McKie, B. G., Tiegs, S. D., Woodward, G., Elosegi, A., Dobson, M., Fleituch, T., & Graça, M. A. (2016). Litter decomposition as an indicator of stream ecosystem functioning at local-to-continental scales: Insights from the European RivFunction project. *Advances in Ecological Research, 55*, 99–182. Elsevier.

Cornejo, A., Pérez, J., Alonso, A., López-Rojo, N., & Boyero, L. (2020). A common fungicide impairs stream ecosystem functioning through effects on aquatic hyphomycetes and detritivorous caddisflies. *Journal of Environmental Management, 263*, 110425.

Davic, R. D. (2003). Linking keystone species and functional groups: A new operational definition of the keystone species concept. *Conservation Ecology, 7*.

Dodds, W. K. (2007). Trophic state, eutrophication and nutrient criteria in streams. *Trends in Ecology & Evolution, 22*, 669–676.

Encalada, A., Calles, J., Ferreira, V., Canhoto, C., & Graça, M. A. S. (2010). Riparian land use and the relationship between the benthos and litter decomposition in tropical montane forest streams. *Freshwater Biology, 55*, 1719–1733.

Fernandes, I., Duarte, S., Cássio, F., & Pascoal, C. (2015). Plant litter diversity affects invertebrate shredder activity and the quality of fine particulate organic matter in streams. *Marine & Freshwater Research, 66*, 449–458.

Ferreira, V., Encalada, A. C., & Graça, M. A. S. (2012). Effects of litter diversity on decomposition and biological colonization of submerged litter in temperate and tropical streams. *Freshwater Science, 31*, 945–962.

Ferreira, V., Gulis, V., & Graça, M. A. S. (2006). Whole-stream nitrate addition affects litter decomposition and associated fungi but not invertebrates. *Oecologia, 149*, 718–729.

Ferreira, V., Larrañaga, A., Gulis, V., Basaguren, A., Elosegi, A., Graça, M. A., & Pozo, J. (2015). The effects of eucalypt plantations on plant litter decomposition and macroinvertebrate communities in Iberian streams. *Forest Ecology and Management, 335*, 129–138.

Fidalgo, M. L., Ferreira, C., & Sampaio, A. (2013). Assessment of the preferences of red swamp crayfish (*Procambarus clarkii*) fed with riparian tree leaves: A microcosm study. *International Review of Hydrobiology, 98*, 183–190.

Folke, C., Holling, C. S., & Perrings, C. (1996). Biological diversity, ecosystems, and the human scale. *Ecological Applications, 6*, 1018–1024.

Foucreau, N., Puijalon, S., Hervant, F., & Piscart, C. (2013). Effect of leaf litter characteristics on leaf conditioning and on consumption by *Gammarus pulex*. *Freshwater Biology, 58*, 1672–1681.

França, J. S., Gregório, R. S., de Paula, J. D. A., Júnior, J. F. G., Ferreira, F. A., & Callisto, M. (2009). Composition and dynamics of allochthonous organic matter inputs and benthic stock in a Brazilian stream. *Marine & Freshwater Research, 60*, 990–998.

Franche, C., Lindström, K., & Elmerich, C. (2008). Nitrogen-fixing bacteria associated with leguminous and non-leguminous plants. *Plant and Soil, 321*, 35–59.

Gessner, M. O., & Chauvet, E. (2002). A case for using litter breakdown to assess functional stream integrity. *Ecological Applications, 12,* 498–510.

Graça, M. A., & Cressa, C. (2010). Leaf quality of some tropical and temperate tree species as food resource for stream shredders. *International Review of Hydrobiology, 95,* 27–41.

Graça, M. A., Ferreira, V., Canhoto, C., Encalada, A. C., Guerrero-Bolaño, F., Wantzen, K. M., & Boyero, L. (2015). A conceptual model of litter breakdown in low order streams. *International Review of Hydrobiology, 100,* 1–12.

Gulis, V., Ferreira, V., & Graça, M. A. S. (2006). Stimulation of leaf litter decomposition and associated fungi and invertebrates by moderate eutrophication: Implications for stream assessment. *Freshwater Biology, 51,* 1655–1669.

Haapala, A., Muotka, T., & Markkola, A. (2001). Breakdown and macroinvertebrate and fungal colonization of alder, birch, and willow leaves in a boreal forest stream. *Journal of the North American Benthological Society, 20,* 395–407.

Higdon, J. (2002). Functionally dominant herbivores as keystone species. *Conservation Ecology, 6.*

Hladyz, S., Gessner, M. O., Giller, P. S., Pozo, J., & Woodward, G. (2009). Resource quality and stoichiometric constraints on stream ecosystem functioning. *Freshwater Biology, 54,* 957–970.

Hladyz, S., Tiegs, S. D., Gessner, M. O., Giller, P. S., Risnoveanu, G., Preda, E., Nistorescu, M., Schindler, M., & Woodward, G. (2010). Leaf-litter breakdown in pasture and deciduous woodland streams: A comparison among three European regions. *Freshwater Biology, 55,* 1916–1929.

Jabiol, J., & Chauvet, E. (2012). Fungi are involved in the effects of litter mixtures on consumption by shredders. *Freshwater Biology, 57,* 1667–1677.

Kominoski, J. S., Marczak, L. B., & Richardson, J. S. (2011). Riparian forest composition affects stream litter decomposition despite similar microbial and invertebrate communities. *Ecology, 92,* 151–159.

Landeira-Dabarca, A., Pérez, J., Graça, M. A., & Boyero, L. (2019). Joint effects of temperature and litter quality on detritivore-mediated breakdown in streams. *Aquatic Sciences, 81,* 1–10.

Larrañaga, A., Basaguren, A., & Pozo, J. (2014a). Resource quality controls detritivore consumption, growth, survival and body condition recovery of reproducing females. *Marine & Freshwater Research, 65,* 910–917.

Larrañaga, A., de Guzmán, I., & Solagaistua, L. (2020). A small supply of high quality detritus stimulates the consumption of low quality materials, but creates subtle effects on the performance of the consumer. *Science of the Total Environment, 726*, 138397.

Larrañaga, S., Larrañaga, A., Basaguren, A., Elosegi, A., & Pozo, J. (2014b). Effects of exotic eucalypt plantations on organic matter processing in Iberian streams. *International Review of Hydrobiology, 99,* 363–372.

Lecerf, A., & Chauvet, E. (2008). Intraspecific variability in leaf traits strongly affects alder leaf decomposition in a stream. *Basic and Applied Ecology, 9,* 598–605.

Lecerf, A., Dobson, M., Dang, C. K., & Chauvet, E. (2005). Riparian plant species loss alters trophic dynamics in detritus-based stream ecosystems. *Oecologia, 146,* 432–442.

Little, C. J., & Altermatt, F. (2018). Species turnover and invasion of dominant freshwater invertebrates alter biodiversity–ecosystem-function relationship. *Ecological Monographs, 88,* 461–480.

Loh, J., & Wackernagel, M. (2004). *Living planet report 2004.* WWF-World Wide Fund for Nature.

López-Rojo, N., Martínez, A., Pérez, J., Basaguren, A., Pozo, J., & Boyero, L. (2018). Leaf traits drive plant diversity effects on litter decomposition and FPOM production in streams. *PLoS One, 13,* e0198243.

López-Rojo, N., Pozo, J., Pérez, J., Basaguren, A., Martínez, A., Tonin, A. M., Correa-Araneda, F., & Boyero, L. (2019). Plant diversity loss affects stream ecosystem multifunctionality. *Ecology, 100*, e02847.

López-Rojo, N., Pérez, J., Pozo, J., Basaguren, A., Apodaka-Etxebarria, U., Correa-Araneda, F., & Boyero, L. (2020). Shifts in key leaf litter traits can predict effects of plant diversity loss on decomposition in streams. *Ecosystems.*

Loreau, M., & Hector, A. (2001). Partitioning selection and complementarity in biodiversity experiments. *Nature, 412,* 72.

Martínez, A., Larrañaga, A., Pérez, J., Descals, E., Basaguren, A., & Pozo, J. (2013). Effects of pine plantations on structural and functional attributes of forested streams. *Forest Ecology and Management, 310,* 147–155.

Martínez, A., Monroy, S., Pérez, J., Larrañaga, A., Basaguren, A., Molinero, J. & Pozo, J. (2016). In-stream litter decomposition along an altitudinal gradient: Does substrate quality matter? *Hydrobiologia, 766,* 17–28.

Martínez, A., Pérez, J., Molinero, J., Sagarduy, M., & Pozo, J. (2015). Effects of flow scarcity on leaf-litter processing under oceanic climate conditions in calcareous streams. *Science of the Total Environment, 503,* 251–257.

Mills, L. S., Soulé, M. E., & Doak, D. F. (1993). The keystone-species concept in ecology and conservation. *BioScience, 43,* 219–224.

Monroy, S., Menendez, M., Basaguren, A., Pérez, J., Elosegi, A., & Pozo, J. (2016). Drought and detritivores determine leaf litter decomposition in calcareous streams of the Ebro catchment (Spain). *Science of the Total Environment, 573,* 1450–1459.

Paine, R. T. (1966). Food web complexity and species diversity. *The American Naturalist, 100,* 65–75.

Paine, R. T. (1969). The Pisaster-Tegula interaction: Prey patches, predator food preference, and intertidal community structure. *Ecology, 50,* 950–961.

Pérez, J., Correa-Araneda, F., López-Rojo, N., Basaguren, A., & Boyero, L. (2021). Extreme temperature events alter stream ecosystem functioning. *Ecological Indicators, 121,* 106984.

Pérez, J., Galan, J., Descals, E., & Pozo, J. (2014). Effects of fungal inocula and habitat conditions on alder and eucalyptus leaf litter decomposition in streams of northern Spain. *Microbial Ecology, 67,* 245–255.

Piccolo, J. J., & Wipfli, M. S. (2002). Does red alder (*Alnus rubra*) in upland riparian forests elevate macroinvertebrate and detritus export from headwater streams to downstream habitats in southeastern Alaska? *Canadian Journal of Fisheries and Aquatic Sciences, 59,* 503–513.

Pozo, J., Casas, J., Menéndez, M., Mollá, S., Arostegui, I., Basaguren, A., Casado, C., Descals, E., García-Avilés, J., González, J. M., Larrañaga, A., López, E., Lusi, M., Moya, O., Pérez, J., Riera, T., Roblas, N., & Salinas, M. J. (2011). Leaf-litter decomposition in headwater streams: A comparison of the process among four climatic regions. *Journal of the North American Benthological Society, 30,* 935–950.

Riipinen, M. P., Fleituch, T., Hladyz, S., Woodward, G., Giller, P., & Dobson, M. (2010). Invertebrate community structure and ecosystem functioning in European conifer plantation streams. *Freshwater Biology, 55,* 346–359.

Rohlf, D. J. (1991). Six biological reasons why the Endangered Species Act doesn't work—And what to do about it. *Conservation Biology, 5,* 273–282.

Santonja, M., Pellan, L., & Piscart, C. (2018). Macroinvertebrate identity mediates the effects of litter quality and microbial conditioning on leaf litter recycling in temperate streams. *Ecology and Evolution, 8,* 2542–2553.

Silvester, W. B. (1977). Dinitrogen fixation by plant associations excluding legumes [Nitrogen]. *Treatise on Dinitrogen Fixation.*

Solagaistua, L., Elosegi, A., & Larrañaga, A. (2019). Consumption and performance responses of the amphipod *Echinogammarus berilloni* change during laboratory incubation. *Annales de Limnologie-International Journal of Limnology* (p. 25). EDP Sciences.

Swan, C. M., & Kominoski, J. S. (2012). Biodiversity and ecosystem function of decomposition. *eLS*. John Wiley & Sons.

Swan, C. M., & Palmer, M. A. (2004). Leaf diversity alters litter breakdown in a piedmont stream. *Journal of the North American Benthological Society, 23,* 15–28.

Swan, C. M., & Palmer, M. A. (2006). Composition of speciose leaf litter alters stream detritivore growth, feeding activity and leaf breakdown. *Oecologia, 147,* 469–478.

Tonin, A. M., Boyero, L., Monroy, S., Basaguren, A., Pérez, J., Pearson, R. G., Cardinale, B. J., Gonçalves, J. F., & Pozo, J. (2017a) Stream nitrogen concentration, but not plant N-fixing capacity, modulates litter diversity effects on decomposition. *Functional Ecology, 31,* 1471–1481.
Tonin, A. M., Gonçalves, J. F., Bambi, P., Couceiro, S. R. M., Feitoza, L. A. M., Fontana, L. E., Hamada, N., Hepp, L. U., Lezan-Kowalczuk, V. G., Leite, G. F. M., Lemes-Silva, A. L., Lisboa, L. K., Loureiro, R. C., Martins, R. T., Medeiros, A. O., Morais, P. B., Moretto, Y., Oliveria, P. C. A., Pereira, E. B., … Boyero, L. (2017b). Plant litter dynamics in the forest-stream interface: Precipitation is a major control across tropical biomes. *Scientific Reports, 7,* 10799.
van der Heijden, G. M., Schnitzer, S. A., Powers, J. S., & Phillips, O. L. (2013). Liana impacts on carbon cycling, storage and sequestration in tropical forests. *Biotropica, 45,* 682–692.
Villanueva, V. D., Albariño, R., & Canhoto, C. (2011). Detritivores feeding on poor quality food are more sensitive to increased temperatures. *Hydrobiologia, 678,* 155–165.
Wipfli, M. S., & Musslewhite, J. (2004). Density of red alder (*Alnus rubra*) in headwaters influences invertebrate and detritus subsidies to downstream fish habitats in Alaska. *Hydrobiologia, 520,* 153–163.
Woodward, G. (2009). Biodiversity, ecosystem functioning and food webs in fresh waters: Assembling the jigsaw puzzle. *Freshwater Biology, 54,* 2171–2187.
Woodward, G., Gessner, M. O., Giller, P. S., Gullis, V., Hladyz, H., Lecerf, A., Malmqvist, B., McKie, B. G., Tiegs, S. D., Cariss, H., Dobson, M., Elosegi, A., Ferreira, V., Graça, M. A. S., Fleituch, T., Lacoursière, J. O., Nistorescu, M., Pozo, J., Risnoveanu, G., … Chauvet, E. (2012). Continental-scale effects of nutrient pollution on stream ecosystem functioning. *Science, 336,* 1438–1440.
Zubrod, J. P., Englert, D., Wolfram, J., Rosenfeldt, R. R., Feckler, A., Bundschuh, R., Seitz, F., Konschak, M., Baudy, P., & Lüderwald, S. (2017). Long-term effects of fungicides on leaf-associated microorganisms and shredder populations—An artificial stream study. *Environmental Toxicology and Chemistry, 36,* 2178–2189.

Chapter 9
Linking Microbial Decomposer Diversity to Plant Litter Decomposition and Associated Processes in Streams

Cláudia Pascoal, Isabel Fernandes, Sahadevan Seena, Michael Danger, Verónica Ferreira, and Fernanda Cássio

Abstract The physiology, biochemistry and diversity of aquatic microbial decomposers have been largely investigated in low-order streams. However, some aspects still need further attention to better ascertain how microbial decomposer diversity can ensure ecosystem processes and services, particularly under the challenges posed by global environmental change. Aquatic microbial decomposers play a key role in processing plant litter in streams by degrading the most recalcitrant compounds and facilitating nutrient and energy transfer to higher trophic levels. Among microbial decomposers, fungi, particularly aquatic hyphomycetes, play a fundamental role at the early stages of plant litter decomposition, while the relevance of bacteria increases at the late stage of the decomposition. High-throughput sequencing and metagenomic techniques open new avenues towards a more comprehensive understanding of microbial decomposer ecology. This chapter provides a general overview

C. Pascoal (✉) · I. Fernandes · F. Cássio
Centre of Molecular and Environmental Biology (CBMA), Department of Biology, University of Minho, Campus de Gualtar, Braga, Portugal
e-mail: cpascoal@bio.uminho.pt

Institute of Science and Innovation for Bio-Sustainability (IB-S), University of Minho, Campus de Gualtar, Braga, Portugal

I. Fernandes
e-mail: isabelrodriguesfernandes@bio.uminho.pt

F. Cássio
e-mail: fcassio@bio.uminho.pt

S. Seena · V. Ferreira
MARE – Marine and Environmental Sciences Centre, Department of Life Sciences, University of Coimbra, Coimbra, Portugal

M. Danger
Laboratoire Interdisciplinaire Des Environnements Continentaux, Université de Lorraine – Centre National de la Recherche Scientifique, Metz, France
e-mail: michael.danger@univ-lorraine.fr

LTSER France, Zone Atelier du Bassin de La Moselle, Vandoeuvre-lès-Nancy, France

V. Ferreira
e-mail: veronica@ci.uc.pt

C. M. Swan et al. (eds.), *The Ecology of Plant Litter Decomposition in Stream Ecosystems*, https://doi.org/10.1007/978-3-030-72854-0_9

of aquatic microbial diversity and activity on decomposing plant litter. Attention will be paid to the relationships between microbial diversity and their ecological functions under the major threats posed by the ongoing global environmental change to provide the response patterns of microbial decomposers to maintain nutrient and energy fluxes in streams.

9.1 An Introduction to Microbial Decomposers in Freshwaters

In forest streams, filamentous fungi, mainly aquatic hyphomycetes, are the key microbial decomposers of plant litter entering streams from the riparian vegetation (Bärlocher, 1992; Gessner & Chauvet, 1994; Hieber & Gessner, 2002; Pascoal & Cássio, 2004). The role of bacteria becomes evident once leaf litter has been partially broken down by fungi (Baldy et al., 1995; Pascoal et al., 2005). The diversity of aquatic hyphomycetes associated with decomposing litter evaluated from spore morphology is well documented, particularly in Europe, Asia and North America (Duarte et al., 2016), but the identity of bacterial plant litter decomposers has been rarely investigated either through cultivable taxa or by counting of morphotypes (Baldy et al., 2002; Suberkropp & Klug, 1976). Moreover, many microbial decomposers are not cultivable, impairing their identification (Bärlocher, 2007, 2010). On the contrary, molecular methods do not depend on the reproductive status of microbes or pure cultures and have the potential to holistically assess microbial decomposer diversity (Bärlocher, 2007, 2010). Molecular methods are being developed allowing disentangling the contribution of each species to the overall community composition. Combining this information with microbial activity measurements might open new research avenues for disentangling the respective roles of different microbial taxa in the decomposition process and to further clarify the missing links between microbial decomposers diversity and the processes they drive in freshwater ecosystems. Indeed, microbes play a key role in biogeochemical cycles, ensuring several ecosystem functions and services, but the impacts of biodiversity losses in microbial communities associated with decomposing plant detritus in freshwaters have been overlooked.

9.2 Profiling Microbial Decomposers to Unravel Microbial Diversity and Functions in Freshwaters

A major challenge to understand the role of microbial decomposers in ecosystem processes is to accurately detect their identity and activity in trophic relationships. The development of molecular approaches either to identify species or their functions in decomposing leaves has had a profound impact on unraveling the diversity

(Baschien et al., 2013; Duarte et al., 2015; Seena et al., 2019) and ecology of microbial decomposers (Andrade et al., 2016; Fernandes et al., 2011; Hayer et al., 2016).

9.2.1 Identification of Aquatic Hyphomycetes

Species identification and quantification is a crucial step in biodiversity assessment. Traditionally, aquatic hyphomycetes are identified microscopically based on the size and shape of asexually produced spores (conidia), typically tetraradiate or filiform (Gulis et al., 2020) (Fig. 9.1). Aquatic hyphomycetes conidia can be found in foam (Pascoal et al., 2005; Sridhar & Bärlocher, 1994), suspended in stream water (Bärlocher, 2000; Bärlocher & Graça, 2002; Pascoal et al., 2005), or released from decomposing leaves (Fernandes et al., 2015; Pascoal & Cássio, 2004). However, the identification of aquatic hyphomycetes based on conidium morphology has drawbacks: some conidia are morphologically similar demanding taxonomic expertise and fungal sporulation is achieved only under particular conditions (Bärlocher, 2007). DNA barcoding offers an opportunity to identify aquatic hyphomycetes accurately. This technique uses short DNA sequences linked to individual species (Letourneau et al., 2010; Seena et al., 2010). The internal transcribed spacer (ITS) gene region is considered to be the most relevant barcode for identifying aquatic hyphomycetes (Seena et al., 2010). Approximately 26% of the described aquatic hyphomycetes species are connected to their ITS barcode in databases (Duarte et al., 2014). However, the major hurdle in DNA barcoding of aquatic hyphomycetes is the low number of DNA sequences derived from voucher specimens thereby lacking reliable sequences in the public database (e.g., National Center for Biotechnology Information—NCBI 2009–2019), which often leads to misidentifications. In a recent study to identify fungi through DNA barcoding using the basic GenBank local alignment search tool program (BLAST), around 30% error was detected in linking the number of ITS sequences to taxon (Hofstetter et al., 2019). Nevertheless, DNA based approaches open opportunities to circumvent the current constrains when assessing aquatic hyphomycete diversity in streams.

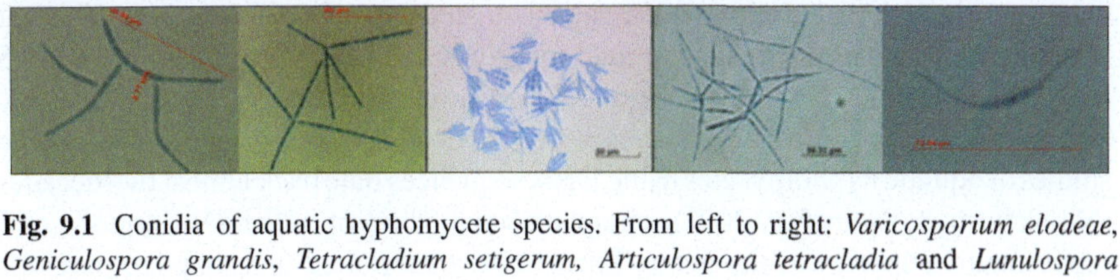

Fig. 9.1 Conidia of aquatic hyphomycete species. From left to right: *Varicosporium elodeae*, *Geniculospora grandis*, *Tetracladium setigerum*, *Articulospora tetracladia* and *Lunulospora curvula*

9.2.2 Genetic diversity

Genetic diversity, which includes inter and intraspecific variability, is a critical feature of natural populations and is considered crucial for ensuring ecosystem functions and stability in a changing environment (Duarte et al., 2019; Fernandes et al., 2011; Seena et al., 2019). Genetic diversity was first explored within the population of two dominant sporulating aquatic hyphomycete species, *Neonectria lugdunensis* (former *Heliscus lugdunensis*) and *Articulospora tetracladia*, suggesting a high degree of genetic variation within the strains from a single foam patch as determined by random amplified polymorphic DNA (RAPD) analyses (Peláez et al., 1996). Later, the population genetic structure of *Tetrachaetum elegans*, a common aquatic hyphomycete found during the initial stages of leaf litter decomposition, was explored by amplified fragment length polymorphism (AFLP) multilocus fingerprints; significant genetic differentiation of this fungus was found within a limited geographic area (Laitung et al., 2004). Furthermore, genetic variation of *T. elegans*, evaluated via RAPD analysis, suggested substrate preferences among the genotypes of aquatic hyphomycetes (Charcosset & Gardes, 1999). Another aquatic fungus, *Tetracladium marchalianum*, was genotyped at eight microsatellite loci and genotypic diversity of this fungus was found to be very large and well connected at local scales (Anderson & Shearer, 2011). Thus, it can be speculated that genetic differentiation is mainly controlled by the distances between streams and dispersal barriers. Conversely, genotypes of *A. tetracladia*, discriminated through ITS barcodes, were generally geographically widespread regardless of sampling time, sites or substrates (Seena et al., 2012). This suggests that the drivers that lead to intraspecific diversity of aquatic fungi and the links to their ecological functions are largely unknown.

9.2.3 Phylogeny and Diversity

Because morphological variations of conidia evolved as a consequence of selective pressure, they do not relate to their phylogeny (Belliveau & Bärlocher, 2005; Seena et al., 2018). Therefore, the strategy to investigate the phylogenetic relationship between microbes is based on DNA sequence analyses of stable markers. Nuclear rDNA sections are reputed to be optimal for the study of fungal phylogeny, e.g., 18S rDNA, ITS and 28S rDNA (Bärlocher, 2007). However, to improve the phylogenetic signal, multi loci comparisons are advocated (Duarte et al., 2013). Phylogenetic studies of aquatic hyphomycetes using DNA sequences date back almost two decades (Nikolcheva & Bärlocher, 2002): complete sequencing of 18S rDNA regions of 5 representative species of the genus *Tetracladium* supported that they are part of a monophyletic group, as suggested by morphology-based taxonomy. The comparison based on 18S rDNA and other rDNA sequences confirmed the polyphyletic origin of aquatic hyphomycetes (Baschien et al., 2006, 2013; Belliveau & Bärlocher, 2005; Campbell et al., 2006, 2009). Moreover, phylogenetic studies allow the establishment of anamorphs and teleomorphs connections of aquatic hyphomycetes (e.g., anamorph *Jaculispora submersa* and teleomorph *Classicula fluitans*; Bauer et al., 2003), provide evidence for their occurrence in several ecological niches (Seena & Monroy, 2016) and show relatives of terrestrial origin (Baschien et al., 2013).

9.2.4 Leaf Litter Associated Microbial Communities

Molecular characterization of the fungal community on decomposing leaves in streams began with terminal restriction fragment length polymorphism (T-RFLP) and denaturing gradient gel electrophoresis (DGGE) targeting a section of 18S rRNA (Nikolcheva et al., 2003). It was concluded that species richness and community evenness decreased with the advancement of litter decomposition, probably due to an increase in substrates homogeneity and a decrease in the nutritional value of the substrate. Moreover, ITS primers with enhanced specificity for particular fungal groups, revealed strikingly high fungal diversity on leaf litter compared to that found based on spore morphology (Nikolcheva & Bärlocher, 2004). Additionally, these analyses showed a clear fungal succession pattern during plant litter decomposition: the role of terrestrial fungi tends to decrease after few days of plant litter entering in streams, while the contribution of aquatic hyphomycetes progressively increases.

The diversity of microbial decomposers in specific habitats and the contribution of different microbial groups to plant litter decomposition have been also addressed. Clone libraries were successfully used to determine fungal diversity in hyporheic zones (Bärlocher et al., 2008) and on decomposing leaf litter (Seena et al., 2008). The relative amount of DNA of Archaea, Bacteria and Fungi during decomposition of leaf litter was assessed for the first-time using quantitative polymerase chain reaction (qPCR) (Manerkar et al., 2008). Moreover, Das and collaborators (2007) were able to uncover the phylotypes associated with actinomycetes on decomposing leaves. The dynamics of yeast populations associated with decomposing leaves have rarely been investigated (but see Sampaio et al., 2004, 2007). Further research has addressed shifts in the structure of fungal, bacterial and ciliate communities associated with plant litter of different species alone or in mixtures using DGGE (Fernandes et al., 2013), but most studies have focused on assessing the shifts in fungal and bacterial communities under exposure to several anthropogenic stressors (e.g., metals, Duarte et al., 2008; eutrophication, Duarte et al., 2009; nanoparticles, Pradhan et al., 2011; Chapters 16, 17, 18).

More recently, metabarcoding, an outcome of the fusion of DNA barcoding with high-throughput or next-generation sequencing (HTS or NGS), has become one of the most relevant tools in biodiversity assessment. By using this method, the genetic diversity of microbial communities can be acquired at a high-resolution scale (e.g., Duarte et al., 2015; Seena et al., 2019a). Pyrosequencing was used to characterize fungal communities on leaves and particulate organic matter by targeting the 18S ribosomal gene region (Wurzbacher et al., 2015) or the ITS gene region (Duarte et al., 2015). Both studies revealed a very high fungal diversity even under eutrophication

(Duarte et al., 2015). Further, Duarte et al. (2017) suggested that larger leaf areas tended to harbor a more diverse and active fungal community, as revealed by Illumina MiSeq analysis based on ITS region from RNA.

9.2.5 Microbial Biomass Accrual and Reproduction

Traditionally, fungal activity has been assessed by measuring biomass build-up and reproduction on decomposing leaves. Fungal biomass on leaf litter is difficult to determine since fungal hyphae penetrate the substrate they are decomposing. To overcome this problem, ergosterol, a fungal membrane constituent (not present in bacteria or plant material) that suffers rapid oxidation after cell death, has been used as a measure of living fungal biomass on leaves (Gessner & Newell, 2002). However, some studies showed that ergosterol could persist at appreciable concentrations and for a considerable time in the absence of living fungi (Mille-Lindblom et al., 2004). Moreover, the mycelial ergosterol concentration can vary with the fungal species (Gessner & Chauvet, 1993), their nutritional status (Charcosset & Chauvet, 2001), growth phase (Barajas-Aceves et al., 2002) and/or the presence of stressors (e.g., fungicide zineb, Barajas-Aceves et al., 2002; fungicide tebuconazole, Baudy et al., 2020). In addition, ergosterol is unable to distinguish individual species biomass in a complex community. Despite these limitations, ergosterol is still the most widely used fungal biomass indicator today (e.g., Bergmann & Graça, 2020; Pimentão et al., 2020).

In aquatic hyphomycetes, a significant proportion of carbon is used in the formation of conidia, which are released from fungi on decomposing leaves (Gessner & Chauvet, 1997; Suberkropp, 1991). Sporulation by aquatic hyphomycetes can be induced by incubating pure cultures or naturally-colonized plant substrates in deionized water or filtered stream water under aeration (Bärlocher, 2020). By identifying and counting these conidia, sporulation rates can be estimated, and so the reproductive potential of individual fungal species (Bärlocher, 1982, 2009). Fungal reproduction is very sensitive to a panoply of abiotic and biotic factors, such as nutrient concentration in water (Abelho & Graça, 2006; Gulis & Suberkropp, 2004), temperature (Chauvet & Suberkropp, 1998; Fernandes et al., 2009; Suberkropp, 1984), pollutants (Pereira et al., 2016; Pradhan et al., 2011; Sridhar & Bärlocher, 2011), oxygen concentration (Medeiros et al., 2009), current velocity (Ferreira & Graça, 2006), litter quality (Fernandes et al., 2012; Pérez et al., 2014), or fungal species competitive interactions (Treton et al., 2004). Fungal reproduction is frequently one of the most sensitive parameters to stressors (e.g., metals, Duarte et al., 2008; eutrophication, Duarte et al., 2009; nanomaterials, Pradhan et al., 2011), making aquatic hyphomycetes potential bioindicators of anthropogenic stress (Solé et al., 2008).

Since bacteria are also important players in the decomposition of plant litter in streams and rivers (Baldy et al., 1995; Duarte et al., 2010), it is important to accurately assess their abundance and biomass. The commonly used approach to obtain such estimates is to detach bacterial from plant litter by sonication, pass the bacterial suspension through a membrane filter, stain the trapped cells with a fluorescent dye and count them under an epifluorescence microscope (Buesing & Gessner, 2020). This technique presents limitations, namely observer bias, requiring strictly standardized counting procedures and thorough cross-calibration among individuals to ensure reproducible results, and it is extremely time-consuming (Frossard et al., 2016). An alternative to epifluorescence microscopy is flow cytometry (Frossard et al., 2016). This technique has been successfully used to assess bacterial biomass associated with decomposing plant litter after chronic exposure to silver nanoparticles (Batista et al., 2020; Tlili et al., 2017) and to compare the effects of litter quality and litter standing stocks on microbial biomass associated with decomposing litter (Frossard et al., 2013). Nevertheless, information on cell size and shape, which can be obtained by epifluorescence microscopy, is mostly lost when using flow cytometry (Frossard et al., 2016).

A limitation of the above described methods to asses microbial biomass accrual is the difficulty in discriminating between living and non-living microbial biomass. A more dynamic measure of microbial activity can be given by estimating fungal or bacterial productivity, from incorporation rates of radioactive labelled acetate or leucine into ergosterol or protein, respectively (Baldy et al., 1995; Pascoal & Cássio, 2004; Suberkropp et al., 2020), that reflect the specific fungal or bacterial growth rate on leaf litter. This has permitted us to conclude that fungal production represents more than 90% of the total microbial production (Baldy et al., 1995; Pascoal & Cássio, 2004), supporting a major role of fungi in leaf litter decomposition in streams, at least during the first stages of the process.

ATP is present in living cells and disappears rapidly in dead cells. The ease of ATP measurement has fostered its use as an indicator of living and active microbial biomass associated with decomposing leaves in streams (Abelho, 2020, 2009; Sales et al., 2015; Suberkropp, 1991). A limitation of this technique is the inability to discriminate between microbial communities (e.g., fungi, bacteria, protists).

9.2.6 Catabolic Reactions and Enzymatic Activity

Community respiration is a measure of biological activity, reflecting the microbial use of organic matter and, therefore, the functional significance of microbes in decomposition (Graça & Abelho, 2020). Respiration rates of microorganisms associated with decomposing leaves are generally determined by measuring oxygen consumption (e.g., Carlisle & Clements, 2005; Stelzer et al., 2003). Alternatively, microbial respiration can be measured using the MicroResp™ method (Tlili et al., 2017), a colorimetric assay based on the color change of a pH indicator dye caused by the release of CO_2 by heterotrophic communities.

Another measure of microbial activity on plant litter can be obtained by determining extracellular enzymatic activities (Romaní et al., 2006). This can be done by incubating decomposing plant litter with fluorescent-linked artificial substrates, specific for each enzyme (Romaní et al., 2006). In a study comparing growth and patterns of degradative enzymes expressed by communities of bacteria and fungi grown separately and in coexistence, enzyme activities were in general low when bacteria grew alone, and the activity of key enzymes in the degradation of lignin and cellulose was undetectable, while fungi growing alone had a high capacity for the decomposition of lignin, cellulose and hemicelluloses (Romaní et al., 2006). Most studies, however, measure the overall enzyme activity that does not discriminate fungal and bacterial contribution (Artigas et al., 2012; Mora-Gómez et al., 2020).

9.2.7 Discriminating Individual Species Performances Within Communities

Discriminating the growth of individual fungal or bacterial species within a community is still a challenge but is critical to better understand the role of microbial decomposers diversity in maintaining ecological processes. Monoclonal antibodies were used in several assays, such as the enzyme-linked immunosorbent assay (ELISA), which allows the identification and quantification of mycelium, and immunofluorescence (IMF) for visualization of mycelium on leaf material (Bermingham et al., 1995, 2001). This allowed the identification and biomass quantification of *Anguillospora longissima*, *Alatospora acuminata*, *T. marchalianum* and *N. lugdunensis* (Bermingham et al., 1995, 2001).

Fluorescent in situ hybridization (FISH) can be an alternative to monoclonal antibodies. It uses short DNA sequences provided with a fluorescent tag complementary to the *taxon* of interest (Baschien et al., 2001; McArthur et al., 2001). However, the autofluorescence of hyphae and colonized substrates, and the weak probe-conferred signals due to low probe permeabilization may lead to confounding results (Baschien, 2003).

DNA-based quantitative stable isotope probing (qSIP) can identify growing microorganisms in environmental samples (Hayer et al., 2016; Hungate et al., 2015). Briefly, organisms are exposed to an isotopically-labelled substrate and those that assimilate the substrate incorporate the isotope into their DNA, increasing their buoyant density (Hayer et al., 2016). The heavy DNA can be separated from non-labelled DNA through isopycnic centrifugation and analyzed, using metabarcoding and qPCR, to identify growing microorganisms (Hayer et al., 2016). Using this technique, Hayer and collaborators (2016) showed that a large proportion of the bacterial taxa associated with decomposing leaf litter grew slowly, and several less abundant taxa were highly frequent, indicating that rare organisms may be important for the decomposition of leaf litter in streams.

The relative intensity of each band (phylotype) in DGGE gels can give semi-quantitative estimates of the relative biomass of each fungal species within the community (Nikolcheva & Bärlocher, 2005). However, caution is needed since techniques relying on PCR like DGGE, can be biased because of preferential amplification (Kanagawa, 2003). PCR can be limited by inhibitors of the polymerase reaction, reagent limitation, or accumulation of pyrophosphate molecules, being the reaction no longer at an exponential rate and so generating more products in some reactions than in others (Ginzinger, 2002).

The qPCR technique allows measurements of the amount of PCR product when the reaction is still at the exponential phase, by determining a fluorescence signal threshold at which all samples can be compared (Ginzinger, 2002). The fractional number of PCR cycles required to generate enough fluorescent signal to reach the threshold (Ct value) is directly proportionate to the amount of starting template (Ginzinger, 2002), allowing accurate quantification of species DNA. The qPCR was successfully used to discriminate the relative abundance of specific microbial groups, Archaea, Bacteria, and Fungi on leaves decomposing in streams (Manerkar et al., 2008) or bacterial and archaeal ammonia oxidizers and nitrogen-fixing bacteria in leaf and sediment samples (Rico et al., 2014). Also, qPCR allowed explaining the putative mechanisms underlying biodiversity effects on leaf decomposition under stress by discriminating the contribution of different aquatic hyphomycete ecotypes to the total fungal biomass produced in multicultures (Fernandes et al., 2011). This approach allowed evidencing the importance of species traits in modulating biodiversity effects under stress (Fernandes et al., 2011). The expression profiles of the functional marker gene also helped to discriminate ecotypes from polluted and nonpolluted habitats (Seena et al., 2020).

Recently, TaqMan® probe-based qPCR assays targeting aquatic hyphomycete species common in temperate regions were designed and validated to detect and quantify species within communities (Baudy et al., 2019; Feckler et al., 2017). In addition, qPCR-obtained DNA levels showed a positive correlation with ergosterol concentrations, confirming that DNA levels are a suitable species-specific biomass proxy (Baudy et al., 2019). However, aquatic hyphomycetes DNA concentrations were found to vary upon exposure to the fungicide tebuconazole (Baudy et al., 2020). This highlights the need to further develop and test this technique to assess its reliability to accurately disentangle single species contribution to the overall community biomass. Nevertheless, molecular techniques open new avenues to gain deeper insights into the ecological role of aquatic hyphomycetes and other microorganisms in freshwaters and to address biodiversity and ecosystem functioning relationships (Baudy et al., 2019; Fernandes et al., 2011).

9.3 Microbial Metabolism and Stoichiometry

Microbial decomposers use the substrate they are inhabiting (e.g., leaf litter) leading to large changes in resource chemical quality through time accompanied by a progressive disappearance of their growth substrate (Suberkropp et al., 1976). Moreover, aquatic microbial communities can acquire nutrients and carbon both from their decomposing substrate and from the water, either in dissolved organic or mineral forms (Cheever et al., 2012). Therefore, the microbial decomposer´s metabolism is largely dependent on detrital resources availability and quality and on the physical and chemical characteristics of their surrounding environment.

9.3.1 Carbon Quality and Priming Effect on Litter Decomposition

While most stoichiometric studies have considered carbon as a single pool, carbon quality has long been acknowledged as a factor modulating leaf litter decomposition. The abundance of lignin, in particular, has been shown as a major factor regulating microbial decomposition (Fernandes et al., 2012; Melillo et al., 1982). For example, the nutrient demand of microorganisms for ensuring leaf litter decomposition depends largely upon carbon recalcitrance, indicating that in nature, nutrient limitations of microbial growth might occur at different nutrient levels depending on the carbon quality of litter (Sinsabaugh & Follstad Shah, 2011). Considering carbon recalcitrance into stoichiometric investigation would permit us to refine stoichiometric predictions about microbial decomposer´s activity and the consequences for ecosystem functioning (including carbon respiration and nutrient remineralization). In particular, splitting leaf litter carbon into distinct pools of different quality permits to consider that, in some specific cases, leaf litter decomposers might be limited by the availability of labile carbon. Stimulatory effects of labile carbon inputs (e.g., through the release of algal exudates) on leaf litter decomposition have been shown in several studies (Danger et al., 2013; Kuehn et al., 2014; Pope et al., 2020), a process called aquatic priming effect (Guenet et al., 2010). These labile carbon additions can also result in opposite trends, i.e., reductions in leaf litter decomposition due to shifts in carbon substrate use (Halvorson et al., 2019). To date, the influence of microbial community structure on priming effect occurrence and intensity has not been deeply investigated. Yet, microbial diversity involving different microorganisms with distinct traits seems to control this phenomenon in soils (Fontaine & Barot, 2005). Thus, a better understanding of leaf litter decomposition budgets should include information on the links between microbial diversity and the relative availability of different carbon sources in water and leaf litter.

9.3.2 Microbial Leaf Litter Decomposition Budgets

Microbial leaf litter decomposition budgets confirm that aquatic fungi are the main microbial contributors to leaf mass loss in freshwaters (Baldy et al., 1995; Komínková et al., 2000; Suberkropp, 1991). Laboratory studies have shown that fungal growth efficiency (i.e., the proportion of leaf organic matter assimilated by fungi channelled for biomass and conidial production) ranged from 24 to 46% (35% on average), with the remaining assimilated matter being respired. Based on such growth efficiency values, aquatic fungi can account for 42 to 65% of the overall carbon loss from leaf litter of diverse deciduous tree species during elevated fungal activity (Baldy et al., 1995; Gessner & Chauvet, 1997). Nevertheless, such budgets might be largely impacted by environmental conditions, such as nutrient availability (Gulis & Suberkropp, 2003; Pascoal & Cássio, 2004; Pascoal et al., 2005).

Since leaf litter is generally a nutrient-poor but carbon-rich substrate, microbial decomposers' activity is often considered as nutrient limited in stream ecosystems (Cross et al., 2005; Enríquez et al., 1993; Güsewell & Gessner, 2009). Elevated nutrient availabilities have thus long been associated with fast microbial litter decomposition. For example, litter rich in nitrogen (N) or phosphorus (P) has been shown to decompose faster than litter exhibiting lower N or P concentrations (Hladyz et al., 2009; Melillo et al., 1982). Similarly, most laboratory (e.g., Gulis et al., 2017) or field studies on nutrient enrichment have shown a stimulatory effect of dissolved N and/or P concentrations (see Ferreira et al., 2015 for a meta-analysis; Rosemond et al., 2015). Overall microbial decomposer activity increases asymptotically (Michaelis–Menten kinetics) with N or P concentrations ($0.09 < \text{N-NO3} < 3.5 \text{ mg L}^{-1}$) (Ferreira et al., 2006; Gulis et al., 2006; Fernandes et al., 2014). This stimulatory effect of nutrients is modulated by litter recalcitrance, with an increase in dissolved N from 0.1 to 3.0 mg N L^{-1} accelerating the decomposition of lignin-poor litter (e.g., < 10% of lignin, 2.9 × increase) more strongly than that of litter rich in lignin (e.g., > 15% of lignin, 1.4 × increase) (Jabiol et al., 2019). The response to nutrient enrichment might also differ between groups of microbial decomposers. For example, fungal biomass, but not necessarily diversity, tends to increase from the most oligotrophic to moderate eutrophic streams and decrease under hypertrophic conditions (Duarte et al., 2009; Dunck et al., 2015; Pascoal & Cássio, 2004; Pereira et al., 2016). In contrast, bacterial biomass increases monotonically from oligotrophic to hypertrophic streams (Duarte et al., 2009; Pascoal et al., 2005). Moreover, higher temperatures appear to reduce litter recalcitrance (lignin) effects on microbes that mediated litter decomposition (Fernandes et al., 2012). Future research on microbial diversity responses to nutrients (see Sect. 9.5. below) might permit to identify nutrient preferences of the different taxa, and ultimately help to predict the responses of leaf litter decomposition to nutrient variations in ecosystems.

9.3.3 Microbial Stoichiometry and Carbon-Use Efficiency

One convenient way to predict microbial decomposer´s responses to nutrient availability is the ecological stoichiometry framework. Ecological stoichiometry specifically investigates consumers elemental requirements (generally focusing on their C:N:P ratios) and the relative availability of the same elements in their resources (see Chapter 3, for further details). The intensity of elemental imbalance permits to estimate which element limits consumers' activity as well as the functional consequences in terms of nutrient recycling (Sterner & Elser, 2002). To optimally decompose leaf litter, microorganisms must balance their carbon and nutrient acquisition with their stoichiometric requirements. Both bacteria and fungi have been shown to be non-homeostatic, at least at the community level (Danger et al., 2008), i.e., microbial biomass exhibit variable elemental composition (Danger et al., 2016). While stoichiometric data are still scarce for aquatic fungi, the few available results suggest differences between the elemental plasticity of different fungal strains (Danger & Chauvet, 2013). For example, in P-depleted liquid cultures, *Lemonniera terrestris* reached higher C:P ratios than *Articulospora tetracladia* and *Tricladium chaetocladium* (Danger & Chauvet, 2013). However, on average, microbial decomposers have lower and less variable carbon-to-nutrient ratios than what can be found in the substrates they colonize (Cleveland & Liptzin, 2007; Manzoni et al., 2010; Sardans et al., 2012). In addition, bacteria have lower carbon-to-nutrient ratios than fungi, which renders bacteria more susceptible to nutrient limitations than fungi. Conversely, fungi generally have higher carbon demand and are able to decompose detritus with higher carbon-to-nutrient ratios than bacteria (Keiblinger et al., 2010).

Carbon Use Efficiency (CUE) corresponds to the ratio of microbial growth to total microbial carbon assimilation (i.e., including microbial respiration) and gives information on the conversion efficiency of detritus into microbial biomass as well as indications on the potential carbon storage in ecosystems (Sinsabaugh et al., 2013). CUE can be modeled from the C:N:P ratios of decomposing material (Manzoni et al., 2010). The application of such models demonstrated that decomposers could adapt to low-nutrient conditions by reducing their CUE.

An alternative way to investigate nutrient effects on microbial activity is to search for optimal stoichiometric ratios, i.e., nutrient ratios that will maximize their development and/or activity. For other groups, like metazoans, these ratios can be approached for each taxon by evaluating their Threshold Elemental Ratios (TER) that correspond to the ratios at which limitation shifts from one element to another (Frost et al., 2006). The large immobilization capacities of microorganisms complexify this approach. Immobilization allows microorganisms to balance their nutrient and carbon requirements even for extremely nutrient-poor substrates (Sinsabaugh et al., 2013). Several studies investigated microbial stoichiometric requirements in terrestrial contexts (e.g., Mooshammer et al., 2014), far less in aquatic ecosystems. For aquatic microbial communities, dissolved N:P ratios maximizing cellulose decomposition varied from 1.7 to 45 depending on the overall nutrient supply (Güsewell & Gessner, 2009). The next step for better understanding the mechanisms at play would involve investigating microbial taxa independently; however, most approaches are generally carried out with complex microbial communities (often including fungi and bacteria).

9.4 Substrate Diversity and Quality for Microbial Decomposers

Microbial decomposers are likely to be sensitive to the benthic litter standing stock's characteristics, which reflects the type of riparian vegetation. It could be anticipated that higher diversity of riparian trees (major contributors to litter inputs in forest streams), and consequently, of benthic organic matter, would support a higher diversity of aquatic microbes ('niche complementarity hypothesis'). Positive correlations have been found between aquatic hyphomycete species richness in stream water and benthic leaf litter species richness (Laitung & Chauvet, 2005; Lecerf et al., 2005) or riparian tree species richness (Ferreira et al., 2016; Rajashekhar & Kaveriappa, 2003) in undisturbed streams. Similar positive relationships have been found between aquatic hyphomycete conidium concentration in stream water and benthic leaf litter species richness and amount (Ferreira et al., 2016; Laitung et al., 2002).

Human-mediated decreases in riparian tree diversity may have adverse effects on aquatic hyphomycete diversity. Aquatic hyphomycete species richness associated with decomposing oak (*Quercus robur*) leaves was lower in streams flowing through commercial beech (*Fagus sylvatica*) forests than in streams flowing through mixed deciduous forests in southwestern France (Lecerf et al., 2005). In central Portugal, aquatic hyphomycete species richness in stream water and associated with decomposing alder (*Alnus glutinosa*) and oak leaves was lower in streams in eucalypt (*Eucalyptus globulus*) monocultures than in streams in mixed deciduous forests (Bärlocher & Graça, 2002; Ferreira et al., 2006). In the Azores islands, aquatic hyphomycete species richness associated with decomposing holly (*Ilex perado*) leaves was lower in streams in cryptomeria (*Crypromeria japonica*) monocultures than in streams with native laurel forests (Ferreira et al., 2017). However, the replacement of native forests by tree plantations not always led to decreases in aquatic hyphomycete species richness if a riparian strip of native tree species is maintained or native tree species are allowed to grow within the plantation, thus ensuring a diverse litter input to streams. In northern Spain, aquatic hyphomycete species richness associated with decomposing alder and pine (*Pinus radiata*) leaves did not significantly differ between streams in deciduous forests and pine plantations, where a riparian strip of native species is maintained (Martínez et al., 2013). Also, in northern Spain, the humid climate allows for the development of an understory of native species in eucalypt plantations; thus, aquatic hyphomycete species richness associated with decomposing leaves did not significantly differ between streams in deciduous forests and those in plantations (Chauvet et al., 1997), or it was even higher in the latter streams (Ferreira et al., 2006).

It could also be expected that higher amounts of benthic litter standing stocks could support higher aquatic hyphomycete species richness and biomass ('productivity hypothesis'). Indeed, higher species richness and reproductive output (i.e., conidium production) have been found where the amount of benthic organic matter is high (Ferreira et al., 2016; Laitung et al., 2002).

Aquatic hyphomycete species richness and community composition also vary with the identity and type of decomposing organic matter since different leaf species and plant parts (e.g., leaves *vs.* woody substrates) differ in physical and chemical characteristics. Although aquatic hyphomycete species are not apparently excluded from any particular substrate, they seem to have substrate preferences (Canhoto & Graça, 1996; Chauvet et al., 1997; Gulis, 2001; Ferreira et al., 2006; Ferreira & Graça, 2016) likely driven by differences in toughness, nutrient concentration and concentration of structural and secondary compounds among substrates. In a given stream, aquatic hyphomycete species richness is generally lower on woody substrates and conifer needles than on deciduous leaves (Ferreira et al., 2006, 2017; Gonçalves et al., 2007; Martínez et al., 2013), which can be attributed to the higher toughness and lignin:nutrients ratio of the former litter types than of leaves. Differences in the aquatic hyphomycete species richness and community composition may also occur among deciduous leaf species, although these are generally smaller than between leaves and woody substrates (Canhoto & Graça, 1996; Ferreira & Graça, 2016; Gonçalves et al., 2013; Gulis, 2001).

Plant-litter decomposition also varies with the substrate diversity: a compilation of manipulative studies showed that 44% of litter mixtures decomposed faster than predicted from the sum of single litter species and 39% of litter mixtures decomposed slower than expected from individual species decomposition (Lecerf et al., 2011). Moreover, effects of leaf-litter quality and diversity on stream ecosystem functioning may vary with the environmental context: synergistic effects of leaf species number on leaf decomposition were found in oligotrophic but not in eutrophic streams (Lima-Fernandes et al., 2015). This suggests that oligotrophic streams are more dependent on the number of leaf species than eutrophic streams. If so, riparian plant diversity should be preserved in oligotrophic systems to maintain leaf-litter decomposition. On the other hand, the positive effects of leaf-litter quality (leaf N) on leaf-litter decomposition were strengthened by moderate increases in nutrient concentrations in the stream water (Lima-Fernandes et al., 2015), suggesting that leaf-litter decomposition depends more on the quality than the number of leaf species in eutrophic streams. These findings support that eutrophication modulates leaf diversity effects on leaf decomposition with potential implications for ecosystem management.

9.5 Microbial Diversity and Litter Decomposition Under Global Change

Many factors regulating microbial diversity and their ecological functions are currently affected by human activities. Such impacts can occur at large spatial scales and represent an important component of human-induced global change. Examples include increases in the concentration of dissolved nutrients in stream water, as a result of atmospheric nitrogen deposition and agriculture; increase in water temperature resulting from global warming, removal of riparian vegetation and urbanization;

impoverishment and homogenization of the riparian vegetation, as a result from tree monocultures and invasions by exotic species; water stress as a result from increases in human needs for freshwater and decreases in precipitation; and contamination by pharmaceuticals, agrochemicals, healthcare products, nanoparticles and plastics. This non-exhaustive list of changes has been addressed in other chapters of this book. Here, we will focus on the effects of increases in the concentration of dissolved nutrients and water temperature, two of the most widespread environmental changes (Woodward et al., 2012; IPCC, 2014).

The number of aquatic hyphomycete species in stream water and associated with decomposing leaf litter on a given sampling date is generally higher under moderate nutrient enrichment (Ferreira et al., 2006; Gulis & Suberkropp, 2003, 2004), which can be explained by the increased availability of resources (i.e., nutrients) not yet confounded by deleterious changes (e.g., sedimentation, pesticides, hypoxia) ('productivity hypothesis' and 'intermediate stress hypothesis'). Nutrient enrichment of stream water can also result in changes in the relative contribution of aquatic hyphomycete species to conidial production, although it is difficult to associate a given species to a given nutrient status across studies (Artigas et al., 2008; Ferreira et al., 2006; Gulis & Suberkropp, 2003, 2004). The reproductive potential of individual fungal species (based on conidia numbers) may also change with nutrient enrichment, with some species being over benefited by increased resource availability (Gulis & Suberkropp, 2003). Since aquatic hyphomycete species have different decomposing capability and palatability to shredders, changes in species richness and relative abundance may affect litter decomposition. Indeed, litter decomposition is generally higher under moderate nutrient enrichment (Ferreira et al., 2015; Gulis & Suberkropp, 2003; Gulis et al., 2019; Rosemond et al., 2015; Woodward et al., 2012), but this is more likely attributed to stimulated microbial activities than to changes in aquatic hyphomycete species richness or community composition (Ferreira, 2020; Chapter 16).

In laboratory experiments, increases in temperature by 5ºC (with maximum temperature ≤ 15ºC) led to changes in aquatic hyphomycete community structure associated with decomposing alder leaves (Dang et al., 2009; Ferreira & Chauvet, 2011a, b). On the contrary, an increase in temperature from 16ºC to 24ºC generally led to decreases in aquatic hyphomycete species richness and shifts in species dominance (Fernandes et al., 2012). Similarly, Gonçalves and collaborators (2013) found higher aquatic hyphomycete species richness associated with alder and oak leaves at 10ºC than at colder (5ºC) or warmer (15ºC and 20ºC) temperatures. Also, increases in temperature induced changes in the species dominance pattern. Moreover, aquatic hyphomycete community structure responded to temperature more strongly on alder than on oak; however, they were affected first by litter species (resource quality) and then by temperature (Gonçalves et al., 2013).

In a manipulative experiment in a springbrook in a mixed forest in Canada, a 4ºC temperature increase led to a rise in aquatic hyphomycete species richness on leaf litter, but not on other substrate types (e.g., needles, grass and wood) (Bärlocher et al., 2008). In a similar experiment in a forest stream in central Portugal, a ~ 3ºC temperature increase did not change aquatic hyphomycetes species richness or community structure associated with oak and chestnut (*Castanea sativa*) leaves during winter (Ferreira et al., 2015). Additionally, season (spring *vs.* autumn) played a greater role in structuring aquatic hyphomycete communities associated with oak leaves than experimental warming (Duarte et al., 2016). This suggests that the major factor structuring aquatic hyphomycete communities may not be temperature, but other seasonally changing factors, such as substrate availability (Gossiaux et al., 2019). However, on the global scale, water temperature has been described as the prime factor ruling aquatic hyphomycete community composition associated with alder leaf litter, independently of biogeographic realms (Seena et al., 2019a).

Different effects of warming on aquatic hyphomycete communities may partially depend on community composition. Aquatic hyphomycete species have distinct thermal preferences, with thermal optimum generally between 10–30°C across species (reviewed by Ferreira et al., 2014). Effects may also depend on the ambient water temperature and the magnitude of the increase. Increases in water temperature when the ambient temperature is already high (>15°C) may have more severe effects than when the ambient temperature is low (<15°C) since species may be already near their thermal optimum (Ferreira et al., 2014).

Increases in temperature have been found to stimulate litter decomposition (Amani et al., 2019; Boyero et al., 2011), but, as for nutrient enrichment, effects of warming are likely mediated by stimulation of microbial activities than by changes in species richness and community structure (Ferreira, 2020).

9.6 Functional Consequences of Microbial Biodiversity Loss

Over the last two centuries, the intensification of human activities on our planet has led to a massive species extinction (Chapin III et al., 2000), with freshwaters being one of the most endangered ecosystems (Dudgeon et al., 2006). This has motivated a bloom of studies on the relationship between biodiversity and ecosystem functioning (BEF, Hooper et al., 2012). BEF research has focused primarily on terrestrial ecosystems, while aquatic ecosystems have received increased attention only over the last decade; BEF studies focusing on microbial communities are even scarcer (Daam et al., 2019; Pascoal & Cássio, 2008). Traditionally, species richness has been used as a biodiversity measure in BEF studies targeting microbial decomposer communities (e.g., Bärlocher & Corkum, 2003; Duarte et al., 2006; Pascoal et al., 2010). Studies in which aquatic hyphomycete species were manipulated point to a positive relationship between fungal diversity and leaf decomposition (Bärlocher & Corkum, 2003; Duarte et al., 2006; Fernandes et al., 2011; Pascoal et al., 2010; Raviraja et al., 2006; Treton et al., 2004). Conversely, other studies failed to detect the effects of fungal diversity on leaf mass loss, pointing to considerable functional redundancy among fungi (Andrade et al., 2016; Dang et al., 2005; Geraldes et al., 2012). However, such studies showed that higher diversity decreases the variability of process rates, probably increasing ecosystem stability (portfolio effect or statistical average effect, see Doak et al., 1998). Fungal diversity had positive effects on other microbial functions such as fungal reproduction (Geraldes et al., 2012) or biomass build-up (Andrade et al., 2016), despite having no effect on leaf decomposition, supporting the importance of considering multiple functions when addressing BEF relationships. In addition, diversity effects of aquatic fungi have also been attributed to species identity, indicating that certain species´ traits may have a greater impact on ecosystem processes than species diversity *per se* (reviewed in Pascoal & Cássio, 2008). Actually, the structure and function of leaf-associated microorganisms can be decoupled under anthropogenic pressure: microbially-mediated leaf litter decomposition remaining stable, increasing or exhibiting a U-shaped response as structural metrics (e.g., taxonomic diversity) change gradually (Feckler & Bundschuh, 2020).

Few studies have addressed other biodiversity measures, like intraspecific diversity considering species background (polluted *vs.* unpolluted stream) (Duarte et al., 2019; Fernandes et al., 2011) or genetic diversity based on species genetic divergence of the ITS1-5.8S-ITS2 rRNA genes (Andrade et al., 2016). Fernandes et al. (2011) demonstrated that positive diversity effects were maintained under metal stress when an ecotype (fungal strain from one species which is adapted to specific environmental conditions) of the worldwide distributed species *A. tetracladia* from a metal-polluted site was incorporated in the assemblage, but these effects were lost when it was replaced by the other ecotype from an unpolluted site. Species or strains that have a redundant role in an ecological process may exhibit noticeable traits when exposed to different environmental contexts (Fernandes et al., 2011). Different environmental contexts have been shown to modulate the impacts of fungal diversity on litter decomposition or decomposer´s activity (nutrients, Bärlocher & Corkum, 2003; metal stress, Pascoal et al., 2010; Fernandes et al., 2011; Duarte et al., 2019; warming, Geraldes et al., 2012).

Diversity effects can result from mechanisms of complementarity and/or selection, whose relative contribution can be quantified if individual species performances in multicultures are determined (partitioning model, Loreau & Hector, 2001). However, most studies examining the relationships between microbial diversity and ecological processes have been limited by difficulties in tracking individual species performances within assemblages (e.g., Bärlocher & Corkum, 2003; Duarte et al., 2006). For instance, Fernandes et al. (2011) successfully determined the contribution of each fungal species to the total biomass produced in multicultures by qPCR. In the absence of metal, positive diversity effects were observed for fungal biomass and leaf decomposition as a result of species complementarity; but, under metal stress, the dominance effect maintained positive diversity effects in assemblages containing the ecotype from the metal-polluted site (Fernandes et al., 2011).

9.7 Outlook

Although the role microbial decomposers play in aquatic detritus-based food webs is acknowledged, there are still gaps that need to be further addressed. Freshwater systems harbor high microbial diversity (Schloss et al., 2016), but data are still scarce compared to other organisms (Balian et al., 2008; Debroas et al., 2017). This lack of knowledge about microbial diversity has limited our deeper understanding of the role that microbial decomposers play in ensuring several ecosystem services, such as nutrient recycling, water purification and carbon sequestration (Ducklow, 2008).

Several gaps that need to be clarified on the relationships between biodiversity and ecosystem functioning in aquatic systems have been identified (Daam et al., 2019) and should be considered when envisaging the functional role of aquatic microbial decomposer´s diversity, namely: (i) looking at multiple ecosystem functions, (ii) studying the role of rare species and focusing on realistic species losses; (iii) integrating different biodiversity metrics (intraspecific diversity, genetic diversity, phylogenetic diversity), (iv) integrating various trophic components (taxonomic groups, trophic composition, trophic interactions), (v) testing different environmental conditions, (vi) targeting larger spatial and temporal scales, (vii) integrating trait-based approaches, and (viii) applying ecological modelling to BEF relationships.

Nowadays, relevant advances to unravel microbial diversity are associated with metabarcoding or environmental DNA techniques. DNA sequencing of environmental samples is increasing in public databases (e.g., NCBI), which facilitate the knowledge of microbial diversity. Moreover, omics are a relevant approach to discover functional traits that can help to explain microbial functional diversity. Further links have to be established between taxonomic, genetic and functional diversity, particularly if we aim to understand how microbial decomposer communities and plant litter decomposition respond to multiple threats derived from global change (Fig. 9.2), such as warming, drought events, eutrophication, persistent and emergent contaminants. Some data already exist on the role of fungal ecotypes to maintain ecological functions under stress (e.g., metal stress, Fernandes et al., 2011) and the biological mechanisms underlying stress responses based on proteomics (e.g., metal nanoparticles, Barros et al., 2020). These data suggest that microbial populations and communities can adapt to stressors following the pollution-induced community tolerance (Tlili et al., 2016). It is conceivable that in nature, these microbes may be able to evolve at relatively short times due to their high replication time. So, they can be crucial to ensure multiple ecosystem functions and services under the ongoing global change (Fig. 9.2).

Another important issue to be further addressed is how fungal assemblages on leaf litter are a food resource for invertebrate detritivores. Beyond the ability of fungi to degrade complex carbon sources from plant litter to easier assimilable food sources due to their enzymatic capabilities (Romaní et al., 2006), it has been claimed that invertebrate shredders have a preference for leaf litter colonized by certain species of fungi (e.g., Arsuffi & Suberkropp, 1984, 1985, 1986). However, no clear explanations have been provided so far for such evidence. Increasing the knowledge on fungal traits

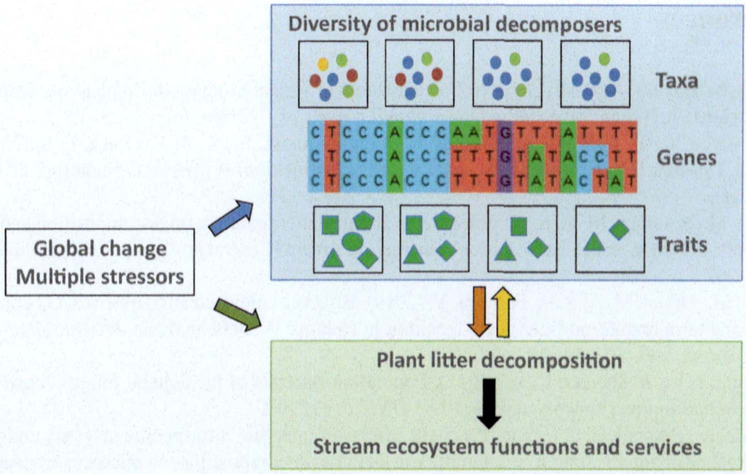

Fig. 9.2 Diagram with possible effects of multiple stressors on different dimensions of diversity (taxonomic, genetic and functional diversity) of microbial decomposers and plant litter decomposition with impacts on stream ecosystem functioning and services. Blue arrow, direct effects of stressors on diversity; green arrow, direct effects of stressors on functions; orange arrow, effects on functions mediated by changes in diversity; yellow arrow; effects on diversity mediated by changes in functions; black arrow, impacts on ecosystem services

could help to elucidate these aspects related to the availability of certain nutrients, such as nitrogen, vitamins or fatty acids (Arce Funck et al., 2015).

How fungal community assembly on plant litter in streams is another aspect that is poorly understood but can be important to ensure plant litter decomposition, particularly under stressful abiotic and biotic conditions. It is recognized that there is a colonization succession with some species of fungi appearing to be early colonizers, while others appear to occur later (e.g., Sridhar et al., 2009). However, the relative importance of stochastic versus deterministic processes in microbial community assembly has been poorly investigated (Chase & Myers, 2011; Stegen et al., 2012). Although stochastic processes, in which communities would be randomly assembled through birth–death, drift, and speciation (neutral theory), are believed to play a role in shaping community structure, most studies focus on deterministic processes considering the selection imposed by biotic interactions and environmental filtering (Vellend, 2010). Anyway, if key early colonizers are lost, it may compromise species succession and, consequently, plant litter decomposition. Therefore, powerful predictive models will contribute to better understand microbial community dynamics and the interactions between plant litter and the micro- and macro-organisms that shape stream ecosystem functioning under global change.

References

Abelho, M. (2009). ATP and ergosterol as indicators of fungal biomass during leaf decomposition in streams: A comparative study. *International Review of Hydrobiology, 94*, 3–15.

Abelho, M. (2020). ATP as a measure of microbial biomass. In M. A. S. Graça, F. Bärlocher, & M. O. Gessner (Eds.), *Methods to study litter decomposition: A practical guide* (pp. 291–299). Springer.

Abelho, M., & Graça, M. A. S. (2006). Effects of nutrient enrichment on decomposition and fungal colonization of sweet chestnut leaves in an Iberian stream (Central portugal). *Hydrobiologia, 560*, 239–247.

Amani, M., Graça, M. A. S., & Ferreira, V. (2019). Effects of elevated atmospheric CO2 concentration and temperature on litter decomposition in streams: A meta-analysis. *International Review of Hydrobiology, 104*, 14–25.

Anderson, J. L., & Shearer, C. A. (2011). Population genetics of the aquatic fungus *Tetracladium marchalianum* over space and time. *PLoS ONE, 6*, e15908.

Andrade, R., Pascoal, C. & Cássio, F. (2016) Effects of inter and intraspecific diversity and genetic divergence of aquatic fungal communities on leaf litter decomposition—a microcosm experiment. *FEMS Microbiology Ecology, 92*, 1–18.

Arce Funck, J., Bec, A., Perrière, F., Felten, V., & Danger, M. (2015). Aquatic hyphomycetes: A potential source of polyunsaturated fatty acids in detritus-based stream food webs. *Fungal Ecology, 13*, 205–210.

Arsuffi, T. L., & Suberkropp, K. (1984). Leaf processing capabilities of aquatic hyphomycetes—Interspecific differences and influence on shredder feeding preferences. *Oikos, 42*, 144–154.

Arsuffi, T. L., & Suberkropp, K. (1985). Selective feeding by stream caddisfly (Trichoptera) detritivores on leaves with fungal-colonized patches. *Oikos, 45*, 50–58.

Arsuffi, T. L., & Suberkropp, K. (1986). Growth of two stream caddisflies (Trichoptera) on leaves colonized by different fungal species. *Journal of the North American Benthological Society, 5*, 297–305.

Artigas, J., Majerholc, J., Foulquier, A., Margoum, C., Volat, B., Neyra, M., & Pesce, S. (2012). Effects of the fungicide tebuconazole on microbial capacities for litter breakdown in streams. *Aquatic Toxicology, 122*, 197–205.

Artigas, J., Romaní, A. M., & Sabater, S. (2008). Effect of nutrients on the sporulation and diversity of aquatic hyphomycetes on submerged substrata in a Mediterranean stream. *Aquatic Botany, 88*, 32–38.

Baldy, V., Chauvet, E., Charcosset, J.-Y., & Gessner, M. O. (2002). Microbial dynamics associated with leaves decomposing in the mainstem and floodplain pond of a large river. *Aquatic Microbial Ecology, 28*, 25–36.

Baldy, V., Gessner, M. O., & Chauvet, E. (1995). Bacteria, fungi and the breakdown of leaf litter in a large river. *Oikos, 74*, 93–102.

Balian, E. V., Segers, H., Lévèque, C., & Martens, K. (2008). The freshwater animal diversity assessment: An overview of the results. *Hydrobiologia, 595*, 627–637.

Barajas-Aceves, M., Hassan, M., Tinoco, R., & Vazquez-Duhalt, R. (2002). Effect of pollutants on the ergosterol content as indicator of fungal biomass. *Journal of Microbiological Methods, 50*, 227–236.

Bärlocher, F. (1982). Conidium production from leaves and needles in four streams. *Canadian Journal of Botany, 60*, 1487–1494.

Bärlocher, F. (1992). *The ecology of aquatic hyphomycetes*. Springer-Verlag.

Bärlocher, F. (2000). Water-borne conidia of aquatic hyphomycetes: Seasonal and yearly patterns in Catamaran Brook, New Brunswick, Canada. *Canadian Journal of Botany-Revue Canadienne De Botanique, 78*, 157–167.

Bärlocher, F. (2007). Molecular approaches applied to aquatic hyphomycetes. *Fungal Biology Reviews, 21*, 19–24.

Bärlocher, F. (2009). Reproduction and dispersal in aquatic hyphomycetes. *Mycoscience, 50*, 3–8.

Bärlocher, F. (2010). Molecular approaches promise a deeper and broader understanding of the evolutionary ecology of aquatic hyphomycetes. *Journal of the North American Benthological Society, 29*, 1027–1041.

Bärlocher, F. (2020). Sporulation by aquatic hyphomycetes. In F. Bärlocher, M. O. Gessner, & M. A. S. Graça (Eds.), *Methods to study litter decomposition*. (pp. 241–245). Springer.

Bärlocher, F., & Corkum, M. (2003). Nutrient enrichment overwhelms diversity effects in leaf decomposition by stream fungi. *Oikos, 101*, 247–252.

Bärlocher, F., & Graça, M. A. S. (2002). Exotic riparian vegetation lowers fungal diversity but not leaf decomposition in Portuguese streams. *Freshwater Biology, 47*, 1123–1135.

Bärlocher, F., Seena, S., Wilson, K. P., & Williams, D. D. (2008). Raised water temperature lowers diversity of hyporheic aquatic hyphomycetes. *Freshwater Biology, 53*, 368–379.

Barros, D., Pradhan, A., Pascoal, C., & Cássio, F. (2020). Proteomic responses to silver nanoparticles vary with the fungal ecotype. *Science of the Total Environment, 704*, 135385.

Baschien, C. (2003). *Development an evaluation of rRNA targeted in situ probes and phylogenetic relationships of freshwater fungi* (PhD thesis). Technischen Universität Berlin.

Baschien, C., Manz, W., Neu, T. R., & Szewzyk, U. (2001). Fluorescence in situ hybridization of freshwater fungi. *International Review of Hydrobiology, 86*, 371–381.

Baschien, C., Marvanova, L., & Szewzyk, U. (2006). Phylogeny of selected aquatic hyphomycetes based on morphological and molecular data. *Nova Hedwigia, 83*, 311–352.

Baschien, C., Tsui, C.K.-M., Gulis, V., Szewzyk, U., & Marvanová, L. (2013). The molecular phylogeny of aquatic hyphomycetes with affinity to the Leotiomycetes. *Fungal Biology, 117*, 660–672.

Batista, D., Tlili, A., Gessner, M. O., Pascoal, C., & Cássio, F. (2020). Nanosilver impacts on aquatic microbial decomposers and litter decomposition assessed as pollution-induced community tolerance (PICT). *Environmental Science: Nano, 7*, 2130–2139.

Baudy, P., Konschak, M., Sakpal, H., Baschien, C., Schulz, R., Bundschuh, M. & Zubrod, J.P. (2020). The fungicide tebuconazole confounds concentrations of molecular biomarkers estimating fungal biomass. *Bulletin of Environmental Contamination and Toxicology*, in press.

Baudy, P., Zubrod, J. P., Röder, N., Baschien, C., Feckler, A., Schulz, R., & Bundschuh, M. (2019). A glance into the black box: Novel species-specific quantitative real-time PCR assays to disentangle aquatic hyphomycete community composition. *Fungal Ecology, 42*, 100858.

Bauer, R., Begerow, D., Oberwinkler, F., & Marvanová, L. (2003). Classicula: The teleomorph of *Naiadella fluitans*. *Mycologia, 95*, 756–764.

Belliveau, M. J. R., & Bärlocher, F. (2005). Molecular evidence confirms multiple origins of aquatic hyphomycetes. *Mycological Research, 109*, 1407–1417.

Bergmann, M., & Graça, M. A. S. (2020). Uranium affects growth, sporulation, biomass and leaf-litter decomposition by aquatic hyphomycetes. *Limnetica, 39*, 141–154.

Bermingham, S., Dewey, F. M., Fisher, P. J., & Maltby, L. (2001). Use of a monoclonal antibody-based immunoassay for the detection and quantification of *Heliscus lugdunensis* colonizing alder leaves and roots. *Microbial Ecology, 42*, 506–512.

Bermingham, S., Dewey, F. M., & Maltby, L. (1995). Development of a monoclonal antibody-based immunoassay for the detection and quantification of *Anguillospora longissima* colonizing leaf material. *Applied and Environmental Microbiology, 61*, 2606–2613.

Bermingham, S., Maltby, L., & Dewey, F. M. (1996). Monoclonal antibodies as tools to quantify mycelium of aquatic hyphomycetes. *New Phytologist, 132*, 593–601.

Bermingham, S., Maltby, L., & Dewey, F. M. (1997). Use of immunoassays for the study of natural assemblages of aquatic hyphomycetes. *Microbial Ecology, 33*, 223–229.

Boyero, L., Pearson, R. G., Gessner, M. O., Barmuta, L. A., Ferreira, V., Graça, M. A. S., Dudgeon, D., Boulton, A. J., Callisto, M., Chauvet, E., Helson, J. E., Bruder, A., Albariño, R. J., Yule, C. M., Arunachalam, M., Davies, J. N., Figueroa, R., Flecker, A. S., Ramírez, A., … West, D. C. (2011). A global experiment suggests climate warming will not accelerate litter decomposition in streams but might reduce carbon sequestration. *Ecology Letters, 14*, 289–294.

Buesing, N., & Gessner, M. O. (2020). Bacterial abundance and biomass determination in plant litter by epifluorescence microscopy. In F. Bärlocher, M. O. Gessner, & M. A. S. Graça (Eds.), *Methods to study litter decomposition* (pp. 265–273). Springer.

Campbell, J., Marvanová, L., & Gulis, V. (2009). Evolutionary relationships between aquatic anamorphs and teleomorphs: *Tricladium* and *Varicosporium*. *Mycological Research, 113*, 1322–1334.

Campbell, J., Shearer, C., & Marvanova, L. (2006). Evolutionary relationships among aquatic anamorphs and teleomorphs: *Lemonniera, Margaritispora*, and *Goniopila*. *Mycological Research, 110*, 1025–1033.

Canhoto, C., & Graça, M. A. S. (1996). Decomposition of *Eucalyptus globulus* leaves and three native leaf species (*Alnus glutinosa, Castanea sativa* and *Quercus faginea*) in a Portuguese low order stream. *Hydrobiologia, 333*, 79–85.

Carlisle, D. M., & Clements, W. H. (2005). Leaf litter breakdown, microbial respiration and shredder production in metal-polluted streams. *Freshwater Biology, 50*, 380–390.

Chapin, F. S., III., Zavaleta, E. S., Eviner, V. T., Naylor, R. L., Vitousek, P. M., Reynolds, H. L., Hooper, D. U., Lavorel, S., Sala, O. E., Hobbie, S. E., Mack, M. C., & Diaz, S. (2000). Consequences of changing biodiversity. *Nature, 405*, 234–242.

Charcosset, J.-Y., & Chauvet, E. (2001). Effect of culture conditions on ergosterol as an indicator of biomass in the aquatic hyphomycetes. *Applied and Environmental Microbiology, 67*, 2051–2055.

Charcosset, J. Y., & Gardes, M. (1999). Infraspecific genetic diversity and substrate preference in the aquatic hyphomycete *Tetrachaetum elegans*. *Mycological Research, 103*, 736–742.

Chase, J. M., & Myers, J. A. (2011). Disentangling the importance of ecological niches from stochastic processes across scales. *Philosophical Transactions of the Royal Society B: Biological Sciences, 366*, 2351–2363.

Chauvet, E., Fabre, E., Elosegui, A., & Pozo, J. (1997). The impact of eucalypt on the leaf-associated aquatic hyphomycetes in Spanish streams. *Canadian Journal of Botany-Revue Canadienne De Botanique, 75*, 880–887.

Chauvet, E., & Suberkropp, K. (1998). Temperature and sporulation of aquatic hyphomycetes. *Applied and Environmental Microbiology, 64*, 1522–1525.

Cheever, B. M., Kratzer, E. B., & Webster, J. R. (2012). Immobilization and mineralization of N and P by heterotrophic microbes during leaf decomposition. *Freshwater Science, 31*, 133–147.

Cleveland, C., & Liptzin, D. (2007). C:N: P stoichiometry in soil: Is there a "Redfield ratio" for the microbial biomass? *Biogeochemistry, 85*, 235–252.

Cross, W. F., Benstead, J. P., Frost, P. C., & Thomas, S. A. (2005). Ecological stoichiometry in freshwater benthic systems: Recent progress and perspectives. *Freshwater Biology, 50*, 1895–1912.

Daam, M. A., Teixeira, H., Lillebø, A. I., & Nogueira, A. J. A. (2019). Establishing causal links between aquatic biodiversity and ecosystem functioning: Status and research needs. *Science of the Total Environment, 656*, 1145–1156.

Dang, C. K., Chauvet, E., & Gessner, M. O. (2005). Magnitude and variability of process rates in fungal diversity-litter decomposition relationships. *Ecology Letters, 8*, 1129–1137.

Dang, C. K., Schindler, M., Chauvet, E., & Gessner, M. O. (2009). Temperature oscillation coupled with fungal community shifts can modulate warming effects on litter decomposition. *Ecology, 90*, 122–131.

Danger, M., Cornut, J., Chauvet, E., Chavez, P., Elger, A., & Lecerf, A. (2013). Benthic algae stimulate leaf litter decomposition in detritus-based headwater streams: A case of aquatic priming effect? *Ecology, 94*, 1604–1613.

Danger, M., Daufresne, T., Lucas, F., Pissard, S., & Lacroix, G. (2008). Does Liebig's law of the minimum scale up from species to communities? *Oikos, 117*, 1741–1751.

Danger, M., Gessner, M. O., & Bärlocher, F. (2016). Ecological stoichiometry of aquatic fungi: Current knowledge and perspectives. *Fungal Ecology, 19*, 100–111.

Das, M., Royer, T. V., & Leff, L. G. (2007). Diversity of fungi, bacteria, and actinomycetes on leaves decomposing in a stream. *Applied and Environmental Microbiology, 73*, 756–767.

Debroas, D., Domaizon, I., Humbert, J.-F., Jardillier, L., Lepère, C., Oudart, A. & Taïb, N. (2017) Overview of freshwater microbial eukaryotes diversity: A first analysis of publicly available metabarcoding data. *FEMS Microbiology Ecology, 93.*

Doak, D. F., Bigger, D., Harding, E. K., Marvier, M. A., O'Malley, R. E., & Thomson, D. (1998). The statistical inevitability of stability-diversity relationships in community ecology. *the American Naturalist, 151*, 264–276.

Duarte, S., Antunes, B., Trabulo, J., Seena, S., Cássio, F., & Pascoal, C. (2019). Intraspecific diversity affects stress response and the ecological performance of a cosmopolitan aquatic fungus. *Fungal Ecology, 41*, 218–223.

Duarte, S., Bärlocher, F., Cássio, F., & Pascoal, C. (2014). Current status of DNA barcoding of aquatic hyphomycetes. *Sydowia, 66*, 191–202.

Duarte, S., Bärlocher, F., Pascoal, C., & Cássio, F. (2016). Biogeography of aquatic hyphomycetes: Current knowledge and future perspectives. *Fungal Ecology, 19*, 169–181.

Duarte, S., Bärlocher, F., Trabulo, J., Cássio, F., & Pascoal, C. (2015). Stream-dwelling fungal decomposer communities along a gradient of eutrophication unraveled by 454 pyrosequencing. *Fungal Diversity, 70*, 127–148.

Duarte, S., Cássio, F., Ferreira, V., Canhoto, C., & Pascoal, C. (2016). Seasonal variability may affect microbial decomposers and leaf decomposition more than warming in streams. *Microbial Ecology, 72*, 263–276.

Duarte, S., Cássio, F., Pascoal, C., & Bärlocher, F. (2017). Taxa-area relationship of aquatic fungi on deciduous leaves. *PLoS ONE, 12*, e0181545.

Duarte, S., Pascoal, C., Alves, A., Correia, A., & Cássio, F. (2008). Copper and zinc mixtures induce shifts in microbial communities and reduce leaf litter decomposition in streams. *Freshwater Biology, 53*, 91–101.

Duarte, S., Pascoal, C., Alves, A., Correia, A., & Cássio, F. (2010). Assessing the dynamic of microbial communities during leaf decomposition in a low-order stream by microscopic and molecular techniques. *Microbiological Research, 165*, 351–362.

Duarte, S., Pascoal, C., Cássio, F., & Bärlocher, F. (2006). Aquatic hyphomycete diversity and identity affect leaf litter decomposition in microcosms. *Oecologia, 147*, 658–666.

Duarte, S., Pascoal, C., Garabetian, F., Cássio, F., & Charcosset, J.-Y. (2009). Microbial decomposer communities are mainly structured by trophic status in circumneutral and alkaline streams. *Applied and Environmental Microbiology, 75*, 6211–6221.

Duarte, S., Seena, S., Bärlocher, F., Pascoal, C., & Cássio, F. (2013). A decade's perspective on the impact of DNA sequencing on aquatic hyphomycete research. *Fungal Biology Reviews, 27*, 19–24.

Ducklow, H. (2008). Microbial services: Challenges for microbial ecologists in a changing world. *Aquatic Microbial Ecology, 53*, 13–19.

Dudgeon, D., Arthington, A. H., Gessner, M. O., Kawabata, Z. I., Knowler, D. J., Leveque, C., Naiman, R. J., Prieur-Richard, A. H., Soto, D., Stiassny, M. L. J., & Sullivan, C. A. (2006). Freshwater biodiversity: Importance, threats, status and conservation challenges. *Biological Reviews, 81*, 163–182.

Dunck, B., Lima-Fernandes, E., Cássio, F., Cunha, A., Rodrigues, L., & Pascoal, C. (2015). Responses of primary production, leaf litter decomposition and associated communities to stream eutrophication. *Environmental Pollution, 202*, 32–40.

Enríquez, S., Duarte, C. M., & Sand-Jensen, K. (1993). Patterns in decomposition rates among photosynthetic organisms: The importance of detritus C:N: P content. *Oecologia, 94*, 457–471.

Feckler, A. & Bundschuh, M. (2020) Decoupled structure and function of leaf-associated microorganisms under anthropogenic pressure: potential hurdles for environmental monitoring. *Freshwater Science, 39*, 652–664.

Feckler, A., Schrimpf, A., Bundschuh, M., Bärlocher, F., Baudy, P., Cornut, J., & Schulz, R. (2017). Quantitative real-time PCR as a promising tool for the detection and quantification of leaf-associated fungal species—A proof-of-concept using *Alatospora pulchella. PLoS ONE, 12*, e0174634.

Fernandes, I., Duarte, S., Cássio, F., & Pascoal, C. (2013). Effects of riparian plant diversity loss on aquatic microbial decomposers become more pronounced with increasing time. *Microbial Ecology, 66*, 763–772.

Fernandes, I., Pascoal, C., & Cássio, F. (2011). Intraspecific traits change biodiversity effects on ecosystem functioning under metal stress. *Oecologia, 166*, 1019–1028.

Fernandes, I., Pascoal, C., Guimarães, H., Pinto, R., Sousa, I., & Cássio, F. (2012). Higher temperature reduces the effects of litter quality on decomposition by aquatic fungi. *Freshwater Biology, 57*, 2306–2317.

Fernandes, I., Pereira, A., Trabulo, J., Pascoal, C., Cássio, F., & Duarte, S. (2015). Microscopy- or DNA-based analyses: Which methodology gives a truer picture of stream-dwelling decomposer fungal diversity? *Fungal Ecology, 18*, 130–134.

Fernandes, I., Seena, S., Pascoal, C., & Cássio, F. (2014). Elevated temperature may intensify the positive effects of nutrients on microbial decomposition in streams. *Freshwater Biology, 59*, 2390–2399.

Fernandes, I., Uzun, B., Pascoal, C., & Cássio, F. (2009). Responses of aquatic fungal communities on leaf litter to temperature-change events. *International Review of Hydrobiology, 94*, 410–418.

Ferreira, V. (2020). Impact of climate change on aquatic hyphomycetes. *Climate change and microbial ecology: current research and future trends* (Ed. J. Marxsen). Norfolk, UK: Caister Academic Press.

Ferreira, V., Castagneyrol, B., Koricheva, J., Gulis, V., Chauvet, E., & Graça, M. A. S. (2015). A meta-analysis of the effects of nutrient enrichment on litter decomposition in streams. *Biological Reviews, 90*, 669–688.

Ferreira, V., Castela, J., Rosa, P., Tonin, A. M., Boyero, L., & Graça, M. A. S. (2016). Aquatic hyphomycetes, benthic macroinvertebrates and leaf litter decomposition in streams naturally differing in riparian vegetation. *Aquatic Ecology, 50*, 711–725.

Ferreira, V., & Chauvet, E. (2011a). Future increase in temperature more than decrease in litter quality can affect microbial litter decomposition in streams. *Oecologia, 167*, 279–291.

Ferreira, V., & Chauvet, E. (2011b). Synergistic effects of water temperature and dissolved nutrients on litter decomposition and associated fungi. *Global Change Biology, 17*, 551–564.

Ferreira, V., Chauvet, E., & Canhoto, C. (2015). Effects of experimental warming, litter species, and presence of macroinvertebrates on litter decomposition and associated decomposers in a temperate mountain stream. *Canadian Journal of Fisheries and Aquatic Sciences, 72*, 206–216.

Ferreira, V., Elosegi, A., Gulis, V., Pozo, J., & Graça, M. A. S. (2006). *Eucalyptus* plantations affect fungal communities associated with leaf-litter decomposition in Iberian streams. *Archiv Fur Hydrobiologie, 166*, 467–490.

Ferreira, V., Faustino, H., Raposeiro, P. M., & Gonçalves, V. (2017). Replacement of native forests by conifer plantations affects fungal decomposer community structure but not litter decomposition in Atlantic island streams. *Forest Ecology and Management, 389*, 323–330.

Ferreira, V., & Graça, M. A. S. (2006). Do invertebrate activity and current velocity affect fungal assemblage structure in leaves? *International Review of Hydrobiology, 91*, 1–14.

Ferreira, V., & Graça, M. A. S. (2016). Effects of whole-stream nitrogen enrichment and litter species mixing on litter decomposition and associated fungi. *Limnologica, 58*, 69–77.

Ferreira, V., Gulis, V., & Graça, M. A. S. (2006). Whole-stream nitrate addition affects litter decomposition and associated fungi but not invertebrates. *Oecologia, 149*, 718–729.

Ferreira, V., Gulis, V., Pascoal, C. & Graça, M.A.S. (2014) Stream pollution and fungi. *Freshwater Fungi and Fungus-like Organisms. De Gruyter Series: Marine and Freshwater Botany* (Eds., G. Jones, K. Hyde & K.-L. Pang, pp. 389–412). Berlin, Germany: De Gruyter.

Fontaine, S., & Barot, S. (2005). Size and functional diversity of microbe populations control plant persistence and long-term soil carbon accumulation. *Ecology Letters, 8*, 1075–1087.

Frossard, A., Gerull, L., Mutz, M., & Gessner, M. O. (2013). Litter supply as a driver of microbial activity and community structure on decomposing leaves: A test in experimental streams. *Applied and Environmental Microbiology, 79*, 4965.

Frossard, A., Hammes, F., & Gessner, M. O. (2016). Flow cytometric assessment of bacterial abundance in soils, sediments and sludge. *Frontiers in Microbiology, 7*, 903.

Frost, P. C., Benstead, J. P., Cross, W. F., Hillebrand, H., Larson, J. H., Xenopoulos, M. A., & Yoshida, T. (2006). Threshold elemental ratios of carbon and phosphorus in aquatic consumers. *Ecology Letters, 9*, 774–779.

Geraldes, P., Pascoal, C., & Cássio, F. (2012). Effects of increased temperature and aquatic fungal diversity on litter decomposition. *Fungal Ecology, 5*, 734–740.

Gessner, M. O., & Chauvet, E. (1993). Ergosterol-to-biomass conversion factors for aquatic hyphomycetes. *Applied and Environmental Microbiology, 59*, 502–507.

Gessner, M. O., & Chauvet, E. (1994). Importance of stream microfungi in controlling breakdown rates of leaf-litter. *Ecology, 75*, 1807–1817.

Gessner, M. O., & Chauvet, E. (1997). Growth and production of aquatic hyphomycetes in decomposing leaf litter. *Limnology and Oceanography, 42*, 496–505.

Gessner, M. O., & Newell, S. Y. (2002). Biomass, growth rate, and production of filamentous fungi in plant litter. In H. J. Christon (Ed.), *Manual of environmental microbiology.* (pp. 390–408). ASM Press.

Ginzinger, D. G. (2002). Gene quantification using real-time quantitative PCR: An emerging technology hits the mainstream. *Experimental Hematology, 30*, 503–512.

Gonçalves, A. L., Gama, A. M., Ferreira, V., Graça, M. A. S., & Canhoto, C. (2007). The breakdown of blue gum (*Eucalyptus globulus* Labill.) bark in a portuguese stream. *Fundamental and Applied Limnology / Archiv Für Hydrobiologie, 168*, 307–315

Gonçalves, A. L., Graça, M. A. S., & Canhoto, C. (2013). The effect of temperature on leaf decomposition and diversity of associated aquatic hyphomycetes depends on the substrate. *Fungal Ecology, 6*, 546–553.

Gossiaux, A., Jabiol, J., Poupin, P., Chauvet, E., & Guérold, F. (2019). Seasonal variations overwhelm temperature effects on microbial processes in headwater streams: Insights from a temperate thermal spring. *Aquatic Sciences, 81*, 30.

Graça, M. A. S., & Abelho, M. (2020). Respiration of litter-associated microbes and invertebrates. In F. Bärlocher, M. O. Gessner, & M. A. S. Graça (Eds.), *Methods to study litter decomposition* (pp. 301–308). Springer.

Guenet, B., Danger, M., Abbadie, L., & Lacroix, G. (2010). Priming effect: Bridging the gap between terrestrial and aquatic ecology. *Ecology, 91*, 2850–2861.

Gulis, V. (2001). Are there any substrate preferences in aquatic hyphomycetes? *Mycological Research, 105*, 1088–1093.

Gulis, V., Ferreira, V., & Graça, M. A. S. (2006). Stimulation of leaf litter decomposition and associated fungi and invertebrates by moderate eutrophication: Implications for stream assessment. *Freshwater Biology, 51*, 1655–1669.

Gulis, V., Kuehn, K. A., Schoettle, L. N., Leach, D., Benstead, J. P., & Rosemond, A. D. (2017). Changes in nutrient stoichiometry, elemental homeostasis and growth rate of aquatic litter-associated fungi in response to inorganic nutrient supply. *The ISME Journal, 11*, 2729–2739.

Gulis, V., Marvanová, L., & Descals, E. (2020). An illustrated key to the common temperate species of aquatic hyphomycetes. In F. Bärlocher, M. O. Gessner, & M. A. S. Graça (Eds.), *Methods to study litter decomposition* (pp. 223–239). Springer.

Gulis, V., Su, R. & Kuehn, K.A. (2019). Fungal decomposers in freshwater environments. *The structure and function of aquatic microbial communities. Advances in Environmental Microbiology* (Ed., C. Hurst, pp. 121–155). Cham: Springer.

Gulis, V., & Suberkropp, K. (2003). Leaf litter decomposition and microbial activity in nutrient-enriched and unaltered reaches of a headwater stream. *Freshwater Biology, 48*, 123–134.

Gulis, V., & Suberkropp, K. (2004). Effects of whole-stream nutrient enrichment on the concentration and abundance of aquatic hyphomycete conidia in transport. *Mycologia, 96*, 57–65.

Güsewell, S., & Gessner, M. O. (2009). N : P ratios influence litter decomposition and colonization by fungi and bacteria in microcosms. *Functional Ecology, 23*, 211–219.

Halvorson, H. M., Francoeur, S. N., Findlay, R. H., & Kuehn, K. A. (2019). Algal-mediated priming effects on the ecological stoichiometry of leaf litter decomposition: A meta-analysis. *Frontiers in Earth Science, 7*, 76.

Hayer, M., Schwartz, E., Marks, J. C., Koch, B. J., Morrissey, E. M., Schuettenberg, A. A., & Hungate, B. A. (2016). Identification of growing bacteria during litter decomposition in freshwater through quantitative stable isotope probing. *Environmental Microbiology Reports, 8*, 975–982.

Hieber, M., & Gessner, M. O. (2002). Contribution of stream detrivores, fungi, and bacteria to leaf breakdown based on biomass estimates. *Ecology, 83*, 1026–1038.

Hladyz, S., Gessner, M. O., Giller, P. S., Pozo, J., & Woodward, G. (2009). Resource quality and stoichiometric constraints on stream ecosystem functioning. *Freshwater Biology, 54*, 957–970.

Hofstetter, V., Buyck, B., Eyssartier, G., Schnee, S., & Gindro, K. (2019). The unbearable lightness of sequenced-based identification. *Fungal Diversity, 96*, 243–284.

Hooper, D. U., Adair, E. C., Cardinale, B. J., Byrnes, J. E. K., Hungate, B. A., Matulich, K. L., Gonzalez, A., Duffy, J. E., Gamfeldt, L., & O'Connor, M. I. (2012). A global synthesis reveals biodiversity loss as a major driver of ecosystem change. *Nature, 486*, 105–108.

Hungate, B. A., Mau, R. L., Schwartz, E., Caporaso, J. G., Dijkstra, P., van Gestel, N., Koch, B. J., Liu, C. M., McHugh, T. A., Marks, J. C., Morrissey, E. M., & Price, L. B. (2015). Quantitative microbial ecology through stable isotope probing. *Applied and Environmental Microbiology, 81*, 7570–7581.

IPCC. (2014). *Climate change 2014: Synthesis report*. Contribution of working groups I, II and III to the fifth assessment report of the Intergovernmental Panel on Climate Change (Eds., Core Writing Team, R.K. Pachauri & L.A. Meyer). Geneva, Switzerland: IPCC.

Jabiol, J., Lecerf, A., Lamothe, S., Gessner, M. O., & Chauvet, E. (2019). Litter quality modulates effects of dissolved nitrogen on leaf decomposition by stream microbial communities. *Microbial Ecology, 77*, 959–966.

Kanagawa, T. (2003). Bias and artifacts in multitemplate polymerase chain reactions (PCR). *Journal of Bioscience and Bioengineering, 96*, 317–323.

Keiblinger, K. M., Hall, E. K., Wanek, W., Szukics, U., Hämmerle, I., Ellersdorfer, G., Böck, S., Strauss, J., Sterflinger, K., Richter, A., & Zechmeister-Boltenstern, S. (2010). The effect of resource quantity and resource stoichiometry on microbial carbon-use-efficiency. *FEMS Microbiology Ecology, 73*, 430–440.

Komínková, D., Kuehn, K. A., Büsing, N., Steiner, D., & Gessner, M. O. (2000). Microbial biomass, growth, and respiration associated with submerged litter of *Phragmites australis* decomposing in a littoral reed stand of a large lake. *Aquatic Microbial Ecology, 22*, 271–282.

Kuehn, K. A., Francoeur, S. N., Findlay, R. H., & Neely, R. K. (2014). Priming in the microbial landscape: Periphytic algal stimulation of litter-associated microbial decomposers. *Ecology, 95*, 749–762.

Laitung, B., & Chauvet, E. (2005). Vegetation diversity increases species richness of leaf-decaying fungal communities in woodland streams. *Archiv Fur Hydrobiologie, 164*, 217–235.

Laitung, B., Chauvet, E., Feau, N., Feve, K., Chikhi, L., & Gardes, M. (2004). Genetic diversity in *Tetrachaetum elegans*, a mitosporic aquatic fungus. *Molecular Ecology, 13*, 1679–1692.

Laitung, B., Pretty, J. L., Chauvet, E., & Dobson, M. (2002). Response of aquatic hyphomycete communities to enhanced stream retention in areas impacted by commercial forestry. *Freshwater Biology, 47*, 313–323.

Lecerf, A., Dobson, M., Dang, C. K., & Chauvet, E. (2005). Riparian plant species loss alters trophic dynamics in detritus-based stream ecosystems. *Oecologia, 146*, 432–442.

Lecerf, A., Marie, G., Kominoski, J. S., LeRoy, C. J., Bernadet, C., & Swan, C. M. (2011). Incubation time, functional litter diversity, and habitat characteristics predict litter-mixing effects on decomposition. *Ecology, 92*, 160–169.

Letourneau, A., Seena, S., Marvanová, L., & Bärlocher, F. (2010). Potential use of barcoding to identify aquatic hyphomycetes. *Fungal Diversity, 40*, 51–64.

Lima-Fernandes, E., Fernandes, I., Geraldes, P., Pereira, A., Cássio, F., & Pascoal, C. (2015). Eutrophication modulates plant-litter diversity effects on litter decomposition in streams. *Freshwater Science, 34*, 31–41.

Loreau, M., & Hector, A. (2001). Partitioning selection and complementarity in biodiversity experiments. *Nature, 412*, 72–76.

Manerkar, M. A., Seena, S., & Bärlocher, F. (2008). Q-RT-PCR for assessing archaea, bacteria, and fungi during leaf decomposition in a stream. *Microbial Ecology, 56*, 467–473.

Manzoni, S., Trofymow, J. A., Jackson, R. B., & Porporato, A. (2010). Stoichiometric controls on carbon, nitrogen, and phosphorus dynamics in decomposing litter. *Ecological Monographs, 80*, 89–106.

Martínez, A., Larrañaga, A., Pérez, J., Descals, E., Basaguren, A., & Pozo, J. (2013). Effects of pine plantations on structural and functional attributes of forested streams. *Forest Ecology and Management, 310*, 147–155.

McArthur, F. A., Baerlocher, M. O., MacLean, N. A. B., Hiltz, M. D., & Bärlocher, F. (2001). Asking probing questions: Can fluorescent in situ hybridization identify and localise aquatic hyphomycetes on leaf litter? *International Review of Hydrobiology, 86*, 429–438.

Medeiros, A. O., Pascoal, C., & Graça, M. A. S. (2009). Diversity and activity of aquatic fungi under low oxygen conditions. *Freshwater Biology, 54*, 142–149.

Melillo, J. M., Aber, J. D., & Muratore, J. F. (1982). Nitrogen and lignin control of hardwood leaf litter decomposition dynamics. *Ecology, 63*, 621–626.

Mille-Lindblom, C., von Wachenfeldt, E., & Tranvik, L. J. (2004). Ergosterol as a measure of living fungal biomass: Persistence in environmental samples after fungal death. *Journal of Microbiological Methods, 59*, 253–262.

Mooshammer, M., Wanek, W., Zechmeister-Boltenstern, S., & Richter, A. (2014). Stoichiometric imbalances between terrestrial decomposer communities and their resources: Mechanisms and implications of microbial adaptations to their resources. *Frontiers in Microbiology, 5*, 22–22.

Mora-Gómez, J., Boix, D., Duarte, S., Cássio, F., Pascoal, C., Elosegi, A., & Romaní, A. M. (2020). Legacy of summer drought on autumnal leaf litter processing in a temporary mediterranean stream. *Ecosystems, 23*, 989–1003.

Nikolcheva, L. G., & Bärlocher, F. (2002). Phylogeny of *Tetracladium* based on 18S rDNA. *Czech Mycology, 53*, 285–295.

Nikolcheva, L. G., & Bärlocher, F. (2004). Taxon-specific fungal primers reveal unexpectedly high diversity during leaf decomposition in a stream. *Mycological Progress, 3*, 41–49.

Nikolcheva, L. G., & Bärlocher, F. (2005). Seasonal and substrate preferences of fungi colonizing leaves in streams: Traditional versus molecular evidence. *Environmental Microbiology, 7*, 270–280.

Nikolcheva, L. G., Cockshutt, A. M., & Bärlocher, F. (2003). Determining diversity of freshwater fungi on decaying leaves: Comparison of traditional and molecular approaches. *Applied and Environmental Microbiology, 69*, 2548–2554.

Pascoal, C., & Cássio, F. (2004). Contribution of fungi and bacteria to leaf litter decomposition in a polluted river. *Applied and Environmental Microbiology, 70*, 5266–5273.

Pascoal, C., & Cássio, F. (2008). Linking fungal diversity to the functioning of freshwater ecosystems. In K. R. Sridhar, F. Bärlocher, & K. D. Hyde (Eds.), *Novel techniques and ideas in mycology* (pp. 1–15). Fungal Diversity Press.

Pascoal, C., Cássio, F., Marcotegui, A., Sanz, B., & Gomes, P. (2005). Role of fungi, bacteria, and invertebrates in leaf litter breakdown in a polluted river. *Journal of the North American Benthological Society, 24*, 784–797.

Pascoal, C., Cássio, F., Nikolcheva, L., & Bärlocher, F. (2010). Realized fungal diversity increases functional stability of leaf litter decomposition under zinc stress. *Microbial Ecology, 59*, 84–93.

Pascoal, C., Marvanová, L., & Cássio, F. (2005). Aquatic hyphomycete diversity in streams of Northwest Portugal. *Fungal Diversity, 19*, 109–128.

Peláez, F., Platas, G., Collado, J., & Díez, M. T. (1996). Infraspecific variation in two species of aquatic hyphomycetes assessed by RAPD analysis. *Mycological Research, 100*, 831–837.

Pereira, A., Geraldes, P., Lima-Fernandes, E., Fernandes, I., Cássio, F., & Pascoal, C. (2016). Structural and functional measures of leaf-associated invertebrates and fungi as predictors of stream eutrophication. *Ecological Indicators, 69*, 648–656.

Pérez, J., Galán, J., Descals, E., & Pozo, J. (2014). Effects of fungal inocula and habitat conditions on alder and eucalyptus leaf litter decomposition in streams of Northern Spain. *Microbial Ecology, 67*, 245–255.

Pimentão, A. R., Pascoal, C., Castro, B. B., & Cássio, F. (2020). Fungistatic effect of agrochemical and pharmaceutical fungicides on non-target aquatic decomposers does not translate into decreased fungi- or invertebrate-mediated decomposition. *Science of the Total Environment, 712*, 135676.

Pope, C. A., Halvorson, H. M., Findlay, R. H., Francoeur, S. N., & Kuehn, K. A. (2020). Light and temperature mediate algal stimulation of heterotrophic activity on decomposing leaf litter. *Freshwater Biology, 65*, 1210–1222.

Pradhan, A., Seena, S., Pascoal, C., & Cássio, F. (2011). Can metal nanoparticles be a threat to microbial decomposers of plant litter in streams? *Microbial Ecology, 62*, 58–68.

Rajashekhar, M., & Kaveriappa, K. M. (2003). Diversity of aquatic hyphomycetes in the aquatic ecosystems of the Western Ghats of India. *Hydrobiologia, 501*, 167–177.

Raviraja, N. S., Nikolcheva, L. G., & Barlocher, F. (2006). Fungal growth and leaf decomposition are affected by amount and type of inoculum and by external nutrients. *Sydowia, 58*, 91–104.

Rico, A., Dimitrov, M. R., Van Wijngaarden, R. P. A., Satapornvanit, K., Smidt, H., & Van den Brink, P. J. (2014). Effects of the antibiotic enrofloxacin on the ecology of tropical eutrophic freshwater microcosms. *Aquatic Toxicology, 147*, 92–104.

Romaní, A. M., Fischer, H., Mille-Lindblom, C., & Tranvik, L. J. (2006). Interactions of bacteria and fungi on decomposing litter: Differential extracellular enzyme activities. *Ecology, 87*, 2559–2569.

Rosemond, A. D., Benstead, J. P., Bumpers, P. M., Gulis, V., Kominoski, J. S., Manning, D. W. P., Suberkropp, K., & Wallace, J. B. (2015). Experimental nutrient additions accelerate terrestrial carbon loss from stream ecosystems. *Science, 347*, 1142–1145.

Sales, M. A., Gonçalves, J. F., Dahora, J. S., & Medeiros, A. O. (2015). Influence of leaf quality in microbial decomposition in a headwater stream in the Brazilian Cerrado: A 1-year study. *Microbial Ecology, 69*, 84–94.

Sampaio, A., Cortes, R., & Leão, C. (2004). Yeast and macroinvertebrate communities associated with leaf litter decomposition in a second order stream. *International Review of Hydrobiology, 89*, 453–466.

Sampaio, A., Sampaio, J. P., & Leão, C. (2007). Dynamics of yeast populations recovered from decaying leaves in a nonpolluted stream: A 2-year study on the effects of leaf litter type and decomposition time. *FEMS Yeast Research, 7*, 595–603.

Sardans, J., Rivas-Ubach, A., & Peñuelas, J. (2012). The elemental stoichiometry of aquatic and terrestrial ecosystems and its relationships with organismic lifestyle and ecosystem structure and function: A review and perspectives. *Biogeochemistry, 111*, 1–39.

Schloss, P.D., Girard, R.A., Martin, T., Edwards, J. & Thrash, J.C. (2016) Status of the archaeal and bacterial census: an update. *mBio, 7*, e00201–00216.

Seena, S., Bärlocher, F., Sobral, O., Gessner, M. O., Dudgeon, D., McKie, B. G., Chauvet, E., Boyero, L., Ferreira, V., Frainer, A., Bruder, A., Matthaei, C. D., Fenoglio, S., Sridhar, K. R., Albariño, R. J., Douglas, M. M., Encalada, A. C., Garcia, E., Ghate, S. D., … Graça, M. A. S. (2019). Biodiversity of leaf litter fungi in streams along a latitudinal gradient. *Science of the Total Environment, 661*, 306–315.

Seena, S., Duarte, S., Pascoal, C., & Cássio, F. (2012). Intraspecific variation of the aquatic fungus *Articulospora tetracladia*: An ubiquitous perspective. *PLoS ONE, 7*, e35884.

Seena, S., Graça, D., Bartels, A., & Cornut, J. (2019). Does nanosized plastic affect aquatic fungal litter decomposition? *Fungal Ecology, 39*, 388–392.

Seena, S., Marvanová, L., Letourneau, A., & Bärlocher, F. (2018). *Articulospora* – Phylogeny vs morphology. *Fungal Biology, 122*, 965–976.

Seena, S., & Monroy, S. (2016). Preliminary insights into the evolutionary relationships of aquatic hyphomycetes and endophytic fungi. *Fungal Ecology, 19*, 128–134.

Seena, S., Pascoal, C., Marvanova, L., & Cassio, F. (2010). DNA barcoding of fungi: A case study using ITS sequences for identifying aquatic hyphomycete species. *Fungal Diversity, 44*, 77–87.

Seena, S., Sobral, O., & Cano, A. (2020). Metabolomic, functional, and ecologic responses of the common freshwater fungus *Neonectria lugdunensis* to mine drainage stress. *Science of the Total Environment, 718*, 137359.

Seena, S., Wynberg, N., & Bärlocher, F. (2008). Fungal diversity during leaf decomposition in a stream assessed through clone libraries. *Fungal Diversity, 30*, 1–14.

Sinsabaugh, R. L., & Follstad Shah, J. J. (2011). Ecoenzymatic stoichiometry of recalcitrant organic matter decomposition: The growth rate hypothesis in reverse. *Biogeochemistry, 102*, 31–43.

Sinsabaugh, R. L., Manzoni, S., Moorhead, D. L., & Richter, A. (2013). Carbon use efficiency of microbial communities: Stoichiometry, methodology and modelling. *Ecology Letters, 16*, 930–939.

Solé, M., Fetzer, I., Wennrich, R., Sridhar, K. R., Harms, H., & Krauss, G. (2008). Aquatic hyphomycete communities as potential bioindicators for assessing anthropogenic stress. *Science of the Total Environment, 389*, 557–565.

Sridhar, K. R., & Bärlocher, F. (1994). Viability of aquatic hyphomycete conidia in foam. *Canadian Journal of Botany, 72*, 106–110.

Sridhar, K. R., & Bärlocher, F. (2011). Reproduction of aquatic hyphomycetes at low concentrations of Ca2+, Zn2+, Cu2+, and Cd2+. *Environmental Toxicology and Chemistry, 30*, 2868–2873.

Sridhar, K. R., Duarte, S., Cássio, F., & Pascoal, C. (2009). The role of early fungal colonizers in leaf-litter decomposition in Portuguese streams impacted by agricultural runoff. *International Review of Hydrobiology, 94*, 399–409.

Stegen, J. C., Lin, X., Konopka, A. E., & Fredrickson, J. K. (2012). Stochastic and deterministic assembly processes in subsurface microbial communities. *The ISME Journal, 6*, 1653–1664.

Stelzer, R. S., Heffernan, J., & Likens, G. E. (2003). The influence of dissolved nutrients and particulate organic matter quality on microbial respiration and biomass in a forest stream. *Freshwater Biology, 48*, 1925–1937.

Sterner, R. W., & Elser, J. J. (2002). Stochiometry and Homeostasis. In R. W. Sterner & J. J. Elser (Eds.), *Ecological Stoichiometry: The Biology of Elements from Molecules to the Biosphere.* (pp. 2–43). Princeton University Press.

Suberkropp, K. (1984). Effect of temperature on seasonal occurrence of aquatic hyphomycetes. *Transactions of the British Mycological Society, 82*, 53–62.

Suberkropp, K. (1991). Relationships between growth and sporulation of aquatic hyphomycetes on decomposing leaf litter. *Mycological Research, 95*, 843–850.

Suberkropp, K., Gessner, M. O., & Kuehn, K. A. (2020). Fungal growth rates and production. In F. Bärlocher, M. O. Gessner, & M. A. S. Graça (Eds.), *Methods to study litter decomposition.* (pp. 257–264). Springer.

Suberkropp, K., Godshalk, G. L., & Klug, M. J. (1976). Changes in the chemical composition of leaves during processing in a woodland stream. *Ecology, 57*, 720–727.

Suberkropp, K., & Klug, M. J. (1976). Fungi and bacteria associated with leaves during processing in a woodland stream. *Ecology, 57*, 707–719.

Tlili, A., Berard, A., Blanck, H., Bouchez, A., Cássio, F., Eriksson, K. M., Morin, S., Montuelle, B., Navarro, E., Pascoal, C., Pesce, S., Schmitt-Jansen, M., & Behra, R. (2016). Pollution-induced community tolerance (PICT): Towards an ecologically relevant risk assessment of chemicals in aquatic systems. *Freshwater Biology, 61*, 2141–2151.

Tlili, A., Jabiol, J., Behra, R., Gil-Allué, C., & Gessner, M. O. (2017). Chronic exposure effects of silver nanoparticles on stream microbial decomposer communities and ecosystem functions. *Environmental Science & Technology, 51*, 2447–2455.

Treton, C., Chauvet, E., & Charcosset, J.-Y. (2004). Competitive interaction between two aquatic hyphomycete species and increase in leaf litter breakdown. *Microbial Ecology, 48*, 439–446.

Vellend, B. M. (2010). Conceptual synthesis in community ecology. *The Quarterly Review of Biology, 85*, 183–206.

Woodward, G., Gessner, M. O., Giller, P. S., Gulis, V., Hladyz, S., Lecerf, A., Malmqvist, B., McKie, B. G., Tiegs, S. D., Cariss, H., Dobson, M., Elosegi, A., Ferreira, V., Graça, M. A. S., Fleituch, T., Lacoursière, J. O., Nistorescu, M., Pozo, J., Risnoveanu, G., … Chauvet, E. (2012). Continental-scale effects of nutrient pollution on stream ecosystem functioning. *Science, 336*, 1438–1440.

Wurzbacher, C., Grimmett, I. J. & Bärlocher, F. (2015) Metabarcoding-based fungal diversity on coarse and fine particulate organic matter in a first-order stream in Nova Scotia, Canada. *F1000Research, 4*, 1378–1378.

Chapter 10
The Role of Macroinvertebrates on Plant Litter Decomposition in Streams

Micael Jonsson and Ryan A. Sponseller

Abstract Macroinvertebrate detritivores (i.e., shredders) in freshwaters are often a main driver of decomposition rates of terrestrial plant litter. Yet, the extent to which shredders drive this process depends on the specific functional traits and species present in the shredder community, which in turn are determined by the broader species pool, as well as a range of local environmental conditions, such as pH, substrate characteristics, water chemistry, water temperature, and current velocity. Projected global change will modify several of these environmental conditions, with potential consequences for litter decomposition rates and overall carbon cycling in freshwaters. In this chapter, we describe how a range of freshwater environmental conditions determines the presence of certain species (i.e., functional traits) and the characteristics of shredder communities (i.e., species composition and richness). We then discuss how these characteristics in turn may influence interactions among shredders, and between shredders and other freshwater organisms, to determine their influence on litter decomposition in streams.

10.1 Introduction

Litter-associated macroinvertebrates (i.e., shredders) are represented by a range of species, which are mainly insects in the orders Diptera, Plectoptera, and Trichoptera, but also include some crustaceans and molluscs that can locally occur in high densities and, as opposed to insect shredders, often have fully aquatic life cycles. Early studies on terrestrial plant litter and shredders in freshwater systems found clear positive associations between standing litter stock and shredder abundance (Anderson & Sedell, 1979; Cummins et al., 1973; Short et al., 1980). Yet, it was not until the landmark experiment by Wallace et al. (1982) that the direct role of shredders for

M. Jonsson (✉) · R. A. Sponseller
Department of Ecology and Environmental Science, Umeå University, Umeå, Sweden
e-mail: micael.jonsson@umu.se

R. A. Sponseller
e-mail: ryan.sponseller@umu.se

C. M. Swan et al. (eds.), *The Ecology of Plant Litter Decomposition in Stream Ecosystems*, https://doi.org/10.1007/978-3-030-72854-0_10

plant litter decomposition was confirmed. Here, application of insecticides to an entire headwater reach resulted in a large reduction in shredder abundance, which dramatically reduced conversion of terrestrial leaf litter (CPOM; coarse particulate organic matter) to small particles (FPOM; fine particulate organic matter) and downstream transport of FPOM (Wallace et al., 1982). During this time, parallel energetic studies also revealed that invertebrate shredders have very low assimilation efficiency (Cummins & Klug, 1979; Golladay et al., 1983; McDiffet, 1970), and that their secondary production and respiration contribute little to the broader ecosystem energy budget (Fisher & Likens, 1973). Collectively, this research suggested that the role of shredders in the litter decomposition process is mainly in the conversion, via fragmentation, of CPOM to FPOM. This functional role is nevertheless critical, as it facilitates overall decomposition of terrestrially derived plant material (Cummins et al., 1989; Mulholland et al., 1985; Villanueva et al., 2012; Webster & Benfield, 1986), increases the availability of litter-based resources to other freshwater organisms (Cummins et al., 1973; Wallace & Webster, 1996), and underpins longitudinal connectivity in river systems (e.g., via FPOM transport; Wallace et al., 1982). As such, studies during this period set the stage for an actively developing and exciting research field over the coming decades (Graça, 2001; Marks, 2019; Tank et al., 2010).

Freshwater environmental conditions interact with regional pool of available species to determine which species (i.e., functional traits) and community characteristics (i.e., species composition and richness) are present or absent locally at any one site (Bonada et al., 2007; Jonsson et al., 2017; Poff, 1997; Poff et al., 2006). These characteristics in turn influence the rate at which the macroinvertebrate shredder community decomposes litter (Dangles & Malmqvist, 2004; Gessner et al., 2010; Jonsson & Malmqvist, 2000; McKie et al., 2008). Hence, altered environmental conditions will likely modify litter decomposition rates via changes in shredder community composition, with consequences for the role that shredders have for overall litter turnover rates. In this chapter, we will describe how certain traits may be present or absent in (or differ in abundance among) shredder communities due to variation in local environmental conditions. We will then go on to describe how such variation in community characteristics may regulate litter processing rates and trophic links between shredders and other freshwater functional feeding groups. As a synthesis, we present possible scenarios as to how predicted global change (i.e., changes in climate and land use) can affect litter decomposition in fresh waters via impacts on shredder communities, and will do so for tropical, temperate, boreal, and Arctic biomes.

All types of freshwater systems may contain macroinvertebrate shredder species, but their role is greatest in ecosystems that receive substantial seasonal inputs of terrestrial (e.g., riparian) plant material relative to the area of aquatic habitat, which is mostly in small to mid-sized streams surrounded by well-developed deciduous, riparian vegetation. Thus, while other types of riverine systems, as well as lakes and ponds, can receive terrestrial litter input and therefore may house shredders, we focus here on macroinvertebrate shredders and litter decomposition in small to mid-sized streams that are forested, which also represent the type of freshwater systems where most research relevant to this topic has been carried out.

10.2 Macroinvertebrate Shredder Functional Traits

Each species that can be classified as a freshwater macroinvertebrate shredder exhibits unique functional traits, or rather a set of traits, that make it more or less likely to exist under certain environmental conditions (Poff et al., 2006), and that determine its role in the litter decomposition process. In general, as all species classified as shredders per definition feed, at least partially, on plant litter, and their life cycle is often intimately tied to seasonal pulses in litter resource availability. In temperate and boreal systems, this means that most shredders time their presence and growth as larvae with autumn leaf senescence and subsequent peaks in litter input and increases in standing stocks (Richardson, 1991; Wallace et al., 1999). However, while this is true for insect shredders, other shredder/detritivorous taxa, such as crustaceans and gastropods, that are present throughout the year, are less responsive to seasonal variation in litter availability, but instead show a high level of feeding plasticity by foraging also on other types of food resources (MacNeil et al., 1997; Moore, 1975).

Plasticity in feeding traits can, however, also be found among insect shredders. In particular, the strategy to shred plant material for food can be mixed with scraping surfaces or collecting FPOM (Cummins & Klug, 1979). The level of plasticity, or the extent to which shredder species use another feeding strategy than shredding, can change with development (i.e., ontogenetic diet shifts: Feminella & Stewart, 1986; Tierno de Figueroa & López-Rodriguez, 2019), or with variation in water chemistry (e.g., pH: Dangles, 2002; Ledger & Hildrew, 2000). Further, it is possible that the inherently low quality of plant litter, and the additionally, successively (seasonally) diminishing quality of litter standing stocks (Chauvet, 1987; Gessner & Chauvet, 1994), promote a diet that includes also higher-quality, autochthonous resources, such as algae (Brett et al., 2017; Jonsson & Stenroth, 2016; Moore, 1975) as well as predation (Dangles, 2002). The reality of these dietary choices complicates the use of traditional, and overly simplistic, functional feeding group designations (Mihuc, 1997). Further, to understand the role of macroinvertebrates for litter decomposition, and how this role may be altered under changing environmental conditions, this potential flexibility in resource use has to be considered.

Shredders also exhibit traits that are directly related to variation in abiotic conditions, such as water chemistry (e.g., pH and nutrient concentrations), water temperature, current velocity, and bottom substrate complexity and grain size. Thus, depending on the local abiotic conditions, different shredder communities are found (Jonsson et al., 2017; Malmqvist & Mäki, 1994; Poff, 1997), and it is therefore to some extent possible to predict community characteristics in a particular freshwater habitat, based on prevailing, local abiotic conditions. For example, strong environmental filters are exerted directly by pH and nutrient concentrations (i.e., level of eutrophication) and, thus, indirectly by land use and land cover that shape water chemistry (e.g., Jonsson et al., 2017). Across a gradient in pH, euholognathan stoneflies tend to dominate in more acidic streams, while trichopterans and dipterans are less common, and crustaceans are very scarce (Dangles & Guérold, 1999). Conversely, in streams of higher pH, stonefly abundance is often lower, and crustaceans and other acid-sensitive species are more abundant (Dangles & Guérold, 1999; Griffith & Perry, 1993). Across a gradient in eutrophication, a similar—but opposite—change in community composition is typically observed (Woodiwiss, 1964). This is because stoneflies in general are sensitive to the low oxygen levels resulting from organic pollution (e.g., from agricultural runoff) and the subsequent high microbial oxygen consumption (Hilsenhoff, 1988). Moreover, some Trichoptera groups are fairly tolerant to low oxygen levels, and crustaceans tolerate, and often dominate, under these conditions (Metcalfe, 1994).

High oxygen (O_2) demand results in stoneflies and some other taxa being more abundant and species rich in colder waters at higher latitudes (and altitudes) when compared to more southern (and/or lowland) streams with higher temperatures, where O_2 saturation is often lower (Verberk et al., 2011). High water velocity promotes oxygenation, and is therefore an environment where more O_2 demanding species can be found, but can also in itself create habitats that are suitable for some (i.e., rheophilic) taxa and an obstacle to others, shaping communities across a gradient from slow- to fast-flowing water (Hart & Finelli, 1999). However, the impact of water velocity on a shredder community can interact with bottom substrate type and complexity (Huryn & Wallace, 1987). For example, high substrate complexity or large grain sizes may moderate potentially adverse effects of a fast current by creating refugia of lower current velocities (Franken et al., 2006). Because of this, and because substrates form the main living space for benthic macroinvertebrate communities, bottom substrate characteristics are important determinants of shredder community composition (Reice, 1980; Sponseller & Benfield, 2001; Williams & Mundie, 1978). In addition, bottom substrate characteristics influence stream retentiveness of terrestrial plant litter input (Ehrman & Lamberti, 1992; Lepori & Malmqvist, 2005), a pre-requisite for whether a rich and abundant shredder community can be found or not (Haapala et al., 2003; Richardson, 1991; Wallace et al., 1999).

10.3 Inter- and Intraspecific Interactions

The low assimilation efficiency of shredders reflects the inherently low quality of plant detritus as a food source. Although some studies have shown that detritivores can assimilate up to 40% of ingested plant biomass, others have found that the conversion rate of ingested leaf litter to shredder biomass more often is $\leq$ 20% (Golladay et al., 1983; McDiffet, 1970). Due to the low resource quality of leaf detritus, shredders are highly dependent on microbial colonization on and within the leaf tissue, as the microbes (primarily fungi) improve the nutritional quality to shredders (Bärlocher, 1985; Cummins & Klug, 1979). Accordingly, studies have shown that microbial colonization of leaf litter increases shredder assimilation efficiency considerably (Cummins & Klug, 1979; Golladay et al., 1983). Nevertheless, the generally low assimilation efficiency also means that feces produced are quite similar to the original detrital resource in terms of nutrient content. Frass and feces from shredder leaf consumption can therefore serve as an important pre-processed food resource to other shredders, collectors, and filter feeders (Dieterich et al., 1997; Grafius & Anderson, 1979; Jonsson & Malmqvist, 2005; Patrick, 2013; Short & Maslin, 1977). Thus, interactions between shredders and leaf-associated microbes are critically important for shredder secondary production, the availability and quality of litter-based resources to other freshwater organisms, and, thus, for overall plant litter processing in streams.

Interactions among shredder individuals within a shredder community may amplify or reduce their impacts on litter decomposition rates. For example, rates of litter decomposition have been found to decrease with increasing shredder density, due to strong interference competition (Jonsson & Malmqvist, 2003). In natural systems, such effects are likely absent initially when resources are abundant soon after leaf senescence, but may become increasingly apparent as the litter resource gradually is fragmented and consumed (Jonsson, 2006). Moreover, due to different species utilizing separate niches (i.e., 'niche complementarity'), competition is often weaker among species than within species (Loreau & Hector, 2001). Thus, total amount of interference competition may be lower in a species-rich community compared to in a community that consists of only one or a few species total (Gessner et al., 2010; Jonsson & Malmqvist, 2000, 2003, but see McKie et al., 2009). Hence, if shredder species are lost, litter decomposition rates may decrease despite compensatory increases in the abundance of remaining species, due to overall increased levels of interference (i.e., resource) competition (Jonsson & Malmqvist, 2000, 2003). Changes in shredder species richness can therefore alter their role as drivers of litter decomposition rates.

Different feeding modes, such as scraping the leaf surface to selectively consume fungal biomass, or ingesting pieces of the leaf matrix together with fungal biomass (Bloor, 2011), are a key aspect of niche complementary among invertebrate shredders. For example, isopods and stoneflies have mouthparts that are more suitable for scraping surfaces than biting bits off a leaf, as many trichopterans do (Graça et al., 1993; Jonsson et al., 2002). Such differences in feeding behavior may create situations of apparent niche complementarity, or cases where facilitation among species occur (Giller et al., 2004). Hence, a higher number of shredder species should, on average, result in higher decomposition rates (Gessner et al., 2010). These differences in feeding modes among distantly related shredder taxa are likely also the mechanistic explanation as to why mixing litter from different plant species may increase decomposition rates (Santonja et al., 2020; Swan & Palmer, 2006; Tonin et al., 2018). However, more subtle niche complementarity—whatever it may be—among closely related species (e.g., within the same family or genus) can also result in higher per-capita litter processing rates in mixed communities than for single species, if it lowers competition or promotes facilitative interactions (Jonsson & Malmqvist, 2000, 2003; McKie et al., 2008).

Changes in decomposition rates caused by a change in the shredder community will likely have consequences also for other organisms, such as litter-associated microbes (via nutrient excretion; Mulholland et al., 1985; Villanueva et al., 2012), filter feeders and collectors (via particle production; Dieterich et al., 1997; Grafius & Anderson, 1979; Jonsson & Malmqvist, 2005; Patrick, 2013; Short & Maslin,

1977, but see Heard & Richardson, 1995; Jonsson et al., 2018), and predators (via prey availability; Peckarsky, 1982). However, how a change in shredder community composition influences other freshwater organisms via altered litter processing rates has rarely been studied (but see Jonsson & Malmqvist, 2005; Patrick, 2013). Moreover, despite several studies showing that shredder species richness is important for rates of litter decomposition, there is ample evidence that the presence of particular shredder species, rather than a change in species richness per se, sometimes can be at least as important for rates of litter decomposition (Boyero et al., 2014; Dudgeon & Gao, 2010; Perkins et al., 2010; Santonja et al., 2018), indicating that dominant functional traits rather than shredder diversity per se (i.e., the 'mass ratio hypothesis'; Grime, 1998) determines litter mass loss (Creed et al., 2009; Stoker et al., 2017). Hence, future research should consider the importance of dominant traits in shredder communities rather than merely species richness, how environmentally induced variation in these trait values results in altered rates of decomposition, and what consequences this has for microbes and other invertebrate guilds.

10.4 Impacts of Global Change on Litter Decomposition via Effect on Invertebrate Shredders

In the face of current and future global change, freshwaters are among the most threatened ecosystems. In addition to potential direct and indirect effects of predicted climate change (IPCC, 2007; Moss et al., 2009; Settele et al., 2014), a long list of other anthropogenic changes, including different types of land uses, will continue to impact freshwater systems and their biodiversity in many ways (Dudgeon et al., 2006). These impacts will alter the rates of ecosystem processes, and in many cases lead to impaired ecosystem functioning (Dudgeon, 2010). Below, we explore how different types of global change may impact rates of litter decomposition via influences on invertebrate shredders (see also Table 10.1).

10.4.1 Warming

Global air temperature is expected to increase by 2–5 °C by the end of the twenty-first century, mainly due to effects of greenhouse gas emissions from human activities (IPCC, 2007). However, these warming trends will not be uniform globally. Instead, northern regions (i.e., the boreal and Arctic) are predicted to experience the greatest future temperature change, whereas considerably smaller increases may be observed in the tropics (IPCC, 2007; Settele et al., 2014). While stream water temperatures are regulated by a complex set of drivers (groundwater, shading, etc.), there is reason to expect that warmer air temperatures will increase water temperature, at least for some portion of the year (Morrill et al., 2005; Webb & Nobilis,

Table 10.1 Hypothesized effects of global change on shredder communities, and subsequent changes in plant litter decomposition rates due to altered importance of the shredder community, in four different biomes, each with expected biome-specific global changes. The hypothesized change ('Effects on decomposition rates') is based on the overview in the above text and cited literature within that text, and the presumed changed importance of shredders plant litter decomposition in fresh waters from before to after the impact of global change. The number of '+' or '–' represents the hypothesized strength of the effect

Biome	*Global change*	*Effect on freshwater system*	*Effect on the shredder community*	*Effect on decomposition rates*
Tundra	Warming	Increased water temperature	Reduced psychrophile species	– –
			Increased metabolic rates	+
	Shrubification	Increased litter input	Increased shredder abundance	+ +
Boreal forest	Warming	Increased water temperature	Reduced psychrophile species	– –
		Increased drought occurrence	Reduced shredder biomass and richness	–
	Precipitation	Increased flood stochasticity	Reduced shredder biomass and richness	–
			Removal of litter input	– –
		Increased N and dissolved C	Increased microbial biomass	+ +
	Forestry	Reduced litter quality	Reduced shredder biomass and richness	– – –
Temperate forest	Warming	Increased anoxic conditions	Reduced shredder biomass and richness	– – –
			Reduced microbial biomass	– –
		Increased drought occurrence	Reduced shredder biomass and richness	– – –
			Disconnect between land and water	– – –
	Precipitation	Increased flash floods	Reduced shredder biomass and richness	– – –
			Removal of litter input	– – –

(continued)

Table 10.1 (continued)

Biome	*Global change*	*Effect on freshwater system*	*Effect on the shredder community*	*Effect on decomposition rates*
	Agriculture	Reduced flows	Reduced shredder biomass and richness	– – –
			Disconnect between land and water	– – –
Tropical rain forest	Warming	Increased water temperature	Increased metabolic rates	+
	Precipitation	Reduced flows	Reduced shredder biomass and richness	–
			Disconnect between land and water	–
	Agriculture	Reduced flows	Reduced shredder biomass and richness	–
			Disconnect between land and water	– –

2007). As water temperature is a strong environmental filter that determines macroinvertebrate community composition (Jacobsen et al., 1997), such changes are likely to impact the distribution of freshwater organisms, their interactions, and thus the processes they carry out (Settele et al., 2014). Hence, warming-induced changes in shredder community composition will likely alter intra- and interspecific interactions, including the presence and strengths of facilitation and effects of niche complementarity, and interactions among different types of organisms that are associated with litter processing or products thereof. For example, warming has been shown to weaken facilitation between macroinvertebrate and microbial decomposers, presumably via increased metabolic demands and reduced nutrient excretion by the macroinvertebrates (Bernabé et al., 2018).

Warming of freshwaters will inevitably have the largest adverse impacts on cold-loving (i.e., psychrophile) macroinvertebrate species at higher latitudes, and if these species are important shredders, also the litter decomposition process will be severely affected (e.g., Perkins et al., 2010). As increases in temperature are likely to be greatest in high-latitude ecosystems, which often have species-poor communities dominated by only a few species that are adapted to colder conditions (e.g., stoneflies; Irons et al., 1994; Jacobsen et al., 1997; Li & Dudgeon, 2009; Masese et al., 2014), the impact of warming on the role of shredders for litter decomposition may be most pronounced in these systems (Table 10.1). Hence, in northern regions, the major effect of warming will likely be a changed shredder community composition due to taxon-specific temperature preferences and responses in metabolic rates to warming in relation to available resources (i.e., starvation; Perkins et al., 2010;

Sweeney, 1978; Sweeney & Schnack, 1977). The impact on litter processing may, however, be alleviated if more southern shredder species expand their ranges northward to fill vacant niches. Yet, the extent to which such species replacement can take place will depend on other environmental filters (e.g., local pH), geographic barriers, the rate at which warming occurs, as well as the migratory ability of the southern, more thermophilic species (Bilton et al., 2001). Moreover, climate-change induced alterations of riparian vegetation may counteract adverse effects of warming or even promote the role of shredders for litter decomposition in streams (Jonsson & Canhoto, 2017; Wondzell et al., 2019).

10.4.2 Climate-Induced Changes in Vegetation

Warming will also gradually change the terrestrial plant community composition (e.g., conifers will be replaced by broadleaf species; e.g., Walther et al., 2002) and functional trait representation, which in itself will further alter productivity on land as well as the quantity and quality of litter supplied to fresh waters during leaf senescence (Kominoski et al., 2013). Further, as tree species differ in phenology of leaf senescence (Dixon, 1976; Eckstein et al., 1999), and as phenology is coupled with litter quality (Campanella & Bertiller, 2008; Niinemets & Tamm, 2005), a gradual change in plant community composition in response to warming will also alter the temporal resource availability to shredders, and the quality of those resources (Jonsson & Canhoto, 2017). The most dramatic warming-induced shifts in vegetation are predicted to occur at high latitudes (i.e., in the tundra) and altitudes (i.e., above the current tree line), as trees will expand into these previously open areas, or, conversely, in regions that become too warm and dry for trees to persist (Table 10.1; Chen et al., 2011; Walther et al., 2002; Zhang et al., 2013). In the former situation, an increased shredder abundance, and thus an increased role of shredders for plant litter decomposition, may be expected, as the availability of litter resources will increase, whereas in the latter situation, shredders likely are lost, or severely reduced in abundance, reducing their role for plant litter processing.

Changes in terrestrial net primary productivity (NPP) are also expected in response to warming, especially in northern regions (i.e., boreal and Arctic) and at high elevations (e.g., Gao et al., 2013), as this is where temperature increases will be the greatest (IPCC, 2007; Settele et al., 2014). Mean annual temperature and NPP are generally positively correlated (Huston & Wolverton, 2009), so increased NPP as a consequence of climate warming is expected. However, lower latitudes may experience reduced NPP due to increasingly dry conditions caused by higher temperatures (IPCC, 2007; Settele et al., 2014; Walther et al., 2002). These potential changes in terrestrial NPP are important because this is tightly coupled to leaf litter production (Wardle et al., 2003), and therefore with amount of terrestrial plant litter that freshwater systems receive from the riparian zone, but also to shading that may counteract effects of warming on water temperature (Wondzell et al., 2019).

More subtle changes in litter quality may also be caused by warming, because in sufficiently warm and wet environments, where resources are abundant, plant strategies involve growing in height to escape intra- and interspecific competition for light (Hautier et al., 2009). This strategy requires allocation of resources to biomass production, and therefore results in lower investment into secondary compounds (i.e., defense against herbivores; Bazzaz et al., 1987; Coley, 1988), which then increases the quality (i.e., palatability) of the litter produced. However, contrastingly, higher concentrations of atmospheric carbon dioxide (CO_2), which is one main agent behind climate warming, and subsequent greater CO_2 uptake by vegetation, may result in poorer litter quality as the carbon (C)-to-nitrogen (N) ratio, as well as concentrations of lignin and phenolics, increase (Norby et al., 2001; Stiling & Cornelissen, 2007). In addition to differences in litter quality having strong effects on decomposition rates (Heal et al., 1997; Lidman et al., 2017; Ostrofsky, 1997), changes in litter quality may also exacerbate stochiometric mismatches between shredders and the litter resources (Norby et al., 2001; Tuchman et al., 2002). Such a change would have immediate consequences for freshwater secondary production, especially in combination with increased metabolic demands due to higher water temperature (Perkins et al., 2010; Sweeney, 1978; Sweeney & Schnack, 1977).

10.4.3 Direct and Indirect Effects of Changed Precipitation

Similar to the effects of increasing air temperature, precipitation patterns will change unevenly across the globe. Current models suggest that some regions will experience greater annual rainfall with positive effects on terrestrial NPP, whereas other areas are expected to experience lesser amounts with more severe and prolonged droughts and adverse effects on terrestrial NPP (Table 10.1; IPCC, 2007; Settele et al., 2014; Walther et al., 2002). Such effects on terrestrial NPP will in themselves affect the role of macroinvertebrates for plant litter decomposition in fresh waters, via changes in litter input quantity and quality (see 10.4.2). In addition, studies suggest that precipitation drives litter input dynamics to fresh waters in the tropics, whereas temperature in itself is a more important driver at higher latitudes (Tonin et al., 2017). However, greater stochasticity in precipitation, in terms of both amounts and frequency (Pendergrass et al., 2017), and thus frequency of floods and droughts in freshwater systems, is expected in a warmer climate (Trenberth, 2011). Hence, besides affecting NPP, altered precipitation may affect the role of shredders for litter processing by regulating litter input dynamics, and via impacts on the frequency and magnitude of floods, runoff of dissolved organic matter, and frequency and length of droughts.

Changes in both magnitude and frequency of floods due to climate change will influence freshwater macroinvertebrate community composition, and most invertebrate groups will exhibit a reduced abundance in response to altered flow regimes (Kakouei et al., 2018; McMullen & Lytle, 2012). In addition, spates due to extreme rainfall events can drive the exports of organic C and nutrients from freshwater systems, especially by reducing the retention of CPOM (Giling et al., 2015). Thus, if flooding events occur during or soon after leaf senescence, these will influence the spatial distribution of litter resources, with consequences for invertebrate shredders and other functional feeding groups that to some extent depend on this resource or products from the decomposition process (i.e., collectors and filter feeders). Such effects of high flow may be moderated by substrate characteristics that increase flow heterogeneity and promote litter retention, or be exacerbated by past human activities that have resulted in reduced structural complexity and increased channelization of streams (Ehrman & Lamberti, 1992; Koljonen et al., 2012; Lepori & Malmqvist, 2005).

Even moderately reduced flows may have large impacts on shredder-mediated litter decomposition rates, if they result in increased distance between riparian vegetation and the water body, reducing litter input and in-stream litter availability (Arroita et al., 2015; Giling et al., 2015), and thus shredder abundance (Richardson, 1991; Wallace et al., 1999). In the event of drastically reduced flows (i.e., extensive and prolonged droughts), invertebrate shredders (as well as other freshwater organisms) can be extremely vulnerable, as habitat conditions (e.g., oxygen levels and temperature) progressively deteriorate and habitats disappear (Bonada et al., 2007; Herbst et al., 2018). This habitat deterioration, in turn, will affect litter decomposition rates in streams where shredders are important actors in that process (Leberfinger et al., 2010; Monroy et al., 2016). It is important to note, however, that droughts (as well as spates) can have very different effects on the role of shredders for plant litter decomposition, depending on when during the year they occur and how they overlap with certain developmental stages of the locally important shredder species.

Runoff of dissolved organic matter (DOM), including critical nutrients (i.e., C, N, and phosphorus [P]), can stimulate litter-associated microbial biomass and activity in freshwater systems (Emilson et al., 2017), and thus increase the palatability of terrestrial plant litter and the rate at which it is decomposed by invertebrate shredders (sensu Heal et al., 1997; Rosemond et al., 2015). In regions where increased precipitation is expected as a consequence of climate change, freshwater systems will likely receive increased amounts of DOM from terrestrial runoff (Christensen et al., 2001; Larsen et al., 2011). In warmer regions, where increasingly dry conditions will have adverse effects of terrestrial vegetation, inputs of DOM may become more sporadic but of higher magnitude following rare, extreme rain episodes (Table 10.1; Alpert et al., 2002; Nunes et al., 2009). In addition to quantitative changes, DOM runoff may also change qualitatively, as a consequence of changed soil nutrient availability following climate-change induced alterations in plant physiology and community composition, and, thus, plant litter chemistry (Bazzaz et al., 1987; Niinemets & Tamm, 2005). Such qualitative changes may, as for quantitative changes, influence litter-associated microbial communities with consequences for litter palatability and the rate at which shredders decompose the plant litter.

Runoff of terrestrial organic matter may also reduce pH, which can be tightly coupled to concentrations of DOM (i.e., organic acidity). Thus, as opposed to positive effects of DOM via increased microbial biomass and litter palatability, runoff may also result in reduced shredder contributions to litter decomposition, and reduced overall decomposition rates, if important acid-sensitive shredder (and microbial) species are lost (Petrin et al., 2007; Schmera et al., 2013). However, in the boreal region, where precipitation and subsequent runoff are predicted to increase the most, many important shredder taxa are naturally tolerant to low pH (Dangles et al., 2004), so effects of DOM on shredder-mediated litter processing via changed shredder communities should be small in the north. However, besides bringing nutrients, or lowering the pH, runoff may also bring sediment other chemicals from land to water. The effect of such environmental change on the role of shredders for the litter decomposition process will depend on the causal agent (e.g., type of human activity) and the way in which catchment characteristics are altered (see 10.4.5).

10.4.4 Fire and Strong Winds

Climate change is expected to increase frequencies of forest fires and strong winds (Seidl et al., 2017). If large areas of a catchment are disturbed by either storm felling or fire, increased runoff (Verkaik et al., 2013) and subsequent effects on macroinvertebrate communities (Minshall, 2003) may persist until the forest has recovered. At smaller scales, fire and wind disturbance can result in increased inputs of dead wood, which in turn could alter water flow and promote retention of plant litter with positive effects on shredder abundance and their importance for litter decomposition. However, both fire and wind can also remove riparian vegetation and open the canopy, and thus change the resource base and the dominant functional feeding groups present in the freshwater system (Vannote et al., 1980). However, this effect is likely to be transient, provided that secondary succession proceeds in the absence of disturbance (Stone & Wallace, 1998). In fact, fire and wind disturbance may promote shredder abundance and shredder-mediated decomposition, if it allows for regeneration of early-successional deciduous vegetation, which produces higher-quality litter in a seasonal manner, as opposed to the often dominant late-successional, coniferous species.

10.4.5 Human Activities

Multiple types of land use will influence freshwater systems and their shredder communities in different ways (Table 10.1). Forestry, and in particular large-scale clear-cutting, affects vegetation and thus runoff of DOM (i.e., nutrients and pH) in a similar way as do large-scale disturbances (see 10.4.4), and may therefore mimic effects of forest secondary succession on freshwater invertebrate communities and plant litter decomposition. However, additional impacts on soils (i.e., damage from forestry machines and soil scarification to promote seedling growth and survival) create novel disturbance regimes, resulting in, for example, increased sediment inputs, with often adverse effects on freshwater invertebrate communities and litter decomposition rates (Gurtz & Wallace, 1984; Lecerf & Richardson, 2010).

Forestry also, typically, transforms the tree community composition, in favor of species (e.g., conifers and Eucalyptus) that produce lower-quality litter to freshwater systems (Ferreira et al., 2016; Laudon et al., 2011). Such changes in riparian vegetation is an important determinant of the presence of shredders; a reduced litter input quality will lessen the role of shredders in the decomposition process (Raposeiro et al., 2018), as shredders contribute more when litter is of higher quality, whereas microbes are more important for the decomposition of lower-quality litter (Hieber & Gessner, 2002; Raposeiro et al., 2018).

Effects of forestry may, however, also be small. In fact, macroinvertebrate abundance has in some cases been found to be higher in streams impacted by forestry, suggesting that other environmental filters, such as pH, override the impact of forestry on macroinvertebrate communities (Liljaniemi et al., 2002). This may be especially true for stream environments that are characterized by strongly limiting conditions in temperature and/or nutrients (i.e., in the boreal region, e.g., Lidman et al., 2017). Moreover, as suggested above, forestry may, in the absence of fire disturbance, emulate some beneficial aspects of natural disturbances, by creating young deciduous riparian vegetation that provide high-quality litter input to fresh waters and thus promote the abundance of shredders (Liljaniemi et al., 2002; McKie & Malmqvist, 2009). In any case, equivocal effects of forestry on freshwater macroinvertebrates may be due to the level of effects being mediated by other conditions, such as substrate type (Gurtz & Wallace, 1984), and will certainly differ among different management strategies. Thus, it is somewhat difficult to draw general conclusions as to how forestry affects the importance of invertebrate shredders for litter decomposition in freshwaters; these effects are likely transient in time and highly context dependent (Ferreira et al., 2016).

Agriculture can, in several ways, have large impacts on freshwater systems and their macroinvertebrate communities, and due to a growing human population, agricultural activities and their associated impacts are predicted to increase (Dudgeon et al., 2006; Laurance et al., 2014; Moss, 2008). When land is cultivated for agricultural purposes, there may be a complete removal of riparian vegetation, resulting in a more autochthonous resource base with subsequent changes in macroinvertebrate community composition (Allan, 2004; Vannote et al., 1980). Alternatively, there is a modified riparian plant community composition, which alters quantity and quality of litter input to streams (Stenroth et al., 2015), and thus likely the role of shredders for litter decomposition. However, one of the more dramatic impacts of agriculture is the runoff of nutrients, which stimulates microbial biomass and microbially mediated litter decomposition (Gulis & Suberkropp, 2003; Woodward, 2012), but thereby likely also litter palatability and shredder-mediated decomposition (Bärlocher, 1985; Cummins & Klug, 1979). However, this potentially positive effect on shredders may be counteracted, because reduced oxygen levels due to increased microbial activity have adverse impacts on some important shredder species, such as stoneflies, which therefore typically are absent in streams impacted by agriculture (Hilsenhoff, 1988;

Stenroth et al., 2015). Accordingly, effects of nutrient enrichment on litter decomposition have been found to be stronger in colder regions, suggesting that the initial importance of macroinvertebrates (higher in colder regions), and effects of nutrient enrichment on these, determine effects of nutrient enrichment on plant litter decomposition (Ferreira et al., 2014). Agricultural activities often also result in runoff of directly harmful substances, such as pesticides (Cooper, 1993; Willis & McDowell, 1982). These substances may influence the role of invertebrate shredders for litter decomposition, either by reducing litter palatability via impacts on the microbial community (i.e., microbial conditioning; Bärlocher, 1985; Cummins & Klug, 1979; Jonsson et al., 2015) or by directly affecting the shredder community (Liess & von der Ohe, 2005).

A reduction in microbial litter conditioning, independent of cause, may require compensatory feeding by shredders to maintain growth (Bärlocher, 1985; Cummins & Klug, 1979; Flores et al., 2014). Hence, despite adverse impacts on the microbial community, the importance of shredders for plant litter decomposition may *increase*. Conversely, the importance of shredders may *decrease* despite positive effects on the microbial community, e.g., via nutrient input, if consumption of less litter is required to sustain shredder growth (Zubrod et al., 2015). Hence, human-induced impacts on litter-associated microbial communities can either decrease or increase the role of shredders in the decomposition process, but the above described, unexpected effects are likely transient, as longer-term effects on per-capita feeding activity will act on shredder abundance. Accordingly, no effects on shredder activity from agricultural pesticides, despite reduced microbially mediated litter decomposition (Rasmussen et al., 2012), or compensatory feeding due to lower litter quality (Flores et al., 2014), will likely eventually result in reduced shredder contribution to the decomposition process (Bärlocher, 1985; Cummins & Klug, 1979).

Human activities also result in voluntary or involuntary introduction of nonnative species (Ricciardi, 2007), and these species may become invasive with potentially large impacts on native organisms and the processes they mediate (Ricciardi & Cohen, 2007; Mueller & Hellmann, 2008). With regard to plant litter decomposition in fresh waters, it is not well studied how introduced and invasive plant (Dangles et al., 2002) or shredder species may influence the role of shredders. Invasive crayfish are, however, a good example of how massive the effects of species introductions can be. Besides the signal crayfish (*Aphanomyces astaci*) being a carrier of the crayfish plague and, thus, reducing (or completely removing) populations of native crayfish (Strand et al., 2014), introduced and invasive crayfish species may impact litter decomposition and other processes in complex ways (Jackson et al., 2014; Turley et al., 2017). Thus, this area of global-change effects on litter decomposition in fresh waters definitely needs more research.

Besides the potentially large impact of each of the above presented global changes on freshwater macroinvertebrates and litter decomposition, freshwater systems are often influenced by several types of disturbances simultaneously (Dudgeon et al., 2006). Hence, it is difficult to predict consequences of global change in natural systems based on studies of isolated disturbance types (Jackson et al., 2016). Moreover, the ongoing loss of freshwater biodiversity (Dudgeon et al., 2006) may weaken the resistance and resilience of fresh waters to disturbances (i.e., the insurance hypothesis; Yachi & Loreau, 1999). For example, effects of an invasive terrestrial plant on shredder-mediated litter decomposition may differ depending on the diversity and composition of the shredder community feeding on litter from that plant (Dangles et al., 2002). Nonetheless, the large environmental variability that is inherent in many freshwater systems may have increased the tolerance of these systems to multiple disturbances, compared to more stable aquatic environments, such as marine systems (Jackson et al., 2016).

10.5 Conclusion

In summary, many types of global change have the potential to modify terrestrial and freshwater environmental conditions that will have consequences for shredder communities and their role as drivers of litter decomposition and overall organic matter dynamics in streams. These impacts are very likely to differ across biomes. Indeed, even the same type of global change, e.g., warming, will likely have different implications for freshwater systems depending on biome, resulting in different effects—in terms of magnitude and/or direction—on the role of shredders for rates of leaf litter decomposition (Table 10.1). For example, increases in water temperature due to climate change are expected to be much higher at northern latitudes than in the tropics (IPCC, 2007; Settele et al., 2014), resulting in losses of important psychrophile shredder species in the north, whereas tropical communities may remain intact. On the other hand, as a result of climate change, terrestrial vegetation may become more abundant in the north and at higher latitudes (Chen et al., 2011; Walther et al., 2002; Zhang et al., 2013), resulting in increased shading of streams and more terrestrial plant litter input; this may reduce water temperatures and promote shredder abundance and thus strengthen their importance as drivers of the decomposition process (e.g., Lagrue et al., 2011; Wondzell et al., 2019).

Overall, while global change may result in a weakened or strengthened role of macroinvertebrate shredders for plant litter decomposition in fresh waters, we hypothesize that the effect of warming will be small in the tropics, in part due to relatively low importance of shredders (as opposed to microbes) for litter decomposition in this biome (Li & Dudgeon, 2009), either negative or positive in the tundra and boreal regions, and the strongest—and only negative—in the temperate region, due to increased habitat fragmentation and deteriorating environment (Bonada et al., 2007; Herbst et al., 2018), and disconnected land–water systems (Arroita et al., 2015; Giling et al., 2015), resulting from greater drought frequencies from global warming and intensified human water use (Table 10.1). Nevertheless, how the role of shredders for organic matter processing in streams will be altered by current and future climate change is immensely difficult to predict. Moreover, other types of global-change drivers, such as land use, may also show biome-specific effects, but not in the same way as climate change. For example, effects of deforestation on plant litter availability and the shredder community may be more pronounced in the tropics than in northern regions, but this has yet not been comparatively studied. Therefore, future research on organic matter processing and C cycling in streams must consider the potentially altered role of shredders under changed environmental conditions, but at the same time also realize that alterations in this role will differ among different types of global change, and be specific depending on the biome that is studied (Table 10.1).

References

Allan, J. D. (2004). Landscapes and riverscapes: The influence of land use on stream ecosystems. *Annual Review of Ecology, Evolution, and Systematics, 35*, 257–284.

Alpert, P., Ben-Gai, T., Baharad, A., Benjamini, Y., Yekutieli, D., Colacino, M., Diodato, L., Ramis, C., Homar, V., Romero, R., Michaelides, S., & Manes, A. (2002). The paradoxical increase of Mediterranean extreme daily rainfall in spite of decrease in total values. *Geophysical Research Letters, 29*, 1536.

Anderson, N. H., & Sedell, J. R. (1979). Detritus processing by macroinvertebrates in stream ecosystems. *Annual Review of Entomology, 24*, 351–377.

Arroita, M., Aristi, I., Díez, J., Martinez, M., Oyarzun, G., & Elosegi, A. (2015). Impact of water abstraction on storage and breakdown of coarse organic matter in mountain streams. *Science of the Total Environment, 503–504*, 233–240.

Bärlocher, F. (1985). The role of fungi in the nutrition of stream invertebrates. *Botanical Journal of the Linnean Society, 91*, 83–94.

Bazzaz, F. A., Chiariello, N. R., Coley, P. D., & Pitelka, L. F. (1987). Allocating resources to reproduction and defense. *BioScience, 37*, 58–67.

Bernabé, T. N., de Omena, P. M., dos Santos, V. P., de Siqueira, V. M., de Oliveira, V. M., & Romero, G. Q. (2018). Warming weakens facilitative interactions between decomposers and detritivores, and modifies freshwater ecosystem functioning. *Global Change Biology, 24*, 3170–3186.

Bilton, D. T., Freeland, J. R., & Okamura, B. (2001). Dispersal in freshwater invertebrates. *Annual Review of Ecology and Systematics, 32*, 159–181.

Blooe, M. C. (2011). Dietary preference of *Gammarus pulex* and *Asellus aquaticus* during laboratory breeding programme for ecotoxicological studies. *International Journal of Zoology, 2011*, 294394.

Bonada, N., Rieradevall, M., & Prat, N. (2007). Macroinvertebrate community structure and biological traits related to flow permanence in a Mediterranean river network. *Hydrobiologia, 589*, 91–106.

Boyero, L., Cardinale, B.J., Bastian, M. & Pearson, R.G. (2014) Biotic vs. abiotic control of decomposition: a comparison of the effects of simulated extinctions and changes in temperature. *PLoS ONE, 9*, e87426.

Brett, M. T., Bunn, S. E., Chandra, S., Galloway, A. W. E., Guo, F., Kainz, M. J., Kankaala, P., Lau, D. C. P., Moulton, T. P., Power, M. E., Rasmussen, J. B., Taipale, S. J., Thorp, J. H., & Wehr, J. D. (2017). How important are terrestrial organic carbon inputs for secondary production in freshwater ecosystems? *Freshwater Biology, 62*, 833–853.

Campanella, M. V., & Bertiller, M. B. (2008). Plant phenology, leaf traits and leaf litterfall of contrasting life forms in the arid Patagonian Monte, Argentina. *Journal of Vegetation Science, 19*, 75–85.

Chauvet, E. (1987). Changes in the chemical composition of alder, poplar and willow leaves during decomposition in a river. *Hydrobiologia, 148*, 35–44.

Chen, I.-C., Hill, J. K., Ohlemüller, R., Roy, D. B., & Thomas, C. D. (2011). Rapid range shifts of species associated with high levels of climate warming. *Science, 333*, 1024–1026.

Christensen, J. H., Räisänen, J., Iversen, T., Bjørge, D., Christensen, O. B., & Rummukainen, M. (2001). A synthesis of regional climate change simulations—A Scandinavian perspective. *Geophysical Research Letters, 28*, 1003–1006.

Coley, P. D. (1988). Effects of plant growth rate and leaf lifetime on the amount and type of anti-herbivore defense. *Oecologia, 74*, 531–536.

Cooper, C. M. (1993). Biological effects of agriculturally derived surface water pollutants on aquatic systems—A review. *Journal of Environmental Quality, 22*, 402–408.

Creed, R. P., Cherry, R. P., Pflaum, J. R., & Wood, C. J. (2009). Dominant species can produce a negative relationship between species diversity and ecosystem function. *Oikos, 118*, 723–732.

Cuffney, T. F., Wallace, J. B., & Lugthart, G. J. (1990). Experimental evidence quantifying the role of benthic invertebrates in organic matter dynamics of headwater streams. *Freshwater Biology, 23*, 281–299.

Cummins, K. W., & Klug, M. J. (1979). Feeding ecology of stream invertebrates. *Annual Review of Ecology and Systematics, 10*, 147–142.

Cummins, K. W., Petersen, R. C., Howard, F. O., Wuycheck, J. C., & Holt, V. I. (1973). Utilization of leaf litter by stream detritivores. *Ecology, 54*, 336–345.

Cummins, K. W., Wilzbach, M. A., Gates, D. M., Perry, J. B., & Taliaferro, W. B. (1989). Shredders and riparian vegetation. *BioScience, 39*, 24–30.

Dangles, O. (2002). Functional plasticity of benthic macroinvertebrates: Implications for trophic dynamics in acid streams. *Canadian Journal of Fisheries and Aquatic Sciences, 59*, 1563–1573.

Dangles, O., & Guérold, F. (1999). Impact of headwater stream acidification on the trophic structure of macroinvertebrate communities. *International Review of Hydrobiology, 84*, 287–297.

Dangles, O., Jonsson, M., & Malmqvist, B. (2002). The importance of detritivore species diversity for maintaining stream ecosystem functioning following the invasion of a riparian plant. *Biological Invasions, 4*, 441–446.

Dangles, O., & Malmqvist, B. (2004). Species richness-decomposition relationships depend on species dominance. *Ecology Letters, 7*, 395–401.

Dangles, O., Malmqvist, B., & Laudon, H. (2004). Naturally acid freshwater ecosystems are diverse and functional: Evidence form boreal streams. *Oikos, 104*, 149–155.

Dieterich, M., Anderson, N. H., & Anderson, T. M. (1997). Shredder-collector interactions in temporary streams of western Oregon. *Freshwater Biology, 38*, 387–393.

Dixon, K. R. (1976). Analysis of seasonal leaf fall in north temperate deciduous forests. *Oikos, 27*, 300–306.

Dudgeon, D. (2010). Prospects for sustaining freshwater biodiversity in the 21st century: Linking ecosystem structure and function. *Current Opinion in Environmental Sustainability, 2*, 422–430.

Dudgeon, D., Arthington, A. H., Gessner, M. O., Kawabata, Z.-I., Knowler, D. J., Lévêque, C., Naiman, R. J., Prieur-Richard, A.-H., Soto, D., Stiassny, M. L. J., & Sullivan, C. A. (2006). Freshwater biodiversity: Importance, threats, status and conservation challenges. *Biological Reviews, 81*, 163–182.

Dudgeon, D., & Gao, B. W. (2010). Biodiversity and ecosystem functioning in a species-poor guild: A test using tropical stream detritivores. *Hydrobiologia, 652*, 329–336.

Eckstein, R. L., Karlsson, P. S., & Weih, M. (1999). Leaf life span and nutrient resorption as determinants of plant nutrient conservation in temperate-arctic regions. *New Phytologist, 143*, 177–189.

Ehrman, T. P., & Lamberti, G. A. (1992). Hydraulic and particulate matter retention in a 3rd-order Indiana stream. *Journal of the North American Benthological Society, 11*, 341–349.

Emilson, C. E., Kreutzweiser, D. P., Gunn, J. M., & Mykytczuk, N. C. S. (2017). Leaf-litter microbial communities in boreal streams linked to forest and wetland sources of dissolved organic carbon. *Ecosphere, 8*, e01678.

Feminella, J. W., & Stewart, K. W. (1986). Diet and predation by three leaf-associated stoneflies (Plecoptera) in an Arkansas mountain stream. *Freshwater Biology, 16*, 521–538.

Ferreira, V., Castagneyrol, B., Koricheva, J., Gulis, V., Chauvet, E., & Graça, M. A. S. (2014). A meta-analysis of the effects of nutrient enrichment on litter decomposition in streams. *Biological Reviews, 90*, 669–688.

Ferreira, V., Koricheva, J., Pozo, J., & Graça, M. A. S. (2016). A meta-analysis on the effects of changes in the composition of native forests on litter decomposition in streams. *Forest Ecology and Management, 364*, 27–38.

Fisher, S. G., & Likens, G. E. (1973). Energy flow in Bear Brook, New Hampshire: An integrative approach to stream ecosystem metabolism. *Ecological Monographs, 43*, 421–439.

Flores, L., Larrañaga, A., & Elosegi, A. (2014). Compensatory feeding of a stream detritivore alleviates the effects of poor food quality when enough food is supplied. *Freshwater Science, 33*, 134–141.

Franken, R. J. M., Batten, S., Beijer, J. A. J., Gardeniers, J. J. P., Scheffer, M., & Peeters, E. (2006). Effects of interstitial refugia and current velocity on growth of the amphipod *Gammarus pulex* Linnaeus. *Journal of the North American Benthological Society, 25*, 656–663.

Gao, Y., Zhou, X., Wang, Q., Wang, C., Zhan, Z., Chen, L., Yan, J., & Qu, R. (2013). Vegetation net primary productivity and its response to climate change during 2001–2008 in the Tibetan Plateau. *Science of the Total Environment, 444*, 356–362.

Gessner, M. O., & Chauvet, E. (1994). Importance of stream microfungi in controlling breakdown rates of leaf litter. *Ecology, 75*, 1807–1817.

Gessner, M. O., Swan, C. M., Dang, C. K., McKie, B. G., Bardgett, R. D., Wall, D. H., & Hättenschwiler, S. (2010). Diversity meets decomposition. *Trends in Ecology and Evolution, 25*, 372–380.

Giling, D. P., Mac Nally, R., & Thompson, R. M. (2015). How might cross-system subsidies in riverine networks be affected by altered flow variability? *Ecosystems, 18*, 1151–1164.

Giller, P. S., Hillebrand, H., Berninger, U.-G., Gessner, M. O., Hawkins, S., Inchausti, P., Inglis, C., Leslie, H., Malmqvist, B., Monaghan, M. T., Morin, P. J., & O'Mullan, G. (2004). Biodiversity effects on ecosystem functioning: Emerging issues and their experimental test in aquatic environments. *Oikos, 104*, 423–436.

Golladay, S. W., Webster, J. R., & Benfield, E. F. (1983). Factors affecting food utilization by a leaf shredding aquatic insect: Leaf species and conditioning time. *Holarctic Ecology, 6*, 157–162.

Graça, M. A. S. (2001). The role of invertebrates on leaf litter decomposition in streams – a review. *International Review of Hydrobiology, 86*, 383–393.

Graça, M. A. S., Maltby, L., & Calow, P. (1993). Importance of fungi in the diet of *Gammarus pulex* and *Asellus aquaticus* II. Effects on growth, reproduction and physiology. *Oecologia, 96*, 304–309.

Grafius, E., & Anderson, N. H. (1979). Population dynamics, bioenergetics, and the role of *Lepidostoma quercina* Ross (Trichoptera: Lepidostomatidae) in an Oregon woodland stream. *Ecology, 60*, 433–441.

Griffith, M. B., & Perry, S. A. (1993). Colonization and processing of leaf-litter by macroinvertebrate shredders in streams of contrasting pH. *Freshwater Biology, 30*, 93–103.

Grime, J. P. (1998). Benefits of plant diversity to ecosystems: Immediate, filter and founder effects. *Journal of Ecology, 86*, 902–910.

Gulis, V., & Suberkropp, K. (2003). Leaf litter decomposition and microbial activity in nutrient-enriched and unaltered reaches of a headwater stream. *Freshwater Biology, 48*, 123–134.

Gurtz, M. E., & Wallace, J. B. (1984). Substrate-mediated response of stream invertebrates to disturbance. *Ecology, 65*, 1556–1569.

Hautier, Y., Niklaus, P. A., & Hector, A. (2009). Competition for light causes plant biodiversity loss after eutrophication. *Science, 324*, 636–638.

Haapala, A., Moutka, T., & Laasonen, P. (2003). Distribution of benthic macroinvertebrates and lead litter in relation to streambed retentivity: Implications for headwater stream restoration. *Boreal Environmental Research, 8*, 19–30.

Hart, D. D., & Finelli, C. M. (1999). Physical-biological coupling in streams: The pervasive effects of flow on benthic organisms. *Annual Review of Ecology and Systematics, 30*, 363–395.

Heal, O. W., Anderson, J. M., & Swift, M. J. (1997). Plant litter quality and decomposition: A historical overview. In G. Cadish & K. E. Giller (Eds.), *Driven by nature: Plant litter quality and decomposition* (pp. 3–30). CAB International.

Heard, S. B., & Richardson, J. S. (1995). Shredder-collector facilitation in stream detrital food webs: Is there enough evidence? *Oikos, 72*, 359–366.

Heino, J., Muotka, T., & Paavola, R. (2003). Determinants of macroinvertebrate diversity in headwater streams: Regional and local influences. *Journal of Animal Ecology, 72*, 425–434.

Herbst, D. B., Cooper, S. D., Medhurst, R. B., Wiseman, S. W., & Hunsaker, C. T. (2018). Drought ecohydrology alters the structure and function of benthic invertebrate communities in mountain streams. *Freshwater Biology, 64*, 886–902.

Hieber, M., & Gessner, M. O. (2002). Contribution of stream detritivores, fungi, and bacteria to leaf breakdown based on biomass estimates. *Ecology, 83*, 1026–1038.

Hilsenhoff, W. L. (1988). Rapid field assessment of organic pollution with a family-level biotic index. *Journal of the North American Benthological Society, 7*, 65–68.

Huryn, A. D., & Wallace, J. B. (1987). The exopterygote insect community of a mountain stream in North Caroline, USA: Life histories, production, and functional structure. *International Journal of Freshwater Entomology, 9*, 229–251.

Huston, M. A., & Wolverton, S. (2009). The global distribution of net primary production: Resolving the paradox. *Ecological Monographs, 79*, 343–377.

IPCC. (2007). *Climate change 2007: The physical science basis.* Contributions of Working Group I to the Fourth Assessment Report of the Intergovernmental Panel on Climate Change (Eds., S. Solomon, D. Qin, M. Manning, Z. Chen, M. Marquis, K.B. Averyt, M. Tignor and H.L. Miller, p. 996). Cambridge, UK, New York, NY, USA: Cambridge University Press.

Irons, J. G., III., Oswood, M. W., Stout, R. J., & Pringle, C. M. (1994). Latitudinal patterns in leaf litter breakdown: Is temperature really important? *Freshwater Biology, 31*, 401–411.

Jackson, M. C., Jones, T., Milligan, M., Sheath, D., Taylor, J., Ellis, A., England, J., & Grey, J. (2014). Niche differentiation among invasive crayfish and their impacts on ecosystem structure and functioning. *Freshwater Biology, 59*, 1123–1135.

Jackson, M. C., Loewen, C. J. G., Vinebrooke, R. D., & Chimimba, C. T. (2016). Net effects of multiple stressors in freshwater ecosystems: A meta-analysis. *Global Change Biology, 22*, 180–189.

Jacobsen, D., Schultz, R., & Enclada, A. (1997). Structure and diversity of stream invertebrate assemblages: The influence of temperature with altitude and latitude. *Freshwater Biology, 38*, 247–261.

Jonsson, M. (2006). Species richness effects on ecosystem functioning increase with time in an ephemeral resource system. *Acta Oecologica, 29*, 72–77.

Jonsson, M., Burrows, R. M., Lidman, J., Fältström, E., Laudon, H., & Sponseller, R. A. (2017). Land use influences macroinvertebrate community composition in boreal headwaters through altered stream conditions. *Ambio, 46*, 311–323.

Jonsson, M., & Canhoto, C. (2017). Climate change and freshwater invertebrates: Their role in reciprocal freshwater-terrestrial resource fluxes. In S. N. Johnson & T. H. Jones (Eds.), *Global climate change and terrestrial invertebrates* (pp. 274–294). Chichester, UK.

Jonsson, M., Dangles, O., Malmqvist, B., & Guérold, F. (2002). Simulating species loss following perturbation: Assessing the effects on process rates. *Proceedings of the Royal Society of London B, 269*, 1047–1052.

Jonsson, M., Ershammar, E., Fick, J., Brodin, T., & Klaminder, J. (2015). Effects of an antihistamine on carbon and nutrient recycling in streams. *Science of the Total Environment, 538*, 240–245.

Jonsson, M., & Malmqvist, B. (2000). Ecosystem process rates increases with animal species richness: Evidence from leaf-eating, aquatic insects. *Oikos, 89*, 519–523.

Jonsson, M., & Malmqvist, B. (2003). Mechanisms behind positive diversity effects on ecosystem functioning: Testing the facilitation and interference hypotheses. *Oecologia, 134*, 554–559.

Jonsson, M., & Malmqvist, B. (2005). Species richness and composition effects in a detrital processing chain. *Journal of the North American Benthological Society, 24*, 798–806.

Jonsson, M., Polvi, L. E., Sponseller, R. A., & Stenroth, K. (2018). Catchment properties predict autochthony in stream filter feeders. *Hydrobiologia, 815*, 83–95.
Jonsson, M., & Stenroth, K. (2016). True autochthony and allochthony in aquatic-terrestrial resource fluxes along a landuse gradient. *Freshwater Science, 35*, 882–894.
Kakouei, K., Kiesel, J., Domisch, S., Irving, K. S., Jähnig, S. C., & Kail, J. (2018). Projected effects of climate-change-induced alterations on stream macroinvertebrate abundances. *Ecology and Evolution, 8*, 3393–3409.
Koljonen, S., Louhi, P., Mäki-Petäys, A., Huusko, A., & Moutka, T. (2012). Quantifying the effects of in-stream habitat structure and discharge on leaf retention: Implications for stream restoration. *Freshwater Science, 31*, 1121–1130.
Kominoski, J. S., Follstad Shah, J. J., Canhoto, C., Fischer, D. G., Giling, D. P., & Gonz.lez, E., Griffiths, N.A., Larrañaga, A., LeRoy, C.J., Mineau, M.M., McElarney, Y.R., Shirley, S.M., Swan, C.M. & Tiegs, S.D. (2013). Forecasting functional implications of global changes in riparian plant communities. *Frontiers in Ecology and the Environment, 11*, 423–432.
Lagrue, C., Kominoski, J. S., Danger, M., Baudoin, J.-M., Lamothe, S., Lambrigot, D., & Lecerf, A. (2011). Experimental shading alters leaf litter breakdown in streams of contrasting riparian canopy cover. *Freshwater Biology, 56*, 2059–2069.
Larsen, S., Andersen, T., & Hessen, D. O. (2011). Climate change predicted to cause severe increase of organic carbon in lakes. *Global Change Biology, 17*, 1186–1192.
Laudon, H., Sponseller, R. A., Lucas, R. W., Futter, M. N., Egnell, G., Bishop, K., Ågren, A., Ring, E., & Högberg, P. (2011). Consequences of more intensive forestry for the sustainable management of forest soils and waters. *Forests, 2*, 243–260.
Laurance, W. F., Sayer, J., & Cassman, K. G. (2014). Agricultural expansion and its impacts on tropical nature. *Trends in Ecology & Evolution, 29*, 107–116.
Leberfinger, K., Bohman, I., & Herrmann, J. (2010). Drought impact on stream detritivores: Experimental effects on leaf litter breakdown and life cycles. *Hydrobiologia, 652*, 247–254.
Lecerf, A., & Richardson, J. S. (2010). Litter decomposition can detect effects of high and moderate levels of forest disturbance on stream conditions. *Forest Ecology and Management, 259*, 2433–2443.
Ledger, M. E., & Hildrew, A. G. (2000). Herbivory in an acid stream. *Freshwater Biology, 43*, 545–556.
Lepori, F., & Malmqvist, B. (2005). Effects of stream restoration on ecosystem functioning: Detritus retentiveness and decomposition. *Journal of Applied Ecology, 42*, 228–238.
Li, A. O., & Dudgeon, D. (2009). Shredders: Species richness, abundance, and role in litter breakdown in tropical Hong Kong streams. *Journal of the North American Benthological Society, 28*, 167–180.
Lidman, J., Jonsson, M., Burrows, R. M., Bundschuh, M., & Sponseller, R. A. (2017). Composition of riparian litter input regulates organic matter decomposition: Implications for headwater stream functioning in a managed forest landscape. *Ecology and Evolution, 7*, 1068–1077.
Liess, M., & von der Ohe, P. C. (2005). Analyzing effects of pesticides on invertebrate communities in streams. *Environmental Toxicology and Chemistry, 24*, 954–965.
Liljaniemi, P., Vuori, K.-M., Ilyashuk, B., & Luotonen, H. (2002). Habitat characteristics and macroinvertebrate assemblages in boreal forest streams: Relations to catchment silvicultural activities. *Hydrobiologia, 474*, 239–251.
Loreau, M., & Hector, A. (2001). Partitioning selection and complementarity in biodiversity experiments. *Nature, 412*, 72–76.
MacNeil, C., Dick, J. T. A., & Elwood, R. W. (1997). The trophic ecology of freshwater *Gammarus* spp. (Crustacea: Amphipoda): Problems and perspectives concerning the functional feeding group concept. *Biological Reviews, 72*, 349–364.
Malmqvist, B., & Mäki, M. (1994). Benthic macroinvertebrate assemblages in north Swedish streams: Environmental relationships. *Ecography, 17*, 9–16.
Marks, J.C. (2019) Revisiting the fates of dead leaves that fall into streams. *Annual Review of Ecology and Systematics, 50*, 24.1–24.22.

Masese, F. O., Kitaka, N., Kipkemboi, J., Gettel, G. M., Irvine, K., & McClain, M. E. (2014). Macroinvertebrate functional feeding groups in Kenyan highland streams: Evidence for a diverse shredder guild. *Freshwater Science, 33*, 435–450.

McDiffet, W. F. (1970). Transformation of energy by a stream detritivore, *Pteronarcys scotti* (Plecoptera). *Ecology, 51*, 975–988.

McKie, B. G., & Malmqvist, B. (2009). Assessing ecosystem functioning in streams affected by forest management: Increased leaf decomposition occurs without changes to the composition of benthic assemblages. *Freshwater Biology, 54*, 2086–2100.

McKie, B. G., Schindler, M., Gessner, M. O., & Malmqvist, B. (2009). Placing biodiversity and ecosystem functioning in context: Environmental perturbations and the effects of species richness in a stream field experiment. *Oecologia, 160*, 757–770.

McKie, B. G., Woodward, G., Hladyz, S., Nistorescu, M., Preda, E., Popescu, C., Giller, P. S., & Malmqvist, B. (2008). Ecosystem functioning in stream assemblages from different regions: Contrasting responses to variation in detritivore richness, evenness and density. *Journal of Animal Ecology, 77*, 495–504.

McMullen, L. E., & Lytle, D. A. (2012). Quantifying invertebrate resistance to floods: A global-scale meta-analysis. *Ecological Applications, 22*, 2164–2175.

Metcalfe, J. L. (1994). Biological water quality assessment of rivers: Use of macroinvertebrate communities. In P. Calow & G. E. Petts (Eds.), *The river handbook: Hydrological and ecological principles.* (pp. 144–170). Blackwell Scientific.

Mihuc, T. B. (1997). The functional trophic role of lotic primary consumers: Generalist versus specialist strategies. *Freshwater Biology, 37*, 455–462.

Minshall, G. W. (2003). Responses of stream benthic macroinvertebrates to fire. *Forest Ecology and Management, 178*, 155–161.

Monroy, S., Menédez, M., Basaguren, A., Pérez, J., Elosegi, A., & Pozo, J. (2016). Drought and detritivores determine leaf litter decomposition in calcareous streams of the Ebro catchment (Spain). *Science of the Total Environment, 573*, 1450–1459.

Moore, J. W. (1975). The role of algae in the diet of *Asellus aquaticus* L. and *Gammarus pulex* L. *Journal of Animal Ecology, 44*, 719–730.

Morrill, J., Bales, R., & Conklin, M. (2005). Estimating stream temperature from air temperature: Implications for future water quality. *Journal of Environmental Engineering, 131*, 139–146.

Moss, B. (2008). Water pollution by agriculture. *Philosophical Transactions of the Royal Society B, 363*, 659–666.

Moss, B., Hering, D., Green, A. J., Aidoud, A., Becares, E., Beklioglu, M., Bennion, H., Boix, D., Brucet, S., Carvalho, L., Clement, B., Davidson, T., Declerck, S., Dobson, M., van Donk, E., Dudley, B., Feuchtmayr, H., Friberg, N., Grenouillet, G., … Weyhenmeyer, G. A. (2009). Climate change and the future of freshwater biodiversity in Europe: A primer for policy-makers. *Freshwater Reviews, 2*, 103–130.

Mueller, J. M., & Hellmann, J. J. (2008). An assessment of invasion risk from assisted migration. *Conservation Biology, 22*, 562–567.

Mulholland, P. J., Elwood, J. W., Newbold, J. D., & Ferren, L. A. (1985). Effect of a leaf-shredding invertebrate on organic matter dynamics and phosphorus spiraling in heterotrophic laboratory systems. *Oecologia, 66*, 199–206.

Niinemets, Ü., & Tamm, U. (2005). Species differences in timing of leaf fall and foliage chemistry modify nutrient resorption efficiency in deciduous temperate forest stands. *Tree Physiology, 25*, 1001–1014.

Norby, R. J., Cotrufo, M. F., Ineson, P., O'Neill, E. G., & Canadell, J. G. (2001). Elevated CO_2, litter chemistry, and decomposition: A synthesis. *Oecologia, 127*, 153–165.

Nunes, J. P., Seixas, J., Keizer, J. J., & Ferreira, A. J. D. (2009). Sensitivity of runoff and soil erosion to climate change in two Mediterranean watersheds. Part I: Model parameterization and evaluation. *Hydrological Processes, 23*, 1202–1211.

Ostrofsky, M. L. (1997). The relationship between chemical characteristics of autumn-shed leaves and aquatic processing rates. *Journal of the North American Benthological Society, 16*, 750–759.

Patrick, C. J. (2013). The effect of shredder community composition on the production and quality of fine particulate organic matter. *Freshwater Science, 32*, 1026–1035.

Peckarsky, B. L. (1982). Aquatic insect predator–prey relations. *BioScience, 32*, 261–266.

Pendergrass, A. G., Knutti, R., Lehner, F., Deser, C., & Sanderson, B. M. (2017). Precipitation variability increases in a warmer climate. *Scientific Reports, 7*, 17966.

Perkins, D. M., McKie, B. G., Malmqvist, B., Gilmour, S. G., Reiss, J., & Woodward, G. (2010). Environmental warming and biodiversity-ecosystem functioning in freshwater microcosms: Partitioning the effects of species identity, richness and metabolism. *Advances in Ecological Research, 43*, 178–201.

Petersen, R. C., & Cummins, K. W. (1974). Leaf processing in a woodland stream. *Freshwater Biology, 4*, 343–368.

Petrin, Z., McKie, B., Buffam, I., Laudon, H., & Malmqvist, B. (2007). Landscape-controlled chemistry variation affects communities and ecosystem function in headwater streams. *Canadian Journal of Fisheries and Aquatic Sciences, 64*, 1563–1572.

Poff, N. L. (1997). Landscape filters and species traits: Towards mechanistic understanding and prediction in stream ecology. *Journal of the North American Benthological Society, 16*, 391–409.

Poff, N. L., Olden, J. D., Vieira, N. K. M., Finn, D. S., Simmons, M. P., & Kondratieff, B. C. (2006). Functional trait niches of North American lotic insects: Trait-based ecological applications in light of phylogenetic relationships. *Journal of the North American Benthological Society, 25*, 730–755.

Raposeiro, P. M., Ferreira, V., Gea, G., & Gonçalves, V. (2018). Contribution of aquatic shredders to leaf litter decomposition in Atlantic island streams depends on shredder density and litter quality. *Marine and Freshwater Research, 69*, 1432–1439.

Rasmussen, J. J., Wiberg-Larsen, P., Baatrup-Pedersen, A., Monberg, R. J., & Kronvang, B. (2012). Impacts of pesticides and natural stressors on leaf litter decomposition in agricultural streams. *Science of the Total Environment, 416*, 148–155.

Reice, S. R. (1980). The role of substratum in benthic macroinvertebrate microdistribution and litter decomposition in a woodland stream. *Ecology, 61*, 580–590.

Ricciardi, A. (2007). Are modern biological invasions an unprecedented form of global change? *Conservation Biology, 21*, 329–336.

Ricciardi, A., & Cohen, J. (2007). The invasiveness of an introduced species does not predict its impact. *Biological Invasions, 9*, 309–315.

Richardson, J. S. (1991). Seasonal food limitation of detritivores in a montane stream: An experimental test. *Ecology, 72*, 873–887.

Rosemond, A. D., Benstead, J. P., Bumpers, P. M., Gulis, V., Kominoski, J. S., Manning, D. W. P., Suberkropp, K., & Wallace, J. B. (2015). Experimental nutrient additions accelerate terrestrial carbon loss from stream ecosystems. *Science, 347*, 1142–1145.

Santonja, M., Pellan, L., & Piscart, C. (2018). Macroinvertebrate identity mediates the effects of litter quality and microbial conditioning on litter recycling in temperate streams. *Ecology and Evolution, 8*, 2542–2553.

Santonja, M., Rodríguez-Pérez, H., Le Bris, N., & Piscart, C. (2020). Leaf nutrients and macroinvertebrates control litter mixing effects on decomposition in temperate streams. *Ecosystems, 23*, 400–416.

Schmera, D., Erös, T., & Heino, J. (2013). Habitat filtering determines spatial variation of macroinvertebrate community traits in northern headwater streams. *Community Ecology, 14*, 77–88.

Seidl, R., Thom, D., Kautz, M., Martin-Benito, D., Peltoniemi, M., Vacchiano, G., Wild, J., Ascoli, D., Petr, M., Honkaniemi, J., Lexer, M. J., Trotsiuk, V., Mairota, P., Svoboda, M., Fabrika, M., Nagel, T. A., & Reyer, C. P. O. (2017). Forest disturbances under climate change. *Nature Climate Change, 7*, 395–402.

Settele, J., Scholes, R., Betts, R., Bunn, S., Leadley, P., Nepstad, D., Overpeck, J. T., & Taboada, M. A. (2014). Terrestrial and inland water systems. In C. B. Field, V. R. Barros, D. J. Dokken, K. J. Mach, M. D. Mastrandrea, T. E. Bilir, M. Chatterjee, K. L. Ebi, Y. O. Estrada, R. C. Genova,

B. Girma, E. S. Kissel, A. N. Levy, S. MacCracken, P. R. Mastrandrea, & L. L. White (Eds.), *Climate change 2014: Impacts, adaptation, and vulnerability. Part A: Global and sectoral aspects*. Contribution of Working Group II to the Fifth Assessment Report of the Intergovernmental Panel on Climate Change (pp. 271–359). Cambridge University Press.

Short, R. A., Canton, S. P., & Ward, J. V. (1980). Detrital processing and associated macroinvertebrates in a Colorado mountain stream. *Ecology, 61*, 727–732.

Short, R. A., & Maslin, P. E. (1977). Processing of leaf litter by a stream detritivore: Effect on nutrient availability to collectors. *Ecology, 58*, 935–938.

Sponseller, R. A., & Benfield, E. F. (2001). Influences of land use on leaf breakdown in southern Appalachian headwater streams: A multiple-scale analysis. *Journal of North American Benthological Society, 20*, 44–59.

Stenroth, K., Polvi, L. E., Fältström, E., & Jonsson, M. (2015). Land-use effects on terrestrial consumers through changed size structure of aquatic insects. *Freshwater Biology, 60*, 136–149.

Stiling, P., & Cornelissen, T. (2007). How does elevated carbon dioxide (CO_2) affect plant–herbivore interactions? A field experiment and meta-analysis of CO2-mediated changes on plant chemistry and herbivore performance. *Global Change Biology, 13*, 1823–1842.

Stoker, D., Flakner, A. J., Murray, K. M., Lang, A. K., Barnum, T. R., Hepinstall-Cymerman, J., Conroy, M. J., Cooper, R. J., & Pringle, C. M. (2017). Decomposition of terrestrial resource subsidies in headwater streams: Does consumer diversity matter? *Ecosphere, 8*, e01868.

Stone, M. K., & Wallace, J. B. (1998). Long-term recovery of a mountain stream from clearcut logging: The effects of forest succession on benthic invertebrate community structure. *Freshwater Biology, 39*, 151–169.

Strand, D. A., Jussila, J., Johnsen, S. I., Viljamaa-Dirks, S., Edsman, L., Wiik-Nielsen, J., Viljugrein, H., Engdahl, F., & Vrålstad, T. (2014). Detection of crayfish plague spores in large freshwater systems. *Journal of Applied Ecology, 51*, 544–553.

Swan, C. M., & Palmer, M. A. (2006). Preferential feeding by an aquatic consumer mediates non-additive decomposition of speciose leaf litter. *Oecologia, 149*, 107–114.

Sweeney, B. W. (1978). Bioenergetic and developmental response of a mayfly to thermal variation. *Limnology and Oceanography, 23*, 461–477.

Sweeney, B. W., & Schnack, J. A. (1977). Egg development, growth, and metabolism of *Sigara alternata* (Say) (Hemiptera: Corixidae) in fluctuating thermal environments. *Ecology, 56*, 265–277.

Tank, J. L., Rosi-Marshall, E. J., Griffiths, N. A., Entrekin, S. A., & Stephen, M. L. (2010). A review if allochthonous organic matter dynamics and metabolism in streams. *Journal of the North American Benthological Society, 29*, 118–146.

Tierno de Figueroa, J. M., & López-Rodriguez, M. J. (2019). Trophic ecology of Plecoptera (Insecta): A review. *the European Zoological Journal, 86*, 79–102.

Tonin, A. M., Gonçalves, J. F., Jr., Bambi, P., Couceiro, S. R. M., Feitoza, L. A. M., Fontana, L. E., Hamada, N., Hepp, L. U., Lezan-Kowalczuk, V. G., & LEite, G.F.M., Lemes-Silva, A.L., Lisboa, L.K., Loureiro, R.C., Martins, R.T., Medeiros, A.O., Morais, P.B., Moretto, Y., Oliveria, P.C.A., Pereira, E.B., Ferreira, L.P., Pérez, J., Petrucio, M.M., Reis, D.F., Rezende, R.S., Roque, N., Santos, L.E.P., Siegloch, A.E., Tonello, G. & Boyero, L. (2017). Plant litter dynamics in the forest-stream interface: Precipitation is a major control across tropical biomes. *Scientific Reports, 7*, 10799.

Tonin, A. M., Pozo, J., Monroy, S., Basaguren, A., Pérez, J., Gonçalves, J. F., Jr., Pearson, R. G., Cardinale, B. J., & Boyero, L. (2018). Interactions between large and small detritivores influence how biodiversity impacts litter decomposition. *Journal of Animal Ecology, 87*, 1465–1474.

Trenberth, K. E. (2011). Changes in precipitation with climate change. *Climate Research, 47*, 123–138.

Tuchman, N. C., Wetzel, R. G., Rier, S. T., Wahtera, K. A., & Teeri, J. A. (2002). Elevated atmospheric CO_2 lowers leaf litter nutritional quality for stream ecosystem food webs. *Global Change Biology, 8*, 163–170.

Turley, M. D., Bilotta, G. S., Gasparrini, A., Sera, F., Mathers, K. L., Humpheryes, I., & England, J. (2017). The effects of non-native signal crayfish (Pacifastacus leniusculus) on fine sediment and sediment-biomonitoring. *Science of the Total Environment, 601–602*, 186–193.
Vannote, R. L., Minshall, G. W., Cummins, K. W., Sedell, J. R., & Cushing, C. E. (1980). The river continuum concept. *Canadian Journal of Fisheries and Aquatic Sciences, 37*, 130–137.
Verberk, W. C., Bilton, D. T., Calosi, P., & Spicer, J. I. (2011). Oxygen supply in aquatic ectotherms: Partial pressure and solubility together explain biodiversity and size patterns. *Ecology, 92*, 1565–1572.
Verkaik, I., Rieradevall, M., Cooper, S. D., Melack, J. M., Dudley, T. L., & Prat, N. (2013). Fire as a disturbance in Mediterranean climate streams. *Hydrobiologia, 719*, 352–382.
Villanueva, V. D., Albariño, R., & Canhoto, C. (2012). Positive effect of shredders on microbial biomass and decomposition in stream microcosms. *Freshwater Biology, 57*, 2504–2513.
Wallace, J. B., Eggert, S. L., Meyer, J. L., & Webster, J. R. (1997). Multiple trophic levels of a forest stream linked to terrestrial litter inputs. *Science, 277*, 102–104.
Wallace, J. B., Eggert, S. L., Meyer, J. L., & Webster, J. R. (1999). Effects of resource limitation on a detrital-based ecosystem. *Ecological Monographs, 69*, 409–422.
Wallace, J. B., & Webster, J. R. (1996). The role of macroinvertebrates in stream ecosystem function. *Annual Review of Ecology and Systematics, 41*, 115–139.
Wallace, J. B., Webster, J. R., & Cuffney, T. F. (1982). Stream detritus dynamics: Regulation by invertebrate consumers. *Oecologia, 53*, 197–200.
Walther, G.-R., Post, E., Convey, P., Menzel, A., Parmesan, C., Beebee, T. J. C., Fromentin, J.-M., Hoegh-Guldberg, O., & Bairlein, F. (2002). Ecological responses to recent climate change. *Nature, 416*, 389–395.
Wardle, D. A., Hörnberg, G., Zackrisson, O., Kalela-Brundin, M., & Coomes, D. A. (2003). Long-term effects of wildfire on ecosystem properties across an island area gradient. *Science, 300*, 972–975.
Webb, B. W., & Nobilis, F. (2007). Long term changes in river temperature and the influence of climatic and hydrologic factors. *Hydrological Sciences Journal, 52*, 74–85.
Webster, J. R., & Benfield, E. F. (1986). Vascular plant breakdown in freshwater ecosystems. *Annual Review of Ecology and Systematics, 17*, 567–594.
Williams, D. D., & Mundie, J. H. (1978). Substrate size selection by stream invertebrates and the influence of sand. *Limnology and Oceanography, 23*, 1030–1033.
Willis, G. H., & McDowell, L. L. (1982). Pesticides in agricultural runoff and their effects on downstream water quality. *Environmental Toxicology and Chemistry, 1*, 267–279.
Wondzell, S. M., Diabat, M., & Haggerty, R. (2019). What matters most: Are future stream temperatures more sensitive to changing air temperatures, discharge, or riparian vegetation? *Journal of the American Water Resources Association, 55*, 116–132.
Woodiwiss, F. S. (1964). The biological system of stream classification used by the Trent River Board. *Chemistry and Industry, 14*, 443–447.
Woodward, G., Gessner, M. O., Giller, P. S., Gulis, V., Hladyz, S., Lecerf, A., Malmqvist, B., McKie, B. G., Tiegs, S. D., Cariss, H., Dobson, M., Elosegi, A., Ferreira, V., Graça, M. A. S., Fleituch, T., Lacoursière, J. O., Nistorescu, M., Pozo, J., Risnoveanu, G., …, Chauvet, E. (2012). Continental-scale effects of nutrient pollution on stream ecosystem functioning. *Science, 336*, 1438–1440.
Yachi, S., & Loreau, M. (1999). Biodiversity and ecosystem productivity in a fluctuating environment: The insurance hypothesis. *Proceedings of the National Academy of Sciences USA, 96*, 1463–1468.
Zhang, W., Miller, P. A., Smith, B., Wania, R., Koenigk, T., & Döscher, R. (2013). Tundra shrubification and tree-line advance amplify arctic climate warming: Results from an individual-based dynamic vegetation model. *Environmental Research Letters, 8*, 034023.
Zubrod, J. P., Englert, D., Rosenfeldt, R. R., Wolfram, J., Lüderwald, S., Wallace, D., Schnetzer, N., Schulz, R., & Bundschuh, M. (2015). The relative importance of diet-related and waterbourne effects of copper for a leaf-shredding invertebrate. *Environmental Pollution, 205*, 16–22.

Chapter 11
The Role of Protozoans and Microscopically Small Metazoans in Aquatic Plant Litter Decomposition

Julia Reiss

Abstract Protozoans and microscopically small invertebrates are omnipresent in aquatic systems but, arguably, under-studied, and this is true when it comes to one of the best described ecosystem processes in fresh water—leaf litter decomposition. However, over the last ten years research has been emerging that points to the important role of these microbes (such as ciliates and micro-crustaceans) as consumers of leaf prokaryotes and fungi. In fact, a decomposing leaf can be viewed as a micro-world of a complex food web with tiny organisms interacting with each other—fuelled by energy and nutrients from the leaf and the water interface. Evidence from the latter comes from biofilm studies that highlight the need to incorporate protozoans and micro-metazoans into our understanding of plant litter decomposition. Here, publications on laboratory and field studies are summarised, to show that protozoans and micro-metazoans are not only present on decomposing detritus, but that they can influence leaf litter decomposition. After presenting this empirical evidence, I conclude with some theoretical and more abstract thoughts on why these microbes have an important role in energy and nutrient cycling that links leaves as a substrate for biofilm with surrounding compartments such as the benthic zone and the open water.

11.1 Decomposing Leaves as 'Micro-Worlds'

Most school children will be able to tell you that leaves are decomposed by 'microbes' and some will even know that bacteria and fungi are two main groups that will contribute to the decay of dead plants. Clearly, decades of research on decomposition (Gessner et al., 2010) have 'paid off' in the sense that microbes are considered key in providing some of the most important ecosystem processes on this planet. However, ecologists working on microscopically small invertebrates (i.e., micro-metazoans),

---

J. Reiss (✉)
Department of Life Sciences, Whitelands College, Roehampton University, London SW15 4JD, UK
e-mail: julia.reiss@roehampton.ac.uk

C. M. Swan et al. (eds.), *The Ecology of Plant Litter Decomposition in Stream Ecosystems*, https://doi.org/10.1007/978-3-030-72854-0_11

and those working on protozoans, will tell you that the organisms they study have been largely ignored when it comes to providing a bigger picture of how energy and nutrients are recycled in ecosystems and this includes leaf litter decomposition in fresh waters. Protozoans and metazoans that are invisible to the naked eye are very biodiverse, omnipresent and abundant—so why then are they omitted from much leaf litter decomposition research? There are probably two main answers here: firstly, studying these microbes is time consuming field- and microscope work and requires expert taxonomic knowledge (this argument is not new (Robertson et al., 2000a)) but still holds true to some extent even after decades of ever advancing research on these groups (e.g., Peralta-Maraver et al., 2019; Poff et al., 1993; Reiss & Schmid-Araya, 2008). Secondly, assessing their role in leaf litter decomposition involves complicated field- and laboratory procedures, because their effects on leaf litter decomposition are probably largely indirect and often counterintuitive. Most protozoans and micro-metazoans will not feed on leaf material directly but on other microbes and influence leaf litter decomposition via top-down control of decomposers (this can stimulate the abundance of their prey even more), nutrient recycling and bioturbation.

In fact, a decomposing leaf is a micro-world (Fig. 11.1) that is an excellent representation of the complex interactions among microbes within, and on, biofilm and an example for 'small worlds that give big answers' (i.e., model systems that allow developing or testing of some general ecological concepts (Altermatt et al., 2015; Reiss, Forster, et al., 2010)). Microscopically small organisms in this micro-world

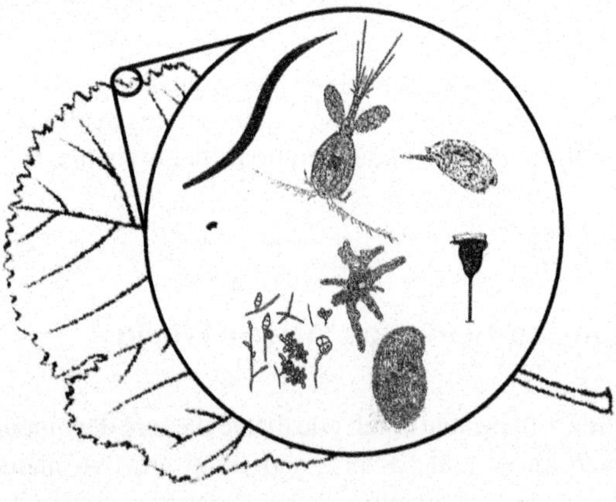

Fig. 11.1 A decomposing leaf as a 'micro-world' for microbes in, and on, leaf biofilm. Fungi and bacteria decompose the leaf and are a food sources for micro-grazers (protozoans and micro-metazoans). In tandem with single-celled algae their biomass and exudates (e.g., exopolymeric substances) create a three-dimensional environment on the surface of decomposing leaves. Tiny animal-like species dwell within—and above—this matrix; and they belong to the protozoans (flagellates, ciliates and amoebae) and micro-metazoans (shown here are nematodes, copepods and rotifers). Also shown are viruses as especially bacteriophages are omnipresent in water. Food chains in this 'micro-world' represent energy links from the leaf (and nutrients from the surrounding water), to the basal decomposers (bacteria and fungi) and tiny primary and secondary consumers

include viruses, prokaryotes (archaea and bacteria), protists (single-celled eukaryotes such as single-celled fungi, algae and protozoans), as well as multicellular fungi and microscopic metazoans (such as nematodes, rotifers, tardigrades, gastrotrichs and micro-crustaceans) (Reiss, 2018).

Within this staggering amount of microbiota, there are two main groups that have animal characteristics: protozoans and micro-metazoans (Reiss, 2018). Protozoans are microscopically small, heterotrophic, single-celled eukaryotes and include species that belong to the flagellates, ciliates or amoebae (Reiss, 2018). The term 'micro-metazoan' is not used widely in the ecological literature and instead 'meiofauna', 'micro- or meso-zooplankton' or 'micro-invertebrate' are established terminology. In this chapter, I am using the term 'micro-metazoan' for two reasons: firstly, meiobenthologists do not unanimously agree on a definition for meiofauna (Robertson et al., 2000b) and, further, base it on sieve sizes (which is not important when it comes to 'leaf litter decomposition methods'). Secondly, decomposing leaves are exposed to the benthic—and the open water zone alike and leaves therefore represent a link between benthic, epibenthic and pelagic food chains and communities.

Indeed, terminological problems do not end here—although protozoans have animal features they are not always included in definitions of 'animal' or 'invertebrate' but at the same time can be included in terms such as 'micro-invertebrate', 'micrograzer' (Weitere et al., 2018), 'micro-animal' or 'micro-fauna'. I settled on 'protozoans' and 'micro-metazoans' in this chapter and together I view them as the microscopic counterpart to 'macro-invertebrates' whose importance is widely established in the leaf litter decomposition literature (see Chapter 10).

Finally, ideas on what constitutes leaf litter 'biofilm' depend on the area of research (e.g., biomedical *vs.* ecological) and in this chapter it is used as defined by (Sigee, 2005) as an assemblage 'of microorganisms occurring at a physical (e.g., water/solid) interface, typically present within a layer of extracellular polysaccharide' (p. 484, Sigee, 2005). Fungi have a special position here as organisms that can be present both in the biofilm and the leaf tissue matrix at the same time.

Terminological issues aside, research that focusses on the role of protozoans and micro-metazoans in leaf litter decomposition is now emerging (e.g., Ribblett et al., 2005; Risse-Buhl et al., 2012) and the purpose of this chapter is to place animal-like microbes (protozoans and micro-metazoans) into the context of energy flow and leaf litter decomposition, while summarising and synthesizing the status quo of what we know about the role of these microbes in leaf litter decomposition (as established by laboratory and field research).

11.2 Protozoans and Micro-metazoans Are Omnipresent in Aquatic Systems and Part of the Food Web

Decomposing leaves provide a substrate for microbe biofilms and the concept of the 'microbial colonization and degradation' stage that 'conditions' the leaf is famous in leaf litter decomposition literature (Abelho, 2001). Indeed, bacterial and fungal spores might be present on the leaves before they fall into the stream but even a sterile leaf will undergo rapid microbial colonisation and quickly show a well-established biofilm—typically after 2 weeks according to Abelho (2001) (depending of course on a plethora of factors such as water temperature and chemistry, see Chapter 2). Much 'microbial colonisation' research has focused on bacteria and fungi and their relative importance (see Abelho, 2001; Gessner et al., 2010) because these groups enhance detrital decomposition directly by macerating leaves and transforming leaf carbon (and other nutrients) into living bacterial/fungal biomass. The latter has been found to increase the palatability of detritus to macrofauna (Abelho, 2001). However, natural biofilms are typically much more complex than these 'simple' food chains suggest. When bacteria and fungi are present, so are their primary consumers (Weitere et al., 2018)—many different species of flagellates, ciliates and amoebae; and tiny metazoans such as rotifers (Fig. 11.1). When light is present, single celled algae will be an important component of biofilms and attract micro-grazers that in turn invite microscopically small predators (Weitere et al., 2018). Ciliates, for example, are an omnipresent group in fresh water and a key component of secondary production in the micro-world (Reiss, 2018; Reiss & Schmid-Araya, 2008, 2010). Ciliates can also be found in high abundances on detrital biofilm (Peralta-Maraver et al., 2019; Ribblett et al., 2005) and solid surface biofilm. For example, Norf et al. (2009) found up to 600 ind./cm^2 of biofilm ciliates on microscope slides exposed to water of the River Rhine. Indeed, as a rule of thumb, this glass slide biofilm will show a rich micro-community consisting of multiple trophic groups after only 1 day of exposure and include protozoans and micro-metazoans after 5 days (Arndt et al., 2003). In a pivotal review, Weitere and colleagues (2018) suggest a framework that places freshwater biofilms into a wider concept of energy flow in aquatic systems, and importantly this framework highlights the crucial ecological role of biofilm-dwelling protozoans and micro-metazoans. The complexity of benthic-pelagic coupling, biofilm maturing and microbe interactions is broken down into three interlinking topics: (1) coupling of the planktonic and benthic food webs, (2) the 'horizontal food-quality axis' (changes in the basal resources over time that are associated with increasing foraging and nutritional costs for the protozoans and micro-metazoans) and (3) vertical food web complexity and food chain length (protozoans and micro-metazoans are part of a complex food web with several trophic levels). An example that illustrates the latter point are food chains that involve epibenthic—and microscopically small—copepods (such as *Eucyclops* spp.) that prey on bacterivorous ciliates in biofilm (Reiss & Schmid-Araya, 2011) and in turn represent prey for juvenile fish (like many micro-crustaceans [Dineen & Robertson, 2010]). Indeed the link between benthic micro-metazoans (meiofauna) and higher trophic levels has been thoroughly reviewed by

Ptatscheck et al. (2020). Weitere et al. (2018) give plenty of detailed genus-genus interactions between biofilm basal groups (bacteria, fungi and algae) and biofilm dwellers, compiled from the literature. For instance, Aphelenchoidida nematodes have mouth parts that allow them to pierce the hyphae of biofilm fungi and feed on them (Weitere et al., 2018). As strong argument for the link between decomposing plant leaf litter and populations of microscopically small metazoans was made by Majdi and Traunspurger (2017) who showed that isotopic signature of grazing nematodes and copepods is closely linked to leaf litter, especially in autumn (Majdi & Traunspurger, 2017). Needless to say that trophic interactions in these micro-worlds have to be painstakingly established via time consuming methods (often microscope observations) and these are summarised in helpful reviews (Majdi et al., 2019; Weitere et al., 2018). These reviews, and the aquatic biofilm literature, point towards an important role of protozoans and micro-metazoans in detrital processing—including leaf litter decomposition.

11.3 Is Identification Key?

I would argue that incorporating protozoans and micro-metazoans into leaf litter decomposition studies is mainly hampered by the 'tradition' in the 'field of decomposition' rather than methodology. For most decomposition studies, a main aim will be to estimate the potential of all decomposers to contribute to the process (i.e., their metabolic power, see Sect. 11.5) and sometimes a rather 'crude' approach will suffice, such as measuring respiration in the light and dark (e.g., via oxygen probes in a laboratory set-up, such sensor dishes [SDR SensorDish®]) to get an estimate for autotroph *vs.* heterotroph activity. The latter estimate can give an indication about the presence and role of animal-like heterotrophs (protozoans and micro-metazoans) on leaf biofilm given biomass of other heterotrophs (bacteria and fungi) is estimated at the same time (e.g., stain and count of bacteria and ergosterol method for fungi). A straightforward and cost-effective way is to identify and count protozoans and micro-metazoans per area of leaf surface. An overview of the main taxonomic keys and counting methods can be found in (Reiss & Schmid-Araya, 2008) and (Reiss, 2018) but a beginner will find the book 'Das Leben im Wassertropfen' (Streble & Krauter, 2001) and the website 'plingfactory' (Plewka, 2020) very useful and be able to identify to group level using the drawings and microscope photos respectively. The main laboratory method is to use a counting chamber (e.g., Fuchs-Rosenthal counting chamber) under a light microscope using 100 times and 400 times magnification (e.g., Reiss et al., 2019).

Lastly, molecular ecology allows using tools that help non-specialists to identify bacteria, fungi and algae in biofilm (Sigee, 2018) and indeed sequencing and eDNA techniques are ever advancing and these methods have already been used in soil and plankton studies for protozoans (Seppey et al., 2017) and micro-metazoans (Yang & Zhang, 2020) respectively. Taken together, these three approaches (measuring heterotrophic activity, microscope identification and molecular ecology) will hopefully contribute to shedding light into the 'black box' (that currently is a good metaphor for the role of protozoans and micro-metazoans in leaf litter decomposition).

11.4 Do Protozoans and Micro-metazoans Play a Role in Leaf Litter Decomposition? What Is the Evidence?

Reviews synthesising the vast field of plant leaf litter decomposition (Abelho, 2001; Gessner et al., 2010) largely ignore protozoans and micro-metazoans, especially in conceptual figures for leaf litter decomposition. However, a study by Majdi and Traunspurger (2017) demonstrates that micro-metazoans (meiofaunal nematodes, harpacticoid copepods and chironomids) derive a large amount of their energy from leaf litter (compared to filamentous algae, macrophytes and FPOM) and the authors speculate that this is mainly due to feeding on microbes growing on the leaves (Majdi & Traunspurger, 2017). The purpose of this chapter is to incorporate protozoans and micro-metazoans into the existing leaf litter decomposition frameworks (e.g., Abelho, 2001; Canhoto & Graça, 2008; Gessner et al., 2010) and it seems obvious to first address the question: how abundant are protozoans and micro-metazoans on decomposing leaves? Evidence comes from a hand-full of field studies (Table 11.1) that have counted and identified protozoans and micro-metazoans on leaves or decomposing material (Albertoni et al., 2020; Brüchner-Hüttemann et al., 2019; Franco et al., 1998; Peralta-Maraver et al., 2019).

A field study by Peralta-Maraver and colleagues (2019) showed the relative importance of prokaryotes, protozoans and micro-metazoans on decomposition (Fig. 11.2). They exposed cotton strips and two types of commercially available tea bags in 30 UK streams (both in the benthic and hyporheic zone). These were retrieved from the streams after 29 to 61 days, depending on flood conditions in the stream. Bacteria were counted via flow cytometry and their activity was assessed with the Biolog EcoPlate System (Biolog Inc.). Flagellates, ciliates and micro-metazoans were counted under a stereomicroscope. All groups were abundant in- and on the decomposition bags, with up to 450 (± 36SE) and 15 (± 6SE) ind./bag for protozoans and micro-metazoans respectively. In fact, the biomass of protists and micro-metazoans was highly correlated with decomposition (Peralta-Maraver, 2019; Fig. 11.2).

A field study by Franco et al. (1998) followed the colonization and succession of ciliates on different leaf types submerged in a stream over the course of five weeks. On a weekly basis, ciliates present on the leaves were identified and assigned to groups based on their feeding type. The authors found that all feeding groups were present on all substrates and 20–35 ciliate species co-existed within a few squared centimetres.

Following on from this we can now speculate about the potential roles of protozoans and micro-metazoans when it comes to leaf litter decomposition. From an

Table 11.1 Studies that have investigated the presence of protozoans and micro-metazoans on plant litter and/or role of these groups on detrital decomposition

	Authors	Study and organisms	Main results
Field studies	Brüchner-Hüttemann et al. (2019)	Natural leaf biofilm community on submerged leaves (bacteria, protists) was monitored in a stream for 13 months	Ciliates made up the highest percentage of total annual biomass in sediment and on dead wood and leaf litter, whereas on macrophytes, bacteria had the highest percentage of total annual biomass. On all substrates, ciliates had the highest share of secondary production
	Peralta-Maraver et al. (2019)	Bundles consisting of cotton strips and tea bags were exposed in 30 streams. Presence of bacteria, protists and micro-metazoans was recorded. There was one sampling occasion	Biomass of all the studied biotic groups was positively related to the decomposition coefficients. Biomass of protists was the same as bacteria and micro-metazoans. Alpha-diversity of micro-metazoans also had a positive influence on leaf litter decomposition
	Franco et al. (1998)	Natural ciliate community on submerged leaves in a stream was observed over 2 weeks	20–35 ciliate species co-existed and the majority were filter feeders
	Majdi and Traunspurger (2017)	Meio- and macrofauna, filamentous algae, macrophytes, leaf litter and FPOM were sampled in the Furlbach, a small lowland stream in both spring and autumn. $\delta 13C$ and δ $15N$ isotopic signatures were established	Compared to the other food sources, leaf litter was the dominant source of carbon for grazing nematodes and harpacticoid copepods in autumn and also a main source of carbon for chironomids

(continued)

Table 11.1 (continued)

	Authors	Study and organisms	Main results
	Albertoni et al. (2020)	Decomposing macrophytes in a lake and presence of bacteria, fungi and micro-metazoans was monitored over 18 days	Nematodes, micro-crustaceans (ostracods, copepods, cladocerans), turbellarians and chironomids were present on decomposing plants
Laboratory studies	Santschi et al. (2018)	Alder, beech, poplar and oak leaves and natural microbial decomposer community consisting of bacteria, fungi and protists were kept at 10 °C, for 72 days	Microbial communities, with a reduced functional and trophic complexity, showed a small but significant overall reduction in decomposition rates compared to communities with the naturally complete functional and trophic complexity
	Risse-Buhl et al. (2015)	The experiment was performed with alder leaves, fungus *Heliscus lugdunensis* and a multi-species bacterial assemblage in the presence of the ciliate *T. pyriformis*; at 18 °C	The fungus and the multi-species bacterial assemblage was significantly enhanced in the presence of *T. pyriformis* after 7 days of incubation
	Ribblett et al. (2005)	The experiment was performed with elder leaves, bacteria, the ciliate *Dexiostoma campyla* and the flagellate, *Spumella* sp.; at 10 °C	The presence of protists increased leaf litter decomposition and decreased bacterial abundances
	Chambord et al. (2017)	The experiment was performed with alder leaves, fungi and the cladoceran *C. sphaericus* and copepod *Cyclops bohater*, with and without *Gammarus pulex* and *Sericostoma personatum*; at 10 °C	The presence of microcrustaceans enhanced leaf mass loss by 62 and 22% in treatments with fungi or trichopteran, respectively

(continued)

Table 11.1 (continued)

	Authors	Study and organisms	Main results
	Unpublished data from Flores et al. (2016)	The experiment was performed with alder leaves, the copepod *Cyclops viridis*, *Asellus aquaticus* and *Gammarus pulex*; at 15 °C	Leaf mass loss in microcosms with *Cyclops* was the same as for *Asellus* or *Gammarus* (because metabolic capacity had been accounted for) despite the fact that *Cyclops* feeds on biofilm and not the leaf material

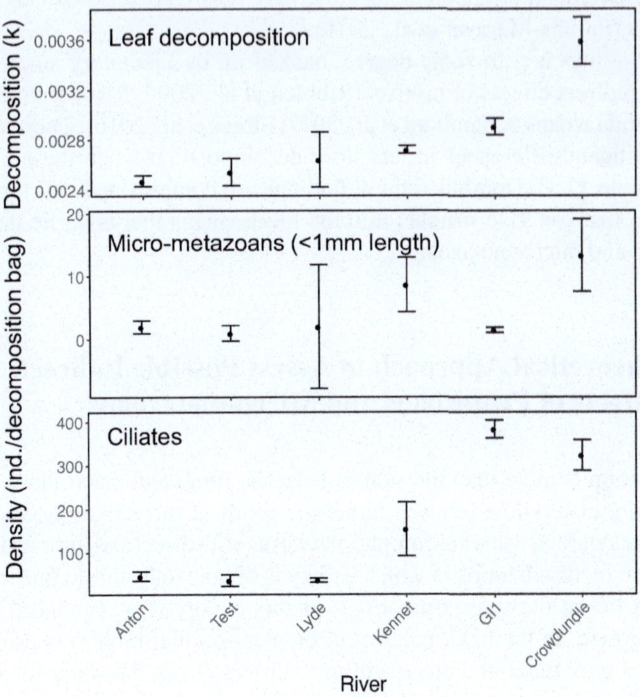

Fig. 11.2 Examples for data collected in the study by Peralta-Maraver et al. (2019) that exposed 'decomposition bags' (consisting of bundled cotton strips, green tea and roibush tea bags) in 30 streams across the UK and retrieved them after 30 days. Shown here are data from some of these streams (selected out of 30 because micro-metazoans were present). The abundance of ciliates (mean ± SE from 3 replicates, lowest panel) was substantial in some of the decomposition bags (e.g., river Crowdudle > 300 ciliates were in the cotton strip and tea bag bundle). The micro-metazoans (meiofauna) did not exceed 15 individuals in a sample (middle panel). There is a trend for higher decomposition (most upper panel) when micrograzer abundance is high. However, not shown here, are the much more significant correlations between protozoan and microscopic metazoan biomass and decomposition rates of all three substrates (see Peralta-Maraver et al., 2019)

ecosystem functioning perspective, we might even ask if their presence speeds leaf litter decomposition up. Biofilm literature has highlighted the resistance of bacterial biofilms to grazing within the microbial food web and established that the main effect of protozoans and micro-metazoans is on biofilm phenotype rather than biofilm biomass. However, aquatic groups working on natural river biofilms argue that protozoans and micro-metazoans have direct feeding effects on biofilm (Arndt et al., 2003; Weitere et al., 2018). Again, the study by Peralta-Maraver et al. (2019) possibly provides the strongest evidence for a link between micro-grazers and decomposition to date (Fig. 11.2). This study showed a general pattern for both the benthic and hyporheic zone: decomposition rates were positively correlated with the biomass of each of the 4 organismal groups studied (prokaryotes, flagellates, ciliates and micro-metazoans). In addition, α-diversity of micro-metazoans and metabolic diversity of prokaryotes were important predictors that were positively related to decomposition coefficients (Peralta-Maraver et al., 2019).

These findings are, to some degree, backed up by laboratory studies that have addressed indirect effects of protists (Ribblett et al., 2005; Risse-Buhl et al., 2015) and micro-metazoans (Chambord et al., 2017; Flores et al., 2016). These studies have found significant differences in leaf litter decomposition when these groups were present (Table 11.1). Combined these findings are complex and call for theoretical approaches that can give insights into the mechanisms that underlie the effects of protozoans and micro-metazoans.

11.5 Theoretical Approach to Assess Possible Indirect Effects of Protozoans and Micro-metazoans

Decomposition is more than the sum of bacteria, fungi and macrofauna feeding on decomposing material and indeed rather the result of more multifaceted food web interactions. All food webs are complex networks with direct and indirect interactions and intricate feedback loops in which energy is of course lost from one trophic level to the next but at the same time a part of this energy (e.g., biomass) is recycled and transformed. In the field, renewal of biomass on leaf biofilm is probably quite high, as bacteria, fungi and algae will use nutrients from the water interface—leaf biofilm does not represent a closed system. Contributing to this biomass are (epi-) benthic protozoans, micro-metazoans and macro-invertebrates that are all temporary residents on leaves and their community composition and biomass will change over time.

Still, principles established in ecological theory and research apply when we consider the role of protozoans and micro-metazoans in leaf litter decomposition and three main 'theoretical' concepts should be considered here and will be discussed in turn:

1. Secondary production (and therefore energy uptake from the leaf by heterotrophs) is invariant of body size within trophic level and decreases with

body size across trophic levels (Huryn & Benke, 2007); within closed systems (closed food webs).

2. The presence of protozoans and micro-metazoans can have counterintuitive effects on leaf litter decomposition because their grazing on bacteria and fungi can have a variety of effects: it can decrease the biomasses of the latter groups, it can free one group from competition and it can also enhance metabolic power of biofilms.
3. Leaves are substrates for micro-worlds but are not the only energy and nutrient source as they represent a link between different habitats—in other words, bottom-up effects (e.g., via food quality and quantity for bacteria) are also driven by abiotic factors in the open water and the fact that the biofilm is accessible for both benthic and pelagic micro-consumers. Weitere et al. (2018) call this 'across-habitat subsidization of resources and consumers'.

Of course, many other ecological principles apply (such as the role of micro-invertebrate biodiversity effects on leaf litter decomposition [Reiss et al., 2009, 2011]), but the focus of this chapter is to distil some aspects of ecological theory that specifically apply to leaf litter decomposition by protozoans and micro-metazoans. Ideally, this chapter should be read in conjunction with Chapters 7 and 9 where more theoretical concepts are explored.

First, I will discuss the metabolic power of assemblages composed of microscopically small decomposers and their consumers. The message of this section is simple: microscopically small organisms, including protozoans and micro-metazoans, have a substantial combined biomass and, compared to macroscopic fauna, they can assimilate the same or greater amounts of energy from decomposing leaves. The relationship between production and biomass (the P/B ratio; it indicates how often biomass is turned over per year) has been well studied for aquatic macro-invertebrates (Huryn & Benke, 2007) and, to a lesser extent, for microbes (Marxsen, 2006), including protozoans and micro-metazoans (Reiss & Schmid-Araya, 2010). We know that small organisms have higher P/B ratios compared to larger ones (Huryn & Benke, 2007) because their generation times are shorter and biomass is 'renewed' more frequently throughout the year. When considering communities or assemblages of a wide range of organisms, including invertebrates, ecological theory predicts that secondary production of heterotrophs is invariant of body mass within trophic levels (Fig. 11.3; Brown et al., 2004). In other words, this means that the standing stock of biomass of protozoans and micro-metazoans might be smaller than that of macro-invertebrates, but this biomass is turned over more quickly (higher P/B value) and therefore the metabolic power represented by each group (i.e., protozoans, micro-metazoans and macrofauna) can be very similar. This means that populations (or assemblages) of small organisms, over time, will assimilate the same amount of energy as larger organisms, given they can reach the same biomass as larger organisms. As a rule of thumb, carbon uptake per unit time (e.g., gC/year) is higher for the same biomass of small organisms compared to larger ones in closed systems (Fig. 11.3). To make exact predictions of how much energy is fluxed from decomposing leaves to a given subset of the protozoan, micro-metazoan and macro-decomposer assemblage, it is

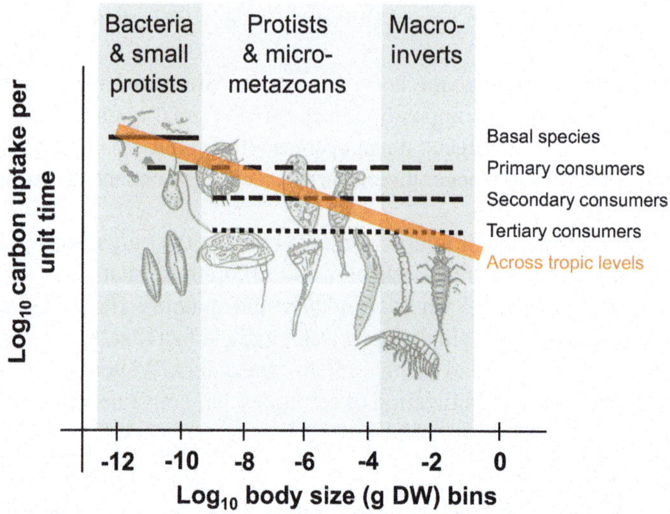

Fig. 11.3 Carbon uptake per unit time (e.g., gC/year) is higher for small organisms compared to larger ones if they have the same biomass and feed in closed systems (e.g., in a mesocosm). This concept figure demonstrates production (which can serve as a proxy of carbon/energy uptake, indicated with lines here) of body size bins within an assemblage feeding on leaf carbon and leaf biofilm. The diagram has simplified assumptions: that there are 4 trophic levels in the leaf decomposer assemblage, that energy is lost from one trophic level to the next in the same percentage, and that all size groups operate within the same metabolic constrains

necessary to consider trophic levels (Fig. 11.3). Across trophic levels, the relationship between body mass and production is negative because energy is lost from one trophic level to the next (Fig. 11.3; Brown et al., 2004; Huryn & Benke, 2007).

We can safely assume that three trophic levels will be present within both micro- and macroinvertebrates, as primary, secondary and tertiary consumers are found in both groups (Majdi & Traunspurger, 2017; Tachet et al., 2010; Weitere et al., 2018). Hence, in theory, production within and across the groups scales negatively with body mass (Fig. 11.3) because energy is lost between trophic levels. Regarding the latter, energy transfer calculations often assume a fixed predator to prey body mass ratio (e.g., one order of magnitude on log axes, Brown et al., 2004). However (and here comes the caveat), microscopically small organisms operate quite differently compared to larger ones. To give two examples here: the way metabolism scales with body mass is very different for prokaryotes, protists and metazoans (DeLong et al., 2010) and also predator:prey body mass ratios are very different indeed. To give an example, we can consider the species rich ciliates—a group in which body size can vary by 4 orders of magnitude (Fig. 11.3). Body mass (in theory) scales with metabolic rate (of size bins) to the power of 1 on logarithmic axes (DeLong et al., 2010) which is different to metazoans (scaling exponent is often close to 0.75 for this group; DeLong et al., 2010); and many predatory ciliates will feed on prey that is the same size as themselves (Foissner & Berger, 1996). Obviously controlled laboratory experiments offer an excellent approach to rigorously study these aspects of energy transfer from the leaf to the protozoans and micro-metazoans. Readers who are new to allometric theories and plan laboratory set-ups with decomposing leaves are directed to publications on allometric scaling principles (Brown et al., 2004; DeLong et al., 2010; Huryn & Benke, 2007; Reiss, Forster, et al., 2010) and approaches to laboratory studies (Chapter 20, Altermatt et al., 2015; Reiss, Bailey, et al., 2010).

Another theoretical concept concerns the relative importance of top-down effects and stimulation of bacterial biofilm and/or fungal biomass, harking back to the concept of grazing optimization and nutrient cycling in theoretical ecology (De Mazancourt et al., 1998). Micro-grazers potentially impact bacteria and fungi by a range of mechanisms, such as feeding effects (e.g., feeding on less active bacteria and food selection or preferential feeding on bacteria that frees fungi from competition), and also by bioturbation and waste products that stimulate prokaryote and fungal growth. To give an example, we can picture a bacterial biofilm growing on a leaf that is grazed by flagellates, ciliates, rotifers and nematodes. Intuitively, these bacterivorous protists and micro-metzoans should reduce bacterial abundances and, consequently, lower decomposition rates. However, empirical studies have often found the opposite result: an increase in decomposition rates in the presence of bacterivorous protozoans and micro-metazoans (Ribblett et al., 2005). Ecological studies on biofilm grazing by protozoans and micro-metazoans can give answers in this regard (Arndt et al., 2003; Risse-Buhl et al., 2012; Weitere et al., 2018) but we can also branch out into the biomedical literature that is very advanced when it comes to detailed, micro-scale, findings regarding how pathogenic bacterial biofilms gain higher resistance and activity. For example, bacteria grow on metabolites excreted by other bacterial species (this is called cross-feeding (Adamowicz et al., 2018)). Transferred to natural biofilms this means that high bacterial diversity could result in high biofilm activity and biomass. Because micro-grazers can feed selectively on particular bacteria (Boenigk et al., 2002; Wey et al., 2008) they could both induce or inhibit bacterial diversity, depending on which micro-grazers are present. Bacterivorous protists have been found to stimulate a high turnover of bacteria (prokaryotes are kept in the accelerating phase of their logarithmic growth (Fenchel & Jørgensen, 1977)) leading to a better physical state and higher metabolic capacity of the bacterial community and this can consequently enhance decomposition. In 'biofilm-macrofauna' ecology (e.g., diatoms grazed by snails), it is generally thought that herbivores can stimulate (algae) biofilm growth by decreasing biofilm thickness, suppressing dead cells, and rendering it more "functional" (Burgmer et al., 2010; Liess & Hillebrand, 2004). Important in the context of leaf litter decomposition is that some micro-metazoans (especially nematodes) can feed on fungi (Weitere et al., 2018) and could hence play a disproportionally large role in structuring leaf biofilm.

To add yet another dimension to the control of bacteria and fungi, we have to mention trophic cascades within the microbial food web (Wey et al., 2008) and across-habitat subsidization of resources and consumers (Weitere et al., 2018). The balance between top-down control and stimulating effects is further modified by the presence of multiple trophic levels. For example, Wey et al. (2008) showed that single bacteria (in contrast to colony forming ones) are released from grazing pressure by flagellates when predatory ciliates are present, possibly through a trophic cascade. Other micro-predators include copepod species and often these are epi-benthic or planktonic species that will feed on biofilms. Weitere et al. (2018) show that up to 4 trophic levels exist in micro-worlds and propose that in terms of flux of matter and energy, the large number of trophic levels suggests substantial energy loss and re-cycling within biofilms.

Across-habitat subsidization of resources is a common phenomenon that can have considerable effects on the productivity and dynamics of the receiving community (Pace et al., 2004; Perkins et al., 2021; Wallace et al., 2015). It is commonly accepted in biofilm research that the majority of dissolved nutrients and organic carbon taken up by biofilm algae and bacteria originates from the water column. However, the subsidization at the level of benthic micro-consumers has received less attention (but see Chapter 3). By existing at the interface between substratum and plankton, biofilm-dwelling organisms can potentially utilize both suspended and surface-associated resources.

11.6 Synthesis and Where Do We Go from Here

The overall conclusion from this chapter must be that the role of protozoans and micro-metazoans is understudied in general and especially in plant litter decomposition research. Laboratory experiments could be the way forward and hopefully this chapter has shown that it is important to know the identity of the microbe species that are present on the leaves—and their abundances and activities (see Reiss, 2018; Reiss & Schmid-Araya, 2008). Further, we need to have an understanding of some ecological theories that can explain how energy and nutrients from the leaves can fuel micro-worlds. Important in this regard are publications on food web links (Majdi et al., 2019; Ptatscheck et al., 2020; Weitere et al., 2018) and those on theoretical principles (e.g., Altermatt et al., 2015; Prosser et al., 2007; Reiss, Forster, et al., 2010; Reiss & Schmid-Araya, 2010).

References

Abelho, M. (2001). From litterfall to breakdown in streams: A review. *The Scientific World Journal, 1*, 656–680. https://doi.org/10.1007/978-94-011-0729-7.

Adamowicz, E. M., Flynn, J., Hunter, R. C., & Harcombe, W. R. (2018). Cross-feeding modulates antibiotic tolerance in bacterial communities. *The ISME Journal, 12*(11), 2723–2735.

Albertoni, E. F., de Moraes, A. L., Guimarães, P. S., & Palma-Silva, C. (2020). Invertebrates and microbiota associated with aquatic macrophyte degradation in a shallow lake in southern Brazil. *Acta Brasiliensis, 4*(1), 38–44.

Altermatt, F., Fronhofer, E. A., Garnier, A., Giometto, A., Hammes, F., Klecka, J., Legrand, D., Mächler, E., Massie, T. M., Pennekamp, F., Plebani, M., Pontarp, M., Schtickzelle, N., Thuillier, V., Petchey, O. L. (2015). Big answers from small worlds: A user's guide for protist microcosms as a model system in ecology and evolution. *Methods in Ecology and Evolution, 6*(2), 218–231.

Arndt, H., Schmidt-Denter, K., Auer, B., & Weitere, M. (2003). Protozoans and biofilms. In W. E. Krumbein, D. M. Paterson, & G. A. Zavarzin (Eds.), Fossil and recent biofilms (pp. 161–179). Springer, Dordrecht. https://doi.org/10.1007/978-94-017-0193-8.

Boenigk, J., Matz, C., Jürgens, K., & Arndt, H. (2002). Food concentration-dependent regulation of food selectivity of interception-feeding bacterivorous nanoflagellates. *Aquatic Microbial Ecology, 27*(2), 195–202. https://doi.org/10.3354/ame027195.

Brown, J. H., Gillooly, J. F., Allen, A. P., Savage, V. M., & West, G. B. (2004). Toward a metabolic theory of ecology. *Ecology, 85*(7), 1771–1789. https://doi.org/10.1890/03-9000.

Brüchner-Hüttemann, H., Ptatscheck, C., & Traunspurger, W. (2019). Unicellular organisms in benthic habitats on different substrates in a first-order stream and their contribution to secondary production. *Aquatic Microbial Ecology, 83*(1), 49–63.

Burgmer, T., Reiss, J., Wickham, S., & Hillebrand, H. (2010). Effects of snail grazers and light on the benthic microbial food web in periphyton communities. *Aquatic Microbial Ecology*. https://doi.org/10.3354/ame01449.

Canhoto, C., & Graça, M. A. S. (2008). Interactions between fungi and stream invertebrates: Back to the future. *Novel Techniques and Ideas in Mycology. Fungal Diversity Research Series, 20*, 205–325.

Chambord, S., Tackx, M., Chauvet, E., Escolar, G., & Colas, F. (2017). Two microcrustaceans affect microbial and macroinvertebrate-driven litter breakdown. *Freshwater Biology, 62*(3), 530–543.

De Mazancourt, C., Loreau, M., & Abbadie, L. (1998). Grazing optimization and nutrient cycling: When do herbivores enhance plant production? *Ecology, 79*(7), 2242–2252. https://doi.org/10.1890/0012-9658(1998)079[2242:GOANCW]2.0.CO;2.

DeLong, J. P., Okie, J. G., Moses, M. E., Sibly, R. M., & Brown, J. H. (2010). Shifts in metabolic scaling, production, and efficiency across major evolutionary transitions of life. *Proceedings of the National Academy of Sciences, 107*(29), 12941–12945.

Dineen, G., & Robertson, A. L. (2010). Subtle top-down control of a freshwater meiofaunal assemblage by juvenile fish. *Freshwater Biology, 55*(9), 1818–1830.

Fenchel T. M., Jørgensen B. B. (1977). Detritus Food Chains of Aquatic Ecosystems: The Role of Bacteria. In M. Alexander (Eds). Advances in Microbial Ecology, vol 1. Springer, Boston, MA. https://doi.org/10.1007/978-1-4615-8219-9_1.

Flores, L., Bailey, R. A., Elosegi, A., Larrañaga, A., & Reiss, J. (2016). Habitat complexity in aquatic microcosms affects processes driven by detritivores. *PLoS ONE, 11*(11), 1–15. https://doi.org/10.1371/journal.pone.0165065.

Foissner, W., & Berger, H. (1996). A user-friendly guide to the ciliates (Protozoa, Ciliophora) commonly used by hydrobiologists as bioindicators in rivers, lakes, and waste waters, with notes on their ecology. *Freshwater Biology, 35*(2), 375–482.

Franco, C., Esteban, G. F., & Téllez, C. (1998). Colonization and succession of ciliated protozoa associated with submerged leaves in a river. *Limnologica, 28*(3), 275–283.

Gessner, M. O., Swan, C. M., Dang, C. K., McKie, B. G., Bardgett, R. D., Wall, D. H., & Hättenschwiler, S. (2010). Diversity meets decomposition. *Trends in Ecology and Evolution, 25*(6), 372–380. https://doi.org/10.1016/j.tree.2010.01.010.

Huryn, A. D., & Benke, A. C. (2007). Relationship between biomass turnover and body size for stream communities. In A. G. Hildrew, D. Raffaelli, & R. Edmons-Brown (Eds.), *Body size: The structure and function of aquatic ecosystems*. (pp. 55–76). Cambridge University Press.

Liess, A., & Hillebrand, H. (2004). Invited review: Direct and indirect effects in herbivore–periphyton interactions. *Archiv Für Hydrobiologie, 159*(4), 433–453.

Majdi, N., Schmid-Araya, J. M., & Traunspurger, W. (2019). Examining the diet of meiofauna: A critical review of methodologies. *Hydrobiologia, 847*, 1–18.

Majdi, N., & Traunspurger, W. (2017). Leaf fall affects the isotopic niches of meiofauna and macrofauna in a stream food web. *Food Webs, 10*, 5–14. https://doi.org/10.1016/j.fooweb.2017.01.002.

Marxsen, J. (2006). Bacterial production in the carbon flow of a central European stream, the Breitenbach. *Freshwater Biology, 51*(10), 1838–1861. https://doi.org/10.1111/j.1365-2427.2006.01620.x.

Norf, H., Arndt, H., & Weitere, M. (2009). Responses of biofilm-dwelling ciliate communities to planktonic and benthic resource enrichment. *Microbial Ecology, 57*(4), 687–700.

Pace, M. L., Cole, J. J., Carpenter, S. R., Kitchell, J. F., Hodgson, J. R., Van de Bogert, M. C., … Bastviken, D. (2004). Whole-lake carbon-13 additions reveal terrestrial support of aquatic food webs. *Nature, 427*(6971), 240–243.

Perkins, D. M., Durance, I., Jackson, M., Jones, J., Lauridsen, Rasmus B., Layer-Dobra, K., Reiss, J., & Thompson, Murray S. A., Woodward, G. (2021). Systematic variation in food web body-size structure linked to external subsidies. *Biology Letters, 17*, 20200798. https://doi.org/10.1098/rsbl.2020.0798.

Peralta-Maraver, I., Perkins, D. M., Thompson, M. S. A., Fussmann, K., Reiss, J., & Robertson, A. L. (2019). Comparing biotic drivers of litter breakdown across stream compartments. *Journal of Animal Ecology, 88*(8), 1146–1157. https://doi.org/10.1111/1365-2656.13000.

Plewka, M. (2020). Pling—Life in water. PlingFactory website. Retrieved November 5, 2020, from https://www.plingfactory.de/

Poff, N. L., Palmer, M. A., Angermeier, P. L., Vadas, R. L., Hakenkamp, C. C., Bely, A., Arensburger, P., Martin, A. P. (1993). Size structure of the metazoan community in a Piedmont stream. *Oecologia, 95*(2), 202–209.

Prosser, J. I., Bohannan, B. J. M., Curtis, T. P., Ellis, R. J., Firestone, M. K., Freckleton, R. P., Green, J. L., Green, L. E., Killham, K., Lennon, J. J., Osborn, A. M., Solan, van der Gast, M., C. J., & Young, J. P. W. (2007). The role of ecological theory in microbial ecology. *Nature Reviews Microbiology, 5*(5), 384–392.

Ptatscheck, C., Brüchner-Hüttemann, H., Kreuzinger-Janik, B., Weber, S., & Traunspurger, W. (2020). Are meiofauna a standard meal for macroinvertebrates and juvenile fish? *Hydrobiologia, 847*(12), 2755–2778. https://doi.org/10.1007/s10750-020-04189-y.

Reiss, J. (2018). Microorganisms 2: Viruses, prokaryotes, fungi, protozoans, and microscopic metazoans. In J. Hughes (Ed.), Freshwater Ecology and Conservation (pp. 157–172). Oxford University Press. https://doi.org/10.1093/oso/9780198766384.003.0008.

Reiss, J., Forster, J., Cássio, F., Pascoal, C., Stewart, R., & Hirst, A. G. (2010). When microscopic organisms inform general ecological theory. *Advances in Ecological Research, 43*(45–85). https://doi.org/10.1016/B978-0-12-385005-8.00004-6.

Reiss, J., Bailey, R. A., Perkins, D. M., Pluchinotta, A., & Woodward, G. (2011). Testing effects of consumer richness, evenness and body size on ecosystem functioning. *Journal of Animal Ecology, 80*(6), 1145–1154. https://doi.org/10.1111/j.1365-2656.2011.01857.x.

Reiss, J., Bridle, J. R., Montoya, J. M., & Woodward, G. (2009). Emerging horizons in biodiversity and ecosystem functioning research. *Trends in Ecology and Evolution, 24*(9), 505–514. https://doi.org/10.1016/j.tree.2009.03.018.

Reiss, J., Perkins, D. M., Fussmann, K. E., Krause, S., Canhoto, C., Romeijn, P., & Robertson, A. L. (2019). Groundwater flooding: Ecosystem structure following an extreme recharge event. *Science of the Total Environment, 652*, 1252–1260. https://doi.org/10.1016/j.scitotenv.2018.10.216.

Reiss, J., & Schmid-Araya, J. M. (2008). Existing in plenty: Abundance, biomass and diversity of ciliates and meiofauna in small streams. *Freshwater Biology, 53*(4), 652–668. https://doi.org/10.1111/j.1365-2427.2007.01907.x.

Reiss, J., & Schmid-Araya, J. M. (2010). Life history allometries and production of small fauna. *Ecology, 91*(2), 497–507. https://doi.org/10.1890/08-1248.1.

Reiss, J., & Schmid-Araya, J. M. (2011). Feeding response of a benthic copepod to ciliate prey type, prey concentration and habitat complexity. *Freshwater Biology, 56*(8), 1519–1530. https://doi.org/10.1111/j.1365-2427.2011.02590.x.

Reiss, J., Bailey, R. A., Cássio, F., Woodward, G., & Pascoal, C. (2010). Assessing the Contribution of Micro-Organisms and Macrofauna to Biodiversity–Ecosystem Functioning Relationships in Freshwater Microcosms. *Advances in Ecological Research, 43,* 151–1760. https://doi.org/10.1016/B978-0-12-385005-8.00004-6.

Ribblett, S. G., Palmer, M. A., & Wayne Coats, D. (2005). The importance of bacterivorous protists in the decomposition of stream leaf litter. *Freshwater Biology, 50*(3), 516–526.

Risse-Buhl, U., Karsubke, M., Schlief, J., Baschien, C., Weitere, M., & Mutz, M. (2012). Aquatic protists modulate the microbial activity associated with mineral surfaces and leaf litter. *Aquatic Microbial Ecology, 66*(2), 133–147.

Risse-Buhl, U., Schlief, J., & Mutz, M. (2015). Phagotrophic protists are a key component of microbial communities processing leaf litter under contrasting oxic conditions. *Freshwater Biology, 60*(11), 2310–2322.

Robertson, A. L., Rundle, S. D., & Schmid-Araya, J. M. (2000a). Putting the meio-into stream ecology: Current findings and future directions for lotic meiofaunal research. *Freshwater Biology, 44*(1), 177–183.

Robertson, A. L., Rundle, S. D., & Schmid-Araya, J. M. (2000b). An introduction to a special issue on lotic meiofauna. *Freshwater Biology, 44*(1), 1–3.

Santschi, F., Gounand, I., Harvey, E., & Altermatt, F. (2018). Leaf litter diversity and structure of microbial decomposer communities modulate litter decomposition in aquatic systems. *Functional Ecology, 32*(2), 522–532.

Seppey, C. V. W., Singer, D., Dumack, K., Fournier, B., Belbahri, L., Mitchell, E. A. D., & Lara, E. (2017). Distribution patterns of soil microbial eukaryotes suggests widespread algivory by phagotrophic protists as an alternative pathway for nutrient cycling. *Soil Biology and Biochemistry, 112*, 68–76. https://doi.org/10.1016/j.soilbio.2017.05.002.

Sigee, D. C. (2005). *Freshwater Microbiology: Biodiversity and Dynamic Interactions of Microorganisms in the Aquatic Environment*. Wiley Online Library.

Sigee, D. C. (2018). Microorganisms 1. In J. Hughes (Ed.), Freshwater Ecology and Conservation: Approaches and Techniques (pp. 133–156). https://doi.org/10.1093/oso/9780198766384.003.0007.

Streble, H., & Krauter, D. (2001). Das Leben im Wassertropfen. *Kosmos*, 21.

Tachet, H., Richoux, P., Bournaud, M., & Usseglio-Polatera, P. (2010). *Invertébrés d'eau douce: Systématique, biologie, écologie*. CNRS Editions.

Wallace, J. B., Eggert, S. L., Meyer, J. L., & Webster, J. R. (2015). Stream invertebrate productivity linked to forest subsidies: 37 stream-years of reference and experimental data. *Ecology, 96*(5), 1213–1228.

Weitere, M., Erken, M., Majdi, N., Arndt, H., Norf, H., Reinshagen, M., Traunspurger, W., Walterscheid, A., Wey, J. K. (2018). The food web perspective on aquatic biofilms. *Ecological Monographs, 88*(4), 543–559.

Wey, J. K., Scherwass, A., Norf, H., Arndt, H., & Weitere, M. (2008). Effects of protozoan grazing within river biofilms under semi-natural conditions. *Aquatic Microbial Ecology, 52*(3), 283–296.

Yang, J., & Zhang, X. (2020). eDNA metabarcoding in zooplankton improves the ecological status assessment of aquatic ecosystems. *Environment International, 134*, 105230. https://doi.org/10.1016/j.envint.2019.105230.

Part III
Global Change and Plant Litter Decomposition

Chapter 12
Individual and Interacting Effects of Elevated CO_2, Warming, and Hydrologic Intensification on Leaf Litter Decomposition in Streams

Jennifer J. Follstad Shah

Abstract Elevated atmospheric carbon dioxide (CO_2) concentration, rising water temperatures, and intensification of hydrologic regimes are characteristics of global climate change that affect rates of leaf litter decomposition in lotic ecosystems. Predicted effects of elevated atmospheric CO_2 concentration and elevated temperature negate each other. However, a recent meta-analysis found no effect of elevated atmospheric CO_2 on decomposition, while temperature consistently stimulates rates of decomposition. The sensitivity of litter decomposition to shifts in temperature is less clear due to methodological differences between studies calculating the apparent activation energy of decomposition and the multitude of biotic and abiotic variables that enhance or mitigate the effect of temperature. Both floods and droughts are becoming more frequent features of hydrologic regimes, but spatial and temporal variation in hydrologic intensification adds further challenge for predicting how climate change will alter decomposition rates. Despite these complexities, it is clear that altered rates of litter decomposition have cascading influence on the global C budget and food web dynamics. Thus, improved understanding of the individual and interactive effects of elevated atmospheric CO_2 concentration, warming, and shifting hydrology on decomposition rates remains a vital research need.

12.1 Predicted Individual Effects of Elevated Atmospheric CO_2 Concentration, Warming, and Hydrologic Intensification on Leaf Litter Decomposition

The decomposition of leaf litter is an integrative ecosystem process that fuels aquatic food webs and links biogeochemical cycles (Gessner et al., 1999; Minshall et al., 1983; Wallace et al., 1997). The decomposition process involves leaching of dissolved

J. J. F. Shah (✉)
Environmental and Sustainability Studies Program & Department of Geography, University of Utah, Salt Lake City, Utah, USA
e-mail: Jennifer.shah@envst.utah.edu

C. M. Swan et al. (eds.), *The Ecology of Plant Litter Decomposition in Stream Ecosystems*, https://doi.org/10.1007/978-3-030-72854-0_12

constituents, degradation by bacteria and fungi, feeding by invertebrate detritivores, and physical fragmentation, all of which are mediated by the chemistry and physical structure of leaves and environmental factors (Gessner et al., 1999; Graça, 2001; Webster & Benfield, 1986). Elevated atmospheric carbon dioxide (CO_2) concentration, rising water temperatures, and intensification of hydrologic regimes are central features of the climate crisis that affect rates of decomposition in lotic ecosystems through changes to decomposer metabolic rates and community composition, plant productivity and leaf chemistry, and the supply of water and organic matter.

Human activities have caused global atmospheric CO_2 concentration to increase from 280 ppm at the start of the Industrial Revolution to over 400 ppm (IPCC, 2014). Elevated CO_2 induces terrestrial plants to be more productive through higher rates of photosynthesis (Finzi et al., 2001; Stiling & Cornelissen, 2007), when nutrients or moisture are not limiting (Norby et al., 2010; Perry et al., 2012). In turn, increased terrestrial primary productivity can lead to greater detrital inputs to aquatic ecosystems that support the process of decomposition. However, greater availability of CO_2 also results in leaf litter with higher C:nutrient ratios and higher concentrations of polyphenolic compounds, condensed tannins, and lignin (Ferreira & Chauvet, 2011; Martins, Rezende et al., 2017; Monroy et al., 2016; Rier et al., 2002, 2005; Tuchman et al., 2002), factors that often suppress rates of decomposition (Ardón & Pringle, 2008; Lecerf & Chauvet, 2008; LeRoy et al., 2007; Martínez et al., 2014; Ostrofsky, 1997). Moreover, elevated atmospheric CO_2 concentration promotes more acidic conditions in streams and rivers with poor buffering capacity. Litter decomposition proceeds more slowly with declining pH because acidic conditions suppress microbial and detritivore activity (Boyero et al., 2016; Young et al., 2008). Overall, elevated atmospheric CO_2 concentration is predicted to inhibit rates of litter decomposition (Amani et al., 2019; Kominoski & Rosemond, 2012; Fig. 12.1a).

Mean global air temperature has risen by approximately 1.0 °C above pre-industrial levels as a result of elevated CO_2 concentration (IPCC, 2018) and increases in water temperature have been observed in many streams and rivers (Kaushal et al.,

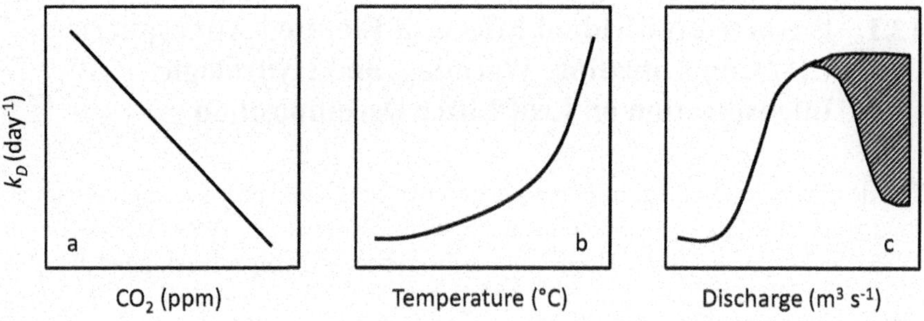

Fig. 12.1 Theoretical effects of CO_2 concentration (a), temperature (b), and water discharge (c) on leaf litter decomposition in streams and rivers. Shading in (c) denotes flushing of organic matter standing stock during high flow conditions

2010). Temperature is a key environmental factor that influences the rate at which organic matter is processed (Tank et al., 2010; Webster & Benfield, 1986; Young et al., 2008). Elevated temperature, within species tolerance limits, increases litter decomposition by stimulating metabolic rates of extant bacteria, fungi, and invertebrate decomposers (Ferreira & Chauvet, 2011; Ferreira, Chauvet et al., 2014; Ferreira et al., 2010; Flury & Gessner, 2011; Moghadam & Zimmer, 2016; Rajashekhar & Kaveriappa, 2000). In addition, higher temperatures typically favor smaller bodied organisms (Atkinson, 1994; James, 1970), which have faster metabolic rates than larger bodied organisms (Gillooly et al., 2001). Higher temperature also promotes leaching of organic matter, which can result in faster litter decomposition by removing recalcitrant compounds (Rier et al., 2005). Litter decomposition is predicted to increase exponentially with temperature (Boyero, Pearson, Gessner et al., 2011; Follstad Shah et al., 2017; Fig. 12.1b), based on first principles of thermodynamics (see *Theory* below). However, temperature and other extrinsic (e.g., resource availability, stream flow) or intrinsic (e.g., leaf litter quality) factors may co-vary, potentially dampening or enhancing the effect of temperature on litter decomposition rates (Fig. 12.2).

Global precipitation patterns are shifting in response to elevated atmospheric and marine temperature (IPCC, 2014), leading to altered flow regimes in streams and rivers (Hattermann et al., 2017; IPCC, 2008). Average annual runoff is expected to increase in high latitudes and the wet tropics, and to decrease at mid-latitudes and in dry tropical regions (IPCC, 2014; Rodell et al., 2018). Snowmelt-fed regions will likely exhibit consistently earlier peak flows, except in limited areas where increases in precipitation are expected to augment snow accumulation (IPCC, 2014). Changes to flow regimes are more varied in other areas (Gosling et al., 2017; IPCC, 2014). Yet, intensification of hydrologic events, in terms of both floods and droughts, is expected in many regions as climate patterns shift (Grimm et al., 2013; IPCC, 2008). Greater frequency of punctuated high flow events will increase the delivery of sediments, organic matter, and contaminants (Grimm et al., 2013). Heavy flow events can lead to faster decomposition of organic matter through physical fragmentation (Paul et al., 2006), unless flows bury litter standing stock with fine sediment, transport toxins and pharmaceuticals that suppress the metabolism of aquatic decomposers, or deplete individual stream reaches of organic substrates by flushing aquatic primary producers and detritus downstream (Roberts et al., 2007; Rosi-Marshall et al., 2013; Young et al., 2008). In consequence, increased stream flashiness can affect rates of litter decomposition in a non-linear fashion.

Increased prevalence of streams and rivers with intermittent or ephemeral flow, particularly in regions with reduced rainfall and greater reliance on water storage in reservoirs, represents the other extreme of hydrologic intensification (Acuña et al., 2014; Datry et al., 2018). In short, the hydrological regimes of intermittent rivers and ephemeral streams are characterized by alternating flowing, non-flowing and dry phases (del Campo et al., 2020). Litter accumulation, processing, and downstream transport vary in importance amongst phases, resulting in high temporal and spatial variability of litter decomposition. Gradual re-wetting events can stimulate high rates of litter processing *in situ*, whereas pulsed high flow re-wetting events can result in export of accumulated leaf litter. Greater variability of hydrologic flow in

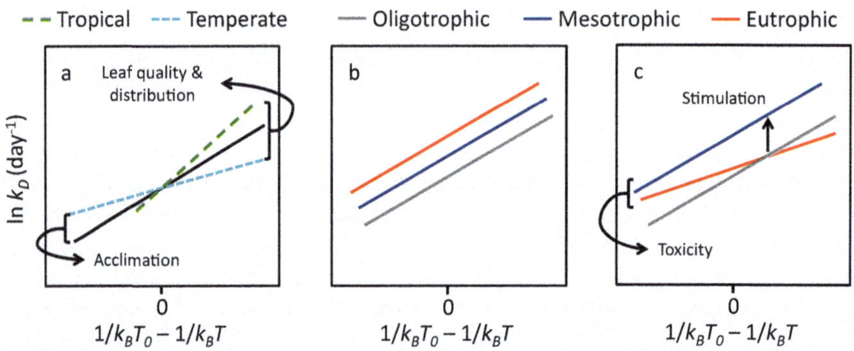

Fig. 12.2 Example mechanisms that influence the temperature dependence of leaf litter decomposition, depicted in the form of Arrhenius plots that visualize the relationship described in Eq. 12.2. The x-axis in (a–c) is normalized inverse temperature, where k_B is Boltzmann's constant (8.62 × 10^{-5} eV K^{-1}), T is temperature in Kelvin (K), and T_0 is a standard temperature. This normalization centers the data at 0 on the x-axis and allows for interpretation of the intercept at the standard temperature. Temperatures cooler than the standard temperature are plotted to the left of 0, warmer temperatures to the right of 0. The y-axis is the natural logarithm of the decomposition coefficient (k_D, day^{-1}). The slope of the relationship quantifies the apparent activation energy (E_a, eV). The solid black line in (a) represents the predicted value of E_a, based on Metabolic Scaling Theory (0.65 eV; Brown et al., 2004). This value is very similar to the Ea of cellulose strip decay in streams (0.68 eV; Tiegs et al., 2019). The colored dashed lines in (a) depict values of E_a observed in temperate (turquoise; 0.27 eV) and tropical (green; 0.75 eV) biomes (Follstad Shah et al., 2017). In temperate biomes, rates may be higher than predicted in cooler water due to acclimation of microbes and detritvores to seasonal pulses of allochthonous c inputs, but lower than predicted in warmer water due to the supply of labile plant litter that decomposes rapidly. The colored lines in (b, c) represent streams and rivers categorized by trophic status: oligotrophic (gray), mesotrophic (blue), and eutrophic (red). Increasing nutrient availability can stimulate rates of decomposition (Ferreira, Chauvet et al., 2014; Rosemond et al., 2015), as illustrated by shifts in intercepts amongst systems with different trophic status (b, c). If rates increase similarly across the thermal spectrum, the E_a may remain constant amongst trophic groups (b; Jabiol et al., 2020). However, nutrient toxicity in eutrophic systems can suppress rates of decomposition, especially when aquatic organisms are near thermal maxima (Woodward et al., 2012). This phenomenon could induce shifts in both the value of E_a (i.e., slope) and the rate of decomposition at the standard temperature (i.e., intercept) in eutrophic systems (c). More details are described in the text (see *Modulation of temperature sensitivity by biotic & abiotic factors)*

river networks increases the relative importance of reservoirs for processing organic matter, as a greater fraction of detritus is transported in pulse events to these receiving water bodies (Acuña & Tockner, 2010). The effects of hydrologic intensification on litter decomposition are not unidirectional, unlike the effects of elevated CO_2 and temperature, but may be characterized as a unimodal pattern, where absence of flow severely restricts decomposition, rates increase towards a metabolic maximum as moisture becomes non-limiting, and high flow events flush organic matter from the system (Fig. 12.1c).

Here, I summarize studies that have tested predictions illustrated in Fig. 12.1 by quantifying 1) the effect size of elevated atmospheric CO_2 concentration and temperature on litter decomposition relative to controls, and 2) the temperature dependence of litter decomposition. I also assess how elevated atmospheric CO_2 concentration, elevated temperature, altered hydrologic regimes, and other major global changes (e.g., eutrophication) may interact to affect litter decomposition in streams and rivers. Finally, I describe how changes to organic matter processing in lotic ecosystems may influence the global C cycle and aquatic food webs.

12.2 Effect Size of Elevated Atmospheric CO_2 Concentration and Warming on Litter Decomposition

12.2.1 Elevated Atmospheric CO_2 Concentration

Amani et al. (2019) conducted a meta-analysis to assess the effects of elevated temperature, elevated CO_2 concentration, and their interaction on litter decomposition rates in streams and rivers. Elevated CO_2 concentration had no effect on decomposition rates relative to controls, contrary to the authors' prediction (Fig. 12.1a), perhaps because differences in leaf quality induced by elevated CO_2 are more subtle relative to the wide variation observed amongst different plant genera. In addition, leaching of leaf litter can mitigate differences in leaf quality induced by elevated CO_2 (Ferreira et al., 2010; Rier et al., 2005), conditioning by colonizing microbial communities (Ferreira et al., 2010), or the diversity of decomposers present (Rier et al., 2005). Amani et al. (2019) also found that the combined effects of elevated atmospheric CO_2 concentration and elevated temperature had no effect on decomposition rates relative to controls, perhaps due to the opposing influence of these factors on decomposition (Fig. 12.1a,b). However, the sample sizes used for comparisons were small (<20 paired control-treatment observations), suggesting their results are not broadly generalizable. Other studies have indicated that temperature has a greater influence on leaf litter decomposition relative to elevated atmospheric CO_2 concentration (Ferreira & Chauvet, 2011; Martins, Melo et al., 2017). Amani et al. (2019) concluded that additional studies are required to assess the role of elevated atmospheric CO_2 concentration and its interaction with elevated temperature on litter decomposition. Specifically, they urged that individual studies manipulate both CO_2 and temperature, utilize fast and slow decomposing litter types, and assess the role of both microbes and detritrivores. They also recommended that more of these studies be carried out in the tropics.

12.2.2 Elevated Temperature

The meta-analysis conducted by Amani et al. (2019) showed that elevated temperature (+1 °C or more) increased rates of litter decomposition with an effect size (Hedge's *g;* Hedges et al., 1999) of 1.20 (95% CI: 0.96–1.43). Stimulation of decomposition rates with elevated temperature was observed across natural streams (both correlative [n = 71] and manipulative studies [n = 20]) and laboratory experiments (n = 57), as well as along latitudinal gradients (n = 22), with an effect size > 1 in these four study types. Elevated temperature stimulated decomposition rates to a lesser extent (effect size < 1) along altitudinal gradients (n = 32). The effect of elevated temperature was large (> 0.80) for total litter decomposition (mediated by both microbes and detritivores) in all scenarios, whereas microbial-driven litter decomposition was strongly stimulated only in manipulative field studies conducted in an oligotrophic, low temperature stream (mean temperature of 2.8 °C) using *Quercus* leaves (a slow decomposing litter type). Elevated temperature augmented decomposition rates of leaf litter from some genera, but not others, and depended on study type. For example, decomposition of both *Alnus* (a fast decomposing litter) and *Quercus* significantly increased in laboratory studies (effect size > 1), but not along altitudinal gradients (non-significant effect size). However, in all but one study type, the sample size used to calculate genus-specific effect sizes was small (≤13), prohibiting broad generalization. The meta-analysis by Amani et al. (2019) demonstrates that elevated temperature generally promotes faster rates of leaf litter decomposition in freshwater ecosystems, but the type of leaf litter present and the degree to which microbes and detritivores contribute to leaf litter processing may modulate its effect.

12.3 Quantifying the Temperature Dependence of Litter Decomposition

12.3.1 Theory

Increased capacity to predict changes to ecosystem process rates with shifts in temperature is needed to better assess the effect of global warming on organic matter decomposition. Metabolic Scaling Theory purports that ecosystem flux rates reflect the combined metabolic rates of individuals within the ecosystem, and can thus be predicted from allometric and thermal scaling relationships (Brown et al., 2004; Enquist et al., 2003). The theory assumes that resources are in steady state supply (Brown et al., 2004). Under such conditions, allometric scaling relationships for individual metabolic rate and the density of organisms in a population using a common resource are predicted to be the inverse of one another, leading to mass invariant community energy flux rates (Damuth, 1981; Enquist & Niklas, 2001). This phenomenon, called the 'energy equivalence rule', suggests ecosystem flux rates should be invariant with biomass standing stock, yet still dependent on shifts in temperature (Enquist et al., 2003). If so, then the magnitude of change in litter decomposition rates with a given change in temperature (i.e., the temperature dependence) can be quantified as an activation energy, using the linearized form of the Arrhenius equation (Arrhenius, 1915):

$$\text{In } k_D = \text{In } r_0 - E \times \frac{1}{k_B T} \tag{12.1}$$

where k_D is the litter decomposition rate coefficient (day^{-1}), r_0 is a normalization constant, E is the activation energy (eV; 1 eV = 1.6×10^{-19} J, or 96 kJ mol^{-1}), k_B is Boltzmann's constant (8.62×10^{-5} eV K^{-1}), and T is temperature in Kelvin (K).

Typically, temperature data are centered using a normalization, such that values of 0 on the x-axis represent rates at a standard temperature (Allen et al., 2005; Demars et al., 2016; Follstad Shah et al., 2017; Yvon-Durocher et al., 2012):

$$\text{In } k_D = \text{In } r_0 - E \times (\frac{1}{k_B T_0} - \frac{1}{k_B T}) \tag{12.2}$$

where T_0 is the standard water temperature (e.g., the average observed temperature in K). Empirical estimates of the activation energy are considered *apparent* (denoted as E_a) rather than *intrinsic* (E) values, because it is difficult to isolate the effect of temperature from the multitude of abiotic (e.g., stream discharge) and biotic (e.g., leaf litter quality) factors that influence rates of decomposition and the interaction of these factors with temperature (Fig. 12.2). The values reported herein represent *apparent* estimates of temperature dependence. Ecosystem C flux rates should reflect the average temperature dependence of organismal metabolism (Enquist et al., 2003), found to be ~ 0.65 eV (Gillooly et al., 2001; Table 12.1). Empirical data show that the average E_a of a multitude of thermal biological response rates associated with freshwater organisms is 0.77 eV (95% CI: 0.64–0.91 eV; Dell et al., 2011; Table 12.1).

12.3.2 Results from Past Studies

Several studies have quantified the temperature sensitivity of organic matter decomposition in aquatic habitats, providing estimates that range from 0.34 to 0.68 eV for microbe-mediated decomposition and 0 to 0.34 eV for total decomposition (Boyero, Pearson, Gessner et al., 2011; Follstad Shah et al., 2017; Tiegs et al., 2019; Fig. 12.2; Table 12.2). Boyero, Pearson, Gessner et al. (2011) conducted a litter decomposition experiment, using *Alnus glutinosa* leaves decaying in fine (0.5 mm) and coarse (10 mm) mesh bags, at 24 sites spanning absolute latitudes ranging from 0.37° to 47.80°. Their experiment lasted 56 days and coincided with the dry season in the tropics and autumn in temperate biomes. The E_a of litter decomposition (0.41 ± 0.21 eV) mediated by microbes differed from the E_a of total decomposition, which

Table 12.1 Comparison of the apparent activation energy (E_a, eV) of metabolic rates by taxonomic group. 'NA' denotes where data were not available

Category	Metabolic process	Taxonomic group	Sample size	E_a		Citation
			(#)	(eV)	95% CI (eV)	
Plants						
	Respiration[a]	Plants	67	0.66	NA	Gillooly et al. (2001)
	Varied	Plants	20	0.57	0.50–0.62	Dell et al. (2011)
Microbes						
	Respiration[a]	Unicells	30	0.76	NA	Gillooly et al. (2001)
	Varied	Unicells	12	0.51	0.32–0.68	Dell et al. (2011)
	Varied	Fungi	4	0.95	0.75–1.15	Dell et al. (2011)
	Respiration	Bacteria	205	0.59	NA	Lopez Urritia and Moran (2007)
	Respiration	Microbes	48	1.00	NA	Jabiol et al. (2020)
	Respiration[a]	Fungi	48	0.55	NA	Jabiol et al. (2020)
	Production	Bacteria	851	0.58	NA	Lopez Urritia and Moran (2007)
	Production	Bacteria	353	0.50	NA	Lopez Urritia and Moran (2007)
	Production	Bacteria	190	0.42	NA	Lopez Urritia and Moran (2007)
	Production	Bacteria	50	0.86	0.56–1.17	Sinsabaugh and Follstad Shah (2010)
Invertebrates						
	Respiration[a]	Invertebrates	20	0.79	NA	Gillooly et al. (2001)

(continued)

Table 12.1 (continued)

Category	Metabolic process	Taxonomic group	Sample size	E_a		Citation
			(#)	(eV)	95% CI (eV)	
	Varied	Invertebrates	81	0.87	0.76–0.95	Dell et al. (2011)
	Varied	Insects	128	0.63	0.47–0.78	Dell et al. (2011)
	Respiration	*Ephemeroptera*	NA	0.75–1.38[b]	NA	Shah et al. (2019)
	Respiration	*Ephemeroptera*	NA	0.56–0.84[b]	NA	Shah et al. (2019)
	Respiration	*Plecoptera*	NA	0.59–1.14[b]	NA	Shah et al. (2019)
	Respiration	*Plecoptera*	NA	1.23–1.24[b]	NA	Shah et al. (2019)
	Respiration	*Gammarus*	48	1.15	NA	Jabiol et al. (2020)
	Respiration	*Potamophylax*	48	0.99	NA	Jabiol et al. (2020)
	Respiration	*Sericostoma*	48	0.55	NA	Jabiol et al. (2020)
Vertebrates						
	Respiration[a]	Fish	113	0.43	NA	Gillooly et al. (2001)
	Respiration[a]	Amphibians	64	0.50	NA	Gillooly et al. (2001)
	Respiration[a]	Reptiles	105	0.76	NA	Gillooly et al. (2001)
	Respiration[a]	Bird & mammals	142	0.78	NA	Gillooly et al. (2001)
	Varied	Vertebrate	127	0.51	0.36–0.67	Dell et al. (2011)
Ecosystem						
	Varied	Freshwater organisms	89	0.77	0.64–0.91	Dell et al. (2011)
	Varied	Marine organisms	78	0.62	0.52–0.68	Dell et al. (2011)
	Varied	Terrestrial organisms	205	0.64	0.57–0.68	Dell et al. (2011)

(continued)

Table 12.1 (continued)

Category	Metabolic process	Taxonomic group	Sample size	E_a		Citation
			(#)	(eV)	95% CI (eV)	
Trophic Group						
	Varied	Producer	24	0.54	0.44–0.64	Dell et al. (2011)
	Varied	Detritivore	4	0.46	0.00–0.97	Dell et al. (2011)
	Varied	Herbivore	50	0.83	0.66–0.99	Dell et al. (2011)
	Varied	Omnivore	100	0.63	0.53–0.73	Dell et al. (2011)
	Varied	Carnivore	181	0.61	0.56–0.67	Dell et al. (2011)

[a]Mass-specific respiration rate
[b]Converted from reported range of Q_{10} values at temperatures considered not stressful to the organisms studied. No average value was reported

was invariant with respect to temperature (Table 12.2). Follstad Shah et al. (2017) synthesized a dataset of 169 studies conducted between absolute latitudes 0° to 60°, comprised of 1,025 observations (169 from fine mesh [$\leq$ 1 mm], 856 from coarse mesh [> 1 mm]) of litter decomposition for 85 plant genera. These authors found a common value of E_a for litter decomposition mediated by microbes and total decomposition (0.34 $\pm$ 0.04 eV; Table 12.1). However, they also found that the E_a varied across temperate (0.27 $\pm$ 0.05 eV, 95% CI: 0.18–0.37 eV) and tropical (0.75 $\pm$ 0.13 eV, 95% CI: 0.50–1.01 eV) biomes and amongst twelve plant genera (Fig. 12.2). Tiegs et al. (2019) coordinated a global-scale standardized assay in which cotton strips were deployed for 30 days in 514 streams spanning 140° latitude, with representation from all of Earth's biomes. The temperature dependence of cotton strip decay, measured as the loss of tensile strength, was 0.68 eV. All of these studies relied on mean temperatures reported during decomposition experiments, which is an important caveat. A recent modeling effort shows that the use of central tendencies of temperature can underestimate the value of E_a as compared to calculations including variation in temperature over the course of leaf incubation (Tomczyk et al., 2020).

The range of E_a estimates for litter decomposition in aquatic habitats is generally lower than E_a values predicted by Metabolic Scaling Theory, as well as empirically derived estimates of E_a for decomposition of plant roots, shoots, and large woody debris in terrestrial habitats, and short-term rates of soil respiration, biofilm respiration, and aquatic ecosystem respiration (Table 12.2). However, a study of benthic respiration with small samples size found its temperature sensitivity to be similar to the observed range of E_a estimates for litter decomposition (Table 12.2).

Table 12.2 Comparison of the apparent activation energy (E_a, eV) of ecosystem processes. 'NA' denotes where data were not available

Process & habitat	Type	Method	Sample Size	E_a			
			(#)	(eV)	95% CI (eV)	SE	Citation
Decomposition							
Streams	Leaf litter[a]	Field experiment	22	0.41	NA	0.21	Boyero, Pearson, Gessner et al. (2011)
Streams	Leaf litter[b]	Field experiment	22	0.00	NA	NA	Boyero, Pearson, Gessner et al. (2011)
Streams & rivers	Leaf litter[a]	Data synthesis	169	0.37	0.19–0.56	0.09	Follstad Shah et al. (2017)
Streams & rivers	Leaf litter[b]	Data synthesis	856	0.33	0.25–0.40	0.04	Follstad Shah et al. (2017)
Streams & rivers	Leaf litter[c]	Data synthesis	1025	0.34	0.27–0.40	0.04	Follstad Shah et al. (2017)
Aquatic mesocosms	Leaf litter[b]	Lab experiment	48	0.56	NA	0.53	Jabiol et al. (2020)
Aquatic mesocosms	Leaf litter[c]	Lab experiment	192	0.12	NA	0.31	Jabiol et al. (2020)
Streams	Cotton strip	Field experiment	360	0.68	NA	NA	Tiegs et al. (2019)
Terrestrial	Cotton strip	Field experiment	346	0.40	NA	NA	Tiegs et al. (2019)
Terrestrial	Root	Data synthesis	48	0.75	0.44–1.06	NA	Allen et al. (2005)
Terrestrial	Plant litter	Lab experiment	75	0.53–0.92[g]	NA	NA	Craine et al. (2010)
Terrestrial	Wood	Field experiment	2016	0.50	0.48–0.52	NA	Hu et al. (2018, 2020)
Primary production							
Aquatic mesocosms	Net	Field experiment	131	0.41	0.32–0.50	NA	Yvon-Durocher et al. (2010)
Aquatic mesocosms	Gross	Field experiment	131	0.45	0.38–0.53	NA	Yvon-Durocher et al. (2010)
Geothermal streams	Gross	Field experiment	13	0.54	NA	0.24	Demars et al. (2011)

(continued)

Table 12.2 (continued)

Process & habitat	Type	Method	Sample Size	E_a			
			(#)	(eV)	95% CI (eV)	SE	Citation
Geothermal streams	Gross	Field experiment	39	0.50	0.35–0.65	0.07	Demars et al. (2016)
Streams	Gross	Field experiment[e]	222	1.15	NA	0.16	Demars et al. (2016)
Streams	Gross[d]	Data synthesis[e]	222	0.92	NA	0.16	Demars et al. (2016)
Streams	Gross	Field experiment	292	0.71[h]	NA	NA	Song et al. (2018)
Terrestrial & wetland	Net	Data synthesis[e]	1599	0.00	NA	NA	Kerkhoff et al. (2005)
Respiration							
Biofilms	Biofilm	Lab experiment	94	0.54	NA	0.12	Acuña et al. (2008)
Geothermal streams	Benthic	Field experiment	13	0.66	0.21–1.11	NA	Perkins et al. (2012)
Sediment	Benthic	Lab experiment	13	0.47	0.31–0.63	NA	Perkins et al. (2012)
Soils	Soil	Data synthesis	174	0.65	0.60–0.70	NA	Allen et al. (2005)
Soils	Soil	Data synthesis[f]	133	0.41	0.28–0.54	NA	Allen et al. (2005)
Soils	Soil	Lab experiment	420	0.85	NA	NA	Craine et al. (2010)
Soils	Soil	Lab experiment[f]	420	0.93	NA	NA	Craine et al. (2010)
Soils	Soil	Data synthesis	206	0.54–1.53[g]	NA	NA	Craine et al. (2010)
Streams	Ecosystem	Field experiment[e]	18	0.63	NA	0.17	Valett et al. (2008)
Streams & rivers	Ecosystem	Data synthesis[e]	222	0.59	NA	0.10	Demars et al. (2016)
Streams & rivers	Ecosystem[d]	Data synthesis[e]	222	0.44	NA	0.10	Demars et al. (2016)
Streams	Ecosystem	Field experiment	292	0.70	NA	NA	Song et al. (2018)
Aquatic mesocosms	Ecosystem	Field experiment	131	0.62	0.55–0.69	NA	Yvon-Durocher et al. (2010)

(continued)

Table 12.2 (continued)

Process & habitat	Type	Method	Sample Size	E_a			
			(#)	(eV)	95% CI (eV)	SE	Citation
Geothermal streams	Ecosystem	Field experiment	13	0.67	0.17–1.17	0.23	Demars et al. (2011)
Forest canopy	Ecosystem	Data synthesis	8999	0.62	NA	NA	Yvon-Durocher et al. (2012)
Non-forest canopy	Ecosystem	Data synthesis	3271	0.70[h]	NA	NA	Yvon-Durocher et al. (2012)
Soils	Ecosystem	Data synthesis	4160	0.65	NA	NA	Yvon-Durocher et al. (2012)
Estuarine pelagic	Ecosystem	Data synthesis	1018	0.59	NA	NA	Yvon-Durocher et al. (2012)
Estuarine benthic	Ecosystem	Data synthesis	443	0.63	NA	NA	Yvon-Durocher et al. (2012)
Lake pelagic	Ecosystem	Data synthesis	3666	0.63	NA	NA	Yvon-Durocher et al. (2012)
Lake benthic	Ecosystem	Data synthesis	428	0.55	NA	NA	Yvon-Durocher et al. (2012)
Rivers	Ecosystem	Data synthesis	154	0.58	NA	NA	Yvon-Durocher et al. (2012)
Ocean microbial	Ecosystem	Data synthesis	438	0.57	NA	NA	Yvon-Durocher et al. (2012)

[a]Mediated by microbes alone
[b]Mediated by both microbes and detritivores
[c]Pooled value across microbial and total decomposition
[d]Discharge-corrected rates
[e]Rates measured over one season. Other rates were measured on a short-term basis, unless noted otherwise
[f]Rates measured over one year. Other rates were measured on a short-term basis, unless noted otherwise
[g]Range of reported values. No average value was reported
[h]Median value

12.3.3 *Modulation of Temperature Sensitivity by Biotic and Abiotic Factors*

The results of studies quantifying the temperature dependence of litter decomposition vary, potentially, due to differences in methodology (i.e., standardized experiment *vs.* data synthesis, sample size, timing and duration of assays, and organic substrates assessed). However, factors that can affect the E_a of decomposition or co-vary with temperature must also be considered. Interactions of these factors with temperature may also help to explain the observed differences in the E_a of litter decomposition relative to the temperature sensitivities of other types of organic matter utilization. Some of these factors include the biogeography of detritivores and plant genera, thermal acclimation capacity of biota, variation in litter quality, and variation in key environmental controls (Fig. 12.2). Each of these factors is explored in more detail below.

Three studies have shown that temperature-corrected rates (i.e., reported per degree-day) of total decomposition increase with absolute latitude (Boyero, Pearson, Gessner et al., 2011; Follstad Shah et al., 2017; Irons et al., 1994), but the mechanisms driving this trend are not yet clear. Temperate streams generally have greater density, relative abundance, and diversity of detritivores relative to streams in the tropics (Boyero, Pearson, Dudgeon et al., 2011). Boyero, Pearson, Gessner et al. (2011) hypothesized that greater densities of detritivores in temperate biomes relative to the tropics contributed to elevated temperature-corrected rates of litter decomposition at higher latitudes, thereby dampening the E_a. In other words, greater densities of metabolically active detritivores can lead to faster rates of decomposition, despite lower temperature. Detritivore density data within *Alnus* leaf packs at half of the sites studied by Boyero, Pearson, Gessner et al. (2011), where such data were available, supported this hypothesis ($\log_e$ detritivore density $= 1.36 + 0.10 \cdot$ latitude, $r^2 = 0.50$, $P = 0.015$, $n = 11$). In contrast, Follstad Shah et al. (2017) did not find a correlation between detritivore density and absolute latitude for the subset of observations within their global database that had the appropriate data to analyze ($P > 0.05$, $n = 61$). Irons et al. (1994) provided two alternative hypotheses for the observed positive correlation between temperature-corrected decomposition and absolute latitude. First, they proposed that microbes may be less metabolically active in colder water relative to detritivores. If so, the rate of temperature-corrected microbe-mediated decomposition should decrease with absolute latitude. Second, they proposed that detritivores may be more important than microbes to decomposition in temperate zones. If so, temperature-corrected decomposition rate at higher latitudes should be greater for detritivores than microbes. Results from Boyero, Pearson, Gessner et al. (2011) and Follstad Shah et al. (2017) provided mixed support for these hypotheses. The former study found that temperature-corrected, microbe-mediated decomposition was invariant with respect to absolute latitude. In contrast, the latter study found a positive correlation between temperature-corrected, microbe-mediated decomposition and absolute latitude, suggesting that microbes also compensate for lower temperatures in temperate biomes. In both studies, rates of total decomposition at higher latitudes were greater than rates of microbe-mediated decomposition. However, new approaches have been developed to isolate the effect of detritivores from total decomposition rates (Lecerf, 2017; Woodward et al., 2012). Application of these approaches would allow for more robust testing of these hypotheses, and help determine whether these mechanisms influence the E_a of litter decomposition.

Many streams and rivers receive seasonal pulses of organic matter inputs from adjacent terrestrial habitats, particularly in temperate biomes. Yet, organic substrates derived from both autochthonous and allochthonous sources can be scarce in heavily shaded streams at the peak of the growing season, coincident with higher water temperature in temperate biomes (Roberts et al., 2007). Based on these observations, Follstad Shah et al. (2017) suggested that selection pressures on aquatic organisms that utilize pulsed litter inputs at low temperatures could have led to physiological adaptations (e.g., properties of enzymes and maximum growth rates associated with microbial communities; Bradford, 2013; Wallenstein et al., 2010); aquatic community compositions (Dang et al., 2009; Findlay et al., 2008; Friberg et al., 2009; Handa et al., 2014; Martínez et al., 2014; Strickland et al., 2015); functional redundancies (Nelson et al., 2020); and trophic interactions (Rall et al., 2010) that facilitate high activity at low temperature and rapid litter exploitation (Benstead & Huryn, 2011). Evolutionary adaptations within both microbial and detritivore communities that allow for utilization of pulsed resources at colder temperatures would promote the observed relationship between temperature-corrected decomposition rates and absolute latitude. Thermal acclimation via evolutionary adaption also helps to explain the weaker temperature sensitivity of total decomposition observed in temperate biomes relative to the tropics (Follstad Shah et al., 2017). This 'thermal acclimation hypothesis' is somewhat supported by a recent study using controlled metabolic chambers set at non-stressful temperatures (7.5 °C and 15 °C; Shah et al., 2019), which showed that the respiration rate of Ephemeroptera species collected from temperate streams have a weaker temperature sensitivity than congeners from tropical streams (Table 12.2). However, this pattern did not hold for Plecoptera species. Many studies have been conducted to quantify the temperature sensitivity of metabolic rates for microbes, invertebrates, and fish (Table 12.1). However, few have assessed the temperature sensitivity of metabolic rates for aquatic organisms collected from different biomes. Further experimentation or synthesis of existing data is required to better test the thermal acclimation hypothesis.

Variation in light availability and seasonal pulses of organic matter inputs from terrestrial vegetation to streams and rivers in temperate biomes violate the Metabolic Scaling Theory assumption of steady state resource supply. More constant litter inputs to tropical streams and rivers relative to temperate biomes may explain why the temperature sensitivity of total decomposition in the tropics observed by Follstad Shah et al. (2017) was similar to the canonical value predicted by Metabolic Scaling Theory, while the value of E_a was much lower for leaf litter processed within streams and rivers from temperate biomes. However, some studies suggest resource pulses may augment the temperature sensitivity of organic matter utilization. Estimates of E_a range from 0.58 to 0.70 eV for ecosystem respiration in streams and rivers calculated over short timescales (Demars et al., 2011, 2016; Song et al., 2018; Valett et al., 2008; Yvon-Durocher et al., 2012; Table 12.2). Most of estimates were made under steady state conditions (i.e., when rates of gross primary production [GPP] and ecosystem respiration [ER] were correlated within systems; Demars et al., 2016; Song et al., 2018). Over annual timescales, Yvon-Durocher et al. (2012) found that ecosystem respiration had greater values of E_a (0.57–1.08 eV) relative to terrestrial ecosystems (0–0.42 eV), possibly due to a stronger influence of allochthonous C subsidies relative to autochthonous primary production in aquatic ecosystems. In Walker Branch, a heavily studied temperate spring-fed stream in North America, GPP and ER are correlated over annual timescales (Roberts et al., 2007), yet neither is related to temperature in a way predicted by thermodynamics due to shifting resource availability (Fig. 12.3). Hence, calculating the E_a for individual systems or particular seasons in temperate systems can be confounded by shifting light and C supply. The same concept applies to estimates of E_a for litter decomposition derived

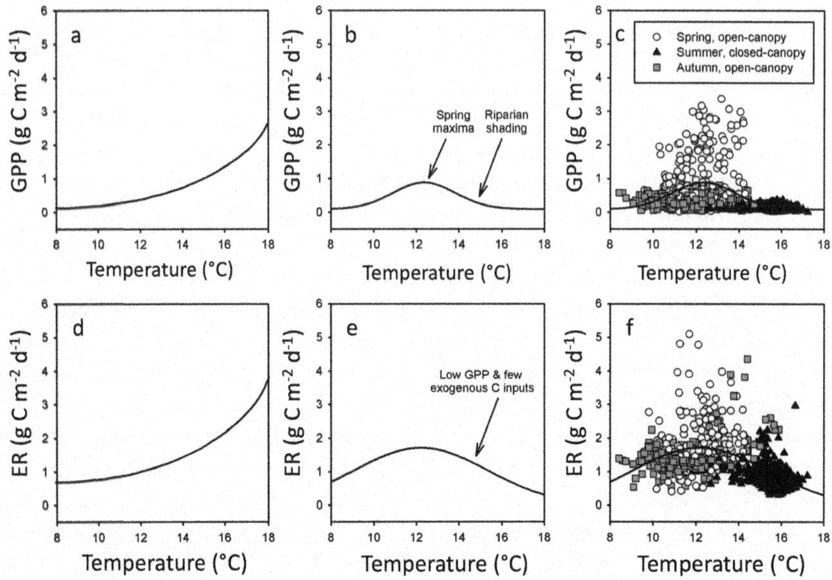

Fig. 12.3 Theoretical (a, d) and empirical (b, c, e, f) relationships between temperature and rates of gross primary production (a–c) and ecosystem respiration (d–f) in Walker Branch, Tennessee, USA. Data from Roberts et al. (2007). Symbols denote different rates in different seasons and forest canopy conditions

from individual systems. For example, Griffiths and Tiegs (2016) found the E_a of litter decomposition to range from 3.3 to 6.3 eV for *Acer rubrum*, *Liriodendron tulipifera*, and *Quercus alba* decomposing within Walker Branch. However, this study was conducted over a narrow temperature range (mean daily difference of ≤ 1.1 °C between sites) in autumn. In addition, summer conditions at this site are marked by low rates of GPP (Fig. 12.3) and organic matter standing stocks (Roberts et al., 2007). Release from substrate limitation may have thus induced the extremely strong E_a observed.

Evolutionary history and adaption to climate shape plant traits, in addition to traits associated with microbes and aquatic invertebrates. Plant traits are strong determinants of litter decomposition rates in aquatic ecosystems (Ardón & Pringle, 2008; Gessner & Chauvet, 1994; Martínez et al., 2014; Ostrofsky, 1997). An experimental study conducted along latitudinal gradients using litter mixtures (Boyero et al., 2016) and a large data synthesis of single species decay rates (LeRoy et al., 2020) found that phylogeny, which influences plant traits (Cornwell et al., 2014), is a better predictor of decomposition rate than climate in streams and rivers. Similarly, plant traits are a stronger driver of leaf and wood decomposition than climate in terrestrial systems (Cornwell et al., 2008; Hu et al., 2018). Follstad Shah et al. (2017) found that variance in the E_a of genus-specific litter decomposition rates could be explained by plant traits, including leaf %N and ratios of C:N, lignin:N, and lignin:P. Lower values of E_a were associated with higher quality litter, similar to studies of litter and soil organic matter processing conducted in terrestrial systems (Conant, Drijber et al., 2008; Conant, Steinweg et al., 2008; Fierer et al., 2005; Hobbie, 1996; Ramirez et al., 2012; Wang et al., 2019). Higher values of E_a are expected for the decomposition of recalcitrant litter because microbial conditioning facilitates detritivore consumption of structurally complex litter (Gessner & Chauvet, 1994; Suberkropp, 1992; Wright & Covich, 2005). Enzymatic reactions required by microbes to metabolize complex, low-quality macromolecules have higher apparent activation energies than enzymatic reactions that metabolize chemically simpler leaf constituents (Bosatta & Ågren, 1999; Conant et al., 2011; Wagai et al., 2013). For example, enzymes expressed to degrade lignocellulose and polyphenols have temperature sensitivities ranging from 0.45 to 0.56 eV, while the E_a of polysaccharide hydrolysis and nutrient mineralization ranges from 0.31to 0.41 eV (Sinsabaugh & Follstad Shah, 2012; Wang et al., 2012).

Plant biogeography may influence the temperature sensitivity of litter decomposition due to co-variation with plant traits. Faster decomposing genera are typically found at higher, colder latitudes, while slower decomposing genera are typically found at lower, warmer altitudes (Boyero et al., 2017). However, Zhang et al. (2019) found no difference in litter decomposition when rates were categorized amongst three biomes, likely due to a wide variation of plant traits within each biome. Correlations between plant traits and values of E_a observed by Follstad Shah et al. (2017) were strongly influenced by the inclusion of *Alnus*, the sole plant capable of N-fixation within the analyses. Furthermore, no difference was found in the E_a of litter decomposition when all 85 plant genera were categorized as 'fast', 'medium', or 'slow' decomposing leaves (common slope of 0.23 ± 0.03 eV, 95% CI: 0.18–0.29 eV), based on the distinctions established by Peterson and Cummins (1974). The database compiled by Follstad Shah et al. (2017) was biased towards experiments using *Alnus* and *Quercus* leaves, together representing 38.4% of total decomposition rate observations. The mean E_a across the twelve plant genera for which temperature sensitivity could be quantified was 0.64 eV. The weighted mean, accounting for the number of observations per genus, was 0.40 eV (Fig. 12.4). This weighted mean value was close to the E_a of total decomposition calculated across all plant genera (0.33 eV; Table 12.2). *Alnus* and *Quercus* are two plant genera broadly distributed throughout temperate biomes. Genus-specific values of E_a for *Alnus* and *Quercus* calculated by Follstad Shah et al. (2017) were weaker than the other ten plant genera (0.16 eV and 0.32 eV, respectively; Fig. 12.4). The distribution of *Alnus* and *Quercus*, combined with their relatively weak sensitivity to temperature, likely influenced the difference in biome-specific estimates of E_a.

Abiotic factors can influence the apparent temperature sensitivity of litter decomposition, in addition to biotic factors. Tiegs et al. (2019) removed the influence of leaf quality on decomposition by using a standard substrate to show that environmental variation, such as nutrient availability and differences in pH, leads to distinct signatures of cotton strip decay amongst biomes. They also found that the E_a of cotton strip decay was stronger in streams (0.60 eV) relative to the surfaces of riparian soils (0.40 eV), and attributed the difference to variation in moisture between habitats. Biotic factors still played a role in this study. Microbial activity contributes to

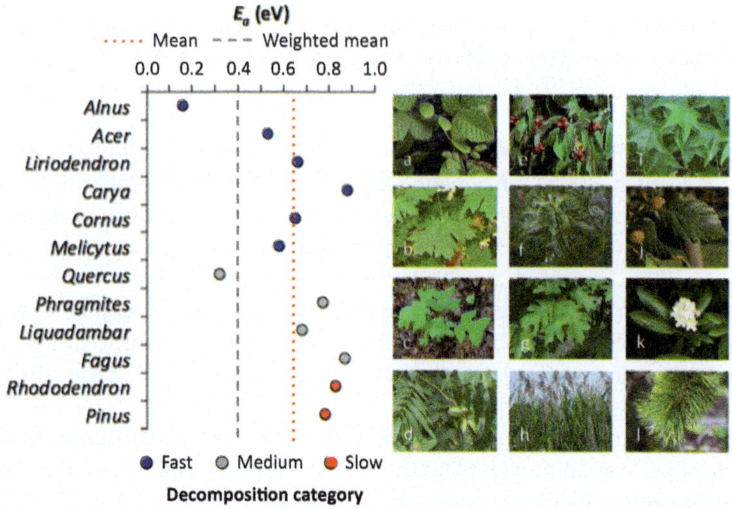

Fig. 12.4 Comparison of the genus-specific apparent activation energy (E_a, eV) amongst twelve plant genera (sample size; panel photo): *Alnus* (224; a), *Acer* (68; b), *Liriodendron* (23; c), *Carya* (14; d), *Cornus* (12; e), *Melicytus* (10; f), *Quercus* (105; g), *Phragmites* (23; h), *Liquadambar* (22; i), *Fagus* (14; j), *Rhododendron* (21; k), *Pinus* (13; l). The mean (0.64 eV) and weighted mean (0.40 eV) genus-specific E_a values are denoted by the red dotted line and gray dashed line, respectively. Colored symbols represent categories of leaf litter decomposition coefficients (k_D) associated with each genus, based on Peterson and Cummins (1974): fast ($k_D > 0.0100$ day^{-1}), medium (k_D = 0.0050–0.0100 day^{-1}), slow ($k_D < 0.0050$ day^{-1}). Photos were obtained from various websites, including but not limited to the USDA Plants Database (https://plants.sc.egov.usda.gov/java/), iNaturalist (https://www.inaturalist.org/), Wikipedia (https://en.wikipedia.org/), and Wikimedia (https://commons.wikimedia.org/). Photo credits with copyrights: J.S. Peterson (b); J. McMillan (c); T. Rodd (d); W.S. Justice (e); Rudolph89 (f); A. Huster (g); W. Mark and J. Reimer (i); P. Rothrock (j); S. McDougal (l)

cotton strip decay, in addition to environmental factors (Colas et al., 2019). The E_a for cotton strip decay in streams was similar to the canonical value predicted by Metabolic Scaling Theory (0.65 eV; Brown et al., 2004) and within the range of E_a values observed for bacterial and fungal metabolic rates (0.42–0.95 eV; Table 12.1). In addition, it was closer to the E_a of lignocellulose degradation (0.45–0.54 eV) than the E_a of polysaccharide hydrolysis and nutrient mineralization (0.31–0.41 eV; Sinsabaugh & Follstad Shah, 2012; Wang et al., 2012), reflecting that cotton strips are largely comprised of cellulose. The E_a for cotton strip decay in streams was stronger than the E_a for microbe-mediated litter decomposition (Boyero, Pearson, Gessner et al., 2011; Follstad Shah et al., 2017; Table 12.2), supporting the idea that variation in leaf quality has a strong influence on the temperature dependence of litter decomposition.

Eutrophication is an abiotic global change affecting streams and rivers (Galloway et al., 2008; Manning et al., 2020), but few studies have examined the interactive effects of temperature and nutrient supply on aquatic ecosystem processes (Cross et al., 2014). Moderate levels of nutrient availability generally stimulate rates of leaf litter decomposition at the scale of leaf packs (Ferreira, Castagneyrol et al., 2014; Rosemond et al., 2015; Woodward et al., 2012; Young et al., 2008), particularly when neither N nor P is limiting (Duarte et al., 2009; Kominoski et al., 2015). However, decomposition rates generally decline in hypertrophic streams due to declines in microbial diversity and biomass (Duarte et al., 2009), shifts in macroinvertebrate community composition (Woodward et al., 2012), stoichiometric imbalance between detritus and consumers (Tonin et al., 2017), or effects of toxins (e.g., high ammonia concentrations or loading of pesticides and pharmaceuticals coincident with nutrient inputs) on the physiology and abundance of aquatic organisms (Duarte et al., 2009; Fernandes et al., 2014; Lecerf et al., 2006; Rosi-Marshall et al., 2013; Schäfer et al., 2007, 2012). The limited number of small-scale studies that have experimentally manipulated both temperature and nutrients within mesocosms or stream-side channels indicate that interaction effects between these two factors are inconsistent. In some cases, additive or synergistic effects of temperature and nutrients have led to faster rates of microbe-mediated and total decomposition for *Alder glutinosa*, *Melicytus ramiflorus*, and *Betula pendula* (Martínez et al., 2014; Moghadam & Zimmer, 2016; Piggott et al., 2015). However, Fernandes et al. (2014) found that elevated temperature stimulated decomposition rates of *Alder glutinosa* and *Quercus robur* mediated by microbes only at low N concentrations (< 0.1 mg L^{-1}). Other studies show that the interaction of temperature and nutrients had no effect on total decomposition rates of *Melicytus ramiflorus* (Piggott et al., 2012) or the decomposition of *Corylus* sp. mediated by microbes or detritivores (Gossiaux et al., 2020; Jabiol et al., 2020). The disparate results amongst these studies suggest that interaction effects of temperature and nutrients on litter decomposition may vary by the dominant taxonomic group processing leaves and characteristics of nutrient gradients in streams and rivers. Only one study to date has assessed whether nutrient supply modulates the apparent temperature dependence of litter decomposition. Jabiol et al. (2020) found that four N concentrations ranging from 0.71–2.81 mg N L^{-1} had no effect on the E_a of litter decomposition mediated by microbes or detritivores. Additional study is needed to determine if this pattern holds over a broader range of nutrient availability and spatial scales.

12.4 Interactions Between Elevated CO_2, Elevated Temperature, and Altered Hydrologic Flow on Litter Decomposition Mediated by Microbes and Detritivores

The combined effects of elevated atmospheric CO_2 concentration, elevated temperature, and hydrologic intensification may vary in different regions of the world due to shifts in vegetation and variation in hydrologic response to climate change. Figure 12.5 illustrates how these factors may interact within tundra, temperate rainforest, desert, and tropical rainforest biomes under future climate scenarios. In the

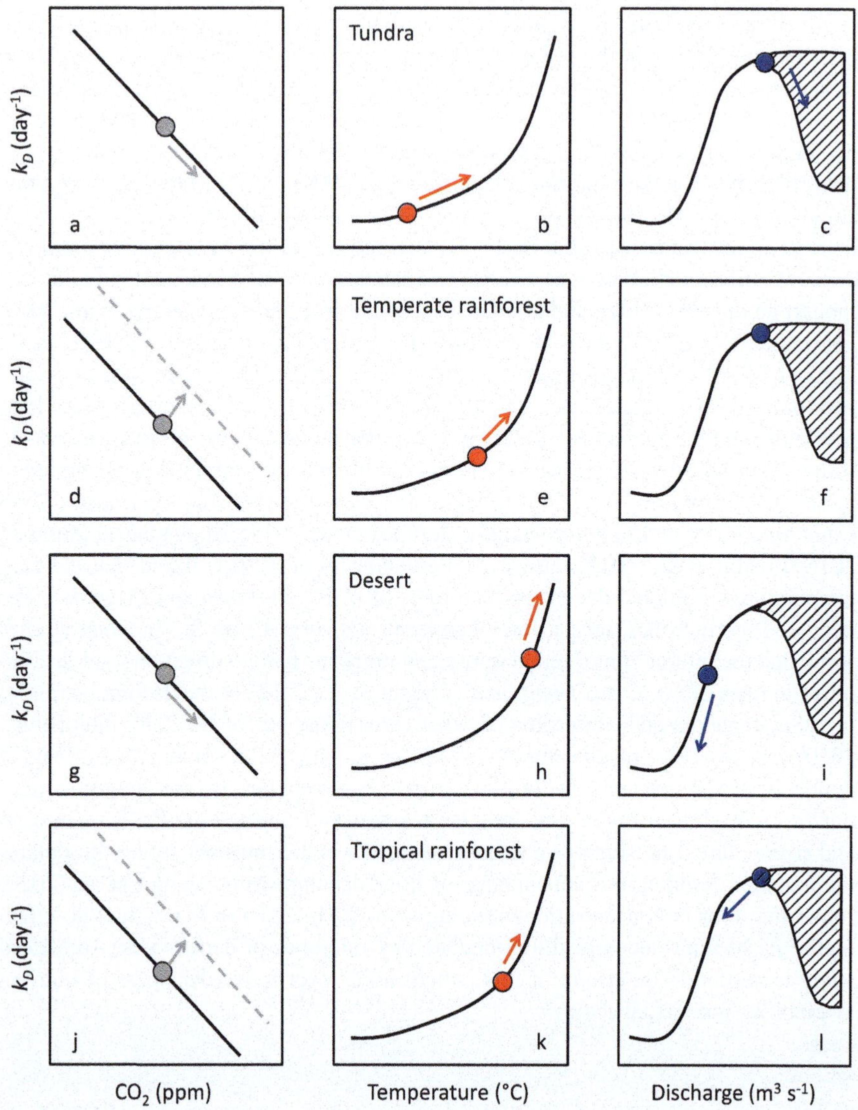

Fig. 12.5 Conceptual figure summarizing potential differences in climate change effects on leaf litter decomposition in tundra (a–c), temperate rainforest (d–f), desert (g–i), and tropical rainforest (j–l) biomes. Symbols represent current conditions, while arrows indicate direction of predicted change in each biome. Gray dashed lines in (d, j) denote more labile leaf litter with elevated CO_2 concentration due to shifts in vegetation. Shaded areas in (c, f, i, l) denote high magnitude hydrologic flow that flushes organic matter from the system

tundra, warmer temperatures and increased precipitation are predicted to stimulate productivity and promote taller vegetation, albeit within narrower riparian zones (Nilsson et al., 2013). However, elevated atmospheric CO_2 concentration may result in lower litter quality and more acidic surface water. Warmer, wetter conditions and potentially greater allochthonous inputs to aquatic ecosystems should favor faster rates of organic matter processing. However, the magnitude of these rates could be modulated or negated by lower litter quality, declining pH, and variable flow in areas of permafrost, decreasing discharge in non-permafrost areas, and glacial outburst flooding (Nilsson et al., 2015). Vegetation in the temperate rainforest of the U.S. Pacific Northwest is shifting to greater dominance of deciduous trees relative to conifers in response to warming (Ball et al., 2010; Kominoski et al., 2013), despite relatively constant water supply (Rodell et al., 2018). Deciduous plants have higher litter quality than conifers and may be more productive with higher CO_2 availability, warmer temperature, and adequate moisture. These conditions should promote faster rates of leaf litter decomposition (Kominoski et al., 2011). In semi-arid to arid western North America, combined effects of elevated temperature, greater frequency and intensity of drought, and river regulation are expected to favor herbaceous (e.g., *Bromus tectorum*) and non-native drought-tolerant (e.g., *Tamarix*) species over native, early-successional tree species (e.g., *Populus* and *Salix*; Perry et al., 2012; Reynolds & Shafroth, 2017). In addition, plant productivity is predicted to decline as higher water demand limits photosynthetic capacity (Perry et al., 2012). Decomposition rates may increase because streams and rivers in this region will be warmer, receive fewer allochthonous inputs of organic matter (Bailey et al., 2001), and be buffered from elevated CO_2 by their alkaline nature. However, rates may be suppressed by greater probability of intermittent flow (Perry et al., 2020). Large portions of the Brazilian rainforest are expected to shift to savannah as annual precipitation and rainfall variability decline (Ciemer et al., 2019). Temperature is predicted to stay relatively constant (Ciemer et al., 2019), yet small shifts can have a large influence on decomposition given strong temperature sensitivity in the tropics. For example, a roughly 10% increase in litter decomposition rate requires only a 1 °C rise in water temperature in the tropical systems but a 4 °C rise in temperate systems, given differences in the E_a observed in these biomes (Follstad Shah et al., 2017). In some cases, grasses do not have lower quality litter when grown at elevated atmospheric CO_2 concentration (Monroy et al., 2016). Hence, decomposition rates may significantly increase with lower allochthonous inputs of litter and warmer temperature, if streams remain perennial. These scenarios suggest that the response of organic matter decomposition to multiple interacting factors associated with climate change will be context specific and coupled to responses of adjacent riparian vegetation.

The relative contribution of microbes and detritivores to litter decomposition may shift in response to elevated atmospheric CO_2 concentration, warming, and hydrologic intensification. The responses of these taxa also may vary amongst biomes. Microbial activity diminishes to a lesser degree than detritivore activity on more recalcitrant leaf litter amongst plant genera (Hieber & Gessner, 2002; Martins, Melo et al., 2017). Similar patterns may hold for increased leaf recalcitrance associated with elevated atmospheric CO_2 concentration, which has been correlated to declines in detritivore activity (Martins, Melo et al., 2017; Tuchman et al., 2002), but not fungal biomass or activity (Ferreira et al., 2010; Ferreira & Chauvet, 2011; Martins, Melo et al., 2017). However, bacterial production on leaves grown under elevated CO_2 concentration has declined in some studies (e.g., Tuchman et al., 2002) but remained unchanged in other studies (e.g., Rier et al., 2005). Experimental results indicate that microbial activity becomes relatively more important with elevated temperature as compared to detritivore activity (Bärlocher et al., 2008; Boyero, Pearson, Gessner et al., 2011; Jabiol et al., 2020; Martins, Melo et al., 2017; O'Gorman et al., 2012). Elevated temperature generally favors smaller bodied metazoans ('James' Rule'; James, 1970), due to oxygen demands and different thermal sensitivities in growth and development rate (or 'temperature-size rule'; Atkinson, 1994). However, the temperature-size rule does not apply to unicellular organisms whose body size remains invariant with temperature (Forster et al., 2012). Taxonomic differences in temperature-body size scaling relationships can alter the temperature dependence of ecosystem processes mediated by microbes and detritivores, and subsequent energy flow through food webs. Finally, stream meiofauna and detritivore population body size-abundance scaling relationships can deviate from the energy equivalence rule (Schmid et al., 2000, 2020) due to stochastic hydrophysical processes (i.e., the frequency of high flow events; Schmid et al., 2020) and adaptations to elevated temperature (O'Gorman et al., 2012, 2017). Thus, shifts in size spectra should be considered when making predictions about detritivore-mediated litter decomposition response to climate change.

12.5 Significance of Leaf Litter Decomposition Responses to Climate Change

12.5.1 Global C Budget

Clearly, more studies are needed to better understand the effects of biotic and abiotic factors on the temperature sensitivity of litter decomposition in order to make predictions about how rates will change with shifts in temperature. However, the studies to date indicate that elevated temperature ranging from 1 °C to 4 °C will increase rates between 5.0–21.4% based on an E_a of 0.34 eV and 10.3–47.4% based on an E_a of 0.68 eV, assuming a standard water temperature of 10 °C. Litter decomposition dominated by microbial activity converts a sizeable fraction of organic matter to CO_2, while detritivores generate large amounts of fine particulate organic C due to low assimilation efficiencies (Baldy et al., 2007; Ward et al., 1994). Similarity in the temperature sensitivity of litter decomposition mediated by microbes and total decomposition observed by Follstad Shah et al. (2017) suggests that the fractions of gaseous C loss and particulate C transport attributed to litter decomposition will not significantly change over broad scales as temperatures rise. Yet, if temperature sensitivity does indeed vary with respect to taxonomic groups or their response to different genera of leaf litter (Boyero, Pearson, Gessner et al., 2011; Ferreira & Canhoto, 2015; Jabiol et al., 2020), then the balance between these two processes may shift in the future. These differences may be more pronounced at higher latitudes, where detritivore diversity and abundance is greater than in the tropics (Boyero, Pearson, Dudgeon et al., 2011).

Organic matter catabolism fuels heterotrophic metabolism in detrital food webs, and is thus a major component of aquatic ecosystem respiration. CO_2 produced by heterotrophic metabolism contributes about 28% of CO_2 evasion from streams and rivers under current climate conditions (Hotchkiss et al., 2015). The influence of streams and rivers on global C cycling depends upon ecosystem-level rates of both GPP and ER (Battin et al., 2008; Demars et al., 2011, 2016; Song et al., 2018). A recent data synthesis of stream metabolism studies conducted in summer (Demars et al., 2016) and models utilizing stream metabolism data collected amongst six biomes (Song et al., 2018) found that the E_a of GPP and ER is of similar magnitude (Table 12.2). Demars et al. (2016) showed that net ecosystem production (NEP; i.e., the balance between GPP and ER) is invariant with respect to temperature and inferred that CO_2 emissions from lotic ecosystems should not increase with warming when GPP and ER are tightly coupled. However, Song et al. (2018) found a rise of 1 °C in water temperature leads to a 23.6% decline in NEP because of differential responses in the ratio of GPP/ER to warming in high temperature streams with greater ratios of GPP/ER on a daily basis relative to low temperature streams with lower ratios of GPP/ER. Consequently, these authors estimated that warming will result in an increase of approximately 0.02 Pg year^{-1} from streams similar in size to the systems studied. However, dominant sources of C to streams and stream size (including seasonal desiccation) are important factors controlling the magnitude of CO_2 emitted from lotic ecosystems (Hotchkiss et al., 2015; Keller et al., 2020). It is still unclear if the E_a of NEP in streams and rivers varies in relation to these factors, both of which are changing with alteration of global climate (del Campo et al., 2020; Gosling et al., 2017; Hattermann et al., 2017; Kominoski et al., 2020; Mcdonough et al., 2020).

12.5.2 *Food Webs*

Altered rates of decomposition in response to factors associated with climate change is consequential to detrital food webs, but also influenced by aquatic community responses to elevated atmospheric CO_2 concentration, warming, and hydrologic intensification (Marks, 2019; O'Gorman et al., 2012, 2017). Leaf litter can be rapidly exploited by the increased metabolic demand induced by warming, particularly in systems with limited allochthonous C inputs (Roberts et al., 2007; Rosemond et al., 2015). Increases in extreme hydrologic events (floods and droughts) and decreases in retentive structures (e.g., large woody debris) from shifts in vegetation may reduce the amount of terrestrial detritus retained in stream ecosystems and support less detritivore secondary production (del Campo et al., 2020; Kominoski et al., 2020; Tank et al., 2010). Depletion of basal resources or changes to the quality of detritus can have cascading effects on higher trophic levels throughout aquatic food webs, manifesting as changes to biotic richness, production, nutrient cycling, and whole stream metabolism (Dudgeon et al., 2006; Estes et al., 2011; Poff et al., 2007). Similarly, loss of species or specific functional groups can significantly suppress rates of leaf litter decomposition (Bärlocher et al., 2008; Boyero et al., 2012; Handa et al., 2014; Stewart et al., 2013; Tonin et al., 2018) and limit energy flow through the food web (Graça et al., 2002; Wallace et al., 1999). However, evidence from litter-poor geothermal streams in the Hengill region of Iceland shows that concomitant shifts in warming and nutrient supply can support higher basal resource (i.e., diatom) production that allow for greater body size and biomass of higher trophic groups through altered community composition (O'Gorman et al., 2012, 2017; Nelson et al., 2017, 2020). These responses promoted patterns of litter decomposition and whole-stream metabolism temperature dependence consistent with more litter-rich systems (Demars et al., 2011, 2016; Friberg et al., 2009), suggesting maintenance of ecosystem stability despite shifts in body size-abundance scaling relationships and community composition. Thus, resource supply and its interaction with other factors of global change are important considerations affecting basal food web pathways and how they support food web dynamics.

12.6 Conclusions

It is evident that climate change factors are affecting rates of litter decomposition in streams and rivers, and subsequently influencing lotic ecosystem C budgets and food web dynamics. Elevated temperature has been more widely tested and has the most consistent effect on litter decomposition, as compared to elevated atmospheric CO_2 concentration and hydrologic intensification. However, there still exists a need to better understand each of these factors in isolation, as well as their interactions. Effects of interactions between these variables are difficult to predict across broad scales due to regional differences in climate and biogeography. No studies exist that test all three factors simultaneously relative to controls and examine the responses of both microbes and detritivores. Most studies have been conducted at small scale (e.g., laboratory experiments or short-term warming of single systems) or have used space for time substitutions (e.g., latitudinal gradients or comparisons amongst closely situated geothermal systems). Additional long-term studies of change *in situ* and simulation models are required to parse individual and interacting effects under realistic climate change conditions (O'Gorman et al., 2012). The open nature of river networks and the connectedness of streams and rivers to adjacent riparian zones requires that fluctuation in resource supply must be considered in tandem to changes in atmospheric CO_2 concentration, temperature, and hydrologic regimes.

References

Acuña, V., Datry, T., Marshall, J., Barceló, D., Dahm, C. N., Ginebreda, A., McGregor, G., Sabater, S., Tockner, K., & Palmer, M. A. (2014). Why should we care about temporary waterways? *Science, 343*(6175), 1080–1081. https://doi.org/10.1126/science.1246666

Acuña, V., & Tockner, K. (2010). The effects of alterations in temperature and flow regime on organic carbon dynamics in Mediterranean river networks. *Global Change Biology, 16*(9), 2638–2650. https://doi.org/10.1111/j.1365-2486.2010.02170.x

Acuña, V., Wolf, A., Uehlinger, U., & Tockner, K. (2008). Temperature dependence of stream benthic respiration in an Alpine river network under global warming. *Freshwater Biology, 53*(10), 2076–2088. https://doi.org/10.1111/j.1365-2427.2008.02028.x

Allen, A. P., Gillooly, J. F., & Brown, J. H. (2005). Linking the global carbon cycle to individual metabolism. *Functional Ecology, 19*(2), 202–213. https://doi.org/10.1111/j.1365-2435.2005.00952.x

Amani, M., Graça, M. A. S., & Ferreira, V. (2019). Effects of elevated atmospheric CO_2 concentration and temperature on litter decomposition in streams: A meta-analysis. *International Review of Hydrobiology, 104*(1–2), 14–25. https://doi.org/10.1002/iroh.201801965

Ardón, M., & Pringle, C. M. (2008). Do secondary compounds inhibit microbial- and insect-mediated leaf breakdown in a tropical rainforest stream, Costa Rica? *Oecologia, 155*, 311–323. https://doi.org/10.1007/s00442-007-0913-x

Arrhenius, S. (1915). *Quantitative Laws in Biological Chemistry.* https://doi.org/10.5962/bhl.title.4661

Atkinson, D. (1994). Temperature and organism size—A biological law for ectotherms? In *Advances in ecological research* (Vol. 25, pp. 1–58). https://doi.org/10.1016/S0065-2504(08)60212-3

Bailey, J. K., Schweitzer, J. A., & Whitham, T. G. (2001). Salt cedar negatively affects biodiversity of aquatic macroinvertebrates. *Wetlands, 21*(3), 442–447. https://doi.org/10.1672/0277-5212(2001)021[0442:SCNABO]2.0.CO;2

Baldy, V., Gobert, V., Guerold, F., Chauvet, E., Lambrigot, D., & Charcosset, J. Y. (2007). Leaf litter breakdown budgets in streams of various trophic status: Effects of dissolved inorganic nutrients on microorganisms and invertebrates. *Freshwater Biology, 52*(7), 1322–1335. https://doi.org/10.1111/j.1365-2427.2007.01768.x

Ball, B. A., Kominoski, J. S., Adams, H. E., Jones, S. E., Kane, E. S., Loecke, T. D., Mahaney, W. M., Martina, J. P., Prather, C. M., Robinson, T. M. P., & Solomon, C. T. (2010). Direct and terrestrial vegetation-mediated effects of environmental change on aquatic ecosystem processes. *BioScience, 60*(8), 590–601. https://doi.org/10.1525/bio.2010.60.8.5

Battin, T., Kaplan, L., Findlay, S., Hopkinson, C. S., Martí, E., Packman, A. I., Newbold, J. D., & Sabater, F. (2008). Biophysical controls on organic carbon fluxes in fluvial networks. *Nature Geoscience, 1*, 95–100. https://doi.org/10.1038/ngeo101

Bärlocher, F., Seena, S., Wilson, K. P., & Dudley Williams, D. (2008). Raised water temperature lowers diversity of hyporheic aquatic hyphomycetes. *Freshwater Biology, 53*(2), 368–379. https://doi.org/10.1111/j.1365-2427.2007.01899.x

Benstead, J. P., & Huryn, A. D. (2011). Extreme seasonality of litter breakdown in an arctic spring-fed stream is driven by shredder phenology, not temperature. *Freshwater Biology, 56*, 2034–2044. https://doi.org/10.1111/j.1365-2427.2011.02635.x

Bosatta, E., & Ågren, G. I. (1999). Soil organic matter quality interpreted thermodynamically. *Soil Biology and Biochemistry, 31*(13), 1889–1891. https://doi.org/10.1016/S0038-0717(99)00105-4

Boyero, L., Graça, M. A. S., Tonin, A. M., Pérez, J., Swafford, A. J., Ferreira, V., … Pearson, R. G. (2017). Riparian plant litter quality increases with latitude. *Scientific Reports, 7*(1), 1–10. https://doi.org/10.1038/s41598-017-10640-3

Boyero, L., Pearson, R. G., Dudgeon, D., Ferreira, V., Graça, M. A. S., Gessner, M. O., … Barmuta, L. A. (2012). Global patterns of stream detritivore distribution: Implications for biodiversity loss in changing climates. *Global Ecology and Biogeography, 21*(2), 134–141. https://doi.org/10.1111/j.1466-8238.2011.00673.x

Boyero, L., Pearson, R. G., Dudgeon, D., Graça, M. A. S., Gessner, M. O., Albariño, R. J., Ferreira, V., Yule, C. M., Boulton, A. J., Arunachalam, M., Callisto, M., Chauvet, E., Ramírez, A., Chará, J., Moretti, M. S., Gonçalves, J. F., Helson, J. E., Chará-Serna, A. M., Encalada, A. C., … Pringle, C. M. (2011). Global distribution of a key trophic guild contrasts with common latitudinal diversity patterns. *Ecology, 92*(9), 1839–1848. https://doi.org/10.1890/10-2244.1

Boyero, L., Pearson, R. G., Gessner, M. O., Barmuta, L. A., Ferreira, V., Graça, M. A. S., Dudgeon, D., Boulton, A. J., Callisto, M., Chauvet, E., Helson, J. E., Bruder, A., Albariño, R. J., Yule, C. M., Arunachalam, M., Davies, J. N., Figueroa, R., Flecker, A. S., Ramírez, A., … West, D. C. (2011). A global experiment suggests climate warming will not accelerate litter decomposition in streams but might reduce carbon sequestration. *Ecology Letters, 14*(3), 289–294. https://doi.org/10.1111/j.1461-0248.2010.01578.x

Boyero, L., Pearson, R. G., Hui, C., Gessner, M. O., Pérez, J., Alexandrou, M. A., Graça, M. A. S., Cardinale, B. J., Albariño, R. J., Arunachalam, M., Barmuta, L. A., Boulton, A. J., Bruder, A., Callisto, M., Chauvet, E., Death, R. G., Dudgeon, D., Encalada, A. C., Ferreira, V., … Yule, C. M. (2016). Biotic and abiotic variables influencing plant litter breakdown in streams: a global study. *Proceedings of the Royal Society B, 283*, 20152664. https://doi.org/10.1098/rspb.2015.2664

Bradford, M. A. (2013). Thermal adaptation of decomposer communities in warming soils. *Frontiers in Microbiology, 4*, 1–16. https://doi.org/10.3389/fmicb.2013.00333

Brown, J. H., Gillooly, J. F., Allen, A. P., Savage, V. M., & West, G. B. (2004). Toward a metabolic theory of ecology. *Ecology, 85*(7), 1771–1789. https://doi.org/10.1890/03-9000@10.1002/(ISSN)1939-9170

Ciemer, C., Boers, N., Hirota, M., Kurths, J., Müller-Hansen, F., Oliveira, R. S., & Winkelmann, R. (2019). Higher resilience to climatic disturbances in tropical vegetation exposed to more variable rainfall. *Nature Geoscience, 12*, 174–179. https://doi.org/10.1038/s41561-019-0312-z

Colas, F., Woodward, G., Burdon, F. J., Guérold, F., Chauvet, E., Cornut, J., Cébron, A., Clivot, H., Danger, M., Danner, M. C., Pagnout, C., & Tiegs, S. D. (2019). Towards a simple global-standard bioassay for a key ecosystem process: Organic-matter decomposition using cotton strips. *Ecological Indicators, 106*, 105466. https://doi.org/10.1016/j.ecolind.2019.105466

Conant, R. T., Drijber, R. A., Haddix, M. L., Parton, W. J., Paul, E. A., Plante, A. F., Six, J., & Steinweg, J. M. (2008). Sensitivity of organic matter decomposition to warming varies with its quality. *Global Change Biology, 14*, 868–877. https://doi.org/10.1111/j.1365-2486.2008.01541.x

Conant, R. T., Ryan, M. G., Ågren, G. I., Birge, H. E., Davidson, E. A., Eliasson, P. E., Evans, S. E., Frey, S. D., Giardina, C. P., Hopkins, F. M., Hyvönen, R., Kirschbaum, M. U. F., Lavallee, J. M., Leifeld, J., Parton, W. J., Steinweg, J. M., Wallenstein, M. D., Wetterstedt, J. Å. M., & Bradford, M. A. (2011). Temperature and soil organic matter decomposition rates – synthesis of current knowledge and a way forward. *Global Change Biology, 17*, 3392–3404. https://doi.org/10.1111/j.1365-2486.2011.02496.x

Conant, R. T., Steinweg, J. M., Haddix, M. L., Paul, E. A., Plante, A. F., & Six, J. (2008). Experimental warming shows that decomposition temperature sensitivity increases with soil organic matter recalcitrance. *Ecology, 89*(9), 2384–2391

Cornwell, W. K., Cornelissen, J. H. C., Amatangelo, K., Dorrepaal, E., Eviner, V. T., Godoy, O., … Westoby, M. (2008). Plant species traits are the predominant control on litter decomposition rates within biomes worldwide. *Ecology Letters, 11*(10), 1065–1071. https://doi.org/10.1111/j.1461-0248.2008.01219.x

Cornwell, W. K., Westoby, M., Falster, D. S., Fitzjohn, R. G., O'Meara, B. C., Pennell, M. W., McGlinn, D. J., Eastman, J. M., Moles, A. T., Reich, P. B., Tank, D. C., Wright, I. J., Aarssen, L., Beaulieu, J. M., Kooyman, R. M., Leishman, M. R., Miller, E. T., Niinemets, Ü., Oleksyn, J., & Zanne, A. E. (2014). Functional distinctiveness of major plant lineages. *Journal of Ecology, 102*, 345–356. https://doi.org/10.1111/1365-2745.12208

Craine, J. M., Fierer, N., & McLauchlan, K. K. (2010). Widespread coupling between the rate and temperature sensitivity of organic matter decay. *Nature Geoscience, 3*, 854–857. https://doi.org/10.1038/NGEO1009

Cross, W. F., Hood, J. M., Benstead, J. P., Huryn, A. D., & Nelson, D. (2014). Interactions between temperature and nutrients across levels of ecological organization. *Global Change Biology*, gcb.12809. https://doi.org/10.1111/gcb.12809

Damuth, J. (1981). Population density and body size in mammals. *Nature, 290*(23), 699–700

Dang, C. K., Schindler, M., & Chauvet, E. (2009). Temperature oscillation coupled with fungal community shifts can modulate warming effects on litter decomposition. *Ecology, 90*(1), 122–131

Datry, T., Foulquier, A., Corti, R., von Schiller, D., Tockner, K., Mendoza-Lera, C., Clément, J. C., Gessner, M. O., Moleón, M., Stubbington, R., Gücker, B., Albariño, R., Allen, D. C., Altermatt, F., Arce, M. I., Arnon, S., Banas, D., Banegas-Medina, A., Beller, E., & Zoppini, A. (2018). A global analysis of terrestrial plant dynamics in non-perennial waterways. *Nature Geoscience*. https://doi.org/10.1038/s41561-018-0134-4

del Campo, R., Foulquier, A., Singer, G., & Datry, T. (2020). Leaf litter decomposition in intermittent rivers and ephemeral streams. In C. M. Swan, L. Boyero, & C. Canhoto (Eds.), *Leaf Litter Decomposition in Freshwater Ecosystems*. Springer.

Dell, A. I., Pawar, S., & Savage, V. M. (2011). Systematic variation in the temperature dependence of physiological and ecological traits. *Proceedings of the National Academy of Sciences of the United States of America, 108*(26), 10591–10596. https://doi.org/10.1073/pnas.1015178108

Demars, B. O. L., Gíslason, G. M., Ólafsson, J. S., Manson, J. R., Friberg, N., Hood, J. M., Thompson, J. J. D., Freitag, T. E. (2016). Impact of warming on CO_2 emissions from streams countered by aquatic photosynthesis. *Nature Geoscience, 9*, 758–761. https://doi.org/10.1038/NGEO2807

Demars, B. O. L., Manson, J. R., Ólafsson, J. S., Gíslason, G. M., Gudmundsdöttir, R., Woodward, G., Reiss, J., Pichler, D. E., Rasmussen, J. J., & Friberg, N. (2011). Temperature and the metabolic balance of streams. *Freshwater Biology*. https://doi.org/10.1111/j.1365-2427.2010.02554.x

Duarte, S., Pascoal, C., Garabétian, F., Cássio, F., & Charcosset, J. Y. (2009). Microbial decomposer communities are mainly structured by trophic status in circumneutral and alkaline streams. *Applied and Environmental Microbiology, 75*(19), 6211–6221. https://doi.org/10.1128/AEM.00971-09

Dudgeon, D., Arthington, A. H., Gessner, M. O., Kawabata, Z. I., Knowler, D. J., Lévêque, C., Naiman, R. J., Prieur-Richard, A.-H., Soto, D., Stiassny, M. L. J., & Sullivan, C. A. (2006). Freshwater biodiversity: Importance, threats, status and conservation challenges. *Biological Reviews of the Cambridge Philosophical Society, 81*, 163–182. https://doi.org/10.1017/S1464793105006950

Enquist, B. J., Economo, E. P., Huxman, T. E., Allen, A. P., Ignace, D. D., & Gillooly, J. F. (2003). Scaling metabolism from organisms to ecosystems. *Nature, 423*(6940), 639–642. https://doi.org/10.1038/nature01671

Enquist, B. J., & Niklas, K. J. (2001). Invariant scaling relations across tree-dominated communities. *Nature, 410*(6829), 655–660. https://doi.org/10.1038/35070500

Estes, J. A., Terborgh, J., Brashares, J. S., Power, M. E., Berger, J., Bond, W. J., Carpenter, S. R., Essington, T. E., Holt, R. D., Jackson, J. B. C., Marquis, R. J., Oksanen, L., Oksanen, T., Paine, R. T., Pikitch, E. K., Ripple, W. J., Sandin, S. A., Scheffer, M., Schoener, T. W., Shurin, J. B., & Wardle, D. A. (2011, July 15). Trophic downgrading of planet earth. *Science, 333*, 301–306. https://doi.org/10.1126/science.1205106

Fernandes, I., Seena, S., Pascoal, C., & Cássio, F. (2014). Elevated temperature may intensify the positive effects of nutrients on microbial decomposition in streams. *Freshwater Biology, 59*(11), 2390–2399. https://doi.org/10.1111/fwb.12445

Ferreira, V., & Canhoto, C. (2015). Future increase in temperature may stimulate litter decomposition in temperate mountain streams: Evidence from a stream manipulation experiment. *Freshwater Biology, 60*(5), 881–892. https://doi.org/10.1111/fwb.12539

Ferreira, V., Castagneyrol, B., Koricheva, J., Gulis, V., Chauvet, E., & Graça, M. A. S. (2014). A meta-analysis of the effects of nutrient enrichment on litter decomposition in streams. *Biological Reviews*, brv.12125. https://doi.org/10.1111/brv.12125

Ferreira, V., & Chauvet, E. (2011). Future increase in temperature more than decrease in litter quality can affect microbial litter decomposition in streams. *Oecologia, 167*, 279–291. https://doi.org/10.1007/s00442-011-1976-2

Ferreira, V., Chauvet, E., & Canhoto, C. (2014). Effects of experimental warming, litter species, and presence of macroinvertebrates on litter decomposition and associated decomposers in a temperate mountain stream. *Canadian Journal of Fisheries and Aquatic Sciences, 72*(2), 206–216. https://doi.org/10.1139/CJFAS-2014-0119

Ferreira, V., Gonçalves, A. L., Godbold, D. L., & Canhoto, C. (2010). Effect of increased atmospheric CO_2 on the performance of an aquatic detritivore through changes in water temperature and litter quality. *Global Change Biology, 16*(12), 3284–3296. https://doi.org/10.1111/j.1365-2486.2009.02153.x

Fierer, N., Craine, J. M., McLaughlin, K., & Schimel, J. P. (2005). Litter quality and the temperature sensitivity of decomposition. *Ecology, 86*(2), 320–326

Findlay, R. H., Yeates, C., Hullar, M. A. J., Stahl, D. A., & Kaplan, L. A. (2008). Biome-level biogeography of streambed microbiota. *Applied and Environmental Microbiology, 74*(10), 3014–3021. https://doi.org/10.1128/AEM.01809-07

Finzi, A. C., Allen, A. S., DeLucia, E. H., Ellsworth, D. S., & Schlesinger, W. H. (2001). Forest litter production, chemistry, and decomposition following tow years of free-air CO_2 enrichment. *Ecology, 82*(2), 470–484. https://doi.org/10.1890/0012-9658(2001)082[0470:FLPCAD]2.0.CO;2

Flury, S., & Gessner, M. O. (2011). Experimentally simulated global warming and nitrogen enrichment effects on microbial litter decomposers in a marsh. *Applied and Environmental Microbiology, 77*(3), 803–809. https://doi.org/10.1128/AEM.01527-10

Follstad Shah, J. J., Kominoski, J. S., Ardón, M., Dodds, W. K., Gessner, M. O., Griffiths, N. A., Hawkins, C. P., Johnson, S. L., Lecerf, A., LeRoy, C. J., Manning, D. W. P., Rosemond, A. D., Sinsabaugh, R. L., Swan, C. M., Webster, J. R., & Zeglin, L. H. (2017). Global synthesis of the temperature sensitivity of leaf litter breakdown in streams and rivers. *Global Change Biology*, gcb.13609. https://doi.org/10.1111/gcb.13609

Forster, J., Hirst, A. G., & Atkinson, D. (2012). Warming-induced reductions in body size are greater in aquatic than terrestrial species. *Proceedings of the National Academy of Sciences of the United States of America, 109*(47), 19310–19314. https://doi.org/10.1073/pnas.1210460109

Friberg, N., Dybkjaer, J. B., Olafsson, G. M., Gislason, S. E., Larsen, T. L., & Lauridsen. (2009). Relationships between structure and function in streams contrasting in temperature. *Freshwater Biology, 54*, 2051–2068. https://doi.org/10.1111/j.1365-2427.2009.02234.x

Galloway, J. N., Townsend, A. R., Erisman, J. W., Bekunda, M., Cai, Z., Freney, J. R., Martinelli, L. A., Seitzinger, S. P., & Sutton, M. A. (2008). Transformation of the nitrogen cycle: Recent trends, questions, and potential solutions. *Science, 320*(5878), 889–892. https://doi.org/10.1126/science.1136674

Gessner, M. O., & Chauvet, E. (1994). Importance of stream microfungi in controlling breakdown rates of leaf litter. *Ecology, 75*(6), 1807–1817. https://doi.org/10.2307/1939639

Gessner, M. O., Chauvet, E., & Dobson, M. (1999). A perspective on leaf litter breakdown in streams. *Oikos, 85*(2), 377–384

Gillooly, J. F., Brown, J. H., West, G. B., Savage, V. M., & Charnov, E. L. (2001). Effects of size and temperature on metabolic rate. *Science, 293*(5538), 2248–2251. https://doi.org/10.1126/science.1061967

Gosling, S. N., Zaherpour, J., Mount, N. J., Hattermann, F. F., Dankers, R., Arheimer, B., Breuer, L., Ding, J., Haddeland, I., Kumar, R., Kundu, D., Liu, J., Griensven, A. v., Veldkamp, T. I. E., Vetter, T., Wang, X., & Zhang, X. (2017). A comparison of changes in river runoff from multiple global and catchment-scale hydrological models under global warming scenarios of 1 °C, 2 °C and 3 °C. *Climatic Change, 141*, 577–595. https://doi.org/10.1007/s10584-016-1773-3

Gossiaux, A., Rollin, M., Guérold, F., Felten, V., Laviale, M., Bachelet, Q., Poupin, P., Chauvet, E., Bec, A., & Danger, M. (2020). Temperature and nutrient effects on the relative importance of brown and green pathways for stream ecosystem functioning: A mesocosm approach. *Freshwater Biology*, fwb.13474. https://doi.org/10.1111/fwb.13474

Graça, M. A. S. (2001). The role of invertebrates on leaf litter decomposition in streams - A review. *International Review of Hydrobiology, 86*(4–5), 383–393. https://doi.org/10.1002/1522-2632(200107)86:4/5%3c383::AID-IROH383%3e3.0.CO;2-D

Graça, M. A. S., Pozo, J., Canhoto, C., & Elosegi, A. (2002). Effects of Eucalyptus plantations on detritus, decomposers, and detritivores in streams. *TheScientificWorldJournal, 2*, 1173–1185. https://doi.org/10.1100/tsw.2002.193

Griffiths, N. A., & Tiegs, S. D. (2016). Organic-matter decomposition along a temperature gradient in a forested headwater stream. *Freshwater Science, 35*(2). https://doi.org/10.1086/685657.

Grimm, N. B., Chapin III, F. S., Bierwagen, B., Gonzalez, P., Groffman, P. M., Luo, Y., Melton, F., Nadelhoffer, K., Pairis, A., Raymond, P. A., Schimel, J., & Williamson, C. E. (2013). The impacts of climate change on ecosystem structure and function in a nutshell. *Frontiers in Ecology and the Environment, 11*(9), 474–482. https://doi.org/10.1890/120282

Handa, I. T., Aerts, R., Berendse, F., Berg, M. P., Bruder, A., Butenschoen, O., Chauvet, E., Gessner, M. O., Jabiol, J., Makkonen, M., McKie, B. G., Malmqvist, B., Peeters, E. T. H. M., Scheu, S., Schmid, B., Ruijven, J. v., Vos, V. C. A., & Hättenschwiler, S. (2014). Consequences of biodiversity loss for litter. *Nature, 509*, 210–221. https://doi.org/10.1038/nature13247

Hattermann, F. F., Krysanova, V., Gosling, S. N., Dankers, R., Daggupati, P., Donnelly, C., Flörke, M., Huang, S., Motovilov, Y., Buda, S., Yang, T., Müller, C., Leng, G., Tang, Q., Portmann, F. T., Hagemann, S., Gerten, D., Wada, Y., Masaki, Y., Alemayehu, T., Satoh, Y., & Samaniego, L. (2017). Cross-scale intercomparison of climate change impacts simulated by regional and global hydrological models in eleven large river basins. *Climatic Change, 141*, 561–576. https://doi.org/10.1007/s10584-016-1829-4

Hedges, L. V., Gurevitch, J., & Curtis, P. S. (1999). The meta-analysis of response ratios in experimental ecology. *Ecology, 80*(4), 1150–1156. https://doi.org/10.1890/0012-9658(1999)080[1150:TMAORR]2.0.CO;2

Hieber, M., & Gessner, M. O. (2002). Contribution of stream detritivores, fungi, and bacteria to leaf breakdown based on biomass estimates. *Ecology, 83*(4), 1026–1038

Hobbie, S. E. (1996). Temperature and plant species control over litter decomposition in Alaskan tundra. *Ecological Monographs, 66*(4), 503–522. https://doi.org/10.2307/2963492

Hotchkiss, E. R., Hall Jr, R. O., Sponseller, R. A., Butman, D., Klaminder, J., Laudon, H., & Karlsson, J. (2015). Sources of and processes controlling CO_2 emissions change with the size of streams and rivers. *Nature Geoscience, 8*, NGEO2507. https://doi.org/10.1038/NGEO2507

Hu, Z., Michaletz, S. T., Johnson, D. J., McDowell, N. G., Huang, Z., Zhou, X., & Xu, C. (2018). Traits drive global wood decomposition rates more than climate. *Global Change Biology, 24*(11), 5259–5269. https://doi.org/10.1111/gcb.14357

Hu, Z., Chen, H. Y. H., Yue, C., Ying, X., Shao, J., Zhou, G., Wang, J., Wang, M., Xia, J., Li, Y., Zhou, X., & Michaletz, S. (2020). Traits mediate drought effects on wood carbon fluxes. *Global Change Biology, 26*, 3429–3442. https://doi.org/10.1111/gcb.15088

IPCC. (2008). Climate change and water. In B. C. Bates, Z. W. Kundzewicz, S. Wu, & J. P. Palutikof (Eds.), *Technical paper of the intergovernmental panel on climate change*. https://doi.org/10.1093/nq/s10-IX.235.507-c

IPCC. (2014). Climate change 2014: Synthesis report. In C. W. Team, R. K. Pachauri, & L. A. Meyer (Eds.), *Contribution of working groups I, II, and III to the Fifth assessment report of the intergovernmental panel on climate change*. https://doi.org/10.1016/S0022-0248(00)00575-3

IPCC. (2018). Summary for policymakers. In T. W. V. Masson-Delmotte, P. Zhai, H. O. Pörtner, D. RobertV. Masson-Delmotte, P. Zhai, H. O. Pörtner, D. Roberts, J. Skea, P. R. Shukla, A. Pirani, W. Moufouma-Okia, C. Péan, R. Pidcock, S. Connors, J. B. R. Matthews, Y. Chen, X. Zhou, M. I. Gomis, & E. Lonnoy (Eds.), *Global warming of 1.5°C. An IPCC Special Report on the impacts of global warming of 1.5°C above pre-industrial levels and related global greenhouse gas emission pathways, in the context of strengthening the global response to the threat of climate change* (p. 32). Retrieved from https://www.ipcc.ch/sr15/

Irons, J. G., III., Oswood, M. W., Stout, R. J., & Pringle, C. M. (1994). Latitudinal pattems in leaf litter breakdown: Is temperature really important? *Freshwater Biology, 32*, 401–411

Isaac, N. J. B., Storch, D., & Carbone, C. (2013). The paradox of energy equivalence. *Global Ecology and Biogeography, 22*, 1–5. https://doi.org/10.1111/j.1466-8238.2012.00782.x

Jabiol, J., Gossiaux, A., Lecerf, A., Rota, T., Guérold, F., Danger, M., Poupin, P., Gilbert, F., & Chauvet, E. (2020). Variable temperature effects between heterotrophic stream processes and organisms. *Freshwater Biology*, fwb.13520. https://doi.org/10.1111/fwb.13520

James, F. C. (1970). Geographic size variation in birds and its relationship to climate. *Ecology, 51*(3), 365–390

Kaushal, S. S., Likens, G. E., Jaworski, N. A., Pace, M. L., Sides, A. M., Seekell, D., Belt, K. T., Secor, D. H., & Wingate, R. L. (2010). Rising stream and river temperatures in the United States. *Frontiers in Ecology and the Environment, 8*(9), 461–466. https://doi.org/10.1890/090037

Keller, P. S., Catalán, N., von Schiller, D. Grossart, H.-P., Koschorreck, M., Obrador, B., ... & Marcé, R. (2020). Global CO2 emissions from dry inland waters share common drivers across ecosystems. *Nature Communications 11*, 2126. https://doi.org/10.1038/s41467-020-15929-y

Kerkhoff, A. J., Enquist, B. J., Elser, J. J., & Fagan, W. F. (2005). Plant allometry, stoichiometry and the temperature-dependence of primary productivity. *Global Ecology and Biogeography, 14*, 585–598. https://doi.org/10.1111/j.1466-822x.2005.00187.x

Kominoski, J. S., Chapman, S. K., Dodds, W. K., Follstad Shah, J. J., & Richardson, J. S. (2020). Understanding the causes and consequences of global changes in riparian vegetation throughout inland and coastal waters. In C. M. Swan, L. Boyero, & C. Canhoto (Eds.), *Leaf litter decomposition in freshwater ecosystems*. Springer.

Kominoski, J. S., Rosemond, A. D., Benstead, J. P., Gulis, V., Maerz, J. C., & Manning, D. W. P. (2015). Low-to-moderate nitrogen and phosphorus concentrations accelerate microbially driven litter breakdown rates. *Ecological Applications, 25*(3), 856–865. https://doi.org/10.1890/14-1113.1

Kominoski, J. S., Follstad Shah, J. J., Canhoto, C., Fischer, D. G., Giling, D. P., González, E., Griffiths, N. A., Larrañaga, A., LeRoy, C. J., Mineau, M. M., McElarney, Y. R., Shirley, S. M., Swan, C. M., & Tiegs, S. D. (2013). Forecasting functional implications of global changes in riparian plant communities. *Frontiers in Ecology and the Environment, 11*(8), 423–432. https://doi.org/10.1890/120056

Kominoski, J. S., Marczak, L. B., & Richardson, J. S. (2011). Riparian forest composition affects stream litter decomposition despite similar microbial and invertebrate communities. *Ecology, 92*(1), 151–159. https://doi.org/10.1890/10-0028.1

Kominoski, J. S., & Rosemond, A. D. (2012). Conservation from the bottom up: Forecasting effects of global change on dynamics of organic matter and management needs for river networks. *Freshwater Science, 31*(1), 51–68. https://doi.org/10.1899/10-160.1

Lecerf, A. (2017). Methods for estimating the effect of litterbag mesh size on decomposition. *Ecological Modelling, 362*, 65–68. https://doi.org/10.1016/j.ecolmodel.2017.08.011

Lecerf, A., & Chauvet, E. (2008). Intraspecific variability in leaf traits strongly affects alder leaf decomposition in a stream. *Basic and Applied Ecology, 9*, 598–605. https://doi.org/10.1016/j.baae.2007.11.003

Lecerf, A., Usseglio-Polatera, P., Lambrigot, D., Bracht, B., & Chauvet, E. (2006). Assessment of functional integrity of eutrophic streams using litter breakdown and benthic macroinvertebrates. *Archiv Für Hydrobiologie*, 105–126. https://doi.org/10.1127/0003-9136/2006/0165-0105

LeRoy, C. J., Hipp, A. L., Lueders, K., Follstad Shah, J. J., Kominoski, J. S., Ardón, M., Dodds, W. K., Gessner, M. O., Griffiths, N. A., Lecerf, A., Manning, D. W. P., Sinsabaugh, R. L., & Webster, J. R. (2020). Plant phylogenetic history explains in-stream decomposition at a global scale. *Journal of Ecology, 108*(1), 17–35. https://doi.org/10.1111/1365-2745.13262

LeRoy, C. J., Whitham, T. G., Wooley, S. C., & Marks, J. C. (2007). Within-species variation in foliar chemistry influences leaf-litter decomposition in a Utah river. *Journal of the North American Benthological Society, 26*(3), 426–438. https://doi.org/10.1899/06-113.1

Lopez-Urrutia, Á., & Moran, X. A. G. (2007). Resource limitation of bacterial production distorts the temperature dependence of oceanic carbon cycling. *Ecology, 88*(4), 817–822

Manning, D. W. P., Rosemond, A. D., Gulis, V., & Ferreira, V. (2020). The multiple effects of eutrophication on decomposition rates. In C. M. Swan, L. Boyero, & C. Canhoto (Eds.), *Leaf litter decomposition in freshwater ecosystems*. Springer.

Marks, J. C. (2019). Revisiting the fates of dead leaves that fall into streams. *Annual Review of Ecology, Evolution, and Systematics, 50*, 24.1–24.22.

Martínez, A., Larrañaga, A., Pérez, J., Descals, E., & Pozo, J. (2014). Temperature affects leaf litter decomposition in low-order forest streams: Field and microcosm approaches. *FEMS Microbiology Ecology, 87*, 257–267. https://doi.org/10.1111/1574-6941.12221

Martins, R. T., Melo, A. S., Gonçalves, J. F., Jr., Campos, C. M., & Hamada, N. (2017). Effects of climate change on leaf breakdown by microorganisms and the shredder Phylloicus elektoros (Trichoptera: Calamoceratidae). *Hydrobiologia, 789*, 31–44. https://doi.org/10.1007/s10750-016-2689-7

Martins, R. T., Rezende, R. D. S., Gonçalves, J. F., Jr., Lopes, A., Piedade, M. T. F., de Cavalcante, H. L., & Hamada, N. (2017). Effects of increasing temperature and CO_2 on quality of litter, shredders, and microorganisms in Amazonian aquatic systems. *PLoS ONE, 12*(11), e0188791. https://doi.org/10.1371/journal.pone.018879

Mcdonough, L. K., Santos, I. R., Andersen, M. S., Carroll, D. M. O., Rutlidge, H., Meredith, K., Oudone, P., Bridgeman, J., Gooddy, D. C., Sorensen, J. P. R., Lapworth, D. J., MacDonald, A. M., Ward, J., & Baker, A. (2020). Changes in global groundwater organic carbon driven by climate change and urbanization. *Nature Communications* (11), 1279. https://doi.org/10.1038/s41467-020-14946-1

Minshall, G. W., Petersen, R. C., Cummins, K. W., Bott, T. L., Sedell, J. R., Cushing, C. E., Bott, T. L., Sedell, J. R., Cushing, C. E., & Vannote, R. L. (1983). Interbiome comparison of stream ecosystem dynamics. *Ecological Monographs, 53*(1), 1–25.

Moghadam, F. S., & Zimmer, M. (2016). Effects of warming, nutrient enrichment and detritivore presence on litter breakdown and associated microbial decomposers in a simulated temperate woodland creek. *Hydrobiologia, 770*(1), 243–256. https://doi.org/10.1007/s10750-015-2596-3

Monroy, S., Miranda-Apodaca, J., Pérez-López, U., Mena-Petite, A., Elosegi, A., Muñoz-Rueda, A., & Pozo, J. (2016). Elevated atmospheric CO2 interacts with drought and competition to produce complex results in plant quality and subsequent microbial aquatic decomposition. *Canadian Journal of Fisheries and Aquatic Sciences, 73*, 1770–1776. https://doi.org/10.1139/cjfas-2016-0034

Nelson, D., Benstead, J. P., Huryn, A. D., Cross, W. F., Hood, J. M., Johnson, P. W., Junker, J. R., Gíslason, G. M. and Ólafsson, J. S. (2017). Shifts in community size structure drive temperature invariance of secondary production in a stream-warming experiment. *Ecology, 98*, 1797–1806. https://doi.org/10.1002/ecy.1857

Nelson, D., Benstead, J. P., Huryn, A. D., Cross, W. F., Hood, J. M., Johnson, P. W., Junker, J. R., Gíslason, G. M., & Ólafsson, J. S. (2020). Thermal niche diversity and trophic redundancy drive neutral effects of warming on energy flux through a stream food web. *Ecology, 101*(4), e02952. https://doi.org/10.1002/ecy.2952

Nilsson, C., Jansson, R., Kuglerová, L., Lind, L., & Ström, L. (2013). Boreal riparian vegetation under climate change. *Ecosystems, 16*(3), 401–410. https://doi.org/10.1007/s10021-012-9622-3

Nilsson, C., Polvi, L. E., & Lind, L. (2015). Extreme events in streams and rivers in arctic and subarctic regions in an uncertain future. *Freshwater Biology, 60*(12), 2535–2546. https://doi.org/10.1111/fwb.12477

Norby, R. J., Warren, J. M., Iversen, C. M., Medlyn, B. E., & McMurtrie, R. E. (2010). CO_2 enhancement of forest productivity constrained by limited nitrogen availability. *Proceedings of the National Academy of Sciences of the United States of America, 107*(45), 19368–19373. https://doi.org/10.1073/pnas.1006463107

O'Gorman, E. J., Pichler, D. E., Adams, G., Benstead, J. P., Cohen, H., Craig, N., Cross, W. F., Demars, B. O. L., Friberg, N., Gíslason, G. M., Gudmundsdóttir, R., Hawczak, A., Hood,

J. M., Hudson, L. N., Johansson, L., Johansson, M. P., Junker, J. R., Laurila, A., Manson, J. R., … Woodward, G. (2012). Impacts of warming on the structure and functioning of aquatic communities. individual- to ecosystem-level responses. In *Advances in ecological research* (Vol. 47). https://doi.org/10.1016/B978-0-12-398315-2.00002-8

O'Gorman, E. J., Zhao, L., Pichler, D. E., Adams, G., Friberg, N., Rall, B. C., Seeney, A., Zhang, H., Reuman, D. C., & Woodward, G. (2017). Unexpected changes in community size structure in a natural warming experiment. *Nature Climate Change*, NCLIMATE3368. https://doi.org/10.1038/NCLIMATE3368

Ostrofsky, M. L. (1997). Relationship between chemical characteristics of autumn-shed leaves and aquatic processing rates. *Journal of the North American Benthological Society, 16*(4), 750–759

Paul, M. J., Meyer, J. L., & Couch, C. A. (2006). Leaf breakdown in streams differing in catchment land use. *Freshwater Biology, 51*, 1684–1695. https://doi.org/10.1111/j.1365-2427.2006.01612.x

Perkins, D. M., Yvon-Durocher, G., Demars, B. O. L., Reiss, J., Pichler, D. E., Friberg, N., Trimmer, M., & Woodward, G. (2012). Consistent temperature dependence of respiration across ecoystems contrasting in thermal history. *Global Change Biology, 18*, 1300–1311. https://doi.org/10.1111/j.1365-2486.2011.02597.x

Perry, L. G., Andersen, D. C., Reynolds, L. V., Nelson, S. M., & Shafroth, P. B. (2012). Vulnerability of riparian ecosystems to elevated CO_2 and climate change in arid and semiarid western North America. *Global Change Biology, 18*(3), 821–842. https://doi.org/10.1111/j.1365-2486.2011.02588.x

Perry, L. G., Shafroth, P. B., Hay, L. E., Markstrom, S. L., & Bock, A. R. (2020). Projected warming disrupts the synchrony of riparian seed release and snowmelt streamflow. *New Phytologist, 225*(2), 693–712. https://doi.org/10.1111/nph.16191

Peterson, R., & Cummins, K. (1974). Leaf processing in a woodland stream. *Freshwater Biology, 4*, 343–368. https://doi.org/10.1111/j.1365-2427.1974.tb00103.x

Piggott, J. J., Lange, K., Townsend, C. R., & Matthaei, C. D. (2012). Multiple stressors in agricultural streams: A mesocosm study of interactions among raised water temperature, sediment addition and nutrient enrichment. *PLoS ONE, 7*(11). https://doi.org/10.1371/journal.pone.0049873

Piggott, J. J., Niyogi, D. K., Townsend, C. R., & Matthaei, C. D. (2015). Multiple stressors and stream ecosystem functioning: Climate warming and agricultural stressors interact to affect processing of organic matter. *Journal of Applied Ecology, 52*(5), 1126–1134. https://doi.org/10.1111/1365-2664.12480

Poff, N. L. R., Olden, J. D., Merritt, D. M., & Pepin, D. M. (2007). Homogenization of regional river dynamics by dams and global biodiversity implications. *Proceedings of the National Academy of Sciences of the United States of America, 104*(14), 5732–5737. https://doi.org/10.1073/pnas.0609812104

Rajashekhar, M., & Kaveriappa, K. M. (2000). Effects of temperature and light on growth and sporulation of aquatic hyphomycetes. *Hydrobiologia, 441*, 149–153. https://doi.org/10.1023/A:1017591109362

Rall, B. C., Vucic-Pestic, O., Ehnes, R. B., Emmerson, M., & Brose, U. (2010). Temperature, predator – prey interaction strength and population stability. *Global Change, 16*, 2145–2157. https://doi.org/10.1111/j.1365-2486.2009.02124.x

Ramirez, K. S., Craine, J. M., & Fierer, N. (2012). Consistent effects of nitrogen amendments on soil microbial communities and processes across biomes. *Global Change Biology, 18*(6), 1918–1927. https://doi.org/10.1111/j.1365-2486.2012.02639.x

Reynolds, L. V., & Shafroth, P. B. (2017). Riparian plant composition along hydrologic gradients in a dryland river basin and implications for a warming climate. *Ecohydrology, 10*(6), 1–13. https://doi.org/10.1002/eco.1864

Rier, S. T., Tuchman, N. C., & Wetzel, R. G. (2005). Chemical changes to leaf litter from trees grown under elevated CO_2 and the implications for microbial utilization in a stream ecosystem. *Canadian Journal of Fisheries and Aquatic Sciences, 62*, 185–194. https://doi.org/10.1139/F04-148

Rier, S. T., Tuchman, N. C., Wetzel, R. G., & Teeri, J. A. (2002). Elevated-CO_2-induced changes in the chemistry of quaking aspen (*Populus tremuloides* Michaux) leaf litter: Subsequent mass loss and microbial response in a stream ecosystem. *Journal of the North American Benthological Society, 21*(1), 16–27. https://doi.org/10.2307/1468296

Roberts, B. J., Mulholland, P. J., & Hill, W. R. (2007). Multiple scales of temporal variability in ecosystem metabolism rates: Results from 2 years of continuous monitoring in a forested headwater stream. *Ecosystems, 10*(4), 588–606. https://doi.org/10.1007/s10021-007-9059-2

Rodell, M., Famiglietti, J. S., Wiese, D. N., Reager, J. T., Beaudoing, H. K., Landerer, F. W., & Lo, M.-H. (2018). Emerging trends in global freshwater availability. *Nature, 557*, 651–659. https://doi.org/10.1038/s41586-018-0123-1

Rosemond, A. D., Benstead, J. P., Bumpers, P. M., Gulis, V., Kominoski, J. S., Manning, D. W. P., Suberkropp, K., & Wallace, J. B. (2015). Experimental nutrient additions accelerate terrestrial carbon loss from stream ecosystems. *Science, 347*(6226), 1142–1145. https://doi.org/10.1126/science.aaa1958

Rosi-Marshall, E. J., Kincaid, D. W., Bechtold, H. A., Royer, T. V., Rojas, M., & Kelly, J. J. (2013). Pharmaceuticals suppress algal growth and microbial respiration and alter bacterial communities in stream biofilms. *Ecological Applications, 23*(3), 583–593. https://doi.org/10.1890/12-0491.1

Schäfer, R. B., Von Der Ohe, P. C., Rasmussen, J., Kefford, B. J., Beketov, M. A., Schulz, R., & Liess, M. (2012). Thresholds for the effects of pesticides on invertebrate communities and leaf breakdown in stream ecosystems. *Environmental Science and Technology, 46*(9), 5134–5142. https://doi.org/10.1021/es2039882

Schäfer, R. B., Caquet, T., Siimes, K., Mueller, R., Lagadic, L., & Liess, M. (2007). Effects of pesticides on community structure and ecosystem functions in agricultural streams of three biogeographical regions in Europe. *Science of the Total Environment, 382*, 272–285. https://doi.org/10.1016/j.scitotenv.2007.04.040

Schmid, P. E., Schmid-Araya, J. M., & Tokeshi, M. (2020). The scaling of biomass variance across trophic levels in stream species communities: A macroecological approach. *Hydrobiologia*. https://doi.org/10.1007/s10750-020-04239-5

Schmid, P. E., Tokeshi, M., & Schmid-Araya, J. M. (2000). Relation between population density and body size in stream communities. *Science, 289*, 1557–1561

Shah, A., Havird, J. C., Encalada, A. C., Flecker, A. S., Funk, C. W., Guayasamin, J. M., Guayasamin, J. M., Kondratieff, B. C., Poff, N. L., Thomas, S. A., Zamudio, K. R., & Ghalambor, C. K. (2019). Temperature-dependence of metabolic rate in tropical and temperate aquatic insects : Support for the Climate Variability Hypothesis in mayflies but not stoneflies. *Pre-Print*. https://doi.org/10.1101/2019.12.25.888578

Sinsabaugh, R. L., & Follstad Shah, J. J. (2010). Integrating resource utilization and temperature in metabolic scaling of riverine bacterial production. *Ecology, 91*(5), 1455–1465. https://doi.org/10.1890/08-2192.1

Sinsabaugh, R. L., & Follstad Shah, J. J. (2012). Ecoenzymatic stoichiometry and ecological theory. *Annual Review of Ecology, Evolution, and Systematics, 43*(1), 313–343. https://doi.org/10.1146/annurev-ecolsys-071112-124414

Song, C., Dodds, W. K., Rüegg, J., Argerich, A., Baker, C. L., Bowden, W. B., Douglas, M. M., Farrell, K. J., Flinn, M. B., Garcia, E. A., Helton, A. M., Harms, T. K., Jia, S., Jones, J. B., Koenig, L. E., Kominoski, J. S., McDowell, W. H., McMaster, D., Parker, S. P., … Whiles, M. R. (2018). Continental-scale decrease in net productivity in streams due to climate warming. *Nature Geoscience, 11*, 415–420. https://doi.org/10.1038/s41561-018-0125-5

Stewart, B. A., Close, P. G., Cook, P. A., & Davies, P. M. (2013). Upper thermal tolerances of key taxonomic groups of stream invertebrates. *Hydrobiologia, 718*, 131–140. https://doi.org/10.1007/s10750-013-1611-9

Stiling, P., & Cornelissen, T. (2007). How does elevated carbon dioxide (CO_2) affect plant-herbivore interactions? A field experiment and meta-analysis of CO_2- mediated changes on plant chemistry and herbivore performance. *Global Change Biology, 13*(9), 1823–1842. https://doi.org/10.1111/j.1365-2486.2007.01392.x

Strickland, M. S., Keiser, A. D., & Bradford, M. A. (2015). Climate history shapes contemporary leaf litter decomposition. *Biogeochemistry, 122*, 165–174. https://doi.org/10.1007/s10533-014-0065-0

Suberkropp, K. (1992). Interactions with invertebrates. In F. Bärlocher (Ed.), *Ecological studies, Vol. 4: The ecology of aquatic hyphomycetes* (pp. 118–134). https://doi.org/10.1007/978-3-642-76855-2_6

Tank, J. L., Rosi-Marshall, E. J., Griffiths, N. A., Entrekin, S. A., & Stephen, M. L. (2010). A review of allochthonous organic matter dynamics and metabolism in streams. *Journal of the North American Benthological Society, 29*(1), 118–146. https://doi.org/10.1899/08-170.1

Tiegs, S. D., Costello, D. M., Isken, M. W., Woodward, G., Mcintyre, P. B., Gessner, M. O., Tiegs, S. D., Costello, D. M., Isken, M. W., Woodward, G., McIntyre, P. B., Gessner, M. O., Chauvet, E., Griffiths, N. A., Flecker, A. S., Acuña, V., Albariño, R., Allen, D. C., Alonso, C., Andino, P., Arango, C., Aroviita, J., Barbosa, M. V. M, Barmuta, L. A., Baxter, C. V., …s Zwart, J. A. (2019). Global patterns and drivers of ecosystem functioning in rivers and riparian zones. *Science Advances, 5*, eeav0486. https://doi.org/10.1126/sciadv.aav0486

Tomczyk, N. J., Rosemond, A. D., Bumpers, P. M., Cummins, C. S., Wenger, S. J., & Benstead, J. P. (2020). Ignoring temperature variation leads to underestimation of the temperature sensitivity of plant litter decomposition. *Ecosphere, 11*(2), e03050. https://doi.org/10.1002/ecs2.3050

Tonin, A. M., Boyero, L., Monroy, S., Basaguren, A., Pérez, J., Pearson, R. G., Cardinale, B. J., Gonçalves, J. F., & Pozo, J. (2017). Stream nitrogen concentration, but not plant N-fixing capacity, modulates litter diversity effects on decomposition. *Functional Ecology, 31*(7), 1471–1481. https://doi.org/10.1111/1365-2435.12837

Tonin, A. M., Pozo, J., Monroy, S., Basaguren, A., Pérez, J., Gonçalves, J. F., Pearson, R., Cardinale, B. J., & Boyero, L. (2018). Interactions between large and small detritivores influence how biodiversity impacts litter decomposition. *Journal of Animal Ecology, 87*(5), 1465–1474. https://doi.org/10.1111/1365-2656.12876

Tuchman, N. C., Wetzel, R. G., Rier, S. T., Wahtera, K. A., & Teeri, J. A. (2002). Elevated atmospheric CO_2 lowers leaf litter nutritional quality for stream ecosystem food webs. *Global Change Biology, 8*, 163–170

Valett, H. M., Thomas, S. A., Mulhulland, P. J., Webster, J. R., Dahm, C. N., Fellows, C. S., Crenshaw, C. L., & Peterson, C. G. (2008). Endogenous and exogenous control of ecosystem function: N cycling in headwater streams. *Ecology, 89*(12), 3515–3527.

Wagai, R., Kishimoto-Mo, A. W., Yonemura, S., Shirato, Y., Hiradate, S., & Yagasaki, Y. (2013). Linking temperature sensitivity of soil organic matter decomposition to its molecular structure, accessibility, and microbial physiology. *Global Change Biology, 19*(4), 1114–1125. https://doi.org/10.1111/gcb.12112

Wallace, J. B., Eggert, S. L., Meyer, J. L., & Webster, J. R. (1997). Multiple trophic levels of a forest stream linked to terrestrial litter inputs. *Science, 102*(277). https://doi.org/10.1126/science.277.5322.102

Wallace, J. B., Eggert, S. L., Meyer, J. L., & Webster, J. R. (1999). Effects of resource limitation on a detrital-based ecosystem. *Ecological Monographs, 69*(January), 409–442

Wallenstein, M., Allison, S. D., Ernakovich, J., Steinweg, J. M., & Sinsabaugh, R. (2010). Controls on the temperature sensitivity of soil enzymes: A key driver of in situ enzyme activity rates. In G. Shukla & A. Varma (Eds.), *Soil Enzymology* (pp. 245–258). https://doi.org/10.1007/978-3-642-14225-3_13

Wang, G., Post, W. M., Mayes, M. A., Frerichs, J. T., & Sindhu, J. (2012). Parameter estimation for models of ligninolytic and cellulolytic enzyme kinetics. *Soil Biology and Biochemistry, 48*, 28–38. https://doi.org/10.1016/j.soilbio.2012.01.011

Wang, Q., Zhao, X., Chen, L., Yang, Q., Chen, S., & Zhang, W. (2019). Global synthesis of temperature sensitivity of soil organic carbon decomposition: Latitudinal patterns and mechanisms. *Functional Ecology, 33*(3), 514–523. https://doi.org/10.1111/1365-2435.13256

Ward, G. M., Ward, A. K., Dahm, C. N., & Aumen, N. G. (1994). Origin and formation of organic and inorganic particles in aquatic system. In R. S. Wotton (Ed.), *The Biology of Particles in Aquatic Systems.* (pp. 45–73). Lewis Publishers.

Webster, J. R., & Benfield, E. F. (1986). Vascular plant breakdown in freshwater ecosystems. *Annual Review of Ecology and Systematics, 17*, 567–594

Woodward, G., Gessner, M. O., Giller, P. S., Gulis, V., Hladyz, S., Lecerf, A., Malmqvist, B., McKie, B. G., Tiegs, S. D., Cariss, H., Dobson, M., Elosegi, A., Ferreira, V., Graça, M. A. S., Fleituch, T., Lacoursiere, J. O., Nistorescu, M., Pozo, J., Risnoveanu, G., Schindler, M., Vadineanu, A., Vought, L. B.-M., & Chauvet, E. (2012). Continental-scale effects of nutrient pollution on stream ecosystem functioning. *Science, 336*(6087), 1438–1440. https://doi.org/10.1126/science.1219534

Wright, M. S., & Covich, A. P. (2005). Relative importance of bacteria and fungi in a tropical headwater stream: Leaf decomposition and invertebrate feeding preference. *Microbial Ecology, 49*(4), 536–546. https://doi.org/10.1007/s00248-004-0052-4

Young, R. G., Matthaei, C. D., & Townsend, C. R. (2008). Organic matter breakdown and ecosystem metabolism: Functional indicators for assessing river ecosystem health. *Journal of the North American Benthological Society, 27*(3), 605–625. https://doi.org/10.1899/07-121.1

Yvon-Durocher, G., Caffrey, J. M., Cescatti, A., Dossena, M., Giorgio, P. Del, Gasol, J. M., Montoya, J. M., Pumpanen, J., Staehr, P. A., Trimmer, M., Woodward, G., & Allen, A. P. (2012). Reconciling the temperature dependence of respiration across timescales and ecosystem types. *Nature, 487*(7408), 472–476. https://doi.org/10.1038/nature11205

Yvon-Durocher, G., Iwan Jones, J., Trimmer, M., Woodward, G., & Montoya, J. M. (2010). Warming alters the metabolic balance of ecosystems. *Philosophical Transactions of the Royal Society B, 365*, 2117–2126. https://doi.org/10.1098/rstb.2010.0038

Zhang, M., Cheng, X., Geng, Q., Shi, Z., Luo, Y., & Xu, X. (2019). Leaf litter traits predominantly control litter decomposition in streams worldwide. *Global Ecology and Biogeography*, geb.12966. https://doi.org/10.1111/geb.12966

Chapter 13
Causes and Consequences of Changes in Riparian Vegetation for Plant Litter Decomposition Throughout River Networks

John S. Kominoski, Samantha K. Chapman, Walter K. Dodds, Jennifer J. Follstad Shah, and John S. Richardson

Abstract Riparian ecosystems occupy land-water interfaces along upland-to-lowland and coastal gradients of river networks. Global changes in riparian vegetation alter the types and processing of organic matter at these interfaces and throughout river networks. Dominant pathways of structural changes in riparian vegetation are associated with (i) temperature increases and changes in precipitation and hydrology, (ii) range expansion/contraction of native and non-native species, (iii) altered land-use for agriculture/forest plantations and harvesting, and urban development, (iv) shifts in disturbance regimes, such as fire, disease, pest outbreaks, and storms, and (v) saltwater intrusion. Widespread changes in riparian vegetation alter above and below-ground carbon (C) stores and shift the relative proportion of algal and detrital basal resources in aquatic ecosystems. Global changes in riparian vegetation likely shift the sources and sinks of organic matter along river networks from upland headwaters to lowland rivers and coastal wetlands. Climate and global changes are expanding and contracting continental vegetation species ranges while sea-level rise and saltwater

J. S. Kominoski (✉)
Department of Biological Sciences, Institute of Environment, Florida International University, Miami, FL 33199, USA
e-mail: jkominos@fiu.edu

S. K. Chapman
Department of Biology, Villanova University, Villanova, PA 19085, USA
e-mail: samantha.chapman@villanova.edu

W. K. Dodds
Division of Biology, Kansas State University, Manhattan, KS 66502, USA
e-mail: wkdodds@ksu.edu

J. J. Follstad Shah
Environmental & Sustainability Studies Program, Department of Geography, University of Utah, Salt Lake City, UT 84112, USA
e-mail: jennifer.shah@envst.utah.edu

J. S. Richardson
Department of Forest and Conservation Sciences, University of British Columbia, Vancouver, BC V6T 1Z4, Canada
e-mail: john.richardson@ubc.ca

C. M. Swan et al. (eds.), *The Ecology of Plant Litter Decomposition in Stream Ecosystems*, https://doi.org/10.1007/978-3-030-72854-0_13

intrusion are transgressing coastal ecosystems landward. Understanding the general pathways and functional consequences of changes in riparian vegetation is vital to conserving ecosystem functions and services throughout continental river networks and coastal wetlands that are supported by organic matter processing.

13.1 Riparia & River Networks

Most aquatic ecosystems rely on allochthonous energy produced in riparian and terrestrial ecosystems (Cebrian & Lartigue, 2004). Riparian ecosystems contain unique species adapted to flooding, drought, erosion and deposition, which collectively result in riparian areas as being control points of biogeochemical cycling and organic matter processing (Bernhardt et al., 2017; McClain et al., 2003; Naiman et al., 2010; Sabo et al., 2005). Riparian and wetland vegetation drive the quantity, quality, and timing of organic matter in aquatic ecosystems (Batzer & Sharitz, 2014; Kominoski & Rosemond, 2012; Meyer et al., 1998), mediating the influence of surface waters on global carbon (C) budgets. Climate and land-use changes drive the transformation, export, and fate of organic matter from local habitats to entire river networks (Benda et al., 2004). However, the composition of riparian communities is in flux on a global scale, resulting in significant changes to aquatic ecosystem processes and services (González et al., 2017; Kominoski et al., 2013).

The river continuum (Vannote et al., 1980), serial discontinuity (Stanford & Ward, 2001; Ward & Stanford, 1983), and flood pulse concepts (Junk et al., 1989) provided conceptual frameworks that stimulated decades of research assessing the longitudinal and lateral connectivity of organic matter from small streams to large rivers to floodplains and wetlands. This research has demonstrated the importance of organic matter to energy and food web dynamics across spatial scales (e.g., Hall & Meyer 1998; Minshall et al., 1983; Thorp & Delong, 1994). Aquatic ecosystems are both connected and disconnected along hydrologic flow paths, influencing recipient and donor-controlled ecosystems through delivery of organic matter (Ball et al., 2010). Aquatic organisms amongst Earth's biomes have evolved over millennia to utilize organic matter inputs, whether delivered in seasonal pulses or supplied continuously over annual cycles (Benstead & Huryn, 2011; Yeung et al., 2019). This connection between organic matter input regimes and evolutionary adaptations likely has profound impacts on the quantity of energy and nutrients stored, transformed, or transported within river networks.

Despite the evolution of these frameworks towards understanding the river network, we know very little about the integration of organic matter processes over time and through multiple components of river networks, such as lakes, reservoirs, floodplains and wetlands. The role of inland aquatic ecosystems in processing terrestrial organic matter is critically large relative to the spatial extent that surface waters cover the globe (Battin et al., 2009). Wetlands, especially coastal wetlands, are foundational to the global storage of carbon, given that they retain and bury massive amounts of organic and inorganic matter relative to their land area (Chmura et al.,

2003, McCleod et al., 2011). River networks actively fix, store, transform, and transport carbon (Cole et al., 2007), and stream litter decomposition processes integrate at network scales (Fig. 13.1). At local scales, riparian vegetation influences litter processing, organic carbon availability, and aquatic ecosystem services (Fig. 13.1). Further, threshold responses or unforeseen consequences caused by environmental

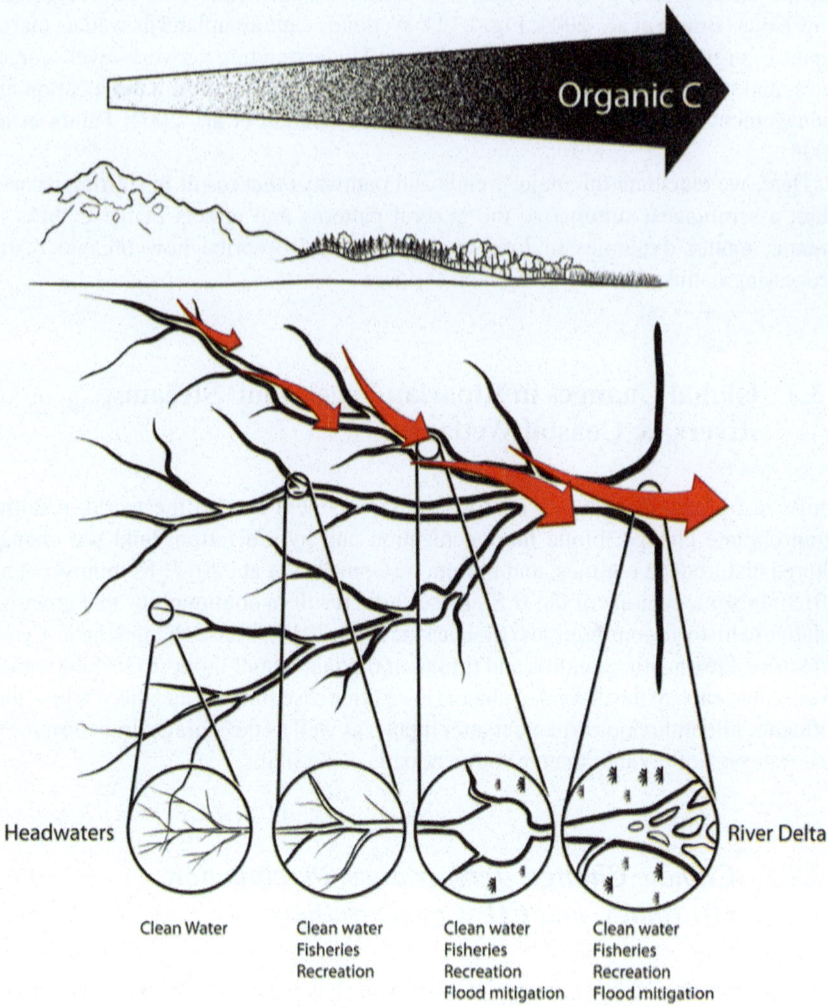

Fig. 13.1 Conceptual figure depicting a river network demonstrating how spatial heterogeneity in riparian and wetland ecosystems influences the distribution and processing of organic matter and ecosystem services. River networks integrate the sources, fluxes, and transformations of organic matter that vary along multiple flow paths from headwaters to the ocean that collectively influence ecosystem services. Copyright E. E. Nixon

change accumulating in separate sub-basins within larger river networks may occur downstream after convergence of multiple flow paths (e.g., Ward & Stanford, 1983).

Quantification of organic matter fate in river networks that includes interior and coastal wetlands is needed, as aquatic ecosystems receive and distribute organic matter from ecosystems to which they are coupled (Aufdenkampe et al., 2011). Carbon dynamics and ecosystem services at any given location in a river network depend on upstream processes that accumulate along hierarchical and conjoining flow paths (Benda et al., 2004; Fig. 13.1). Wetlands capture upland as well as marine inputs of organic matter, increasing retention. Understanding network-level sources, fates, and transformations of organic matter will aid in identifying the location and management of freshwater ecosystem services (Benda et al., 2004; Peters et al., 2008).

Here, we elucidate the major trends and pathways that result in shifting riparian plant assemblages, summarize the general patterns and effects of these shifts on organic matter dynamics in lotic ecosystems, and describe how organic matter processing is linked to key ecosystem services.

13.2 Global Changes in Riparian Vegetation: Streams, Rivers, & Coastal Wetlands

Shifts in riparian community composition are evident around the world, resulting from climate change, biotic homogenization and hybridization, land use change, altered disturbance regimes, and pollution (González et al., 2017; Kominoski et al., 2013). In some regions of the U.S., these shifts result in communities that are novel relative to historic communities (Macfarlane et al., 2017). These changes have myriad effects on the quantity, quality, and timing of organic matter inputs to surface waters. We explore each of the drivers of change in riparian assemblages and discuss how they influence allochthonous organic matter inputs, as well as the storage, transformation, and transport of organic matter within aquatic ecosystems.

13.2.1 Climate Change: Temperature, Precipitation, Hydrology, and CO_2 Concentrations

The biomes that freshwaters are embedded in have strong influences on riparian vegetation (Fig. 13.2; Dodds et al., 2015, 2019). When the amount of precipitation is less than the potential evapotranspiration, the riparian zone can shift from trees to grasses and shrubs. Intermittent habitats in extremely dry areas can have almost no woody riparian vegetation, through stream channels and wetlands are often wetter areas likely to support a greater plant biomass. In very cold areas (high altitude or latitude), the development of woody vegetation may be inhibited by climate severity

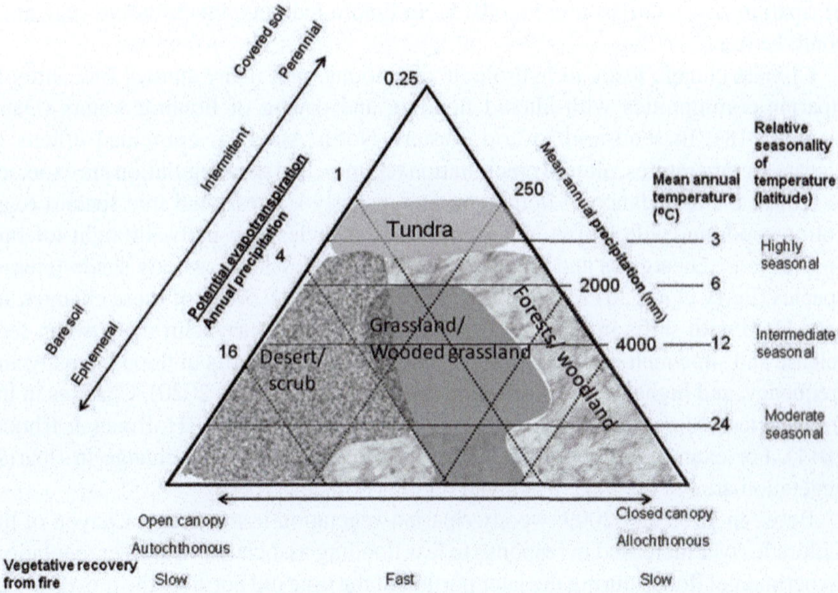

Fig. 13.2 Climate zones, vegetation types, and relationship to stream characteristics (modified from Holdridge [1947]). Image from Dodds et al. (2015)

allowing only low-stature vegetation in riparian zones. Woody vegetation can have indirect influences on decomposition as well. In restoring Australian rivers to mitigate temperature increases, a 10% increase in riparian cover leads to a 1 degree Centigrade lowering of temperature (Davies, 2010).

Within areas where precipitation is, at least seasonally, unable to meet demands for plant growth, the effects on riparian vegetation may be strong. Climate change will increase climate variability, and intense dry periods could alter riparian community composition. For example, climate change and lower fire frequency and intensity may be responsible for riparian woody expansion into tallgrass prairie streams (Veach et al., 2015) and this expansion has substantial influence on stream food webs and community composition (Riley & Dodds, 2012; Vandermyde & Whiles, 2015; Veach et al., 2015), water chemistry, and whole-stream metabolic rates (Larson et al., 2018). In similarly dry Mediterranean habitats, climate change is expected to inhibit early successional riparian communities slowing regeneration of these communities in response to disturbances such as floods (Rivaes et al., 2013).

Increased temperatures associated with climate change will alter Arctic and high altitude riparian vegetation. For example, alterations in ice breakup and scour will change riparian communities (Prowse et al., 2006). Warmer temperatures will allow higher stature and greater biomass of riparian vegetation to move to higher latitudes and altitudes. In the Boreal zone, increased temperature is predicted to allow invasive species to take hold, and lead to shifts in community structure as well as narrowing of riparian zones (Nilsson et al., 2013). In tundra habitats, this invasive vegetation could be trees.

Climate change leads to hydrologic alterations, and many studies link shifts in riparian communities with altered flooding and drying of flowing waters (Datry et al., 2018). In semi-arid to arid western North America, combined effects of elevated temperatures, altered precipitation regimes, and river regulation are expected to reduce the abundance of dominant, native, early-successional tree species (e.g., cottonwoods and willows) and favor herbaceous species, non-native drought-tolerant species (e.g., cheatgrass and tamarisk), and late-successional, woody shade-tolerant species (Perry et al., 2012; Reynolds & Shafroth, 2017). Some of these changes are associated with shifts in seeding phenology that result in asynchrony between seed release and snowmelt runoff or monsoonal precipitation, shifts in flood intensity and frequency, and higher plant water demand (Perry et al., 2012, 2020). Changes in the frequencies of large floods are also important in other regions (Hoffman & Rhode, 2011). For example, global climate models predict up to 27% change in riparian vegetation area in the Lake Michigan region (Primack, 2000).

Between 1965 and 2009, woody riparian vegetation in the Grand Canyon of the Colorado river increased in response to less flooding associated with river regulation. Experimental floods during the later portion of the time did not slow the movement of riparian vegetation to colonize and stabilize sand bars (Sankey et al., 2015). Increased flooding (flood augmentation) in Rocky Mountains (upper Arkansas River basin) decreased riparian wetland vegetation cover by 10% (Dominick & O'Neill, 1998). The North Platte River narrowed and more dense riparian tree cover developed in response to decreased spring flooding, however the upper Missouri River did not demonstrate similar responses (Johnson, 1998).

13.2.2 Native and Non-native Plant Species Changes

Shifts in native and non-native riparian plant species can have predictable effects on aquatic ecosystem structure and function (Kominoski et al., 2013). Invasive riparian plant species such as salt cedar (*Tamarix* spp.) Russian olive (*Elaeagnus angustifolia*), Japanese knotweed (*Fallopia japonica*) and others may displace native species and alter leaf litter quality and ecosystem functions in inland streams and rivers worldwide (e.g., Kominoski et al., 2013; Lecerf & Chauvet, 2008) (Fig. 13.3). The phylogeny of riparian plant species strongly influences leaf litter decomposition in adjacent aquatic ecosystems (LeRoy et al., 2020). A number of studies have shown that differences in litter quality are linked to genotypes of individual trees, resulting in variable rates of decomposition rates in streams (Marks, 2019). LeRoy et al. (2006) showed that different genotypes of poplar and their hybrids had very different decomposition rates in streams, and a similar result was seen in experimental ponds (Crutsinger et al., 2014). Jackrel and Wooton (2014) demonstrated using an experiment of reciprocal transplants of leaves that communities of detritivores showed adaptations to leaf litter derived from the local genotypes of red alder. Phylogeny

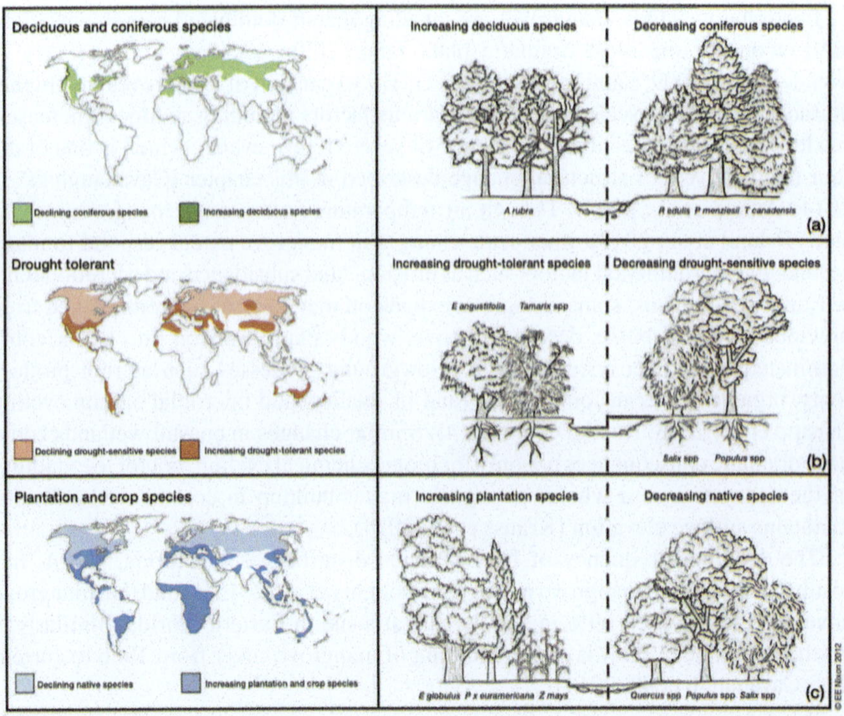

Fig. 13.3 Three global examples of broad-scale changes in riparian plant species composition: **a** shifts in coniferous and deciduous tree species, **b** increases in drought-tolerant species, and **c** global distribution of plantation and crop species. Distribution of increasing and decreasing tree species for each of the three examples are shown separately on each map. **a** *Pinus and Tsuga* species are declining throughout the Northern Hemisphere. These species are being replaced by deciduous species, such as *Alnus* species in North America. **b** *Populus* and *Salix* species are declining throughout the Northern Hemisphere, whereas drought-tolerant specious such as *Tamarix* specious (native to Eurasia) and *Elaeagnus angustifolia* (native from the Middle East to central Asia) are gaining in North America, Europe, South America, Asia, and North Africa. *Acer negundo (*native to North America) is increasing in Europe, and Salix (native to Europe) is invading Australia and New Zealand. **c** *Populus* and *Salix* species are declining throughout the Northern Hemisphere, with the exception of *Populus* hybrids grown in plantations in North America and Europe. Native riparian vegetation throughout North America, Europe, Asia, Australia, and South America are declining as land is used for plantation and crop species, such as *Eucalyptus* species and *Zea mays,* which are being planted globally. Illustrations of increasing and decreasing species emphasize how shifts in plant species composition have structural and functional consequences for riparian and aquatic ecosystems. Image from Kominoski et al. (2013)

also plays a role with regards to the rate at which litter mixtures decay in streams and rivers. Boyero et al. (2016) found that lower phylogenetic distance amongst leaf species decaying together in 24 streams along a latitudinal gradient promoted greater rates of mass loss in temperate biomes but slower rates of mass loss in the tropics.

In coastal wetlands, mangroves are invading marsh-dominated ecosystems globally, in one of the most dramatic plant range shifts occurring today (Perry & Mendelssohn, 2009; Saintilan et al., 2014). The expansion of mangroves into higher latitudes on a global scale is driven by various factors including sediment increases, sea level rise, and a declining frequency of severe freeze events, which is one of the four major drivers of structural change described in this chapter (Cavanaugh et al., 2014; Osland et al., 2013). Though air temperatures often drive mangrove expansion (Osland et al., 2013), finer-scale changes in mangrove extent respond to many secondary environmental factors such as erosion, land subsidence and accretion (Giri & Long, 2014). Thus, some mangrove expansion may be due to re-emergence from previous populations. As described above, woody plant invasion into herbaceous-dominated upland ecosystems alters belowground processes such as root productivity, organic matter decomposition (and its quality) and microbial carbon cycling (Knapp et al., 2008; Rundel et al., 2014). Similar changes in coastal wetlands could have dramatic consequences not only for biogeochemical cycling, but for the viability of the ecosystem as a whole, as organic matter buildup in coastal wetlands also maintains surface elevation (Krauss et al., 2014).

The declining frequency of freeze-related disturbances in Florida, U.S.A. has resulted in increased mangrove extent, and Doughty et al. (2015) found that mangrove coverage increased by 69% in just 7 years at a site in Eastern Florida. Similarly, a northeastern site in Florida saw a doubling of mangrove cover from 1986 to current day (Cavanaugh et al., 2014).

13.2.3 Agriculture and Forest Harvesting

Shifts in the composition of riparian contributions of leaf litter can result from natural disturbances, human disturbances, and long-term changes in land use. These events often create space for fast-growing, disturbance-dependent trees, some of which have undefended and high-quality leaf litter, such as red alder (*Alnus rubra*) and yellow poplar (*Liriodendron tulipifera*). These species depend on primary succession to establish, but are often unable to persist at a site without further disturbance, so over a time span of 60 to 80 years these trees are replaced by later successional species. Some similar successional changes can occur due to forest harvesting (e.g., Kominoski et al., 2011). The types of leaf litter inputs from riparian vegetation has a profound impact on the resulting stream community, particularly the composition of the detritivores trophic level (Kominoski et al., 2013).

In many parts of the world, there has been extensive planting of non-native species as forest crop trees. Some of the trees most commonly planted outside of their native ranges are eucalypts (*Eucalyptus globulus* and *E. nitens*) and radiata pine (*Pinus radiata*) (Ferreira et al., 2019; García et al., 2012). Ferreira et al. (2019) demonstrated that decomposition rates of same leaf species in streams flowing through *Eucalyptus* plantations are on average 23% lower than for streams in native forests, mostly due to reduction in macroinvertebrate densities. In that study, the magnitude of the inhibition of decomposition varied geographically, with greater effect sizes in more temperate regions, attributed to the greater influence of macroinvertebrates in subtropical regions (Ferreira et al., 2019). In addition to *Eucalyptus* plantations for fiber, other trees, such as palm oil and rubber trees are planted close to streamsides. In Malaysia, there were very small differences in macroinvertebrate assemblages on leaf packs draining palm oil forests versus native forest (Chellaiah & Yule, 2018).

Many tree species planted for fiber production in riparian habitats have leaf litter of low quality. In particular, conifers (pines, firs, etc.) and eucalypts are considered as low-quality due to being well defended physically and chemically against breakdown, having low N:C, and possibly chemical defenses (Graça & Cressa, 2010). Pine plantations in the UK have been a common target for afforestation, however, pine litter decomposes less than 20% as quickly as birch leaves, despite similar detritivore assemblages, and suggests planting pines or other conifers alongside streams could lead to reductions in stream productivity (Collen et al., 2004). In other parts of the world, North American trees such as Douglas-fir and Sitka spruce have been planted, often right up to streamside (e.g., Gee & Smith, 1997), although the impacts of these plantings on instream decomposition are not clear. Kominoski et al. (2011) showed that leaf litter decomposition rates were lower in streams flowing through conifer forests than streams with a larger component of angiosperms, primarily red alder). The overall productivity and yield of particulate detritus from catchments in Alaska with a high component of alder in riparian areas were much higher than streams draining mostly conifer stands (Wipfli & Musselwhite, 2004). These forestry related shifts in streamside species composition and consequent inputs will affect decomposition rates, either through types of inputs or catchment characteristics.

Forest harvesting, and other land-use, can impact instream decomposition rates. In some studies forest harvest has resulted in decreased rates of decomposition (e.g., Kreutzweiser et al., 2008; Lecerf & Richardson, 2010), whereas other studies have documented increased rates from forestry (Benfield et al., 2001, McKie & Malmqvist, 2009). In each of these studies there was little evidence for why decomposition rates changed, particularly in opposite directions in different regions. These impacts of forestry on decomposition rates occur whether or not riparian buffers are retained during harvesting, and the mechanisms for these changes remains an open research question.

Agricultural crops may be grown right up to stream edges in some jurisdictions. This would certainly alter leaf litter composition, and probably reduce input rates. An additional consideration is that in some parts of the world, crop plants have been genetically modified, and their leaf litter may include anti-herbivory chemicals. In much of the USA corn has been so-modified to include *Bacillus thuringiensis* (Bt) genes to reduce insect damage. The leaf litter from this corn has been shown to slow the growth rate of a detritivorous caddisfly larva, *Lepidostoma liba*, although there were no other large-scale effects on stream communities from such studies (Chambers et al., 2010). Conversion of forest to pasture has several effects on instream processes, including the particular types of leaf litter inputs. In three streams in Ecuador, the contrast between reaches in forest and pasture showed lower decomposition rates of a standardized leaf litter in pasture reaches, largely attributed to the absence of a single species of shredding caddisfly larvae from pasture streams (Encalada et al., 2010). Young et al. (1994) found rates of leaf litter decomposition were higher in streams draining more intensively managed pastures, which they attributed to higher nutrient yield, particularly Nitrogen, from pastures.

13.2.4 Urbanization

The urban stream syndrome represents the suite of alterations to stream ecosystems that accompany urbanization (e.g., Booth et al., 2016; Paul & Meyer, 2001). These include increased peaks in flows, increased nutrients and contaminants, warmer peak temperatures and rapid shifts, channel simplification, altered riparian vegetation, etc. All of these aspects can influence litter decomposition rates (Young et al., 2008). Litter decomposition rates are often different in urban streams from those in comparable forested sites, due to a host of potential influences. For example, in urban streams in Puerto Rico, decomposition rates were only 30% or lower than rates in forested streams, attributed to loss of consumer species in urban streams (Classen-Rodriguez et al., 2019). Similarly, breakdown rates in urban streams in Brazil were about half that of rates in forested streams, considered to be due to shifts in decomposer assemblages, especially loss of detritivorous invertebrates (Martins et al., 2015). Decomposition of litter in a stream in Malaysia was nearly twice as rapid in urban reaches than in forested sections, presumably due to nutrient enrichment (Yule et al., 2015). Breakdown rates in streams in Maine were higher in more urbanized catchments, and was attributed to elevated concentrations of nutrients along the rural to urban gradient (Huryn et al., 2002). Rates of breakdown were almost four times higher in urban streams in the Southeast USA compared to forest streams (Paul et al., 2006). The influence of urbanization is large, but the direction and magnitude are highly variable, and depends on the suite of processes altered in any given urban setting.

13.3 Impacts of Altered Litter Decomposition Throughout River Networks

13.3.1 Land-Use Change Impacts

Organic matter source and bioavailability are closely linked to human land use activities. Global reductions in terrestrial C are attributed to increasing human co-option of terrestrial gross primary production (Running, 2012). For example, wetland drainage and riparian land clearing remove terrestrial organic matter storage, and widespread replacement of native vegetation with agricultural crops for direct and indirect human consumption reduce carbon availability for ecosystem functions and services (Allan, 2004; Running, 2012). Projected increases in cultivated land area coupled with increased nutrient mobilization (MEA, 2005) will reduce the amount of terrestrial organic matter loading (Running, 2012) and increase in situ aquatic organic matter production, which is more bioavailable. This shift towards more readily available organic matter will influence secondary production and energy flow paths within aquatic food webs from headwaters to downstream lakes and rivers (Griffiths et al., 2009). The distribution of dissolved organic carbon in the Ipswich River, Massachusetts demonstrates how organic matter processes throughout a river network potentially contribute to nutrient regulating and water supply ecosystem services (Stewart et al., 2011). In agricultural regions, the interaction of increased nutrient inputs, reduced terrestrial carbon inputs, and reduced light limitation explain why most carbon export from the Mississippi River Basin is due to aquatic primary production (Shih et al., 2010).

13.3.2 Climate Change and Eutrophication Impacts

Global changes in soil and water temperatures, hydrologic variability, and nutrient availability will increase rates of ecosystem metabolism and alter organic matter export (Acuña & Tockner, 2010). Elevated temperature increases rates of leaf litter decomposition in streams and rivers (Follstad Shah et al., 2017). The magnitude of this increase is similar, whether decay is mediated by microbes alone or due to the combined effects of microbes and detritivores (Follstad Shah et al., 2017). However, this thermal response of detritivores may not scale to the level of river networks, as decomposition rates at this scale depend on the availability of terrestrial organic matter inputs. For example, some of the land use changes (described above) reduce inputs of terrestrial organic matter inputs to streams and rivers. In addition, net primary production may be diminished in water-stressed riparian plant communities, despite elevated CO_2 concentrations that promote higher rates of photosynthesis and growth and greater water use efficiency (Perry et al., 2013). These studies imply that changes to riparian community organic matter production can modulate predicted effects of climate change at the scale of river networks. Reservoirs alter upstream-downstream organic matter linkages via increased storage of organic matter within reservoirs (Vörösmarty et al., 2003) and enhanced loss of carbon as CO_2 and methane (CH_4) emissions (Kominoski & Rosemond, 2012; Tranvik et al., 2009). Streams with lower terrestrial organic matter inputs and lower standing stocks of benthic organic matter, as well as downstream reaches of river networks that have lower quality organic matter, are likely to remove less nutrients (Barnes et al., 2012; Taylor & Townsend 2010), potentially contributing to denitrification efficiency loss (Mulholland et al., 2008). Reductions in standing stocks of organic matter could result in lower secondary production of food webs that support downstream fisheries (Cross et al., 2006).

Elevated temperature and moderate eutrophication are shifting streams and rivers towards greater heterotrophy, resulting in more emissions of CO_2. Whole-stream metabolism studies along natural thermal gradients show that ecosystem respiration increases in response to rising temperature to a greater extent than gross primary production, resulting in declines in net ecosystem production (Demars et al., 2011). Moderate nutrient enrichment stimulates organic matter decomposition rates in streams and rivers (Woodward et al., 2012) and at the network scale results in reduced terrestrial C residence time (Rosemond et al., 2015). In temperate biomes, terrestrial C losses can exceed instream C production resulting in greater net heterotrophy (Rosemond et al., 2015). Again, however, this pattern is dependent on terrestrial C supply concomitant with heterotrophic demand for C in eutrophic systems.

Model simulations show that flow regime alterations have a greater effect on organic C dynamics within river networks than altered thermal regimes, but this effect is most pronounced in headwater streams relative to mainstem rivers (Acuña & Tockner, 2010). More floods and longer droughts are predicted to reduce the amount of organic C processed within the river network due to reduced rates of respiration and increased C export (Acuña & Tockner, 2010). Some of the increased export of C in Mediterranean systems may also be due to drought-induced phenological shifts in leaf litter senescence, followed by winter flood events (Acuña et al., 2007).

Climate change further poses unprecedented effects on coastal regions by causing saltwater intrusion into vulnerable ecosystems through accelerating rates of sea-level rise, changes in the hydrologic cycle and temperature regime, and potentially the increasing strength and frequency of storms (Farfan et al., 2014; Herbert et al., 2015, Nicholls & Cazenave, 2010; Osland et al., 2016). Direct effects of climate change on coastal ecosystems can result in loss of wetland area and ecosystem function if and when rates of sea-level rise exceed the natural capacity of foundation species in wetlands to adapt (Charles et al., 2019; Saha et al., 2011; Wilson et al., 2018). Low-lying coastal ecosystems are periodically influenced by storms and continuously influenced by sea-level rise and saltwater intrusion (Dessu et al., 2018; Herbert et al., 2015; Osland et al., 2016). Our collective understanding is that saltwater intrusion generally decreases C storage and increases nutrient export in coastal wetlands (Ardón et al., 2013, 2016; Charles et al., 2019; Herbert et al., 2015, 2018; Neubauer, 2013; Wilson et al., 2018).

13.3.3 Impacts of Altered Hydrologic Connectivity

Ecosystems are becoming increasingly more or less connected through globalization, fragmentation, and climate change. Connectivity—the flow of organisms, water, materials, and ecological processes across landscapes (Taylor et al., 1993)—can be used to understand how to better manage and restore threatened and declining ecosystems (Haddad et al., 2015; Kominoski et al., 2019; Pringle, 2001). The extent and health of coastal ecosystems are declining worldwide (Dahl & Stedman, 2013; Nicholls et al., 2007), so understanding how changes in hydrologic connectivity influence the structure and function of these threatened ecosystems is paramount (Sheaves, 2009). Storage and accumulation rates of carbon and nutrients in coastal wetlands are sensitive to underlying topographic gradients that influence hydrologic connectivity to these sources, as well as temporal changes in connectivity that can be both directional (e.g., sea-level rise) and episodic (e.g., hurricanes, droughts, floods). Therefore, hydrologic connectivity can influence biogeochemical processes, including net primary productivity and organic matter mineralization (Bouillon et al., 2008; Castañeda-Moya et al., 2013; Koch et al., 2012; Noe et al., 2001).

Changes in hydrology associated with urbanization and river regulation influence the sources and fates of organic matter in river networks. Urbanization increases impervious cover, reducing infiltration rates and increasing the tendency of flash floods as well as increasing organic matter via septic and sewage inputs (Paul & Meyer, 2001; Walsh et al., 2005). Therefore, urban watersheds are likely to have less continuous sources of soil carbon due to reduced infiltration rates, but potentially more C point sources associated with engineered open spaces (Aitkenhead-Peterson et al., 2009) or instream production and wastewater inputs (Newcomer et al., 2012). Urban watersheds may also exhibit increased retention and burial of C in reservoirs (Vörösmarty et al., 2003). All of these factors ultimately alter the quantity and bioavailability of organic matter in urbanized waterways. River regulation alters the hydrologic regimes, but also the distribution of lotic and lentic surface waters in river networks. Acuña and Tockner (2010) showed that the majority of organic C inputs to river networks is processed within reservoirs, which altered C dynamics in river reaches below impoundments. Reservoirs alter upstream-downstream organic matter linkages via increased storage of organic matter within reservoirs (Vörösmarty et al., 2003) and enhanced loss of carbon as CO_2 and methane (CH_4) emissions (Kominoski & Rosemond 2012; Tranvik et al., 2009).

13.3.4 Impacts on Ecosystem Services

Detrital organic matter is a critical supporting component of many ecosystem functions, and the production, storage, transformation, and transport of organic matter are spatially and temporally dynamic throughout watersheds and river networks (Fig. 13.1; Hall & Meyer 1998; Minshall et al., 1983; Thorp & Delong, 1994). The ability for science to link biophysical processes of litter, to societal values of ecosystem services requires a mechanistic understanding of how these functions and services are related and linked spatially and temporally (Bennett et al., 2009; Daily et al., 2009; Rosemond et al., 2015). A critical knowledge gap is understanding how these ecosystem services are driven by network-level organic matter dynamics, which is required given dynamic and heterogeneous changes in transport, sources, and processing of litter from various locations throughout river networks. The relative importance of organic matter sources changes longitudinally within river networks (Vannote et al., 1980). The relative support of the basal resources supporting aquatic food webs from particulate and dissolved terrestrial and aquatic sources is highly variable throughout river networks and estuaries (Cawley et al., 2014; Sobczak et al., 2002). Worldwide construction of dams for hydropower generation increases the accumulation and burial of organic matter (Downing et al., 2008) that can lead to sources of CO_2 and methane (Kominoski & Rosemond, 2012; St Louis et al., 2000; Tranvik et al., 2009). The cycling of nitrogen and phosphorus are closely linked to the availability of detrital carbon (Kominoski et al., 2018; Rosemond et al., 2015; Seitzinger et al., 2006; Taylor & Townsend, 2010) and thus the amount of nutrients retained on land rather than exported downstream where eutrophication and hypoxia of coastal waters has caused collapse of estuaries (NRC, 2000; Rabalais et al., 2002). Organic matter primarily contributes to regulating and supporting ecosystem services via its control of ecosystem function through microbial food webs and biogeochemical processes that are foundational to many—if not all—aquatic ecosystem services.

In coastal wetlands, mangrove encroachment can increase ecosystem C storage (Doughty et al., 2015; Kelleway et al., 2016), but the magnitude of these C storage differences depends on environmental setting (Yando et al., 2016). The majority of mangrove carbon storage increases are often due to aboveground biomass (Charles et al., 2020; Doughty et al., 2015) but some studies show that soil carbon can also increase either over longer time scales (Kelleway et al., 2016) or in some cases rapidly (Simpson et al., 2019). Using experimental mangrove removals, Guo et al. (2017) found that mangrove cover was positive related to soil organic carbon content. Increases in soil carbon storage when mangroves invade into marshes are likely driven by an increase in root growth (Fig. 13.4; Coldren et al., 2019). Although organic matter decomposition is an important driver of the blue carbon storage in wetlands and of surface elevation, less is known about how decomposition changes when mangroves encroach into salt marshes (but see Charles et al., 2020). Decomposition rates in coastal wetlands depends in part on litter quality, but also are controlled by the oxygen availability in oils, as anoxic conditions limit enzyme reactions that control organic matter breakdown (Chapman et al., 2019). It's possible that mangrove encroachment into salt marshes could alter both the litter quality and the oxygen availability. Perry and Mendelssohn (2009) found no difference in decomposition rates in plots where mangroves had encroached into the marsh as compared to those still dominated by the salt marsh grass *Spartina alterniflora*. However, Charles et al. (2020) found that mangrove litter decomposed much more slowly than the dominant salt marsh plant *Batis martima* in coastal Texas sites where mangroves are encroaching. Taken together, these findings suggest that changes in organic matter decomposition with mangrove encroachment may depend on the species composition of the salt marsh.

The ability of aquatic ecosystems to support ecosystem services for society depends on the timing, quantity, and source of organic matter inputs. Global environmental changes (e.g., climate, land-use, and hydrology) influence the quantity, source, and processing rates of organic matter (Kominoski & Rosemond 2012; Tank et al., 2010), which challenge our ability to maintain and sustain aquatic ecosystem services (Fig. 13.1 and examples above). The magnitude and speed of these global changes lends urgency to better quantify organic matter dynamics that support aquatic ecosystem services at network scales. The variability in organic matter sources and transformations across different aquatic ecosystems (streams, lakes, wetlands) will result in retention, production and transport dynamics that vary over space and time.

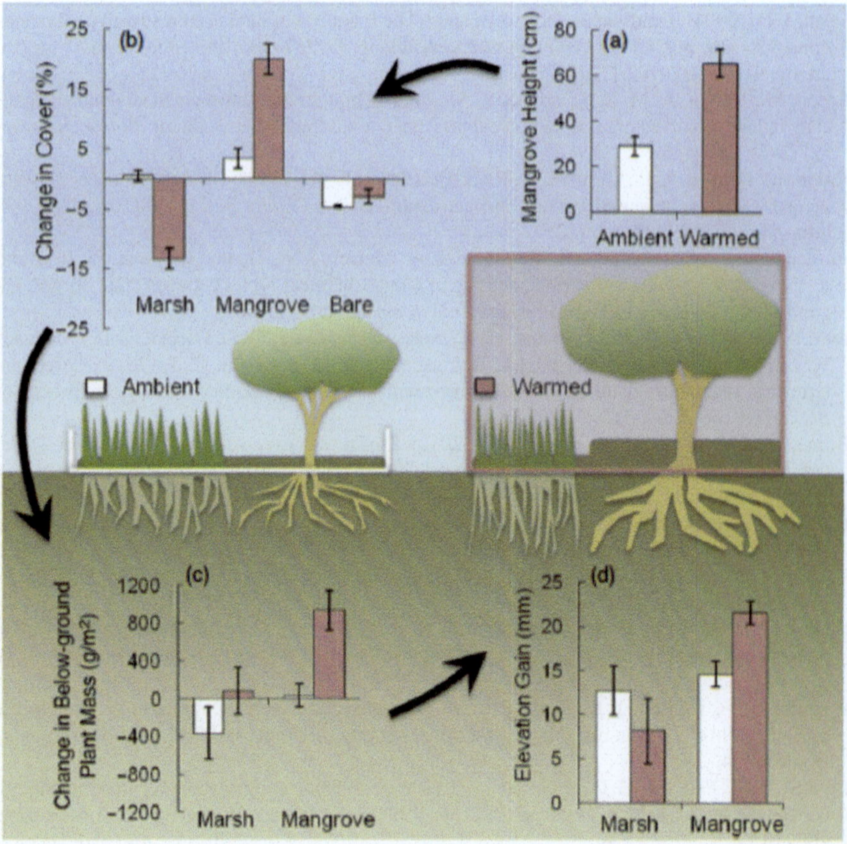

Fig. 13.4 Response of mangroves and salt marshes to warming conditions: **a** mangrove height (cm), where chronic warming accelerates vertical growth, **b** percent change in cover, where areal expansions of mangrove shade out salt marsh plants, **c** change in below-ground plant mass (g/m^2), where increased root productivity and areal coverage of individual mangroves result in greater below-ground growth, and **d** change in elevation (mm), which is largely determined by changes in below-ground biomass. Warming treatments included ambient (control) and warming chamber. Data are means ± SE. Image from Coldron et al. (2019)

References

Acuña, V., Giorgi, A., Muñoz, I., Sabater, F. & Sabater, S. (2007). Meteorological and riparian influences on organic matter dynamics in a forested Mediterranean stream. *Journal of the North American Benthological Society, 26*(1), 54–69.

Acuña, V., & Tockner, K. (2010). The effects of alterations in temperature and flow regime on organic carbon dynamics in Mediterranean river networks. *Global Change Biology, 16*(9), 2638–2650. https://doi.org/10.1111/j.1365-2486.2010.02170.x.

Aitkenhead-Peterson, J. A., Steele, M. K., Nahar, N., & Santhy, K. (2009). Dissolved organic carbon and nitrogen in urban and rural watersheds of south-central Texas: Land use and land management influences. *Biogeochemistry, 96*(1), 119–129. https://doi.org/10.1007/s10533-009-9348-2.

Allan, J. D. (2004). Landscapes and riverscapes: The influence of land use on stream ecosystems. *Annual Review of Ecology Evolution and Systematics, 35,* 257–284. https://doi.org/10.1146/annurev.ecolsys.35.120202.110122.

Ardón, M., Helton, A. M. & Bernhardt, E. S. (2016). Drought and saltwater incursion synergistically reduce dissolved organic carbon export from coastal freshwater wetlands. *Biogeochemistry, 127*(2–3), 411–426.

Ardón, M., Morse, J. L., Colman, B. P., & Bernhardt, E. S. (2013). Drought-induced saltwater incursion leads to increased wetland nitrogen export. *Global Change Biology, 19*(10), 2976–2985. https://doi.org/10.1111/gcb.12287.

Aufdenkampe, A. K., Mayorga, E., Raymond, P. A., Melack, J. M., Doney, S. C., Alin, S. R., Aalto, R. E., & Yoo, K. (2011). Riverine coupling of biogeochemical cycles between land, oceans, and atmosphere. *Frontiers in Ecology and the Environment, 9*(1), 53–60.

Ball, B. A., Kominoski, J. S., Adams, H. E., Jones, S. E., Kane, E. S., Loecke, T. D., Mahaney, W. M., Martina, J. P., Prather, C. M., Robinson, T. M., & Solomon, C. T. (2010). Direct and terrestrial vegetation-mediated effects of environmental change on aquatic ecosystem processes. *BioScience, 60,* 590–601.

Barnes, R. T., Smith, R. L., & Aiken, G. R. (2012). Linkages between denitrification and dissolved organic matter quality, Boulder Creek watershed, Colorado. *Journal of Geophysical Research: Biogeosciences, 117*(G1).

Battin, T. J., Luyssaert, S., Kaplan, L. A., Aufdenkampe, A. K., Richter, A. & Tranvik, L. J. (2009). The boundless carbon cycle. *Nature GeoScience, 2*: 598–600.

Batzer, D. P., & Sharitz, R. R. (Eds.). (2014). *Ecology of freshwater and Estuarine Wetlands.* University of California Press.

Benda, L., Poff, N. L., Miller, D., Dunne, T., Reeves, G., Pess, G., & Pollock, M. (2004). The network dynamics hypothesis: How channel networks structure riverine habitats. *BioScience, 54,* 413–427.

Benfield, E. F., Webster, J. R., Tank, J. L., & Hutchens, J. J. (2001). Long-term patterns in leaf breakdown in streams in response to watershed logging. *International Review of Hydrobiology, 86,* 467–474.

Bennett, E. M., Peterson, G. D., & Gordon, L. J. (2009). Understanding relationships among multiple ecosystem services. *Ecology Letters, 12*(12), 1394–1404. https://doi.org/10.1111/j.1461-0248.2009.01387.x.

Benstead, J. P., & Huryn, A. D. (2011). Extreme seasonality of litter breakdown in an arctic spring-fed stream is driven by shredder phenology, not temperature. *Freshwater Biology, 56*(10), 2034–2044. https://doi.org/10.1111/j.1365-2427.2011.02635.x.

Bernhardt, E. S., Blaszczak, J. R., Ficken, C. D., Fork, M. L., Kaiser, K. E., & Seybold, E. C. (2017). Control points in ecosystems: Moving beyond the hot spot hot moment concept. *Ecosystems, 20*(4), 665–682.

Booth, D. B., Roy, A. H., Smith, B., & Capps, K. A. (2016). Global perspectives on the urban stream syndrome. *Freshwater Science, 35,* 412–420.

Bouillon, S., Borges, A. V., Castañeda-Moya, E., Diele, K., Dittmar, T., Duke, N. C., Kristensen, E., Lee, S. Y., Marchand, C., Middelburg, J. J., Rivera-Monroy, V. H., Smith III, T. J., & Twilley, R. R. (2008). Mangrove production and carbon sinks: A revision of global budget estimates. *Global Biogeochemical Cycles, 22*(2). https://doi.org/10.1029/2007gb003052

Boyero, L., Pearson, R. G., Hui, C., Gessner, M. O., Pérez, J., Alexandrou, M. A., Graça, M. A. S., Cardinale, B. J., Albariño, R. J., Arunachalam, M., Barmuta, L. A., Boulton, A. J., Bruder, A., Callisto, M., Chauvet, E., Death, R. G., Dudgeon, D., Encalada, A. C., Ferreira, V., … Yule, C. M. (2016). Biotic and abiotic variables influencing plant litter breakdown in streams: A global study. *Proceedings of the Royal Society B: Biological Sciences, 283*(1829). https://doi.org/10.1098/rspb.2015.2664

Castañeda-Moya, E., Twilley, R. R., & Rivera-Monroy, V. H. (2013). Allocation of biomass and net primary productivity of mangrove forests along environmental gradients in the Florida Coastal Everglades, USA. *Forest Ecology and Management, 307,* 226–241.

Cavanaugh, K. C., Kellner, J. R., Forde, A. J., Gruner, D. S., Parker, J. D., Rodriguez, W., et al. (2014). Poleward expansion of mangroves is a threshold response to decreased frequency of extreme cold events. *Proceedings of the National Academy of Sciences*. https://doi.org/10.1073/pnas.1315800111.

Cawley, K. M., Yamashita, Y., Maie, N., & Jaffe, R. (2014). Using optical properties to quantify fringe mangrove inputs to the dissolved organic matter (DOM) pool in a subtropical estuary. *Estuaries and Coasts, 37,* 399–410.

Cebrian, J., & Lartigue, J. (2004). Patterns of herbivory and decomposition in aquatic and terrestrial ecosystems. *Ecological Monographs, 74*(2), 237–259. https://doi.org/10.1890/03-4019.

Chambers, C. P., Whiles, M. R., Rosi-Marshall, E. J., Tank, J. L., Royer, T. V., Griffiths, N. A., Evans-White, M. A., & Stojak, A. R. (2010). Responses of stream macroinvertebrates to Bt maize leaf detritus. *Ecological Applications, 20*(7), 1949–1960. https://doi.org/10.1890/09-0598.1

Chapman, S. K., Hayes, M. A., Kelly, B., & Langley, J. A. (2019). Exploring the oxygen sensitivity of wetland soil carbon mineralization. *Biology Letters, 15*(1). https://doi.org/10.1098/rsbl.2018.0407

Charles, S. P., Kominoski, J. S., Armitage, A. R., Guo, H., Weaver, C. A., & Pennings, S. C. (2020). Quantifying how changing mangrove cover affects ecosystem carbon storage in coastal wetlands. *Ecology, 101*(2). https://doi.org/10.1002/ecy.2916

Charles, S. P., Kominoski., J. S. Troxler, T. G., Gaiser, E. E., Servais, S., Wilson, B. J., Davis, S. E., Sklar, F. H., Coronado-Molina, C., Madden, C. J., & Kelly, S. (2019). Experimental saltwater intrusion drives rapid soil elevation and carbon loss in freshwater and brackish Everglades marshes. *Estuaries and Coasts, 42,* 1868–1881.

Chellaiah, D., & Yule, C. M. (2018). Riparian buffers mitigate impacts of oil palm plantations on aquatic macroinvertebrate community structure in tropical streams of Borneo. *Ecological Indicators, 95,* 53–62.

Chmura, G. L., Anisfeld, S. C., Cahoon, D. R., & Lynch, J. C. (2003). Global carbon sequestration in tidal, saline wetland soils. *Global Biogeochemical Cycles, 17*(4).

Classen-Rodríguez, L., Gutiérrez-Fonseca, P. E., & Ramírez, A. (2019). Leaf litter decomposition and macroinvertebrate assemblages along an urban stream gradient in Puerto Rico. *Biotropica, 51*(5), 641–651. https://doi.org/10.1111/btp.12685.

Coldren, G. A., Langley, J. A., Feller, I. C., & Chapman, S. K. (2019). Warming accelerates mangrove expansion and surface elevation gain in a subtropical wetland. *Journal of Ecology, 107,* 79–90.

Cole J. J., Prairie Y. T., Caraco N. F., McDowell, W. H., Tranvik, L. J., Striegl, R. G., Duarte, C. M., Kortelainen, P., Downing, J. A., Middelburg, J. J., & Melack, J. (2007). Plumbing the global carbon cycle: Integrating inland waters into the terrestrial carbon budget. *Ecosystems, 10,* 172–185.

Collen, P., Keay, E. J., & Morrison, B. R. S. (2004). Processing of pine (*Pinus sylvestris*) and birch (*Betula pubescens*) leaf material in a small river system in the northern Cairngorms, Scotland. In *European geosciences union* (Vol. 8). Retrieved from https://hal.archives-ouvertes.fr/hal-00304945

Cross, W. F., Wallace, J. B., Rosemond, A. D., & Eggert, S. L. (2006). Whole-system nutrient enrichment increases secondary production in a detritus-based ecosystem. *Ecology, 87,* 1556–1565.

Crutsinger, G. M., Rudman, S. M., Rodriguez-Cabal, M. A., McKown, A. D., Sato, T., Macdonald, A. M., Heavyside, J., Geraldes, A., Hart, E. M., LeRoy, C. J., & El-Sabaawi, R. W. (2014). Testing a "genes-to-ecosystems" approach to understanding aquatic-terrestrial linkages. *Molecular Ecology, 23*(23), 5888–5903. https://doi.org/10.1111/mec.12931

Dahl, T., & Stedman, S. (2013). *Status and trends of wetlands in the coastal watersheds of the Conterminous United States 2004 to 2009.*

Daily, G. C., Polasky, S., Goldstein, J., Kareiva, P. M., Mooney, H. A., Pejchar, L., Ricketts, T. H., Salzman, J., Shallenberger, R. (2009). Ecosystem services in decision making: time to deliver. *Frontiers in Ecology and the Environment, 7,* 21–28.

Datry, T., Boulton, A. J., Bonada, N., Fritz, K., Leigh, C., Sauquet, E., Tockner, K., Hugueny, B., & Dahm, C. N. (2018). Flow intermittence and ecosystem services in rivers of the Anthropocene. *Journal of Applied Ecology, 55,* 353–364.

Davies, P. M. (2010). Climate change implications for river restoration in global biodiversity hotspots. *Restoration Ecology, 18,* 261–268.

Demars, B. O. L., Manson, J. R., Ólafsson, J. S., Gíslason, G. M., & Friberg, N. (2011). Stream hydraulics and temperature determine the metabolism of geothermal Icelandic streams. *Knowledge and Management of Aquatic Ecosystems, 402,* 5. https://doi.org/10.1051/kmae/2011046.

Dessu, S. B., Price, R. M., Troxler, T. G., & Kominoski, J. S. (2018). Effects of sea-level rise and freshwater management on long-term water levels and water quality in the Florida Coastal Everglades. *Journal of Environmental Management, 211,* 164–176.

Dodds, W. K., Bruckerhoff, L., Batzer, D., Schechner, A., Pennock C., Renner, E., Tromboni, F., Bigham, K., & Grieger, S. (2019). The freshwater biome gradient framework: Predicting macroscale properties based on latitude, altitude, and precipitation. *Ecosphere, 10,* e02786.

Dodds, W. K., Gido, K., Whiles, M. R., Daniels, M. D., & Grudzinski, B. P. (2015). The stream biome gradient concept: Factors controlling lotic systems across broad biogeographic scales. *Freshwater Science, 34,* 1–19. https://doi.org/10.1086/679756.

Dominick, D. S., & O'Neill, M. P. (1998). Effects of flow augmentation on stream channel morphology and riparian vegetation: Upper Arkansas River Basin, Colorado. *Wetlands, 18,* 591–607.

Doughty, C. D., Langley, J. A., Walker, W. C., Schaub, R., & Chapman, S. K. (2015). Mangrove range expansion increases coastal wetland carbon storage. *Estuaries and Coasts, 39,* 385–396.

Downing, J. A., Cole, J. J., Middelburg, J. J., Striegl, R. G., Duarte, C. M., Kortelainen, P., Prairie, Y. T., & Laube, K. A. (2008). Sediment organic carbon burial in agriculturally eutrophic impoundments over the last century. *Global Biogeochemical Cycles, 22*(1). https://doi.org/10.1029/2006gb002854

Encalada, A. C., CAlles, J., Ferreira, V., Canhoto, C. M., & Graça, M. A. S. (2010). Riparian land use and the relationship between the benthos and litter decomposition in tropical montane streams. *Freshwater Biology, 55*(8), 1719–1733. https://doi.org/10.1111/j.1365-2427.2010.02406.x

Farfan, L., D'Sa, E., Liu, K., & Rivera-Monroy, V. H. (2014). Tropical cyclone impacts on coastal regions: The case of the Yucatan and the Baja California Peninsulas, Mexico. *Estuaries and Coasts, 37,* 1388–1402.

Ferreira, V., Boyero, L., Calvo, C., Correa, F., Figueroa, R., Gonçalves Jr., J. F., Goyenola, G., Graça, M. A. S., Hepp, L. U., Kariuki, S., López-Rodríguez, A., Mazzeo, N., M'Erimba, C., Monroy, S., Peil, A., Pozo, J., Rezende, R., & Teixeira-de-Mello, F. (2019). A global assessment of the effects of *Eucalyptus* plantations on stream ecosystem functioning. *Ecosystems, 22,* 629–642.

Ferreira, V., Castela, J., Rosa, P., Tonin, A. M., Boyero, L., & Graca, M. A. S. (2016). Aquatic hyphomycetes, benthic macroinvertebrates and leaf litter decomposition in streams naturally differing in riparian vegetation. *Aquatic Ecology, 50,* 711–725.

Follstad Shah, J. J., Kominoski, J. S., Ardón, M., Dodds, W. K., Gessner, M. O., Griffiths, N. A., Hawkins, C. P., Johnson, S. L., Lecerf, A., LeRoy, C. J., & Manning, D. W. (2017). Global synthesis of the temperature sensitivity of leaf litter breakdown in streams and rivers. *Global Change Biology, 23*(8), 3064–3075.

García, L., Richardson, J. S., & Pardo, I. (2012). Leaf quality influences invertebrate colonization and drift in a temperate rainforest stream. *Canadian Journal of Fisheries and Aquatic Sciences, 69*(10), 1663–1673. https://doi.org/10.1139/F2012-090.

Gee, J. H. R., & Smith, B. D. (1997). Benthic invertebrates in the headwaters of the Wye and Severn: Effects of forestry and clearfelling. *Hydrological and Earth Systems Science, 1,* 549–556.

Giri, C. P., & Long, J. (2014). Mangrove reemergence in the northernmost range limit of eastern Florida. *Proceedings of the National Academy of Sciences of the United States of America, 111,* E1447–E1448.

González, E., Felipe-Lucia, M. R., Bourgeois, B., Boz, B., Nilsson, C., Palmer, G. Sher, A. A. (2017). Integrative conservation of riparian zones. *Biological Conservation, 211*, 20–29.

Graça, M. A. S., & Cressa, C. (2010). Leaf quality of some tropical and temperate tree species as food resource for stream shredders. *International Review of Hydrobiology, 95*(1), 27–41. https://doi.org/10.1002/iroh.200911173.

Griffiths, N. A., Tank, J. L., Royer, T. V., Rosi-Marshall, E. J., Whiles, M. R., Chambers, C. P., Frauendorf, T. C., & Evans-White, M. A. (2009). Rapid decomposition of maize detritus in agricultural headwater streams. *Ecological Applications, 19*(1), 133–142. https://doi.org/10.1890/07-1876.1

Guo, H., Weaver, C., Charles, S. P., Whitt, A., Dastidar, S., D'Odorico, P., Fuentes, J. D., Kominoski, J. S., Armitage, A. R., & Pennings, S. C. (2017). Coastal regime shifts: Rapid responses of coastal wetlands to changes in mangrove cover. *Ecology, 98*(3), 762–772. https://doi.org/10.1002/ecy.1698

Haddad, N. M., Brudvig, L. A., Clobert, J., Davies, K. F., Gonzalez, A., Holt, R. D., Lovejoy, T. E., Sexton, J. O., Austin, M. P., Collins, C. D., Cook, W. M., Damschen, E. I., Ewers, R. M., Foster, B. L., Jenkins, C. N., King, A. J., Laurance, W. F., Levey, D. J., Margules, C. R., Melbourne, B. A., ... Townshend, J. R. (2015). Habitat fragmentation and its lasting impact on Earth's ecosystems. *Science Advances, 1*, e1500052.

Hall, R. O., Jr., & Meyer, J. L. (1998). The trophic significance of bacteria in a detritus-based stream food web. *Ecology, 79*, 1995–2012.

Herbert, E. R., Boon, P., Burgin, A. J., Neubauer, S. C., Franklin, R. B., Ardón, M., Hopfensperger, K. N., Lamers, L. P., & Gell, P. (2015). A global perspective on wetland salinization: Ecological consequences of a growing threat to freshwater wetlands. *Ecosphere, 6*, 1–43.

Herbert, E. R., Schubauer-Berigan, J., & Craft, C. B. (2018). Differential effects of chronic and acute simulated seawater intrusion on tidal freshwater marsh carbon cycling. *Biogeochemistry, 138*, 137–154.

Hoffman, M. T., & Rohde, R. F. (2011). Rivers through time: Historical changes in the riparian vegetation of the semi-arid, winter rainfall region of South Africa in response to climate and land use. *Journal of the History of Biology, 44*, 59–80.

Holdridge, L. R. (1947). Determination of world plant formations from simple climatic data. *Science, 105*(2727), 367–368. https://doi.org/10.1126/science.105.2727.367.

Huryn, A. D., Butz Huryn, V. M., Arbuckle, C. J., & Tsomides, L. (2002). Catchment land-use, macroinvertebrates and detritus processing in headwater streams: Taxonomic richness versus function. *Freshwater Biology, 47*(3), 401–415. https://doi.org/10.1046/j.1365-2427.2002.00812.x.

Jackrel, S. L., & Wootton, J. T. (2014). Local adaptation of stream communities to intraspecific variation in a terrestrial ecosystem subsidy. *Ecology, 95*(1), 37–43. https://doi.org/10.1890/13-0804.1.

Johnson, W. C. (1998). Adjustment of riparian vegetation to river regulation in the Great Plains, USA. *Wetlands, 18*, 608–618.

Junk, W. J., Bayley, P. B., & Sparks, R. E. (1989). The flood pulse concept in river-floodplain systems. *Canadian Special Publication of Fisheries and Aquatic Sciences, 106*, 110–127.

Kelleway, J. J., Saintilan, N., Macreadie, P. I., Skilbeck, C. G., Zawadzki, A., & Ralph, P. J. (2016). Seventy years of continuous encroachment substantially increases "blue carbon" capacity as mangroves replace intertidal salt marshes. *Global Change Biology, 22*(3), 1097–1109. https://doi.org/10.1111/gcb.13158.

Knapp, A. K., Briggs, J. M., Collins, S. L., Archer, S. R., Bret-Harte, M. S., Ewers, B. E., Peters, D. P., Young, D. R., Shaver, G. R., Pendall, E., & Cleary, M. B. (2008). Shrub encroachment in North American grasslands: Shifts in growth form dominance rapidly alters control of ecosystem carbon inputs. *Global Change Biology, 14*(3), 615–623. https://doi.org/10.1111/j.1365-2486.2007.01512.x

Koch, G., Childers, D. L., Staehr, P. A., Price, R. M., Davis, S. E., & Gaiser, E. E. (2012). Hydrological conditions control P loading and aquatic metabolism in an oligotrophic, subtropical estuary. *Estuaries and Coasts, 35,* 292–307.

Kominoski, J. S., Follstad Shah, J. J., Canhoto, C., Fischer, D. G., Giling, D. P., González, E., Griffiths, N. A., Larrañaga, A., LeRoy, C. J., Mineau, M. M., McElarney, Y. R., Shirley, S. M., Swan, C. M., & Tiegs, S. D. (2013). Forecasting functional implications of global changes in riparian plant communities. *Frontiers in Ecology and the Environment, 11*(8), 26. https://doi.org/10.1890/120056

Kominoski, J. S., Marczak, L. B., & Richardson, J. S. (2011). Riparian forest composition affects stream litter decomposition despite similar microbial and invertebrate communities. *Ecology, 92,* 151–159.

Kominoski J. S., Rehage, J., Anderson, W., Boucek, R., Briceno, H. O., Bush, M. R., Dreschel, T. W., Heithaus, M. R., Jaffe, R., Larsen, L., Matich, P., McVoy, C., Rosenblatt, A. E., & Troxler, T. (2019). Ecosystem fragmentation and connectivity-legacies and future implications of a restored Everglades. In D. L. Childers, E. Gaiser, & L. Ogden (Eds.), *The dynamics of social-ecological transformation in the South Florida landscape* (pp. 73–96). Oxford University Press.

Kominoski, J. S., & Rosemond, A. D. (2012). Conservation from the bottom up: Forecasting effects of global change on dynamics of organic matter and management needs for river networks. *Freshwater Science, 31,* 51–68.

Kominoski, J. S., Rosemond, A. D., Benstead, J. P., Gulis, V., & Manning, D. W. P. (2018). Experimental nitrogen and phosphorus additions increase rates of stream ecosystem respiration and carbon loss. *Limnology and Oceanography, 63*(1), 22–36. https://doi.org/10.1002/lno.10610.

Krauss, K. W., McKee, K. L., Lovelock, C. E., Cahoon, D. R., Saintilan, N., Reef, R., & Chen, L. (2014). How mangrove forests adjust to rising sea level. *New Phytologist, 202*(1), 19–34.

Kreutzweiser, D. P., Good, K. P., Capell, S. S., & Holmes, S. B. (2008). Leaf-litter decomposition and macroinvertebrate communities in boreal forest streams linked to upland logging disturbance. *Journal of the North American Benthological Society, 27*(1), 1–15. https://doi.org/10.1899/07-034R.1.

Larson, D. M., Dodds, W. K., & Veach, A. M. (2018). Removal of woody riparian vegetation substantially altered a stream ecosystem in an otherwise undisturbed grassland watershed. *Ecosystems, 22,* 64–76.

Lecerf, A., & Chauvet, E. (2008). Intraspecific variability in leaf traits strongly affects alder leaf decomposition in a stream. *Basic and Applied Ecology, 9*(5), 598–605.

Lecerf, A., & Richardson, J. S. (2010). Litter decomposition can detect effects of high and moderate levels of forest disturbance on stream condition. *Forest Ecology and Management, 259,* 2433–2443.

LeRoy, C. J., Hipp, A. L., Lueders, K., Follstad Shah, J. J., Kominoski, J. S., Ardón, M., Dodds, W. K., Gessner, M. O., Griffiths, N. A., Lecerf, A., Manning, D. W. P., Sinsabaugh, R. L., & Webster, J. R. (2020). Plant phylogenetic history explains in-stream decomposition at a global scale. *Journal of Ecology, 108*(1), 17–35. https://doi.org/10.1111/1365-2745.13262

LeRoy, C. J., Whitham, T. G., Keim, P., & Marks, J. C. (2006). Plant genes link forests and streams. *Ecology, 87*(1), 255–261. https://doi.org/10.1890/05-0159.

Macfarlane, W. W., Gilbert, J. T., Jensen, M. L., Gilbert, J. D., Hough-Snee, N., McHugh, P. A., Wheaton, J. M. & Bennett, S. N. (2017). Riparian vegetation as an indicator of riparian condition: Detecting departures from historic condition across the North American West. *Journal of Environmental Management, 202,* 447–460.

Marks, J. C. (2019). Revisiting the fates of dead leaves that fall into streams. *Annual Review of Ecology Evolution and Systematics, 50,* 547–568.

Martins, R. T., Melo, A. S., Gonçalves, J. F., Jr., & Hamada, N. (2015). Leaf-litter breakdown in urban streams of Central Amazonia: Direct and indirect effects of physical, chemical, and biological factors. *Freshwater Science, 34,* 716–726.

McClain, M. E., Boyer, E. W., Dent, C. L., Gergel, S. E., Grimm, N. B., Groffman, P. M., Hart, S. C., Harvey, J. W., Johnston, C. A., Mayorga, E. & McDowell, W. H. (2003). Biogeochemical hot spots and hot moments at the interface of terrestrial and aquatic ecosystems. *Ecosystems, 6*, 301–312.

McKie, B. G., & Malmqvist, B. (2009). Assessing ecosystem functioning in streams affected by forest management: Increased leaf decomposition occurs without changes to the composition of benthic assemblages. *Freshwater Biology, 54*(10), 2086–2100. https://doi.org/10.1111/j.1365-2427.2008.02150.x.

Mcleod, E., Chmura, G. L., Bouillon, S., Salm, R., Björk, M., Duarte, C. M., Lovelock, C. E., Schlesinger, W. H., & Silliman, B. R. (2011). A blueprint for blue carbon: Toward an improved understanding of the role of vegetated coastal habitats in sequestering CO_2 *Frontiers in Ecology and Evolution, 9*, 552–560.

Meyer, J. L., Wallace, J. B., & Eggert, S. L. (1998). Leaf litter as a source of dissolved organic carbon in streams. *Ecosystems, 1*(3), 240–249. https://doi.org/10.1007/s100219900019.

Millenium Ecosystem Assessment. (2005). *Ecosystems and human well-being: The assessment series (four volumes and summary)*. Washington, DC: Island Press.

Minshall G. W., Peterson R. C., Cummins K. W., Bott, T. L., Sedell, J. R., Cushing, C. E., & Vannote, R. L. (1983). Interbiome comparison of stream ecosystem dynamics. *Ecological Monographs, 53*, 1–25.

Mulholland, P. J., Helton, A. M., Poole, G. C., Hall, R. O., Hamilton, S. K., Peterson, B. J., Tank, J. L., Ashkenas, L. R., Cooper, L. W., Dahm, C. N., & Dodds, W. K. (2008). Stream denitrification across biomes and its response to anthropogenic nitrate loading. *Nature, 452*(7184), 202–205.

Naiman, R., Decamps, H., & McClain, M. (2010). *Riparia: Ecology, conservation, and management of streamside communities*. Elsevier.

National Research Council (NRC). (2000). *Clean coastal waters: Understanding and reducing the effects of nutrient pollution*. Washington, DC: National Academy Press.

Neubauer, S. C. (2013). Ecosystem responses of a tidal freshwater marsh experiencing saltwater intrusion and altered hydrology. *Estuaries and Coasts, 36*(3), 491–507.

Newcomer, T. A., Kaushal, S. S., Mayer, P. M., Shields, A. R., Canuel, E. A., Groffman, P. M., & Gold, A. J. (2012). Influence of natural and novel organic carbon sources on denitrification in forest, degraded urban, and restored streams. *Ecological Monographs, 82*(4), 449–466. https://doi.org/10.1890/12-0458.1

Nicholls, R. J., Wong, P. P., Burkett, V., Codignotto, J., Hay, J., McLean, R., Ragoonaden, S., Woodroffe, C. D., Abuodha, P. A. O., Arblaster, J. & Brown, B. (2007). Coastal systems and low-lying areas. In Parry, M., Parry, M. L., Canziani, O., Palutikof, J., Van der Linden, P. & Hanson, C (Eds.), *Climate change 2007-impacts, adaptation and vulnerability: Working group II contribution to the fourth assessment report of the IPCC* (Vol. 4). Cambridge University Press.

Nicholls, R. J., & Cazenave, A. (2010). Sea-level rise and its impact on coastal zones. *Science, 328*, 1517–1520.

Nilsson, C., Jansson, R., Kuglerová, L., Lind, L., & Ström, L. (2013). Boreal riparian vegetation under climate change. *Ecosystems, 16*, 401–410.

Noe, G., Childers, D. L., & Jones, R. D. (2001). Phosphorus biogeochemistry and the impacts of phosphorus enrichment: Why are the Everglades so unique? *Ecosystems, 4*, 603–624.

Osland, M. J., Enwright, N., Day, R. H., & Doyle, T. W. (2013). Winter climate change and coastal wetland foundation species: Salt marshes versus mangrove forests in the southeastern US. *Global Change Biology, 19*, 1482–1494.

Osland, M. J., Enwright, N. M., Day, R. H., Gabler, C. A., Stagg, C. L., & Grace, J. B. (2016). Beyond just sea-level rise: Considering macroclimatic drivers within coastal wetland vulnerability assessments to climate change. *Global Change Biology, 22*, 1–11.

Paul, M. J., & Meyer, J. L. (2001). Streams in the urban landscape. *Annual Review of Ecology and Systematics, 32*, 333–365. https://doi.org/10.1146/annurev.ecolsys.32.081501.114040.

Paul, M. J., Meyer, J. L., & Couch, C. A. (2006). Leaf breakdown in streams differing in catchment land use. *Freshwater Biology, 51*(9), 1684–1695. https://doi.org/10.1111/j.1365-2427.2006.01612.x.

Perry, C. L., & Mendelssohn, I. A. (2009). Ecosystem effects of expanding populations of *Avicennia germinans* in a Louisiana salt marsh. *Wetlands, 29*(1), 396–406. https://doi.org/10.1672/08-100.1.

Perry, L. G., Andersen, D. C., Reynolds, L. V., Nelson, S. M., & Shafroth, P. B. (2012, March 1). Vulnerability of riparian ecosystems to elevated CO_2 and climate change in arid and semiarid western North America. *Global Change Biology, 18*, 821–842. https://doi.org/10.1111/j.1365-2486.2011.02588.x

Perry, L. G., Shafroth, P. B., Blumenthal, D. M., Morgan, J. A., & LeCain, D. R. (2013). Elevated CO2 does not offset greater water stress predicted under climate change for native and exotic riparian plants. *New Phytologist, 197*(2), 532–543.

Perry, L. G., Shafroth, P. B., Hay, L. E., Markstrom, S. L., & Bock, A. R. (2020). Projected warming disrupts the synchrony of riparian seed release and snowmelt streamflow. *New Phytologist, 225*(2), 693–712. https://doi.org/10.1111/nph.16191.

Peters D. P. C., Groffman P. M., Nadelhoffer K. J., Grimm, N. B., Collins, S. L., Michener, W. K., & Huston, M. A. (2008). Living in an increasingly connected world: A framework for continental-scale environmental science. *Frontiers in Ecology and the Environment, 6*, 229–237.

Primack, A. G. (2000). Simulation of climate-change effects on riparian vegetation in the Pere Marquette River, Michigan. *Wetlands, 20*, 538–547.

Pringle, C. M. (2001). Hydrologic connectivity and the management of biological reserves: A global perspective. *Ecological Applications, 11*, 981–998.

Prowse, T. D., Wrona, F. J., Reist, J. D., Gibson, J. J., Hobbie, J. E., Lévesque, L. M., & Vincent, W. F. (2006). Climate change effects on hydroecology of Arctic freshwater ecosystems. *AMBIO: A Journal of the Human Environment, 35*, 347–359.

Rabalais, N. N., Turner, R. E., & Wiseman, W. J. (2002). Gulf of Mexico hypoxia, a.k.a. "The Dead Zone". *Annual Reviews in Ecology and Systematics, 33*, 235–263.

Reynolds, L. V., & Shafroth, P. B. (2017). Riparian plant composition along hydrologic gradients in a dryland river basin and implications for a warming climate. *Ecohydrology, 10*(6). https://doi.org/10.1002/eco.1864

Riley, A. J., & Dodds, W. K. (2012). The expansion of woody riparian vegetation, and subsequent stream restoration, influences the metabolism of prairie streams. *Freshwater Biology, 57*(6), 1138–1150. https://doi.org/10.1111/j.1365-2427.2012.02778.x.

Rivaes, R., Rodríguez-González, P. M., Albuquerque, A., Pinheiro, A. N., Egger, G., & Ferreira, M. T. (2013). Riparian vegetation responses to altered flow regimes driven by climate change in Mediterranean rivers. *Ecohydrology, 6*(3), 413–424. https://doi.org/10.1002/eco.1287.

Rosemond, A. D., Benstead, J. P., Bumpers, P. M., Gulis, V., Kominoski, J. S., Manning, D. W. P., Suberkropp, K., & Wallace, J. B. (2015). Experimental nutrient additions accelerate terrestrial carbon loss from stream ecosystems. *Science.Sciencemag.Org*. https://doi.org/10.1126/science.aaa1934

Rundel, P. W., Dickie, I. A., & Richardson, D. M. (2014). Tree invasions into treeless areas: Mechanisms and ecosystem processes. *Biological Invasions, 16*, 663–675.

Running, S. W. (2012). A measurable planetary boundary for the biosphere. *Science, 337*, 1458–1459. https://doi.org/10.1126/science.1227620.

Sabo, J. L., Sponseller, R., Dixon, M., Gade, K., Harms, T., Heffernan, J., Jani, A., Katz, G., Soykan, C., Watts, J., & Welter, J. (2005). Riparian zones increase regional species richness by harboring different, not more, species. *Ecology, 86*(1), 56–62. https://doi.org/10.1890/04-0668

Saha A. K., Saha, S., Sadle, J., Jiang, J., Ross, M. S., Price, R. M., Sternberg, L. S. L., & Wendelberger, K. S. (2011). Sea level rise and South Florida coastal forests. *Climate Change, 107*, 81–108.

Saintilan, N., Wilson, N. C., Rogers, K., Rajkaran, A., & Krauss, K. W. (2014). Mangrove expansion and salt marsh decline at mangrove poleward limits. *Global Change Biology, 20*(1), 147–157. https://doi.org/10.1111/gcb.12341.

Sankey, J. B., Ralston, B. E., Grams, P. E., Schmidt, J. C., & Cagney, L. E. (2015). Riparian vegetation, Colorado River, and climate: Five decades of spatiotemporal dynamics in the Grand

Canyon with river regulation. *Journal of Geophysical Research G: Biogeosciences, 120*(8), 1532–1547. https://doi.org/10.1002/2015JG002991

Seitzinger, S. P., Harrison, J. A., Böhlke, J. K., Bouwman, A. F., Lowrance, R., Peterson, B., Tobias, C., & Van Drecht, G. (2006). Denitrification across landscapes and waterscapes: a synthesis. *Ecological Applications, 16*, 2064–2090.

Simpson, L., Stein, C., Osborne, T., & Hydrobiologia, I. F. (2019). Mangroves dramatically increase carbon storage after 3 years of encroachment. *Springer, 834*(1), 13–26. https://doi.org/10.1007/s10750-019-3905-z

Sheaves, M. (2009). Consequences of ecological connectivity: The coastal ecosystem mosaic. *Marine Ecology Progress Series, 391*, 107–115.

Shih J. -S., Alexander R. B., Smith R. A., et al. (2010). An initial SPARROW model of land use and in-stream controls on total organic carbon in streams of the coterminous United States. U.S. Geological Survey Open-File Report 2010–1276, 22 p.

Sobczak, W. V., Cloern, J. E., Jassby, A. D., & Müller-Solger, A. B. (2002). Bioavailability of organic matter in a highly disturbed estuary: the role of detrital and algal resources. *Proceedings of the National Academy of Sciences, 99*, 8101–8105.

Stanford, J. A., & Ward, J. V. (2001). Revisiting the serial discontinuity concept. *Regulated Rivers: Research & Management, 17*(4–5), 303–310. https://doi.org/10.1002/rrr.659.

Stewart R. J., Wollheim W. M., Gooseff M. N., Briggs, M. A., Jacobs, J. M., Peterson, B. J., & Hopkinson, C. S. (2011). Separation of river network-scale nitrogen removal among the main channel and two transient storage compartments. *Water Resources Research, 47*. https://doi.org/10.1029/2010WR009896

St Louis, V., Kelly, C., Duchemin, É., Rudd, J., & Rosenberg, D. (2000). Reservoir surfaces as sources of greenhouse gases to the atmosphere: A global estimate. *BioScience, 50*, 766–775.

Tank J. L., Rosi-Marshall E. J., Griffiths N. A., Entrekin, S. A., & Stephen, M. L. (2010). A review of allochthonous organic matter dynamics and metabolism in streams. *Journal of the North American Benthological Society, 29*, 118–146.

Taylor, P. D., Fahrig, L., Henein, K., & Merriam, G. (1993). Connectivity is a vital element of landscape structure. *Oikos, 68*, 571–572.

Taylor, P., & Townsend, A. (2010). Stoichiometric control of organic carbon–nitrate relationships from soils to the sea. *Nature, 464*, 1178–1181.

Thorp, J. H., & Delong, M. D. (1994). The riverine productivity model: An heuristic view of carbon sources and organic processing in large river ecosystems. *Oikos, 70*, 305–308.

Tranvik, L. J., Downing, J. A., Cotner, J. B., Loiselle, S. A., Striegl, R. G., Ballatore, T. J., Dillon, P., Finlay, K., Fortino, K., Knoll, L. B., Kortelainen, P. L., Kutser, T., Larsen, S., Laurion, I., Leech, D. M., McCallister, S. L., McKnight, D. M., Melack, J. M., Erin Overholt, … Weyhenmeyer, G. A. (2009). Lakes and reservoirs as regulators of carbon cycling and climate. *Limnology and Oceanography, 54*(6 PART 2), 2298–2314. https://doi.org/10.4319/lo.2009.54.6_part_2.2298

Vandermyde, J. M., & Whiles, M. R. (2015). Effects of experimental forest removal on macroinvertebrate production and functional structure in tallgrass prairie streams. *Freshwater Science, 34*, 519–534.

Vannote, R. L., Minshall, G. W., Cummins, K. W., Sedell, J. R., & Cushing, C. E. (1980). The river continuum concept. *Canadian Journal of Fisheries and Aquatic Sciences, 37*(1), 130–137. https://doi.org/10.1139/f80-017.

Veach, A. M., Dodds, W. K., & Jumpponen A. (2015). Woody plant encroachment, and its removal, impact bacterial and fungal communities across stream and terrestrial habitats in a tallgrass prairie ecosystem. *FEMS Microbiology Ecology, 91*, fiv109.

Vörösmarty, C. J., Meybeck, M., Fekete, B., Sharmad, K., Greena, P., & Syvitskie, J. P. M. (2003). Anthropogenic sediment retention: Major global impact from registered river impoundments. *Global Planet Change, 39*, 169–90.

Walsh, C. J., Roy, A. H., Feminella, J. W., Cottingham, P. D., Groffman, P. M. & Morgan, R. P. (2005). The urban stream syndrome: current knowledge and the search for a cure. *Journal of the North American Benthological Society, 24*(3), 706–723.

Ward, J. V., & Stanford, J. A. (1983). Serial discontinuity concept of lotic ecosystems. *Dynamics of lotic systems* (pp. 29–42). Ann Arbor MI: Ann Arbor Science.

Wilson, B. J., Servais, S. M., Charles, S. P., Davis, S. E., Gaiser, E. E., Kominoski, J. S., Richards, J. H., & Troxler, T. G. (2018). Declines in plant productivity drive carbon loss from brackish coastal wetland mesocosms exposed to saltwater intrusion. *Estuaries and Coasts, 41*, 2147–2158.

Wipfli, M. S., & Musselwhite, J. (2004). Density of red alder (*Alnus rubra*) in headwaters influences invertebrate and detritus subsidies to downstream fish habitats in Alaska. *Hydrobiologia, 520,* 153–163.

Wollheim, W. M., Pellerin, B. A., Vörösmarty, C. J., & Hopkinson, C. S. (2005). N retention in urbanizing headwater catchments. *Ecosystems, 8,* 871–884.

Wollheim, W. M., Vörösmarty, C. J., Bouwman, A. F., Green, P., Harrison, J., Linder, E., Peterson, B. J., Seitzinger, S. P., & Syvitski, J. P. M. (2008). Global N removal by freshwater aquatic systems using a spatially distributed, within-basin approach. *Global Biogeochemical Cycles, 22*(2). https://doi.org/10.1029/2007gb002963

Woodward, G., Gessner, M. O., Giller, P. S., Gulis, V., Hladyz, S., Lecerf, A., Malmqvist, B., McKie, B. G., Tiegs, S. D., Cariss, H. & Dobson, M. (2012). Continental-scale effects of nutrient pollution on stream ecosystem functioning. *Science, 336*(6087), 1438–1440.

Yando, E. S., Osland, M. J., Willis, J. M., Day, R. H., Krauss, K. W., & Hester, M. W. (2016). Salt marsh-mangrove ecotones: Using structural gradients to investigate the effects of woody plant encroachment on plant–soil interactions and ecosystem carbon pools. *Journal of Ecology, 104,* 1020–1031.

Yeung, A. C. Y., Kreutzweiser, D. P., & Richardson, J. S. (2019). Stronger effects of litter origin on the processing of conifer than broadleaf leaves: A test of home-field advantage of stream litter breakdown. *Freshwater Biology, 64*(10), 1755–1768. https://doi.org/10.1111/fwb.13367.

Young, R. G., Huryn, A. D., & Townsend, C. R. (1994). Effects of agricultural development on processing of tussock leaf litter in high country New Zealand streams. *Freshwater Biology, 32*(2), 413–427. https://doi.org/10.1111/j.1365-2427.1994.tb01136.x.

Young, R. G., Matthaei, C. D., & Townsend, C. R. (2008). Organic matter breakdown and ecosystem metabolism: Functional indicators for assessing river ecosystem health. *Journal of the North American Benthological Society, 27*(3), 605–625. https://doi.org/10.1899/07-121.1.

Yule, C. M., Gan, J. Y., Jinggut, T., & Lee, K. V. (2015). Urbanization affects food webs and leaf-litter decomposition in a tropical stream in Malaysia. *Freshwater Science, 34,* 702–715. https://doi.org/10.1086/681252.

Chapter 14
Effects of Exotic Tree Plantations on Plant Litter Decomposition in Streams

Aitor Larrañaga, Aingeru Martínez, Ricardo Albariño, J. Jesús Casas, Verónica Ferreira, and Romina Principe

Abstract The need for tree-derived industrial products is causing an increase in the land surface covered by fast-growing monoculture plantations throughout the world. Species planted are selected mostly prioritizing their rapid growth, with less consideration to minimizing the negative environmental effects they create. Among the various ecosystems that can be negatively affected by plantations, streams are among the most impacted, as they strongly depend on dead organic matter from the surrounding vegetation. Changes in land use in favour of monocultures can have large consequences on stream biodiversity and functioning since they can lead to alterations in the diversity, quantity, quality and timing of litter inputs. Here, we review the literature dealing with the effects of plantations on litter decomposition in streams, with special focus on eucalyptus and conifer plantations, which have been more thoroughly studied than other planted tree species. The effects of plantations on litter decomposition in streams have a degree of regional specificity, and depend on the composition of detritivores, litter characteristics (of native and exotic species) and local environmental conditions. There is, nonetheless, a need for further research

A. Larrañaga (✉)
University of the Basque Country, UPV/EHU, Bilbao, Spain
e-mail: aitor.larranagaa@ehu.eus

A. Martínez
Centre for Functional Ecology, University of Coimbra, Coimbra, Portugal
e-mail: aingeru.martinez@ehu.eus

R. Albariño
INIBIOMA, National University of Comahue - CONICET, Bariloche, Argentina
e-mail: ralbarin@comahue-conicet.gob.ar

J. J. Casas
CAESCG, University of Almería, Almería, Spain
e-mail: jjcasas@ual.es

V. Ferreira
MARE – Marine and Environmental Sciences Centre, Department of Life Sciences, University of Coimbra, Coimbra, Portugal
e-mail: veronica@ci.uc.pt

R. Principe
ICBIA, National University of Río Cuarto – CONICET, Río Cuarto, Argentina

C. M. Swan et al. (eds.), *The Ecology of Plant Litter Decomposition in Stream Ecosystems*, https://doi.org/10.1007/978-3-030-72854-0_14

describing the effects of specific planted exotic species and for more studies out of Europe and North America.

14.1 Introduction

Trees form dense forests in many parts of the world, but their diversity varies widely depending on their degree of human intervention, from primary growth forests (where humans have had a negligible effect) to the most intensively managed plantations (where tree growth is maximized in detriment of diversity; see Box 14.1). The historical human use of wood and the woodland conversion to agriculture have led to a gradual but profound modification of landscapes and the reduction in global forest cover. Thus, today's natural forests are comprised mostly of modified natural forests (74%) as opposed to primary (i.e., pristine) forests (26%) (FAO, 2015). Also, with the growing global population and demand for forest resources, planted forests are becoming an increasingly important part of the human-modified landscape, alongside urban, agricultural, and natural forest land cover types (FAO, 2018). Globally, natural forests cover 3695 million ha (Fig. 14.1) and are decreasing at an annual rate of 0.24% (FAO, 2015). On the other hand, planted forests cover around 291 million ha, accounting for 7% of the global forest area (Fig. 14.1). The largest area of planted forests is found in the temperate domain, accounting for 150 million ha, followed by the tropical and boreal domains with almost 60 million ha each. Asia has 44.4% of the world's planted forests, followed by Europe (28.6%), Central and North America (15%), Africa (5.5%), South America (2.4%) and Oceania (1.5%) (FAO, 2015; Fig. 14.1). Moreover, planted land surface area is increasing globally at an average annual rate of 1.84%, ranging from 1.1% in Europe to 2.5% in North and Central America (FAO, 2015; Fig. 14.1). Land cover by planted exotic tree species constitutes 25% of planted forests worldwide (FAO, 2015), with dissimilar representation around the globe. Plantations of exotic species dominate in the southern hemisphere, where they represent 88% of the total planted surface in South America, 75% in Oceania, 31% in Africa, and 42% in the Caribbean (Payn et al., 2015). In contrast, continents with more planted forest cover have a lower proportion of exotic plantations, from 25% in eastern Asia to 1% in North America (Payn et al., 2015).

Box 14.1 Glossary box for forest definitions modified from FAO (2006b)

Forest: Land spanning more than 0.5 ha with trees higher than 5 m and a canopy cover of more than 10%, or trees able to reach these thresholds in situ. It does not include land that is predominantly under agricultural or urban land use. Includes native and introduced tree species growing naturally or human-assisted.

Natural forest: A forest composed of indigenous/native trees.

Primary forest: Forest of native species, where there are no clear visible indications of human activities and the ecological processes are not significantly disturbed.

Modified natural forest: Forest of naturally regenerated native species where there are clearly visible indications of human activities. Today most natural forest cover belongs globally to this category.

Planted forest: Forest in which trees, both native and introduced species, have been established through planting or seeding. Includes all stands established through planting or seeding. Includes the subcategories semi-natural forests and plantations.

Semi-natural forest: Forest of native species, established through planting, seeding or assisted natural regeneration, such as thinning or fertilization.

Plantation: Forest of usually exotic, but also native, species established through planting or seeding mainly for production of wood or non-wood goods.

The majority of planted forest area (three quarters) is grown for productive purposes (i.e., production of wood, fibre, fuel or non-wood forest products), while just one quarter is intended for protective purposes (e.g., rehabilitation of degraded lands, combating desertification or protection of soil and water) (FAO, 2006a). Meanwhile, afforestation, the process where new forests are planted across tree-less land (i.e., grasslands and abandoned agricultural lands), has become one of the most technologically simple methods of removing carbon dioxide from the atmosphere and storing it as biomass, while controlling water erosion and dust storms, reducing

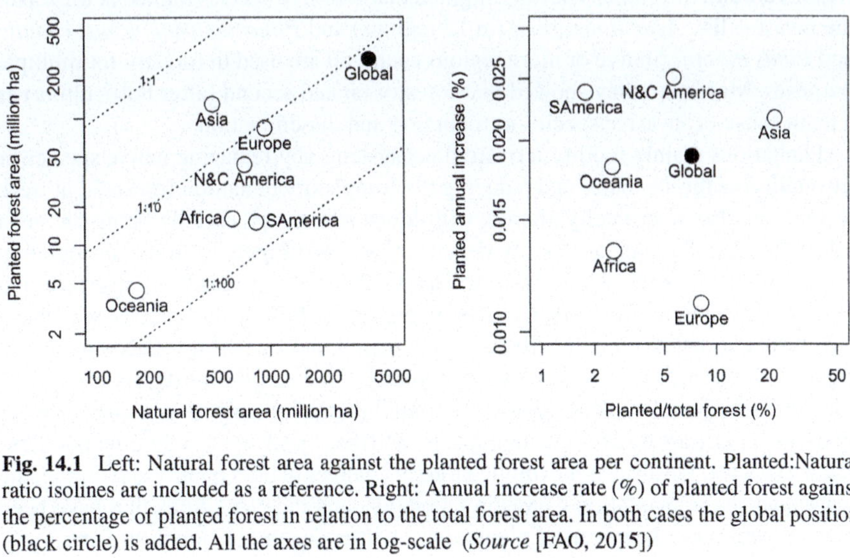

Fig. 14.1 Left: Natural forest area against the planted forest area per continent. Planted:Natural ratio isolines are included as a reference. Right: Annual increase rate (%) of planted forest against the percentage of planted forest in relation to the total forest area. In both cases the global position (black circle) is added. All the axes are in log-scale (*Source* [FAO, 2015])

river sedimentation, and mitigating small floods. Still, plantations (including both afforestation and reforestation practices) raise concerns about their potential negative effects on ecosystems. For example, plantations can alter nutrient cycling and catchment hydrology, with negative consequences on soil and water quality and quantity (Farley et al., 2008; Jackson et al., 2005; Mátyás & Sun, 2014), or even become a source for invasion when individuals are dispersed from plantations to other areas (Richardson & Rejmánek, 2011).

Exotic tree species have a long history of being used in forestry, mostly because selected species have improved productivity compared with that of native species (e.g., Elfving et al., 2001; Morris et al., 2011). The higher growth rates of selected exotic species, compared with those of native species, can indeed result in shorter rotations (Salmón Rivera et al., 2016). The species used in plantations differ among regions, with conifers dominating colder regions and broadleaves dominating warmer regions; overall, conifers account for 52% of total plantations, with broadleaves representing 37% and the remaining 11% being unspecified (FAO, 2006a). In order of importance, the main coniferous genera by cover area are *Pinus*, *Cunninghamia*, *Larix*, *Picea* and *Cryptomeria* while the main broadleaf genera are *Eucalyptus*, *Populus*, *Acacia* and *Tectona* (Fig. 14.2). Exotic planted woody species may become invasive if they expand naturally beyond plantations (Hayson & Murphy, 2003). Effectively, exotic woody species invading native forests were found to cover 79 million ha worldwide in 2010, with an estimated annual increase rate of 11.3% considering the period 1990–2010 (FAO, 2015). In a study on forestry tree invasiveness, Hayson and Murphy (2003) found that 282 out of 458 species exclusively used in forestry had become naturalised and invasive, with most species belonging to the genera *Acacia*, *Eucalyptus*, *Larix*, *Picea*, *Pinus*, *Populus* and *Tectona*. More recently, Richardson and Rejmánek (2011) pointed to the genus *Acacia* (32 spp.) and *Pinus* (22 spp.) among the most widespread invasive exotic tree species in 15 regions around the globe. They highlighted that four tree species ranked at the top of the invasive list: *Acacia mearnsii* (in 12 regions) and *Pinus pinaster*, *Pinus radiata* and *Pinus elliottii* (in five or more regions each); all are used in forestry for multiple purposes. Moreover, they showed that forestry ranked second (after horticulture) as a main cause of invasive species introduction and dissemination.

Plantations mainly modify terrestrial ecosystems (by replacing native species or by modifying the attributes and reducing the complexity of natural forests), but fresh waters can also be critically altered, with forest streams potentially being the most affected given that they are highly dependent on litter inputs from the surrounding terrestrial ecosystems, which in turn vary with forest composition, structure and production (Fausch et al., 2010; Whiles & Wallace, 1997). In this chapter we review the current knowledge on the effects of tree plantations on leaf litter decomposition capacity of streams. We focus on two case studies, which are the most prolific in scientific research, namely on eucalyptus and conifer plantations. We summarize the main findings of existing research on these species and then briefly review the effects of less-known species. We discuss the implications of changes in litter quality and timing, the consequences of plantation harvesting methods, and the procedures that have been proven to mitigate the effects of plantations on litter decomposition in streams.

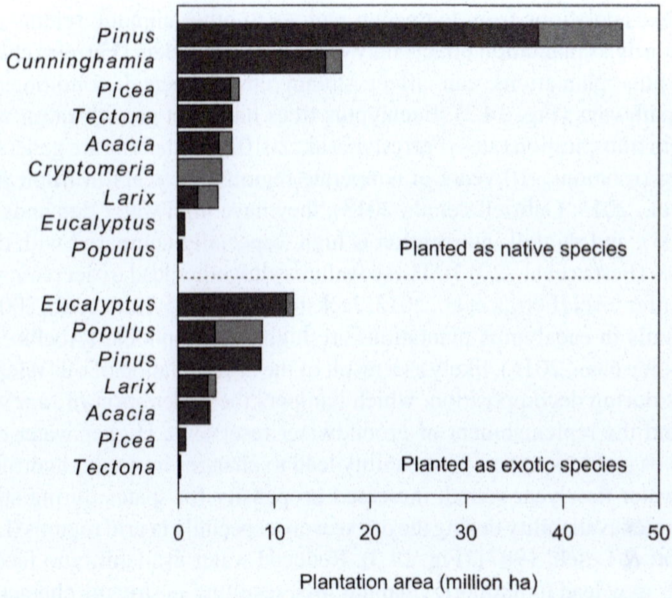

Fig. 14.2 Plantation area of the most common genera in selected countries of the world with the largest areas of planted forest (Algeria, Argentina, Australia, Belgium, Brazil, Bulgaria, Chile, China, Croatia, Czech Republic, Finland, France, India, Indonesia, Iran, Italy, Japan, Latvia, Lithuania, Malaysia, Myanmar, Netherlands, New Zealand, Norway, Philippines, Poland, Slovakia, Slovenia, South Africa, Sudan, Sweden, Turkey, Ukraine, United Kingdom, United States, Vietnam). Plantation areas for production (dark grey) and protection (light grey) goals are shown in the bars (*Source* [FAO, 2006a])

14.2 Case Studies

14.2.1 Eucalyptus Plantations (Fig. 14.3)

Eucalyptus is the most planted tree genus out of its natural range worldwide. In 2009, eucalyptus plantations covered already >20 million ha worldwide, mostly between 35°S and 35°N (Iglesias-Trabado et al., 2009). Three countries alone (Brazil, India and China) contribute with >50% of total eucalyptus plantation area, but they are present in >95 countries (Iglesias-Trabado et al., 2009), often also in stream banks. Still, most studies addressing the effects of eucalyptus plantations on litter decomposition in streams have been performed in central Portugal and northern Spain. In these two locations, the climate is warm temperate (drier in central Portugal and more humid in northern Spain), native forests are characterized by mixed deciduous broadleaf species dominated by *Quercus* spp. and *Castanea sativa*, and the species used in plantations is *Eucalyptus globulus* (Ferreira et al., 2016). Therefore, the current knowledge of this topic is geographically biased and so must be our revision. Information from other regions will be included whenever possible, but in its absence, extrapolations from temperate regions to other climatic realms should be made carefully as plantation effects may be context dependent (Ferreira et al., 2019).

Eucalyptus plantations can affect stream environmental conditions through multiple pathways (Fig. 14.3). Eucalyptus trees have fast growth rates, with rapid increases in transpiration rates (Forrester et al., 2010). Plantations are generally dense and young (rotation: ~10 years in temperate regions, ~7 years in tropical regions; Ferraz et al., 2013; Gabrielle et al., 2013), they have high water demands (Jackson et al., 2005), and rainfall interception is high, especially compared with deciduous tree species (Ferraz et al., 2013). These features altogether lead to decreases in runoff and in aquifer level (Ferraz et al., 2013; Jackson et al., 2005; Lara et al., 2009). Additionally, soils in eucalyptus plantations are highly hydrophobic (Abelho & Graça, 1996; Walden et al., 2015), likely as a result of the accumulation of oils released from the leaves during decomposition, which hampers the penetration of rain water into the soil and the replenishment of groundwater reservoirs. Higher water consumption by trees and loss of soil permeability lead to changes in stream hydrology, with reduced water flow year round, increased propensity for spates during storms and reduced water availability during the dry season, especially in arid regions (Lara et al., 2009; Scott & Lesch, 1997) (Fig. 14.3). Reduced water availability to feed streams in summer may lead to temporary habitat loss, resulting in streams characterised by isolated pools or completely dry (Canhoto & Laranjeira, 2007). Under these conditions, there is reduced solute dilution capacity, which results in toxic conditions for

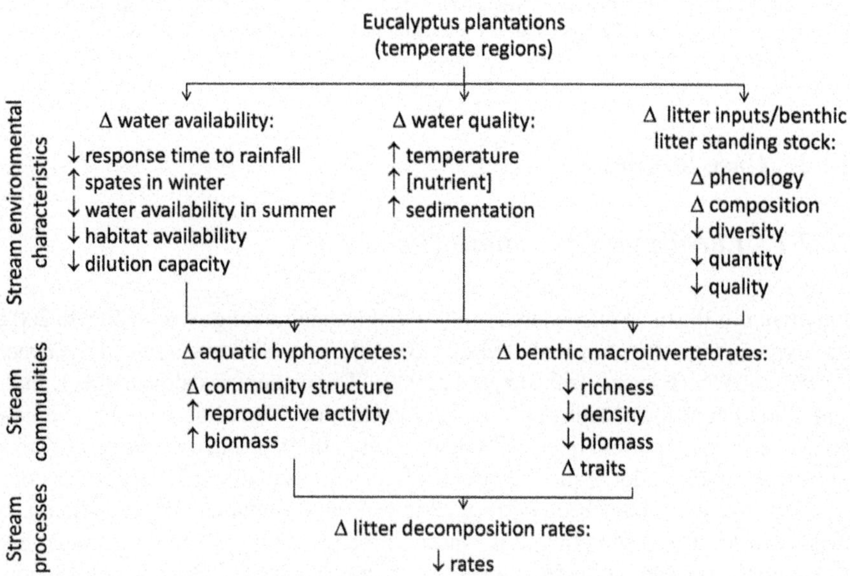

Fig. 14.3 Conceptual diagram of the effects of eucalyptus plantations on stream environmental characteristics, communities and processes in temperate regions, which are characterized by native forests composed of mixed broadleaf deciduous tree species and seasonality in litter fall. Δ indicates change, ↑ indicates increase and ↓ indicates decrease

stream biota due to the accumulation of polyphenolic compounds leached from eucalyptus leaves, reduced pH and reduced dissolved oxygen concentration (Canhoto & Laranjeira, 2007; Canhoto et al., 2013). Water quality may be reduced in streams flowing through eucalyptus plantations also as a result of increases in water temperature if single-stemmed eucalyptus trees with a small crown on the top (the typical tree habit in plantations) replace wide-canopy native tree species (e.g., *Alnus glutinosa*, *Salix* spp.) (Cordero-Rivera et al., 2017). There may also be an increase in dissolved nutrient concentrations due to the use of fertilizers in plantations (Hopmans & Bren, 2007) and soil erosion, especially during and after plantation and harvesting, resulting in increased fine sediment inputs to streams (Siegloch et al., 2014).

Eucalyptus plantations can also alter considerably litter inputs to streams (Fig. 14.3). These changes can be in terms of timing of litter inputs to streams, with inputs peaking during autumn in streams flowing through deciduous forests, while litter inputs are more homogeneously distributed year round or peaking in summer due to water stress under eucalyptus plantations (Abelho & Graça, 1996; Canhoto et al., 2002; Molinero & Pozo, 2003, 2004, 2006; Pozo et al., 1997). The annual quantity of litter inputs to eucalyptus streams may be similar to (Abelho & Graça, 1996) or lower than (Molinero & Pozo, 2003, 2004, 2006; Pozo et al., 1997) litter inputs to streams under native forests. However, the relative composition of litter inputs to streams is strongly altered under eucalyptus plantations. Thus, litter accumulated in eucalyptus streams has a higher fraction of twigs and bark (Molinero & Pozo, 2004; Pozo et al., 1997). Plant species richness and identity in riparian ecotones under plantations is shifted, resulting in litter inputs dominated by the recalcitrant eucalyptus litter (Graça et al., 2002; Molinero & Pozo, 2003, 2006; Pozo et al., 1997). As a result, the quality of litter inputs to eucalyptus streams is decreased (Molinero & Pozo, 2004, 2006; Pozo et al., 1997). Since litter inputs occur mostly during low flow conditions and litter is more recalcitrant (i.e., woody material and eucalyptus leaves), benthic litter standing stock is larger than in streams flowing through native forests (Molinero & Pozo, 2003, 2004; Pozo et al., 1997).

The changes in stream environmental characteristics described above can affect stream communities (Fig. 14.3). The community structure of aquatic hyphomycetes differs between stream types (Bärlocher & Graça, 2002; Ferreira et al., 2006). Species richness is lower in streams flowing through eucalyptus plantations than in those under native forests in central Portugal (Bärlocher & Graça, 2002; Ferreira et al., 2006), while it shows the opposite pattern (Ferreira et al., 2006) or it does not differ between stream types in northern Spain (Chauvet et al., 1997). In general, conidia concentration in the water (Bärlocher & Graça, 2002) and sporulation rates by aquatic hyphomycetes on *A. glutinosa* and *Quercus robur* litter incubated in both stream types does not differ (Ferreira et al., 2006), but maximum sporulation rates can be higher in eucalyptus streams. Fungal biomass is also higher under eucalyptus plantations in Portugal, likely caused by higher water temperature (Ferreira et al., 2006), but not in Spain (Diez et al., 2002; Molinero et al., 1996; Pozo et al., 1998). The stronger effects in Portugal compared to Spain are likely due to the more arid climate in the former region that leads to stronger decreases in water availability and changes in litter inputs in eucalyptus streams (Ferreira et al., 2006).

Regarding benthic macroinvertebrates, there is generally lower taxa richness, density and biomass of total macroinvertebrates and shredders in eucalyptus streams than in streams flowing through deciduous forests (Abelho & Graça, 1996; Cordero-Rivera et al., 2017; Ferreira et al., 2015; Larrañaga et al., 2006, 2009a, 2009b; Monroy et al., 2017) (Fig. 14.3). Negative effects are especially strong for large shredder taxa (e.g., amphipods of the genus *Echinogammarus* and caddisflies of the family Limnephilidae), which are more negatively affected by the poor quality of eucalyptus litter (Larrañaga et al., 2009a; Monroy et al., 2017). Assemblages of detritivore macroinvertebrates in streams flowing through eucalyptus plantations are dominated by taxa that feed on fine particulate organic matter and algae (since litter is recalcitrant in these streams), and are mostly small bodied, with short life cycles, reduced voltinism and active dispersal (likely an adaptation to summer streamflow reduction and drought risk) (Larrañaga et al., 2009a, 2009b; Monroy et al., 2017).

The above-mentioned changes in abiotic and biotic factors may alter litter decomposition in streams (Fig. 14.3). However, it seems that the magnitude and the direction of the effect depend on specific local factors (Chauvet et al., 2016; Ferreira et al., 2016; Graça et al., 2002). A recent meta-analysis based on 92 comparisons between streams flowing through eucalyptus plantations and streams flowing through native deciduous forests reported in 10 published studies found an overall significant inhibition of 22% in litter decomposition rate in streams flowing through plantations when compared with streams flowing through deciduous forests (Ferreira et al., 2016). The magnitude of the effect varies, however, with plantation extent, resource type, litter identity and type of community involved (microbes only or microbes plus macroinvertebrates). The effects of eucalyptus plantations on litter decomposition are significant when eucalyptus plantation extends into the riparian area (inhibition of 27%) but not when plantations keep a native riparian buffer, highlighting the relevance of maintaining riparian ecotones intact to mitigating plantation effects (Ferreira et al., 2016). When considering resource type, significant effects are observed only for leaf litter decomposition (inhibition of 26%) but not for substrates that included woody components, likely due to the larger role played by invertebrates in the decomposition of leaves compared to that of woody substrates (Ferreira et al., 2016). When considering just leaf litter decomposition, significant effects were found when macroinvertebrates had access to the leaves (inhibition by 36%) but not in microbial-driven leaf decomposition. This result suggests that there is functional redundancy among microbial communities and that plantation effects on litter decomposition are mediated through changes in macroinvertebrate communities in temperate regions (Ferreira et al., 2016). Finally, litter quality resulting from contrasting leaf traits also moderates the effects of eucalyptus plantations on leaf decomposition with stronger inhibition being found on more palatable leaf species (inhibition of 51% for alder, *A. glutinosa,* 27% for oak, *Q. robur*, and no effect for eucalyptus, *Eucalyptus globulus*). This again suggests a stronger role of macroinvertebrates in mediating plantation effects on leaf litter decomposition (Ferreira et al., 2016). The re-analysis of a data subset considering only studies addressing the effects of eucalyptus plantation on alder and oak leaf litter decomposition in coarse and fine mesh bags fully illustrates these last results (Fig. 14.4).

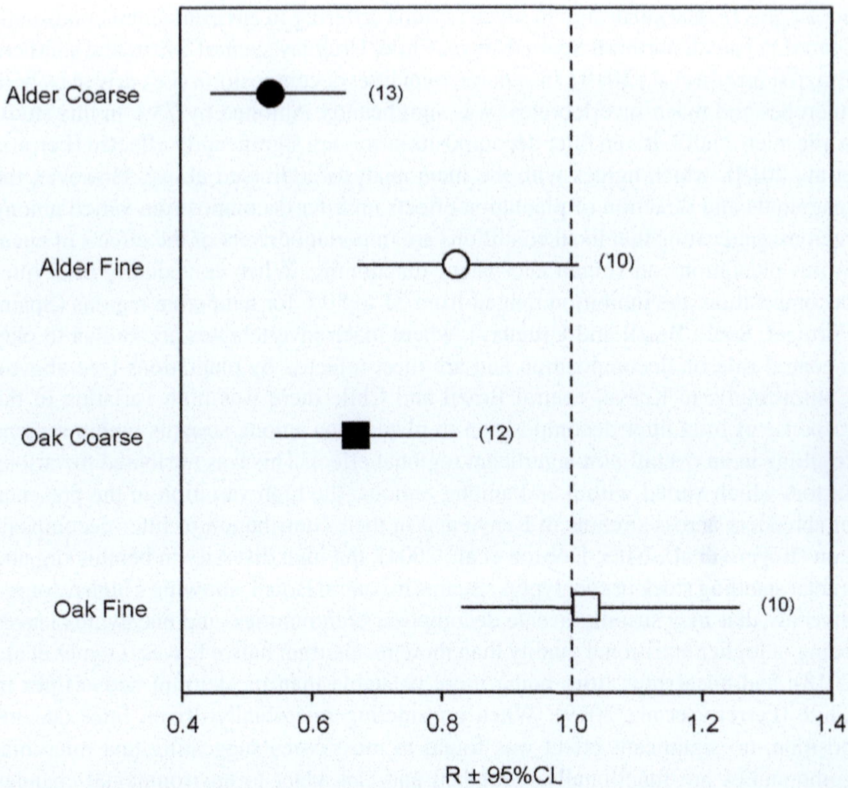

Fig. 14.4 Effect of eucalyptus (*Eucalyptus globulus*) plantations on alder (*Alnus glutinosa*) and oak (*Quercus robur*) leaf litter decomposition (k,/dd) in streams in the Iberian Peninsula, when enclosed in coarse and fine mesh bags. $R = k_{\text{Eucalyptus stream}}/k_{\text{Reference stream}}$ (total n = 45 from 7 published studies; data subset from Ferreira et al., 2016). R = 1 (dashed line) indicates no effect of eucalyptus plantations while R < 1 indicates inhibition of k in eucalyptus streams. Effects are significant when the 95%CI (confidence interval) does not include 1 and treatments significantly differ when their 95% CI do not overlap. Values in parenthesis indicate sample size (i.e., number of comparisons between eucalyptus and reference streams). Global R: 0.73 (95% CI: 0.63–0.84). Missing effects sizes were detected to the left of the global R, with the new estimate being lower (Global R: 0.63; 95% CI: 0.54–0.73) indicating that the dataset used is conservative. Eucalyptus plantations inhibit k by an average of 27% (16–37%), but the effect is stronger for coarse than for fine mesh bags (Q_M (df = 1) = 11.68, $p < 0.001$). The effect also tends to be stronger for alder than for oak, especially in fine mesh bags, but no significant effect of litter species was found (Q_M (df = 1) = 2.92, $p = 0.087$)

The reported effects of eucalyptus plantations on litter decomposition in the Iberian Peninsula may differ from those in other regions because of differences in climate, type of native vegetation, eucalyptus species used in plantations and relative contribution of macroinvertebrates and microbes to litter decomposition, including their identity. A recent coordinated experiment evaluated the effects of eucalyptus plantations on *A. glutinosa* litter decomposition following a paired native forest *vs.*

eucalyptus plantation design in seven regions differing in environmental conditions: central Portugal, northern Spain, Kenya, Chile, Uruguay, central Brazil and southern Brazil (Ferreira et al., 2019). In general, total litter decomposition (i.e., driven by both microbes and macroinvertebrates) was significantly inhibited by 23% in this study, while microbially-driven litter decomposition was not significantly affected (Ferreira et al., 2019), which agrees with the meta-analysis addressed above. However, the magnitude and direction of plantation effects on litter decomposition varied among regions, suggesting that local conditions are important drivers of the effects of eucalyptus plantations on stream ecosystem functioning. When considering total litter decomposition, the inhibition ranged from 31 to 50% for temperate regions (Spain, Portugal, South Brazil and Uruguay), where macroinvertebrates are known to play a central role on decomposition and are most affected by plantations (see above). Contrastingly, in Kenya, central Brazil and Chile there was high variation in the response of total litter decomposition to plantations among streams within regions resulting in an overall non-significant regional effect. This was attributed to various factors which varied within and among regions: the high variation in the presence of shredders across streams in Kenya and in their contribution to litter decomposition (Boyero et al., 2015; Dobson et al., 2004); the high diversity in benthic organic matter standing stock in eucalyptus streams in central Brazil, allowing a high resource diversity that may sustain diverse decomposer communities, and eucalyptus leaves being of higher nutritional quality than more recalcitrant native leaves (Gomes et al., 2018); and eucalyptus litter being more palatable than recalcitrant native litter in Chile (Ferreira et al., 2019). When considering microbially-driven litter decomposition, no significant effect was found in most cases suggesting that microbial communities are functionally redundant and can adapt to environmental changes created by plantations more easily. However, stimulation of microbial litter decomposition by 32–110% was found in Uruguay and central Brazil and inhibition by 48% in Kenya, suggesting that microbial functional redundancy cannot be generalized across regions (Ferreira et al., 2019). The high solar irradiation and primary production in unshaded eucalyptus streams in Uruguay may allow a priming effect by which microbial decomposers benefit from labile carbon originating from primary producers (Danger et al., 2013; Kuehn et al., 2014) while the high diversity in benthic organic matter standing stock in eucalyptus streams in central Brazil may allow a higher diversity of decomposers and microbial inoculum (Ferreira et al., 2019).

To sum up, the effects of eucalyptus plantations on litter decomposition depend on the region (with stronger effects in regions where macroinvertebrates play an important role on decomposition), extent of plantations (with stronger effects if eucalyptus are planted in riparian areas), resource type (decomposition of leaves being more affected than that of woody substrates), litter species (decomposition of high quality leaves being more affected than that of low quality leaves), and the type of decomposer community involved (decomposition mediated by both microbes and macroinvertebrates being more affected than microbial-driven decomposition). Thus, stronger negative effects of eucalyptus plantations can then be expected for streams naturally receiving high-quality litter and where shredders are abundant.

14.2.2 Conifer Plantations (Fig. 14.5)

Conifer plantations are widespread in temperate and cold regions at both hemispheres. Nevertheless, the majority of studies tackling the effects of conifer plantations, mostly evergreen species (though *Larix* is deciduous), in streams have been carried out in temperate climates, where deciduous broadleaf species grow naturally. Consequently, as above, this review will inevitably offer a biased view of the effects of these plantations on litter dynamics in streams. Contrary to eucalyptus studies, which are based on a single genus (*Eucalyptus*), with most focusing on *E. globulus*, studies on conifers have focused on different species from the genera *Abies*, *Cryptomeria*, *Larix*, *Picea* and *Pinus*. Thus, even if many traits are shared among conifer species, the variability of effects reviewed here is larger than that of eucalyptus effects. Moreover, conifer species are in some cases planted within their natural distribution range or biome, but they are included here because they are planted as monospecific dense stands and/or are intensively managed, thus causing similar harmful effects to those of their exotic counterparts.

Similar to eucalyptus plantations, conifer plantations have been associated with an increase in soil hydrophobicity compared to deciduous forests, which has been linked to increases in surface runoff during rainfall (Butzen et al., 2015). A larger evapotranspiration of conifers compared to deciduous forests has been proven in many studies after the seminal work of Swank and Douglass (1974) and has been shown to reduce annual stream flow, especially in the dry season (Jackson et al., 2005; Jobbágy et al., 2013). Nevertheless, other studies have shown that this effect is context-dependent across the world (Komatsu et al., 2007, 2011). Conifer plantations have been related to water acidification in the UK (Ormerod et al., 1993). However, the majority of studies in other regions show that this forestry activity does not alter stream water physicochemical properties (Martínez et al., 2013, 2016; Riipinen et al., 2010; Swank & Crossley, 1988; Thompson & Townsend, 2004; Woodall & Wallace, 1972).

Most conifer species planted are evergreen, and thus create important differences in the timing of litter inputs to streams compared with streams flowing through deciduous forests where trees shed leaves in autumn (Inoue et al., 2012; Martínez et al., 2016). Consequently, while in streams flowing through deciduous forests there is a peak of benthic storage of coarse particulate organic matter (CPOM) in autumn, litter storage in conifer streams shows less variation throughout the year (Inoue et al., 2012; Martínez et al., 2016). Nevertheless, CPOM annual budgets are often similar between both stream types (Martínez et al., 2013, 2016; Miserendino & Masi, 2010; Riipinen et al., 2009). However, given the lower concentration of nutrients (both nitrogen and phosphorus) in needles than in leaves of many deciduous tree species (Casas et al., 2013; Martínez et al., 2013), streams flowing through conifer plantations store lower amounts of nutrients in benthic CPOM compared to deciduous forest streams (Martínez et al., 2016) (Fig. 14.5).

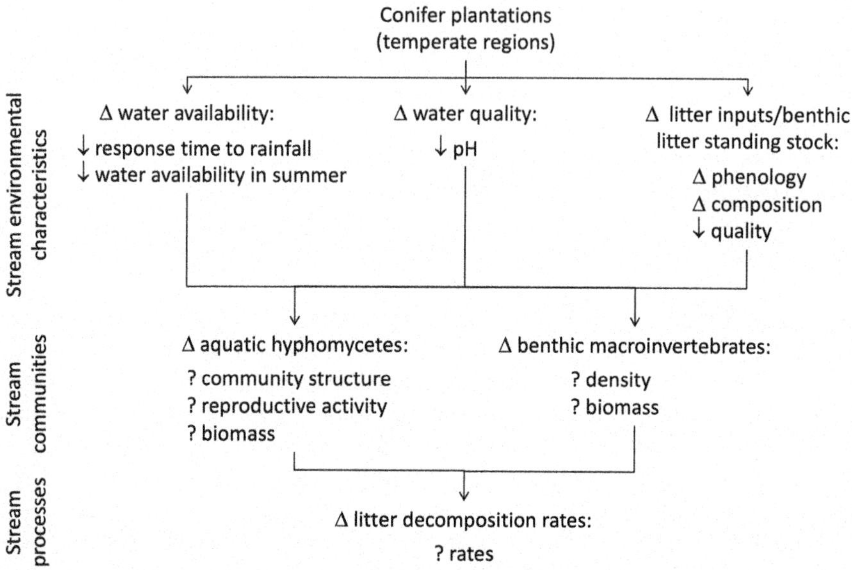

Fig. 14.5 Conceptual diagram of the effects of conifer plantations on stream environmental characteristics, communities and processes in temperate regions, which are characterized by native forests composed of mixed broadleaf deciduous tree species and seasonality in litter fall. Δ indicates change, ↑ indicates increase, ↓ indicates decrease and ? indicates contradictory results

Conifer tree species produce litter that is poor in nutrients and tough (Casas et al., 2013; Martínez et al., 2013), so aquatic communities can be energetically challenged in places where conifers are dominant (Fig. 14.5). Efforts to elucidate the effects of conifer plantations on aquatic communities have focused mainly on invertebrates, with information about microbial decomposers being scarce. Two studies addressed the effects of conifer plantations on aquatic hyphomycetes associated with decomposing litter. In northern Spain, Martínez et al. (2013) did not find a clear alteration in aquatic hyphomycete sporulation rates and community structure in streams flowing through *Pinus radiata* plantations compared to those flowing through deciduous forests. In contrast, Ferreira et al. (2017) found distinct aquatic hyphomycete community structure between streams flowing through *Cryptomeria japonica* plantations and streams flowing through native laurel forests in Azores (Fig. 14.5). Regarding the effects on macroinvertebrates, density and biomass of total invertebrates have been found to be lower (Friberg, 1997; Whiles & Wallace, 1997) or similar (Martínez et al., 2013, 2016; Monroy et al., 2017) in streams flowing through pine plantations compared to those under deciduous forests. Similarly, effects on shredders vary, with various metrics such as abundance and biomass being reported as higher (Friberg, 1997; Riipinen et al., 2009, 2010) or lower (Martínez et al., 2013, 2016; Miserendino & Masi, 2010; Riipinen et al., 2010; Whiles & Wallace, 1997) in streams flowing through conifer plantations. Moreover, specific taxa have been shown to respond differently to conifer plantations; while Friberg (1997) and Whiles and Wallace (1997) found higher abundance of small shredders such as *Leuctra* in streams under conifer plantations, Martínez et al. (2016) and Monroy et al. (2017) reported a size-specific response to pine plantations, with lower density of small detritivores under this type of streams.

A highly consistent result is that conifer needles are processed slower than most deciduous leaves (Albariño & Balseiro, 2002; Hisabae et al., 2011; Martínez et al., 2013; Whiles & Wallace, 1997), due to their low quality (i.e., high toughness and low nutrient concentration) (Casas et al., 2013; Martínez et al., 2013). However, the effect of conifer plantations on litter decomposition capacity of streams is unclear (Fig. 14.6). Martínez et al. (2013) found lower alder litter decomposition rates in streams under conifer plantations. However, other studies showed faster decomposition rates for broadleaves (Riipinen et al., 2009; Whiles & Wallace, 1997) and needles (Whiles & Wallace, 1997) in conifer streams compared to native forest streams. Moreover, Riipinen et al. (2010) did not find differences in litter decomposition rates between streams flowing through broadleaf deciduous forests and conifer plantations when pH was fitted as covariate. Finally, Ferreira et al. (2017) did not find differences in decomposition rates of broadleaves and needles when comparing streams under native evergreen laurel forests and conifer plantations.

A special case is when natural grasslands are converted to plantations with no maintenance of intact riparian bands, as it occurs in mountain grasslands of central Argentina afforested extensively with the conifer *Pinus elliottii*. Grassland afforestation has profound effects in stream ecosystems, as trees growing in riparian ecotones reduce sunlight incidence and water temperature regimes (Cibils-Martina et al., 2017). Planted trees also feed streams with large amounts of CPOM, both woody and needle litter, thus shifting availability of basal food web resources from primary producers to allochthonous organic matter (Principe et al., 2015). As a result, afforested streams have strongly changed community structure and lower total invertebrate abundance and richness (Márquez et al., 2015). Interestingly, needle litter decomposes at similar rates in plantation and grassland streams, indicating that large storage of needle litter in afforested streams is not easily transformed and transferred along food webs (Márquez et al., 2017; Principe et al., 2015).

14.3 Other Planted Species and Management of Plantations

The scientific literature about the effects of eucalyptus and conifer plantations on litter decomposition in streams is steadily growing. Although these studies do not cover all the biomes and the entire latitudinal gradient where eucalyptus and conifer plantations have been established, the current knowledge can already be used with high confidence for predicting large-scale effects of eucalyptus and conifer plantations on streams worldwide. For other planted species, the number of studies comparing litter decomposition in plantation and native forest streams is scarce. A few studies show that litter decomposition rates in streams are reduced under intensive monospecific beech (*Fagus sylvatica*) plantations compared to traditionally managed forests with

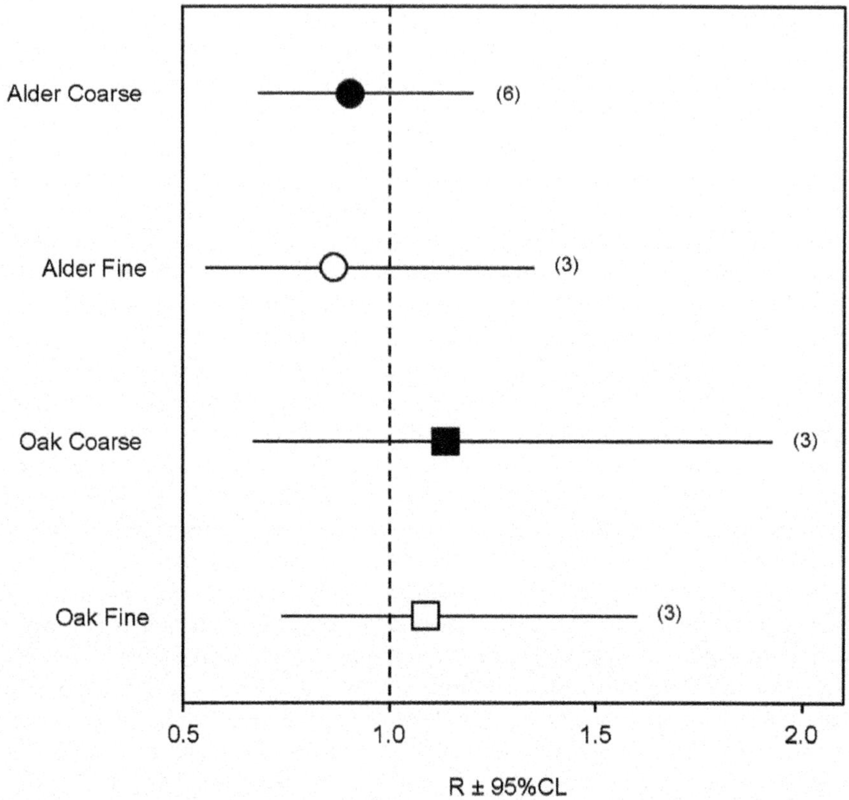

Fig. 14.6 Effect of conifer plantations on alder (*Alnus glutinosa*) and oak (*Quercus robur*) leaf litter decomposition (*k*,/dd) in streams, when enclosed in coarse and fine mesh bags. $R = k_{\text{Conifer stream}}/k_{\text{Reference stream}}$ (total n = 15 from 2 published studies; data subset from Ferreira et al., 2016). R = 1 (dashed line) indicates no effect of conifer plantations, R < 1 indicates inhibition and R > 1 indicates stimulation of *k* in conifer streams. Effects are significant when the 95%CI does not include 1 and treatments significantly differ when their 95%CI do not overlap. Values in parenthesis indicate sample size (i.e., number of comparisons between conifer and reference streams). Global R: 0.96 (95% CI: 0.77–1.21). Missing effects sizes were detected to the left of the global R, with the new estimate being lower (Global R: 0.87; 95% CI: 0.70–1.08) but it does not change the result that was not-significant

high tree diversity (Ferreira et al., 2016; Hladyz et al., 2011; Lecerf & Chauvet, 2008; Lecerf et al., 2005). In contrast, litter decomposition rates are stimulated in streams flowing through plane (*Platanus hybrida*) riparian plantations than in streams flowing through native vegetation (Menéndez et al., 2013). Finally, a recent study in Malaysia found faster litter decomposition of both the native *Macaranga* sp. and the exotic oil palm (*Elaeis guineensis*) in streams flowing through oil palm plantations than in streams surrounded by native vegetation; accelerated litter decomposition was mainly driven by microbes, which were boosted by nutrient inputs from plantation fertilizers (Chellaiah & Yule, 2018a).

Without specific studies about the effects of other plantations, we can only infer their effects from the biology, ecology and traits of the planted species. Secondary compounds produced by plants force evolutionary adaptations in consumers in order to cope with that toxicity and gain an advantage over consumers who lack the same adaptation (Futuyma & Agrawal, 2009). This suggests that exotic species used in plantations may have a larger effect than planted native species on stream detritivore feeding and growth, which is more likely if exotic species differ deeply from native species in leaf litter traits. As studies accumulate, it is more evident that leaf traits override the effect of the origin of plant species on decomposition (Davis et al., 2011; Kennedy & El-Sabaawi, 2017). Litter traits such as the concentration of nutrients, structural compounds and toxic chemicals, and toughness are key for understanding the effects of plantations on litter decomposition. However, as the different traits might have synergistic or antagonistic effects on decomposition, predicting the consequences of plantations on stream nutrient cycling only by analysing litter characteristics is unattainable (also, plantations lead to other environmental changes unrelated to litter characteristics). Although some detritivores are able to compensate for the low quality of some resources by increasing their consumption (Flores et al., 2014) and resource preferences do not always correlate with growth and survival of consumers (Larrañaga et al., 2014; Lewis et al., 2017), consumption of litter, or, by extension, litter decomposition rates, might be a good approach to partially predict the effect of plantations on ecosystem level litter decomposition. Litter decomposition rates of worldwide planted exotic species span a wide range (Fig. 14.7), with the slowest (plane and *Cryptomeria japonica*, for total and microbial decomposition, respectively) and the fastest (poplar *Populus* × *canadensis* and eucalyptus) values falling within the range found for native species (Fig. 14.7). Moreover, decomposition rates of dominant deciduous forest species such as oak (*Quercus* spp.) or beech (*Fagus* sp.), which form old and well-preserved forests across Europe, are similar to those of species used for intensive plantations such as *E. globulus* or *Pinus* spp. (Fig. 14.7). Nevertheless, high tree diversity in native forests provides streams with a large range of litter traits, which make communities in these streams more productive and diverse than those under monoculture plantations. Similarly, the presence of a native species riparian buffer in streams flowing through plantations usually helps mitigating the direct effects of plantations on aquatic communities and on ecosystem processes, such as litter decomposition. For instance, riparian trees, which are adapted to floods, may offer resources of higher quality, such as alder (*Alnus* spp.) litter, which can help sustaining aquatic food webs in streams flowing through plantations with otherwise low quality litter available. Not only riparian buffers strips, but patches of well-preserved native forests in other places of the catchment can also create a similar effect. For example, macroinvertebrate communities in stream reaches surrounded by eucalyptus plantations, but with a large cover of native deciduous vegetation in the catchment were more similar to stream reaches totally surrounded by deciduous native vegetation than to those completely surrounded by eucalyptus plantations (Larrañaga et al., 2009a). Similarly, riparian buffers have been observed to preserve the structure of aquatic communities in streams in intensively managed oil palm plantations (Chellaiah & Yule, 2018b). As expected, decomposition rates are higher

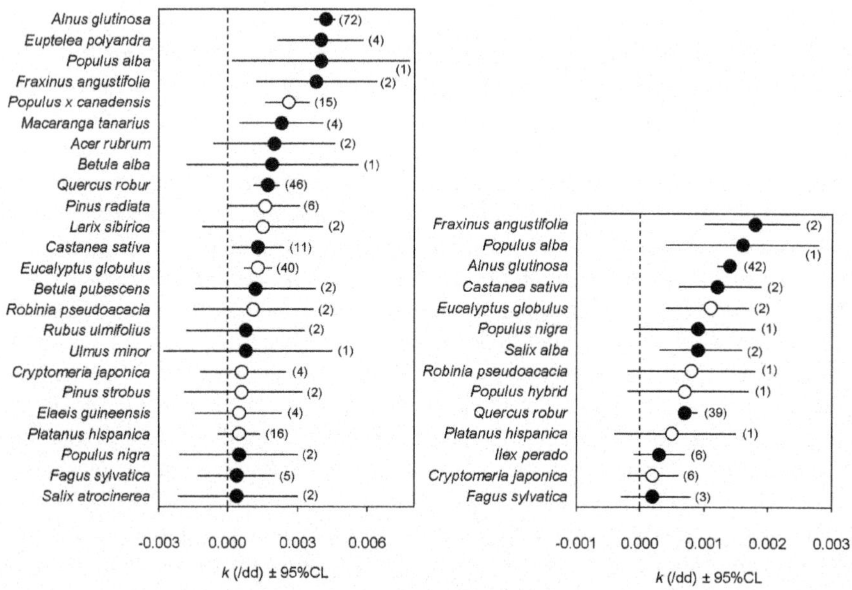

Fig. 14.7 Leaf litter decomposition rates (*k*, dd) of native (black circles) and exotic (white circles) tree species in streams in coarse (left) and fine mesh bags (right) (total n = 357 trials from 30 published studies) addressing effects of exotic plantations on leaf litter decomposition in streams by comparing rates of different species at the same location and/or by comparing rates of the same species in different locations (only studies that report rates for each location separately are considered; species considered native or exotic as defined by the authors). *k* values were calculated including streams flowing through native forests and through plantations together, but litter decomposition in coarse and fine mesh bags were kept separated. *k* values are non-significant when 95% CI includes 0 (dashed line). Values in parenthesis indicate sample size (i.e., number of *k* values). Large 95%CI are generally due to small sample size and interpretation needs to be made with caution. Studies used to build this plot marked with an asterisk (*) in the reference section

in locations where riparian buffers are preserved compared to locations with riparian zones altered by thinning or invaded by exotic species (Casotti et al., 2015). Likewise, litter decomposition rates were similar between streams with native vegetation and streams under eucalyptus plantations maintaining its riparian native vegetation (Ferreira et al., 2016). Beyond providing high quality litter to streams, riparian buffers help reducing nutrient and sediment inputs (de Souza et al., 2013; Jones et al., 2001) and increasing shading and mitigating temperature fluctuations (Broadmeadow et al., 2011; read Feld et al., 2018 for a recent review on riparian buffers). Nonetheless, the characteristics of the riparian buffers needed to preserve decomposition rates remain poorly known. In this line, Lecerf and Richardson (2010) observed that litter decomposition in streams with 10–30 metres wide riparian buffers in native conifer forest catchments that had been harvested 8 years before still differed in decomposition from streams flowing through preserved forest.

In addition to the negative effects attributed to most plantations as trees grow, harvesting is by far the single event with the largest impact on the structure and functioning of streams. Flow regime is disrupted (Martin et al., 2000), inputs of fine sediments are incremented (Kreutzweiser & Capell, 2001), inputs of organic matter are suddenly augmented during the forestry activities and then reduced (Santiago et al., 2011) and biota is consequently altered (Banks et al., 2007). Road density built for harvesting increases sediment inputs and deposition and slows down in situ litter decomposition (Erdozain et al., 2018). Contrastingly, litter decomposition can be stimulated by the release of soil nutrients from clearcut practices (McKie & Malmqvist, 2009) and by changing plant composition in the riparian zones (Kominoski et al., 2011). Best management practices (BMPs) can reduce the effect of forest harvesting on streams (McBroom et al., 2008; Smolders et al., 2018), but even following BMPs logging is able to reduce decomposition (Kreutzweiser et al., 2008), as BMPs do not totally mitigate the changes originated. As expected, the partial felling of trees (i.e., thinning) instead of large scale clearcuts reduces the impact of plantations on stream communities (Quinn et al., 2004) and litter decomposition (Lecerf & Richardson, 2010). Regardless of changes induced by clearcutting, they are long lasting; more than a decade is usually needed for a full recovery back to reference conditions. For instance, Stone and Wallace (1998) reported stream macroinvertebrate communities had recovered to reference conditions only 16 years after catchment clearcutting. Similarly, effects of harvesting on litter decomposition rates can last more than a decade (Griffith & Perry, 1991; Guevara et al., 2015; Webster et al., 2014; Yeung et al., 2017). In plantations where the harvesting cycle is shorter than the time needed for the recovery of stream communities and decomposition rates, we might anticipate an accumulation of effects of multiple harvesting cycles, although this has not been proven yet.

When to harvest a plantation is a paramount decision in order to maximize the production of resources. As trees develop the biomass produced per year decreases, thus, short harvesting cycles are optimal from an economic point of view, albeit they can become the main culprit of the impact that some planted species have on aquatic systems. Not all planted species are managed equally and there is a link between their productivity and the tree species cultivated. Hengeveld et al. (2012) summarized the applicability of different plantation tree species to different forest management approaches in Europe (Fig. 14.8). These authors consider five categories of forestry, from the most intensive, based on short rotations, to the most natural, i.e., natural reserves. The genera most suited for intensive forestry, and thus least suitable for conservation purposes, were *Eucalyptus*, *Robinia* and *Pseudotsuga*, exotic species in Europe (Fig. 14.8). These three genera are considered totally inadequate for nature reserves in Europe, with *Eucalyptus* not even adequate for close-to-nature forests. On the other end *Carpinus*, *Betula*, *Fraxinus*, *Alnus* or *Fagus* are considered of high value for nature reserves, but inadequate for short rotation forestry (Fig. 14.8). This analysis emphasizes that traits of particular tree species can constrain their use more to production or to conservation. Nevertheless, most species show a relatively high applicability (>20%) to at least three management approaches, which shows that beside the selection of species, the intensity of plantation exploitation can be adapted either to maximize production or to minimize environmental impacts. In this sense, the previously described negative effects of native beech plantations on freshwater

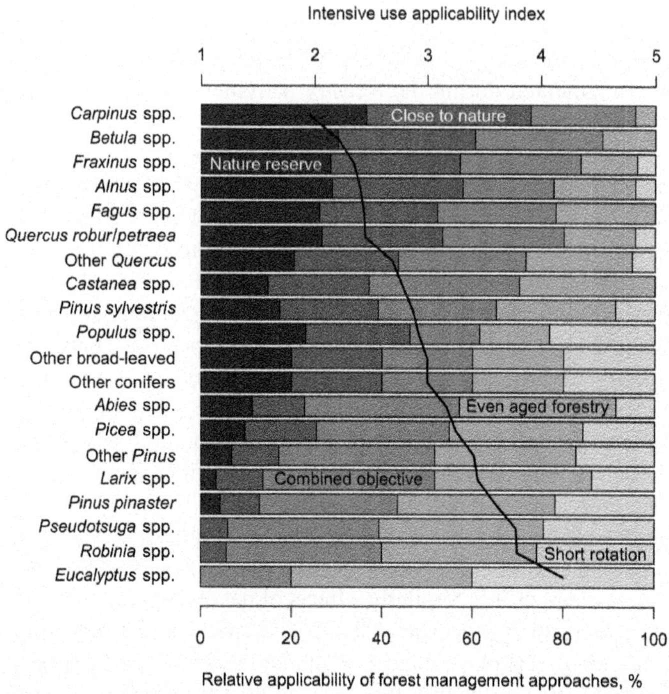

Fig. 14.8 Relative applicability of different tree genus/species for different forest management approaches in Europe, from nature reserves to short rotation forestry. The transversal line compiles the relative applicability into a single index (i.e., Intensive use applicability index; scale in top X axis) considering the five forest management approaches as numbers, from Nature reserve (1) to Short rotation forestry (5) and weighing them by the relative applicability percentages (e.g., *Eucalyptus* spp.: (3 * 20% + 4 * 40% + 5 * 40%)/100 = 4.2). Species are ordered following the applicability to intensive use (less to more, from top to bottom) (*Source* [Hengeveld et al., 2012])

litter decomposition (Ferreira et al., 2016; Hladyz et al., 2011a, 2011b; Lecerf et al., 2005), illustrates the relevance of *how* forestry is implemented beside *which* is the species selected for plantations.

14.4 Concluding Remarks

From this review we can distil the following ideas:

1. Plantations can alter litter decomposition in stream ecosystems by means of changes in hydrology, water quality, leaf litter inputs (including quantity, quality, timing) and biota.
2. Some planted tree species have received special scientific attention (eucalyptus and conifers), with most of the studies carried out in Europe and North America.

Studies on other tree species and regions are needed to allow drawing more general conclusions.

3. The decomposability of the leaf litter produced by planted species is similar to that of many species dominating native forests. Deficiency or absence of appropriate riparian buffers strongly contribute to the deleterious effect of plantations on stream decomposition capacity.
4. Although planted tree species can display intrinsic economic and environmental values, they can always be managed in a range of different ways to find a trade-off between productivity and conservation.
5. There are proven measures that help mitigate the impacts of plantations on litter decomposition in freshwaters (conserving riparian buffers and following Best Management Practices, for instance). Unfortunately, they are yet to be implemented in many places of the world.

References

The ones used to create Fig. 14.7 have an asterisk at the end, "*".

Abelho, M., & Graça, M. A. S. (1996). Effects of eucalyptus afforestation on leaf litter dynamics and macroinvertebrate community structure of streams in Central Portugal. *Hydrobiologia, 324*(3), 195–204. https://doi.org/10.1007/bf00016391 (*).

Albariño, R. J., & Balseiro, E. G. (2002). Leaf litter breakdown in Patagonian streams: Native versus exotic trees and the effect of invertebrate size. *Aquatic Conservation: Marine and Freshwater Ecosystems, 12*(2), 181–192. https://doi.org/10.1002/aqc.511.

Baldy, V., Gessner, M. O., & Chauvet, E. (1995). Bacteria, fungi and the breakdown of leaf litter in a large river. *Oikos, 74*(1), 93–102. https://doi.org/10.2307/3545678 (*).

Banks, J. L., Li, J., & Herlihy, A. T. (2007). Influence of clearcut logging, flow duration, and season on emergent aquatic insects in headwater streams of the Central Oregon Coast Range. *Journal of the North American Benthological Society, 26*(4), 620–632. https://doi.org/10.1899/06-104.1.

Bärlocher, F., & Graça, M. A. S. (2002). Exotic riparian vegetation lowers fungal diversity but not leaf decomposition in Portuguese streams. *Freshwater Biology, 47*(6), 1123–1135. https://doi.org/10.1046/j.1365-2427.2002.00836.x (*).

Boyero, L., Pearson, R. G., Gessner, M. O., Dudgeon, D., Ramírez, A., Yule, C. M., Callisto, M., Pringle, C. M., Encalada, A. C., Arunachalam, M., Mathooko, J., Helson, J. E., Rincón, J., Bruder, A., Cornejo, A., Flecker, A. S., Mathuriau, C., M'Erimba, C., Gonçalves Jr, J. F.,… Jinggut, T. (2015). Leaf-litter breakdown in tropical streams: Is variability the norm? *Freshwater Science, 34*(2), 759–769. https://doi.org/10.1086/681093.

Broadmeadow, S. B., Jones, J. G., Langford, T. E. L. L., Shaw, P. J., & Nisbet, T. R. (2011). The influence of riparian shade on lowland stream water temperatures in southern England and their viability for brown trout. *River Research and Applications, 27*(2), 226–237. https://doi.org/10.1002/rra.1354.

Brown, J. H., Rosenzweig, M. L., Gardener, M. R., Carroll, S. P., Thompson, K., Pickett, S. T. A., Stromberg, J. C., Del Tredici, P., Suding, K. N., Ehrenfeld, J. G., Grime, J. P., Mascaro, J., & Briggs, J. C. (2011). Don't judge species on their origins. *Nature, 474*(7350), 153–154. https://doi.org/10.1038/474153a.

Butzen, V., Seeger, M., Marruedo, A., de Jonge, L., Wengel, R., Ries, J. B., & Casper, M. C. (2015). Water repellency under coniferous and deciduous forest—Experimental assessment and impact on overland flow. *CATENA, 133,* 255–265. https://doi.org/10.1016/j.catena.2015.05.022.

Canhoto, C., Bärlocher, F., & Graça, M. A. S. (2002). The effects of *Eucalyptus globulus* oils on fungal enzymatic activity. *Archiv fur Hydrobiologie, 154*(1), 121–132. https://doi.org/10.1127/archiv-hydrobiol/154/2002/121.

Canhoto, C., Calapez, R., Gonçalves, A. L., & Moreira-Santos, M. (2013). Effects of *Eucalyptus* leachates and oxygen on leaf-litter processing by fungi and stream invertebrates. *Freshwater Science, 32*(2), 411–424. https://doi.org/10.1899/12-062.1.

Canhoto, C., & Laranjeira, C. (2007). Leachates of *Eucalyptus globulus* in intermittent streams affect water parameters and invertebrates. *International Review of Hydrobiology, 92*(2), 173–182. https://doi.org/10.1002/iroh.200510956.

Casas, J. J., Larrañaga, A., Menéndez, M., Pozo, J., Basaguren, A., Martínez, A., Pérez, J., González, J. M., Mollá, S., Casado, C., Descals, E., Roblas, N., López-González, J. A., & Valenzuela, J. L. (2013). Leaf litter decomposition of native and introduced tree species of contrasting quality in headwater streams: How does the regional setting matter? *Science of the Total Environment, 458–460*, 197–208. https://doi.org/10.1016/j.scitotenv.2013.04.004.

Casotti, C. G., Kiffer, W. P., Costa, L. C., Rangel, J. V., Casagrande, L. C., & Moretti, M. S. (2015). Assessing the importance of riparian zones conservation for leaf decomposition in streams. *Natureza e Conservação, 13*(2), 178–182. https://doi.org/10.1016/j.ncon.2015.11.011.

Chauvet, E., Fabre, E., Elósegui, A., & Pozo, J. (1997). The impact of eucalypt on the leaf-associated aquatic hyphomycetes in Spanish streams. *Canadian Journal of Botany, 75*(6), 880–887. https://doi.org/10.1139/b97-097.

Chauvet, E., Ferreira, V., Giller, P. S., McKie, B. G., Tiegs, S. D., Woodward, G., Elosegi, A., Dobson, M., Fleituch, T., Graça, M.A.S., Gulis, V., Hladyz, S., Lacoursière, J. O., Lecerf, A., Pozo, J., Preda, E., Riipinen, M., Rîşnoveanu, G., & Gessner, M. O. (2016). Litter decomposition as an indicator of stream ecosystem functioning at local-to-continental scales: Insights from the European RivFunction project. *Advances in Ecological Research, 55*, 99–182. https://doi.org/10.1016/bs.aecr.2016.08.006.

Chellaiah, D., & Yule, C. M. (2018a). Litter decomposition is driven by microbes and is more influenced by litter quality than environmental conditions in oil palm streams with different riparian types. *Aquatic Sciences, 80*(4), 1–11. https://doi.org/10.1007/s00027-018-0595-y.

Chellaiah, D., & Yule, C. M. (2018b). Riparian buffers mitigate impacts of oil palm plantations on aquatic macroinvertebrate community structure in tropical streams of Borneo. *Ecological Indicators, 95*(1), 53–62. https://doi.org/10.1016/j.ecolind.2018.07.025.

Cibils-Martina, L., Márquez, J., Principe, R., Gari, N., & Albariño, R. (2017). Pine afforestation affects key primary producers in mountain grassland streams in Córdoba, Argentina. *New Zealand Journal of Marine and Freshwater Research.* https://doi.org/10.1080/00288330.2017.1329746.

Cordero-Rivera, A., Martínez Álvarez, A., & Álvarez, M. (2017). Eucalypt plantations reduce the diversity of macroinvertebrates in small forested streams. *Animal Biodiversity and Conservation, 40*(1), 87–97. https://doi.org/10.32800/abc.2017.40.0087.

Danger, M., Cornut, J., Chauvet, E., Chavez, P., Elger, A., & Lecerf, A. (2013). Benthic algae stimulate leaf litter decomposition in detritus-based headwater streams: A case of aquatic priming effect? *Ecology, 94*(7), 1604–1613. https://doi.org/10.1890/12-0606.1.

de Souza, A. L. T. D., Fonseca, D. G., Libório, R. A., & Tanaka, M. O. (2013). Influence of riparian vegetation and forest structure on the water quality of rural low-order streams in SE brazil. *Forest Ecology and Management, 298*, 12–18. https://doi.org/10.1016/j.foreco.2013.02.022.

Diez, J., Elosegi, A., Chauvet, E., & Pozo, J. (2002). Breakdown of wood in the Agüera stream. *Freshwater Biology, 47*(11), 2205–2215. https://doi.org/10.1046/j.1365-2427.2002.00965.x.

Dobson, M., Mathooko, J. M., Ndegwa, F. K., & M'Erimba, C. (2004). Leaf litter processing rates in a Kenyan highland stream, the Njoro River. *Hydrobiologia, 519*(1–3), 207–210. https://doi.org/10.1023/b:hydr.0000026592.50734.ea.

Elfving, B., Ericsson, T., & Rosvall, O. (2001). The introduction of lodgepole pine for wood production in Sweden—A review. *Forest Ecology and Management, 141*(1–2), 15–29. https://doi.org/10.1016/s0378-1127(00)00485-0.

Erdozain, M., Kidd, K., Kreutzweiser, D. P., & Sibley, P. (2018). Linking stream ecosystem integrity to catchment and reach conditions in an intensively managed forest landscape. *Ecosphere, 9*(5). https://doi.org/10.1002/ecs2.2278.

FAO. (2006a). *Global planted forests thematic study: Results and analysis, by A. Del Lungo, J. Ball and J. Carle. Planted forests and trees working paper 38.* Rome, Italy.

FAO. (2006b). *Responsible management of planted forests: Voluntary guidelines. Planted forests and trees working paper 37/E.* Rome, Italy.

FAO. (2015). *Global forest resources assessment 2015: How are the world's forests changing?* Rome, Italy.

FAO. (2018). *The state of the world's forests 2018—Forest pathways to sustainable development.* Rome, Italy.

Farley, K. A., Piñeiro, G., Palmer, S. M., Jobbágy, E. G., & Jackson, R. B. (2008). Stream acidification and base cation losses with grassland afforestation. *Water Resources Research, 44*(7), 1–11. https://doi.org/10.1029/2007wr006659.

Fausch, K. D., Baxter, C. V., & Murakami, M. (2010). Multiple stressors in north temperate streams: Lessons from linked forest-stream ecosystems in northern Japan. *Freshwater Biology, 55*(1), 120–134. https://doi.org/10.1111/j.1365-2427.2009.02378.x.

Feld, C. K., Fernandes, M. R., Ferreira, M. T., Hering, D., Ormerod, S. J., Venohr, M., & Gutiérrez-Cánovas, H. (2018). Evaluating riparian solutions to multiple stressor problems in river ecosystems—A conceptual study. *Water Research, 139,* 381–394. https://doi.org/10.1016/j.watres.2018.04.014.

Ferraz, S. F. B., Lima, W. de P., & Rodrigues, C. B. (2013). Managing forest plantation landscapes for water conservation. *Forest Ecology and Management, 301,* 58–66. https://doi.org/10.1016/j.foreco.2012.10.015.

Ferreira, V., Boyero, L., Calvo, C., Correa, F., Figueroa, R., Gonçalves, J. F., Goyenola, G., Grac, M. A. S., Hepp, L. U., Kariuki, S., López-Rodríguez, A., Mazzeo, N., M'Erimba, C., Monroy, S., Peil, A., Pozo, J., Rezende, R., & Teixeira-de-Mello, F. (2019). A global assessment of the effects of eucalyptus plantations on stream ecosystem functioning. *Ecosystems, 22*(3), 629–642. https://doi.org/10.1007/s10021-018-0292-7.

Ferreira, V., Elosegi, A., Gulis, V., Pozo, J., & Graça, M. A. S. (2006). Eucalyptus plantations affect fungal communities associated with leaf-litter decomposition in Iberian streams. *Archiv Für Hydrobiologie, 166*(4), 467–490. https://doi.org/10.1127/0003-9136/2006/0166-0467 (*).

Ferreira, V., Faustino, H., Raposeiro, P. M., & Gonçalves, V. (2017). Replacement of native forests by conifer plantations affects fungal decomposer community structure but not litter decomposition in Atlantic island streams. *Forest Ecology and Management, 389,* 323–330. https://doi.org/10.1016/j.foreco.2017.01.004.

Ferreira, V., Koricheva, J., Pozo, J., & Graça, M. A. S. (2016). A meta-analysis on the effects of changes in the composition of native forests on litter decomposition in streams. *Forest Ecology and Management, 364,* 27–38. https://doi.org/10.1016/j.foreco.2016.01.002.

Ferreira, V., Larrañaga, A., Gulis, V., Basaguren, A., Elosegi, A., Graça, M. A. S., & Pozo, J. (2015). The effects of eucalypt plantations on plant litter decomposition and macroinvertebrate communities in Iberian streams. *Forest Ecology and Management, 335,* 129–138. https://doi.org/10.1016/j.foreco.2014.09.013 (*).

Flores, L., Larrañaga, A., & Elosegi, A. (2014). Compensatory feeding of a stream detritivore alleviates the effects of poor food quality when enough food is supplied. *Freshwater Science, 33*(1), 134–141. https://doi.org/10.1086/674578.

Forrester, D. I., Collopy, J. J., & Morris, J. D. (2010). Transpiration along an age series of *Eucalyptus globulus* plantations in southeastern Australia. *Forest Ecology and Management, 259*(9), 1754–1760. https://doi.org/10.1016/j.foreco.2009.04.023.

Friberg, N. (1997). Benthic invertebrate communities in six Danish forest streams: Impact of forest type on structure and function. *Ecography, 20*(1), 19–28. https://doi.org/10.1111/j.1600-0587.1997.tb00343.x.

Futuyma, D. J., & Agrawal, A. A. (2009). Macroevolution and the biological diversity of plants and herbivores. *Proceedings of the National Academy of Sciences of the United States of America, 106,* 18054–18061. https://doi.org/10.1073/pnas.0904106106.

Gabrielle, B., Nguyen The, N., Maupu, P., & Vial, E. (2013). Life cycle assessment of eucalyptus short rotation coppices for bioenergy production in southern France. *GCB Bioenergy, 5*(1), 30–42. https://doi.org/10.1111/gcbb.12008.

Gomes, P. P., Ferreira, V., Tonin, A. M., Medeiros, A. O., & Júnior, J. F. G. (2018). Combined effects of dissolved nutrients and oxygen on plant litter decomposition and associated fungal communities. *Microbial Ecology, 75*(4), 854–862. https://doi.org/10.1007/s00248-017-1099-3.

Graça, M. A. S., Pozo, J., Canhoto, C., & Elosegi, A. (2002). Effects of *Eucalyptus* plantations on detritus, decomposers, and detritivores in streams. *The Scientific World Journal, 2,* 1173–1185. https://doi.org/10.1100/tsw.2002.193.

Griffith, M. B., & Perry, S. A. (1991). Leaf pack processing in two Appalachian mountain streams draining catchments with different management histories. *Hydrobiologia, 220*(3), 247–254. https://doi.org/10.1007/bf00006580.

Guevara, G., Godoy, R., Boeckx, P., Jara, C., & Oyarzún, C. (2015). Effects of riparian forest management on Chilean mountain in-stream characteristics. *Ecohydrology and Hydrobiology, 15*(3), 160–170. https://doi.org/10.1016/j.ecohyd.2015.07.003.

Hayson, K. A., & Murphy, S. (2003). *The status of invasiveness of forest tree species outside their natural habitat: A global review and discussion paper. Forest health and biosecurity working paper FBS/3E. Forestry Department. FAO.* Rome, Italy.

Hengeveld, G. M., Didion, M., van den Wyngaert, I., Clerkx, A. P. P. M. S., Schelhaas, M.-J., & Nabuurs, G.-J. (2012). A forest management map of European forests. *Ecology and Society, 17*(4). https://doi.org/10.5751/es-05149-170453.

Hisabae, M., Sone, S., & Inoue, M. (2011). Breakdown and macroinvertebrate colonization of needle and leaf litter in conifer plantation streams in Shikoku, southwestern Japan. *Journal of Forest Research, 16*(2), 108–115. https://doi.org/10.1007/s10310-010-0210-0 (*).

Hladyz, S., Åbjörnsson, K., Chauvet, E., Dobson, M., Elosegi, A., Ferreira, V., Fleituch, T., Gessner, M. O., Giller, P. S., Gulis, V., Hutton, S. A., Lacoursière, J. O., Lamothe, S., Lecerf, A. Malmqvis, B., McKie, B. G., Nistorescu, M., Preda, E., & Woodward, G. (2011a). Stream ecosystem functioning in an agricultural landscape. The importance of terrestrial-aquatic linkages. *Advances in Ecological Research, 44,* 211–276. https://doi.org/10.1016/b978-0-12-374794-5.00004-3 (*).

Hladyz, S., Åbjörnsson, K., Giller, P. S., & Woodward, G. (2011b). Impacts of an aggressive riparian invader on community structure and ecosystem functioning in stream food webs. *Journal of Applied Ecology, 48*(2), 443–452. https://doi.org/10.1111/j.1365-2664.2010.01924.x.

Hopmans, P., & Bren, L. J. (2007). Long-term changes in water quality and solute exports in headwater streams of intensively managed radiata pine and natural eucalypt forest catchments in south-eastern Australia. *Forest Ecology and Management, 253,* 244–261. https://doi.org/10.1016/j.foreco.2007.07.027.

Iglesias-Trabado, G., Carballeira-Tenreiro, R., & Folgueiro-Lozano, J. (2009, October 19). *Eucalyptus universalis*: Global cultivated eucalyptus forests map. Version 1.2. In: *Forestry Consulting's EUCALYPTOLOGICS: Information resources on Eucalyptus cultivation worlwide.* http://gif-forestry.com.

Inoue, M., Shinotou, S., Maruo, Y., & Miyake, Y. (2012). Input, retention, and invertebrate colonization of allochthonous litter in streams bordered by deciduous broadleaved forest, a conifer plantation, and a clear-cut site in southwestern Japan. *Limnology, 13*(2), 207–219. https://doi.org/10.1007/s10201-011-0369-x.

Jackson, R. B., Jobbágy, E. G., Avissar, R., Roy, S. B., Barrett, D. J., Cook, C. W., Farley, K. A., le Maitre, D. C., McCarl, B. A., & B. C. Murray (2005). Trading water for carbon with biological carbon sequestration. *Science, 310*(5756), 1944–1947. https://doi.org/10.1126/science.111928.

Jobbágy, E. G., Acosta, A. M., & Nosetto, M. D. (2013). Rendimiento hídrico en cuencas primarias bajo pastizales y plantaciones de pino de las sierras de Córdoba (Argentina). *Ecologia Austral, 23*(2), 87–96.

Jones, K. B., Neale, A. C., Nash, M. S., Van Remortel, R. D., Wickham, J. D., Riitters, K. H., & O'neill, R. V. (2001). Predicting nutrient and sediment loadings to streams from landscape metrics: A multiple watershed study from the United States Mid-Atlantic Region. *Landscape Ecology, 16*(4), 301–312. https://doi.org/10.1023/a:1011175013278.

Kennedy, K. T. M., & El-Sabaawi, R. W. (2017). A global meta-analysis of exotic versus native leaf decay in stream ecosystems. *Freshwater Biology, 62*(6), 977–989. https://doi.org/10.1111/fwb.12918.

Komatsu, H., Kume, T., & Otsuki, K. (2011). Increasing annual runoff-broadleaf or coniferous forests? *Hydrological Processes, 25*(2), 302–318. https://doi.org/10.1002/hyp.7898.

Komatsu, H., Tanaka, N., & Kume, T. (2007). Do coniferous forests evaporate more water than broad-leaved forests in Japan? *Journal of Hydrology, 336*(3–4), 361–375. https://doi.org/10.1016/j.jhydrol.2007.01.009.

Kominoski, J. S., Marczak, L. B., & Richardson, J. S. (2011). Riparian forest composition affects stream litter decomposition despite similar microbial and invertebrate communities. *Ecology, 92*(1), 151–159. https://doi.org/10.1890/10-0028.1.

Kreutzweiser, D. P., & Capell, S. S. (2001). Fine sediment deposition in streams after selective forest harvesting without riparian buffers. *Canadian Journal of Forest Research, 32*(6), 1108. https://doi.org/10.1139/x02-086.

Kreutzweiser, D. P., Good, K. K. P., Capell, S. S. S. S., & Holmes, S. B. S. (2008). Leaf-litter decomposition and macroinvertebrate communities in boreal forest streams linked to upland logging disturbance. *Journal of the North American Benthological Society, 27*(1), 1–15. https://doi.org/10.1899/07-034r.1.

Kuehn, K. A., Francoeur, S. N., Findlay, R. H., & Neely, R. K. (2014). Priming in the microbial landscape: Periphytic algal stimulation of litter-associated microbial decomposers. *Ecology, 95*(3), 749–762. https://doi.org/10.1890/13-0430.1.

Laćan, I., Rers, V. H., & McBride, J. R. (2010). Similar breakdown rates and benthic macroinvertebrate assemblages on native and *Eucalyptus globulus* leaf litter in Californian streams. *Freshwater Biology, 55*(4), 739–752. https://doi.org/10.1111/j.1365-2427.2009.02312.x (*).

Lara, A., Little, C., Urrutia, R., McPhee, J., Álvarez-Garretón, C., Oyarzún, C., Soto, D., Donoso, P., Nahuelhual, L., Pino, M., & Arismendi, I. (2009). Assessment of ecosystem services as an opportunity for the conservation and management of native forests in Chile. *Forest Ecology and Management, 258*(4), 415–424. https://doi.org/10.1016/j.foreco.2009.01.004.

Larrañaga, A., Basaguren, A., Elosegi, A., & Pozo, J. (2009a). Impacts of *Eucalyptus globulus* plantations on Atlantic streams: Changes in invertebrate density and shredder traits. *Fundamental and Applied Limnology/Archiv Für Hydrobiologie, 175*(2), 151–160. https://doi.org/10.1127/1863-9135/2009/0175-0151.

Larrañaga, A., Basaguren, A., & Pozo, J. (2009b). Impacts of *Eucalyptus globulus* plantations on physiology and population densities of invertebrates inhabiting Iberian atlantic streams. *International Review of Hydrobiology, 94*(4), 497–511. https://doi.org/10.1002/iroh.200811156.

Larrañaga, A., Basaguren, A., & Pozo, J. (2014a). Resource quality controls detritivore consumption, growth, survival and body condition recovery of reproducing females. *Marine & Freshwater Research, 65*(10), 910–917. https://doi.org/10.1071/mf13165.

Larrañaga, A., Larrañaga, S., Basaguren, A., Elosegi, A., & Pozo, J. (2006). Assessing impact of eucalyptus plantations on benthic macroinvertebrate communities by a litter exclusion experiment. *Annals of Limnology-International Journal of Limnology, 42*(1), 1–8. https://doi.org/10.1051/limn/2006002.

Larrañaga, S., Larrañaga, A., Basaguren, A., Elosegi, A., & Pozo, J. (2014b). Effects of exotic eucalypt plantations on organic matter processing in Iberian streams. *International Review of Hydrobiology, 99*(5), 363–372. https://doi.org/10.1002/iroh.201301665 (*).

Lecerf, A., & Chauvet, E. (2008). Diversity and functions of leaf-decaying fungi in human-altered streams. *Freshwater Biology, 53*(8), 1658–1672. https://doi.org/10.1111/j.1365-2427.2008.01986.x.

Lecerf, A., Dobson, M., Dang, C. K., & Chauvet, E. (2005). Riparian plant species loss alters trophic dynamics in detritus-based stream ecosystems. *Oecologia, 146*(3), 432–442. https://doi.org/10.1007/s00442-005-0212-3.

Lecerf, A., & Richardson, J. S. (2010). Litter decomposition can detect effects of high and moderate levels of forest disturbance on stream condition. *Forest Ecology and Management, 259*(12), 2433–2443. https://doi.org/10.1016/j.foreco.2010.03.022.

Lewis, S. E., Piatt, J. J., & Lewis, A. S. L. (2017). Amphipods (*Gammarus pseudolimnaeus*) prefer but are harmed by a diet of non-native leaves. *Freshwater Science, 36*(4), 739–749. https://doi.org/10.1086/694855.

López, E. S., Pardo, I., & Felpeto, N. (2001). Seasonal differences in green leaf breakdown and nutrient content of deciduous and evergreen tree species and grass in a granitic headwater stream. *Hydrobiologia, 464*(1–3), 51–61. https://doi.org/10.1023/a:1013903500888 (*).

Márquez, J. A., Cibils, L., Principe, R. E., & Albariño, R. J. (2015). Stream macroinvertebrate communities change with grassland afforestation in central Argentina. *Limnologica, 53,* 17–25. https://doi.org/10.1016/j.limno.2015.05.002.

Márquez, J. A., Principe, R. E., Cibils Martina, L., & Albariño, R. J. (2017). Pine needle litter acts as habitat but not as food source for stream invertebrates. *International Review of Hydrobiology, 102*(1–2), 29–37. https://doi.org/10.1002/iroh.201601856.

Martin, C. W., Hornbeck, J. W., Likens, G. E., & Buso, D. C. (2000). Impacts of intensive harvesting on hydrology and nutrient dynamics of northern hardwood forests. *Canadian Journal of Fisheries and Aquatic Sciences, 57*(S2), 19–29. https://doi.org/10.1139/cjfas-57-s2-19.

Martínez, A., Larrañaga, A., Miguélez, A., Yvon-Durocher, G., & Pozo, J. (2016). Land use change affects macroinvertebrate community size spectrum in streams: The case of Pinus radiata plantations. *Freshwater Biology, 61*(1). https://doi.org/10.1111/fwb.12680.

Martínez, A., Larrañaga, A., Pérez, J., Descals, E., Basaguren, A., & Pozo, J. (2013). Effects of pine plantations on structural and functional attributes of forested streams. *Forest Ecology and Management, 310,* 147–155. https://doi.org/10.1016/j.foreco.2013.08.024 (*).

Mátyás, C., & Sun, G. (2014). Forests in a water limited world under climate change. *Environmental Research Letters, 9*(8), https://doi.org/10.1088/1748-9326/9/8/085001.

McBroom, M. W., Beasley, R. S., Chang, M., & Ice, G. G. (2008). Storm runoff and sediment losses from forest clearcutting and stand re-establishment with best management practices in East Texas, USA. *Hydrological Processes, 22*(10), 1509–1522. https://doi.org/10.1002/hyp.6703.

McKie, B. G., & Malmqvist, B. (2009). Assessing ecosystem functioning in streams affected by forest management: Increased leaf decomposition occurs without changes to the composition of benthic assemblages. *Freshwater Biology, 54*(10), 2086–2100. https://doi.org/10.1111/j.1365-2427.2008.02150.x.

Menéndez, M., Descals, E., Riera, T., & Moya, O. (2013). Do non-native *Platanus hybrida* riparian plantations affect leaf litter decomposition in streams? *Hydrobiologia, 716*(1), 5–20. https://doi.org/10.1007/s10750-013-1539-0.

Miserendino, M. L., & Masi, C. I. (2010). The effects of land use on environmental features and functional organization of macroinvertebrate communities in Patagonian low order streams. *Ecological Indicators, 10*(2), 311–319. https://doi.org/10.1016/j.ecolind.2009.06.008.

Molinero, J., & Pozo, J. (2003). Balances de hojarasca en dos arroyos forestados: Impacto de las plantaciones de eucalipto en el funcionamiento ecológico de un sistema lótico. *Limnetica, 22*(3–4), 63–70.

Molinero, J., & Pozo, J. (2004). Impact of a eucalyptus (*Eucalyptus globulus* Labill.) plantation on the nutrient content and dynamics of coarse particulate organic matter (CPOM) in a small stream. *Hydrobiologia, 528*(1–3), 143–165. https://doi.org/10.1007/s10750-004-2338-4.

Molinero, J., & Pozo, J. (2006). Organic matter, nitrogen and phosphorus fluxes associated with leaf litter in two small streams with different riparian vegetation: A budget approach. *Archiv fur Hydrobiologie, 166*(3), 363–385. https://doi.org/10.1127/0003-9136/2006/0166-0363.

Molinero, J., Pozo, J., & González, E. (1996). Litter breakdown in streams of the Agüera catchment: Influence of dissolved nutrients and land use. *Freshwater Biology, 36*(3), 745–756. https://doi.org/10.1046/j.1365-2427.1996.00125.x (*).

Monroy, S., Martínez, A., López-Rojo, N., Pérez-Calpe, A. V., Basaguren, A., & Pozo, J. (2017). Structural and functional recovery of macroinvertebrate communities and leaf litter decomposition after a marked drought: Does vegetation type matter? *Science of the Total Environment, 599–600,* 1241–1250. https://doi.org/10.1016/j.scitotenv.2017.05.093.

Morris, T. L., Esler, K. J., Barger, N. N., Jacobs, S. M., & Cramer, M. D. (2011). Ecophysiological traits associated with the competitive ability of invasive Australian acacias. *Diversity and Distributions, 17*(5), 898–910. https://doi.org/10.1111/j.1472-4642.2011.00802.x.

Ormerod, S. J., Rundle, S. D., Lloyd, E. C., & Douglas, A. A. (1993). The influence of riparian management on the habitat structure and macroinvertebrate communities of upland streams draining plantation forests. *The Journal of Applied Ecology, 30*(1), 13. https://doi.org/10.2307/2404266.

Payn, T., Carnus, J.-M., Freer-Smith, P., Kimberley, M., Kollert, W., Liu, S., Orazio, C., Rodriguez, L., Silva, L. N., & Wingfield, M. J. (2015). Changes in planted forests and future global implications. *Forest Ecology and Management, 352,* 57–67. https://doi.org/10.1016/j.foreco.2015.06.021.

Pozo, J. (1993). Leaf litter processing of alder and eucalyptus in the Agüera stream system (northern Spain). I: Chemical changes. *Archiv Für Hydrobiologie, 127*(3), 299–317. (*).

Pozo, J., Basaguren, A., Elósegui, A., Molinero, J., Fabre, E., & Chauvet, E. (1998). Afforestation with *Eucalyptus globulus* and leaf litter decomposition in streams of northern Spain. *Hydrobiologia, 373*(374), 101–109. https://doi.org/10.1007/bf00017570 (*).

Pozo, J., González, E., Díez, J. R., Molinero, J., & Elósegui, A. (1997). Inputs of particulate organic matter to streams with different riparian vegetation. *Journal of the North American Benthological Society, 16*(3), 602–611. https://doi.org/10.2307/1468147.

Principe, R. E., Márquez, J. A., Martina, L. C., Jobbágy, E. G., & Albariño, R. J. (2015). Pine afforestation changes more strongly community structure than ecosystem functioning in grassland mountain streams. *Ecological Indicators, 57,* 366–375. https://doi.org/10.1016/j.ecolind.2015.04.033.

Quinn, J. M., Boothroyd, I. K. G., & Smith, B. J. (2004). Riparian buffers mitigate effects of pine plantation logging on New Zealand streams 2. Invertebrate communities. *Forest Ecology and Management, 191,* 129–146. https://doi.org/10.1016/j.foreco.2003.11.013.

Richardson, D. M., & Rejmánek, M. (2011). Trees and shrubs as invasive alien species—A global review. *Diversity and Distributions, 17*(5), 788–809. https://doi.org/10.1111/j.1472-4642.2011.00782.x.

Riipinen, M. P., Davy-Bowker, J., & Dobson, M. (2009). Comparison of structural and functional stream assessment methods to detect changes in riparian vegetation and water pH. *Freshwater Biology, 54*(10), 2127–2138. https://doi.org/10.1111/j.1365-2427.2008.01964.x.

Riipinen, M. P., Fleituch, T., Hladyz, S., Woodward, G., Giller, P., & Dobson, M. (2010). Invertebrate community structure and ecosystem functioning in European conifer plantation streams. *Freshwater Biology, 55*(2), 346–359. https://doi.org/10.1111/j.1365-2427.2009.02278.x (*).

Salmón Rivera, B., Barrette, M., & Thiffault, N. (2016). Issues and perspectives on the use of exotic species in the sustainable management of Canadian forests. *Reforesta,* (1), 261–280. https://doi.org/10.21750/refor.1.13.13.

Santiago, J., Molinero, J., & Pozo, J. (2011). Impact of timber harvesting on litterfall inputs and benthic coarse particulate organic matter (CPOM) storage in a small stream draining a eucalyptus plantation. *Forest Ecology and Management, 262*(6), 1146–1156. https://doi.org/10.1016/j.foreco.2011.06.018.

Scott, D. F., & Lesch, W. (1997). Streamflow responses to afforestation with *Eucalyptus grandis* and *Pinus patula* and to felling in the Mokobulaan experimental catchments, South Africa. *Journal of Hydrology, 199*(3–4), 360–377. https://doi.org/10.1016/s0022-1694(96)03336-7.

Siegloch, A. E., Suriano, M., Spies, M., & Fonseca-Gessner, A. (2014). Effect of land use on mayfly assemblages structure in neotropical headwater streams. *Anais da Academia Brasileira de Ciências, 86*(4), 1735–1747. https://doi.org/10.1590/0001-3765201420130516.

Smolders, K. E., Rolls, R. J., Boulton, A. J., Webb, A. A., & Sheldon, F. (2018). Effects of selective forest harvesting best management practices on organic matter and invertebrate detritivores in streams draining subtropical eucalypt fores. *Ecological Engineering, 122*(May), 271–285. https://doi.org/10.1016/j.ecoleng.2018.08.010.

Stefánsdóttir, H. M. (2010). *Transport and decomposition of allochthonous litter in Icelandic headwater streams: Effects of forest cover. MS - Thesis* (p. 106) (*).

Stone, M. K., & Wallace, J. B. (1998). Long-term recovery of a mountain stream from clear-cut logging: The effects of forest succession on benthic invertebrate community structure. *Freshwater Biology, 39*(1), 151–169. https://doi.org/10.1046/j.1365-2427.1998.00272.x.

Swank, W. T., & Crossley Jr, D. A. (1988). Summary for policymakers. In Intergovernmental Panel on Climate Change (Ed.), *Climate Change 2013—The Physical Science Basis* (pp. 1–30). https://doi.org/10.1017/cbo9781107415324.004.

Swank, W. T., & Douglass, J. E. (1974). Streamflow greatly reduced by converting deciduous hardwood stands to pine. *Science, 185*(4154), 857–859. https://doi.org/10.1126/science.185.4154.857.

Thompson, R. M., & Townsend, C. R. (2004). Impacts of riparian afforestation on stream biofilms: An exotic forest-native grassland comparison. *New Zealand Journal of Marine and Freshwater Research, 38*(5), 895–902. https://doi.org/10.1080/00288330.2004.9517288.

Walden, L. L., Harper, R. J., Mendham, D. S., Henry, D. J., & Fontaine, J. B. (2015). Eucalyptus reforestation induces soil water repellency. *Soil Research, 53*(2), 168–177. https://doi.org/10.1071/sr13339.

Webster, J. R., Benfield, E. F., Golladay, S. W., & McTammany, M. E. (2014). Recovery of particulate organic matter dynamics in a stream draining a logged watershed. In Wayne T. Swank & J. R. Webster (Eds.), *Long-term response of a forest watershed ecosystem. Clearcutting in the southern Appalachian.* (p. 610). Oxford University Press, 156–176.

Whiles, M. R., & Wallace, J. B. (1997). Leaf litter decomposition and macroinvertebrate communities in headwater streams draining pine and hardwood catchments. *Hydrobiologia, 353,* 107–119. https://doi.org/10.1023/a:1003054827248 (*).

Woodall, W. R., & Wallace, J. B. (1972). The benthic fauna in four small southern Appalachian s treams. *American Midland Naturalist, 88*(2), 393. https://doi.org/10.2307/2424364.

Yeung, A. C. Y., Lecerf, A., & Richardson, J. S. (2017). Assessing the long-term ecological effects of riparian management practices on headwater streams in a coastal temperate rainforest. *Forest Ecology and Management, 384,* 100–109. https://doi.org/10.1016/j.foreco.2016.10.044.

Chapter 15
Salt Modulates Plant Litter Decomposition in Stream Ecosystems

Cristina Canhoto, Felix Bärlocher, Miguel Cañedo-Argüelles, Rosa Gómez, and Ana Lúcia Gonçalves

Abstract Human activities are altering ion composition and concentrations in streams around the globe. This has implications for aquatic biota and for key ecosystem functions such as leaf litter decomposition. In this chapter we present an overview of the causes of stream salinization and its impacts on litter decomposition through changes in the quality and characteristics of leaf litter inputs, the habitat condition and biological communities within the context of global change. Also, salinized streams are put into perspective and compared with naturally saline streams to better understand the potential of salinization to alter biological communities and ecosystem functioning. Finally, we discuss major knowledge gaps and new research lines. Understanding how increased salinity and altered ion concentration

---

The original version of this chapter was revised: The correct co-author name "Rosa Gómez" was updated. The correction to this chapter is available at https://doi.org/10.1007/978-3-030-72854-0_23

C. Canhoto (✉) · A. L. Gonçalves
Centre for Functional Ecology, Department of Life Sciences, University of Coimbra, Calçada Martim de Freitas, 3000-456, Coimbra, Portugal
e-mail: ccanhoto@ci.uc.pt

A. L. Gonçalves
e-mail: aga@ci.uc.pt

F. Bärlocher
Department of Biology, Mt. Allison University, Sackville, NB, Canada
e-mail: fbaerloc@mta.ca

M. Cañedo-Argüelles
Grup de Recerca FEHM (Freshwater Ecology, Hydrology and Management), Departament de Biologia Evolutiva, Ecologia I Ciència Ambientals, Universitat de Barcelona, Avda Diagonal 643, 08028 Barcelona, Spain
e-mail: mcanedo-arguelles@ub.edu

R. Gómez
Department of Ecology and Hydrology, Faculty of Biology, Regional Campus of International Excellence "Campus Mare Nostrum", University of Murcia, Campus de Espinardo, 30100 Murcia, Spain
e-mail: rgomez@um.es

C. M. Swan et al. (eds.), *The Ecology of Plant Litter Decomposition in Stream Ecosystems*, https://doi.org/10.1007/978-3-030-72854-0_15

modulate litter decomposition in streams is urgent, since we remain largely oblivious to the implications for ecosystem functioning and services that sustain human welfare.

15.1 Stream Salinization

Soil salinization (i.e., soil salt content increase) has been long recognized as an environmental problem causing ecosystem degradation and economic losses associated with crop damage (Ghassemi et al., 1995; Thomas & Middleton, 1993). Although the salt that builds up in soil is obviously being transported to surrounding rivers and streams via surface runoff and groundwater (Peck, 1978), the salinization of freshwater ecosystems was largely overlooked during the twentieth century (Cañedo-Argüelles, 2020). Increased salinity of rivers and streams first started to attract public and scientific attention in Australia, mainly due to investigations by Professor William D. Williams (e.g., Williams, 2001; Williams et al., 1991) and the salinization of the Murray-Darling basin, which drains one of the most significant agricultural areas in the country (Jolly et al., 2001). Different researchers pointed out that increased salt concentration (mainly sodium chloride) of the water posed a considerable risk to Australian freshwater biodiversity (Hart et al., 1990; James et al., 2003; Kefford et al., 2003; Nielsen et al., 2003) and led to significant economic (e.g., infrastructure corrosion) and human health (e.g., virus transmission) costs that forced managers and policy-makers to take action (Jardine et al., 2007; Wilson, 2004).

The severe impact of salinization on freshwater ecosystems in Australia was largely perceived as a local phenomenon restricted to arid landscapes (Cañedo-Argüelles, 2020). A notable exception was documented in Germany, where the Werra and Wipper Rivers were severely salinized by potash mining industries (Schulz & Cañedo-Argüelles, 2019). In the first decade of the twenty-first century other causes of salinization emerged, such as road salt application (Kaushal et al., 2005) or coal mining (Pond et al., 2008), and freshwater salinization started being recognized as a global environmental problem. All these activities are leading to an increase in the salt concentration of freshwater ecosystems, but each has a specific ionic signature. For example, sodium and chloride are associated with agriculture and farming (Rengasamy, 2006), whereas sulfate is associated with coal mining (Cormier et al., 2013; Pond et al., 2008). Nowadays, the increase in salinity associated with human activities has been documented for freshwater ecosystems around the world (Cañedo-Argüelles et al., 2016, 2019) such as lakes (Dugan et al., 2017), rivers (Cañedo-Argüelles et al., 2013) and wetlands (Herbert et al., 2015).

Recent calculations show that around one third of the streams in the USA (Kaushal et al., 2018) and Spain (Estévez et al., 2019) are salinized. This is close to the 25% that Williams (2001) estimated for the surface waters of the World. In the future, changes in land use (e.g., increased agriculture and resource extraction) coupled with climate change will very likely exacerbate freshwater salinization. For example, Le et al. (2019) predicted 10–15% increases in mean electrical conductivity of German rivers and streams for the period between 2070 and 2100. For the USA, Olson (2019) predicted an increase in median conductivity of streams and rivers from 0.319 ms/cm to 0.524 ms/cm by the end of the century. Arid areas are probably at greatest risk of salinization due to their geological characteristics and the scarcity of water (Estévez et al., 2019; Kaushal et al., 2018; Olson, 2019). Among them, regions with naturally saline groundwaters are most prone to salinization. In these areas, the replacement of deep-rooted vegetation by crops and pastures leads to rising water tables that can cause drastic salinization of soils and surface waters (Rengasamy, 2006; Williams, 2001). Although this has been mainly shown for Southwestern Australia, it is very likely occurring in many regions of the World (e.g., South Asia) without being reported (Cañedo-Argüelles, 2020).

Despite the general recognition that freshwaters are among the ecosystems most threatened by human activities (Brondizio et al., 2019), and that freshwater salinization will expand and intensify in the near future, the emergence of salinization as a major freshwater problem has been overshadowed by other environmental stressors (Cañedo-Argüelles, 2020). The natural occurrence of salts in freshwaters (e.g., due to rock weathering) may have contributed to this lack of attention by both scientists and managers, particularly in non-arid areas (Gorostiza & Sauri, 2019). However, pressed by global change contexts (Jones & van Vliet, 2018; Velasco et al., 2019), the urgency of establishing guidelines (e.g., setting ion concentration standards) is now on the agenda to protect these systems and the goods and services that they provide (Schuler et al., 2019).

15.2 Stream Ecosystems Are Intimately Linked to Their Surroundings

The importance of mutual exchanges between running waters and their terrestrial surroundings has been recognized since the early decades of the twentieth century, and has become the focus of considerable research effort since the late 1960s (Hynes, 1975). Many streams are basically heterotrophic: they derive most of their energy from the riparian vegetation in the form of leaves, needles, twigs and branches (Allan & Castillo, 2007; Giller & Malmquist, 1999). This detritus, especially in the form of deciduous leaves, is a major source of food for many types of invertebrates (Allan & Castillo, 2007; Cummins & Klug, 1979). It is also a substrate for Archaea, Bacteria and Fungi (Das et al., 2007; Manerkar et al., 2008). Among the latter, aquatic hyphomycetes predominate, but other groups, such as zoosporic fungi and yeasts, occur as well (Bärlocher et al., 2012; Duarte et al., 2013; Seena et al., 2019). Kaushik and Hynes (1968, 1971) published two landmark papers, where they demonstrated the greater involvement of fungi than bacteria in leaf decomposition and observed that detritus-feeding invertebrates (shredders) prefer to consume leaves colonized by fungi. This preference for fungally conditioned leaves has since been confirmed many times (though there are exceptions), and the topic has been reviewed recently by Bärlocher and Sridhar (2014) and Marks (2019). By combining fungal production rates and standing crops of leaf litter, annual fungal production in streams has been estimated to range between 16 and 193 g/m^2 (Gulis et al., 2006,

2019), which is equal to or exceeds bacterial or invertebrate production in comparable streams. These numbers have stimulated a great deal of interest in both fungi and invertebrates, and their individual or joint impact on litter decomposition and food webs, and how these might be impacted by anthropogenic stressors (Bärlocher, 2016; Gessner et al., 2007; Krauss et al., 2011).

15.3 Effects of Stream Salinization on Litter Decomposition

Much of the stream water chemistry is profoundly influenced by the surrounding geology; even during heavy storms considerably less than half the water volume may enter streams via overland flow (Allan & Castillo, 2007). This implies that most of the water entering streams has been in contact with the soil for a considerable length of time. Therefore, most of the salt accumulated in soils (e.g., due to agricultural activities) will reach streams and rivers via surface, subsurface and groundwater flow (Peck, 1978), potentially affecting litter decomposition.

Information on the ecological impacts of salinization on leaf litter decomposition is still fragmentary, particularly considering the predicted geographical expansion of the problem (Kaushal et al., 2018). However, our knowledge has improved in the last decade (Cañedo-Arguelles et al., 2019; Hintz & Relyea, 2019) and indicates that, generally, salinization leads to a decrease in leaf decay rates (Berger et al., 2019; Cañedo-Argüelles et al., 2014, 2019; Canhoto et al., 2017; Gonçalves, Carvalho, et al., 2019; Sauer et al., 2016; Schäfer et al., 2012) or, less commonly, has no effect (Casas et al., 2011; Gonçalves, Simões, et al., 2019; Sauer et al., 2016; Vorste et al., 2019). Leaf degradation impairment seems to result initially from osmotic imbalances in macroinvertebrates (see below), which promote a decrease in total abundance and a taxon-specific decrease in diversity (Kefford et al., 2003). The lethal and sublethal effects of salinization on invertebrates result in microbially dominated decomposition, promoted by a less diverse and highly salt tolerant fungal community (Fig. 15.1).

Fig. 15.1 Leaf decomposition in a gradient of salt-contamination. Increasing concentrations first impair taxonomic and/or functional diversity of shredders through waterborne toxicity and reduced detritus quality. At high salt levels, decomposition is maintained by a more tolerant, but less diverse and abundant, fungal community with much reduced or inexistent sporulation

15.3.1 *Microbial-Mediated Decomposition*

Studies of salinization effects on leaf associated microbial communities (i.e., fungi, bacteria) are scarce and sometimes contradictory. Nonetheless, while little information exists on bacteria, a global pattern has emerged from the published studies: aquatic hyphomycetes are the main drivers of litter decomposition in salt-contaminated streams.

Aquatic Hyphomycetes

Aquatic fungi are highly tolerant of salt. The negative effect of salinity on fungal growth is species-specific, and is usually observed at concentrations $\geq$ 4 g/L NaCl (Canhoto et al., 2017). EC50 (effective concentration resulting in 50% inhibition in growth rate) values of 9 isolates of aquatic hyphomycetes were $\geq$ 7.8 g/L with more than half the species showing values above 16 g/L (Canhoto et al., 2017). But the mechanisms that allow the aquatic hyphomycetes mycelium to grow and cope with high salinity remain hidden in a black box. Several authors indicate the accumulation of osmolytes and modifications of the cell wall and plasma membrane that may resemble mechanisms for coping with desiccation (Canhoto et al., 2017; Gonçalves, Carvalho, et al., 2019; Gonçalves, Simões, et al., 2019; Hooley et al., 2003; Overy et al., 2017; Tamie, 2016). These compounds may help to ensure mycelial structural and functional integrity at high salinity (Araújo et al., 2020; Canhoto et al., 2017; Welsh, 2000). Furthermore, based on evidence from terrestrial systems, fungi may sequester sodium at concentrations of up to 1000 times higher than plants (Kaspari et al., 2009). Such accumulation by the fungal mycelium, exposed to Na-rich salt environments, will help maintaining the cell's turgor, enhancing leaf invasion while increasing the attractiveness of the detritus to invertebrates (Kaspari, 2020; Scharnagl et al., 2017). If confirmed in aquatic hyphomycetes, a differential Na-sequestration or release may explain their species-specific high tolerance to wide salinity ranges.

Leaf associated fungal biomass tends to decrease with salt-contamination (Cañedo-Argüelles et al., 2014; Gonçalves, Carvalho, et al., 2019; Gonçalves, Simões, et al., 2019; Martínez, Barros, et al., 2020; Schäfer et al., 2012). Studies on the effects of NaCl (6 g/L) on oak leaf decomposition indicate biomass reductions by up to 40% (Gonçalves, Carvalho, et al., 2019; Gonçalves, Simões, et al., 2019; Martínez, Barros, et al., 2020). However, mismatches between fungal biomass and mass loss can occur and result in alternative responses, such as an increase in fungal biomass facilitated by higher Na-availability at lower salt concentrations (< 2 g/L NaCl) (Cañedo-Argüelles et al., 2014; Canhoto et al., 2017; Sauer et al., 2016). Fungal biomass responses, resultant mass loss, and connections between the two, may result from (a) species-specific differences in ergosterol (proxy of fungal biomass) concentrations in cell membranes (Charcosset & Chauvet, 2001), (b) changes in ergosterol concentrations in the fungal membrane induced by water salinization (Gonçalves, Carvalho, et al., 2019), (c) distinct degradation efficiency of more tolerant species (Canhoto et al., 2017; Gonçalves, Carvalho, et al., 2019; Gonçalves, Simões, et al., 2019), (d) higher investment in biomass accrual by tolerant, but less effective species, and (e) changed intra- and interspecific interactions within fungal (Gonçalves, Carvalho, et al., 2019) and, eventually, fungal-bacteria communities (Hintz & Relyea, 2019; Vorste et al., 2019).

Fungal respiration decreases along the salinization gradient, although correlations with other fungal descriptive variables are often inconsistent (Canhoto et al., 2017; Connolly et al., 2014; Gonçalves, Simões, et al., 2019; Martínez, Barros, et al., 2020; Swan, 2007). Apparent discrepancies may be based on species-specific metabolic requirements and energetic strategies to guarantee cell integrity, growth, sporulation and organic matter degradation. To date, little information exists on the modification of extracellular enzyme activities by salts (Roache et al., 2006; Servais et al., 2019).

Fungal sporulation is inhibited or suppressed by salt-addition. *Trade-offs* between maintaining cellular integrity and vegetative growth *vs.* conidial production seem to be a common response of aquatic hyphomycetes to salt-stress (Byrne & Jones, 1975; Canhoto et al., 2017; Gonçalves, Simões, et al., 2019; Sridhar & Kaveriappa, 1988). Nonetheless, divergent strategies can be found in strains collected from historically salt-contaminated streams, where the investment in reproduction seems to persist despite high salt concentrations (Gonçalves, Carvalho, et al., 2019). Studies, mainly based on traditional conidia identification, indicate that salinization triggers modifications of fungal community structure (Byrne & Jones, 1975; Cañedo-Argüelles et al., 2014; Canhoto et al., 2017; Gonçalves, Carvalho, et al., 2019; Gonçalves, Simões, et al., 2019; Sauer et al., 2016; Schäfer et al., 2012; Sridhar & Kaveriappa, 1988). Generally, no diversity effect has been observed with increasing salt addition of up to 4 g/L, which points to pervasive functional redundancy among fungal species (Gonçalves, Carvalho, et al., 2019; Pascoal & Cássio, 2008; Pascoal et al., 2005). Nonetheless, a significant reduction of decomposing efficiency tends to be apparent above this concentration (e.g., Canhoto et al., 2017), which may correspond to a reduction to 1–3 sporulating species (Canhoto et al., 2017; Gonçalves, Carvalho, et al., 2019; Gonçalves, Simões, et al., 2019). In extreme cases, the decrease in fungal species richness results in communities with single, highly halotolerant species (e.g., *Flagellospora curta* and *Heliscus lugdunensis*; Canhoto et al., 2017; Gonçalves, Carvalho, et al., 2019). Whether this decrease corresponds to actual species losses remains to be confirmed by molecular techniques—mycelial outgrowth has been suggested as compensating for reduced conidium production (Canhoto et al., 2017).

Bacteria

Despite the pivotal role of fungi on litter decomposition, we cannot rule out the influence of bacteria, on their own and through interactions with fungi, on litter decomposition dynamics. Available information, mostly gathered in lakes (water, sediments) and stream biofilms (Martínez, Gonçalves, et al., 2020; van Gray et al., 2020), indicates that common freshwater bacteria may cope with low salt concentrations (Bordalo, 1993; Sleator & Hill, 2002), while showing high mortality at high salt concentrations (Painchaud et al., 1995; Zhang et al., 2014b). To our knowledge, a single study evaluated bacterial, fungal and macroinvertebrate taxonomic richness associated with leaf litter along a coal-mine induced salinity gradient (25 μS/cm – 1383 μS/cm). Results indicated a modulating effect of salinity on the composition of all leaf-associated communities, which, however, did not translate into changes of decomposition rates (Vorste et al., 2019). The decrease in bacterial richness suggests community shifts from abundant low salt tolerant taxa to fewer more halotolerant and, eventually, halophilic taxa (Vorste et al., 2019). Whether this is true in short-term scenarios of salinization (e.g., pulses) needs to be assessed. In any case, and even considering potential functional stability of the bacterial communities at distinct salinities (Berga et al., 2017; They et al., 2013; Zhang et al., 2014a, 2014b), a thorough investigation is needed to clarify the relative importance of the two microbial groups and their interactions during colonization and degradation of organic matter in salt-contaminated streams.

15.3.2 Invertebrate-Mediated Decomposition

Aquatic macroinvertebrate shredders include mainly insects belonging to the orders Ephemeroptera, Plecoptera and Trichoptera (EPT), Diptera, and Crustaceans (Tachet et al., 2010). These groups present a gradient of tolerance to salinization (Cañedo-Argüelles et al., 2013; Clements & Kotalik, 2016), likely related with the ionic stress and metabolic requirements for osmoregulation (e.g., Kefford et al., 2003; Scheibener et al., 2016). Most of the taxa belonging to the EPT group tend to be extirpated from salinized streams (Kefford et al., 2011), although responses seem to be context-specific (Clements & Kotalik, 2016), and important differences between species (Bonada et al., 2004; Pond, 2010), life-stages (Kefford et al., 2007), and even between populations (Sala et al., 2016), have been reported.

The sensitivity of invertebrates, shredders included, to salinization has been demonstrated in several studies through lethal and sublethal effects—e.g., reduced feeding rates, delayed growth, increased drift (Clements & Kotalik, 2016; Hassell et al., 2006; Kefford et al., 2003). It is still not clear how the changes in the abundance and composition of shredder assemblages affect organic matter processing (Berger et al., 2019), but available evidence suggests that litter decomposition rates by invertebrates are reduced by salinization. For example, Cañedo-Argüelles and co-authors (2014) showed lower decomposition rates of *Populus nigra* leaves subjected to repeated salt pulses (maximum conductivity = 15 mS/cm) related with decreasing EPT abundances in stream mesocosms. Concordantly, *Eucalyptus camaldulensis* decomposition in coarse-mesh bags was negatively correlated with electrical conductivity in streams in southeast Australia (Schäfer et al., 2012). Whether changes in the detritivore community structure towards salt tolerant taxa can be compensated by functional redundancy throughout the salinity gradients is conceivable, but currently unknown. For instance, more tolerant taxa such as crustacea (Kefford et al., 2003), which can be very active detritivores (Bergfur et al., 2007; Piscart et al., 2011), could compensate for the disappearance of more sensitive leaf-consumers (e.g., EPT).

Laboratory studies on the detritus-shredders relationships are limited, but globally support the important modulator effect of salt concentration and ionic composition on the processing efficacy of leaf-consumers (Martínez, Barros, et al., 2020; Tyree et al., 2016; Zalizniak et al., 2006). For example, Tyree et al. (2016) reported that a very small increase of NaCl levels (from 3 to 7 mg/L Na) depressed leaf consumption by *Tipula abdominalis* (Diptera), while a higher concentration (140 mg/L Na) enhanced the decomposer activity of *Lirceus sp.* (Isopoda). However, in the study of Martínez, Barros, et al. (2020), 3 g/L of $CaCl_2$ and 6 g/L of $C_2H_3KO_2$ and NaCl were needed to slow down the feeding activity of *Schizopelex festiva* (Trichoptera). It is generally accepted that shredders preferentially consume and assimilate conditioned over unconditioned leaves and show preferences for certain leaf species, fungi, or combinations of both (Canhoto & Graça, 2008). Nuances promoted by salinization on such feeding behavior are unknown; we may only anticipate that a decrease in fungal diversity and biomass through a salinity gradient (Canhoto et al., 2017) may alter detritus palatability, affecting consumption rates and processing efficacies of the tolerant (less abundant and diverse) shredders. Alternatively, parallel to terrestrial systems, a potential stimulation of litter degradation at intermediate salinities, mediated by the aforementioned Na-rich fungal mycelium, could favor invertebrate fitness, also facilitating the digestion of tannin rich diets (Kaspari, 2020; Scharnagl et al., 2017). Whether any (or both) hypothesis are true, remains to be assessed.

Salinization of forested riparian soils (e.g., runoffs, flooding) may (a) decrease leaf litter quantity and diversity supplied to the stream by eliminating plants or reducing their viability, (b) accelerate senescence and riparian litterfall and/or (c) change plant salt uptake promoting species-specific alterations of chemical leaf-litter traits (Entrekin et al., 2019; Kaspari, 2020). Detritivore extirpation (Ladrera et al., 2017), litter-diversity effects on decomposition and potential mismatches between the timing of the inputs and invertebrate life cycle (Swan & Kominoski, 2012; Voelz & McArthur, 2000) may result from these alterations of riparian soils (Fig. 15.2). In addition, we can expect shifts in leaf quality—e.g., accumulation of Na in leaves from plants exposed to salinization (Entrekin et al., 2019; Parida & Das, 2005)—to directly affect leaf processing by the tolerant biota (Swan & Kominoski, 2012). A single study assessed the importance of leaf quality in salinized streams. It showed that the deleterious effects of water salinization on fungal-mediated litter mass-loss was not due to the quality of the tested riparian subsidies; differences were suggested to occur only in the presence of marked leaf trait dissimilarities (Almeida et al., 2020).

15.4 Factors Modulating Salinization Effects on Litter Decomposition

Litter decomposition in salt-contaminated environments is closely related to the kind of anthropogenic activity affecting the stream, which not only determines the ionic composition of the water (Estévez et al., 2019), but also induces distinct disturbance patterns in space (diffuse or point-source pollution; see also Kaushal et al., 2018) and time (chronic *vs.* pulsed *vs.* ramped; *sensu* Lake, 2000). Two single studies assessed the importance of salt type on litter decomposition in streams (Martínez, Barros, et al., 2020; Sauer et al., 2016). The intensity of the effect, observed in both

Fig. 15.2 Salinized stream (Torrent de Soldevilla, Sallent, Catalonia, Spain). Salt accumulation in the stream bed and riparian degradation are visible (Photo Ruben Ladrera)

decomposers and detritivores, seems to be largely attributable to their physiological tolerance of the water's ionic composition (e.g., Kunz et al., 2013; Martínez, Barros, et al., 2020).

Streams salinization is the primer responsible for triggering multiple changes in water chemical properties (e.g., pH, contaminant mobilization) in a phenomenon termed Freshwater Salinization Syndrome (FSS; Kaushal et al., 2018). In fact, several contaminants related with salinization (e.g., deicers, mining waste) are themselves mixtures of pollutants that may include heavy metals, phosphorus in many forms, nitrogen and cyanide (Fay & Shi, 2012; Kaushal et al., 2019) contributing to the panoply of alterations in water chemistry (Gorostiza & Sauri, 2017). Although no specific relationships have yet been established between water chemical composition and the rate of litter decomposition in salinized streams, this ecosystem-level process holds promise as a useful tool for assessing the functional integrity of multi-impacted saline streams (Chauvet et al., 2016). Its application would be independent of the composition (Kath et al., 2018; Szöcs et al., 2012) and potential interactions within the chemical mixtures (Côté et al., 2016; Kaushal et al., 2018; Schäfer & Piggot, 2018), while connecting the system's response to specific geochemical scenarios.

Scanty information exists on the consequences of short-term (*vs.* chronic) repeated salt additions on litter decomposition. Available information indicates that pulse exposures should generally be less deleterious than chronic exposures, provided that the exposure frequency is low enough to allow community recovery (Cañedo-Argüelles et al., 2014). However, these results should be interpreted with caution.

Two field tests suggest more severe impacts of pulsed (*vs.* press) inputs on macroinvertebrates most likely to affect litter processing (Findlay & Kelly, 2011; Hart et al., 1990).

In the present global environmental context, the most common abiotic factors interacting with stream salinization are, beside pollutants (e.g., pesticides), nutrients, extreme hydrological events (i.e., drought and floods), and increased temperature. Despite their ubiquitous prevalence, limited information exists on the impacts of the co-occurrence of these stressors and salinity on the stream biota and processes they drive (e.g., Verberk et al., 2020). A recent metanalysis by Velasco et al. (2019) indicates that negative, more commonly additive, effects on organismal (invertebrates included) performance may occur in streams, with salinity having an overriding influence. Whether, and how, such responses can propagate to stream processes such as leaf decomposition is still inadequately documented. A few studies have assessed the joint effects of salinity and pesticides (Schäfer et al., 2012; field study), nutrients (Hale & Goffman, 2006; Swan & DePalma, 2012 microcosms and field study, respectively) and drought regimes (Gonçalves, Simões, et al., 2019; microcosm study) on litter decomposition by invertebrates and/or decomposers. All studies point to a pronounced reduction of litter decomposition in the presence of both tested stressors regardless of presence of detritivores. Nonetheless, while additive responses were observed in the field, the effects of ionic increases were not always straightforward and additive in microcosm studies (Swan & DePalma, 2012). Antagonistic (Hale & Goffman, 2006) or synergistic (Gonçalves, Simões, et al., 2019) effects were found as well, particularly at high salt-concentrations. Salinization studies usually consider concomitant rather than sequential exposures; however, this is not always the case, which may affect outcomes (Pallarés et al., 2017). Existing information precludes predictions of the effects of salinization paired with other stressors. This reinforces the need for studies in distinct environmental contexts. The analysis of litter decomposition dynamics in naturally extreme environments (e.g., Gómez et al., 2016; Reice & Herbst, 1982; Steinke & Charles, 1986; see below) may provide a promising approach to predict possible trajectories of salinized streams' responses to multiple stressors.

15.5 Decomposition in Saline Streams

Naturally saline streams and rivers are geographically widespread, but are concentrated in arid and semiarid regions (Williams, 1996), like the Mediterranean area, where geology and climatic features are the main factors determining long-term stream water salinization (Millán et al., 2011). In the Southeast of Spain, one of the most arid area of the Iberian Peninsula, the presence of highly impermeable sedimentary marls, together with the high evaporation rates and low precipitation, result in water courses and soils with a high content of salts (likely calcium sulfate and sodium chloride; Gómez et al., 2005). The range of water salinity varies widely defining hypo (3–20 g/L), meso (> 20–100 g/L) and hypersaline (> 100 g/L) streams (Arribas

Fig. 15.3 Naturally saline stream (Rambla Salada, Fortuna, Murcia, Spain). Salt accumulation and halophytic plants are visible in the riparian area (Photo Rosa Gómez)

et al., 2009). In addition to the stress imposed by high salinities, aquatic organisms have to deal with acute daily and seasonal variation in salt and nutrient concentrations (Boulton & Suter, 1986; Gómez et al., 2017) as result of the highly variable surface discharge in these ecosystems. For example, in Chicamo river (Murcia, Spain), diurnal variation (averaged for the four seasons of the year) for water salinity ranged between 1 and 4 g/L (Vidal-Abarca et al., 2002). In addition, woody perennial shrubs prevail in the riparian areas of these Mediterranean streams (Fig. 15.3), which provides a dominant pool of recalcitrant substrate in the streambed. Under such extreme conditions, detritivores tend to be absent or rare (Gutiérrez-Cánovas et al., 2012).

Most research on the effects of salinity on litter decomposition has focused on freshwaters using saline (coastal and freshwater wetlands; Connolly et al., 2014; Hemminga et al., 1991; Mendelssohn et al., 1999; Quintino et al., 2009; Sangiorgio et al., 2007) or salt-contaminated (micro and mesocosms tests; Cañedo-Argüelles et al., 2013; Canhoto et al., 2017; Roache et al., 2006; van Meter et al., 2012) systems. Comparisons with naturally saline streams need to be made with caution as laboratory or semi-natural approaches give insights into short-term community responses to salinization, while saline stream organisms have been shaped by selection and/or evolution.

Information on organic matter decomposition in naturally saline streams is limited, to the best of our knowledge, to two studies (Gómez et al., 2016; Reice & Herbst, 1982). Results on biologically mediated degradation are in line with studies in streams subjected to salt contamination: salinity decreases both leaf litter (Reice & Herbst,

1982) and wood decomposition rates (Gómez et al., 2016) while presenting negative effects on fungal biomass. Such impairment seems to be mainly related with the energetic costs of haloadaptation by the fungal community (Oren, 1999, 2001; Sinha & Khare, 2014). Nonetheless, it seems that "saline biofilms", developed on woody surfaces, are able to maintain activity even at high salt concentrations. These findings strengthen the need to understand the response of other microbial groups, such as bacteria, in decomposition both in naturally saline and salinized environments.

Although salts in saline streams are non-toxic for both halophiles and halo-tolerant organisms, the effect of Na^+ and Cl^-, and additional divalent ions (e.g., SO_4^{2-}, Ca^{2+} and Mg^{2+}), may influence both the rate and patterns of microbial decomposition in a complex manner (Egglishaw, 1968; Roache et al., 2006; Weston et al., 2006). In fact, in these systems, the biotic degradation of the woody material seems to be supported (or at least facilitated) by an increase in the content of Ca, Mg, and especially of Na in organic matter, which may act as physical and chemical abrasive agents (Blanchette et al., 2002; Schmitz, 1924). Additionally, in saline streams affected by marked water level fluctuations, salt crystals can be formed in desiccated woody debris as a result of water evaporation (Johnson et al., 1992); the consequent mechanical cleavage of this detritus may trigger microbial invasion and wood decay. Further studies are needed to accurately evaluate the importance of such abiotic (*vs.* biotic) processes on organic matter decomposition in these streams.

While experiencing the ongoing increases in temperature, naturally saline streams are also paradoxically affected by dilution stress as result of increased irrigation for agricultural production in their catchment areas (Gutiérrez-Cánovas et al., 2012; Millán et al., 2011). The effects of such perturbation on organic matter decomposition is again unknown, but must be considered for the preservation of the diversity and processes that characterize these peculiar ecosystems.

15.6 Future Directions and Perspectives

Understanding the impacts of salinization on streams functioning is complex because (a) there is an intimate relationship between these streams and their riparian area, frequently also salt-contaminated; (b) stream salinization and salinity encompass diverse origins / ionic compositions supplied to the stream with different temporal and spatial patterns; (c) salt-contamination frequently co-occurs with other stressors; (d) there are no commonly agreed-upon assessment tools. For example, common indices, based on invertebrates, do not clearly reflect salinization (Cañedo-Argüelles et al., 2017). Such wide gaps in our knowledge limit the ability to establish links between salinized ecosystems functioning and ecosystem services. This impairs our ability to take action against salinization and prevent the collapse of freshwater ecosystems, which can compromise human welfare (Cañedo-Argüelles et al., 2016).

Our lack of focus on the basal response of brown webs to salinization precludes a thorough understanding of the responses of the forested stream ecosystems to this stressor. There is a dearth of information on the relative impact of the riparian forest characteristics (inherent and salt-induced) on the aquatic biota's processing capacity and resilience. Also, the mechanistic responses associated with the trilogy leaf litter/decomposers/shredders are still largely unknown. It seems clear, however, that the role of fungi changes somewhat predictably through the salinization gradient: from linking terrestrial litter to the aquatic food chain (pristine streams) to Na accumulators/suppliers (moderately salinized streams) to dominant leaf processors (strongly salinized streams). Nevertheless, the relative importance of fungi and their interactions with bacteria and invertebrates, mechanisms of fungal salt tolerance and how they affect life history trade-offs, must be further scrutinized. Fungal trait-based frameworks (morphological, physiological and biochemical characteristics; Allison, 2012; Treseder & Lennon, 2015) may constitute a valuable new approach to gain insights into the functional capacities of this microbial group, in various spatial and temporal salt-gradients, patterns and multi-stressor scenarios.

Our current understanding on the links between the response of aquatic macroinvertebrate communities to salinization and litter decomposition is very limited. Information on the processing efficiency of leaf-consumers—under waterborne and/or food-induced constraints—and on sublethal endpoints (e.g., development, feeding, reproduction) need to be explored and coupled with information at molecular levels (e.g., use of biomarkers; Cañedo-Argüellles et al., 2016) to anticipate the impacts of salt on litter decomposition and ecosystem function. Finally, consideration of indirect salinization effects on decomposition is required. Competition and predatory interactions may be particularly relevant under warming scenarios as the increase temperature may amplify salt-induced changes of behavioural traits (Bray et al., 2019; Ingram & Burns, 2018; Jackson & Funck, 2019).

The biological and ecological effects of salinization on the detritus-based food webs in streams are the result of several ions that may act per se or in combination. An "ionic ecology" (*sensu* Kaspari, 2020) approach may attract interest to clarify the distinct effect of salts on stream processes, litter decomposition included; particular attention should be devoted to the role of Na on the response of the two main drivers of leaf decomposition—fungi and invertebrate shredders. This is especially relevant not only due to the primordial biological role of this ion on metabolic (temperature-dependent) processes (Kaspari, 2020; Orr & Buchwalter, 2020), but also because of the extensive use of NaCl as deicer agent and its lasting presence in the system (Findlay & Kelly, 2011). Nonetheless, new approaches using other ion mixtures—e.g., those usually found on road deicer alternatives (Fay & Shi, 2012), forest fertilizers—should be considered as part of the endeavour to establish ionic-specific consequences and thresholds for correct and integrated forest-stream management practices. Approaches require the inclusion of distinct contamination scenarios—concentrations, exposure patterns (chronic, pulse, ramp), timings (e.g., season)—in both multi-stressor concomitant and sequential contexts. Such comprehensive list of "needs" acquire a new dimension in urban streams lined with trees (Paul & Meyer, 2001) and non-perennial running waters, whose geographical range and abundance (now comprising over 50% of the global fluvial networks; Arce et al., 2019) are expected to increase due to global warming, human growth and intensified water use.

Intermittency patterns determine fluctuations in the natural (or exacerbated by anthropogenic activities) background salinity of the stream water that follows surface flow contractions/resumptions. Little is known of the relevance of such salinity gradients (usually coupled with altered water temperature, oxygen levels, pH and desiccation) on the decomposition dynamics through the wet-dry cycles. Gaining insights into the relative importance of salt on litter decomposition in forested intermittent streams requires studies in inundated, moist, and dry (potentially "salinized") streambeds, where terrestrial decomposers and detritivores may become involved. In this context, it would be interesting to investigate if community desiccation protective strategies may also confer salt-tolerance (and vice versa) (Pallarés et al., 2017). If true, this could result in higher system resilience in the face of a wider range of salt-concentrations induced by fiercer, more frequent, and abrupt droughts and/or an expansion of secondary salinization.

More inclusive local and large-scale surveys (see Olson, 2019) need to be considered to detect consistent patterns or to draw global conclusions on the effects of salinization on litter decomposition. Developing models linking litter decomposition dynamics and related processes to predicted temperature scenarios, nutrient status and spatial and temporal patterns of salt-contamination (e.g., based on the proximity of roads and urban areas), may help identifying susceptibilities of different systems and geographical areas to salinization, and to prioritize the implementation of protective practices.

Overall, salt pollution of streams is increasing worldwide as water scarcity reduces the dilution capacity of streams (van Vliete et al., 2017), agricultural activities and resource extraction are intensified and expanded (Estévez et al., 2019; Krausmann et al., 2018), and mountain areas become increasingly urbanized and crossed by roads (Kaushal et al., 2005). Thus, understanding how increased salinity and altered ion concentration modulate litter decomposition in streams is urgent; currently, we remain largely oblivious to the implications for ecosystem functioning and services that sustain human well-being.

References

Allan, J. D., & Castillo, M. M. (2007). *Stream ecology*. Dordrecht: Springer. https://doi.org/10.1007/978-1-4020-5583-6.

Allison, S. D. (2012). A trait-based approach for modelling microbial litter decomposition. *Ecology Letters, 15,* 1058–1070. https://doi.org/10.1111/j.1461-0248.201201807.x.

Almeida, E., Martínez, A., Gonçalves, A. L., & Canhoto, C. (2020). Combined effects of freshwater salinization and leaf traits on litter decomposition. *Hydrobiologia, 847,* 3427–3435. https://doi.org/10.1007/s10750-020-04348-1.

Araújo, C. A. S., Ferreira, P. C., Pupin, B., Dias, L. P., Avalos, J., Edwards, J., Hallsworth, J. E., & Rangel, D. E. N. (2020). Osmotolerance as a determinant of microbial ecology: A study of phylogenetically diverse fungi. *Fungal Biology, 124,* 273–288. https://doi.org/10.1016/j.funbio.2019.09.001.

Arce, M. I., Mendoza-Lera, C., Almagro, M., Catalán, N., Romaní, A. M., Martí, E., Marcé, R., Bernal, S., Foulquier, A., Mutz, M., Marcé, R., Zoppini, A., Gionchetta, G., Weigelhofer, G., del

Campo, R., Robinson, C. T., Gilmer, A., Rulik, M., Obrador, B., ... von Schiller, D. (2019). A conceptual framework for understanding the biogeochemistry of dry riverbeds through the lens of soil science. *Earth-Science Reviews, 188,* 441–453. https://doi.org/10.1016/j.earscirev.2018.12.001.

Arribas, P., Gutiérrez-Cánovas, C., Abellán, P., Sánchez-Fernández, D., Picazo, F., Velasco, J., & Millán, A. (2009). Tipificación de los ríos salinos ibéricos. *Revista Ecosistemas, 18*(3), 1–13.

Bärlocher, F. (2016). Aquatic hyphomycetes in a changing environment. *Fungal Ecology, 19,* 14–27. https://doi.org/10.1016/j.funeco.2015.05.005.

Bärlocher, F., & Sridhar, K. R. (2014). Association of animals and fungi in leaf decomposition. In E. B. Gareth Jones, K. D. Hyde, & K. L. Pang (Eds.), *Freshwater Fungi* (pp. 413–441). de Gruyter. https://doi.org/10.1515/9783110333480.v.

Bärlocher, F., Stewart, M., & Ryder, D. S. (2012). Processing of *Eucalyptus viminalis* leaves in Australian streams—Importance of aquatic hyphomycetes and zoosporic fungi. *Fundamental and Applied Limnology, 179,* 305–319. https://doi.org/10.1127/1863-9135/2012/0229.

Berga, M., Zha, Y., Székely, A. J., & Langenheder, S. (2017). Functional and compositional stability of bacterial metacommunities in response to salinity changes. *Frontiers in Microbiology, 8,* 948. https://doi.org/10.3389/fmicb.2017.00948.

Berger, E., Frör, O., & Schäfer, R. B. (2019). Salinity impacts on river ecosystem processes: A critical mini-review. *Philosophical Transactions of the Royal Society B, 374,* 20180010. https://doi.org/10.1098/rstb.2018.0010.

Bergfur, J., Johnson, R. K., Sandin, L., Goedkoop, W., & Nygren, K. (2007). Effects of nutrient enrichment on boreal streams: Invertebrates, fungi and leaf-litter breakdown. *Freshwater Biology, 52*(8), 1618–1633. https://doi.org/10.1111/j.1365-2427.2007.01770.x.

Blanchette, R. A., Held, B. W., & Farrell, R. L. (2002). Defibration of wood in the expedition huts of Antarctica: And unusual deterioration process occurring in the polar environment. *Polar Record, 38,* 313–322. https://doi.org/10.1017/S0032247400018003.

Bonada, N., Zamora-Munoz, C., Rieradevall, M., & Prat, N. (2004). Ecological profiles of caddisfly larvae in Mediterranean streams: Implications for bioassessment methods. *Environmental Pollution, 132*(3), 509–521. https://doi.org/10.1016/j.envpol.2004.05.006.

Bordalo, A. A. (1993). Effects of salinity on bacterioplankton: Field and microcosm experiments. *Journal of Applied Bacteriology, 75*(4), 393–398.

Boulton, A. J., & Suter, P. J. (1986). Ecology of temporary streams—An Australian perspective. In P. De Deckker & W. David Williams (Eds.), *Limnology in Australia* (pp. 313–327). Springer.

Bray, J. P., Reich, J., Nichols, S. J., Kon Kam King, G., Mac Nally, R., Thompson, R., O'Reilly-Nugent, A. & Kefford, B. J. (2019). Biological interactions mediate context and species-specific sensitivities to salinity. *Philosophical Transactions of the Royal Society B, 374,* 20180020. https://doi.org/10.1098/rstb.2018.0020.

Brondizio, E. S., Settele, J., Díaz, S., & Ngo, H. T. (2019). *Global assessment report on biodiversity and ecosystem services of the Intergovernmental Science-Policy Platform on Biodiversity and Ecosystem Services.* IPBES Secretariat.

Byrne, P. J., & Jones, E. B. (1975). Effect of salinity on the reproduction of terrestrial and marine fungi. *Transactions of the British Mycological Society, 65,* 185–200. https://doi.org/10.1016/s0007-1536(75)80002-7.

Cañedo-Argüelles, M. (2020). A review of recent advances and future challenges in freshwater salinization. *Limnetica, 39*(1), 185–211. https://doi.org/10.23818/limn.39.13.

Cañedo-Argüelles, M., Brucet, S., Carrasco, S., Flor-Arnau, N., Ordeix, M., Ponsá, S., & Coring, E. (2017). Effects of potash mining on river ecosystems: An experimental study. *Environmental Pollution, 224,* 759–770. https://doi.org/10.1016/j.envpol.2016.12.072.

Cañedo-Argüelles, M., Bundschuh, M., Gutiérrez-Cánovas, C., Kefford, B. J., Prat, N., Trobajo, R., & Schäfer, R. B. (2014). Effects of repeated salt pulses on ecosystem structure and functions in a stream mesocosm. *Science of The Total Environment, 476,* 634–642. https://doi.org/10.1016/j.scitotenv.2013.12.067.

Cañedo-Argüelles, M., Hawkins, C. P., Kefford, B. J., Schäfer, R. B., Dyack, B. J., Brucet, S., & Coring, E. (2016). Saving freshwater from salts. *Science, 351*(6276), 914–916. https://doi.org/10.1126/science.aad3488.

Cañedo-Argüelles, M., Kefford, B. J., Piscart, C., Prat, N., Schäfer, R. B., & Schulz, C. J. (2013). Salinisation of rivers: An urgent ecological issue. *Environmental Pollution, 173,* 157–167. https://doi.org/10.1016/j.envpol.2012.10.011.

Cañedo-Argüelles, M., Kefford, B., & Schäfer, R. (2019). Salt in freshwaters: Causes, effects, and prospects-introduction to the theme issue. *Philosophical Transactions of the Royal Society B, 374*(1764). https://doi.org/10.1098/rstb.2018.0002.

Canhoto, C., & Graça, M. A. S. (2008). Interactions between fungi (Aquatic Hyphomycetes) and invertebrates. In K. R. Sridhar, F. Bärlocher, & K. D. Hyde (Eds.), *Novel techniques and ideas in mycology* (Chap 14, pp. 205–325). Fungal Diversity Press. (ISSN:16086791).

Canhoto, C., Simões, S., Gonçalves, A. L., Guilhermino, L., & Bärlocher, F. (2017). Stream salinization and fungal mediated leaf decomposition: A microcosm study. *Science of the Total Environment, 599,* 1638–1645. https://doi.org/10.1016/j.scitotenv.2017.05.101.

Casas, J. J., Gessner, M. O., López, D., & Descals, E. (2011). Leaf-litter colonisation, and breakdown in relation to stream typology: Insights from Mediterranean low-order streams. *Freshwater Biology, 56,* 2594–2608. https://doi.org/10.1111/j.1365-2427.2011.02686.x.

Charcosset, J. Y., & Chauvet, E. (2001). Effect of culture conditions on ergosterol as an indicator of biomass in the aquatic hyphomycetes. *Applied and Environmental Microbiology, 67*(5), 2051–2055. https://doi.org/10.1128/AEM.67.5.2051-2055.2001.

Chauvet, E., Ferreira, V., Giller, P. S., McKie, B. G., Tiegs, S. D., Woodward, G., & Gulis, V. (2016). Litter decomposition as an indicator of stream ecosystem functioning at local-to-continental scales: Insights from the European RivFunction project. In *Advances in ecological research* (Vol. 55, pp. 99–182). Academic Press. https://doi.org/10.1016/bs.aecr.2016.08.006.

Clements, W. H., & Kotalik, C. (2016). Effects of major ions on natural benthic communities: An experimental assessment of the US environmental protection agency aquatic life benchmark for conductivity. *Freshwater Science, 35*(1), 126–138. https://doi.org/10.1086/685085.

Connolly, C. T., Sobczak, W. V., & Findlay, S. E. (2014). Salinity effects on *Phragmites* decomposition dynamics among the Hudson River's freshwater tidal wetlands. *Wetlands, 34*(3), 575–582. https://doi.org/10.1007/s13157-014-0526-1.

Cormier, S. M., Suter, G. W., & Zheng, L. (2013). Derivation of a benchmark for freshwater ionic strength. *Environmental Toxicology and Chemistry, 32*(2), 263–271. https://doi.org/10.1002/etc.2064.

Côté, I. M., Darling, E. S., & Brown, C. J. (2016). Interactions among ecosystem stressors and their importance in conservation. *Proceedings of the Royal Society B, 283,* 20152592. https://doi.org/10.1098/rspb.2015.2592.

Cummins, K. W., & Klug, M. J. (1979). Feeding ecology of stream invertebrates. *Annual Review of Ecology and Systematics, 10,* 147–172. https://doi.org/10.1146/annurev.es.10.110179.001051.

Das, M., Royer, T. V., & Leff, L. G. (2007). Diversity of Fungi, Bacteria, and Actinomycetes on leaves decomposing in a stream. *Applied and Environmental Microbiology, 73,* 756–767. https://doi.org/10.1128/AEM.01170-06.

Duarte, S., Seena, S., Bärlocher, F., Pascoal, C., & Cássio, F. (2013). A decade's perspective on the impact of DNA sequencing on aquatic hyphomycete research. *Fungal Biology Reviews, 27,* 19–24. https://doi.org/10.1016/j.fbr.2013.02.003.

Dugan, H. A., Bartlett, S. L., Burke, S. M., Doubek, J. P., Krivak-Tetley, F. E., Skaff, N. K., Summers, J. C., Farrell, K. J., McCullough, I. M., Morales-Williams, A. M., Ouyang, Z., Scordo, F., Hanson, P. C., & Weathers, K. C., … Roberts, D. C. (2017). Salting our freshwater lakes. *Proceedings of the National Academy of Sciences, 114*(17), 4453–4458. https://doi.org/10.1073/pnas.1620211114.

Egglishaw, H. J. (1968). The quantitative relationship between bottom fauna and plant detritus in streams of different calcium concentrations. *Journal of Applied Ecology, 5,* 731–740.

Entrekin, S. A., Clay, N. A., Mogilevski, A., Howard-Parker, B., & Evans-White, M. A. (2019). Multiple riparian–stream connections are predicted to change in response to salinization. *Philosophical Transactions of the Royal Society B, 374*(1764), 20180042. https://doi.org/10.1098/rstb.2018.0042.

Estévez, E., Rodríguez-Castillo, T., González-Ferreras, A. M., Cañedo-Argüelles, M., & Barquín, J. (2019). Drivers of spatio-temporal patterns of salinity in Spanish rivers: A nationwide assessment. *Philosophical Transactions of the Royal Society B, 374*(1764), 20180022. https://doi.org/10.1098/rstb.2018.0022.

Fay, L., & Shi, X. (2012). Environmental impacts of chemicals for snow and ice control: State of the knowledge. *Water, Air, and Soil pollution, 223*(5), 2751–2770. https://doi.org/10.1007/s11270-011-1064-6.

Findlay, S. E., & Kelly, V. R. (2011). Emerging indirect and long-term road salt effects on ecosystems. *Annals of the New York Academy of Sciences, 1223*(1), 58–68. https://doi.org/10.1111/j.1749-6632.2010.05942.x.

Gessner, M. O., Gulis, V., Kuehn, K. A., Chauvet, E., & Suberkropp, K. (2007). Fungal decomposers of plant litter in aquatic ecosystems. In C. P. Kubicek & I. S. Druzhinina (Eds.), *Environmental and microbial relationships*. The Mycota IV. Springer-Verlag. https://doi.org/10.1007/978-3-540-71840-6_17.

Ghassemi, F., Jakeman, A. J., & Nix, H. A. (1995). *Salinisation of land and water resources: Human causes, extent, management, and case studies*. CAB international.

Giller, P. S., & Malmqvist, B. (1999). *The biology of streams and rivers*. Oxford University Press.

Gómez, R., Arce, M. I., Baldwin, D. S., & Dahm, C. N. (2017). Water physicochemistry in intermittent rivers and ephemeral streams. In T. Datry, N. Bonada, & A. Boulton (Eds.), *Intermittent rivers and ephemeral streams* (pp. 109–134). Academic Press.

Gómez, R., Asencio, A. D., Picón, J. M., Del Campo, R., Arce, M. I., del Mar Sánchez-Montoya, M., Suárez, M. L., & Vidal- Abarca, M. R. (2016). The effect of water salinity on wood breakdown in semiarid Mediterranean streams. *Science of the Total Environment*, *541*, 491–501. https://doi.org/10.1016/j.scitotenv.2015.09.040.

Gómez, R., Hurtado, I., Suárez, M. L., & Vidal-Abarca, M. R. (2005). Ramblas in south-east Spain: Threatened and valuable ecosystems. *Aquatic Conservation: Marine and Freshwater Ecosystems, 15*(4), 387–402. https://doi.org/10.1002/aqc.680.

Gonçalves, A. L., Carvalho, A., Bärlocher, F., & Canhoto, C. (2019). Are fungal strains from salinized streams adapted to salt-rich conditions? *Philosophical Transactions of the Royal Society B, 374,* 20180018. https://doi.org/10.1098/rstb.2018.0018.

Gonçalves, A. L., Simões, S., Bärlocher, F., & Canhoto, C. (2019). Leaf litter microbial decomposition in salinized streams under intermittency. *Science of the Total Environment*, *653,* 1204–1212. https://doi.org/10.1016/j.scitotenv.2018.11.050.

Gorostiza, S., & Saurí, D. (2017). Dangerous assemblages: Salts, trihalomethanes and endocrine disruptors in the water palimpsest of the Llobregat River, Catalonia. *Geoforum, 81,* 153–162. https://doi.org/10.1016/j.geoforum.2017.03.005.

Gorostiza, S., & Saurí, D. (2019). Naturalizing pollution: A critical social science view on the link between potash mining and salinization in the Llobregat river basin, northeast Spain. *Philosophical Transactions of the Royal Society B, 374*(1764), 20180006. https://doi.org/10.1098/rstb.2018.0006.

Gulis, V., Kuehn, K. A., & Suberkropp, K. (2006). The role of fungi in carbon and nitrogen cycles in freshwater ecosystems. In G. M. Gadd (Ed.), *Fungi in biogeochemical cycles* (pp. 404–435). Cambridge University Press. https://doi.org/10.1017/cbo9780511550522.018.

Gulis, V., Su, R., & Kuehn, K. A. (2019). Fungal decomposers in freshwater environments. In C. J. Hurst (Ed.), *Advances in environmental microbiology* (pp. 121–155). Springer-Nature. https://doi.org/10.1007/978-3-030-16775-2_5.

Gutiérrez-Cánovas, C., Hernández, J., Millán, A., & Velasco, J. (2012). Impact of chronic and pulse dilution disturbances on metabolism and trophic structure in a saline Mediterranean stream. *Hydrobiologia, 686*(1), 225–239. https://doi.org/10.1007/s10750-012-1004-5.

Hale, R. L., & Groffman, P. M. (2006). Chloride effects on nitrogen dynamics in forested and suburban stream debris dams. *Journal of Environmental Quality, 35,* 2425–2432. https://doi.org/10.2134/jeq2006.0164.

Hart, B. T., Bailey, P., Edwards, R., Hortle, K., James, K., Mcmahon, A., Meredith, C., & Swadling, K. (1990). Effects of salinity on river, stream and wetland ecosystems in Victoria. *Australian Water Resources Council, 24*(9), 1103–1117. https://doi.org/10.1016/0043-1354(90)90173-4.

Hassell, K. L., Kefford, B. J., & Nugegoda, D. (2006). Sub-lethal and chronic salinity tolerances of three freshwater insects: *Cloeon* sp. and *Centroptilum* sp. (Ephemeroptera: Baetidae) and *Chironomus* sp. (Diptera: Chironomidae). *Journal of Experimental Biology, 209*(20), 4024–4032. https://doi.org/10.1242/jeb.02457.

Hemminga, M. A., De Leeuw, J., De Munck, W., & Koutsaal, B. P. (1991). Decomposition in estuarine salt marshes: The effect of soil salinity and soil water content. *Vegetation, 94,* 25–34. https://doi.org/10.1007/BF00044913.

Herbert, E. R., Boon, P., Burgin, A. J., Neubauer, S. C., Franklin, R. B., Ardón, M., Hopfensperger, K. N., Lamers, L. P. M., & Gell, P. (2015). A global perspective on wetland salinization: Ecological consequences of a growing threat to freshwater wetlands. *Ecosphere, 6*(10), 1–43. https://doi.org/10.1890/es14-00534.1.

Hintz, W. D., & Relyea, R. A. (2019). A review of the species, community, and ecosystem impacts of road salt salinisation in fresh waters. *Freshwater Biology, 64,* 1081–1097. https://doi.org/10.1111/fwb.13286.

Hooley, P., Fincham, D. A., Whitehead, M. P., & Clipson, N. J. W. (2003). Fungal osmotolerance. *Advances in Applied Microbiology, 53,* 177–211. https://doi.org/10.1016/s0065-2164(03)53005-2.

Hynes, H. B. N. (1975). The stream and its valley. *Verhandlungen des Internationalen Verein Limnologie, 19,* 1–15. https://doi.org/10.1080/03680770.1974.11896033.

Ingram, T., & Burns, Z. D. (2018). Top-down control by an aquatic invertebrate predator increases with temperature but does not depend on individual behavioral type. *Ecology and Evolution, 8*(16), 8256–8265. https://doi.org/10.1002/ece3.4367.

Jackson, J. K., & Funk, D. H. (2019). Temperature affects acute mayfly responses to elevated salinity: Implications for toxicity of road de-icing salts. *Philosophical Transactions of the Royal Society B, 374,* 20180081. https://doi.org/10.1098/rstb.2018.0081.

James, K. R., Can't, B., & Ryan, T. (2003). Responses of freshwater biota to rising salinity levels and implications for saline water management: A review. *Australian Journal of Botany, 51*(6), 703–713. https://doi.org/10.1071/bt02110.

Jardine, A., Speldewinde, P., Carver, S., & Weinstein, P. (2007). Dryland salinity and ecosystem distress syndrome: Human health implications. *EcoHealth, 4*(1), 10–17. https://doi.org/10.1007/s10393-006-0078-9.

Johnson, B. R., Ibach, R. E., & Baker, A. J. (1992). Effect of saltwater evaporation on trached separation from wood surfaces. *Forest Products Journal, 42,* 57–59.

Jolly, I. D., Williamson, D. R., Gilfedder, M., Walker, G. R., Morton, R., Robinson, G., Jones, H., Zhang, L., Dowling, T. I., Dyce, P., Nandakumar, N., Clarke, N., McNeill, V., & Nathan, R. J. (2001). Historical stream salinity trends and catchment salt balances in the Murray–Darling Basin, Australia. *Marine and Freshwater Research, 52*(1), 53–63. https://doi.org/10.1071/mf00018.

Jones, E., & Van Vliet, M. T. H. (2018). Drought impacts on river salinity in the south- ern US: Implications for water scarcity. *Science of the Total Environment, 644,* 844–853. https://doi.org/10.1016/j.scitotenv.2018.06.373.

Kaspari, M. (2020). The seventh macronutrient: How sodium shortfall ramifies through populations, food webs and ecosystems. *Ecology Letters, 23,* 1153–1168. https://doi.org/10.1111/ele.13517.

Kaspari, M., Yanoviak, S. P., Dudley, R., Yuan, M., & Clay, N. A. (2009). Sodium shortage as a constraint on the carbon cycle in an inland tropical rainforest. *Proceedings of the National Academy of Sciences, 106*(46), 19405–19409. https://doi.org/10.1073/pnas.0906448106.

Kath, J., Thomson, J. R., Thompson, R. M., Kefford, B. J., Dyer, F. J., & Nally, R. M. (2018). Interactions among stressors may be weak: Implications for management of freshwater macroinvertebrate communities. *Diversity and Distributions, 24*(7), 939–950. https://doi.org/10.1111/ddi.12737.

Kaushal, S. S., Groffman, P. M., Likens, G. E., Belt, K. T., Stack, W. P., Kelly, V. R., Band, L. B., & Fisher, G. T. (2005). Increased salinization of fresh water in the northeastern United States. *Proceedings of the National Academy of Science, 102*(38), 13517–13520. https://doi.org/10.1073/pnas.0506414102.

Kaushal, S. S., Likens, G. E., Pace, M. L., Haq, S., Wood, K. L., Galella, J. G., Räike, A., Morel, C., Doody, T. R., Wessel, B., Kortelainen, P., Skinner, V., Utz, R., & Jaworski, N. (2019). Novel 'chemical cocktails' in inland waters are a consequence of the freshwater salinization syndrome. *Philosophical Transactions of the Royal Society B, 374*(1764), 20180017. https://doi.org/10.1098/rstb.2018.0017.

Kaushal, S. S., Likens, G. E., Pace, M. L., Utz, R. M., Haq, S., Gorman, J., & Grese, M. (2018). Freshwater salinization syndrome on a continental scale. *Proceedings of the National Academy of Sciences, 115*(4), E574–E583. https://doi.org/10.1073/pnas.1711234115.

Kaushik, N. K., & Hynes, H. B. N. (1968). Experimental study on the role of autumn-shed leaves in aquatic environments. *Journal of Ecology, 56,* 229–243. https://doi.org/10.2307/2258076.

Kaushik, N. K., & Hynes, H. B. N. (1971). The fate of the dead leaves that fall into streams. *Arch. Hydrobiol. 68,* 465–515.

Kefford, B. J., Marchant, R., Schäfer, R. B., Metzeling, L., Dunlop, J. E., Choy, S. C., & Goonan, P. (2011). The definition of species richness used by species sensitivity distributions approximates observed effects of salinity on stream macroinvertebrates. *Environmental Pollution, 159*(1), 302–310. https://doi.org/10.1016/j.envpol.2010.08.025.

Kefford, B. J., Nugegoda, D., Zalizniak, L., Fields, E. J., & Hassell, K. L. (2007). The salinity tolerance of freshwater macroinvertebrate eggs and hatchlings in comparison to their older life-stages: A diversity of responses. *Aquatic Ecology, 41,* 335–348. https://doi.org/10.1007/s10452-006-9066-y.

Kefford, B. J., Papas, P. J., & Nugegoda, D. (2003). Relative salinity tolerance of macroinvertebrates from the Barwon River, Victoria, Australia. *Marine & Freshwater Research, 54*(6), 755–765. https://doi.org/10.1071/MF02081.

Krausmann, F., Lauk, C., Haas, W., & Wiedenhofer, D. (2018). From resource extraction to outflows of wastes and emissions: The socioeconomic metabolism of the global economy. *Global Environmental Change, 52,* 131–140. https://doi.org/10.1016/j.gloenvcha.2018.07.003.

Krauss, G. J., Sole, M., Krauss, G., Schlosser, D., Wesenberg, D., & Bärlocher, F. (2011). Fungi in freshwaters: Ecology, physiology, and biochemical potential. *FEMS Microbiology Reviews, 35,* 620–651. https://doi.org/10.1111/j.1574-6976.2011.00266.x.

Kunz, J. L., Conley, J. M., Buchwalter, D. B., Norberg-King, T. J., Kemble, N. E., Wang, N., & Ingersoll, C. G. (2013). Use of reconstituted waters to evaluate effects of elevated major ions associated with mountaintop coal mining on freshwater invertebrates. *Environmental Toxicology and Chemistry, 32*(12), 2826–2835. https://doi.org/10.1002/etc.2391.

Ladrera, R., Cañedo-Argüelles, M., & Prat, N. (2017). Impact of potash mining in streams: The Llobregat basin (northeast Spain) as a case study. *Journal of Limnology, 76*(1), 343–354. https://doi.org/10.4081/jlimnol.2016.1525.

Lake, P. S. (2000). Disturbance, patchiness, and diversity in streams. *Journal of the North American Benthological Society, 19,* 573–592. https://doi.org/10.2307/1468118.

Le, T. D. H., Kattwinkel, M., Schützenmeister, K., Olson, J. R., Hawkins, C. P., & Schäfer, R. B. (2019). Predicting current and future background ion concentrations in German surface water under climate change. *Philosophical Transactions of the Royal Society B, 374*(1764), 20180004. https://doi.org/10.1098/rstb.2018.0004.

Manerkar, M. A., Seena, S., & Bärlocher, F. (2008). Q-RT-PCR for assessing archaea, bacteria, and fungi during leaf decomposition in a stream. *Microbial Ecology, 56,* 467–473. https://doi.org/10.1007/s00248-008-9365-z.

Marks, J. C. (2019). Revisiting the fates of dead leaves that fall into streams. *Annual Review of Ecology Evolution and Systematics, 50,* 547–568. https://doi.org/10.1146/annurev-ecolsys-110218-024755.

Martínez, A., Barros, J., Gonçalves, A. L., & Canhoto, C. (2020). Salinization effects on leaf litter decomposition in fresh waters: Does the ionic composition of salt matter? *Freshwater Biology.* https://doi.org/10.1111/fwb.13514.

Martínez, A., Gonçalves, A. L., & Canhoto, C. (2020). Salinization effects on stream biofilm functioning. *Hydrobiologia, 847,* 1453–1459. https://doi.org/10.1007/s10750-020-04199-w.

Mendelssohn, I. A., Sorrell, B. K., Brix, H., Schierup, H. H., Lorenzen, B., & Maltby, E. (1999). Controls on soil cellulose decomposition along a salinity gradient in a *Phragmites australis* wetland in Denmark. *Aquatic Botany, 64,* 381–398. https://doi.org/10.1016/S0304-3770(99)00065-0.

Millán, A., Velasco, J., Gutiérrez-Cánovas, C., Arribas, P., Picazo, F., Sánchez-Fernández, D., & Abellán, P. (2011). Mediterranean saline streams in southeast Spain: What do we know? *Journal of Arid Environments, 75,* 1352–1359. https://doi.org/10.1016/j.jaridenv.2010.12.010.

Nielsen, D. L., Brock, M. A., Rees, G. N., & Baldwin, D. S. (2003). Effects of increasing salinity on freshwater ecosystems in Australia. *Australian Journal of Botany, 51*(6), 655–665. https://doi.org/10.1071/BT02115.

Olson, J. R. (2019). Predicting combined effects of land use and climate change on river and stream salinity. *Philosophical Transactions of the Royal Society B, 374,* 20180005. https://doi.org/10.1098/rstb.2018.0005.

Oren, A. (1999). Bioenergetic aspects of halophilism. *Microbiology and Molecular Biology Reviews, 63,* 334–348.

Oren, A. (2001). The bioenergetic basis for the decrease metabolic diversity at increasing salt concentrations: Implications for the functioning of salt lake ecosystems. *Hydrobiologia, 466,* 61–72. https://doi.org/10.1134/S0026261706010012.

Orr, S. E., & Buchwalter, D. B. (2020). It's all about the fluxes: Temperature influences ion transport and toxicity in aquatic insects. *Aquatic Toxicology, 221,* 105405. https://doi.org/10.1016/j.aquatox.2020.105405.

Overy, D., Correa, H., Roullier, C., Chi, W.-C., Pang, K.-L., Rateb, M., Ebel, R., Shang, Z., Capon, R., Bills, G., & Kerr, R. (2017). Does osmotic stress affect natural product expression in fungi? *Marine Drugs, 15,* 254. https://doi.org/10.3390/md15080254.

Painchaud, J., Therriault, J., & Legendre, L. (1995). Assessment of salinity-related mortality of freshwater bacteria in the saint lawrence estuary. *Applied and Environmental Microbiology, 61,* 205–208. https://doi.org/10.1128/aem.61.1.205-208.1995.

Pallarés, S., Botella-Cruz, M., Arribas, P., Millán, A., & Velasco, J. (2017). Aquatic insects in a multistress environment: Cross-tolerance to salinity and desiccation. *Journal of Experimental Biology, 220*(7), 1277–1286. https://doi.org/10.1242/jeb.152108.

Parida, A. K., & Das, A. B. (2005). Salt tolerance and salinity effects on plants: A review. *Ecotoxicology and Environmental Safety, 60*(3), 324–349. https://doi.org/10.1016/j.ecoenv.2004.06.010.

Pascoal, C., & Cássio, F. (2008). Linking fungal diversity to the functioning of freshwater ecosystems. In K. R. Sridhar, F. Bärlocher, & K. D. Hyde (Eds.), *Novel techniques and ideas in mycology* (pp. 1–18). Fungal Diversity Press.

Pascoal, C., Cássio, F., & Marvanová, L. (2005). Anthropogenic stress may affect aquatic hyphomycete diversity more than leaf decomposition in a low-order stream. *Archiv für Hydrobiologie, 162,* 481–496. https://doi.org/10.1127/0003-9136/2005/0162-0481.

Paul, M. J., & Meyer, J. L. (2001). Streams in the urban landscape. *Annual Review of Ecology and Systematics, 32*(1), 333–365. https://doi.org/10.1146/annurev.ecolsys.32.081501.114040.

Peck, A. J. (1978). Salinization of non-irrigated soils and associated streams: A review. *Soil Research, 16*(2), 157–168. https://doi.org/10.1071/SR9780157.

Piscart, C., Mermillod-Blondin, F., Maazouzi, C., Merigoux, S., & Marmonier, P. (2011). Potential impact of invasive amphipods on leaf litter recycling in aquatic ecosystems. *Biological Invasions, 13*(12), 2861–2868. https://doi.org/10.1007/s10530-011-9969-y.

Pond, G. J. (2010). Patterns of Ephemeroptera taxa loss in Appalachian headwater streams (Kentucky, USA). *Hydrobiologia, 641*(1), 185–201. https://doi.org/10.1007/s10750-009-0081-6.

Pond, G. J., Passmore, M. E., Borsuk, F. A., Reynolds, L., & Rose, C. J. (2008). Downstream effects of mountaintop coal mining: Comparing biological conditions using family-and genus-level macroinvertebrate bioassessment tools. *Journal of the North American Benthological Society, 27*(3), 717–737. https://doi.org/10.1899/08-015.1.

Quintino, V., Sangiorgio, F., Ricardo, F., Mamede, R., Pires, A., Freitas, R., Rodrigues, A. M., & Basset, A. (2009). *In situ* experimental study of reed leaf decomposition along a full salinity gradient. *Estuarine, Coastal and Shelf Science, 85*(3), 497–506. https://doi.org/10.1016/j.ecss.2009.09.016.

Reice, S. R., & Herbst, G. (1982). The role of salinity in decomposition of leaves of Phragmites australis in desert streams. *Journal of Arid Environments, 5*(4), 361–368. https://doi.org/10.1016/S0140-1963(18)31618-5.

Rengasamy, P. (2006). World salinization with emphasis on Australia. *Journal of Experimental Botany, 57*(5), 1017–1023. https://doi.org/10.1093/jxb/erj108.

Roache, M. C., Bailey, P. C., & Boon, P. I. (2006). Effects of salinity on the decay of the freshwater macrophyte, *Triglochin procerum. Aquatic Botany, 84,* 45–52. https://doi.org/10.1016/j.aquabot.2005.07.014.

Sala, M., Faria, M., Sarasúa, I., Barata, C., Bonada, N., Brucet, S., Llenas, L., Ponsá, S., Prat, N., Soares, A. M. V. M., & Cañedo-Arguelles, M. (2016). Chloride and sulphate toxicity to *Hydropsyche exocellata* (Trichoptera, Hydropsychidae): Exploring intraspecific variation and sub-lethal endpoints. *Science of the Total Environment, 566,* 1032–1041. https://doi.org/10.1016/j.scitotenv.2016.05.121.

Sangiorgio, F., Basset, A., Pinna, M., Sabetta, L., Abbiati, M., Ponti, M., Trayanova, A., Minocci, M., Orfanidis, S., Nicolaidou, A., Moncheva, S., Georgescu, L., Dragan, S., Beqiraj, S., Koutsoubas, D., Evagelopoulos, A., & Reizopoulou, S. (2007). Ecosystem processes: Litter breakdown patterns in Mediterranean and Black Sea transitional waters. *Transitional Waters Bulletin, 1*(3), 51–55. https://doi.org/10.1285/i1825229xv1n3p51.

Sauer, F. G., Bundschuh, M., Zubrod, J. P., Schäfer, R. B., Thompson, K., & Kefford, B. J. (2016). Effects of salinity on leaf breakdown: Dryland salinity versus salinity from a coalmine. *Aquatic Toxicology, 177,* 425–432. https://doi.org/10.1016/j.aquatox.2016.06.014.

Schäfer, R. B., Bundschuh, M., Rouch, D. A., Szöcs, E., Peter, C., Pettigrove, V., Schulz, R., Nugegoda, D., & Kefford, B. J. (2012). Effects of pesticide toxicity, salinity and other environmental variables on selected ecosystem functions in streams and the relevance for ecosystem services. *Science of the Total Environment, 415,* 69–78. https://doi.org/10.1016/j.scitotenv.2011.05.063.

Schäfer, R. B., & Piggott, J. J. (2018). Advancing understanding and prediction in multiple stressor research through a mechanistic basis for null models. *Global Change Biology, 24,* 1817–1826. https://doi.org/10.1111/gcb.14073.

Scharnagl, K., Scharnagl, A., von Wettberg, E. (2017). Nature's potato chip: The role of salty fungi in a changing world. *American Journal of Botany, 104,* 1–4. https://doi.org/10.3732/ajb.1700034.

Scheibener, S. A., Richardi, V. S., & Buchwalter, D. B. (2016). Comparative sodium transport patterns provide clues for understanding salinity and metal responses in aquatic insects. *Aquatic Toxicology, 171,* 20–29. https://doi.org/10.1016/j.aquatox.2015.12.006.

Schmitz, H. (1924). Studies in wood decay IV: The effect of sodium carbonate, bicarbonate, sulphate, and chloride on the rate of decay of Douglas Fir Sawdust induced by Lenzites saepiaria fr. With special reference to the effect of alkaline soils on the rate of decay of wood in contact with them. *American Journal of Botany, 11,* 108–121.

Schuler, M. S., Cañedo-Argüelles, M., Hintz, W. D., Dyack, B., Birk, S., & Relyea, R. A. (2019). Regulations are needed to protect freshwater ecosystems from salinization. *Philosophical Transactions of the Royal Society B, 374,* 20180019. https://doi.org/10.1098/rstb.2018.0019.

Schulz, C. J., & Cañedo-Argüelles, M. (2019). Lost in translation: The German literature on freshwater salinization. *Philosophical Transactions of the Royal Society B, 374,* 20180007. https://doi.org/10.1098/rstb.2018.0007.

Seena, S., Bärlocher, F., Sobral, O., Gessner, M. O., Dudgeon, D., McKie, B. G., Chauvet, E., Boyero, L., Ferreira, V., Frainer, A., Bruder, A., Matthaei, C. D., Fenoglio, S., Sridhar, K. R., Albariño, R. J., Douglas, M. M., Encalada, A. C., Garcia, E., Ghate, S. D., … Graça, M. A. S. (2019). Biodiversity of leaf litter fungi in streams along a latitudinal gradient. *Science of the Total Environment, 661,* 306–315. https://doi.org/10.1016/j.scitotenv.2019.01.122.

Servais, S., Kominoski, J. S., Charles, S. P., Gaiser, E. E., Mazzei, V., Troxler, T. G., & Wilson, B. J. (2019). Saltwater intrusion and soil carbon loss: Testing effects of salinity and phosphorus loading on microbial functions in experimental freshwater wetlands. *Geoderma, 337,* 1291–1300. https://doi.org/10.1016/j.geoderma.2018.11.013.

Sinha, R., & Khare, S. K. (2014). Protective role of salt in catalysis and maintaining structure of halophilic proteins against denaturation. *Frontiers in Microbiology, 5,* 165. https://doi.org/10.3389/fmicb.2014.00165.

Sleator, R. D., & Hill, C. (2002). Bacterial osmoadaptation: The role of osmolytes in bacterial stress and virulence. *FEMS Microbiology Reviews, 26*(1), 49–71. https://doi.org/10.1111/j.1574-6976.2002.tb00598.x.

Sridhar, K. R., & Kaveriappa, K. M. (1988). Occurrence and survival of aquatic hyphomycetes in brackish and sea water. *Archiv für Hydrobiologie, 113,* 153–160. https://doi.org/10.1515/botm.1979.22.7.421.

Steinke, T. D., & Charles, L. M. (1986). *In vitro* rates of decomposition of leaves of the mangrove *Bruguiera gymnorrhiza* as affected by temperature and salinity. *South African Journal of Botany, 52*(1), 39–42.

Swan, C. M. (2007). *Assessing the role of road salt runoff on the critical ecological interactions that regulate carbon processing in small, headwater streams in the Chesapeake Bay watershed* (Final Report. MWRRC [Project #2007MD148B]).

Swan, C. M., & DePalma, C. A. (2012). Elevated chloride and consumer presence independently influence processing of stream detritus. *Urban Ecosystems, 15,* 625–635. https://doi.org/10.1007/s11252-011-0210-7.

Swan, C. M., & Kominoski, J. S. (2012). *Biodiversity and Ecosystem Function of Decomposition.* Wiley. https://doi.org/10.1002/9780470015902.a0023601.

Szöcs, E., Kefford, B. J., & Schäfer, R. B. (2012). Is there an interaction of the effects of salinity and pesticides on the community structure of macroinvertebrates? *Science of the Total Environment, 437,* 121–126. https://doi.org/10.1016/j.scitotenv.2012.07.066.

Tachet, H., Richoux, P., Bournaud, M., & Usseglio-Polatera, P. (2010). *Invertébrés d'eau douce: systématique, biologie, écologie* (Vol. 15). CNRS éditions.

Tamie, M. S. A. (2016). Sodium chloride stress induced morphological changes in some halotolerant fungi. *The Egyptian Journal of Hospital Medicine, 62*(1), 109–126. https://doi.org/10.12816/0021420.

They, N., Ferreira, L., Marins, L., & Abreu, P. (2013). Stability of bacterial composition and activity in different salinity waters in the dynamic Patos Lagoon Estuary: Evidence from a Lagrangian-like approach. *Microbial Ecology, 66*(3), 551–562. https://doi.org/10.1007/s00248-013-0259-3.

Thomas, D. S. G., & Middleton, N. J. (1993). Salinization: New perspectives on a major desertification issue. *Journal of Arid Environments, 24*(1), 95–105. https://doi.org/10.1006/jare.1993.1008.

Treseder, K. K., & Lennon, J. T. (2015). Fungal traits that drive ecosystem dynamics on land. *Microbiology and Molecular Biology Reviews, 79,* 243–262. https://doi.org/10.1128/MMBR.00001-15.

Tyree, M., Clay, N., Polaskey, S., & Entrekin, S. (2016). Salt in our streams: Even small sodium additions can have negative effects on detritivores. *Hydrobiologia, 775,* 109–122. https://doi.org/10.1007/s10750-016-2718-6.

van Gray, J. B., Roberto, A. A., & Leff, L. G. (2020). Acute salt stress promotes altered assembly dynamics of nascent freshwater microbial biofilms. *Hydrobiologia, 847,* 2465–2484. https://doi.org/10.1007/s10750-020-04266-2.

van Meter, R. J., Swan, C. M., & Trossen, C. A. (2012). Effects of road deicer (NaCl) and amphibian grazers on detritus processing in pond mesocosms. *Environmental Toxicology and Chemistry, 31,* 2306–2310. https://doi.org/10.1002/etc.1949.

van Vliet, M. T., Flörke, M., & Wada, Y. (2017). Quality matters for water scarcity. *Nature Geoscience, 10*(11), 800–802. https://doi.org/10.1038/ngeo3047.

Velasco, J., Gutiérrez-Cánovas, C., Botella-Cruz, M., Sánchez-Fernández, D., Arribas, P., Carbonell, J. A., Millán, A., & Pallarés, S. (2019). Effects of salinity changes on aquatic organisms in a multiple stressor context. *Philosophical Transactions of the Royal Society B, 374,* 20180011. https://doi.org/10.1098/rstb.2018.0011.

Verberk, W. C., Buchwalter, D. B., & Kefford, B. J. (2020). Energetics as a lens to understanding aquatic insect's responses to changing temperature, dissolved oxygen and salinity regimes. *Current Opinion in Insect Science, 41,* 46–53. https://doi.org/10.1016/j.cois.2020.06.001.

Vidal-Abarca, R., Suárez, L., Gómez, R., Moreno, J. L., & Guerrero, C. (2002). Diel variations in physical and chemical parameters in a semi-arid stream in Spain (Chicamo). *Internationale Vereinigung für Theoretische und Angewandte Limnologie: Verhandlungen, 28,* 1922–2010. https://doi.org/10.1080/03680770.2001.11901890.

Voelz, N. J., & McArthur, J. V. (2000). An exploration of factors influencing lotic insect species richness. *Biodiversity and Conservation, 9,* 1543–1570. https://doi.org/10.1023/A:1008984802844.

Vorste, R. V., Timpano, A. J., Cappellin, C., Badgley, B. D., Zipper, C. E., & Schoenholtz, S. H. (2019). Microbial and macroinvertebrate communities, but not leaf decomposition, change along a mining-induced salinity gradient. *Freshwater Biology, 64,* 671–684. https://doi.org/10.1111/fwb.13253.

Welsh, D. T. (2000). Ecological significance of compatible solute accumulation by micro-organisms: From single cells to global climate. *FEMS Microbiology Reviews, 24,* 263–290. https://doi.org/10.1111/j.1574-6976.2000.tb00542.x.

Weston, N. B., Dixon, R. E., & Joyce, S. B. (2006). Ramifications of increased salinity in tidal freshwater sediments: Geochemistry and microbial pathways of organic matter mineralization. *Journal of Geophysical Research, 111,* G01009. https://doi.org/10.1029/2005JG000071.

Williams, W. D. (1996). The largest, highest and lowest lakes of the world: Saline lakes. *Internationale Vereinigung für Theoretische und Angewandte Limnologie: Verhandlungen, 26,* 61–79.

Williams, W. D. (2001). Anthropogenic salinisation of inland waters. In *Saline lakes* (pp. 329–337). Springer. https://doi.org/10.1023/a:1014598509028.

Williams, W. D., Taaffe, R. G., & Boulton, A. J. (1991). Longitudinal distribution of macroinvertebrates in two rivers subject to salinization. *Hydrobiologia, 210*(1–2), 151–160. https://doi.org/10.1007/BF00014329.

Wilson, S. M. (2004). *Dryland and urban salini- ty costs across the Murray-Darling Basin: An overview & guidelines for identifying and valuing the impacts.* Murray-Darling Basin Commission. Available at: https://www.mdba.gov.au/sites/default/files/archived/mdbc-salinityreports/2070_Dryland_and_urban_salinity_costs_across_MDB.pdf.

Zalizniak, L., Kefford, B. J., & Nugegoda, D. (2006). Is all salinity the same? I. The effect of ionic compositions on the salinity tolerance of five species of freshwater invertebrates. *Marine and Freshwater Research, 57*(1), 75–82. https://doi.org/10.1071/mf05103.

Zhang, L., Gao, G., Tang, X., & Shao, K. (2014a). Can the freshwater bacterial communities shift to the "marine-like" taxa? *Journal of Basic Microbiology, 54,* 1264–1272. https://doi.org/10.1002/jobm.201300818.

Zhang, L., Gao, G., Tang, X., & Shao, K. (2014b). Impacts of different salinities on bacterial biofilm communities in fresh water. *Canadian Journal of Microbiology, 60,* 319–326. https://doi.org/10.1139/cjm-2013-0808.

Chapter 16
Pathways, Mechanisms, and Consequences of Nutrient-Stimulated Plant Litter Decomposition in Streams

David W. P. Manning, Verónica Ferreira, Vladislav Gulis, and Amy D. Rosemond

Abstract Excess nitrogen (N) and phosphorus (P) inputs to streams occur globally, and affect not only stream autotrophs, but also heterotrophic microbes and detrital carbon processing. Detrital carbon, such as leaf litter, supports stream food webs and their connectivity via downstream detritus fluxes. Nutrient enrichment increases litter decomposition rates across multiple scales and trophic levels by stimulating activity of microbial decomposers and enhancing interactions among microbial decomposers, detritivores, and physical abrasion. Nutrient effects on microbial and detritivore-mediated decomposition are typically greater for recalcitrant *vs.* labile litter, especially when coupled to low initial nutrient concentrations. Recent studies and syntheses show that (1) dissolved N and P affect litter by stimulating fungal activity and nutrient immobilization, thus, increasing detrital nutrient content, (2) nutrient effects are greatest with N and P together (*vs.* individually) and when detritivores are present, and (3) ecosystem-level effects of nutrient enrichment can be predicted from small-scale measurements. Despite extensive studies of leaf litter decomposition, its application as a tool to manage nutrient enrichment issues trails comparable tools for autotrophic (i.e., algal) pathways. Thus, better understanding of the consequences of nutrient enrichment on leaf litter and other detrital carbon is important to predict how nutrients will affect stream ecosystem functioning.

D. W. P. Manning (✉)
Department of Biology, University of Nebraska at Omaha, Omaha, NE, USA
e-mail: davidmanning@unomaha.edu

V. Ferreira
MARE—Marine and Environmental Sciences Center, Department of Life Sciences, University of Coimbra, Coimbra, Portugal
e-mail: veronica@ci.uc.pt

V. Gulis
Department of Biology, Coastal Carolina University, Conway, SC, USA
e-mail: vgulis@coastal.edu

A. D. Rosemond
Odum School of Ecology, University of Georgia, Athens, GA, USA
e-mail: rosemond@uga.edu

C. M. Swan et al. (eds.), *The Ecology of Plant Litter Decomposition in Stream Ecosystems*, https://doi.org/10.1007/978-3-030-72854-0_16

16.1 Pathways of Nutrient Enrichment Effects in Streams

Human activities have increased nutrient concentrations in streams and rivers worldwide (Dodds & Smith, 2016; Wurtsbaugh et al., 2019) modifying critical ecosystem functions, including leaf litter decomposition. The effects of excessive nitrogen (N) and phosphorus (P) concentrations have been studied most extensively, because these two elements often co-limit growth and subsequent biological activity of both autotrophic and heterotrophic biota in freshwater ecosystems (Ferreira et al., 2015; Harpole et al., 2011). Nitrogen and P enter streams via both point sources, and diffuse, non-point pollution. Land use changes, fertilizer application, atmospheric deposition (Fowler et al., 2004; Linker et al., 2013), and animal or human wastes (e.g., livestock manure, wastewater) all contribute to increasing nutrients in streams, and each source can yield distinct patterns of nutrient loading and resultant streamwater nutrient concentrations and ratios (Manning et al., 2020; Stets et al., 2020). Excessive concentrations of total N (TN) and total P (TP) are estimated to affect 41 and 46% of total stream length in the United States (US EPA, 2016), and less than 2% of U.S. streams have TP concentrations indicative of reference conditions (TP <10 μg L^{-1}; Stoddard et al., 2016; Fig. 16.1). Other parts of the globe are not immune to this

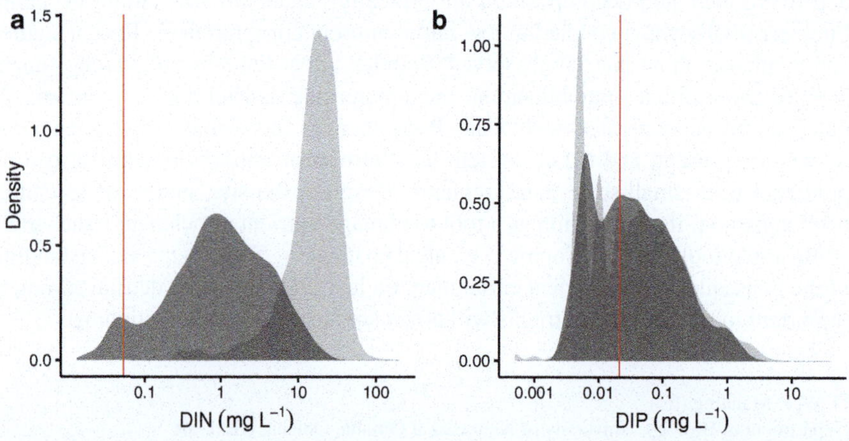

Fig. 16.1 Kernel density plots based on Gaussian kernel density estimators at each x-value representing the distribution of dissolved inorganic N (DIN; **a**) and dissolved inorganic P (DIP; **b**) concentrations observed in publicly available data sets (USA [dark grey areas] and global [light grey areas]). Concentrations generally span several orders of magnitude, and concentrations near the lower end of the spectrum (i.e., near pristine) are rare. We also indicate the mean of half-saturation constants (K_m) for litter decomposition rates reported in the literature for both N and P (red vertical lines in **a** and **b**; Ferreira et al., 2006; Kominoski et al., 2015; Rosemond et al., 2002). About 94 and 100% of concentration values were above the DIN thresholds in the U.S. and global datasets, respectively, compared to 59 and 56% for DIP. Data were collected and made available by the U.S. Geological Survey, as part of the U.S. National Water Quality Assessment (dark grey density curves; n =7,653), and global sites from the Global River Chemistry dataset (GLORICH; Hartmann et al., 2014, 2019) with streamflow <20 m^3 s^{-1} (light grey density curves; n = 14,097 [DIN] and 53,861 [DIP]).

problem; elevated N and P concentrations are pervasive across streams and rivers in Europe, Asia, Oceania, Africa, and Central and South America (McDowell et al., 2020).

The effects of nutrient pollution in streams and rivers can vary spatially and temporally, because of the multiple sources of N and P, and their interactions with the physical attributes and biological activity of streams. For example, nutrient availability may vary according to stream flow regimes controlled by climate and land use, where floods and droughts can episodically deplete, dilute, or elevate nutrient concentrations (Kaushal et al., 2014; Moss et al., 2011; Rose et al., 2018). Similarly, seasonal patterns of nutrient availability may occur because of the timing of fertilizer application, seasonality of riparian or in-stream nutrient uptake, animal migrations, or climatic variables (e.g., interactions among precipitation and wet/dry nutrient deposition; Mulholland & Hill, 1997). The spatial distribution of excessive nutrient concentrations in streams is affected by the prevalence of nutrient sources within watersheds. For instance, the dominant land use in the corn-belt region of the Midwest USA is row-crop agriculture, and streams in this region tend to exhibit higher N concentrations compared to regions with other prevailing types of land use (Hill et al., 2011).

The spatial and temporal variation of stream N and P concentration highlights the need for metrics that can integrate their effects in time and space. The spatial and temporal dynamics of nutrient availability in streams interact with seasonal pulses of terrestrial organic matter inputs that are the base of food webs in most forest streams as well as with the seasonal patterns in activity of plant litter decomposers. Thus, the processes through which this organic matter is broken down offer several integrative metrics that can be used to understand how both acute and chronic nutrient enrichment modify stream ecosystem functions. Specifically, leaf litter decomposition is well suited to quantify the multiple effects of nutrient enrichment in streams because leaf litter from terrestrial sources remains within streams for long periods of time (months to years; Webster & Benfield, 1986), and involves multiple facets of physical, chemical and biological attributes of stream ecosystems (Frainer et al. [Chapter 21 in this volume]).

16.1.1 Nutrient Effects Are Less Understood in Heterotrophic vs. Autotrophic Systems

Some of the most obvious impacts of nutrient pollution in freshwaters include increased algal biomass, blooms of harmful algae and associated hypoxic zones, fish kills, and drinking water contamination (McDowell et al., 2020; Smith et al., 2006). Increased nutrient availability can stimulate ecosystem productivity (i.e., eutrophication), and can increase the importance of primary producers as the energy base for the food web when light is not a limiting factor. This process of anthropogenic eutrophication and the build-up of within-system carbon (C) via increased photosynthesis has justifiably received considerable attention (Conley et al., 2009 and references therein). However, in forest streams, solar irradiation is limited by the riparian vegetation, and aquatic food webs derive most of their C and energy from terrestrial organic matter (i.e., allochthonous C, such as leaf litter, detailed below; Cebrian & Lartigue, 2004; Fisher & Likens, 1973; Moore et al., 2004; Wallace et al., 1997). Nutrient effects on these heterotrophic, "brown" food web pathways have received less attention than autotrophic, "green" food web pathways, particularly for management. For example, efforts to define ecosystem trophic state in streams first focused on relationships between streamwater TN and TP and benthic algae (Dodds, 2006) while more recent work has begun to encompass both green and brown pathways to define ecosystem responses to nutrient enrichment (Arroita et al., 2019; Dodds & Cole, 2007). Whereas nutrient enrichment leads to increased biomass and biomass-specific rates of primary production in autotroph-based ecosystems, it can increase microbial respiration rates, stimulate detritivore activity and result in ecosystem-level C losses in detritus-based systems (Benstead et al., 2009). However, heterotrophic responses to nutrient enrichment, such as decomposition of coarse particulate organic matter (i.e., leaf litter), remain largely absent from conventional strategies to monitor and manage the problems associated with nutrient enrichment of streams and rivers.

Food webs in forest streams are fueled by plant litter (leaves and wood) derived from terrestrial ecosystems (Wallace et al., 1997; Walther & Whiles, 2011). The decomposition of leaf litter is a key ecosystem process that has been studied extensively for several decades (Abelho, 2001; Chauvet et al., 2016; Marks, 2019; Tank et al., 2010; Webster & Benfield, 1986). Moreover, leaf litter decomposition integrates biological activity across multiple trophic levels (microbial decomposers, such as aquatic hyphomycetes, to predators; Gessner et al., 2010) and patterns of patch- or reach-scale decomposition rates can inform models addressing decomposition rates at catchment to river network scales (Rosemond et al., 2015; Webster, 2007). The stimulation of leaf litter decomposition by nutrients results in accelerated loss of stream C altering its availability to stream consumers, and affecting climate feedbacks via increased respiratory loss of leaf litter C to CO_2 (Follstad-Shah [Chapter 12 in this volume]; Manning et al., 2018; Rosemond et al., 2015). As both N and P can limit the growth of key stream biota, and especially those that play an outsized role in leaf litter processing, increased stream nutrient concentrations can lead to rapid turnover of leaf litter in streams, and may ultimately result in reduced C standing stock and availability to in-stream biota. Indeed, a recent meta-analysis of observational and experimental studies suggest that moderate increases of nutrient concentrations can increase leaf litter decomposition rates by 50%, on average (Ferreira et al., 2015). Consistent with this finding, Rosemond et al. (2015) found comparable increases (~50%) for whole-stream leaf litter loss rates from experimentally enriched streams. However, when streams are not nutrient limited (e.g., due to underlying geology or diffuse non-point source pollution), further increases in nutrient concentrations may have no effect on litter decomposition (Baldy et al., 2007; Chadwick & Huryn, 2003). Also, litter decomposition can be inhibited at high nutrient concentrations due to toxic effects of high concentrations of nitrite or ammonia to detritivores (but not microorganisms), or to the concomitant changes in other environmental factors (e.g., decrease in dissolved oxygen concentration or increase in fine sediment load and pesticide or other contaminant concentrations) that may negatively affect both microbial and invertebrate activity (Lecerf et al., 2006; Woodward et al., 2012).

In this chapter, we describe the pathways and explore the mechanisms through which streamwater nutrient enrichment can affect leaf litter decomposition rates. We develop a conceptual model based on current evidence for the effects of nutrient enrichment on leaf litter decomposition driven by microbial decomposers and detritivores. These explorations and conceptualizations are not exhaustive (see Ferreira et al., 2015), but rather focus on both observational and experimental evidence that illustrate the emerging mechanisms of how leaf litter decomposition responds to nutrient enrichment. We also discuss potential interactions of nutrient effects on litter decomposition and other global change stressors, such as rising stream temperatures. We conclude with an overview of consequences for whole ecosystems, including C residence times, and offer perspectives on the need to promote efforts to fully incorporate leaf litter decomposition responses into strategies to monitor and manage nutrient pollution at extensive (i.e., continental) scales.

16.2 Mechanisms of Nutrient Effects on Leaf Litter Decomposition

16.2.1 Microbially Mediated Litter Processing

Microbial communities associated with decaying leaf litter in streams encompass fungi, including so-called aquatic hyphomycetes that are specially adapted to stream environments, as well as bacteria. Fungi dominate these microbial communities in terms of biomass (88–99.9%) and production (up to 627 × higher) while the importance of bacteria in leaf litter decomposition is rather minor (Gessner et al., 2007; Pascoal & Cássio, 2004; Pascoal et al., 2005; Suberkropp et al., 2010; Tant et al., 2013; Weyers & Suberkropp, 1996). For example, fungi contributed 95–99.7% of total microbial biomass and 88–95% of total microbial production on submerged leaf litter in southern Appalachian streams at Coweeta Hydrologic Laboratory, NC, USA (Gulis & Suberkropp, 2003a; Suberkropp et al., 2010). Fungal hyphae are capable of penetrating inside the leaf litter matrix and directly accessing plant polymers while bacteria are restricted to leaf surfaces. In addition, bacteria may rely to a greater extent on dissolved organic C from streamwater than fungi, rather than participating in leaf C processing. While leaf-associated fungi are strongly stimulated by elevated nutrient concentrations in water (see below), bacteria are either only slightly affected or the effect of dissolved nutrients is lacking (Suberkropp et al., 2010). Thus, we will focus on the nutrient effects on litter-associated fungi that drive increases in litter decomposition rates under nutrient enrichment.

In contrast to decomposition in terrestrial ecosystems, fungi associated with submerged leaf litter are capable of obtaining N and P from both the substrate and the water column (Cheever et al., 2013; Suberkropp, 1995). Thus, in streams, decomposition of leaf litter may depend not only on the nutrient content of the substrate but also the availability of N and P from the water column that can modify the activity of microbial decomposers (Suberkropp & Chauvet, 1995). Since plant litter C:N and C:P ratios are considerably higher than those of fungal biomass (Danger & Chauvet, 2013; Grimmett et al., 2013; Gulis et al., 2017), fungi have to alleviate the stoichiometric imbalance by either retaining N and P from leaf litter more efficiently than C or by immobilizing N and P from streamwater. Production of extracellular enzymes to obtain N and P from leaf litter is energetically costly, thus, fungi should preferentially use dissolved inorganic nutrients from the water column. This notion is supported by findings that higher concentrations of dissolved inorganic nutrients in laboratory studies and in whole-stream nutrient addition experiments or due to anthropogenic activities stimulate fungal activity, leading to nutrient immobilization and faster plant litter decomposition (e.g., Ferreira et al., 2006; Gulis, Ferreira et al., 2006; Gulis & Suberkropp, 2003a). Stimulation of fungal activity and plant litter decomposition by inorganic nutrients should be theoretically more pronounced for substrates with high initial C:N and C:P ratios, such as wood or rhododendron (*Rhododendron maximum*) leaves, due to more severe nutrient limitation of microbial activity on these substrates; on the other hand, external nutrients should have a less pronounced effect on leaf litter with initially high N or P content (e.g., leaves of N-fixing alder species). This pattern has been frequently reported in streams (e.g., Ferreira et al., 2006; Gulis et al., 2004, 2006; Gulis & Suberkropp, 2003a; Stelzer et al., 2003), though it can be complicated by variable lignin content of plant litter (Jabiol et al., 2019).

Early studies testing the effects of nutrient addition on microbially driven decomposition of plant litter in streams produced variable results (Elwood et al., 1981; Newbold, Elwood, Schulze, et al., 1983). In the last decades, however, multiple experiments in lab microcosms simulating stream conditions clearly demonstrated stimulation of microbial activity (fungal biomass accrual, growth efficiency, sporulation rate and cumulative spore production, respiration) and leaf litter decomposition by dissolved inorganic nutrients (Ferreira & Chauvet, 2011; Gulis & Suberkropp, 2003a, 2003b; Sridhar & Bärlocher, 1997; Suberkropp, 1998). Short-term whole-stream nutrient addition experiments have also shown positive effects of dissolved nutrients on microbial activity (fungal biomass accrual, sporulation, respiration) and plant litter decomposition rates (Ferreira et al., 2006; Rosemond et al., 2002). Multi-year nutrient enrichments provided additional fine details (Gulis et al., 2004; Gulis & Suberkropp, 2003a; Gulis et al., 2008; Rosemond et al., 2015; Tant et al., 2013, 2015), including uncovering important ecosystem-level consequences of elevated microbial activity and decomposition rates, namely accelerated C loss from the system due to downstream export of fine particulate organic matter (FPOM) and CO_2 evolution (Benstead et al., 2009). Since leaf litter decomposition rates, fungal biomass and sporulation rates in higher-order streams can be similar to those found in headwater streams (Baldy et al., 1995), the importance of aquatic fungi in regulating leaf litter decomposition extends beyond the reaches of headwater streams. Recently, manipulative experiments in streams also addressed the relative importance of dissolved

N and P (Kominoski et al., 2015; Manning et al., 2015, 2016). From the microbial perspective, it appears that nitrate-N has stronger effect on fungal activity and microbial decomposition rates while excess soluble reactive phosphorus (SRP) results in luxury P immobilization by fungi and sharp decreases in leaf litter C:P ratios (Gulis et al., 2017; Manning et al., 2015).

The relationship between dissolved inorganic nutrients and parameters of fungal activity or plant litter decomposition rates can be described in some cases by asymptotic saturation-type models (Ferreira et al., 2006; Gulis, Ferreira, et al., 2006; Gulis, Kuehn et al., 2006; Rosemond et al., 2002). In such models, large increases in fungal activity or decomposition rates occur with relatively small increases in nutrients and at low concentration levels suggesting that microbial nutrient demands can be easily satisfied by moderate nutrient enrichment. The half-saturation constants K_m (concentration at which half of the maximum decomposition rate or activity is reached) for DIN was estimated at <300 $\mu g\ L^{-1}$ and was as low as <20 $\mu g\ L^{-1}$ for SRP.

There are several possible mechanisms, some of them operating concurrently, that translate elevated fungal activity into faster leaf litter decomposition under nutrient addition scenarios (Fig. 16.2):

i. Previous studies have shown that, in general, the greater metabolic energy spent by microbes on acquisition of N and P, such as nutrient mining by extracellular enzymes, the less energy is directed towards the acquisition of C (Gallo et al., 2009; Linkins et al., 1990). Therefore, readily accessible external inorganic N and P should stimulate the activity of extracellular enzymes involved in sequestration of C from plant polymers (Güsewell & Freeman, 2005). Indeed, the activity of beta-glucosidase involved in degradation of cellulose from leaf litter was positively affected by dissolved inorganic N (but not P) availability (Gulis et al., unpublished).
ii. A related mechanism involves nutrient stimulation of the activity of fungal pectin-degrading enzymes that are crucial for litter mass loss resulting in enhanced maceration of leaf litter by releasing whole plant cells as FPOM (Jenkins & Suberkropp, 1995).
iii. Elevated nutrient concentrations stimulate fungal production including conversion of plant C into fungal spores (e.g., Ferreira et al., 2006; Gulis & Suberkropp, 2003a, 2003b, 2003c; Suberkropp et al., 2010) that are released into the current, and fungi can channel up to 80% of production (or 8–12% of leaf mass loss) into sporulation (Suberkropp, 1991).
iv. Elevated nutrients are known to stimulate microbial respiration associated with decaying submerged leaf litter resulting in increased litter C losses as CO_2 to the atmosphere (Benstead et al., 2009; Gulis & Suberkropp, 2003a, 2003b, 2003c; Suberkropp et al., 2010).
v. As leaf litter decomposes, fungal biomass accrual and nutrient immobilization from the water column lead to changes in litter stoichiometry (decreases in C:N and C:P ratios) and an increase in nutritional quality and palatability of decaying leaf litter to detritivores (Bärlocher, 1985; Gessner et al., 2007) that

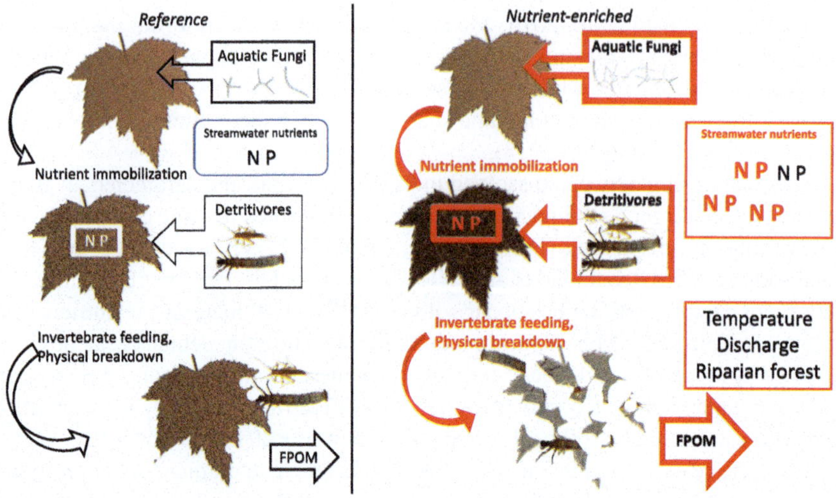

Fig. 16.2 Conceptual diagram depicting leaf litter decomposition in low nutrient streams (left diagram) and effects of nutrient (N, P) enrichment of streamwater on leaf litter decomposition and associated microbial decomposers (right diagram). In general, nutrient enrichment of streamwater with N and P stimulates fungal biomass accrual and activity on leaf litter with the release of C as CO_2 and fine particulate organic matter (FPOM; including fungal spores), both of which occur throughout the decomposition sequence. Fungal-mediated dissolved nutrient immobilization and litter softening lead to changes in key litter attributes, such as decreases in litter C:N and C:P ratios and litter toughness, which result in increased litter palatability for detritivores. Increased detritivore biomass and activity promotes further litter mass loss by incorporation of litter C into secondary production or release as FPOM (i.e., small leaf fragments and feces). Red arrows in the right diagram indicate stimulation with increased nutrient concentrations. Effects of nutrient enrichment on microbes, detritivores and litter decomposition are likely modified by temperature, discharge, and changes in riparian vegetation. Figure modeled after Cummins and Klug (1979) and Marks (2019). Images of *Pycnopsyche* and *Taeniopteryx* from Macroinvertebrates.org (CC BY-NC 4.0). Maple leaf vector images: Tracy Saxby, Integration and Application Network, University of Maryland Center for Environmental Science (ian.umces.edu/imagelibrary/)

in turn stimulate detritivore feeding and furthers leaf litter decomposition (see below).

The ability of aquatic fungi to control and homogenize detrital stoichiometry as plant litter decomposes may have important consequences to stream detritivores, which rely on plant-associated microbial biomass rather than plant material alone as a major source of nutrients (Chung & Suberkropp, 2009a, 2009b; Suberkropp, 1992). Thus, litter-associated fungi are important intermediaries in energy and nutrient transfer in streams while their activity and role in leaf litter decomposition can be modified by inorganic N and P availability.

16.2.2 Detritivore-Mediated Litter Processing

Detritivores colonizing leaf litter promote decomposition via feeding activity and fragmentation and due to using leaf litter to make cases (e.g., case-building caddisflies of the families Calamoceratidae, Lepidostomatidae, Limnephilidae) (Azevedo-Pereira et al., 2006; Moretti et al., 2009; Rincón & Martínez, 2006; Sanpera-Calbet et al., 2009). Detritivores have been shown to contribute up to 63.5% of total leaf litter mass loss in streams (Cornut et al., 2010; Hieber & Gessner, 2002; Taylor & Chauvet, 2014). The fine particles released by shredding detritivore activities (i.e., small leaf fragments and feces) are used by invertebrate collectors (Cummins & Klug, 1979), while shredders and collectors serve as food for predators (e.g., invertebrates, fish) (Flecker & Townsend, 1994; Yule et al., 2010). Thus, detritivores play an important role in mediating energy and matter transfer from the litter to higher trophic levels with the effects of nutrient enrichment modifying invertebrate-dominated food webs and nutrient cycles (Cross et al., 2003, 2006; Davis et al., 2010).

The colonization of submerged leaf litter by microbes, with the accumulation of microbial biomass and litter softening, generally increases susceptibility to physical abrasion in high flows (Manning et al., 2015), and its palatability to invertebrate detritivores (i.e., shredders; reviewed by Bärlocher & Sridhar, 2014; Graça, 2001). Thus, the stimulation of microbial biomass accumulation and activity on submerged leaf litter by increased nutrient availability in streamwater (see above, *Microbially mediated litter processing*) may facilitate and promote detritivore colonization of the litter. For instance, Gulis, Ferreira et al. (2006) found higher macroinvertebrate abundance and taxa richness on oak (*Quercus robur*) and alder (*Alnus glutinosa*) leaves decomposing in nutrient enriched streams (216–2996 μg NO_3-N L^{-1} and 25–56 μg SRP L^{-1}) than in paired reference streams with ambient nutrient concentration (42–483 μg NO_3-N L^{-1} and 3–16 μg SRP L^{-1}) in central Portugal, and Greenwood et al. (2007) found higher detritivore biomass on rhododendron and maple (*Acer rubrum*) leaves decomposing in a stream experimentally enriched with N and P (~400 μg DIN L^{-1} and ~45 μg SRP L^{-1}) than in a reference stream (<30 μg DIN L^{-1} and <10 μg SRP L^{-1}) in the Appalachian Mountains (North Carolina, USA). Ferreira et al. (2006), however, did not find an effect of experimental N enrichment (214–983 μg NO_3-N L^{-1} *vs.* 33–104 μg NO_3-N L^{-1} in reference conditions) on macroinvertebrate abundance associated with decomposing litter in a forest stream in the protected area of Açor Mountain (central Portugal). Effects of nutrient enrichment on benthic detritivores associated with decomposing litter are, thus, likely to be context dependent. The facilitation of detritivore colonization of litter under nutrient enriched conditions may be more pronounced on low-nutrient leaf species (e.g., oak, rhododendron), where microbial activity may be nutrient limited to a greater extent and, therefore, more responsive to dissolved N and P than on nutrient-rich leaf species (e.g., alder, maple), which translates into a stronger stimulation of microbial biomass accrual and nutrient immobilization (Greenwood et al., 2007; Gulis, Ferreira et al., 2006; but see Ardón et al., 2006). Also, detritivore contribution to litter decomposition may be greater under moderate nutrient enrichment compared to reference conditions (Gulis, Ferreira et al., 2006), while it may decrease under high nutrient concentrations (Lecerf et al., 2006; Woodward et al., 2012) due to toxicity or other negative effects of concomitant pollutants.

Correlative and experimental tests of the mechanisms through which increased stream nutrient concentrations affect detritivore communities are becoming more common. Available evidence suggests that interactions among streamwater nutrients, microbial immobilization of nutrients on leaf litter, and subsequent reductions to imbalances between detritivore nutrient demands and resource nutrient content are important (Cornut et al., 2015; Demi et al., 2018; Frainer et al., 2016; Manning et al., 2016). Thus, a useful framework that has advanced our understanding of nutrient enrichment-detritivore interactions is ecological stoichiometry theory (Cross et al., 2005 and references therein; Danger et al. [Chapter 3 in this volume]), which considers the mass balance between consumer nutrient demand (typically in terms of biomass C:nutrient ratio) and nutrients supplied in resources (C:nutrient ratio of leaf litter). In detritus-based ecosystems such as forest streams under reference conditions, detritivore nutrient demands can far exceed nutrients contained in autumn-shed leaf litter, which tends to be nutrient poor (Cross et al., 2003; Hladyz et al., 2009). The consequences of these drastic imbalances between nutrient demand and litter nutrient resources for detritivores likely include combinations of reduced growth, reproduction and survival. Thus, since nutrient enrichment modulates litter nutrient concentrations, these constraints on detritivore populations can be reduced, which may lead to increases in the consumption efficiency, individual mass, body condition, and abundance of key detritivore taxa (Connolly & Pearson, 2013; Danger et al., 2013; Halvorson et al., 2018).

Increased P content of leaf litter after nutrient enrichment appears to be a critical factor for detritivore responses to nutrients in some cases, especially given the relatively high and potentially flexible body P of some detritivore taxa (e.g., *Tallaperla* sp. ~1% P; Cross et al., 2003; Prater et al., 2020), and the importance of dietary P for rapid growth (Demi et al., 2018; Halvorson et al., 2016; Prater et al., 2015), and detritivore fitness (Connolly & Pearson, 2013). In an experimental study of five streams continuously enriched with varying concentrations of N and P, Demi et al. (2019) showed strong positive detritivore community responses, with 30–300% increases in detritivore biomass under nutrient-enriched conditions. The taxa that responded to nutrient enrichment (especially stoneflies of the genera *Allocapnia*, *Leuctra*, and *Tallaperla* and the caddisfly genus *Pycnopsyche*) also showed increased biomass with nutrient-induced decreases in litter C:P ratios. The relationships between experimental nutrient enrichment, reduced litter C:P ratios, and detritivore biomass found by Demi et al. (2019) are also consistent with correlative evidence from a landscape-scale study that dealt with the effects of a streamwater nutrient (TN, TP) and corresponding litter nutrient (C:N, C:P) gradient (Prater et al., 2015). In this study, detritivores with low body C:P ratios (e.g., caddisflies of the genus *Pycnopsyche*) tended to have higher abundance and biomass in streams that exhibited higher P concentrations, and corresponding higher quality leaf litter (lower C:P ratios).

16.2.3 *Comparing the Magnitude of Microbial Decomposer vs. Detritivore Effects on Decomposition*

The relative importance of microbial and detritivore contributions to leaf litter decomposition changes as decomposition progresses (Hieber & Gessner, 2002) with the initial microbial colonization and conditioning of leaf litter being critical for the subsequent consumption by detritivores (Fig. 16.2; see above *Microbially mediated litter processing* and *Detritivore mediated litter processing*). The different traits and functions of microbial decomposers, detritivores, and their interactions can result in different outcomes for their cumulative contributions to leaf litter decomposition under nutrient-enriched conditions. For instance, because microbial decomposers can obtain nutrients from both leaves and continuously renewed dissolved nutrients in flowing water, their responses to elevated dissolved nutrients can be rapid, however, microbial demands will be likely met at relatively low dissolved nutrient concentrations. Detritivores exclusively rely on nutrients from leaf litter (including associated microbial biomass), so their responses to nutrient enrichment are mediated by microbial immobilization of N and P from the water column, especially when freshly fallen leaf litter has low nutrient content. Using field data from an experimental enrichment of a forest stream, combined with microbial assimilation and invertebrate feeding models, Tant et al. (2015) quantified the relative contributions of fungi, bacteria, and detritivores to decomposition rates under reference and nutrient-enriched conditions. Their findings largely confirmed that microbial decomposers contribute 3.9–6.9× more than detritivores at early stages of rhododendron leaf decay (days 0–49) under reference conditions, with contributions of detritivores outweighing those of microbial decomposers at later stages of decay (days 49–108). Under nutrient-enriched conditions, the relative importance of detritivores was greatly increased, so they contributed more to decomposition than microbial decomposers by earlier stages of rhododendron decay (days 7–49) due to an early and relatively high peak of fungal biomass (Tant et al., 2015). This modeling approach to quantify the relative importance of microbial decomposers *vs.* detritivores in response to nutrient enrichment underscores the possibility that detritivore-mediated decomposition may respond strongly to nutrients, particularly when leaf litter is nutrient poor.

Another common method to separate the effects of macroinvertebrates and microorganisms for litter processing is to enclose leaf litter in mesh bags of different mesh size. Typically, mesh aperture <1 mm is sufficient to exclude larger detritivores, allowing for comparisons between litter decomposition mediated by microbial decomposers alone (k_{fine}) *vs.* litter processing by shredding detritivores and microbial decomposers together (k_{coarse} or k_{total}). For example, Gulis, Ferreira et al. (2006) showed that litter type was an important driver of differential response of litter decomposition rates to nutrient enrichment when invertebrates were excluded (k_{fine}) *vs.* present (k_{coarse}). The relative importance of each group of decomposers can be compared by examining the ratio of k_{coarse}/k_{fine} (i.e., decomposition due to microbial and detritivore activity together relative to decomposition due to microbial activity only). In Gulis, Ferreira et al. (2006), k_{coarse}/k_{fine} ratio was 2.54 and 3.59

for alder leaves in reference and nutrient-enriched conditions, respectively, while for slower-decomposing oak, k_{coarse}/k_{fine} was 1.92 and 3.72 in reference *vs.* nutrient-enriched streams. Both cases illustrate that k_{coarse} responded to nutrient enrichment to a greater extent, especially for nutrient-poor oak, demonstrating that nutrient effects on decomposition rates are amplified when detritivores are present. In contrast to this finding, an exhaustive meta-analysis (Ferreira et al., 2015) of nutrient enrichment effects in correlative studies found that litter decomposition rates in coarse and fine mesh bags were stimulated to a comparable degree (26 and 21%, respectively), while in manipulative studies nutrient effects were significant for litter decomposition rates in fine mesh bags (35% increase) but not in coarse mesh bags (Fig. 16.3a). This finding is inconsistent with the prediction that nutrient enrichment should stimulate total decomposition rates more than microbial decomposition rates; however, as noted previously, leaf litter traits, such as C quality and C:nutrient ratios, likely have a modulating effect on the magnitude of microbial and detritivore responses to nutrient enrichment.

Insights from experimental studies and watershed- to regional-scale surveys of litter decomposition rates have been important for gaining a firm understanding of the mechanisms that stimulate leaf litter decomposition in nutrient-rich streams. However, these studies often face inherent shortcomings in terms of the narrow range of nutrient concentrations that are tractable to achieve, in addition to the

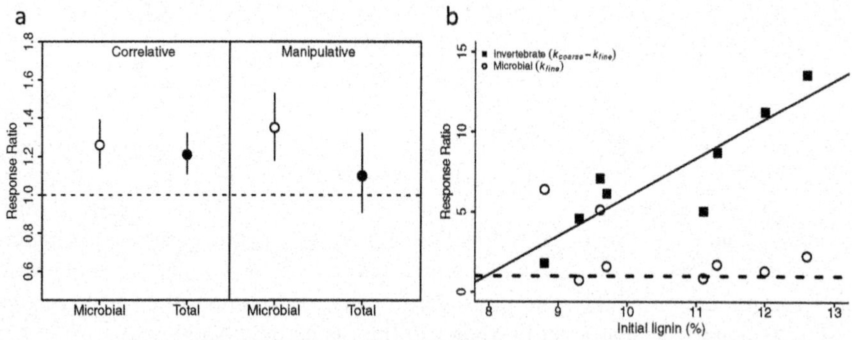

Fig. 16.3 Response ratios ($k_{\text{nutrient-enriched}}/k_{\text{reference}}$) and 95% confidence intervals (CI [vertical bars]) of microbially mediated (i.e., fine mesh) and total (i.e., coarse mesh) decomposition rates from Ferreira et al. (2015) (**a**), and of microbially (i.e., fine mesh) and invertebrate-mediated (i.e., coarse–fine mesh) decomposition rates as a function of initial litter lignin concentration (%) from a multi-year experimental enrichment of 5 streams (Manning et al., 2016) (**b**). The horizontal dashed line in (**a**) and (**b**) indicates a response ratio of 1 ($k_{\text{nutrient-enriched}} = k_{\text{reference}}$). In (**a**), nutrient enrichment stimulated both microbially and total (microbially and invertebrate-mediated) decomposition rates in correlative studies, but only microbially mediated decomposition showed a response in manipulative stream channel studies (the 95% CI around the mean response ratio does not cross 1); response ratios did not differ across treatments (the 95% CIs overlap). In (**b**), invertebrate-mediated decomposition rate response ratios (filled squares) increased as a function of initial lignin concentration (linear regression; $k_{coarse\text{–}fine}$ response ratio = 2.4 × initial lignin – 18.3; R^2=0.75, $P = 0.003$), whereas microbially mediated decomposition response ratios showed no relationship with initial lignin concentration

idiosyncrasies of limited spatial scales (e.g., effects of regional climate or land use). Thus, continental-scale experiments that involve standardized methods and extensive spatial replication offer considerable promise for testing hypotheses related to nutrient enrichment that cannot be addressed by small-scale studies. These continental-scale experiments are especially effective because they reflect the spatial scale of pervasive nutrient enrichment. The pan-European RivFunction project addressed the effects of nutrient enrichment on decomposition rates of oak and alder leaf litter enclosed in both coarse and fine mesh bags that were incubated in 100 streams spanning a broad nutrient gradient (Woodward et al., 2012). Decomposition rates of both oak and alder leaf litter showed a hump-shaped relationship with dissolved nutrient concentration, especially for detritivore-mediated litter decomposition (Woodward et al., 2012). Notably, decomposition rates responded in this way to both dissolved N and P concentrations. Slow leaf litter decomposition in low nutrient streams was likely due to nutrient limitation of microbial activity, while low decomposition rates at extreme nutrient enrichment were likely due to toxic effects to detritivores (e.g., high ammonium or nitrite concentrations) or concomitant changes in other environmental characteristics (e.g., oxygen depletion, smothering by fine sediments, other toxic pollutants). Thus low-to-moderate nutrient enrichment should elicit the strongest response of litter decomposition rates in streams affected by anthropogenic pollution. This generalization was made possible due to the extremely wide ranges of nutrient concentrations that occurred in the 100 study streams across Europe (1–926 μg SRP L^{-1}, 14–21,641 μg DIN L^{-1}). Similar extreme ranges in nutrient concentrations have been documented on other continents, including North and South America, Australia, and Asia (McDowell et al., 2020). However, continental-scale studies investigating nutrient enrichment effects on leaf litter decomposition in these regions are nonexistent, such that it remains unclear whether similar patterns in leaf litter decomposition rates exist across other continents. The sampling bias toward temperate regions is especially important to address given that leaf litter species and benthic communities in tropical regions may respond differently to nutrient enrichment due to the unique phenologies of riparian forests, their phylogenetic context (e.g., LeRoy et al., 2020), and the biogeography of microbial decomposers and detritivores in these catchments (Boyero, Pearson, Dudgeon et al., 2011; Boyero et al., 2015; Seena et al., 2019).

16.2.4 Litter C Quality and C:Nutrient Stoichiometry

Multiple initial characteristics of autumn-shed leaf litter affect their decomposition rates (Enríquez et al., 1993; Webster & Benfield, 1986). Among these characteristics, we will focus on leaf litter C quality, or abundance of structural plant polymers (e.g., lignin), in addition to leaf litter nutrient stoichiometry (i.e., C:nutrient ratios) and their interactions with streamwater nutrient supply. Litter types with high concentration of recalcitrant compounds (e.g., lignin) have been shown to consistently respond to nutrient enrichment to a greater extent than more labile litter types (Ferreira et al.,

2015; Manning et al., 2016), with some exceptions (Ardón et al., 2006). A recent study showed that for standardized C substrates that differed in structural compounds (recalcitrant wood veneers *vs.* labile cellulose sponge), microbial respiration and decomposition were stimulated on both substrates, but decomposition had a greater response to nutrients on recalcitrant wood veneers (Usher et al., 2020).

A key predictor of slower leaf litter decomposition, beyond bulk C:nutrient concentrations, is leaf litter structural and defense compounds that can delay initial colonization by fungi, and impede detritivore consumption. Specifically, structural plant polymers, like lignin and aromatic compounds related to plant defenses (e.g., polyphenols, tannins), are often associated with slower decomposition rates (Ardón et al., 2006; Jabiol et al., 2019; LeRoy & Marks, 2006). Lignin is especially difficult for many microbial decomposers to break down, requiring the ability to produce ligninolytic enzymes (Hendel et al., 2020). Slow-decomposing leaf litter that contains higher amounts of these structural compounds is often also nutrient poor (i.e., high C:nutrient ratios). However, the relative importance of litter C:nutrient concentration *vs.* lignin or polyphenol concentration in modulating the effects of dissolved nutrient enrichment remains understudied. In a microcosm experiment investigating the interactions between litter nutrient and lignin concentration across 38 litter species, Jabiol et al. (2019) found that lignin concentration was a stronger predictor of microbial responses to dissolved nutrient enrichment than initial litter nutrient concentration; leaf litter decomposition rates increased 2.9× for lignin-poor leaf litter *vs.* 1.4× for lignin-rich leaf litter. However, it is well established that microbial activity and decomposition rates of wood that has much higher lignin content than leaf litter (but also lower nutrients) nevertheless respond more strongly to dissolved nutrients than those of leaf litter (Ferreira et al., 2006; Gulis et al., 2004; Stelzer et al., 2003). Thus, it appears that relaxed nutrient limitation due to availability of dissolved nutrients may alleviate limitations specific to structural C compounds as well.

Some detritivore taxa harbor gut microbiota that aid in the digestion of complex C compounds, perhaps reducing the importance of lignin as a barrier to its consumption by detritivores (Canhoto & Graça, 2006). Detritivore consumption of bulk leaf material and non-consumptive uses (i.e., case-building by caddisflies) implies that detritivores may respond differently to the interactive effects of high-lignin litter and nutrient enrichment because of enhanced conditioning that changes the mechanical features of the litter (reduced leaf toughness; Foucreau et al., 2013). To further examine the interplay between leaf litter lignin and nutrient enrichment, we explored responses (response ratio = nutrient-enriched/reference) of 4 leaf types (red maple, tulip poplar [*Liriodendron tulipifera*], chesnut oak [*Quercus prinus*], and rhododendron) with variable initial lignin concentration to nutrient enrichment using data from Manning et al. (2016). Litter was enclosed in litterbags with fine and coarse meshes, to allow for comparison between microbial (k_{fine}) and detritivore-mediated ($k_{coarse}-k_{fine}$) decomposition rates. These data illustrate increasing response magnitude for decomposition rates mediated by detritivores (plus physical abrasion), compared to microbially mediated decomposition, as a function of initial lignin concentration (Fig. 16.3b). This evidence, combined with previous findings that indicate differential effects of lignin on microbial decomposers *vs.* detritivores, suggests that microbial decomposition of lignin-rich litter species will respond to a lesser degree to nutrient enrichment than detritivore-mediated litter decomposition.

As mentioned previously (see *Microbially mediated litter processing*, above), aquatic fungi that colonize leaf litter are able to use both dissolved nutrients and nutrients contained in leaf litter (e.g., Cheever et al., 2013; Pastor et al., 2014; Suberkropp, 1998). This ability can allow fungi to immobilize dissolved inorganic nutrients and alter litter C:nutrient ratios as decomposition progresses (Cheever et al., 2013; Cornut et al., 2015; Gulis et al., 2017). As a result, patterns of nutrient immobilization and the ratio of C:nutrients in decomposing leaf litter are expected to vary through time (from initial to late stages of decomposition), and with litter type, where initially nutrient poor species likely gain disproportionate amounts of nutrients relative to nutrient-rich species (Manzoni et al., 2010; Scott et al., 2013).

Landscape-scale relationships between stream nutrient concentrations and the C:nutrient ratio of leaf litter have been observed, consistent with the prediction that the degree of nutrient immobilization in leaf litter is driven partly by dissolved nutrient availability. Across landscape-scale gradients of stream P concentrations, Scott et al. (2013) observed a negative relationship between litter C:P ratio and increasing stream P. A similar relationship was observed by Prater et al. (2015), with associated consequences for shredding macroinvertebrate communities (see *Detritivore-mediated litter processing*, above). In experimental contexts, fungal-mediated changes to leaf litter C:nutrient stoichiometry have been shown to be an important link between streamwater nutrients and decomposition rates, especially for invertebrate-mediated decomposition (Manning et al., 2015).

16.2.5 Mechanistic Effects of N *vs.* P

The mechanisms of effects of streamwater N *vs.* P via effects on fungal biomass, associated change in litter nutrient content, and effects on detritivores are somewhat similar (Manning et al., 2015), with a key difference. A structural equation analysis was used to discern differences in the effects of N *vs.* P in an experiment in which both nutrients were increased in 5 streams, but with nutrient gradients running in opposite directions (i.e., in high P streams, N was low and in high N streams, P was low). This analysis revealed that dissolved P had effects on decomposition not only through fungal uptake and effects on litter stoichiometry, but that there was additional variance in litter stoichiometry explained by streamwater P concentrations alone, suggesting storage of P when it was available (Fig. 16.4). Nitrogen effects on litter stoichiometry were explained by fungal biomass, with no additional evidence of storage. The ability of fungi to store P but not N, affecting fungal and detrital stoichiometry, has been corroborated in laboratory studies (Gulis et al., 2017). These findings may have implications for management, as fungi may be able to store excess P in relation to its availability in streamwater, which is temporally variable.

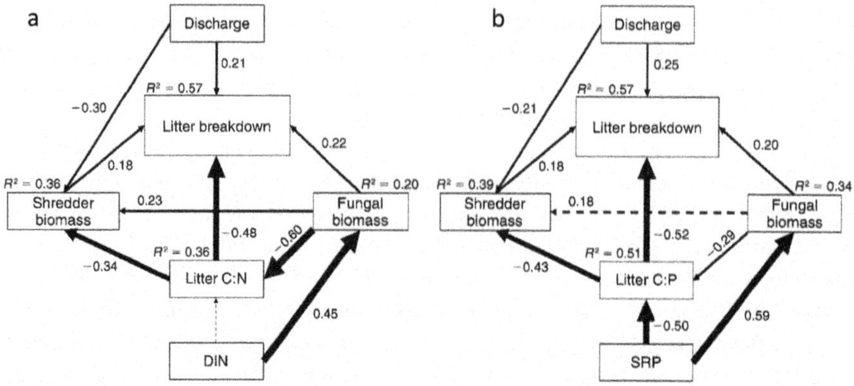

Fig. 16.4 Structural equation models showing relationships among streamwater dissolved inorganic nitrogen (DIN; **a**) and soluble reactive phosphorus (SRP; **b**) concentrations and drivers of leaf litter decomposition rates. Standardized path coefficients are shown, where the sign of the coefficient indicates the direction of the effect between variables. Arrow weights correspond to path coefficients adjusted based on standard deviations, with the strength of the correlations indicated by arrow width (small, medium, and large arrows denote adjusted coefficients <0.30, >0.30 and <0.45, >0.45, respectively). Path coefficients not significantly different from zero are shown with dashed arrows. Figure redrawn from Manning et al. (2015), with permission.

16.3 Consequences of Nutrient Effects on Litter Decomposition for Aquatic Ecosystems

16.3.1 Other Global Change Drivers Interact with Nutrients: Nutrient × Temperature Effects on Leaf Litter Decomposition

The effects of nutrient enrichment on litter decomposition in streams may be modified by concomitant changes in other environmental variables. For instance, nutrient enrichment of streamwater in agricultural contexts can be accompanied by increases in water temperature if the riparian vegetation is removed, which increases the amount of solar irradiation reaching the stream (Gomi et al., 2006; Johnson & Jones, 2000; Kiffney et al., 2003). Also, higher water temperatures in a global warming context can be accompanied by increases in dissolved nutrient concentration because of increased nutrient mineralization and decreased water volume (Moss et al., 2011; Murdoch et al., 2000).

The last decade has witnessed an increase in the number of studies addressing possible interactive effects of nutrient enrichment and increased temperature on litter decomposition. While moderate increases in each factor generally stimulate litter decomposition, their interaction is difficult to predict with synergistic, antagonistic and additive effects reported. For instance, Ferreira and Chauvet (2011) reported synergistic effects between nutrient concentration (low: 1390 μg NO_3-N L^{-1} and

10 μg PO_4-P L^{-1}; high: 13860 μg NO_3-N L^{-1} and 100 μg PO_4-P L^{-1}) and water temperature (5, 10 and 15 °C) on alder leaf litter decomposition in microcosms, with stronger effects of nutrient enrichment in warm than cold water conditions and stronger effects of warming in high than in low nutrient conditions. Fernandes et al. (2014) also found an interaction between the effects of nutrient concentration (0.09–5 mg NO_3-N L^{-1}) and water temperature (12 and 18 °C) on alder and oak leaf litter decomposition in microcosms with stronger effects of temperature at low nutrient concentrations. Piggott et al. (2012) found an interaction between the effects of nutrient concentration and water temperature (ambient and elevated by 1.4 °C) on mahoe (*Melicytus ramiflorus*) leaf litter tensile strength loss in mesocosms, whereas Piggott et al. (2015) found no significant interaction between the effects of nutrient concentration and water temperature (6 levels, 0.7–6 °C above ambient temperature) on seven mahoe litter decomposition measures in mesocosms. Manning et al. (2018) found additive effects of nutrient concentration (82–517 μg DIN L^{-1} and 93–9.7 μg SRP L^{-1}, N and P varied in opposite directions) and water temperature (0.8–19.5 °C) on microbial respiration rates in streams under experimental nutrient enrichment over a seasonal temperature gradient. Leaf litter AFDM-specific respiration rates were 1.24–1.51× higher under nutrient enriched conditions and were positively related with temperature, but no interaction was found between factors (Manning et al., 2018). When fungal biomass-specific respiration rates were considered for leaf litter, no effect of nutrient enrichment was detected, while the effect of temperature was positive.

Studies addressing the interaction between nutrient concentration and water temperature are still scarce and use distinct methodological approaches, which make comparisons and predictions difficult. The distinct types of interaction found may reflect different methods or suggest that other factors may modify the interaction between nutrients and temperature (e.g., litter characteristics, biotic communities). For instance, while both temperature and nutrients are generally predicted to increase microbially mediated respiration and leaf processing (Follstad-Shah [Chapter 12 in this volume], Tiegs et al., 2019), higher temperatures likely inhibit stream detritivores over the short term, or potentially extirpate cold-adapted detritivores over the long term, decreasing litter decomposition rates (Boyero, Pearson, Dudgeon et al., 2011; Boyero et al., 2016). Effects of temperature on growth and activity of aquatic fungi also vary across temperatures commonly found in streams (Dang et al., 2009) and can be further complicated by temperature-driven changes in fungal community structure. Nevertheless, considering possible interaction between nutrient concentrations and water temperature could be critical for water resource managers in the face of climate and land use changes that couple higher temperatures and nutrient enrichment. For instance, low nutrient concentration in an oligotrophic mountain stream likely mitigated the effect of experimental stream warming on litter decomposition (Ferreira & Canhoto, 2015), while low water temperature reduced the potential stimulatory effect of high nutrient concentration on litter decomposition in insular streams (Ferreira et al., 2016).

16.3.2 Nutrient Enrichment Results in Shorter C Residence Time in Streams

Managing nutrient pollution requires consideration of how leaf litter in streams contributes to ecosystem functions that benefit and support ecosystem health and human well-being (i.e., ecosystem services; Frainer et al. [Chapter 21 in this volume], Richardson et al. [Chapter 22 in this volume]); with the understanding that nutrient enrichment may modify the availability (timing, retention and export) of leaf litter resources that fuel stream food webs, as well as stream nutrient uptake rates and export (Newbold, Elwood, O'Neill et al., 1983; Robbins et al., 2019), and feedbacks to the global C cycle and climate change (Follstad-Shah [Chapter 12 in this volume]). Models and observations that target site- to catchment-scale understanding of how nutrients speed the sequence of leaf litter depletion from annual peaks to annual minima could provide several useful benchmarks that link stream leaf litter to its important roles as a driver of other critical stream ecosystem functions. Fortunately, measuring stream 'decomposition potential' via litterbag experiments generally mirrors rates of detritus loss at the stream-reach scale (Rosemond et al., 2015) such that litterbag studies remain a viable and economical option to parameterize models and make predictions about annual depletion of leaf litter standing stocks for any stream of interest with relatively few data points (Fig. 16.5). With litter decomposition rate data in hand, investigators could simulate multiple scenarios, including the effects of nutrient enrichment, and apply them to predict how quickly leaf litter would be processed within a given stream (e.g., time to 50% mass loss [T_{50}], mean residence time, etc.). In addition, estimates of the temporal dynamics of leaf litter standing stocks at stream-reach scale could potentially be used within hierarchical models to predict other ecosystem functions that are coupled to the amount of organic matter in the stream, such as nutrient uptake (Robbins et al., 2019), consumer production (Venarsky et al., 2018; Walther & Whiles, 2011), and ecosystem metabolism (Bernot et al., 2010).

16.4 Management Implications of Nutrient Enrichment Effects on Leaf Litter Decomposition

16.4.1 Litterbag-Scale Measurements Are Predictive of Whole Stream Reach Processes

Nutrient enrichment has been shown to have effects on standing stocks of detrital C at whole stream-reach scales (Rosemond et al., 2015). Because both small-scale litterbag measurements and stream-reach measurements were made concurrently, Rosemond et al. (2015) provide evidence that litterbag measurements accurately predict larger scale processes (Fig. 16.5b). Further, decomposition rates from

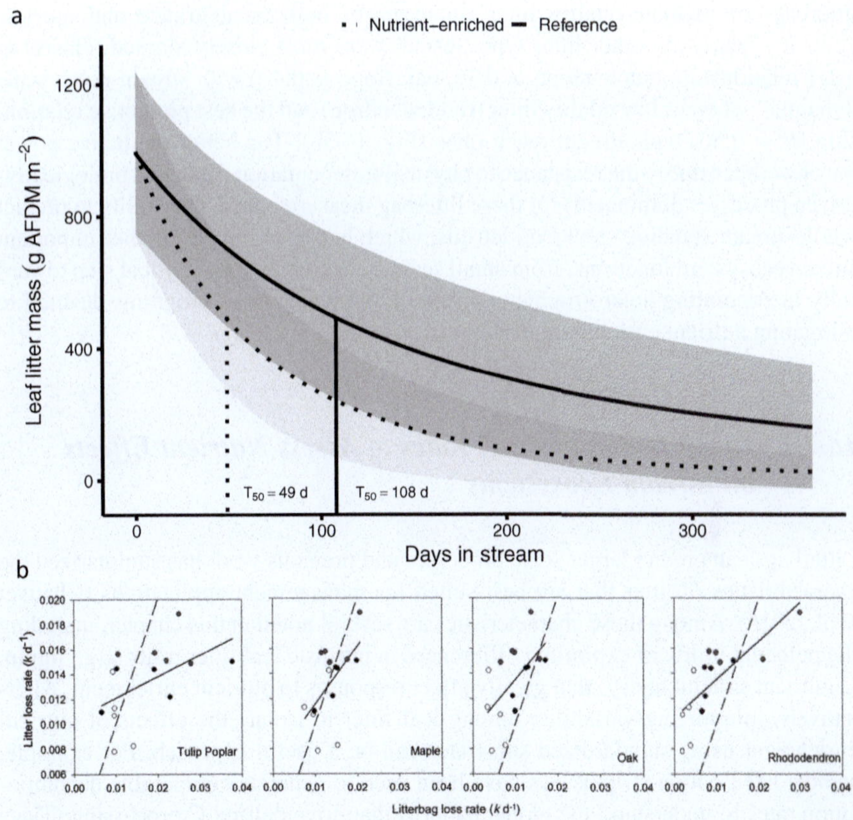

Fig. 16.5 Modeled benthic leaf litter mass (g AFDM m^{-2}) within hypothetical streams under reference and nutrient-enriched conditions (after Rosemond et al., 2015) (**a**). We simulated peak benthic leaf litter mass within the stream from a normal distribution with mean 1000 g AFDM m^{-2} and standard deviation of 250 g AFDM m^{-2}. We then used random draws from decay coefficients observed for 4 leaf litter species (*Acer rubrum* [Maple], *Quercus prinus* [Oak], *Liriodendron tulipifera* [Tulip poplar], *Rhododendron maximum* [Rhododendron]) under reference ($n = 80$) and nutrient-enriched ($n = 160$) conditions from an experimental nutrient enrichment of 5 streams in the Southern Appalachian mountains, North Carolina, USA (Manning et al., 2016). Solid and dotted curved lines indicate the mean leaf litter mass at time t from 1000 simulations under reference and nutrient-enriched conditions, respectively; gray and light gray areas indicate ± 1SD under reference and nutrient-enriched conditions, respectively. Solid and dotted vertical lines show the time to 50% mass remaining (T_{50}) for reference and nutrient-enriched, respectively. On average, T_{50} was reduced by about half with nutrient enrichment (T_{50} = 108 days at reference; T_{50} = 49 days at nutrient-enriched). In (**b**), we show how measuring litter decomposition rates using litterbags for these same four litter species can approximate whole-stream litter loss rates. Whole-stream leaf litter loss rates are shown from corresponding streams and years for the four leaf litter types across the five streams under reference (open circles) and years 1 (gray circles) and 2 (black circles) of experimental N and P enrichment. Litterbag rates were quantified from coarse-mesh bags and represent microbial + detritivore-mediated decomposition. The dashed line in each panel represents a 1:1 relationship, the solid line indicates the linear relationship between litterbag and whole-stream litter loss rates

litterbags of medium-quality litter (in terms of both recalcitrance and nutrient content—relative to other litter types tested) were most closely aligned (closest to a 1:1 relationship; maple slope = 0.46, oak slope = 0.40) with stream-reach scale dynamics, whereas low quality litter (rhododendron) had the best predictive relationship (R^2= 0.56) with stream-reach rates (Fig. 16.5b). The better predictive power may have been due to the resistance to physical fragmentation (such that biologically-driven processes dominated) for these litterbag measurements. The ability to predict whole-stream standing stocks of detritus, which has been linked to other important stream ecosystem functions, from small scale measurements is a critical step toward fully incorporating heterotrophic processes into management programs devoted to mitigating nutrient enrichment in streams.

16.4.2 Using Decomposition Rates to Assess Nutrient Effects on Stream Ecosystems

Litterbags can predict larger scale processes and previous work has summarized the characteristics of litter that are best-suited for management applications (Chauvet et al., 2016). Among these characteristics are several noted in this chapter, including the potential utility of exploiting differences in intrinsic leaf litter traits (e.g., lignin, C:nutrient stoichiometry) that greatly affect responses to nutrient enrichment. Alternatively, minimizing variability among leaf litter to isolate the effects of nutrient enrichment using standardized substrates can be a useful approach. For example, standardized cotton-strip assays have been used to generate comparable decomposition rates to understand the global patterns that drive detrital C processing (Tiegs et al., 2019), but they have yet to be applied to understand the effects of nutrient enrichment at such extensive scales (Colas et al., 2019). Further, litter decomposition responses to nutrient enrichment can often be non-linear, complicating efforts to detect and effectively manage the effects of nutrient pollution in streams (Dodds et al., 2010; Jarvie et al., 2013; Woodward et al., 2012). In general, nutrient enrichment is predicted to increase rates of leaf litter processing according to the asymptotic model (e.g., Monod- or Michaelis-Menten-type relationships; Ferreira et al., 2006; Kominoski et al., 2015; Rosemond et al., 2002), but as mentioned throughout this chapter, certain conditions can result in unchanged or slower leaf litter decomposition than would be predicted based on nutrient concentrations alone (e.g., Royer & Minshall, 2001; Woodward et al., 2012). These challenges may require the use of additional standardized substrates, in tandem with substrates like cotton strips, that are sensitive to both microbial and detritivore-mediated decomposition, in order to adequately quantify their interaction in response to nutrient enrichment. Recent evidence from a study that exploited landscape scale nutrient concentration gradient suggests that wood veneer substrates can be predictive of nutrient enrichment effects, and notably, sensitive to relatively narrow ranges of nutrient concentrations (Usher et al., 2020). While wood veneers are unlike leaf litter in many ways, they have several similar intrinsic characteristics that may make them a suitable substrate for detecting either N or P enrichment effects: they have low nutrient content and high lignin concentrations, are consumed by detritivores (e.g., Eggert & Wallace, 2007), and are resistant to physical abrasion. Thus, the combination of standardized cellulose substrates (such as cotton strips) and standardized substrates that are recalcitrant (e.g., wood veneers) could be a powerful tool for predicting nutrient enrichment effects on leaf litter across landscape-scale gradients that either dampen or enhance interactions among nutrients, microbial decomposers, and stream detritivore communities.

16.5 Conclusions

Nutrient enrichment will continue to threaten freshwater resources that provide for crucial ecological and societal needs. The extent of the problem suggests that many ecosystem functions and the services they provide will be affected, with undesirable consequences for stream ecosystem health. Among the ecosystem functions that occur within streams, leaf litter decomposition is an established metric that is currently and will continue to be affected by nutrient enrichment. Stimulation of heterotrophic pathways via elevated nutrient concentrations has been shown to involve multiple levels of organization, from microorganisms to invertebrates, can affect respiration (CO_2 flux), and ecosystem-scale processing of litter-derived C and nutrients. Collectively, experimental enrichment studies, micro- and mesocosm experiments and landscape-scale studies indicate that (1) dissolved N and P affect litter by stimulating fungal activity and nutrient immobilization, thus, increasing detrital nutrient content, (2) the joint effects of N and P together (compared to N and P alone) as well as the presence of detritivores result in greater effects, and (3) the whole-stream reach effects of nutrient enrichment can be predicted from small scale measurements. With the caveat that watershed land use and ecological context are important (e.g., are nutrients limiting to stream heterotroph growth and activity? Are high nutrient concentrations occurring alongside water-quality issues such as low oxygen and sedimentation?), the combined effects of nutrient enrichment on multiple levels of stream food webs increase litter decomposition rates by ~50% on average, reducing residence time of this important resource that forms the energy base of most stream ecosystems. The importance of terrestrial organic matter, and specifically leaf litter, for stream food webs has been documented in numerous contexts; thus, any changes to the timing of its availability could also modify functions that are mediated by stream biota (e.g., nutrient uptake and retention, secondary production, insect emergence). As our mechanistic understanding of leaf litter decomposition and its responses to nutrient enrichment continues to solidify, future efforts should endeavor to fully incorporate this integrative measure of stream ecosystem functioning into nutrient monitoring and management strategies.

References

Abelho, M. (2001). From litterfall to breakdown in streams: A review. *The Scientific World Journal, 1*, 656–680. https://doi.org/10.1100/tsw.2001.103.

Ardón, M., Stallcup, L. A., & Pringle, C. M. (2006). Does leaf quality mediate the stimulation of leaf breakdown by phosphorus in Neotropical streams? *Freshwater Biology, 51*(4), 618–633. https://doi.org/10.1111/j.1365-2427.2006.01515.x.

Arroita, M., Elosegi, A., & Hall, R. O. (2019). Twenty years of daily metabolism show riverine recovery following sewage abatement. *Limnology and Oceanography, 64*(S1), S77–S92. https://doi.org/10.1002/lno.11053.

Azevedo-Pereira, H. V. S., Graça, M. A. S., & González, J. M. (2006). Life history of Lepidostoma hirtum in an Iberian stream and its role in organic matter processing. *Hydrobiologia, 559*(1), 183–192. https://doi.org/10.1007/s10750-005-1267-1.

Baldy, V., Gessner, M. O., & Chauvet, E. (1995). Bacteria, fungi and the breakdown of leaf litter in a large river. *Oikos, 74*(1), 93–102. JSTOR. https://doi.org/10.2307/3545678.

Baldy, V., Gobert, V., Guerold, F., Chauvet, E., Lambrigot, D., & Charcosset, J.-Y. (2007). Leaf litter breakdown budgets in streams of various trophic status: Effects of dissolved inorganic nutrients on microorganisms and invertebrates. *Freshwater Biology, 52*(7), 1322–1335. https://doi.org/10.1111/j.1365-2427.2007.01768.x.

Bärlocher, F. (1985). The role of fungi in the nutrition of stream invertebrates. *Botanical Journal of the Linnean Society, 91*(1–2), 83–94. https://doi.org/10.1111/j.1095-8339.1985.tb01137.x.

Bärlocher, F., & Sridhar, K. R. (2014). 19. Association of animals and fungi in leaf decomposition. In E. B. G. Jones, K. D. Hyde, & K.-L. Pang (Eds.), *Freshwater Fungi*. De Gruyter. https://doi.org/10.1515/9783110333480.413.

Benstead, J. P., Rosemond, A. D., Cross, W. F., Wallace, J. B., Eggert, S. L., Suberkropp, K., Gulis, V., Greenwood, J. L., & Tant, C. J. (2009). Nutrient enrichment alters storage and fluxes of detritus in a headwater stream ecosystem. *Ecology, 90*(9), 2556–2566. https://doi.org/10.1890/08-0862.1.

Bernot, M. J., Sobota, D. J., Hall, R. O., Mulholland, P. J., Dodds, W. K., Webster, J. R., Tank, J. L., Ashkenas, L. R., Cooper, L. W., Dahm, C. N., Gregory, S. V., Grimm, N. B., Hamilton, S. K., Johnson, S. L., Mcdowell, W. H., Meyer, J. L., Peterson, B., Poole, G. C., Valett, H. M., … Wilson, K. (2010). Inter-regional comparison of land-use effects on stream metabolism. *Freshwater Biology, 55*(9), 1874–1890. https://doi.org/10.1111/j.1365-2427.20.

Boyero, L., Pearson, R. G., Dudgeon, D., Graça, M. A. S., Gessner, M. O., Albariño, R. J., Ferreira, V., Yule, C. M., Boulton, A. J., Arunachalam, M., Callisto, M., Chauvet, E., Ramírez, A., Chará, J., Moretti, M. S., Gonçalves, J. F., Helson, J. E., Chará-Serna, A. M., Encalada, A. C., … Pringle, C. M. (2011). Global distribution of a key trophic guild contrasts with common latitudinal diversity patterns. *Ecology, 92*(9), 1839–1848. https://doi.org/10.1890/10-2244.1.

Boyero, L., Pearson, R. G., Gessner, M. O., Barmuta, L. A., Ferreira, V., Graça, M. A. S., Dudgeon, D., Boulton, A. J., Callisto, M., Chauvet, E., Helson, J. E., Bruder, A., Albariño, R. J., Yule, C. M., Arunachalam, M., Davies, J. N., Figueroa, R., Flecker, A. S., Ramírez, A., … West, D. C. (2011). A global experiment suggests climate warming will not accelerate litter decomposition in streams but might reduce carbon sequestration. *Ecology Letters, 14*(3), 289–294. https://doi.org/10.1111/j.1461-0248.2010.01578.x.

Boyero, L., Pearson, R. G., Hui, C., Gessner, M. O., Pérez, J., Alexandrou, M. A., Graça, M. A. S., Cardinale, B. J., Albariño, R. J., Arunachalam, M., Barmuta, L. A., Boulton, A. J., Bruder, A., Callisto, M., Chauvet, E., Death, R. G., Dudgeon, D., Encalada, A. C., Ferreira, V., … Yule, C. M. (2016). Biotic and abiotic variables influencing plant litter breakdown in streams: A global study. *Proceedings of the Royal Society B: Biological Sciences, 283*(1829), 20152664. https://doi.org/10.1098/rspb.2015.2664.

Boyero, L., Pearson, R. G., Swan, C. M., Hui, C., Albariño, R. J., Arunachalam, M., Callisto, M., Chará, J., Chará-Serna, A. M., Chauvet, E., Cornejo, A., Dudgeon, D., Encalada, A. C., Ferreira, V., Gessner, M. O., Gonçalves, J., Graça, M. A. S., Helson, J. E., Mathooko, J. M., … Yule,

C. M. (2015). Latitudinal gradient of nestedness and its potential drivers in stream detritivores. *Ecography, 38*(9), 949–955. https://doi.org/10.1111/ecog.00982.

Canhoto, C., & Graça, M. A. S. (2006). Digestive tract and leaf processing capacity of the stream invertebrate Tipula lateralis. *Canadian Journal of Zoology, 84*(8), 1087–1095. https://doi.org/10.1139/z06-092.

Cebrian, J., & Lartigue, J. (2004). Patterns of herbivory and decomposition in aquatic and terrestrial ecosystems. *Ecological Monographs, 74*(2), 237–259. https://doi.org/10.1890/03-4019.

Chadwick, M. A., & Huryn, A. D. (2003). Effect of a whole-catchment N addition on stream detritus processing. *Journal of the North American Benthological Society, 22*(2), 194–206. https://doi.org/10.2307/1467992.

Chauvet, E., Ferreira, V., Giller, P. S., McKie, B. G., Tiegs, S. D., Woodward, G., Elosegi, A., Dobson, M., Fleituch, T., Graça, M. A. S., Gulis, V., Hladyz, S., Lacoursière, J. O., Lecerf, A., Pozo, J., Preda, E., Riipinen, M., Rîşnoveanu, G., Vadineanu, A., … Gessner, M. O. (2016). Litter decomposition as an indicator of stream ecosystem functioning at local-to-continental scales: Insights from the European RivFunction project. In A. J. Dumbrell, R. L. Kordas, & G. Woodward (Eds.), *Advances in ecological research* (pp. 99–182). Academic Press. https://doi.org/10.1016/bs.aecr.2016.08.006.

Cheever, B. M., Webster, J. R., Bilger, E. E., & Thomas, S. A. (2013). The relative importance of exogenous and substrate-derived nitrogen for microbial growth during leaf decomposition. *Ecology, 94*(7), 1614–1625. https://doi.org/10.1890/12-1339.1.

Chung, N., & Suberkropp, K. (2009a). Contribution of fungal biomass to the growth of the shredder, Pycnopsyche gentilis (Trichoptera: Limnephilidae). *Freshwater Biology, 54*(11), 2212–2224. https://doi.org/10.1111/j.1365-2427.2009.02260.x.

Chung, N., & Suberkropp, K. (2009b). Effects of aquatic fungi on feeding preferences and bioenergetics of Pycnopsyche gentilis (Trichoptera: Limnephilidae). *Hydrobiologia, 630*(1), 257–269. https://doi.org/10.1007/s10750-009-9820-y.

Colas, F., Woodward, G., Burdon, F. J., Guérold, F., Chauvet, E., Cornut, J., Cébron, A., Clivot, H., Danger, M., Danner, M. C., Pagnout, C., & Tiegs, S. D. (2019). Towards a simple global-standard bioassay for a key ecosystem process: Organic-matter decomposition using cotton strips. *Ecological Indicators, 106*, 105466. https://doi.org/10.1016/j.ecolind.2019.105466.

Conley, D. J., Paerl, H. W., Howarth, R. W., Boesch, D. F., Seitzinger, S. P., Havens, K. E., Lancelot, C., & Likens, G. E. (2009). Ecology: Controlling eutrophication: Nitrogen and phosphorus. *Science, 323*(5917), 1014–1015. https://doi.org/10.1126/science.1167755.

Connolly, N. M., & Pearson, R. G. (2013). Nutrient enrichment of a heterotrophic stream alters leaf litter nutritional quality and shredder physiological condition via the microbial pathway. *Hydrobiologia, 718*(1), 85–92. https://doi.org/10.1007/s10750-013-1605-7.

Cornut, J., Elger, A., Lambrigot, D., Marmonier, P., & Chauvet, E. (2010). Early stages of leaf decomposition are mediated by aquatic fungi in the hyporheic zone of woodland streams. *Freshwater Biology, 55*(12), 2541–2556. https://doi.org/10.1111/j.1365-2427.2010.02483.x.

Cornut, J., Ferreira, V., Gonçalves, A. L., Chauvet, E., & Canhoto, C. (2015). Fungal alteration of the elemental composition of leaf litter affects shredder feeding activity. *Freshwater Biology, 60*(9), 1755–1771. https://doi.org/10.1111/fwb.12606.

Cross, W. F., Benstead, J. P., Frost, P. C., & Thomas, S. A. (2005). Ecological stoichiometry in freshwater benthic systems: Recent progress and perspectives. *Freshwater Biology, 50*(11), 1895–1912. https://doi.org/10.1111/j.1365-2427.2005.01458.x.

Cross, W. F., Benstead, J. P., Rosemond, A. D., & Wallace, J. B. (2003). Consumer-resource stoichiometry in detritus-based streams. *Ecology Letters, 6*(8), 721–732. https://doi.org/10.1046/j.1461-0248.2003.00481.x.

Cross, W. F., Wallace, J. B., Rosemond, A. D., & Eggert, S. L. (2006). Whole-system nutrient enrichment increases secondary production in a detritus-based ecosystem. *Ecology, 87*(6), 1556–1565. https://doi.org/10.1890/0012-9658(2006)87%5b1556:wneisp%5d2.0.co;2.

Cummins, K. W., & Klug, M. J. (1979). Feeding ecology of stream invertebrates. *Annual Review of Ecology and Systematics, 10*(1), 147–172. https://doi.org/10.1146/annurev.es.10.110179.001051.

Dang, C. K., Schindler, M., Chauvet, E., & Gessner, M. O. (2009). Temperature oscillation coupled with fungal community shifts can modulate warming effects on litter decomposition. *Ecology, 90*(1), 122–131. https://doi.org/10.1890/07-1974.1.

Danger, M., & Chauvet, E. (2013). Elemental composition and degree of homeostasis of fungi: Are aquatic hyphomycetes more like metazoans, bacteria or plants? *Fungal Ecology, 6*(5), 453–457. https://doi.org/10.1016/j.funeco.2013.05.007.

Danger, M., Funck, J. A., Devin, S., Heberle, J., & Felten, V. (2013). Phosphorus content in detritus controls life-history traits of a detritivore. *Functional Ecology, 27*(3), 807–815. https://doi.org/10.1111/1365-2435.12079.

Davis, J. M., Rosemond, A. D., Eggert, S. L., Cross, W. F., & Wallace, J. B. (2010). Nutrient enrichment differentially affects body sizes of primary consumers and predators in a detritus-based stream. *Limnology and Oceanography, 55*(6), 2305–2316. https://doi.org/10.4319/lo.2010.55.6.2305.

Demi, L. M., Benstead, J. P., Rosemond, A. D., & Maerz, J. C. (2018). Litter P content drives consumer production in detritus-based streams spanning an experimental N: P gradient. *Ecology, 99*(2), 347–359. https://doi.org/10.1002/ecy.2118.

Demi, L. M., Benstead, J. P., Rosemond, A. D., & Maerz, J. C. (2019). Experimental N and P additions alter stream macroinvertebrate community composition via taxon-level responses to shifts in detrital resource stoichiometry. *Functional Ecology, 33*(5), 855–867. https://doi.org/10.1111/1365-2435.13289.

Dodds, W. K. (2006). Eutrophication and trophic state in rivers and streams. *Limnology and Oceanography, 51,* 671–680. https://doi.org/10.4319/lo.2006.51.1_part_2.0671.

Dodds, W. K., Clements, W. H., Gido, K., Hilderbrand, R. H., & King, R. S. (2010). Thresholds, breakpoints, and nonlinearity in freshwaters as related to management. *Journal of the North American Benthological Society, 29*(3), 988–997. https://doi.org/10.1899/09-148.1.

Dodds, W. K., & Cole, J. J. (2007). Expanding the concept of trophic state in aquatic ecosystems: It's not just the autotrophs. *Aquatic Sciences, 69*(4), 427–439. https://doi.org/10.1007/s00027-007-0922-1.

Dodds, W., & Smith, V. (2016). Nitrogen, phosphorus, and eutrophication in streams. *Inland Waters, 6*(2), 155–164. https://doi.org/10.5268/IW-6.2.909.

Eggert, S. L., & Wallace, J. B. (2007). Wood biofilm as a food resource for stream detritivores. *Limnology and Oceanography, 52*(3), 1239–1245. https://doi.org/10.4319/lo.2007.52.3.1239.

Elwood, J. W., Newbold, J. D., Trimble, A. F., & Stark, R. W. (1981). The limiting role of phosphorus in a woodland stream ecosystem: Effects of P enrichment on leaf decomposition and primary producers. *Ecology, 62*(1), 146–158. https://doi.org/10.2307/1936678.

Enríquez, S., Duarte, C. M., & Sand-Jensen, K. (1993). Patterns in decomposition rates among photosynthetic organisms: The importance of detritus C:N: P content. *Oecologia, 94*(4), 457–471. https://doi.org/10.1007/BF00566960.

Fernandes, I., Seena, S., Pascoal, C., & Cássio, F. (2014). Elevated temperature may intensify the positive effects of nutrients on microbial decomposition in streams. *Freshwater Biology, 59*(11), 2390–2399. https://doi.org/10.1111/fwb.12445.

Ferreira, V., & Canhoto, C. (2015). Future increase in temperature may stimulate litter decomposition in temperate mountain streams: Evidence from a stream manipulation experiment. *Freshwater Biology, 60*(5), 881–892. https://doi.org/10.1111/fwb.12539.

Ferreira, V., Castagneyrol, B., Koricheva, J., Gulis, V., Chauvet, E., & Graça, M. A. S. (2015). A meta-analysis of the effects of nutrient enrichment on litter decomposition in streams. *Biological Reviews, 90*(3), 669–688. https://doi.org/10.1111/brv.12125.

Ferreira, V., & Chauvet, E. (2011). Synergistic effects of water temperature and dissolved nutrients on litter decomposition and associated fungi. *Global Change Biology, 17*(1), 551–564. https://doi.org/10.1111/j.1365-2486.2010.02185.x.

Ferreira, V., Gulis, V., & Graça, M. A. S. (2006). Whole-stream nitrate addition affects litter decomposition and associated fungi but not invertebrates. *Oecologia, 149*(4), 718–729. https://doi.org/10.1007/s00442-006-0478-0.

Ferreira, V., Raposeiro, P. M., Pereira, A., Cruz, A. M., Costa, A. C., Graça, M. A. S., & Gonçalves, V. (2016). Leaf litter decomposition in remote oceanic island streams is driven by microbes and depends on litter quality and environmental conditions. *Freshwater Biology, 61*(5), 783–799. https://doi.org/10.1111/fwb.12749.

Fisher, S. G., & Likens, G. E. (1973). Energy flow in Bear Brook, New Hampshire: An integrative approach to stream ecosystem metabolism. *Ecological Monographs, 43*(4), 421–439. JSTOR. https://doi.org/10.2307/1942301.

Flecker, A. S., & Townsend, C. R. (1994). Community-wide consequences of trout introduction in New Zealand streams. *Ecological Applications, 4*(4), 798–807. https://doi.org/10.2307/1942009.

Foucreau, N., Puijalon, S., Hervant, F., & Piscart, C. (2013). Effect of leaf litter characteristics on leaf conditioning and on consumption by Gammarus pulex. *Freshwater Biology, 58*(8), 1672–1681. https://doi.org/10.1111/fwb.12158.

Fowler, D., O'Donoghue, M., Muller, J. B. A., Smith, R. I., Dragosits, U., Skiba, U., Sutton, M. A., & Brimblecombe, P. (2004). A chronology of nitrogen deposition in the UK between 1900 and 2000. *Water, Air, & Soil Pollution: Focus, 4*(6), 9–23. https://doi.org/10.1007/s11267-004-3009-1.

Frainer, A., Jabiol, J., Gessner, M. O., Bruder, A., Chauvet, E., & McKie, B. G. (2016). Stoichiometric imbalances between detritus and detritivores are related to shifts in ecosystem functioning. *Oikos, 125*(6), 861–871. https://doi.org/10.1111/oik.02687.

Gallo, M. E., Porras-Alfaro, A., Odenbach, K. J., & Sinsabaugh, R. L. (2009). Photoacceleration of plant litter decomposition in an arid environment. *Soil Biology & Biochemistry, 41*(7), 1433–1441. https://doi.org/10.1016/j.soilbio.2009.03.025.

Gessner, M. O., Gulis, V., Kuehn, K., Chauvet, E., & Suberkropp, K.. (Eds.). (2007). Fungal decomposers of plant litter in aquatic ecosystems. In C. P. Kubicek, & I. S. Druzhinina (eds.), *Environmental and microbial relationships* (pp. 301–324). Springer. https://doi.org/10.1007/978-3-540-71840-6_17.

Gessner, M. O., Swan, C. M., Dang, C. K., McKie, B. G., Bardgett, R. D., Wall, D. H., & Hättenschwiler, S. (2010). Diversity meets decomposition. *Trends in Ecology & Evolution, 25*(6), 372–380. https://doi.org/10.1016/j.tree.2010.01.010.

Gomi, T., Moore, R. D., & Dhakal, A. S. (2006). Headwater stream temperature response to clear-cut harvesting with different riparian treatments, coastal British Columbia, Canada. *Water Resources Research, 42*(8). https://doi.org/10.1029/2005wr004162.

Graça, M. A. S. (2001). The role of invertebrates on leaf litter decomposition in streams—A review. *International Review of Hydrobiology, 86*(4–5), 383–393. https://doi.org/10.1002/1522-2632(200107)86:4/5%3c383:AID-IROH383%3e3.0.CO;2-D.

Greenwood, J. L., Rosemond, A. D., Wallace, J. B., Cross, W. F., & Weyers, H. S. (2007). Nutrients stimulate leaf breakdown rates and detritivore biomass: Bottom-up effects via heterotrophic pathways. *Oecologia, 151*(4), 637–649. https://doi.org/10.1007/s00442-006-0609-7.

Grimmett, I. J., Shipp, K. N., Macneil, A., & Bärlocher, F. (2013). Does the growth rate hypothesis apply to aquatic hyphomycetes? *Fungal Ecology, 6*(6), 493–500. https://doi.org/10.1016/j.funeco.2013.08.002.

Gulis, V., Ferreira, V., & Graça, M. A. S. (2006). Stimulation of leaf litter decomposition and associated fungi and invertebrates by moderate eutrophication: Implications for stream assessment. *Freshwater Biology, 51*(9), 1655–1669. https://doi.org/10.1111/j.1365-2427.2006.01615.x.

Gulis, V., Kuehn, K., & Suberkropp, K. (2006). The role of fungi in carbon and nitrogen cycles in freshwater ecosystems. In G. M. Gadd (Ed.), *Fungi in biogeochemical cycles* (pp. 404–435). Cambridge University Press. https://doi.org/10.1017/cbo9780511550522.018.

Gulis, V., Kuehn, K. A., Schoettle, L. N., Leach, D., Benstead, J. P., & Rosemond, A. D. (2017). Changes in nutrient stoichiometry, elemental homeostasis and growth rate of aquatic litter-associated fungi in response to inorganic nutrient supply. *The ISME Journal, 11*(12), 2729–2739. https://doi.org/10.1038/ismej.2017.123.

Gulis, V., Rosemond, A. D., Suberkropp, K., Weyers, H. S., & Benstead, J. P. (2004). Effects of nutrient enrichment on the decomposition of wood and associated microbial activity in streams. *Freshwater Biology, 49*(11), 1437–1447. https://doi.org/10.1111/j.1365-2427.2004.01281.x.

Gulis, V., & Suberkropp, K. (2003a). Leaf litter decomposition and microbial activity in nutrient-enriched and unaltered reaches of a headwater stream. *Freshwater Biology, 48*(1), 123–134. https://doi.org/10.1046/j.1365-2427.2003.00985.x.

Gulis, V., & Suberkropp, K. (2003b). Effect of inorganic nutrients on relative contributions of fungi and bacteria to carbon flow from submerged decomposing leaf litter. *Microbial Ecology, 45*(1), 11–19. https://doi.org/10.1007/s00248-002-1032-1.

Gulis, V., & Suberkropp, K. (2003c). Interactions between stream fungi and bacteria associated with decomposing leaf litter at different levels of nutrient availability. *Aquatic Microbial Ecology, 30,* 149–157.

Gulis, V., Suberkropp, K., & Rosemond, A. D. (2008). Comparison of fungal activities on wood and leaf litter in unaltered and nutrient-enriched headwater streams. *Applied and Environmental Microbiology, 74*(4), 1094–1101. https://doi.org/10.1128/AEM.01903-07.

Güsewell, S., & Freeman, C. (2005). Nutrient limitation and enzyme activities during litter decomposition of nine wetland species in relation to litter N: P ratios. *Functional Ecology, 19*(4), 582–593. https://doi.org/10.1111/j.1365-2435.2005.01002.x.

Halvorson, H. M., Fuller, C. L., Entrekin, S. A., Scott, J. T., & Evans-White, M. A. (2018). Detrital nutrient content and leaf species differentially affect growth and nutritional regulation of detritivores. *Oikos, 127*(10), 1471–1481. https://doi.org/10.1111/oik.05201.

Halvorson, H. M., White, G., Scott, J. T., & Evans-White, M. A. (2016). Dietary and taxonomic controls on incorporation of microbial carbon and phosphorus by detritivorous caddisflies. *Oecologia, 180*(2), 567–579. https://doi.org/10.1007/s00442-015-3464-6.

Harpole, W. S., Ngai, J. T., Cleland, E. E., Seabloom, E. W., Borer, E. T., Bracken, M. E. S., Elser, J., Gruner, D. S., Hillebrand, H., Shurin, J. B., & Smith, J. E. (2011). Nutrient co-limitation of primary producer communities. *Ecology Letters, 14*(9), 852–862. https://doi.org/10.1111/j.1461-0248.2011.01651.x.

Hartmann, J., Lauerwald, R., & Moosdorf, N. (2014). A brief overview of the GLObal RIver Chemistry Database, Glorich. *Procedia Earth and Planetary Science, 10,* 23–27. https://doi.org/10.1016/j.proeps.2014.08.005.

Hartmann, J., Lauerwald, R., & Moosdorf, N. (2019). GLORICH—Global river chemistry database [Data set]. *Supplement to: Hartmann, J et al. (2014): A Brief Overview of the GLObal RIver Chemistry Database, GLORICH. Procedia Earth and Planetary Science, 10,* 23–27. https://doi.org/10.1016/j.Proeps.2014.08.005. PANGAEA. https://doi.org/10.1594/pangaea.902360.

Hendel, B., Sinsabaugh, R. L., & Marxsen, J. (2020). Lignin-degrading enzymes: Phenoloxidase and Peroxidase. In F. Bärlocher, M. O. Gessner, & M. A. S. Graça (Eds.), *Methods to study litter decomposition: A practical guide* (pp. 425–431). Springer International Publishing. https://doi.org/10.1007/978-3-030-30515-4_46.

Hieber, M., & Gessner, M. O. (2002). Contribution of stream detritivores, fungi, and bacteria to leaf breakdown based on biomass estimates. *Ecology, 83*(4), 1026–1038. https://doi.org/10.2307/3071911.

Hill, B. H., Bolgrien, D. W., Herlihy, A. T., Jicha, T. M., & Angradi, T. R. (2011). A synoptic survey of nitrogen and phosphorus in tributary streams and Great Rivers of the upper Mississippi, Missouri, and Ohio River basins. *Water, Air, and Soil pollution, 216*(1), 605–619. https://doi.org/10.1007/s11270-010-0556-0.

Hladyz, S., Gessner, M. O., Giller, P. S., Pozo, J., & Woodward, G. (2009). Resource quality and stoichiometric constraints on stream ecosystem functioning. *Freshwater Biology, 54*(5), 957–970. https://doi.org/10.1111/j.1365-2427.2008.02138.x.

Jabiol, J., Lecerf, A., Lamothe, S., Gessner, M. O., & Chauvet, E. (2019). Litter quality modulates effects of dissolved nitrogen on leaf decomposition by stream microbial communities. *Microbial Ecology, 77*(4), 959–966. https://doi.org/10.1007/s00248-019-01353-3.

Jarvie, H. P., Sharpley, A. N., Withers, P. J. A., Scott, J. T., Haggard, B. E., & Neal, C. (2013). Phosphorus mitigation to control river eutrophication: Murky waters, inconvenient truths, and "postnormal" science. *Journal of Environmental Quality, 42*(2), 295–304. https://doi.org/10.2134/jeq2012.0085.

Jenkins, C. C., & Suberkropp, K. (1995). The influence of water chemistry on the enzymatic degradation of leaves in streams. *Freshwater Biology, 33*(2), 245–253. https://doi.org/10.1111/j.1365-2427.1995.tb01165.x.

Johnson, S. L., & Jones, J. A. (2000). Stream temperature responses to forest harvest and debris flows in western Cascades, Oregon. *Canadian Journal of Fisheries and Aquatic Sciences, 57*(S2), 30–39. https://doi.org/10.1139/f00-109.

Kaushal, S. S., Mayer, P. M., Vidon, P. G., Smith, R. M., Pennino, M. J., Newcomer, T. A., Duan, S., Welty, C., & Belt, K. T. (2014). Land use and climate variability amplify carbon, nutrient, and contaminant pulses: A review with management implications. *Journal of the American Water Resources Association, 50*(3), 585–614. https://doi.org/10.1111/jawr.12204.

Kiffney, P. M., Richardson, J. S., & Bull, J. P. (2003). Responses of periphyton and insects to experimental manipulation of riparian buffer width along forest streams. *Journal of Applied Ecology, 40*(6), 1060–1076. https://doi.org/10.1111/j.1365-2664.2003.00855.x.

Kominoski, J. S., Rosemond, A. D., Benstead, J. P., Gulis, V., Maerz, J. C., & Manning, D. W. P. (2015). Low-to-moderate nitrogen and phosphorus concentrations accelerate microbially driven litter breakdown rates. *Ecological Applications, 25*(3), 856–865. https://doi.org/10.1890/14-1113.1.

Lecerf, A., Usseglio-Polatera, P., Charcosset, J.-Y., Lambrigot, D., Bracht, B., & Chauvet, E. (2006). Assessment of functional integrity of eutrophic streams using litter breakdown and benthic macroinvertebrates. *Archiv Für Hydrobiologie, 165*(1), 105–126. https://doi.org/10.1127/0003-9136/2006/0165-0105.

LeRoy, C. J., Hipp, A. L., Lueders, K., Shah, J. J. F., Kominoski, J. S., Ardón, M., Dodds, W. K., Gessner, M. O., Griffiths, N. A., Lecerf, A., Manning, D. W. P., Sinsabaugh, R. L., & Webster, J. R. (2020). Plant phylogenetic history explains in-stream decomposition at a global scale. *Journal of Ecology, 108*(1), 17–35. https://doi.org/10.1111/1365-2745.13262.

LeRoy, C. J., & Marks, J. C. (2006). Litter quality, stream characteristics and litter diversity influence decomposition rates and macroinvertebrates. *Freshwater Biology, 51*(4), 605–617. https://doi.org/10.1111/j.1365-2427.2006.01512.x.

Linker, L. C., Dennis, R., Shenk, G. W., Batiuk, R. A., Grimm, J., & Wang, P. (2013). Computing atmospheric nutrient loads to the Chesapeake Bay watershed and tidal waters. *Journal of the American Water Resources Association, 49*(5), 1025–1041. https://doi.org/10.1111/jawr.12112.

Linkins, A. E., Sinsabaugh, R. L., Mcclaugherty, C. A., & Melillo, J. M. (1990). Comparison of cellulase activity on decomposing leaves in a hardwood forest and woodland stream. *Soil Biology & Biochemistry, 22*(3), 423–425. https://doi.org/10.1016/0038-0717(90)90123-H.

Manning, D. W. P., Rosemond, A. D., Benstead, J. P., Bumpers, P. M., & Kominoski, J. S. (2020). Transport of N and P in U.S. streams and rivers differs with land use and between dissolved and particulate forms. *Ecological Applications*. https://doi.org/10.1002/eap.2130.

Manning, D. W. P., Rosemond, A. D., Gulis, V., Benstead, J. P., & Kominoski, J. S. (2018). Nutrients and temperature additively increase stream microbial respiration. *Global Change Biology, 24*(1), e233–e247. https://doi.org/10.1111/gcb.13906.

Manning, D. W. P., Rosemond, A. D., Gulis, V., Benstead, J. P., Kominoski, J. S., & Maerz, J. C. (2016). Convergence of detrital stoichiometry predicts thresholds of nutrient-stimulated breakdown in streams. *Ecological Applications, 26*(6), 1745–1757. https://doi.org/10.1890/15-1217.1.

Manning, D. W. P., Rosemond, A. D., Kominoski, J. S., Gulis, V., Benstead, J. P., & Maerz, J. C. (2015). Detrital stoichiometry as a critical nexus for the effects of streamwater nutrients on leaf litter breakdown rates. *Ecology, 96*(8), 2214–2224. https://doi.org/10.1890/14-1582.1.

Manzoni, S., Trofymow, J. A., Jackson, R. B., & Porporato, A. (2010). Stoichiometric controls on carbon, nitrogen, and phosphorus dynamics in decomposing litter. *Ecological Monographs, 80*(1), 89–106. https://doi.org/10.1890/09-0179.1.

Marks, J. C. (2019). Revisiting the fates of dead leaves that fall into streams. *Annual Review of Ecology, Evolution, and Systematics, 50*(1). https://doi.org/10.1146/annurev-ecolsys-110218-024755.

McDowell, R. W., Noble, A., Pletnyakov, P., Haggard, B. E., & Mosley, L. M. (2020). Global mapping of freshwater nutrient enrichment and periphyton growth potential. *Scientific Reports, 10*(1), 3568. https://doi.org/10.1038/s41598-020-60279-w.

Moore, J. C., Berlow, E. L., Coleman, D. C., Ruiter, P. C., de, Dong, Q., Hastings, A., Johnson, N. C., McCann, K. S., Melville, K., Morin, P. J., Nadelhoffer, K., Rosemond, A. D., Post, D. M., Sabo, J. L., Scow, K. M., Vanni, M. J., & Wall, D. H. (2004). Detritus, trophic dynamics and biodiversity. *Ecology Letters, 7*(7), 584–600. https://doi.org/10.1111/j.1461-0248.2004.00606.x.

Moretti, M. S., Loyola, R. D., Becker, B., & Callisto, M. (2009). Leaf abundance and phenolic concentrations codetermine the selection of case-building materials by Phylloicus sp. (Trichoptera, Calamoceratidae). *Hydrobiologia, 630*(1), 199–206. https://doi.org/10.1007/s10750-009-9792-y.

Moss, B., Kosten, S., Meerhoff, M., Battarbee, R. W., Jeppesen, E., Mazzeo, N., Havens, K., Lacerot, G., Liu, Z., de Meester, L., Paerl, H., & Scheffer, M. (2011). Allied attack: Climate change and eutrophication. *Inland Waters, 1*(2), 101–105. https://doi.org/10.5268/IW-1.2.359.

Mulholland, P. J., & Hill, W. R. (1997). Seasonal patterns in streamwater nutrient and dissolved organic carbon concentrations: Separating catchment flow path and in-stream effects. *Water Resources Research, 33*(6), 1297–1306. https://doi.org/10.1029/97WR00490.

Murdoch, P. S., Baron, J. S., & Miller, T. L. (2000). Potential effects of climate change on surface-water quality in North America. *Journal of the American Water Resources Association, 36*(2), 347–366. https://doi.org/10.1111/j.1752-1688.2000.tb04273.x.

Newbold, J. D., Elwood, J. W., O'Neill, R. V., & Sheldon, A. L. (1983). Phosphorus dynamics in a woodland stream ecosystem: A study of nutrient spiralling. *Ecology, 64*(5), 1249–1265. https://doi.org/10.2307/1937833.

Newbold, J. D., Elwood, J. W., Schulze, M. S., Stark, R. W., & Barmeier, J. C. (1983). Continuous ammonium enrichment of a woodland stream: Uptake kinetics, leaf decomposition, and nitrification. *Freshwater Biology, 13*(2), 193–204. https://doi.org/10.1111/j.1365-2427.1983.tb00671.x.

Pascoal, C., & Cássio, F. (2004). Contribution of fungi and bacteria to leaf litter decomposition in a polluted river. *Applied and Environmental Microbiology, 70*(9), 5266–5273. https://doi.org/10.1128/AEM.70.9.5266-5273.2004.

Pascoal, C., Cássio, F., Marcotegui, A., Sanz, B., & Gomes, P. (2005). Role of fungi, bacteria, and invertebrates in leaf litter breakdown in a polluted river. *Journal of the North American Benthological Society, 24*(4), 784–797. https://doi.org/10.1899/05-010.1.

Pastor, A., Compson, Z. G., Dijkstra, P., Riera, J. L., Martí, E., Sabater, F., Hungate, B. A., & Marks, J. C. (2014). Stream carbon and nitrogen supplements during leaf litter decomposition: Contrasting patterns for two foundation species. *Oecologia, 176*(4), 1111–1121. https://doi.org/10.1007/s00442-014-3063-y.

Piggott, J. J., Lange, K., Townsend, C. R., & Matthaei, C. D. (2012). Multiple stressors in agricultural streams: A mesocosm study of interactions among raised water temperature, sediment addition and nutrient enrichment. *PLoS ONE, 7*(11), https://doi.org/10.1371/journal.pone.0049873.

Piggott, J. J., Niyogi, D. K., Townsend, C. R., & Matthaei, C. D. (2015). Multiple stressors and stream ecosystem functioning: Climate warming and agricultural stressors interact to affect processing of organic matter. *Journal of Applied Ecology, 52*(5), 1126–1134. https://doi.org/10.1111/1365-2664.12480.

Prater, C., Bumpers, P. M., Demi, L. M., Rosemond, A. D., & Jeyasingh, P. D. (2020). Differential responses of macroinvertebrate ionomes across experimental N: P gradients in detritus-based headwater streams. *Oecologia*. https://doi.org/10.1007/s00442-020-04720-x.

Prater, C., Norman, E. J., & Evans-White, M. A. (2015). Relationships among nutrient enrichment, detritus quality and quantity, and large-bodied shredding insect community structure. *Hydrobiologia, 753*(1), 219–232. https://doi.org/10.1007/s10750-015-2208-2.

Rincón, J., & Martínez, I. (2006). Food quality and feeding preferences of Phylloicus sp. (Trichoptera: Calamoceratidae). *Journal of the North American Benthological Society, 25*(1), 209–215. https://doi.org/10.1899/0887-3593(2006)25%5b209:fqafpo%5d2.0.co;2.

Robbins, C. J., Matthaeus, W. J., Cook, S. C., Housley, L. M., Robison, S. E., Garbarino, M. A., LeBrun, E. S., Raut, S., Tseng, C.-Y., & King, R. S. (2019). Leaf litter identity alters the timing of lotic nutrient dynamics. *Freshwater Biology*. https://doi.org/10.1111/fwb.13410.

Rose, L. A., Karwan, D. L., & Godsey, S. E. (2018). Concentration–discharge relationships describe solute and sediment mobilization, reaction, and transport at event and longer timescales. *Hydrological Processes, 32*(18), 2829–2844. https://doi.org/10.1002/hyp.13235.

Rosemond, A. D., Benstead, J. P., Bumpers, P. M., Gulis, V., Kominoski, J. S., Manning, D. W. P., Suberkropp, K., & Wallace, J. B. (2015). Experimental nutrient additions accelerate terrestrial carbon loss from stream ecosystems. *Science, 347*(6226), 1142–1145. https://doi.org/10.1126/science.aaa1958.

Rosemond, A. D., Pringle, C. M., Ramírez, A., Paul, M. J., & Meyer, J. L. (2002). Landscape variation in phosphorus concentration and effects on detritus-based tropical streams. *Limnology and Oceanography, 47*(1), 278–289. https://doi.org/10.4319/lo.2002.47.1.0278.

Royer, T. V., & Minshall, G. W. (2001). Effects of nutrient enrichment and leaf quality on the breakdown of leaves in a hardwater stream. *Freshwater Biology, 46*(5), 603–610. https://doi.org/10.1046/j.1365-2427.2001.00694.x.

Sanpera-Calbet, I., Lecerf, A., & Chauvet, E. (2009). Leaf diversity influences in-stream litter decomposition through effects on shredders. *Freshwater Biology, 54*(8), 1671–1682. https://doi.org/10.1111/j.1365-2427.2009.02216.x.

Scott, E. E., Prater, C., Norman, E., Baker, B. C., Evans-White, M., & Scott, J. T. (2013). Leaf-litter stoichiometry is affected by streamwater phosphorus concentrations and litter type. *Freshwater Science, 32*(3), 753–761. https://doi.org/10.1899/12-215.1.

Seena, S., Bärlocher, F., Sobral, O., Gessner, M. O., Dudgeon, D., McKie, B. G., Chauvet, E., Boyero, L., Ferreira, V., Frainer, A., Bruder, A., Matthaei, C. D., Fenoglio, S., Sridhar, K. R., Albariño, R. J., Douglas, M. M., Encalada, A. C., Garcia, E., Ghate, S. D., … Graça, M. A. S. (2019). Biodiversity of leaf litter fungi in streams along a latitudinal gradient. *Science of the Total Environment, 661*, 306–315. https://doi.org/10.1016/j.scitotenv.2019.01.122.

Smith, V. H., Joye, S. B., & Howarth, R. W. (2006). Eutrophication of freshwater and marine ecosystems. *Limnology and Oceanography, 51*, 351–355. https://doi.org/10.4319/lo.2006.51.1_part_2.0351.

Sridhar, K. R., & Bärlocher, F. (1997). Water chemistry and sporulation by aquatic hyphomycetes. *Mycological Research, 101*(5), 591–596. https://doi.org/10.1017/S0953756296003024.

Stelzer, R. S., Heffernan, J., & Likens, G. E. (2003). The influence of dissolved nutrients and particulate organic matter quality on microbial respiration and biomass in a forest stream. *Freshwater Biology, 48*(11), 1925–1937. https://doi.org/10.1046/j.1365-2427.2003.01141.x.

Stets, E. G., Sprague, L. A., Oelsner, G. P., Johnson, H. M., Murphy, J. C., Ryberg, K., Vecchia, A. V., Zuellig, R. E., Falcone, J. A., & Riskin, M. L. (2020). Landscape drivers of dynamic change in water quality of U.S. rivers. *Environmental Science & Technology, 54*(7), 4336–4343. https://doi.org/10.1021/acs.est.9b05344.

Stoddard, J. L., Van Sickle, J., Herlihy, A. T., Brahney, J., Paulsen, S., Peck, D. V., Mitchell, R., & Pollard, A. I. (2016). Continental-scale increase in lake and stream phosphorus: Are oligotrophic systems disappearing in the United States? *Environmental Science & Technology, 50*(7), 3409–3415. https://doi.org/10.1021/acs.est.5b05950.

Suberkropp, K. (1991). Relationships between growth and sporulation of aquatic hyphomycetes on decomposing leaf litter. *Mycological Research, 95*(7), 843–850. https://doi.org/10.1016/S0953-7562(09)80048-8.

Suberkropp, K. (1992). Interactions with invertebrates. In F. Bärlocher (Ed.), *The ecology of aquatic hyphomycetes* (pp. 118–134). Springer. https://doi.org/10.1007/978-3-642-76855-2_6.

Suberkropp, K. (1995). The influence of nutrients on fungal growth, productivity, and sporulation during leaf breakdown in streams. *Canadian Journal of Botany, 73*(S1), 1361–1369. https://doi.org/10.1139/b95-398.

Suberkropp, K. (1998). Effect of dissolved nutrients on two aquatic hyphomycetes growing on leaf litter. *Mycological Research, 102*(8), 998–1002. https://doi.org/10.1017/S0953756297005807.

Suberkropp, K. & Chauvet, E. (1995). Regulation of leaf breakdown by fungi in streams: Influences of water chemistry. *Ecology, 76*(5), 1433–1445. https://doi.org/10.2307/1938146.

Suberkropp, K., Gulis, V., Rosemond, A. D., & Benstead, J. P. (2010). Ecosystem and physiological scales of microbial responses to nutrients in a detritus-based stream: Results of a 5-year continuous enrichment. *Limnology and Oceanography, 55*(1), 149–160. https://doi.org/10.4319/lo.2010.55.1.0149.

Tank, J. L., Rosi-Marshall, E. J., Griffiths, N. A., Entrekin, S. A., & Stephen, M. L. (2010). A review of allochthonous organic matter dynamics and metabolism in streams. *Journal of the North American Benthological Society, 29*(1), 118–146. https://doi.org/10.1899/08-170.1.

Tant, C. J., Rosemond, A. D., & First, M. R. (2013). Stream nutrient enrichment has a greater effect on coarse than on fine benthic organic matter. *Freshwater Science, 32*(4), 1111–1121. https://doi.org/10.1899/12-049.1.

Tant, C. J., Rosemond, A. D., Helton, A. M., & First, M. R. (2015). Nutrient enrichment alters the magnitude and timing of fungal, bacterial, and detritivore contributions to litter breakdown. *Freshwater Science, 34*(4), 1259–1271. https://doi.org/10.1086/683255.

Taylor, B. R., & Chauvet, E. E. (2014). Relative influence of shredders and fungi on leaf litter decomposition along a river altitudinal gradient. *Hydrobiologia, 721*(1), 239–250. https://doi.org/10.1007/s10750-013-1666-7.

Tiegs, S. D., Costello, D. M., Isken, M. W., Woodward, G., McIntyre, P. B., Gessner, M. O., Chauvet, E., Griffiths, N. A., Flecker, A. S., Acuña, V., Albariño, R., Allen, D. C., Alonso, C., Andino, P., Arango, C., Aroviita, J., Barbosa, M. V. M., Barmuta, L. A., Baxter, C. V., … Zwart, J. A. (2019). Global patterns and drivers of ecosystem functioning in rivers and riparian zones. *Science Advances, 5*(1), eaav0486. https://doi.org/10.1126/sciadv.aav0486.

US Environmental Protection Agency. (2016). *National rivers and streams assessment 2008–2009 report* [Reports and Assessments]. https://www.epa.gov/national-aquatic-resource-surveys/national-rivers-and-streams-assessment-2008-2009-report.

Usher, R. L., Wood, J., Bumpers, P. M., Wenger, S. J., & Rosemond, A. D. (2020). Streamwater nutrients stimulate respiration and breakdown of standardized detrital substrates across a landscape gradient: Effects of nitrogen, phosphorus, and carbon quality. *Freshwater Science, 39*(1), 101–114. https://doi.org/10.1086/707598.

Venarsky, M. P., Benstead, J. P., Huryn, A. D., Huntsman, B. M., Edmonds, J. W., Findlay, R. H., & Bruce Wallace, J. (2018). Experimental detritus manipulations unite surface and cave stream ecosystems along a common energy gradient. *Ecosystems, 21*(4), 629–642. https://doi.org/10.1007/s10021-017-0174-4.

Wallace, J. B., Eggert, S. L., Meyer, J. L., & Webster, J. R. (1997). Multiple trophic levels of a forest stream linked to terrestrial litter inputs. *Science, 277*(5322), 102–104. https://doi.org/10.1126/science.277.5322.102.

Walther, D. A., & Whiles, M. R. (2011). Secondary production in a southern Illinois headwater stream: Relationships between organic matter standing stocks and macroinvertebrate productivity. *Journal of the North American Benthological Society, 30*(2), 357–373. https://doi.org/10.1899/10-006.1.

Webster, J. R. (2007). Spiraling down the river continuum: Stream ecology and the U-shaped curve. *Journal of the North American Benthological Society, 26*(3), 375–389. https://doi.org/10.1899/06-095.1.

Webster, J. R., & Benfield, E. F. (1986). Vascular plant breakdown in freshwater ecosystems. *Annual Review of Ecology and Systematics, 17*(1), 567–594. https://doi.org/10.1146/annurev.es.17.110186.003031.

Weyers, H. S., & Suberkropp, K. (1996). Fungal and bacterial production during the breakdown of yellow poplar leaves in 2 Streams. *Journal of the North American Benthological Society, 15*(4), 408–420. https://doi.org/10.2307/1467795.

Woodward, G., Gessner, M. O., Giller, P. S., Gulis, V., Hladyz, S., Lecerf, A., Malmqvist, B., McKie, B. G., Tiegs, S. D., Cariss, H., Dobson, M., Elosegi, A., Ferreira, V., Graça, M. A. S., Fleituch, T., Lacoursière, J. O., Nistorescu, M., Pozo, J., Risnoveanu, G., … Chauvet, E. (2012). Continental-scale effects of nutrient pollution on stream ecosystem functioning. *Science, 336*(6087), 1438–1440. https://doi.org/10.1126/science.1219534.

Wurtsbaugh, W. A., Paerl, H. W., & Dodds, W. K. (2019). Nutrients, eutrophication and harmful algal blooms along the freshwater to marine continuum. *WIREs Water, 6*(5), https://doi.org/10.1002/wat2.1373.

Yule, C. M., Boyero, L., & Marchant, R. (2010). Effects of sediment pollution on food webs in a tropical river (Borneo, Indonesia). *Marine & Freshwater Research, 61*(2), 204. https://doi.org/10.1071/MF09065.

Chapter 17
How Toxicants Influence Organic Matter Decomposition in Streams

Mirco Bundschuh, Alexander Feckler, Ralf B. Schäfer, Ralf Schulz, and Jochen P. Zubrod

Abstract The ecosystem process of organic matter decomposition (OMD) in fresh waters is realised by a complex interaction among different groups of microorganisms (including bacteria and fungi) and detritivorous animals. As a consequence of this multi-level interaction, biotic (e.g., competition, predation) and abiotic (e.g., flow, temperature, toxicants) factors may influence the performance of either (micro)organism group with potential feedback to detritivores, and in turn OMD. In this chapter, we discuss how the abiotic factor toxicants affects OMD. We organised the chapter along a conceptual model that pinpoints groups of organisms and their interactions, which are critical for OMD. We focus on four toxicant classes (i.e., fungicides, antibiotics, insecticides and metals) that specifically affect fungi, bacteria or detritivorous invertebrates or have a broad activity spectrum. We summarize the effects caused by these toxicant classes on relevant groups of organisms, which are often determined by the toxicant mode of action. On this basis, we develop effect pathways leading to alterations in OMD dynamics. Finally, we discuss whether and how these effect pathways may support the interpretation of effect patterns observed

M. Bundschuh (✉) · A. Feckler · R. B. Schäfer · R. Schulz · J. P. Zubrod
iES Landau, Institute for Environmental Sciences, University of Koblenz-Landau, Landau, Germany
e-mail: bundschuh@uni-landau.de

A. Feckler
e-mail: feckler@uni-landau.de

R. B. Schäfer
e-mail: schaefer-ralf@uni-landau.de

R. Schulz
e-mail: schulz@uni-landau.de

J. P. Zubrod
e-mail: zubrod@uni-landau.de

M. Bundschuh · A. Feckler
Department of Aquatic Sciences and Assessment, Swedish University of Agricultural Sciences, Uppsala, Sweden

R. Schulz · J. P. Zubrod
Eußerthal Ecosystem Research Station, University of Koblenz-Landau, Eußerthal, Germany

C. M. Swan et al. (eds.), *The Ecology of Plant Litter Decomposition in Stream Ecosystems*, https://doi.org/10.1007/978-3-030-72854-0_17

under (semi-)field conditions and highlight research gaps we suggest addressing in order to improve understanding and prediction power.

17.1 Introduction

Understanding carbon and nutrient dynamics in ecosystems is one of the most challenging tasks in ecology (Guenet et al., 2010). Carbon and nutrients form a major part of the 100 gigatons of terrestrial plant material produced annually (Gessner et al., 2010), of which a large share (90%) enters the detrital pool as organic material (OM) of variable quality (Guenet et al., 2010). The OM also tightly links terrestrial and aquatic ecosystems as the latter is subsidized with more than 90% of its budget by carbon, nutrients and energy of allochthonous sources (Brett et al., 2017). The energy stored in allochthonous OM is released to the aquatic system through decomposition to fine particulate OM (FPOM) and dissolved OM (DOM) primarily via the activity of microbial decomposers (including bacteria and fungi) and detritivores (Anderson & Sedell, 1979). The latter group mainly comprises amphipods, isopods, plecopterans, and trichopterans (Cummins et al., 1989) and contributes substantially to OM decomposition (OMD). However, depending on the ecosystem studied, the contribution of bacteria and fungi can be as high as 70% and thus exceed the OMD realised by detritivores (Taylor & Chauvet, 2014). Additionally, the colonization of leaves by microorganisms (=conditioning) is of fundamental importance for the OMD process as nutrition of higher trophic levels depends strongly on this conditioning process. This is because microbial conditioning increases the quality of OM for detritivores due to the production of proteins and lipids as well as the degradation of indigestible OM components (Jenkins & Suberkropp, 1995). The finally released FPOM, a high-quality food source (Bundschuh & McKie, 2016), is eventually ingested by filtering and collecting invertebrates (Cummins, 1973). As a consequence, OMD is an important ecosystem process driving element cycling within and across ecosystem boundaries (e.g., via emergence of detritivores with an aquatic and terrestrial life stage).

Table 17.1 Various context parameters of importance for OMD not explicitly covered by the conceptual model in Fig. 17.1

Spatial scale	Parameter(s)	Affected group or parameter
Biome/landscape	Geology and climate	Indirect effects on microbial decomposers and detritivores through effect cascades on hydrology and temperature, which in turn affect water chemistry, leaf quality & quantity, and water turbulence
Watershed	Leaf quantity	Direct effect on detritivores
	Water chemistry	Direct effects on microbial decomposers and detritivores
	Water turbulence	Direct effect on OMD through physical fragmentation

The conditioning of OM is determined by complex interactions among microorganisms. Some authors, for instance, reported antagonistic effects of bacteria on fungi during OM conditioning in streams (Gulis & Suberkropp, 2003; Schneider et al., 2010). On the contrary, contrasting results have been reported for the influence of fungi on bacterial colonization. Gulis and Stephanovich (1999) reported antibiotic properties of some species of aquatic hyphomycetes (a polyphyletic group of fungi), whereas Schneider et al. (2010) observed a stimulation of bacterial growth on leaves in the presence of fungi. The authors suggested that the positive effect of fungal species on bacteria could be driven by the provision of OM degradation products formed by fungal extracellular enzymes. Furthermore, antagonistic interactions among fungal species competing for resources have been reported (e.g., Bärlocher, 1991; Yuen et al., 1999). Despite these complex interactions, it has been suggested that biodiversity of all involved organism groups plays an important role in OMD (Gessner et al., 2010). It is assumed that a higher diversity of aquatic hyphomycetes leads to a higher functional diversity and ultimately higher OMD rates (Costantini & Rossi, 2010). Moreover, the functions provided by species-rich microbial communities may be more stable under increasing levels of stress (Pascoal et al., 2010), which can be explained by the redundancy hypothesis. This hypothesis postulates that the effect of species loss on ecosystem functioning is compensated by species with similar functional traits (Walker, 1992).

Over 350,000 chemicals and mixtures of chemicals have been registered for production and use (Wang et al., 2020). Chemicals can enter freshwater ecosystems as a consequence of their intended application in the field (e.g., deliberate emission of pesticides) or as unintended by-product (e.g., pharmaceutical emission in wastewater discharge) of their use (Schwarzenbach et al., 2006). In the environment, many chemicals (hereafter called toxicants) can exert adverse toxic effects on freshwater organisms and in turn on ecosystem functions. A recent study highlighted that concentrations of organic toxicants exceeded low risk thresholds for approximately half of German rivers, which was similar to the risk exerted by invasive species, although hydromorphological stress and excessive nutrients were the dominant stressors (Schäfer et al., 2016). Increasing evidence suggests that toxicants influence freshwater (micro)organisms and cause a selection of more tolerant strains, phenotypes or species (e.g., Feckler et al., 2018), which secure stable OMD (see as one case study Schäfer et al., 2007). Mechanistic studies on how stressors affect populations and communities at the landscape scale are lacking. This is largely because the identification of mechanisms is typically hampered by the spatio-temporal variability of multiple environmental and biotic conditions that shape populations and communities, as well as by the absence of true replicates. Studies under controlled conditions (laboratory) may allow for the identification of such mechanisms under (semi-)field conditions. However, the sheer amount of organic and inorganic toxicants of natural or anthropogenic origin that are found in freshwater ecosystems and may harm organisms complicates the extrapolation from the laboratory to the field. Toxicants subsume chemicals with diverse modes of action (MoAs), which is defined as the biochemical mechanism through which a chemical influences an organism. They are, moreover, used for different purposes in a range of fields such as medicine, agriculture, life stock and industrial production leading to different exposure profiles in freshwater systems. The exposure profiles can be separated in press (long term) and pulsed (temporary) disturbances. The first is reflected by a constant release of, for instance, municipal wastewater, and the latter by a short-term exposure as a consequence of runoff from roads or agricultural fields. Additionally, the MoA of a chemical determines the group of organisms that responds strongest to it and could shape the effect pathway leading to implications in OMD.

Here we will focus on four groups of toxicants (i.e., fungicides, antibiotics, insecticides and metals) primarily affecting central groups of organisms in OMD. While fungicides and antibiotics are designed to directly interact through different MoAs with fungi (e.g., ergosterol synthesis inhibition) and bacteria (e.g., inhibition of bacterial cell wall synthesis), respectively, insecticides influence the nervous system (e.g., neonicotinoids) of detritivorous invertebrates or their development (moulting inhibition). On the contrary, metals constitute a diverse group comprising various MoAs. For example, some metals have a primarily biocidal MoA and thus affect fungi and bacteria. The identification of effect pathways within and among trophic levels specific to chemical groups will help to understand their potential impacts on ecosystems. The outlined sensitivity differences among the organisms involved in OMD to the various toxicant groups can trigger distinct effect pathways within the food web and in turn the ecosystem process OMD. Against this background, we introduce a conceptual model that highlights the relevance of interactions among OM-associated microorganisms as well as detritivorous invertebrates for OMD under the assumption of pristine (uncontaminated) conditions. Based on this model, we screened the literature for evidence of effect pathways induced by selected toxicant groups (i.e., fungicides, antibiotics, insecticides and metals). We first analyse laboratory studies for evidence for the theoretical effect pathways. Then we interpret these findings in the context of mesocosm and field studies, which are typically characterised by a more complex interaction of multiple environmental factors potentially interfering with such pathways. We conclude by highlighting research gaps to address in order to better understand and predict effects of toxicants on OMD.

17.2 Conceptual Model

The conceptual model visualized in Fig. 17.1 is based on the more general perspective of Graça et al. (2015), where microbial decomposers (including bacteria and fungi) are subsumed as one organism group. Here, we distinguish between these two microbial decomposer groups, because toxicants differ in their MoA and thus the group of (micro)organisms they primarily target. The conceptual model assumes that bacteria and fungi play a central role in the overall OMD either directly (pathway 1) or indirectly through the consumption of OM by detritivores (pathway 2). Please note that the term "pathway" refers to mechanisms covered in the conceptual model; while we use the term "mechanism" in the remaining chapter in a larger context. The

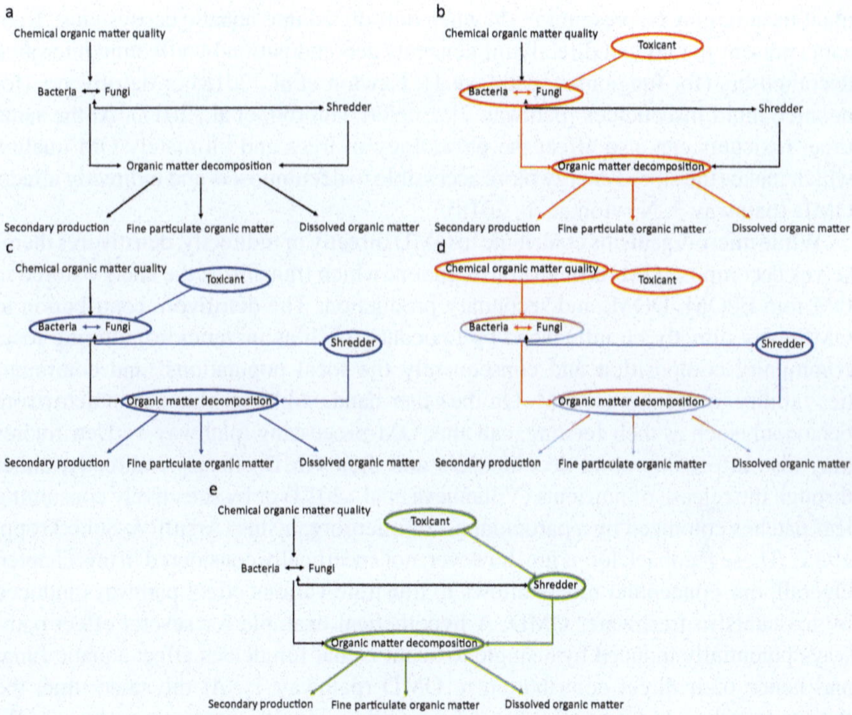

Fig. 17.1 (**a**) Conceptual model showing different groups of organisms that are central to OMD and considered in this chapter. The model was modified after Graça et al. (2015) and forms the basis for the cascade effects or pathways introduced in the section "*Conceptual Model*". The different pathways (1–4) are visualised in parts **b**–**e**. In part (**b**), pathway 1 is highlighted in *red*, namely toxicants (i.e., fungicides and antibiotics) affect microorganisms colonising OM with direct consequences on OMD. Pathway 2 is highlighted in *blue* in part (**c**), where toxicants affect microorganisms colonising OM, which affects detritivores. Through this bottom up directed pathway, OMD is affected. The joint interaction of pathway 1 and 2 is described in pathway 3 (combination of *red* and *blue*, **d**), which is incorporating the toxicants into aquatic systems as part of the OM and affects OMD indirectly. Finally, panel (**e**) highlights in *green* pathway 4 induced by toxicants (such as insecticides) and directly affecting detritivores and ultimately OMD. This model is used to highlight potential cascade effect induced by toxicants with variable modes of action (see following subchapters). Context parameters affecting individual organism groups (in)directly, which are, however, not explicitly considered in the present chapter, are listed in Table 17.1 (cf. Graça et al., 2015)

quality of OM, such as lignin content and C:N:P ratio, influences the ability of microbial decomposers to colonise and decompose OM (García-Palacios et al., 2016). Differences in microbial colonisation affect the nutritious quality of OM as food for detritivores, which could have implications for the organisms' physiology, reproduction and population development, ultimately altering OMD dynamics (pathway 2). Besides OM-inherent properties that determine its colonisation by microorganisms and ultimately OMD, OM may already contain toxicants. Systemic pesticides such as fungicides or insecticides, which are rapidly taken up and translocated within plant tissues, may be present in OM when introduced into aquatic ecosystems. Toxicants present in OM can directly influence its decomposition by affecting microbial decomposers (for fungicides [pathway 1], Newton et al., 2018) or detritivores (for neonicotinoid insecticides [pathway 2], Englert, Zubrod, et al., 2017). At the same time, toxicants can also affect the physiology of trees and ultimately OM quality, which makes the stored energy more accessible to decomposers and indirectly affects OMD (pathway 3, Newton et al., 2018).

While microorganisms contribute to OMD directly or indirectly, detritivores themselves decompose OM through consumption, which transforms the energy stored in OM into FPOM, DOM, and secondary production. The detritivore contribution to OMD may directly be influenced by toxicants (such as insecticides) altering local community composition and consequently the local populations' and communities' ability to decompose OM. On the other hand, sublethal changes in detritivore behaviour, such as their feeding, can alter OM-processing (pathway 4). Detritivores may also affect microbial decomposers and their OMD efficacy indirectly, either through the release of nutrients (Villanueva et al., 2012) or by selectively consuming leaf patches colonized by a particular set of microorganisms (Arsuffi & Suberkropp, 1985). These feedback loops are, however, not specifically considered in this chapter. Overall, our conceptual model allows to illustrate various effect pathways induced by toxicants in freshwater OMD. A hypothetical example for several effect pathways potentially induced by a single toxicant is that fungicides affect aquatic fungi and hence their direct contribution to OMD (pathway 1). At the same time, the fungicide-induced impact on aquatic fungi influences the nutritious quality of OM for detritivores, changing their feeding behaviour with consequences for growth, reproduction and population development. These alterations can feed back on the overall OMD (pathway 2). In a different scenario, the fungicides could also directly affect detritivores, with impacts on their physiology and ultimately OMD (pathway 4). These schematic effect pathways would likely affect the provisioning of FPOM, DOM as well as secondary production (see for a more detailed evaluation of FPOM Bundschuh & McKie, 2016).

17.3 Effect Pathways Induced by Toxicants with Different Target Organisms

17.3.1 Fungicides

Fungicides are considered an essential tool in crop protection against fungal diseases (Strange & Scott, 2005). Additionally, fungicides are used in urban areas in paints and coatings on facades, flat roofs, and basement seals and are applied to public and private spaces (as reviewed by Zubrod, Bundschuh, et al., 2019). Due to their intensive use and moderate to high mobility (Reilly et al., 2012), fungicides enter aquatic ecosystems via point (e.g., discharge from wastewater treatment plants, Kahle et al., 2008) or non-point (e.g., surface runoff, Bereswill et al., 2012) sources. There, they can cause toxic effects in a wide range of nontarget organisms since their MoA interferes with basic biological processes that are not specific to fungi (e.g., cellular respiration, Stenersen, 2004). Consequently, fungicides have been reported to affect OMD via different pathways (Fig. 17.1).

As fungicides are designed to control fungi and fungal-like organisms, OM-associated fungi and their functioning might be affected with potential direct effects on microbially mediated OMD (pathway 1 in Fig. 17.1). Accordingly, Artigas et al. (2012) found a decreased microbial OMD at 33 μg/L of the fungicide tebuconazole in a 6-weeks indoor stream channel experiment. Similarly, in 13-day lasting laboratory microcosm experiments, a significant reduction in OMD was observed for the fungicides azoxystrobin, carbendazim, and cyprodinil (100, 245, and 200 μg/L, respectively, Zubrod, Englert, Feckler, et al., 2015). Effects on microbial decomposition may be related to general reduction in microbial biomass/abundance, enzymatic activity, and shifts in microbial community composition. Bundschuh et al. (2011), for instance, showed a decreasing fungal biomass (up to 45%; as indicated by the proxy molecule ergosterol) with increasing tebuconazole concentrations, while at the highest tested concentration (500 μg/L) also the abundance of OM-associated bacteria was reduced by 33%. Likewise, Zubrod et al. (2011) and Artigas et al. (2012) measured reduced fungal biomass upon exposure to 65 and 33 μg tebuconazole/L, respectively. Moreover, fungicides potentially affect the kinetics of fungal enzymes. Accordingly, Artigas et al. (2012) demonstrated that tebuconazole affects leaf cellulose and hemicellulose but not lignin decomposition. Laccase activity, in contrast, is increased during tebuconazole exposure in the aquatic hyphomycete *Alatospora acuminata*, potentially as a means of detoxification (Artigas et al., 2017).

In contrast to microbially mediated OMD and biomass/abundance, shifts in microbial community composition seem to occur already at low fungicide concentrations (sensu Cornejo et al., 2020). The number of OM-associated fungal species was significantly reduced at a sum concentration of 6 μg/L of a fungicide mixture composed of azoxystrobin, carbendazim, cyprodinil, quinoxyfen, and tebuconazole (Zubrod, Englert, Feckler, et al., 2015). Reported changes in fungal composition reflect an increasing importance of species that are seemingly fungicide-tolerant supporting functional stability (cf., functional redundancy). In this context, *Tetracladium marchalianum* and *Neonectria lugdunensis* (former *Heliscus lugdunensis*) are often dominating fungal communities from Central and Northern Europe when exposed to fungicides (e.g., Bundschuh et al., 2011; Zubrod, Englert, Feckler, et al., 2015). Accordingly, fungal communities collected from agricultural streams had a higher tolerance than those from pristine streams (cf., pollution induced community tolerance concept (PICT), Blanck et al., 1988; Feckler et al., 2018; Gardeström et al., 2016). In line with this observation, Schreiner et al. (2018) documented a functional adaptation of OM-associated microbial communities from originally pristine streams to fungicide stress over 3 colonisation-decomposition cycles. However, to rigorously test for functional redundancy and PICT, future studies need to relate single-species biomass in complex communities through, for instance, quantitative real-time PCR

(Baudy et al., 2019; Feckler et al., 2017) to the species' traits (i.e., decomposition efficiency or enzyme activity).

Given the fungicide-induced changes in fungal community composition, these toxicants have the potential to affect the quality of OM and thus to trigger effects on higher trophic levels (pathway 2 in Fig. 17.1C). Indeed, the fungal species dominating fungicide-exposed communities (see above) are of low palatability to or are even rejected by detritivores such as gammarids (e.g., Arsuffi & Suberkropp, 1989). Accordingly, Bundschuh et al. (2011) showed that the amphipod *Gammarus fossarum* preferred leaf material conditioned under control conditions over leaves conditioned in the presence of 50 and 500 μg tebuconazole/L. In another study, similar food-choice patterns (significant or by trend) of *G. fossarum* were found for a number of other fungicides (azoxystrobin, carbendazim, cyprodinil, quinoxyfen, and tebuconazole) or their mixture. This fungicide mixture induced a comparable response pattern in larvae of the insect *Chaetopteryx villosa* (Konschak et al., 2019). These findings indicate a uniform response to fungicide-induced changes in food quality among major detritivore taxa. It must, however, be noted that fungicide residues adsorbed to the leaves' surface might also contribute to the observed responses (cf. Zubrod, Englert, Feckler, et al., 2015).

Independent of the underlying mechanism causing a lower palatability of fungicide-exposed OM, long-term consumption of such material can impact detritivore energy processing and physiology and thus may result in effects along the food chain (e.g., via an altered FPOM or secondary production; see pathway 2 in Fig. 17.1). When fed over 24 days with leaf material conditioned in the presence of a fungicide mixture at a sum concentration of 62.5 μg/L, *G. fossarum* produced less faeces (i.e., FPOM, Zubrod, Englert, Wolfram, et al., 2015). This indicates an enhanced assimilation of ingested food to compensate for a lower nutritional quality of leaf material (see above) or investments in detoxification mechanisms necessary to cope with dietary fungicide exposure. This compensatory measure was, however, incomplete as gammarid growth (i.e., secondary production) was reduced by approximately 40%. A similar response pattern was found for the isopod *Asellus aquaticus* when exposed to an epoxiconazole concentration of 15 μg/L. Although effects were not statistically significant, the fatty acid composition of asellids was meaningfully affected (Feckler et al., 2016). The change in the fatty acid composition, particularly the reduction in essential fatty acids, suggests important physiological implications potentially interfering with reproduction and consequently with the contribution of this detritivore to OMD. Another potential indirect pathway related to detritivorous invertebrates, despite the lack of related studies, are fungicide-effects mediated via detritivores gut microbiome. Wilson et al. (2014) showed that the prevalence of symbiotic gut fungi of black flies is negatively associated with the flies' fungicide body burden and similar relationships might be observed in other (leaf-shredding) invertebrates. Moreover, the consequences of fungicides being taken up by trees for OM quality as well as species interactions during its decomposition in aquatic systems is hardly assessed (pathway 3 in Fig. 17.1D). To the best of our knowledge only Newton et al. (2018) addressed this question. They reported a higher palatability of leaves for detritivores from fungicide treated trees after conditioning. This suggests that energy resources in leaves are easily accessible for OM-associated microbes, ultimately increasing their growth in addition to leaf palatability for detritivores. Further studies are required to elucidate the mechanistic basis of this finding that largely contrasts with the studies above.

Besides indirect implications, detritivores may also suffer from direct effects of fungicide exposure (pathway 4 in Fig. 17.1E). Apart from potential toxicity related to dietary uptake (see above), waterborne toxicity was described in several studies. For a comprehensive review on such effects on invertebrates (including taxa involved in OMD) and the general scarcity of data on insects compared to crustaceans (see Zubrod, Bundschuh, et al., 2019). Waterborne fungicide toxicity in detritivores was, for instance, shown to affect energy processing and physiological fitness, potentially translating to effects on the wider food web and ecosystem processes including OMD (see Fig. 17.1). Such effects can exacerbate in the presence of additional stressors such as food scarcity (see e.g., for tebuconazole Zubrod et al., 2010). Two long-term laboratory studies, using the fungicide mixture mentioned above or epoxiconazole, indicate direct waterborne effects (pathway 4 in Fig. 17.1C) being generally more relevant than the indirect food-quality related pathway (pathway 2 in Fig. 17.1C, Feckler et al., 2016; Zubrod, Englert, Wolfram, et al., 2015). It must, however, be noted that in natural systems both exposure pathways occur simultaneously, which seems to result in most cases in additive effects. In conclusion, evidence from the laboratory studies suggests that fungicide exposure can affect OMD via all described pathways in the conceptual model (Fig. 17.1), although the direct effect pathways are clearly dominating.

17.3.2 Antibiotics

Antibiotics are extensively applied for human and animal disease prophylaxis and treatment (Bernhardt et al., 2017). Due to incomplete metabolization and degradation, many antibiotics are frequently detected in effluents and downstream of wastewater treatment plants as well as animal production facilities (Janecko et al., 2016). Unlike other pollutants that enter aquatic systems seasonally, such as pesticides, the year-round use of antibiotics results in their continuous release into freshwater environments (Rosi-Marshall & Royer, 2012), where they have the potential to affect non-target organisms and ecosystem functions (Daughton & Ternes, 1999). Studies conducted at the laboratory (microcosm) scale revealed adverse effects of antibiotics on microorganisms (bacteria and fungi) and detritivorous invertebrates, causing alterations in OMD via different pathways as highlighted by the conceptual model in Fig. 17.1.

Antibiotics may modify the interaction between bacteria and fungi (here in particular aquatic hyphomycetes) that colonize terrestrially-derived OM in streams and cause direct effects on microbial OMD (pathway 1 in Fig. 17.1B). Additionally, antibiotics can alter OMD through a cascade effect on detritivorous invertebrates (pathway 2 in Fig. 17.1C). Some evidence for effects on OM-associated microbial communities has been reported by Maul et al. (2006). These authors showed that exposure to the antibiotic ciprofloxacin (100 μg/L) shifted the composition and functioning of the entire OM-associated microbial community (bacteria and fungi), which was indicated by a reduced carbohydrate usage. The structural implications observed by Maul et al. (2006) are in accordance with Bundschuh et al. (2009), who studied the effects of an antibiotic mixture (clarithromycin, erythromycin-H_2O, roxithromycin, sulfa-methoxazole, and trimethoprim) on OM-associated microbes. The OM-associated fungal biomass (indicated by the proxy molecule ergosterol) was increased by ~40% under antibiotic exposure. This may be explained by a modification of the interaction between bacteria and fungi that reduced the competitive pressure on the latter. As no differences in bacterial counts were found between the control and antibiotic treatments (up to 200 μg/L), a shift in the bacterial community composition may be hypothesized as underlying mechanism. Despite these shifts in resource usage and composition of microbes, no effects of antibiotics on microbial OMD were observed (see Bundschuh et al., 2009; Hughes et al., 2016).

Besides direct antibiotic effects on OM-associated microorganisms (pathway 1 in Fig. 17.1B), implications of antibiotics on OMD via the conjectured cascade effect on higher trophic levels (pathway 2 in Fig. 17.1C) have been observed at the laboratory scale. Alterations at the microbial level could theoretically affect the quality of OM as food for detritivorous invertebrates (Bärlocher 1985; Graça et al., 1993). Due to the prime role of detritivores in OMD in temperate streams (Hieber & Gessner, 2002), the consequences of such cascade effects (pathway 2)—leading to an altered OM consumption—are potentially significant. Bundschuh et al. (2009), for instance, observed a food selection behaviour of *G. fossarum* when given the choice to feed on OM (i.e., leaves) microbially colonized under control conditions or under exposure to an antibiotic mixture, with higher consumption rates on the latter (but see Hahn & Schulz, 2007). Given the role of fungi in the diet of detritivores (Suberkropp, 1992), it was suggested that this food selection behaviour was triggered by the higher fungal biomass associated with antibiotic-exposed OM (see above). Furthermore, it was assumed that a shift in the fungal community towards species with a higher palatability for detritivores may have contributed to the higher consumption of antibiotic-treated OM (for examples of similar Cu-induced effects see Zubrod, Feckler et al., 2015). The consumption of more palatable and more nutritious OM should ultimately allow detritivores to grow faster (see for ciprofloxacin Konschak et al., 2020). Accordingly, Bundschuh et al. (2017) found an increased growth of *G. fossarum* at low (2 μg/L) but not at high (200 μg/L) levels (simultaneous dietary and water exposure) of the same antibiotic mixture as applied in Bundschuh et al. (2009). However, energetically costly detoxification mechanisms in gammarids under high antibiotic exposure (Maltby, 1999) could not be compensated by an increased food intake. The effects observed by Bundschuh et al. (2009) for the low exposure scenario (2 μg/L) seem beneficial for OMD through an increased secondary production of detritivores. Over the long term, however, a lowered secondary production or curtailment of life cycles of some detritivores with subsequent consequences for OMD might be observed. This is due to a shortage of OM stocks and therefore food in affected stream sections (cf., Suberkropp et al., 2010), which ultimately may provoke pervasive modifications on OMD dynamics given the key role of detritivores.

Additionally, antibiotics may cause indirect effects on the physiology of detritivorous invertebrates by altering the gut microflora and its efficiency to assimilate nutrients and energy from the ingested food. Such reduced efficiency may eventually affect the growth and reproduction of detritivores and cause adverse effects on OMD over the long term. Although this effect pathway has not been empirically tested yet, it was put forward, for instance, by Bundschuh et al. (2017) as a potential mechanism. The authors assumed that a negative impact on the gut microflora could explain the lack of growth observed for *G. fossarum* exposed to an antibiotic mixture at 200 μg/L (see above). This assumption is supported by a study of Gorokhova et al. (2015) that revealed a lower diversity in the gut microflora of *Daphnia magna* associated with a lower assimilation efficiency when exposed to one of the antibiotics (i.e., trimethoprim) assessed in the study by Bundschuh et al. (2017) at a similar concentration (250 μg/L).

Finally, several studies (e.g., Konschak et al., 2020) indicate effects of antibiotics on OMD by directly affecting detritivorous invertebrates (pathway 4, Fig. 17.1E). However, effect concentrations (e.g., concentration at which the feeding rate of a detritivores was reduced by 50% (EC_{50}): 6.4 mg ciprofloxacin/L) are up to three orders of magnitude above those measured in the environment (e.g., Gracia-Lor et al., 2011). This discrepancy in reported effect concentrations and those measured in the environment may, nevertheless, be biased by the low number of published studies on direct antibiotic effects on detritivores influencing OMD, which calls for further scrutiny. All in all, it is evident from literature that various effect pathways are possible, while pathway 2 seems to be the most likely mechanism inducing changes in OMD as a consequence of antibiotic exposure.

17.3.3 Insecticides

Insecticides, as chemicals used, for instance, in agriculture to control invertebrate pests, regularly occur in surface waters at concentrations known to cause effects in aquatic life (Stehle & Schulz, 2015a; Wolfram et al., 2018). They enter surface waters via various routes, although, due to their physico-chemical properties and use patterns, edge-of-field surface runoff seems to be most important (Dabrowski & Schulz, 2003; Schulz, 2004). Insecticides occur at higher water concentrations in smaller surface water bodies. These smaller water bodies depend, according to the river continuum concept, at the same time on allochthonous input (Vannote et al., 1980). Insecticides are also associated with sediments in larger water bodies (Wolfram et al., 2019). In contrast to other pesticides, insecticides regularly occur in surface waters for rather short time periods spanning a few hours up to a few days (Stehle et al., 2013), which is sufficient to cause negative effects in aquatic ecosystems due to their often fast uptake and presence at their site of action (Tang

& Siegfried, 1995). When following standard risk assessment procedures, insecticides have by far the highest ecotoxicological potential among all pesticide types, with aquatic invertebrates being the most sensitive group of organisms (Malaj et al., 2014; Stehle et al., 2011; Stehle & Schulz, 2015b). Often the single most toxic insecticide within a mixture drives entirely the predicted aquatic toxicity (Wolfram et al., 2019). It is thus likely that effects on OMD are mediated via effects on detritivorous invertebrates (pathway 4; Fig. 17.1E).

Direct mortality-driven insecticide effects on OMD have rarely been reported, while sublethal responses prevail. A reduced survival of stonefly nymphs (*Pteronarcys dorsata*) following aqueous-phase exposure to 96 μg/L imidacloprid, together with a reduced invertebrate-mediated leaf mass loss at exposure levels ≥12 μg/L, were observed in laboratory microcosms, while no impact on microbial decomposition was detectable even at 96 μg imidacloprid/L (Kreutzweiser et al., 2008). Another laboratory study detected a sublethal reduction in leaf mass loss by 25% following 10-d exposure of *Sericostoma vittatum* (Trichoptera) larvae to 2 μg/L chlorantraniliprole (Rodrigues, Bordalo, et al., 2018). Chara-Serna and Richardson (2018) found effects of a field-relevant nominal concentration of 0.3 μg/L chlorpyrifos on invertebrate communities in lab microcosms. Leaf mass loss was reduced by more than 20%, presumably as a result of feeding inhibition. While mortality was unaffected, a tendency for reduced invertebrate biomass was observed. A laboratory study based on a 7-day feeding assay with *Gammarus roeseli* on black alder leaves (*Alnus glutinosa*) found effects already at 5 ng/L lambda-cyhalothrin (Bundschuh, Gergs, et al., 2013). Another pyrethroid, esfenvalerate, led at 0.25 μg/L to feeding inhibition of caddisfly larvae (Rodrigues et al., 2017). As another example of a sublethal effect study, Feckler et al. (2012) used a *G. fossarum* 7-d feeding assay to show a reduced feeding activity at 1 μg/L thiacloprid. Consequently, insecticide exposure can indeed affect OMD negatively by interacting with key detritivores in this ecosystem function (Fig. 17.1E).

All the studies mentioned above looked at aqueous-phase exposure to insecticides, although a food mediated pathway might have been present. Exposure via contaminated food is of importance for lipophilic pyrethroid insecticides that readily partition to organic carbon and thus may adsorb to OM (Bundschuh, Zubrod et al., 2013; Maul et al., 2008). Exposure of beech leaves for 3 h to 0.1 μg/L of the pyrethroid alpha-cypermethrin reduced the shredding activity of invertebrates (Rasmussen, Monberg et al., 2012). There are a number of studies which investigated the effects of contaminated leaves with systemic insecticides on aquatic OMD. A reduced leaf mass loss was, for instance, observed in aquaria receiving imidacloprid-contaminated alder leaves (Englert, Bakanov, et al., 2017) offered as food to the stonefly detritivore *Protonemura* sp. (Lima Fernandes et al., 2019). Contaminated leaves have been obtained through trunk injection of trees or soil drenching. Trunk injection with azadirachtin (neem) to ash trees has been studied by Kreutzweiser et al. (2011), and only a high concentration (6 times above the expected field level) led to effects on microbial decomposition, while no effects on aquatic invertebrates were observed. Similarly, only high concentrations of imidacloprid applied to ash trees led to mortality of two detritivorous insects (*P. dorsata* and *Tipula* sp. [Diptera]) and to the impairment of OMD. However, no effects on microbial OMD were observed (Kreutzweiser et al., 2007). The potential release of imidacloprid from senescent leaves of trees treated via soil drenching about four months before leaf fall into the water was studied by Englert, Bakanov, et al. (2017). The authors modelled, on the basis of simplified assumptions, concentrations of up to 250 ng/L in stream water. The exposure route via ingestion of such leaf material containing systemic insecticides has only been studied rarely (Bundschuh et al., 2019). For the neonicotinoid insecticide thiacloprid, Englert, Zubrod, et al. (2017) have shown, for instance, that dietary exposure caused—relative to the control—similar reductions in gammarids' leaf consumption (~35%) and lipid content (~20%) as observed for the waterborne exposure pathway (30 and 22%). Englert et al. (2018) have, furthermore, documented that leaching and UV irradiation reduces the load of imidacloprid in leaves and thus also the negative effects on the feeding rate of *G. fossarum*.

The potential effects of insecticides on OMD should finally be considered in a wider ecological and environmental context. Organisms exposed to insecticides may have to also face other stressors. Zubrod, Englert et al. (2017) have detected a 2.5-fold increased impairment of the 7-d feeding rate of *G. fossarum* by thiacloprid, when the animals were pre-exposed to wastewater for up to six weeks. Feeding rates of amphipods can be indicative for impacts on the population level (Baird et al., 2007; Maltby et al., 2002), ultimately affecting their contribution to OMD. Moreover, impacts on detritivore-mediated OMD caused by insecticides can be transient and thus potentially leading to an underestimation of negative impacts in more complex (semi-) field conditions (Agatz et al., 2014; Zubrod, Bundschuh et al., 2017).

17.3.4 Metals

Metal pollution in surface waters can originate from various anthropogenic activities such as agriculture, industrial production, and mining (Hogsden & Harding, 2012). Within surface waters, sediments often act as (temporal) sinks for metals, from which they can be dissolved and become biologically available (Chapman et al., 1998). Once they are bioavailable, metals can affect the structural and functional properties of aquatic ecosystems, despite their partly essential nature (Rainbow, 1993). Consequently, metals [including cadmium (Cd), copper (Cu), silver (Ag), uranium (U), and zinc (Zn)] have been shown to adversely affect OMD at the laboratory scale via pathways 1 (direct effects on microorganisms), 2 (cascade effect on detritivorous invertebrates), and 4 (direct effects on detritivorous invertebrates) of the theoretical model (Fig. 17.1B–E). Metal identity and therefore MoA may be important moderators of metal toxicity, as metals may act on different contributors in the OMD process (Fig. 17.1A). A meta-analysis by Ferreira et al. (2016) showed that observed effects were stronger in laboratory than in field studies (see a more detailed discussion in "Evidences from field work"). This difference in effect magnitudes is likely explained by the better control of confounding variables in the former, antagonistic interactions between metals and other environmental variables in the latter, or differences in metal identity/speciation and concentration among studies. However, despite this discrepancy in the effect strengths observed in laboratory and field studies, results gained by laboratory studies can still be seen as a valuable tool for a mechanistic understanding of metal effects on OMD.

Metals with a biocidal (general antimicrobial) MoA, such as Cu (Flemming & Trevors, 1989) and Ag (Silver, 2003) should (theoretically) cause stronger direct effects on bacteria and fungi and lead to structural changes, with conceivable subsequent effects on microbial OMD (pathway 1; Fig. 17.1). Accordingly, a multitude of studies (e.g., Azevedo et al., 2007; Duarte et al., 2008, 2009; Guimaraes-Soares et al., 2007; Zubrod, Feckler et al., 2015) reported structural and functional effects for OM-associated microbes caused by Cu exposure. A common observation (e.g., Duarte et al., 2008, 2009) is a strong adverse Cu effect on the abundance of OM-associated bacteria. The reduced number of bacteria likely reduces the competitive pressure for sympatric fungi (Mille-Lindblom & Tranvik, 2003), resulting in higher fungal biomass, species richness, and partially their reproduction (indicated by an increased number of released spores; see below). As fungi are assumed to be the main drivers for microbial OMD (Hieber & Gessner, 2002), the competitive advantage of fungi under Cu exposure, together with changes in fungal (trait) composition and biomass, may explain the mostly increased OMD rates compared to control treatments (e.g., Zubrod, Feckler et al., 2015). On the contrary, Ag was found to reduce microbial OMD at comparatively low concentrations (~100 μg/L, Funck et al., 2013), likely explained by a reduced fungal biomass and altered community composition (bacteria were not assessed). Finally, Zn can reduce microbial OMD at concentrations as low as 32.7 μg/L (Duarte et al., 2004), probably due to negative effects on the structure and OMD activity of aquatic fungi (e.g., Duarte et al., 2004, 2008, 2009; Fernandes et al., 2009; Medeiros et al., 2010).

As described above for fungicides and antibiotics, shifts in the OM-associated microbial community that trigger changes in the palatability and resource quality of OM may be seen as an underlying mechanism for a cascade effect on detritivores' feeding activity and consequently OMD (pathway 2; Fig. 17.1). For Cu-exposed OM, for instance, increased fungal biomasses (e.g., Zubrod, Feckler et al., 2015) alongside fungal community shifts towards species with a higher palatability for detritivores (Zubrod, Englert, Rosenfeldt et al., 2015) have been observed compared to OM from control treatments, suggesting an increase in food quality (see for instance Arsuffi & Suberkropp, 1989; Bärlocher & Kendrick, 1975; Rong et al., 1995; Suberkropp, 1992). Indeed, OM palatability was positively affected by Cu-exposure and resulted in a higher consumption on Cu-exposed OM by *G. fossarum* compared to control OM (Zubrod, Feckler et al., 2015). The higher consumption of Cu-exposed OM may have been caused by the positive Cu effect on the OM-associated fungal biomass and the ability of detritivores to discriminate between the palatability of different fungal species. Irrespective of the underlying mechanism, the food-choice response of *Gammarus* indicates an altered food quality and/or palatability of Cu-exposed OM. A long-term consumption of leaves conditioned in presence of Cu, however, may have negative consequences for growth and energy reserves of detritivores driven by the dietary exposure towards Cu (Zubrod, Englert, Rosenfeldt et al., 2015). Moreover, and as discussed above for antibiotic-induced changes in the OM palatability, a lowered secondary production or curtailment of life cycles of some detritivores with subsequent consequences for OMD might be observed over the long term because of a shortage of OM stocks and therefore food in stream sections affected by Cu-contamination.

Finally, direct metal effects on detritivorous invertebrates have been shown to reduce OMD (pathway 4; Fig. 17.1E). Cd and U, for instance, caused strong direct effects on the feeding activity, growth, and mortality of detritivores, with metal uptake through feeding on contaminated OM (Abel & Bärlocher, 1988; Campos et al., 2014; Gonçalves et al., 2011), over the water phase (e.g., Abel & Bärlocher, 1988; Pestana et al., 2007), and when exposed to sediment-associated metals (Zubrod, Englert, et al., 2019). On the contrary, only weak Cd and U effects on the structure and functioning of OM-associated microbial communities were found at comparatively high concentrations (i.e., 3.6–4.5 mg Cd/L and 27.8 μg U/L, Batista et al., 2012; Gonçalves et al., 2011; Moreirinha et al., 2011). This indicates that adverse effects of Cd and U on OMD mainly stem from a direct impact on detritivores, whereas effects on microbial decomposers and subsequent indirect effects on detritivores may be of minor importance (pathway 2, Fig. 17.1C). Furthermore, despite its strong biocidal MoA and being an essential metal for invertebrates (Rainbow, 1993), direct effects on detritivores are documented for Cu that reduced OMD. Zubrod, Englert, et al. (2019), for instance, showed that Cu which resorbed from Cu-contaminated sediments (150–650 mg Cu/kg), negatively affected the feeding activity and survival of three sediment-associated detritivores, namely *A. aquaticus, G. fossarum,* and *Hyalella azteca*. Likewise, Zubrod et al. (2014) found 20% decreased feeding rates of *G. fossarum* already at Cu water concentrations between ~35 and 85 μg/L, depending on the applied Cu salt that might have influenced the metal speciation in the medium. In accordance, Rosenfeldt et al. (2015) observed a high mortality as well as significantly decreased feeding activity of and excretion by *G. fossarum* at an environmentally relevant concentration of 40 μg Cu/L. In conclusion, results from laboratory studies at the microcosm scale suggest that metal exposure can affect OMD via three pathways set out in the conceptual model (Fig. 17.1), however, effects on the different OMD-contributors and respective pathways are partially dependent on metal identity.

17.4 Evidence for Cascade Effects from Field and Mesocosm Studies

In this chapter, we review to which extent the effect pathways (Fig. 17.1) identified for different toxicant groups under laboratory conditions match patterns in freshwater ecosystems. We consider field studies or those performed in controlled outdoor experimental systems with multiple organism groups (hereafter called mesocosms). However, studies in these systems typically provide an aggregate of direct and indirect effects of toxicants, which does not allow to distinguish between either pathways. Moreover, the identification of cascade effects is complicated by the fact that additional stressors may mask relationships between toxicant concentrations and ecological responses that have been identified under controlled conditions. Notwithstanding, (semi-)field studies represent a reality check for studies that simplify the environmental context for the sake of establishing causal relationships (Schindler, 1998).

Laboratory studies on fungicides (see previous section) showed that they can affect microbial communities and subsequently microbial OMD (pathway 1). A fraction of these studies demonstrated direct effects on detritivores (pathway 4) but also indirect effects resulting from changes in microbial communities (pathway 2) at environmentally relevant concentrations. In cases where detritivore-driven OMD responded, not only reductions but also increases were found, for example, in response to a higher energy-demand due to detoxification processes. Several field and mesocosm studies confirm that pathway 1 occurs in natural environments. Specifically, mesocosm studies (e.g., Artigas et al., 2012; Cuppen et al., 2000; Dimitrov et al., 2014; Rossi et al., 2018) and field studies (Fernandez et al., 2015; Rasmussen, Wiberg-Larsen et al., 2012) showed a reduction in microbial OMD, typically associated with a microbial community change and a decrease in fungal biomass. With respect to pathway 4, several mesocosm studies provide evidence for the effects of fungicides on invertebrates communities (Cuppen et al., 2000; Daam & Van den Brink, 2010; Halstead et al., 2014; van Wijngaarden et al., 2010, 2014), though at concentrations of 10 s to 100 s μg/L, whereas field concentrations are typically in the lower μg/L range (Schäfer, von der Ohe, et al., 2012). Furthermore, the concentration levels for effects on invertebrates are for most fungicides higher than for fungi and microorganisms (Zubrod, Bundschuh et al., 2019), suggesting that if pathway 4 occurs, it co-occurs with pathway 1, and that pathway 2 is less relevant. We are not aware of a mesocosm or field study where fungicides exclusively affected invertebrates and the detritivore-mediated OMD. A study in French streams found a decrease in detritivore-driven OMD, while microbial OMD was not affected (Brosed et al., 2016). However, fungicides likely played a minor role and co-occurring insecticides and herbicides were the main drivers of pesticide effects. Similarly, a mesocosm study with additional pesticides beside fungicides found exclusive effects on detritivore-OMD, but this is very likely due to insecticides (Auber et al., 2011). Field studies where fungicides played a major role for microbial and invertebrate toxicity rather found exclusive effects on microbial OMD (Rasmussen, Wiberg-Larsen et al., 2012; Fernandez et al., 2015) or effects on both microbial and detritivore OMD (Schäfer, Bundschuh, et al., 2012). Moreover, several mesocosm studies only measured total OMD and found a decrease in response to fungicides (Cuppen et al., 2000; Daam & Van den Brink, 2010; Halstead et al., 2014) or mixtures including fungicides, associated with effects on invertebrate and fungal communities, if monitored. Overall, this suggests that several current-use fungicides likely affect fungal communities and microbial OMD (pathway 1), and in some cases invertebrates and detritivore-driven OMD (combination of pathway 1 and 4). Moreover, the effects are concentration-dependent and this will determine whether any effects occur. A few mesocosm studies found no or negligible effects of suggested environmentally realistic fungicide concentrations on fungal communities and related OMD (Arts et al., 2006; Lin et al., 2012; Talk et al., 2016). However, in the study by Talk et al. (2016) this was attributed to nutrients increasing the tolerance of fungi towards fungicides. Similarly, a field study in agricultural, urban and forest catchments in France suggested that despite effects on fungal biomass the higher microbial decomposition was due to nutrients masking potential toxicant effects (Rossi et al., 2019). This hypothesis of compensation by nutrients of fungicide effects on microorganisms was not supported in an outdoor mesocosm study with detritivores and microbial communities on leaves, where changes in the community composition in response to fungicides and nutrients were observed but did not translate to implications in OMD (Fernandez et al., 2016). Thus, the prediction of specific effect pathways in the field is complicated by additional stressors and the environmental context.

For antibiotics, typically a higher fungal biomass and contrasting effects on detritivores are found in laboratory systems. This suggests a negative direct effect on bacteria leading to a positive indirect effect on fungi and in turn detritivores, although the latter may be compromised by direct negative effects on detritivores. Moreover, changes in the OMD rates were typically minor. Antibiotics have been widely found in field monitoring studies and typically originate from treated or untreated wastewater discharge, effluents from antibiotic manufacturing as well as the use of animal manure (Janecko et al., 2016). These input sources are typically associated with additional stressors such as other pharmaceuticals, pesticides and excessive nutrients, which hamper the attribution of effects to antibiotics. For example, a field study comparing agricultural, forested and urban stream sites, with the highest pharmaceutical concentrations in the urban sites, found a similar microbial OMD rate between the forested and urban sites, but the fungal biomass was reduced, contrasting with the cascade effect from laboratory studies (Rossi et al., 2019). Several further studies focused on the influence of wastewater treatment plants on microbial communities, invertebrate communities and OMD (e.g., Ashauer, 2016; Chonova et al., 2016; Englert et al., 2013; Peschke et al., 2019). The abovementioned co-occurrence of stressors and the associated complexity of biotic and abiotic interactions complicates the comparison of the results from such field studies to those for antibiotics under controlled laboratory conditions. A field study found a reduction of microbial growth and respiration following exposure to a pharmaceutical mixture including antibiotics (Rosi-Marshall et al., 2013). In a tropical mesocosm study, OMD and invertebrate communities were similar between control treatments and those treated with an antibiotic at environmentally relevant concentrations, though the density of certain bacterial groups declined (Rico et al., 2014). Overall, the fact that antibiotics rarely occur in isolation suggests that the cascade effects observed under laboratory conditions are more relevant for mechanistic insight than for prediction of effects in the natural environment. In the field, potential effects may be masked by the action of associated stressors, though a change in bacterial communities can be expected.

For insecticides mainly effect pathway 4 occurs, i.e., an effect on invertebrates resulting in changes in the OMD was detected in laboratory studies. This is in line with results from field and mesocosm studies. Generally, several field studies have demonstrated effects on invertebrates, with typical shifts in community composition from sensitive to tolerant taxa (overviews in Schäfer, 2019; Schulz, 2004). This includes, for example, effects on important detritivore taxa such as amphipods, which were strongly reduced in 22 Argentinian streams (Hunt et al., 2017). It is therefore very likely that the laboratory results translate to the field, although only few studies examined the potential response of ecosystem functioning including OMD. A field study in 19 streams located in rural Germany found a reduced leaf decomposition rate and changes in the invertebrate community (Munze et al., 2015). Two further field studies in Australia and France also found a reduced OMD with an increase in pesticide toxicity, mainly driven by insecticides (Schäfer, Caquet et al., 2007, Schäfer, Bundschuh et al., 2012). The relationship between pesticide toxicity and OMD was remarkably similar in both studies, despite completely different detritivore communities in Australia and France (Schäfer, von der Ohe, et al., 2012). For example, crustaceans play a major role in OMD in France, but are largely irrelevant in Australia. Only one study, where *Gammarus pulex* was the dominant detritivore, found no effect of pesticide toxicity, mainly driven by a carbamate insecticide and a strobilurine fungicide, on OMD in 14 Danish streams. However, the abundance of the dominant detritivore was also not associated with pesticide toxicity (Rasmussen, Wiberg-Larsen et al., 2012).

Several mesocosm studies show that insecticides at environmentally realistic concentrations affect invertebrate communities and reduce detritivore-driven OMD. A realistic spray series including insecticides reduced OMD, most likely due to a reduction in crustacean detritivores (Auber et al., 2011). Another study by Rodrigues, Machado et al. (2018) suggested that the introduction of native crayfish increased the negative effect that chlorantraniliprole had on OMD by detritivores. Imidacloprid exposure at 18 μg/L, which represents the upper end of environmentally-relevant concentrations in a global meta-analysis (Morrissey et al., 2015), reduced detritivore-driven OMD, whereas microbial OMD was not affected (Pestana et al., 2009). This effect was confirmed in an associated bioassay with a stonefly detritivore. However, the approximately 10-fold lower concentration in the same study and in another study (Chara-Serna et al., 2019) had only negligible or no effect on OMD. Similarly, short pulses of this substance had no effect on OMD in a stream mesocosm study with caged crustacean detritivores (Böttger et al., 2013). In a study where several macroinvertebrate species have been exposed to 10.65 ng/L of the pyrethroid lambda-cyhalothrin for 90 min with a subsequent assessment in outdoor stream channels over 10 days, a reduced OMD was observed (Rasmussen et al., 2008). Similarly, the pyrethroid etofenprox reduced the in situ measured feeding rate of *A. aquaticus* on alder leaves in outdoor stream mesocosms at levels of 0.04 μg/L (6-h pulse) and thus at concentrations below those detectable in the field (Wieczorek et al., 2018). By contrast, concentrations of approximately 1 μg/L of the pyrethroid deltamethrin affected the macroinvertebrate community, while the detritivore-mediated OMD increased by up to 55%. This is potentially due to the release of non-insect detritivores from competition with insects and an increased energy demand (Caquet et al., 2007). Similarly, exposure to the insecticide malathion increased OMD, either through reduced predation or increased energy demand associated with detoxification in stream mesocosm communities (Bray et al., 2019). No effects on OMD were found in mesocosms exposed to a potato spraying regime (Arts et al., 2006). However, only coarse mesh bags were used and it was thus not possible to account for microbial decay and to draw conclusions for detritivore-driven OMD. Furthermore, the assessment of OMD took place during weeks without insecticide application (i.e., 4.8 and 120 ng/L of lambda-cyhalothrin, a pyrethroid which degrades fast [DT_{50} < 2 days]), suggesting that potential effects of the spraying regime on OMD were missed.

Overall, these examples suggest that the effect pathway 4 is widely occurring and relevant for insecticides. In a broader context, alpha diversity increases at higher latitudes (Boyero et al., 2012) and many cool adapted taxa are detritivores. Temperate environments are more important for detritivores due to lower temperatures and higher availability of high-quality leaf litter (Boyero et al., 2011). It is thus likely that detritivores in temperate environments, which are largely characterized by extensive agricultural production, are at a wide scale impaired functionally by insecticides as non-target organisms. The effects may not be restricted to aquatic systems, because aquatic-terrestrial coupling through organic matter cycling (Kayler et al., 2019) may also be impacted by pesticides. This has potential far-reaching effects on adjacent riparian ecosystems (Schulz et al., 2015) as also suggested by first field studies (Graf et al., 2019).

The effects of metal contamination on OMD have been frequently studied in both field and mesocosm studies. The laboratory studies (see chapter "Effect pathways induced by toxicants with different target organisms - metals") identified three main effect paths: (1) Stronger effect on detritivores than microorganisms (e.g., for U and Cd), with potential indirect effects on microorganisms and reduction in detritivore OMD (pathway 4, Fig. 17.1E), (2) Up to a certain concentration a stronger effect on bacteria than on fungi and detritivores, where the release of fungi from competition results in higher microbial OMD (e.g., for Cu; pathway 1, Fig. 17.1B). At higher concentrations, invertebrate density and detritivore OMD is reduced (pathway 4, Fig. 17.1E). (3) Stronger effect on microorganisms, resulting in a decrease in microbial OMD (e.g., for Zn and Ag; pathway 1, Fig. 17.1B). A recent meta-analysis has compared the effects of metal contamination on OMD under laboratory and field conditions (Ferreira et al., 2016). The analysis found a stronger effect on total OMD than on microbial OMD, suggesting a lower effect on microbial communities and functional resistance of microorganisms. Generally, the effects were strongest for a mixture of Cu and Zn and higher in laboratory than field studies, suggesting that interactions under real world conditions buffer the effects from metals. However, field studies on metals often deal with mixtures of metals and elevated metal concentrations are often associated with mining, which frequently results in a decreased pH (Niyogi et al., 2001). This complicates the establishment of clear relationships. Notwithstanding, several field and mesocosm studies found stronger effects on invertebrates than on microorganisms for Zn, which contrasts the data available from laboratory studies. In streams affected by mine drainage, invertebrate densities and total OMD responded stronger to Zn concentrations and the pH-driven precipitation of metal oxides than microbial communities, where the response was likely masked by nutrients (Niyogi et al., 2001). A study in an industrial area of Germany showed that microbial OMD still works at extreme concentrations of Zn and Cu where invertebrates have largely been extirpated (Sridhar et al., 2001), although at a slower rate. Furthermore, the impact of elevated Zn concentrations in five streams in the US found a decrease in detritivores, microbial activity and total OMD (Carlisle & Clements, 2005). Several studies examined the effects of Cu, where up to a certain concentration microbial OMD increased in the laboratory. A mesocosm study found a reduction of invertebrate densities and, presumably as a consequence, of total OMD for higher copper treatments. At the same time, leaf-associated fungal biomass, sporulation rates and microbial community composition were similar between treatments (Roussel et al., 2008). Another mesocosm study found only negligible differences between control and different Cu treatments on total OMD (Gardham et al., 2015). In a creek in the US, increasing Cu concentrations lead to a reduction in detritivorous invertebrates and total OMD (Schultheis et al., 1997), although this study does not necessarily contradict the laboratory findings as microbial OMD was not assessed. Similarly, total OMD and invertebrate abundance decreased with Pb concentrations in a study in a lake (Oguma & Klerks, 2015), but microbial OMD was not considered. Overall, the evidence from field and mesocosm studies suggests that direct effects of metals on invertebrates that in turn reduce OMD are more widespread than cases where microbial communities are stronger affected than detritivores. This may be explained by shorter recovery and adaptation times of microbial compared to invertebrate communities (Dahl & Blanck, 1996).

17.5 Conclusions and Perspectives

The information gathered in this chapter highlights that several of the effect pathways detailed in Fig. 17.1 can be considered relevant under field conditions, which often involve more complex exposure regimes to toxicants, further environmental factors (e.g., nutrients), and biological interactions that interfere with the stress response in (eco)systems. More specifically, effect pathways 1 and 4 likely play a considerable role in the field for fungicides and insecticides, respectively. Pathway 2, in contrast, is difficult to disentangle in complex systems but could contribute to the observed effects in natural ecosystems. Toxicant-induced modifications in plants (pathway 3) and the ultimate consequences for OMD and heterotrophic food webs has hardly been considered but could be relevant. Pathway 3 may be particularly relevant either in anthropogenically dominated landscapes where riparian trees may accumulate toxicants or in cases where senescent leaves falling from pesticide-treated deciduous trees enter adjacent water bodies (e.g., orchards). In the following, we suggest various research topics against the background of this chapter that would broaden our understanding of toxicant-induced effects on OMD.

Generally, research gaps exist with respect to better understand and predict effects in real-world ecosystems. Intelligent study designs that link microcosm and field experiments over different relevant gradients of environmental conditions are required to enhance our capacity to predict changes in community composition and related changes in functioning. In this context, microbial and invertebrate OMD also represents an interesting study system to test predictions from ecological theories related to food webs and community assembly. For example, biotic interactions and dispersal could be modified to study their role in community assembly in the context of toxicant effects.

A fundamental uncertainty related to the research on the impact of toxicants on OMD is the question of the transferability of results obtained either in the laboratory or under (semi)field conditions to field conditions. Are the results obtained with one substrate (e.g., alder leaves) representative for effects observed for another substrate (e.g., different leaf species, wood, carcass)? In the case of black alder leaves this may be questioned as this substrate contains high levels of nitrogen that could buffer the negative effects induced by toxicants. Studies suggest that detritivores react in a similar manner on the physiological level (Feckler et al., 2016; Konschak et al., 2019; Zubrod, Englert, Rosenfeldt et al., 2015; Zubrod, Englert, Wolfram et al., 2015) to fungicide-induced changes in OM quality or waterborne exposure. Further evidence is needed to scrutinise the idea of uniformity in response. Additionally, OM-associated microbial communities and their ability to decompose OM varies among sites within and across biogeographic regions (Tiegs et al., 2019), which may influence their susceptibility to toxicants. The only available study suggests that microbial communities along a latitudinal gradient across Europe react in a comparable manner when exposed to fungicides over three colonisation-decomposition cycles (Schreiner et al., 2018). Nonetheless, the study was carried out in microcosms with a limited species number and without potential for recolonization from undisturbed stream sections. Therefore, further collaborative studies within and across biogeographic regions are required to provide more robust conclusions.

OM-associated microbial communities may adapt to toxicant exposure (Schreiner et al., 2018), which results in a higher tolerance after several seasons (Feckler et al., 2018). This adaptation follows the Pollution-Induced Community Tolerance (PICT) concept (Blanck et al., 1988). The PICT concept has mainly been used for phytoplankton and periphyton communities and only sporadically in the context of OMD (Feckler et al., 2018). This discrepancy may be explained by the difficulties to characterise OM-associated microbial communities (particularly the fungal component) regarding both species diversity and the respective species' quantitative contribution to community composition but also functioning. Recent developments in molecular biology can close this gap but still need further development; whereas metabarcoding can inform on community composition (Pawlowski et al., 2018), quantitative real time PCR analyses can provide biomass estimates of individual species or groups of species (Baudy et al., 2019) and ultimately support the testing for PICT and understanding the underlying mechanisms. Besides the challenges related to the assessment of communities, we lack knowledge on the traits of individual species or groups of species, in particular for microorganisms, but also detritivorous invertebrates (e.g., leaf preferences, processing rates). Consequently, changes in OMD resulting from toxicant-induced alterations in community composition can hardly be predicted. Hence, developing and maintaining a trait (enzyme inventory, sensitivity, nutritious quality) database for OM-associated microorganisms should be one of the top priorities of the scientific community to foster the linking of communities with functioning.

Studies assessing the impact of toxicants on plants and the consequences for OMD are largely lacking. The only study we are aware of, highlights that fungicides could improve OM quality, which may be beneficial for detritivores and their population development (Newton et al., 2018) finally increasing OMD. As a consequence of the exploratory nature of this work, the mechanisms leading to these results are unclear. It is, for instance, proposed that fungicides reduced the pressure by fungal diseases, which may have made energy (carbohydrates) available for storage in leaves. These energy resources may have allowed the OM-associated microbial community to increase its performance and ultimately support detritivore growth. Similarly, fungicides may have changed the activity of soil microbial communities that support plant performances and increase OM quality. Overall, further studies on this issue are required, in particular because some fungicides are systemic and may accumulate in relevant concentrations in plant material (Zubrod, Bundschuh et al., 2019).

Finally, the gut microbiome is still a black box in many organisms and in particular for invertebrates (Adamovsky et al., 2018). Changes in the food quality may require some (costly) adaptations in the gut microbiome, with consequences for detritivore growth and reproduction (pathway 2). Similarly, antimicrobial substances such as fungicides, antibiotics, and some metals can potentially change the composition of the gut microbiome with implications for detritivore nutrition (Gorokhova et al., 2015). Consequently, understanding how structural changes in the gut microbiome link to functional changes (e.g., enzyme activity) and ultimately the physiological condition of detritivores is certainly an important step to understand the mechanism by which toxicants affect OMD in a larger context.

It is evident from this review, that our understanding on how toxicants can affect OMD has developed in the last decades meaningfully. At the same time, these developments have informed the scientific community about gaps in various aspects of this research field that need to be addressed to further improve understanding and prediction power.

References

Abel, T., & Bärlocher, F. (1988). Uptake of cadmium by *Gammarus fossarum* (Amphipoda) from food and water. *Journal of Applied Ecology, 25,* 223–231.

Adamovsky, O., Buerger, A. N., Wormington, A. M., Ector, N., Griffitt, R. J., Bisesi, J. H., et al. (2018). The gut microbiome and aquatic toxicology: An emerging concept for environmental health. *Environmental Toxicology and Chemistry, 37,* 2758–2775.

Agatz, A., Ashauer, R., & Brown, C. D. (2014). Imidacloprid perturbs feeding of *Gammarus pulex* at environmentally relevant concentrations. *Environmental Toxicology and Chemistry, 33,* 648–653.

Anderson, N. H., & Sedell, J. R. (1979). Detritus processing by macroinvertebrates in stream ecosystems. *Annual Review of Entomology, 24,* 351–377.

Arsuffi, T. L., & Suberkropp, K. (1985) Selective feeding by stream caddisfly (Trichoptera) detritivores on leaves with fungal-colonized patches *Oikos,* 45, 50–58.

Arsuffi, T. L., & Suberkropp, K. (1989). Selective feeding by shredders on leaf-colonizing stream fungi: comparison of macroinvertebrate taxa. *Oecologia, 79,* 30–37.

Artigas, J., Majerholc, J., Foulquier, A., Margoum, C., Volat, B., Neyra, M., et al. (2012). Effects of the fungicide tebuconazole on microbial capacities for litter breakdown in streams. *Aquatic Toxicology, 122–123,* 197–205.

Artigas, J., Rossi, F., Gerphagnon, M., & Mallet, C. (2017). Sensitivity of laccase activity to the fungicide tebuconazole in decomposing litter. *Science of the Total Environment, 584–585,* 1084–1092.

Arts, G. H. P., Buijse-Bogdan, L. L., Belgers, J. D. M., van Rhenen-Kersten, C. H., van Wijngaarden, R. P. A., Roessink, I., et al. (2006). Ecological impact in ditch mesocosms of simulated spray drift from a crop protection program for potatoes. *Integrated Environmental Assessment and Management, 2,* 105–125.

Ashauer, R. (2016). Post-ozonation in a municipal wastewater treatment plant improves water quality in the receiving stream. *Environmental Science Europe, 28,* 1. https://doi.org/10.1186/s12302-12015-10068-z.

Auber, A., Roucaute, M., Togola, A., & Caquet, T. (2011). Structural and functional effects of conventional and low pesticide input crop-protection programs on benthic macroinvertebrate communities in outdoor pond mesocosms. *Ecotoxicology, 20,* 2042–2055.

Azevedo, M. M., Carvalho, A., Pascoal, C., Rodrigues, F., & Cassio, F. (2007). Responses of antioxidant defenses to Cu and Zn stress in two aquatic fungi. *Science of the Total Environment, 377,* 233–243.

Baird, D. J., Brown, S. S., Lagadic, L., Liess, M., Maltby, L., Moreira-Santos, M., et al. (2007). In situ-based effects measures: Determining the ecological relevance of measured responses. *Integrated Environmental Assessment and Management, 3,* 259–267.

Bärlocher, F. (1985). The role of fungi in the nutrition of stream invertebrates. *Botanical Journal of the Linnean Society, 91,* 83–94.

Bärlocher, F. (1991). Intraspecific hyphal interactions among aquatic hyphomycetes. *Mycologia, 83,* 82–88.

Bärlocher, F., & Kendrick, B. (1975). Assimilation efficiency of *Gammarus pseudolimnaeus* (Amphipoda) feeding on mycelium or autumn-shed leaves. *Oikos, 26,* 55–59.

Batista, D., Pascoal, C., & Cassio, E. (2012). Impacts of warming on aquatic decomposers along a gradient of cadmium stress. *Environmental Pollution, 169,* 35–41.

Baudy, P., Zubrod, J. P., Röder, N., Baschien, C., Feckler, A., Schulz, R., et al. (2019). A glance into the black box: Novel species-specific quantitative real-time PCR assays to disentangle aquatic hyphomycete community composition. *Fungal Ecology, 42,*.

Bereswill, R., Golla, B., Streloke, M., & Schulz, R. (2012). Entry and toxicity of organic pesticides and copper in vineyard streams: Erosion rills jeopardise the efficiency of riparian buffer strips. *Agriculture, Ecosystems & Environment, 146,* 81–92.

Bernhardt, E. S., Rosi-Marshall, E. J., & Gessner, M. O. (2017). Synthetic chemicals as agents of global change. *Frontiers in Ecology and the Environment, 15,* 84–90.

Blanck, H., Wängberg, S.-A., & Morlander, S. (1988). Pollution-induced community tolerance—A new ecotoxicological tool. In J. Cairns & J. R. Pratt (Eds.), *Functional testing of aquatic biota for estimating hazards of chemicals* (pp. 219–230). Philadelphia, PA, USA: American society for testing and materials.

Böttger, R., Feibicke, M., Schaller, J., & Dudel, G. (2013). Effects of low-dosed imidacloprid pulses on the functional role of the caged amphipod *Gammarus roeseli* in stream mesocosms. *Ecotoxicology and Environmental Safety, 93,* 93–100.

Boyero, L., Pearson, R. G., Dudgeon, D., Ferreira, V., Graça, M. A. S., Gessner, M. O., Boulton, A. J., Chauvet, E., Yule, C. M., Albarino, R. J., Ramirez, A., Helson, J. E., Callisto, M., Arunachalam, M., Chara, J., Figueroa, R., Mathooko, J. M., Gonçalves, J. F., Moretti, M.S. et al. (2012). Global patterns of stream detritivore distribution: Implications for biodiversity loss in changing climates. *Global Ecology and Biogeography,* 21, 134–141.

Boyero, L., Pearson, R. G., Dudgeon, D., Graça, M. A. S., Gessner, M. O., Albarino, R. J., Ferreira, V., Yule, C. M., Boulton, A. J., Arunachalam, M., Callisto, M., Chauvet, E., Ramirez, A., Chara, J., Moretti, M. S., Gonçalves, J. F., Helson, J. E., Chara-Serna, A. M., Encalada, A. C. et al. (2011). Global distribution of a key trophic guild contrasts with common latitudinal diversity patterns. *Ecology,* 92, 1839–1848.

Bray, J. P., Nichols, S. J., Keely-Smith, A., Thompson, R., Bhattacharyya, S., Gupta, S., et al. (2019). Stressor dominance and sensitivity-dependent antagonism: Disentangling the freshwater effects of an insecticide among co-occurring agricultural stressors. *Journal of Applied Ecology, 56,* 2020–2033.

Brett, M. T., Bunn, S. E., Chandra, S., Galloway, A. W. E., Guo, F., Kainz, M. J., et al. (2017). How important are terrestrial organic carbon inputs for secondary production in freshwater ecosystems? *Freshwater Biology, 62,* 833–853.

Brosed, M., Lamothe, S., & Chauvet, E. (2016). Litter breakdown for ecosystem integrity assessment also applies to streams affected by pesticides. *Hydrobiologia, 773,* 87–102.

Bundschuh, R., Bundschuh, M., Otto, M., & Schulz, R. (2019). Food-related exposure to systemic pesticides and pesticides from transgenic plants—Evaluation of aquatic test strategies. *Environmental Science Europe, 31,* 87.

Bundschuh, M., Gergs, R., Schadt, S., & Schulz, R. (2013). Do differences in sensitivity between native and invasive amphipods explain their coexistence in Lake Constance? A case study with lambda-cyhalothrin *Chemosphere, 92,* 483–489.

Bundschuh, M., Hahn, T., Gessner, M. O., & Schulz, R. (2009). Antibiotics as a chemical stressor affecting an aquatic decomposer-detritivore system. *Environmental Toxicology and Chemistry, 28,* 197–203.

Bundschuh, M., Hahn, T., Gessner, M. O., & Schulz, R. (2017). Antibiotic mixture effects on growth of the leaf-shredding stream detritivore *Gammarus fossarum. Ecotoxicology, 26,* 547–554.

Bundschuh, M., & McKie, B. G. (2016). An ecological and ecotoxicological perspective on fine particulate organic matter in streams. *Freshwater Biology, 61,* 2063–2074.

Bundschuh, M., Zubrod, J. P., Klemm, P., Elsaesser, D., Stang, C., & Schulz, R. (2013). Effects of peak exposure scenarios on *Gammarus fossarum* using field relevant pesticide mixtures. *Ecotoxicology and Environmental Safety, 95,* 137–143.

Bundschuh, M., Zubrod, J. P., Kosol, S., Maltby, L., Stang, C., Duester, L., et al. (2011). Fungal composition on leaves explains pollutant-mediated indirect effects on amphipod feeding. *Aquatic Toxicology, 104,* 32–37.

Campos, D., Alves, A., Lemos, M. F. L., Correia, A., Soares, A. M. V. M., & Pestana, J. L. T. (2014). Effects of cadmium and resource quality on freshwater detritus processing chains: A microcosm approach with two insect species. *Ecotoxicology, 23,* 830–839.

Caquet, T., Hanson, M. L., Roucaute, M., Graham, D. W., & Lagadic, L. (2007). Influence of isolation on the recovery of pond mesocosms from the application of an insecticide. II. Benthic macroinvertebrate responses. *Environmental Toxicology and Chemistry, 26,* 1280–1290.

Carlisle, D. M., & Clements, W. H. (2005). Leaf litter breakdown, microbial respiration and shredder production in metal polluted streams. *Freshwater Biology, 50,* 380–390.

Chapman, P. M., Wang, F., Janssen, C., Persoone, G., & Allen, H. E. (1998). Ecotoxicology of metals in aquatic sediments: Binding and release, bioavailability, risk assessment, and remediation. *Canadian Journal of Fisheries and Aquatic Sciences, 55,* 2221–2243.

Chara-Serna, A. M., Epele, L. B., Morrissey, C. A., & Richardson, J. S. (2019). Nutrients and sediment modify the impacts of a neonicotinoid insecticide on freshwater community structure and ecosystem functioning. *Science of the Total Environment, 692,* 1291–1303.

Chara-Serna, A. M., & Richardson, J. S. (2018). Chlorpyrifos interacts with other agricultural stressors to alter stream communities in laboratory microcosms. *Ecological Applications, 28,* 162–176.

Chonova, T., Keck, F., Labanowski, J., Montuelle, B., Rimet, F., & Bouchez, A. (2016). Separate treatment of hospital and urban wastewaters: A real scale comparison of effluents and their effect on microbial communities. *Science of the Total Environment, 542,* 965–975.

Cornejo, A., Pérez, J., Alonso, A., Lopez-Rojo, N., Monroy, S., & Boyero, L. (2020). A common fungicide impairs stream ecosystem functioning through effects on aquatic hyphomycetes and detritivorous caddisflies. *Journal of Environmental Management,* 263.

Costantini, M. L., & Rossi, L. (2010). Species diversity and decomposition in laboratory aquatic systems: The role of species interactions. *Freshwater Biology, 55,* 2281–2295.

Cummins, K. W. (1973). Trophic relations of aquatic insects. *Annual Reviews Entomology, 18,* 183–206.

Cummins, K. W., Wilzbach, M. A., Gates, D. M., Perry, J. B., & Taliaferro, W. B. (1989). Shredders and riparian vegetation. *BioScience, 39,* 24–30.

Cuppen, J. G., Van den Brink, P. J., Camps, E., Uil, K. F., & Brock, T. C. (2000). Impact of the fungicide carbendazim in freshwater microcosms. I. Water quality, breakdown of particulate organic matter and responses of macroinvertebrates. *Aquatic Toxicology, 48,* 233–250.

Daam, M. A., & Van den Brink, P. J. (2010). Implications of differences between temperate and tropical freshwater ecosystems for the ecological risk assessment of pesticides. *Ecotoxicology, 19,* 24–37.

Dabrowski, J. M., & Schulz, R. (2003). Predicted and measured levels of azinphosmethyl in the Lourens River, South Africa: Comparison of runoff and spray drift. *Environmental Toxicology and Chemistry, 22,* 494–500.

Dahl, B., & Blanck, H. (1996). Pollution-induced community tolerance (PICT) in periphyton communities established under tri-n-butyltin (TBT) stress in marine microcosms. *Aquatic Toxicology, 34,* 305–325.

Daughton, C. G., & Ternes, T. A. (1999). Pharmaceuticals and personal care products in the environment: Agents of subtle change. *Environmental Health Perspectives, 107,* 907–938.

Dimitrov, M. R., Kosol, S., Smidt, H., Buijse, L., Van den Brink, P. J., Van Wijngaarden, R. P., et al. (2014). Assessing effects of the fungicide tebuconazole to heterotrophic microbes in aquatic microcosms. *Science of the Total Environment, 490,* 1002–1011.

Duarte, S., Pascoal, C., Alves, A., Correia, A., & Cassio, F. (2008). Copper and zinc mixtures induce shifts in microbial communities and reduce leaf litter decomposition in streams. *Freshwater Biology, 53,* 91–101.

Duarte, S., Pascoal, C., & Cassio, F. (2004). Effects of zinc on leaf decomposition by fungi in streams: studies in microcosms. *Microbial Ecology, 48,* 366–374.

Duarte, S., Pascoal, C., & Cassio, F. (2009). Functional stability of stream-dwelling microbial decomposers exposed to copper and zinc stress. *Freshwater Biology, 54,* 1683–1691.

Englert, D., Bakanov, N., Zubrod, J. P., Schulz, R., & Bundschuh, M. (2017). Modeling remobilization of neonicotinoid residues from tree foliage in streams—A relevant exposure pathway in risk assessment? *Environmental Science & Technology, 51,* 1785–1794.

Englert, D., Zubrod, J. P., Pietz, S., Stefani, S., Krauss, M., Schulz, R., & Bundschuh, M. (2017). Relative importance of dietary uptake and waterborne exposure for a leaf-shredding amphipod exposed to thiacloprid-contaminated leaves. *Scientific Reports, 7,* 16182. https://doi.org/10.11038/s41598-16017-16452-16189.

Englert, D., Zubrod, J. P., Neubauer, C., Schulz, R., & Bundschuh, M. (2018). UV-irradiation and leaching in water reduce the toxicity of imidacloprid-contaminated leaves to the aquatic leaf-shredding amphipod *Gammarus fossarum*. *Environmental Pollution, 236,* 119–125.

Englert, D., Zubrod, J. P., Schulz, R., & Bundschuh, M. (2013). Effects of municipal wastewater on aquatic ecosystem structure and function in the receiving stream. *Science of the Total Environment, 454–455,* 401–410.

Feckler, A., Goedkoop, W., Konschak, M., Bundschuh, R., Kenngott, K. G. J., Schulz, R., et al. (2018). History matters: Heterotrophic microbial community structure and function adapt to multiple stressors. *Global Change Biology, 24,* e402–e415.

Feckler, A., Goedkoop, W., Zubrod, J. P., Schulz, R., & Bundschuh, M. (2016). Exposure pathway-dependent effects of the fungicide epoxiconazole on a decomposer-detritivore system. *Science of the Total Environment, 571,* 992–1000.

Feckler, A., Schrimpf, A., Bundschuh, M., Bärlocher, F., Baudy, P., Cornut, J., et al. (2017). Quantitative real-time PCR as a promising tool for the detection and quantification of leaf-associated fungal species—A proof of principle using *Alatospora pulchella. PLoS ONE, 12,*.

Feckler, A., Thielsch, A., Schwenk, K., Schulz, R., & Bundschuh, M. (2012). Differences in the sensitivity among cryptic lineages of the *Gammarus fossarum* complex. *Science of the Total Environment, 439,* 158–164.

Fernandes, I., Duarte, S., Cassio, F., & Pascoal, C. (2009). Mixtures of zinc and phosphate affect leaf litter decomposition by aquatic fungi in streams. *Science of the Total Environment, 407,* 4283–4288.

Fernandez, D., Tummala, M., Schreiner, V. C., Duarte, S., Pascoal, C., Winkelmann, C., et al. (2016). Does nutrient enrichment compensate fungicide effects on litter decomposition and decomposer communities in streams? *Aquatic Toxicology, 174,* 169–178.

Fernandez, D., Voss, K., Bundschuh, M., Zubrod, J. P., & Schäfer, R. B. (2015). Effects of fungicides on decomposer communities and leaf decomposition in vineyard streams. *Science of the Total Environment, 533,* 40–48.

Ferreira, V., Koricheva, J., Duarte, S., Niyogi, D. K., & Guerold, F. (2016). Effects of anthropogenic heavy metal contamination on litter decomposition in streams—A meta-analysis. *Environmental Pollution, 210,* 261–270.

Flemming, C. A., & Trevors, J. T. (1989). Copper toxicity and chemistry in the environment—A review. *Water, Air, and Soil Pollution, 44,* 143–158.

Funck, J. A., Clivot, H., Felten, V., Rousselle, P., Guerold, F., & Danger, M. (2013). Phosphorus availability modulates the toxic effect of silver on aquatic fungi and leaf litter decomposition. *Aquatic Toxicology, 144–145,* 199–207.

García-Palacios, P., McKie, B. G., Handa, I. T., Frainer, A., & Hättenschwiler, S. (2016). The importance of litter traits and decomposers for litter decomposition: A comparison of aquatic and terrestrial ecosystems within and across biomes. *Functional Ecology, 30,* 819–829.

Gardeström, J., Ermold, M., Goedkoop, W., & Mckie, B. G. (2016). Disturbance history influences stressor impacts: Effects of a fungicide and nutrients on microbial diversity and litter decomposition. *Freshwater Biology, 61,* 2171–2184.

Gardham, S., Chariton, A. A., & Hose, G. C. (2015). Direct and indirect effects of copper-contaminated sediments on the functions of model freshwater ecosystems. *Ecotoxicology, 24,* 61–70.

Gessner, M. O., Swan, C. M., Dang, C. K., Mckie, B. G., Bardgett, R. D., Wall, D. H., et al. (2010). Diversity meets decomposition. *Trends in Ecology & Evolution, 25,* 372–380.

Gonçalves, A. L., Lirio, A. V., Pratas, J., & Canhoto, C. (2011). Uranium contaminated water does not affect microbial activity but decreases feeding by the shredder *Sericostoma vittatum. Fundamental and Applied Limnology, 179,* 17–25.

Gorokhova, E., Rivetti, C., Furuhagen, S., Edlund, A., Ek, K., & Breitholtz, M. (2015). Bacteria-mediated effects of antibiotics on *Daphnia* nutrition. *Environmental Science and Technology, 49,* 5779–5787.

Graça, M. A. S., Ferreira, V., Canhoto, C., Encalada, A. C., Guerrero-Bolano, F., Wantzen, K. M., et al. (2015). A conceptual model of litter breakdown in low order streams. *International Review of Hydrobiology, 100,* 1–12.

Graça, M. A. S., Maltby, L., & Calow, P. (1993). Importance of fungi in the diet of *Gammarus pulex* and *Asellus aquaticus* I: Feeding strategies. *Oecologia, 93,* 139–144.

Gracia-Lor, E., Sancho, J. V., & Hernández, F. (2011). Multi-class determination of around 50 pharmaceuticals, including 26 antibiotics, in environmental and wastewater samples by ultra-high performance liquid chromatography-tandem mass spectrometry. *Journal of Chromatography A, 1218,* 2264–2275.

Graf, N., Battes, K. P., Cimpean, M., Dittrich, P., Entling, M. H., Link, M., et al. (2019). Do agricultural pesticides in streams influence riparian spiders? *Science of the Total Environment, 660,* 126–135.

Guenet, B., Danger, M., Abbadie, L., & Lacroix, G. (2010). Priming effect: Bridging the gap between terrestrial and aquatic ecology. *Ecology, 91,* 2850–2861.

Guimaraes-Soares, L., Pascoal, C., & Cassio, F. (2007). Effects of heavy metals on the production of thiol compounds by the aquatic fungi *Fontanospora fusiramosa* and *Flagellospora curta. Ecotoxicology and Environmental Safety, 66,* 36–43.

Gulis, V., & Suberkropp, K. (2003). Interactions between stream fungi and bacteria associated with decomposing leaf litter at different levels of nutrient availability. *Aquatic Microbial Ecology, 30,* 149–157.

Gulis, V. I., & Stephanovich, A. I. (1999). Antibiotic effects of some aquatic hyphomycetes. *Mycological Research, 103,* 111–115.

Hahn, T., & Schulz, R. (2007). Indirect effects of antibiotics in the aquatic environment: A laboratory study on detritivore food selection bahavior. *Human and Ecological Risk Assessment, 13,* 535–542.

Halstead, N. T., McMahon, T. A., Johnson, S. A., Raffel, T. R., Romansic, J. M., Crumrine, P. W., et al. (2014). Community ecology theory predicts the effects of agrochemical mixtures on aquatic biodiversity and ecosystem properties. *Ecology Letters, 17,* 932–941.

Hieber, M., & Gessner, M. O. (2002). Contribution of stream detrivores, fungi, and bacteria to leaf breakdown based on biomass estimation. *Ecology, 83,* 1026–1038.

Hogsden, K. L., & Harding, J. S. (2012). Consequences of acid mine drainage for the structure and function of benthic stream communities: A review. *Freshwater Science, 31,* 108–120.

Hughes, S. R., Kay, P., & Brown, L. E. (2016). Impact of anti-inflammatories, beta-blockers and antibiotics on leaf litter breakdown in freshwaters. *Environmental Science and Pollution Research, 23,* 3956–3962.

Hunt, L., Bonetto, C., Marrochi, N., Scalise, A., Fanelli, S., Liess, M., Lydy, M. J., Chiu, M.-C., & Resh, V. H. (2017) Species at risk (SPEAR) index indicates effects of insecticides on stream invertebrate communities in soy production regions of the Argentine Pampas. *Science of the Total Environment,* 580, 699–709.

Janecko, N., Pokludova, L., Blahova, J., Svobodova, Z., & Literak, I. (2016). Implications of fluoroquinolone contamination for the aquatic environment—A review. *Environmental Toxicology and Chemistry, 35,* 2647–2656.

Jenkins, C. C., & Suberkropp, K. (1995). The influence of water chemistry on the enzymatic degradation of leaves in streams. *Freshwater Biology, 33,* 245–253.

Kahle, M., Buerge, I. J., Hauser, A., Muller, M. D., & Poiger, T. (2008). Azole fungicides: Occurrence and fate in wastewater and surface waters. *Environmental Science and Technology, 42,* 7193–7200.

Kayler, Z. E., Premke, K., Gessler, A., Gessner, M. O., Griebler, C., Hilt, S., Klemedtsson, L., Kuzyakov, Y., Reichstein, M., Siemens, J., Totsche, K. U., Tranvik, L., Wagner, A., Weitere, M., & Grossart, H. P. (2019). Integrating aquatic and terrestrial perspectives to improve insights Into organic matter cycling at the landscape scale. *Frontiers in Earth Science, 7.*

Konschak, M., Zubrod, J. P., Baudy, P., Englert, D., Herrmann, B., Schulz, R., et al. (2019). Waterborne and diet-related effects of inorganic and organic fungicides on the insect leaf shredder *Chaetopteryx villosa* (Trichoptera). *Aquatic Toxicology, 206,* 33–42.

Konschak, M., Zubrod, J. P., Baudy, P., Fink, P., Kenngott, K. G. J., Lüderwald, S., et al. (2020). The importance of diet-related effects of the antibiotic ciprofloxacin on the leaf-shredding invertebrate *Gammarus fossarum* (Crustacea; Amphipoda). *Aquatic Toxicology, 222,*.

Kreutzweiser, D., Good, K., Chartrand, D., Scarr, T., & Thompson, D. (2007). Non-target effects on aquatic decomposer organisms of imidacloprid as a systemic insecticide to control emerald ash borer in riparian trees. *Ecotoxicology and Environmental Safety, 68,* 315–325.

Kreutzweiser, D. P., Good, K. P., Chartrand, D. T., Scarr, T. A., & Thompson, D. G. (2008). Toxicity of the systemic insecticide, imidacloprid, to forest stream insects and microbial communities. *Bulletin of Environmental Contamination and Toxicolgy, 80,* 211–214.

Kreutzweiser, D., Thompson, D., Grimalt, S., Chartrand, D., Good, K., & Scarr, T. (2011). Environmental safety to decomposer invertebrates of azadirachtin (neem) as a systemic insecticide in trees to control emerald ash borer. *Ecotoxicology and Environmental Safety, 74,* 1734–1741.

Lima Fernandes, E., Bundschuh, M., Bakanov, N., Englert, D., SChulz, R., & Schäfer, R. B. (2019). Effects of a systemic pesticide along an aquatic tri-trophic food chain. *Bulletin of Environmental Contamination and Toxicology, 103*, 507–514.

Lin, R., Buijse, L., Dimitrov, M., Dohmen, P., Kosol, S., Maltby, L., et al. (2012). Effects of the fungicide metiram in outdoor freshwater microcosms: Responses of invertebrates, primary producers and microbes. *Ecotoxicology, 21*, 1550–1569.

Malaj, E., von der Ohe, P. C., Grothe, M., Kühne, R., Mondy, C. P., Usseglio-Polatera, P., et al. (2014). Organic chemicals jeopardise freshwater ecosystems health on the continental scale. *Proceedings of the National Academy of Sciences of the United States of America, 111*, 9549–9554.

Maltby, L. (1999). Studying stress: The importance of organism-level responses. *Ecological Applications, 9*, 431–440.

Maltby, L., Clayton, S. A., Wood, R. M., & McLoughlin, N. (2002). Evaluation of the *Gammarus pulex* in situ feeding assay as a biomonitor of water quality: Robustness, responsiveness and relevance. *Environmental Toxicology and Chemistry, 21*, 361–368.

Maul, J. D., Schuler, L. J., Belden, J. B., Whiles, M. R., & Lydy, M. J. (2006). Effects of the antibiotic ciprofloxacin on stream microbial communities and detritivorous macroinvertebrates. *Environmental Toxicology and Chemistry, 25*, 1598–1606.

Maul, J. D., Trimble, A. J., & Lydy, M. J. (2008). Partitioning and matrix-specific toxicity of bifenthrin among sediments and leaf-sourced organic matter. *Environmental Toxicology and Chemistry, 27*, 945–952.

Medeiros, A., Duarte, S., Pascoal, C., Cassio, F., & Graça, M. (2010). Effects of Zn, Fe and Mn on leaf litter breakdown by aquatic fungi: A microcosm study. *International Review of Hydrobiology, 95*, 12–26.

Mille-Lindblom, C., & Tranvik, L. J. (2003). Antagonism between bacteria and fungi on decomposing aquatic plant litter. *Microbial Ecology, 45*, 173–182.

Moreirinha, C., Duarte, S., Pascoal, C., & Cassio, F. (2011). Effects of cadmium and phenanthrene mixtures on aquatic fungi and microbially mediated leaf litter decomposition. *Archives of Environmental Contamination and Toxicology, 61*, 211–219.

Morrissey, C. A., Mineau, P., Devries, J. H., Sanchez-Bayo, F., Liess, M., Cavallaro, M. C., et al. (2015). Neonicotinoid contamination of global surface waters and associated risk to aquatic invertebrates: A review. *Environment International, 74*, 291–303.

Munze, R., Orlinskiy, P., Gunold, R., Paschke, A., Kaske, O., Beketov, M. A., et al. (2015). Pesticide impact on aquatic invertebrates identified with Chemcatcher(R) passive samplers and the SPEAR(pesticides) index. *Science of the Total Environment, 537*, 69–80.

Newton, K., Zubrod, J. P., Englert, D., Lüderwald, S., Schell, T., Baudy, P., et al. (2018). The evil within? Systemic fungicide application in trees enhances litter quality for an aquatic decomposer-detritivore system. *Environmental Pollution, 241*, 549–556.

Niyogi, D. K., Lewis, W. M., & McKnight, D. M. (2001). Litter breakdown in mountain streams affected by mine drainage: Biotic mediation of abiotic controls. *Ecological Applications, 11*, 506–516.

Oguma, A. Y., & Klerks, P. L. (2015). Evidence for mild sediment Pb contamination affecting leaf-litter decomposition in a lake. *Ecotoxicology, 24*, 1322–1329.

Pascoal, C., Cassio, F., Nikolcheva, L., & Barlocher, F. (2010). Realized fungal diversity increases functional stability of leaf litter decomposition under zinc stress. *Microbial Ecology, 59*, 84–93.

Pawlowski, J., Kelly-Quinn, M., Altermatt, F., Apotheloz-Perret-Gentil, L., Beja, P., Boggero, A., Borja, A., Bouchez, A., Cordier, T., Domaizon, I., Feio, M. J., Filipe, A. F., Fornaroli, R., Graf, W., Herder, J., van der Hoorn, B., Iwan Jones, J., Sagova-Mareckova, M. et al. (2018). The future of biotic indices in the ecogenomic era: Integrating (e)DNA metabarcoding in biological assessment of aquatic ecosystems. *Science of the Total Environment, 637–638*, 1295–1310.

Peschke, K., Capowiez, Y., Kohler, H. R., Wurm, K., & Triebskorn, R. (2019). Impact of a wastewater treatment plant upgrade on amphipods and other macroinvertebrates: Individual and community responses. *Frontiers in Environmental Science, 7*.

Pestana, J. L., Re, A., Nogueira, A. J., & Soares, A. M. (2007). Effects of Cadmium and Zinc on the feeding behaviour of two freshwater crustaceans: *Atyaephyra desmarestii* (Decapoda) and *Echinogammarus meridionalis* (Amphipoda). *Chemosphere, 68,* 1556–1562.

Pestana, J. L. T., Alexander, A. C., Culp, J. M., Baird, D. J., Cessna, A. J., & Soares, A. M. V. M. (2009). Structural and functional responses of benthic invertebrates to imidacloprid in outdoor stream mesocosms. *Environmental Pollution, 157,* 2328–2334.

Rainbow, P. S. (1993). The significance of trace metal concentrations in marine invertebrates. In R. Dallinger & P. S. Rainbow (Eds.), *Ecotoxicology of metals in invertebrates* (pp. 3–23). Boca Raton: Lewis Publishers.

Rasmussen, J. J., Friberg, N., & Larsen, S. E. (2008). Impact of lambda-cyhalothrin on a macroinvertebrate assemblage in outdoor experimental channels: Implications for ecosystem functioning. *Aquatic Toxicology, 90,* 228–234.

Rasmussen, J. J., Monberg, R.J., Baattrup-Pedersen, A., Cedergreen, N., Wiberg-Larsen, P., Strobel, B., & Kronvang, B. (2012). Effects of a triazole fungicide and a pyrethroid insecticide on the decomposition of leaves in the presence or absence of macroinvertebrate shredders. *Aquatic Toxicology, 118,* 54–61.

Rasmussen, J. J., Wiberg-Larsen, P., Baattrup-Pedersen, A., Monberg, R. J., & Kronvang, B. (2012). Impacts of pesticides and natural stressors on leaf litter decomposition in agricultural streams. *Science of the Total Environment, 416,* 148–155.

Reilly, T. J., Smalling, K. L., Orlando, J. L., & Kuivila, K. M. (2012). Occurrence of boscalid and other selected fungicides in surface water and groundwater in three targeted use areas in the United States. *Chemosphere, 89,* 228–234.

Rico, A., Dimitrov, M. R., Van Wijngaarden, R. P. A., Satapornvanit, K., Smidt, H., & Van den Brink, P. J. (2014). Effects of the antibiotic enrofloxacin on the ecology of tropical eutrophic freshwater microcosms. *Aquatic Toxicology, 147,* 92–104.

Rodrigues, A. C. M., Bordalo, M.D., Golovko, O., Koba, O., Barata, C., Soares, A. M. V. M., & Pestana, J. L. T. (2018). Combined effects of insecticide exposure and predation risk on freshwater detritivores. *Ecotoxicology, 27,* 794–802.

Rodrigues, A. C. M., Gravato, C., Quintaneiro, C., Bordalo, M. D., Barata, C., Soares, A. M. V. M., et al. (2017). Energetic costs and biochemical biomarkers associated with esfenvalerate exposure in *Sericostoma vittatum*. *Chemosphere, 189,* 445–453.

Rodrigues, A. C. M., Machado, A. L., Bordalo, M. D., Saro, L., Simao, F. C. P., Rocha, R. J. M., Golovko, O., Zlabek, V., Barata, C., Soares, A. M. V. M., & Pestana, J. L. T. (2018). Invasive species mediate insecticide effects on community and ecosystem functioning. *Environmental Science & Technology, 52,* 4889–4900.

Rong, Q., Sridhar, K. R., & Bärlocher, F. (1995). Food selection in three leaf-shredding stream invertebrates. *Hydrobiologia, 316,* 173–181.

Rosenfeldt, R. R., Seitz, F., Zubrod, J. P., Feckler, A., Merkel, T., Lüderwald, S., et al. (2015). Does the presence of titanium dioxide nanoparticles reduce copper toxicity? A factorial approach with the benthic amphipod *Gammarus fossarum*. *Aquatic Toxicology, 165,* 154–159.

Rosi-Marshall, E. J., Kincaid, D. W., Bechtold, H. A., Royer, T. V., Rojas, M., & Kelly, J. J. (2013). Pharmaceuticals suppress algal growth and microbial respiration and alter bacterial communities in stream biofilms. *Ecological Applications, 23,* 583–593.

Rosi-Marshall, E. J., & Royer, T. V. (2012). Pharmaceutical compounds and ecosystem function: An emerging research challenge for aquatic ecologists. *Ecosystems, 15,* 867–880.

Rossi, F., Mallet, C., Portelli, C., Donnadieu, F., Bonnemoy, F., & Artigas, J. (2019). Stimulation or inhibition: Leaf microbial decomposition in streams subjected to complex chemical contamination. *Science of the Total Environment, 648,* 1371–1383.

Rossi, F., Pesce, S., Mallet, C., Margoum, C., Chaumot, A., Masson, M., et al. (2018). Interactive effects of pesticides and nutrients on microbial communities responsible of litter decomposition in streams. *Frontiers in Microbiology, 9,* 2437.

Roussel, H., Chauvet, E., & Bonzom, J. M. (2008). Alteration of leaf decomposition in copper-contaminated freshwater mesocosms. *Environmental Toxicology and Chemistry, 27,* 637–644.

Schäfer, R. B. (2019). Responses of freshwater macroinvertebrates to pesticides: Insights from field studies. *Current Opinion in Environmental Science & Health, 11,* 1–7.

Schäfer, R. B., Bundschuh, M., Rouch, D. A., Szöcs, E., von der Ohe, P. C., Pettigrove, V., Schulz, R., Nugegoda, D., & Kefford, B. (2012). Effects of pesticide toxicity, salinity and other environmental variables on selected ecosystem functions in streams and the relevance for ecosystem services. *Science of the Total Environment, 415*, 69–78.

Schäfer, R. B., Caquet, T., Siimes, K., Mueller, J., Lagadic, L., & Liess, M. (2007). Effects of pesticides on community structure and ecosystem functions in agricultural streams of three biogeographical regions in Europe. *Science of the Total Environment, 382,* 272–285.

Schäfer, R. B., Kuhn, B., Malaj, E., Konig, A., & Gergs, R. (2016). Contribution of organic toxicants to multiple stress in river ecosystems. *Freshwater Biology, 61,* 2116–2128.

Schäfer, R. B., von der Ohe, P. C., Rasmussen, J., Kefford, B. J., Beketov, M. A., Schulz, R., & Liess, M. (2012). Thresholds for the effects of pesticides on invertebrate communities and leaf breakdown in stream ecosystems. *Environmental Science & Technology, 46*, 5134–5142.

Schindler, D. W. (1998). Replication versus realism: The need for ecosystem-scale experiments. *Ecosystems, 1,* 323–334.

Schneider, T., Gerrits, B., Gassmann, R., Schmid, E., Gessner, M. O., Richter, A., et al. (2010). Proteome analysis of fungal and bacterial involvement in leaf litter decomposition. *Proteomics, 10,* 1819–1830.

Schreiner, V. C., Feckler, A., Fernandez, D., Frisch, K., Munoz, K., Szöcs, E., et al. (2018). Similar recovery time of microbial functions from fungicide stress across biogeographical regions. *Scientific Reports, 8,* 17021.

Schultheis, A. S., Sanchez, M., & Hendricks, A. C. (1997). Structural and functional responses of stream insects to copper pollution. *Hydrobiologia, 346,* 85–93.

Schulz, R. (2004). Field studies on exposure, effects, and risk mitigation of aquatic nonpoint-source insecticide pollution: A review. *Journal of Environmental Quality, 33,* 419–448.

Schulz, R., Bundschuh, M., Gergs, R., Brühl, C. A., Diehl, D., Entling, M., et al. (2015). Review on environmental alterations propagating from aquatic to terrestrial ecosystems. *Science of the Total Environment, 538,* 246–261.

Schwarzenbach, R. P., Escher, B. I., Fenner, K., Hofstetter, T. B., Johnson, C. A., von Gunten, U., et al. (2006). The challenge of micropollutants in aquatic systems. *Science, 313,* 1072–1077.

Silver, S. (2003). Bacterial silver resistance: Molecular biology and uses and misuses of silver compounds. *FEMS Microbiology Reviews, 27,* 341–353.

Sridhar, K. R., Krauss, G., Bärlocher, F., Raviraja, N. S., Wennrich, R., Baumbach, R., et al. (2001). Decomposition of alder leaves in two heavy metalpolluted streams in central Germany. *Aquatic Microbial Ecology, 26,* 73–80.

Stehle, S., Elsaesser, D., Gregoire, C., Imfeld, G., Niehaus, E., Passeport, E., et al. (2011). Pesticide risk mitigation by vegetated treatment systems: A meta-analysis. *Journal of Environmental Quality, 40,* 1068–1080.

Stehle, S., Knäbel, A., & Schulz, R. (2013). Probabilistic risk assessment of insecticide concentrations in agricultural surface waters: A critical appraisal. *Environmental Monitoring and Assessment, 185,* 6295–6310.

Stehle, S., & Schulz, R. (2015a). Agricultural insecticides threaten surface waters at the global scale. *Proceedings of the National Academy of Sciences of the United States of America, 112,* 5750–5755.

Stehle, S., & Schulz, R. (2015b). Pesticide authorization in the EU—Environment unprotected? *Environmental Science and Pollution Research, 22,* 19632–19647.

Stenersen, J. (2004). *Chemical pesticides—Mode of action and toxicology.* Boca Raton, Florida, USA: CRC Press LLC.

Strange, R. N., & Scott, P. R. (2005). Plant disease: A threat to global food security. *Annual review of Phytopathology, 43,* 83–116.

Suberkropp, K. (1992). Interactions with invertebrates. In F. Bärlocher (Ed.), *The ecology of aquatic hyphomycetes* (pp. 118–134). Berlin, Germany: Springer Verlag.

Suberkropp, K., Gulis, V., Rosemond, A. D., & Benstead, J. P. (2010). Ecosystem and physiological scales of microbial responses to nutrients in a detritus-based stream: Results of a 5-year continuous enrichment. *Limnology and Oceanography, 55,* 149–160.

Talk, A., Kublik, S., Ulzsa, M., Engel, M., Berghahn, R., Welzl, G., et al. (2016). Effects of multiple but low pesticide loads on aquatic fungal communities colonizing leaf litter. *Journal of Environmental Sciences, 46,* 116–125.

Tang, J.-X., & Siegfried, B. D. (1995). Comparative uptake of a pyrethroid and organophosphate insecticide by selected aquatic insects. *Bulletin of Environmental Contamination and Toxicolgy, 55,* 130–135.

Taylor, B. R., & Chauvet, E. E. (2014). Relative influence of shredders and fungi on leaf litter decomposition along a river altitudinal gradient. *Hydrobiologia, 721,* 239–250.

Tiegs, S. D., Costello, D. M., Isken, M. W., Woodward, G., McIntyre, P.B., Gessner, M. O., Chauvet, E., Griffiths, N. A., Flecker, A. S., Acuna, V., Albarino, R., Allen, D. C., Alonso, C., Andino, P., Arango, C., Aroviita, J., Barbosa, M. V. M., Barmuta, L. A., Baxter, C.V. et al. (2019). Global patterns and drivers of ecosystem functioning in rivers and riparian zones. *Science Advances, 5,* eaav0486.

van Wijngaarden, R. P. A., Arts, G. H. P., Belgers, J. D. M., Boonstra, H., Roessink, I., Schroer, A. F. W., et al. (2010). The species sensitivity distribution approach compared to a microcosm study: A case study with the fungicide fluazinam. *Ecotoxicology and Environmental Safety, 73,* 109–122.

van Wijngaarden, R. P. A., Belgers, D. J. M., Zafar, M. I., Matser, A. M., Boerwinkel, M. C., & Arts, G. H. P. (2014). Chronic aquatic effect assessment for the fungicide azoxystrobin. *Environmental Toxicology and Chemistry, 33,* 2775–2785.

Vannote, R. L., Minshall, G. W., Cummins, K. W., Sedell, J. R., & Cushing, C. E. (1980). The river continuum concept. *Canadian Journal of Fisheries and Aquatic Sciences, 37,* 130–137.

Villanueva, V. D., Albarino, R., & Canhoto, C. (2012). Positive effect of shredders on microbial biomass and decomposition in stream microcosms. *Freshwater Biology, 57,* 2504–2513.

Walker, B. H. (1992). Biodiversity and ecological redundancy. *Conservation Biology, 6,* 18–23.

Wang, Z. Y., Walker, G. W., Muir, D. C. G., & Nagatani-Yoshida, K. (2020). Toward a global understanding of chemical pollution: A first comprehensive analysis of national and regional chemical inventories. *Environmental Science and Technology, 54,* 2575–2584.

Wieczorek, M. V., Bakanov, N., Bilancia, D., Szöcs, E., Stehle, S., Bundschuh, M., et al. (2018). Structural and functional effects of a short-term pyrethroid pulse exposure on invertebrates in outdoor stream mesocosms. *Science of the Total Environment, 610–611,* 810–819.

Wilson, E. R., Smalling, K. L., Reilly, T. J., Gray, E., Bond, L., Steele, L., et al. (2014). Assessing the potential effects of fungicides on nontarget gut fungi (Trichomycetes) and their associated larval black fly hosts. *Journal of the American Water Resources Association, 50,* 420–433.

Wolfram, J., Stehle, S., Bub, S., Petschick, L. L., & Schulz, R. (2018). Meta-analysis of insecticides in United States surface waters: Status and future implications. *Environmental Science and Technology, 52,* 14452–14460.

Wolfram, J., Stehle, S., Bub, S., Petschick, L. L., & Schulz, R. (2019). Insecticide risk in United States surface waters: drivers and spatiotemporal modeling. *Environmental Science and Technology, 53,* 12071–12080.

Yuen, T. K., Hyde, K. D., & Hodgkiss, I. J. (1999). Interspecific interactions among tropical and subtropical freshwater fungi. *Microbial Ecology, 37,* 257–262.

Zubrod, J. P., Baudy, P., Schulz, R., & Bundschuh, M. (2014). Effects of current-use fungicides and their mixtures on the feeding and survival of the key shredder *Gammarus fossarum. Aquatic Toxicology, 150,* 133–143.

Zubrod, J. P., Bundschuh, M., Arts, G., Brühl, C. A., Imfeld, G., Knäbel, A., Payraudeau, S., Rasmussen, J., Rohr, J., Scharmüller, A., Smalling, K., Stehle, S., Schulz, R. & Schäfer, R. (2019) Fungicide—An overlooked pesticide class? *Environmental Science & Technology,* 53, 3347–3365.

Zubrod, J. P., Bundschuh, M., Feckler, A., Englert, D., & Schulz, R. (2011). Ecotoxicological impact of the fungicide tebuconazole on an aquatic decomposer-detritivore system. *Environmental Toxicology and Chemistry, 30,* 2718–2724.

Zubrod, J. P., Bundschuh, M., & Schulz, R. (2010). Effects of subchronic fungicide exposure on the energy processing of *Gammarus fossarum* (Crustacea; Amphipoda). *Ecotoxicology and Environmental Safety, 73,* 1674–1680.

Zubrod, J. P., Bundschuh, R., Englert, D., Rohrberg, M., Wieczorek, M. V., Bakanov, N., Schulz, R., & Bundschuh, M. (2017). Transient effects following peak exposures towards pesticides—An explanation for the unresponsiveness of in situ measured functional variables *Environmental Pollution, 231,* 1393–1397.

Zubrod, J. P., Englert, D., Feckler, A., Koksharova, N., Konschak, M., Bundschuh, R., Schnetzer, N., Englert, K., Schulz, R. & Bundschuh, M. (2015). Does the current fungicide risk assessment provide sufficient protection for key drivers in aquatic ecosystem functioning? *Environmental Science & Technology, 49,* 1173–1181.

Zubrod, J. P., Englert, D., Feckler, A., Rosenfeldt, R. R., Pasternack, H., Hollert, H., Seiler, T.-B., Schulz, R., & Bundschuh, M. (2019). Is *Hyalella azteca* a suitable model leaf-shredding benthic crustacean for testing the toxicity of sediment-associated metals in Europe? *Bulletin of Environmental Contamination and Toxicolgy, 102,* 303–309.

Zubrod, J. P., Englert, D., Lüderwald, S., Poganiuch, S., Schulz, R., & Bundschuh, M. (2017) History matters: Pre-exposure to wastewater enhances pesticide toxicity in invertebrates. *Environmental Science & Technology, 51,* 9280–9287.

Zubrod, J. P., Englert, D., Rosenfeldt, R. R., Wolfram, J., Lüderwald, S., Wallace, D., Schnetzer, N., Schulz, R., & Bundschuh, M. (2015). The relative importance of diet-related and waterborne effects of copper for a leaf-shredding invertebrate. *Environmental Pollution, 205,* 16–22.

Zubrod, J. P., Englert, D., Wolfram, J., Wallace, D., Schnetzer, N., Baudy, P., Konschak, M., Schulz, R., & Bundschuh, M. (2015) Waterborne toxicity and diet-related effects of fungicides in the key leaf shredder *Gammarus fossarum* (Crustacea; Amphipoda) *Aquatic Toxicology, 169,* 105–112.

Zubrod, J. P., Feckler, A., Englert, D., Koksharova, N., Schulz, R., & Bundschuh, M. (2015). Inorganic fungicides as routinely applied in organic and conventional agriculture can increase palatability but reduce microbial decomposition of leaf litter. *Journal of Applied Ecology, 52,* 310–322.

Chapter 18
Effects of Engineered Nanoparticles on Plant Litter Decomposition in Streams

Ahmed Tlili

Abstract Engineered nanoparticles (ENPs) are the basis of novel and most promising technologies of the twenty-first century, and their production and use in commercial products is exponentially growing. Consequently, freshwater ecosystems will increasingly be recipients for these nanomaterials, which comes at a potential environmental risk when sensitive organisms assume important functions. This chapter describes the current knowledge on ENP effects on plant litter decomposition, an important process in streams. The main focus is on microbial decomposer communities, with both their structure and functions being affected upon ENP exposure. The consequences on detritus-based food-webs are also discussed in the light of evidence from published literature and a case study, showing that both waterborne and dietary exposures to ENPs are important pathways. Particular attention is given to the multifaceted nature and complexity of the environmental behaviour of ENPs and the significant risks they pose to freshwater ecosystems. Finally, several knowledge gaps are highlighted throughout the chapter and suggestions for future research directions are provided.

18.1 Engineered Nanoparticles: Benefits, Uncertainties and Inherent Risks

As the introduction rate of novel chemicals in the markets grows, emerging contaminants play an increasingly important role as potential environmental stressors and cause of adverse effects (Gavrilescu et al., 2015). Emerging contaminants are defined as chemicals that currently lack regulations for environmental monitoring. Among these chemicals, engineered nanoparticles (ENPs), defined as manufactured particles smaller than 100 nm, have received particular attention in recent years. Because their unique physicochemical properties have prompted growing economic interest, hundreds of commercial products are on the market already, and prospects for further

A. Tlili (✉)
Department of Environmental Toxicology, Eawag, Swiss Federal Institute of Aquatic Science and Technology, 8600 Dübendorf, Switzerland
e-mail: ahmed.tlili@eawag.ch

C. M. Swan et al. (eds.), *The Ecology of Plant Litter Decomposition in Stream Ecosystems*, https://doi.org/10.1007/978-3-030-72854-0_18

expansion are bright. ENPs are added to a variety of consumer products, from composite materials in electronic devices and computer storage units, to pharmaceuticals, personal care products, food additives, clothes, facade paints and many other applications.

The attractiveness of ENPs, compared to micro- and macro-sized particles of the same material, stems from specific effects at the nanoscale that result from the increased surface to volume ratio (increased chemical reactivity) and the dominance of physical quantum effects, influencing properties like transparency and conductivity. From an economic and material science perspective, application of nanotechnology offers staggering potential and advantages. However, similarly to other potentially good and innovative new technologies, the knowledge about the release of ENPs into the environment, their fate, behaviour, and subsequent impacts on human health and the environment is critical for the development and success of this industry.

Given the lack of well-organized data, mainly due to the absence of regulations for ENP declaration in the products, it is difficult to estimate environmental release (Holden et al., 2014). Detection and quantification of ENPs in natural environments is also challenging, partly because analytical techniques that allow differentiation of ENPs from natural colloids in complex environmental matrices are very scarce (Laux et al., 2018; Wagner et al., 2014). This is particularly true for organic and metal-based ENPs that contain abundant elements in the earth's crust (Hassellov et al., 2008; von der Kammer et al., 2012). Alternatively, mathematical models are used to estimate predicted environmental concentrations (PEC) of ENPs. For example, Gottschalk et al. (2009) used a probabilistic material flow analysis from a life-cycle perspective of ENP-containing products to predict concentrations of ENPs in surface waters for nanosized silver (nanoAg), zinc oxide (nanoZnO), titanium dioxide (nanoTiO_2) and cerium dioxide (nanoCeO_2). Estimates indicated that except for nanoTiO_2 with concentrations of up to about 50 $\mu g\ L^{-1}$, PECs for all other ENPs barely reach one $\mu g\ L^{-1}$.

Despite the importance of mathematical models to provide rough exposure estimates, their output must be considered with care due to the lack of empirical data for validation (Gottschalk et al., 2010, 2013). This notwithstanding, the estimated PECs suggest that ENPs, when considered individually, do not pose an acute risk to aquatic ecosystems at present. However, with markets for ENPs growing, average environmental concentrations estimated today will soon have to be adjusted upwards. Consequently, water bodies receiving wastewater treatment plant effluents or directly discharged wastewater during heavy rain events, might even experience loads and concentrations well above the currently estimated averages. What is more, considerable risks due to chronic exposure to even low levels of ENPs over extended periods cannot be excluded, as would be typical in field situations.

In addition to exposure information, monitoring of ecotoxicological effects is an important aspect when evaluating the risks that ENPs pose to natural environments. Assessing such impacts largely relied on assays in simplified laboratory settings at the single-species level. Yet, it became increasingly recognised that to draw conclusions on potential impacts under relevant environmental conditions, it is essential also to consider more intricate levels of ecological organisation, and to include effects on ecosystem structure and processes (Bundschuh et al., 2016; Gessner & Tlili, 2016).

18.2 ENP Effects on Microbial Decomposers and Leaf Litter Decomposition

Fresh waters are a critical environmental compartment to consider when assessing potential impacts of ENPs, because they are downstream collectors and integrators of processes within their catchments. This includes various types of industry, wastewater treatment plants and other processes related to human activities. Consequently, fresh waters are prime candidates for exposure to ENPs, especially if notable amounts are released from wastewater treatment plants. Most profound effects of ENPs within ecosystems are to be expected when highly sensitive organisms that assume important ecological functions are being negatively affected. Heterotrophic microbial communities play key roles in fresh waters by colonizing and enzymatically decomposing plant litter derived from terrestrial vegetation, a key resource in forest streams and other shallow aquatic ecosystems (Gessner et al., 1999). This litter-decomposition system is dominated by a group of fungi known as aquatic hyphomycetes (Bärlocher, 2009; Gessner & Chauvet, 1994). They interact strongly with litter-colonizing bacteria and litter-consuming invertebrates and produce substantial amounts of biomass that is subsequently channelled to higher trophic levels in the food web, including fish (Jabiol et al., 2013). This ecological importance in ecosystem functioning and the high microbial diversity comprising contrasting life-styles (e.g., filamentous fungi and single-celled bacteria) (Gessner et al., 2010) and sensitivities to contaminants (Pascoal & Cassio, 2004) suggests that the microbes associated with decomposing leaf litter in fresh waters are a suitable model to assess ENP effects on complex ecological systems (Bundschuh et al., 2016; Pascoal et al., 2003). If ENPs affect these fungal and bacterial decomposers, or their trophic interactions with detritivores, then profound consequences ensue for food-web configuration, litter decomposition, and overall energy flow in freshwater ecosystems (Fig. 18.1).

18.2.1 Acute Versus Chronic Exposure to ENPs and Ensuing Effects on Microbial Decomposers

Two major approaches, with various exposure scenarios and endpoints, are applied to examine ENP effects: (i) acute functional responses via short-term inhibition assays and (ii) effects on microbial diversity and community composition or changes in litter decomposition rates and nutrient composition upon chronic exposures. Multiple studies focussing on acute toxicity of ENPs to single aquatic bacterial and fungal species show that toxic concentrations vary depending on the species and endpoint

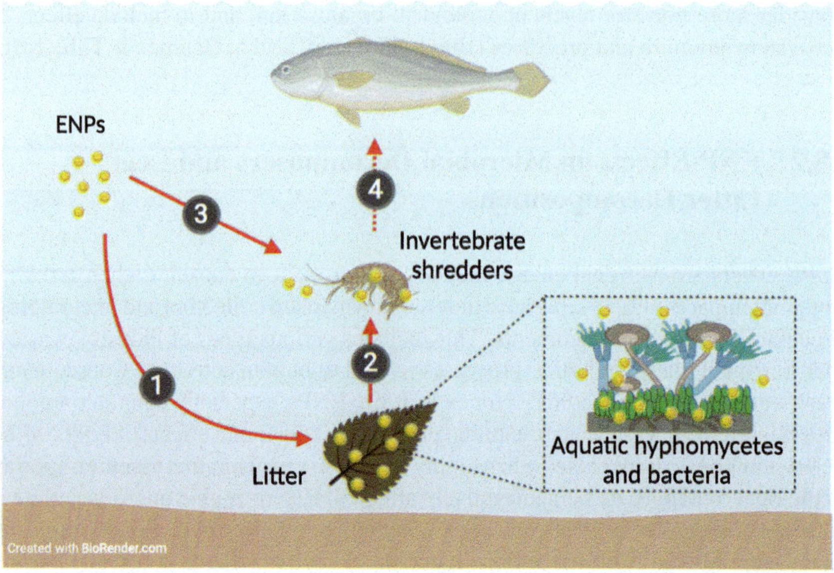

Fig. 18.1 Potential exposure pathways to ENPs in aquatic detrital food-webs via (1) direct sorption of the particles to plant material and associated microbial-decomposers, (2) dietary uptake via contaminated food as well as (3) direct waterborne exposure of invertebrate shredders, and (4) trophic transfer to higher trophic levels in the food-chain, leading to adverse effects

used (Fabrega et al., 2009; Kim et al., 2008; Pradhan, Seena et al., 2015; Seena & Kumar, 2019; Sondi & Salopek-Sondi, 2004). Nevertheless, such information on single organism studies cannot easily inform about what is occurring in a community because they do not account for species interactions. Thus, knowledge of ENP concentration levels that cause acute toxicity at the community level are instrumental for an environmentally relevant risk assessment to determine the magnitude of hazard (Bundschuh et al., 2016). This is achieved by measuring the inhibition of various physiological endpoints in acute short-term assays upon exposure of leaf-associated microorganisms to increasing concentrations of an ENP of interest. Such highly controlled laboratory assays allow deriving specific acute toxicity thresholds, such as NOECs (no observed effect concentrations), LOECs (lowest observed effect concentrations) and EC_{50}s (effective concentrations that cause the 50% effect) at the community level for the tested ENPs.

An essential aspect to consider when performing short-term assays with ENPs and chemicals in general is the selection of the appropriate physiological endpoints (Holden et al., 2016). Ideally, these endpoints should not only reveal the direct acute toxicity of the tested ENPs but also be potential indicators for functional integrity of freshwater ecosystems. Leaf-associated microbial communities ensure a wide range of ecologically relevant functions with associated endpoints that can fulfil such criteria. Among several measured metrics, reduction of acetate-incorporation into ergosterol and sporulation rate of aquatic hyphomycetes, as proxies for fungal growth and reproduction, respectively, as well as bacterial growth emerged as the most sensitive indicators of short-term exposures to ENPs (i.e., from 12 to 24 h). Indeed, inhibitory effects on these parameters occurred at lower concentrations of ENPs than other measured function. This has been reported for various types of ENPs, including nanoCuO (Seena & Kumar, 2019), nanoAg (Batista, Pascoal et al., 2020; Tlili et al., 2016), nanoTiO_2 (Jain et al., 2019) and others. In addition to functional descriptors specifically targeting fungi and bacteria, the reduction of metabolic activity, such as microbial respiration or the activity of extracellular enzymes, such as phosphatase, β-glucosidase and leucine-aminopeptidase, has been reported (Tlili et al., 2016). Overall, these general functional descriptors are less sensitive to ENPs than the specific ones for fungi and bacteria. This can be related to the higher redundancy of such functions within the decomposer communities as they are assumed by both fungal and bacterial communities. Irrespective of their sensitivity, the use of key biological processes such as nutrient acquisition, growth and reproduction of microbial decomposers, allows examining acute toxicity of ENPs and with that establishing protective threshold concentrations for an ecologically relevant risk assessment and environmental management of fresh waters.

Knowledge on the potential impacts of chronic exposure to ENPs remains limited (Holden et al., 2016). This is an important gap given that disruption of decomposer community structure and activities by chronic exposure to ENPs could have large ecosystem consequences, including on food webs, nutrient cycling, and whole-ecosystem metabolism. Nonetheless, multiple studies show that chronic exposure of microbial decomposers, even to low concentrations of ENPs, leads to shifts in leaf-associated microbial communities. For instance, the composition and diversity of aquatic hyphomycete communities changed upon exposures to nanoAg for 25 days (Tlili et al., 2017), nanoCuO for 21 days (Pradhan et al., 2011) and nanoTiO_2 for 50 days (Du et al., 2018) starting at the μg per litter level. Overall, the results of these studies suggest that ENPs can alter species diversity within the community by selecting tolerant species with a high ability to cope with the exerted stress (Du et al., 2019; Pradhan, Seena et al., 2015).

Structural changes in leaf-associated microbial communities, due to chronic exposure to ENPs, were mostly focussing on hyphomycetes via taxonomic analysis of spore morphology. Hence, effects on the diversity and composition of bacterial communities were neglected, despite their important contribution in litter decomposition (Gessner et al., 2010). By benefiting from the development of molecular biology tools, this knowledge gap is increasingly addressed. These tools are based on the comparison of rRNA gene sequences amplified by polymerase chain reactions. In particular, DNA fingerprinting patterns obtained by using various techniques, such as denaturing gradient gel electrophoresis (Batista et al., 2017b), automated ribosomal intergenic spacer analysis (Tlili et al., 2017), and more recently high-throughput sequencing technologies (Riyami et al., 2019; Zhang et al., 2019), allowed comparison of effects on both fungal and bacterial communities. Overall, these studies show that the structure of chronically-exposed bacterial communities to ENPs deviated from the control communities faster than the fungal communities did. One possible reason is that growth within decomposing leaf tissues protects fungi from ENPs more effectively than for bacteria, although high intrinsic tolerance of fungi to ENP stress could also play a role. Furthermore, fungi tend to have longer generation times than many bacteria and hence are less likely capable of responding rapidly to changing conditions.

Moreover, ENP exposure may affect the metabolism of microbial decomposers, reducing their capability to decompose leaf litter (Batista et al., 2017b; Jain et al., 2019; Tlili et al., 2017). For example, Du et al. (2018) reported a positive correlation between the decrease of leaf decomposition rates and activities of eight extracellular enzymes in response to 50 and 500 mg L^{-1} of nanoTiO_2. These enzymes are instrumental for the degradation of plant polymers into small molecules and they play a central role in both the acquisition of carbon, nitrogen and phosphorus for microbial growth and reproduction (e.g., Arnosti et al., 2014) and the cycling of elements at the ecosystem level (Frossard et al., 2012; Lin & Webster, 2014). Multiple studies show that the decrease of extracellular enzyme activities due to chronic exposure to ENPs is accompanied by reduced microbial biomass (Du et al., 2019; Pradhan et al., 2011; Riyami et al., 2019; Zhang et al., 2019). In most cases, more adverse effects on bacterial than on fungal biomass are observed, although the opposite is also reported in response to relatively low nanoAg concentrations (Tlili et al., 2017). These contrasting effects of ENPs on bacterial and fungal biomass point to potential effects of ENPs on species interactions (Gionchetta et al., 2019). If ENP stress affects bacteria more than fungi, or vice-versa, competitive relationships for resources are expected to decrease thus favouring the growth of the less sensitive microbial component.

18.2.2 Toxicity Mechanisms of ENPs in Microbial Decomposers

Irrespective of the duration and concentration of exposure, two major mechanisms underlying effects of ENPs on leaf-associated decomposers and microorganisms in general are suggested: mechanical disruption and physiological damages. The attachment of ENPs to the leaves (e.g., via capture within the extra-polymeric matrix) can provide additional colonisation surface to microorganisms, thereby altering species succession dynamics and ultimately species interactions and community functioning. What is more, as shown for algal cells exposed to nanoCeO_2 (Manier et al., 2013) or to nanoTiO_2 (Hartmann et al., 2010), the surface of fungal and bacterial cells may be covered by ENPs. Consequently, nutrient uptake and interactions between microbial cells and their environment are disrupted through the blockage of biological surfaces.

The potential of ENPs to directly or indirectly induce intracellular formation and accumulation of reactive oxygen species (ROS), which leads to oxidative stress, is considered one of the primary mechanisms causing physiological damages (Petersen & Nelson, 2010). ROS are highly reactive oxygen-containing molecules that are known to cause, among others, mitochondrial and DNA damage, as well as lipid peroxidation (Valko et al., 2005). Induction of oxidative stress has been shown for various types of ENPs, such as nanoCuO (Pradhan, Silva et al., 2016), nanoTiO_2, nanoZnO (Lee & An, 2013), nanoAg (Lapresta-Fernández et al., 2012) and nanoCeO_2 (Zhang et al., 2011). For example, Barros et al. (2019) reported alterations in the proteome of the bacteria *Pseudomonas* sp. due to nanoAg, and notably an increased content of the proteins KatG and AhpC that both play an antioxidant role by scavenging ROS. Similar responses to nanoAg at the proteome level were observed in two fungal ecotypes of *Articulospora tetracladia*, collected from a metal-polluted and a non-polluted stream, although the magnitude of change was higher in the ecotype from the metal-polluted stream (Barros et al., 2020). All of these examples suggest that ENPs can trigger adaptation within populations and communities, leading to more tolerant communities towards ENPs (Batista, Tlili et al., 2020).

18.2.3 What Factors Influence Toxic Effects of ENPs?

Understanding factors affecting the risks posed by ENPs and chemicals in general to food-web interactions and key ecosystem processes has progressed significantly over the past years. For conventional chemicals, it is typically assumed that ecotoxicity is related to the active ingredient mass concentration, which in turn is dependent on its diffusion and equilibrium partitioning. Consequently, risk is characterized using exposure and effects data expressed in terms of mass of the active ingredient per volume or per mass. For ENPs, on the other hand, such an evaluation is not applicable as reflected by the high variation in the outcomes on fate and effects among the studies. An important point that emerges is that environmental fate, bioavailability and toxicity are highly dependent on the nanoparticle chemistry and characteristics, all of which will change as function of the receiving environment (Handy et al., 2008; Navarro et al., 2008).

The inherent physicochemical properties of ENPs, such as particle size, surface charge and coating, crystalline structure, as well as agglomeration state, are key factors to consider when assessing potential effects. For instance, Du et al. (2017) and Pradhan, Geraldes et al. (2016) observed increased toxicity of nanoZnO and nanoCuO, respectively, to microbial decomposers with decreasing particle size, a commonly reported outcome for other types of ENPs and freshwater organisms (Scown et al., 2010; Silva et al., 2014). That the small ENPs exert higher toxicity than large ones is expected, because the ratio between reactive surface area and mass exponentially increases as particle size decreases (Farre et al., 2011; Seitz et al., 2014). However, Batista et al. (2017a) have shown the opposite when comparing the impacts of 100-nm polyvinylpyrrolidone-coated and uncoated 35-nm nanoAg on freshwater microbial decomposers colonizing oak leaves. These seemingly contradictory results can be explained by the high aggregation rate of the uncoated small ENPs during experiments, leading to increased particle size and a lack of effects.

Indeed, metal-based ENPs are usually coated with inorganic or organic compounds, such as citrate, cysteine, carbonate or surfactants. The surface coating is applied in order to selectively change or influence distinct particle properties, such as stability (e.g., coating with charged polymers) or prevention of particle core dissolution (e.g., coating with inorganic layers). Consequently, the functionalization of ENPs via a specific coating can significantly influence their fate, stability and thereby toxicity. Moreover, distinct adverse effects have already been reported for some types of coatings (Angel et al., 2013; Zhao & Wang, 2012). Hence, if a coating for ENPs is applied in a research study the coating has to be studied and assessed as well.

A further layer of complexity is added because ecotoxicological effects of ENPs are highly influenced by multiple environmental factors. The pH, water temperature, dissolved oxygen and sulfide or chloride concentrations are important characteristics of the aquatic environment affecting the fate of ENPs and their toxicity to microbial decomposers (Navarro et al., 2008). For instance, dissolved organic matter (DOM) in natural waters can cover the surface of ENPs and therefore stabilises particle size via steric or electrostatic repulsion (Hall et al., 2009) but also acts as a physical barrier separating nanoparticles and cells (Fabrega et al., 2009). Importantly, this coating by DOM may also limit ion release from ENPs and reduce toxicity (Blinova et al., 2010). Pradhan, Geraldes et al. (2016) observed mitigating effects of humic acids, an important fraction of DOM in fresh waters (Steinberg et al., 2006), for nanoCuO which alone reduced decomposition rate, fungal and bacterial biomass, fungal sporulation and spore diversity associated with decomposing leaves. This reduced toxicity of ENPs in the presence of humic acids was most likely caused by their carboxylic groups that can act as binding sites for ionic metals and thereby decrease metal bioavailability and toxicity (Al-Reasi et al., 2011; Lorenzo et al., 2002). Humic acid macromolecules may also act as a bridge between multivalent cations such as Ca^{2+}, resulting in enhanced nanoparticle aggregation (Chen & Elimelech, 2007).

Because several ENPs are photoactive, sunlight and more specifically UV-radiation significantly influences their environmental toxicity. Due to its redox ability and chemical stability, nanoTiO_2 is among the most frequently used ENPs as a photocatalyst for the breakdown of organic compounds via advanced oxidation processes (Prado et al., 2013). The interaction between nanoTiO_2 and UV-radiation leads to the formation of ROS, mainly the hydroxyl radical (•OH) and superoxide anion (O_2^-), at the surface of the photocatalyst and thus increases its toxicity (Nowotny et al., 2008). One study, however, reported a mitigating effect of UV-radiation for nanoTiO_2 by promoting microbial leaf decomposition. This was explained by enhanced breakdown of the cellulose and lignin in the leaves, through the direct action of the generated ROS, thereby facilitating microbial decomposition of less recalcitrant organic matter (Prado et al., 2013). The enhanced adverse effects on litter decomposition and associated microbial activities and biomass has been also reported for combined exposure to sunlight and ion-releasing ENPs such as nanoZnO (Du et al., 2017, 2019). This observed sunlight-promoted toxicity may be due to increased ion release from the nanoparticles under oxidative conditions following the formation of ROS (Liu & Hurt, 2010). Taken together, all of these examples point to the multifaceted nature of ENP pollution in fresh waters and the complexity of assessing effect outcomes on microbial decomposers and litter decomposition.

18.3 Trophic Transfer of ENPs and Effects on Detrital Food-Web

Decomposing plant material has been identified as a likely sink for ENPs entering fresh waters (e.g., Batista et al., 2017a; Tlili et al., 2017). This can be due to the ENPs being captured within the extra-polymeric substances at the leaf surface or to a direct attachment or sorption to the outer surface of bacterial cells and fungal hyphae (Gil-Allue et al., 2018; Stojak et al., 2011; Tlili et al., 2017). Consequently, risks of ENP exposure are particularly high for aquatic organisms feeding on contaminated leaves. However, although various studies have shown that primary consumers take up and accumulate ENPs from both water and their diet (Bour et al., 2016; Croteau et al., 2011, 2014; Fouqueray et al., 2012; Tangaa et al., 2016), data on trophic transfer of ENPs from contaminated plant material to invertebrate shredders and consequent effects on higher trophic levels is scarce.

Leaf-shredding invertebrates, which rely highly on leaf litter as food source, are widely distributed in fresh waters (Graça, 2001). They constitute an important link in aquatic detrital food webs by transferring carbon and energy from plant litter to higher trophic levels (Jabiol et al., 2013). The responses of aquatic invertebrates to ENPs through dietary uptake are generally assumed to be slower than via direct waterborne exposure (Croteau et al., 2011; Tangaa et al., 2016). Nonetheless, feeding of shredders on contaminated plant material with metal-based ENPs has been reported to also cause severe adverse effects on their fitness and ecological functions (Batista et al., 2017a; Pradhan et al., 2012; Pradhan, Geraldes et al., 2015; see also the case study bellow; Bundschuh et al., 2019). The resulting adverse effects varied from significant reduction of leaf consumption rates and invertebrate growth, reproduction as well as delayed emergence to impairment of the animal neuronal activity and increased mortality. Moreover, once taken up by invertebrate shredders, ENPs could potentially be transferred across ecosystem boundaries, from aquatic to terrestrial food webs (Schulz et al., 2015). For instance, this has been shown for nanoTiO_2 and nanoAu that have been carried by trichopteran caddisflies into their terrestrial life stages during their emergence and after feeding on contaminated food (Bundschuh et al., 2019). Such an outcome implies that terrestrial predators feeding on contaminated aquatic preys may ingest relatively high levels of ENPs but also suffer from alterations in nutritional quality of their prey, as well as their availability. Thus, uptake of ENPs via contaminated food seems to be an important pathway of ENP exposure that needs further considerations (Holden et al., 2016).

Besides uptake via contaminated food and trophic transfer, shifts in leaf-associated microbial communities caused by ENP may also alter food quality for leaf-shredding invertebrates (Zubrod et al., 2011). Freshwater invertebrate shredders prefer to feed on plant litter colonized by microbes, predominantly fungi, whose activities increase plant litter palatability for shredders (Graça, 2001). For instance, Batista et al. (2017a) compared the feeding preference of a common invertebrate shredder, *Limnephilus* sp., on leaves exposed to similar nanoAg concentrations but at different sizes, leading to different leaf colonization by microbes. Results indicated that animals avoided feeding on leaves with lower fungal diversity and biomass. This could therefore have knock-on effects on the entire detritus-based food web (Graça, 2001; Schäfer et al., 2012), although the magnitude of such effects may also rely on the species identity that have been selected by the exposure to ENPs.

18.3.1 Case Study: Relative Importance of Waterborne Exposure to and Dietary Uptake of NanoAg for the Leaf-Shredding Amphipod **Gammarus pulex**[1]

Background & Aims. The opportunities and risks inherent in the widespread use of nanoAg are closely linked to the antimicrobial potency of silver, which appears to be further enhanced when silver occurs in the form of nanoparticles (Sondi & Salopek-Sondi, 2004). NanoAg has been reported to be toxic to most aquatic organisms on all trophic levels, although toxicity varied by several orders of magnitude (i.e., from few nanograms to several milligrams per litter of effective Ag concentrations), depending on the tested organism and experimental setup (Bondarenko et al., 2013). Even though waterborne exposure is considered the most relevant for ENPs, sorption and accumulation of nanoAg to plant material suggest that dietary exposure through the consumption of contaminated food is also an important pathway. The vulnerability of leaf-shredding invertebrates to nanoAg has been demonstrated for both exposure pathways (Al Mahrouqi et al., 2018; Luderwald et al., 2019). Notwithstanding this, studies on their relative contribution to the observed adverse effects are still rare. Therefore, the overarching aim of this case study was to examine the relevance of dietary uptake of and waterborne exposure to nanoAg, or a combination of both exposure pathways, of the leaf shredder *Gammarus pulex*, a widely used amphipod in toxicity testing.

Experimental approach. Discs (14 mm-diameter) from *Populus nigra* L. leaves were enclosed in fine nylon mesh bags and submerged for four weeks in a first-order forest stream located in the north-eastern lowlands of Germany (53°06′46″N, 13°08′43″E). Adult *G. pulex* (6–8 mm body length) were also collected from the same stream two weeks prior to the start of the feeding trials and allowed to acclimatize gradually to the experimental conditions. The microcosms used in the study consisted

---

[1]This case study was performed in collaboration with Dr. Jeremy Jabiol and Prof. Mark O. Gessner (IGB, Department of Experimental Limnology, Germany) and Dr. Renata Behra (Eawag, Department of Environmental Toxicology, Switzerland). Financial support was provided by the Swiss National Science Foundation (SNF, grant no. 200020_134750/1) as part of the National Research Programme NRP 64 on Opportunities and Risks of Nanomaterials.

of polypropylene vessels that were filled with 150 mL of stream water diluted with nanopure water at a ratio of 1:3 to ensure nanoAg stability. Each microcosm was split in two with a 0.5 mm mesh screen. During the feeding trials, one compartment contained two leaf discs and one *G. pulex*, while the second contained only two leaf discs to account for abiotic and microbial leaf mass loss.

The experimental design involved three exposure scenarios to citrate-coated nanoAg suspensions (size 25 ± 13 nm; zeta potential −36.6 ± 3.2 mV in nanopure water; Fig. 18.2). In the waterborne exposure treatments, the animals were first exposed to 100 μg nanoAg L^{-1} during two days, before being placed in the microcosms to feed on non-contaminated leaf discs. To examine the effects of dietary exposure, the gammarids fed on leaf discs that were pre-exposed in nanopure water to 100-μg nanoAg L^{-1} for two days. Finally, a set of microcosms were used to test for the combination of both waterborne and dietary exposure. Additional microcosms were kept as controls with non-contaminated food and water. All microcosms (20 replicates per treatment) were incubated for five days at 15 °C under constant aeration and a 12 h/12 h light/dark cycle (BioSun fluorescent tubes, Radium Lampenwerk GmbH, Germany).

The colloidal stability of 100-μg nanoAg L^{-1} in the exposure media was examined by dynamic light scattering (DLS), using a Zetasizer. Measurements of particle diameter and surface charge were taken immediately (i.e., within 15 min) after addition of the particles and then again after seven days. The total silver concentration in the leaf discs was measured by ICP-MS after acidic digestion (Tlili et al., 2017). Moreover, the silver strongly associated with the leaves and gammarids was also analysed following a washing procedure with cysteine (a strong chelating ligand for silver) (see Tlili et al., 2017 for the detailed procedure). All microcosms were inspected daily for dead animals, which were immediately removed and stored at -20° until

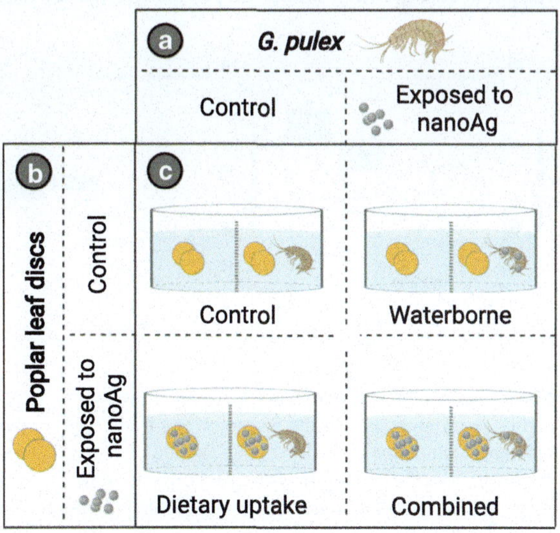

Fig. 18.2 Experimental design of the feeding trials with (**a**) *G. pulex* and (**b**) colonized poplar leaf discs pre-exposed separately to 100 μg L^{-1} nanoAg during two days, and then (**c**) distributed in the microcosms, corresponding to the various exposure scenarios for the feeding trials that lasted for 5 days

analysis. Gammarids' consumption of poplar leaf discs (in mg mg^{-1} DW consumer day^{-1}), corrected for the abiotic and microbial leaf mass loss, was calculated as described in Pradhan et al. (2012).

Results & discussion. Particle characterization in the exposure media showed that nanoAg aggregated only slightly during the experiment (size at t0 days = 50 ± 10 nm and at t10 days = 80 ± 25 nm). The measured low aggregation of the particles is most likely due to their coating with the negatively charged citrate, which was confirmed by the relatively high stability of their surface charge (zeta potential = −16 ± 2 mV at t0 days and t10 days).

Our results clearly show that leaf litter efficiently accumulates silver (Fig. 18.3), which supports the notion that this ecosystem compartment acts as a significant sink for ENPs. Indeed, the largest fraction of accumulated silver was strongly bound to the leaves and ranged from 65 to 100% of the measured total silver concentrations. Moreover, concentrations of silver in the exposure media of all treatments at the end of the feeding trials were below the quantification limit (0.1 nM Ag, data not shown), indicating that the release of silver from the leaves to the water phase was minimal. The shredders that consumed the contaminated leaves for five days assimilated 65 ± 13 and 81 ± 18 μg Ag g^{-1} DW in the dietary and combined exposure treatments, respectively, while exposure via waterborne to nanoAg resulted in at least 10 times less silver in *G. pulex* (6 ± 3 μg Ag g^{-1} DW). These results show that invertebrate shredders are highly vulnerable to nanoAg contamination of decomposing leaves and bioaccumulate the metal when assimilating it with their diet.

The feeding rate of *G. pulex* in control microcosms averaged 0.21 ± 0.04 mg DW mg^{-1} consumer^{-1} day^{-1} and was significantly reduced in all exposure scenarios to nanoAg (i.e., 0.07 ± 0.03, 0.09 ± 0.02 and 0.05 ± 0.02 mg DW mg^{-1} consumer^{-1} day^{-1} in the waterborne, dietary and combined exposure treatments, respectively) (Fig. 18.4). Feeding rates of invertebrate shredders can be influenced by shifts in fungal species, which alters the palatability of leaves offered as food (Zubrod et al.,

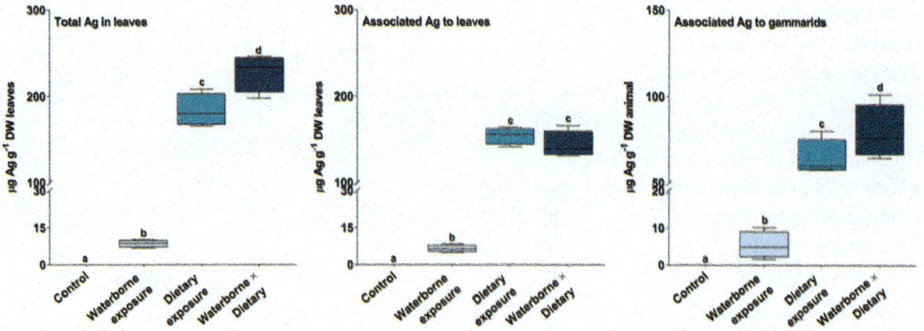

Fig. 18.3 Box plots of the concentrations of total silver in leaves and associated silver to leaves and *G. pulex* at the end of the 5-day feeding trials. The horizontal lines within the boxes indicate medians, boundaries of the boxes indicate the 25th and 75th percentiles, and the whiskers indicate the highest and lowest values ($n = 4$). Different letters denote a significant difference ($p < 0.05$; Tukey's test) between treatments. Silver concentrations in the control treatments were below the quantification limit of 0.1 nM Ag

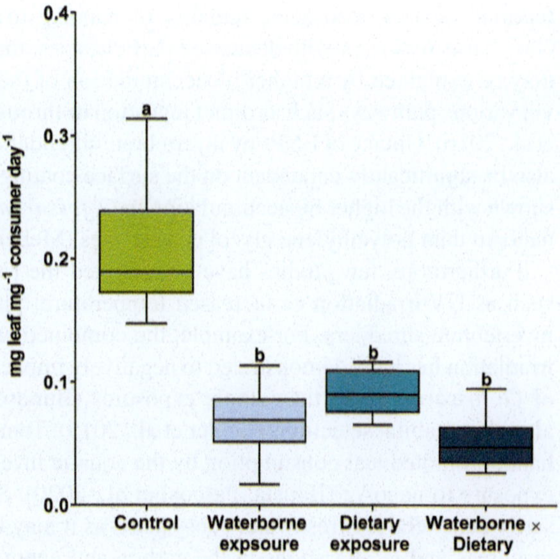

Fig. 18.4 Box plots of the leaf consumption rates by *G. pulex* for 5 days. The horizontal lines within the boxes indicate medians, boundaries of the boxes indicate the 25th and 75th percentiles, and the whiskers indicate the highest and lowest values ($n = 17, 12, 9$ and 9 in the control, waterborne exposure, dietary exposure and waterborne × dietary treatments, respectively). Different letters denote a significant difference ($p < 0.05$; Tukey's test) between treatments

2011). This may not be the case in our study since we have demonstrated previously that under the exact same exposure conditions microbial decomposer communities are not affected (Tlili et al., 2017). Moreover, although significantly reduced in comparison to the control treatments, feeding rates of *G. pulex* did not differ between waterborne and dietary exposure. This is in contradiction to the fact that large differences in silver accumulation were observed between the two exposure pathways. The lack of significance suggests that other mechanisms than alteration of food quality such as interference of silver with food digestion may contribute to the impairment of leaf consumption by *G. pulex*.

In contrast to feeding rates, shredder mortality was clearly correlated to the assimilated silver concentrations in the animal body and differed among the three exposure scenarios. Mortality during the five-day feeding increased from 10% in the control microcosms to 40% in the waterborne exposure treatments, and even more in the dietary and combined exposure treatments, reaching 55 and 60%, respectively. This outcome points to a high sensitivity of *G. pulex* survival to nanoAg, which was exacerbated when exposure occurred via contaminated food. One mechanism could be related to an increased silver ion release after uptake of nanoAg with food, caused by exposure of the particles to acidic conditions in the digestive tract. Overall, our results show that nanoAg can be transferred from one trophic level to another and therefore can be incorporated into food chains. In conclusion, dietary exposure to nanoAg appears to be an important pathway to consider and the related mechanisms underlying adverse effects need further systematic investigation.

Irrespective of the exposure pathway, effects of ENPs on leaf-shredding invertebrates are dependent on multiple factors, particularly the properties of the particles themselves and environmental characteristics. For instance, Mehennaoui et al. (2016)

reported on increased acute lethality of nanoAg to the leaf shredding amphipod Gammarus *fossarum* with decreased particle sizes, ranging from 20 to 200 nm. This may be explained by a higher bioaccumulation of the small ENPs that are taken up via various pathways such as direct ingestion or through the Gammarus gills (Andrei et al., 2016). Uptake of ENPs by invertebrate shredders and therefore toxic effects can also be significantly dependent on the surface coating of the particles, as it has been shown with the higher bioaccumulation in *G. fossarum* of citrate-coated nanoAg and nanoAu than polyethylene glycol coated ones (Mehennaoui et al., 2018).

Furthermore, few studies have investigated the role that environmental factors, such as UV-irradiation or increased temperature, play in the toxicity of ENPs to invertebrate shredders. For example, the combined exposure to nanoTiO_2 and UV-irradiation has been demonstrated to negatively impact the feeding rates and survival of G. *fossarum* more than single exposures (Bundschuh et al., 2011) and also to alter their habitat selection (Feckler et al., 2015). Temperature increase, on the other hand, promoted leaf consumption by the aquatic invertebrate Limnephilus sp. upon exposure to nanoAg (Batista, Pascoal et al., 2020). Such an outcome is not necessarily beneficial for freshwater ecosystems as it may lead to a fast depletion of food resources and to alterations in the carbon and energy fluxes (Schulz et al., 2015). Although further investigations with other types of ENPs and environmental factors are needed, these studies underpin the importance of considering environmental parameters and the unique particle properties during the assessment of the risks posed by ENPs to invertebrate shredders and detritus-based food-webs.

18.4 Conclusions and Outlook

Advances have been made toward characterizing the impact of ENPs on microbial decomposers and the associated functions they ensure in fresh waters over the past years. Clearly, their potency bears significant environmental costs by disrupting key biological processes such as nutrient acquisition, growth and reproduction of microbial decomposers. This has potential consequences for stream food-web configuration, plant litter decomposition, nutrient cycling and the overall flow of energy in fresh waters. Nonetheless, it is clear that the wide range of particle properties together with interactions with environmental factors and water chemistry result in large uncertainties regarding the fate and effects of ENPs in natural ecosystems. Because particle agglomeration is the most important process driving their fate and effects, it is crucial to understand the conditions that promote or prevent this process. Therefore, a systematic characterization of particle size, surface charge, coating, ion release for metal-based ENPs and crystalline structure under relevant experimental conditions is a prerequisite when testing for toxicity. DOM is an important factor affecting the behaviour of ENPs by changing their surface charge or by providing an adsorption surface, and should also be quantified and its composition characterized. Importantly, transformation and aging of the ENPs over time is an additional key to understand their fate and transport in fresh waters, as well as their potential biological effects. In addition, the presence of other organic and inorganic contaminants (i.e., chemical mixtures) can interact with metal- and carbon-based ENPs and impacts toxic outcomes. Such interactions underpin the need for a more adapted environmental risk assessment including mixture toxicity, to account for the role of ENPs.

In general, detailed knowledge about effects of ENPs on levels of ecological organization above the individual is still lacking. Important questions about how ENPs influence species interactions within microbial decomposer communities, as well as with higher trophic levels, and thus their ability to cope with stress remain to be addressed. Assessing and predicting impacts of ENPs on species interactions are not enough. A comprehensive assessment requires the consideration of ecosystem processes, in addition to variables describing the structure of populations and communities (Rosi-Marshall & Royer, 2012). For the decomposition of plant litter, multiple functionally distinct groups of organisms may be involved, which influences the net outcome of ENP exposure on these process rates and further complicates the prediction of adverse effects. Given the difficulty to address these questions with natural communities, it seems reasonable to use a stepwise approach that consists in testing for effects in systems of gradually increasing biological complexity. This can be achieved by manipulating for example fungal diversity under controlled experimental conditions (Dang et al., 2005; Jabiol et al., 2013).

References

Al-Reasi, H. A., Wood, C. M., & Smith, D. S. (2011). Physicochemical and spectroscopic properties of natural organic matter (NOM) from various sources and implications for ameliorative effects on metal toxicity to aquatic biota. *Aquatic Toxicology, 103*, 179–190

Al Mahrouqi, D., Al Riyami, S., & Barry, M. J. (2018). Effects of Zn and Ti Nanoparticles on the Survival and Growth of Sclerophrys arabica Tadpoles in a Two Level Trophic System. *Bulletin of Environment Contamination and Toxicology, 101*, 592–597

Andrei, J., Pain-Devin, S., Felten, V., Devin, S., Giamberini, L., Mehennaoui, K., Cambier, S., Gutleb, A. C., & Guerold, F. (2016). Silver nanoparticles impact the functional role of Gammarus roeseli (Crustacea Amphipoda). *Environmental Pollution, 208*, 608–618

Angel, B. M., Batley, G. E., Jarolimek, C. V., & Rogers, N. J. (2013). The impact of size on the fate and toxicity of nanoparticulate silver in aquatic systems. *Chemosphere, 93*, 359–365

Arnosti, C., Bell, C., Moorhead, D. L., Sinsabaugh, R. L., Steen, A. D., Stromberger, M., Wallenstein, M., & Weintraub, M. N. (2014). Extracellular enzymes in terrestrial, freshwater, and marine environments: Perspectives on system variability and common research needs. *Biogeochemistry, 117*, 5–21

Bärlocher, F. (2009). Reproduction and dispersal in aquatic hyphomycetes. *Mycoscience, 50*, 3–8

Barros, D., Pradhan, A., Mendes, V. M., Manadas, B., Santos, P. M., Pascoal, C., & Cassio, F. (2019). Proteomics and antioxidant enzymes reveal different mechanisms of toxicity induced by ionic and nanoparticulate silver in bacteria. *Environmental Science-Nano, 6*, 1207–1218

Barros, D., Pradhan, A., Pascoal, C., & Cassio, F. (2020). Proteomic responses to silver nanoparticles vary with the fungal ecotype. *Science of the Total Environment, 704*.

Batista, D., Pascoal, C., & Cassio, F. (2017a). How do physicochemical properties influence the toxicity of silver nanoparticles on freshwater decomposers of plant litter in streams? *Ecotoxicology and Environmental Safety, 140*, 148–155

Batista, D., Pascoal, C., & Cassio, F. (2017b). Temperature modulates AgNP impacts on microbial decomposer activity. *Science of the Total Environment, 601–602*, 1324–1332

Batista, D., Pascoal, C., & Cassio, F. (2020). The increase in temperature overwhelms silver nanoparticle effects on the aquatic invertebrate Limnephilus sp. *Environmental Toxicology and Chemistry, 39*, 1429–1437

Batista, D., Tlili, A., Gessner, M. O., Pascoal, C., & Cassio, F. (2020). Nanosilver impacts on aquatic microbial decomposers and litter decomposition assessed as pollution-induced community tolerance (PICT). *Environmental Science-Nano, 7*, 2130–2139

Blinova, I., Ivask, A., Heinlaan, M., Mortimer, M., & Kahru, A. (2010). Ecotoxicity of nanoparticles of CuO and ZnO in natural water. *Environmental Pollution, 158*, 41–47

Bondarenko, O., Juganson, K., Ivask, A., Kasemets, K., Mortimer, M., & Kahru, A. (2013). Toxicity of Ag, CuO and ZnO nanoparticles to selected environmentally relevant test organisms and mammalian cells in vitro: A critical review. *Archives of Toxicology, 87*, 1181–1200

Bour, A., Mouchet, F., Cadarsi, S., Silvestre, J., Verneuil, L., Baqué, D., Chauvet, E., Bonzom, J. M., Pagnout, C., Clivot, H., Fourquaux, I., Tella, M., Auffan, M., Gauthier, L., & Pinelli, E. (2016). Toxicity of CeO_2 nanoparticles on a freshwater experimental trophic chain: A study in environmentally relevant conditions through the use of mesocosms. *Nanotoxicology, 10*, 245–255

Bundschuh, M., Englert, D., Rosenfeldt, R. R., Bundschuh, R., Feckler, A., Luderwald, S., Seitz, F., Zubrod, J. P., & Schulz, R. (2019). Nanoparticles transported from aquatic to terrestrial ecosystems via emerging aquatic insects compromise subsidy quality. *Sci Rep, 9*, 15676

Bundschuh, M., Seitz, F., Rosenfeldt, R. R., & Schulz, R. (2016). Effects of nanoparticles in fresh waters: Risks, mechanisms and interactions. *Freshwater Biology, 61*, 2185–2178

Bundschuh, M., Zubrod, J. P., Englert, D., Seitz, F., Rosenfeldt, R. R., & Schulz, R. (2011). Effects of nano-TiO_2 in combination with ambient UV-irradiation on a leaf shredding amphipod. *Chemosphere, 85*, 1563–1567

Chen, K. L., & Elimelech, M. (2007). Influence of humic acid on the aggregation kinetics of fullerene (C-60) nanoparticles in monovalent and divalent electrolyte solutions. *Journal of Colloid and Interface Science, 309*, 126–134

Croteau, M. N., Misra, S. K., Luoma, S. N., & Valsami-Jones, E. (2011). Silver bioaccumulation dynamics in a freshwater invertebrate after aqueous and dietary exposures to nanosized and ionic Ag. *Environmental Science & Technology, 45*, 6600–6607

Croteau, M. N., Misra, S. K., Luoma, S. N., & Valsami-Jones, E. (2014). Bioaccumulation and toxicity of CuO nanoparticles by a freshwater invertebrate after waterborne and dietborne exposures. *Environmental Science & Technology, 48*, 10929–10937

Dang, C. K., Chauvet, E., & Gessner, M. O. (2005). Magnitude and variability of process rates in fungal diversity-litter decomposition relationships. *Ecology Letters, 8*, 1129–1137

Du, J., Zhang, Y., Cui, M., Yang, J., Lin, Z., & Zhang, H. (2017). Evidence for negative effects of ZnO nanoparticles on leaf litter decomposition in freshwater ecosystems. *Environmental Science: Nano, 4*, 2377–2387

Du, J., Zhang, Y., Guo, R., Meng, F., Gao, Y., Ma, C., & Zhang, H. (2019). Harmful effect of nanoparticles on the functions of freshwater ecosystems: Insight into nanoZnO-polluted stream. *Chemosphere, 214*, 830–838

Du, J., Zhang, Y., Guo, W., Li, N., Gao, C., Cui, M., Lin, Z., Wei, M., & Zhang, H. (2018). Chronic impacts of TiO_2 nanoparticles on Populus nigra L. leaf decomposition in freshwater ecosystem. *Journal of Hazardous Materials, 350*, 121–127

Du, J. J., Zhang, Y. Y., Liu, L. N., Qv, M. X., Lv, Y. N., Yin, Y. F., Zhou, Y. F., Cui, M. H., Zhu, Y. F., & Zhang, H. Z. (2017). Can visible light impact litter decomposition under pollution of ZnO nanoparticles? *Chemosphere, 187*, 368–375

Du, J. J., Zhang, Y. Y., Qv, M. X., Li, K., Yin, X. Y., Sorrell, B. K., Wei, M. B., & Ma, C. (2019). The effects of ZnO nanoparticles on leaf litter decomposition under natural sunlight. *Environmental Science-Nano, 6*, 1180–1188

Fabrega, J., Fawcett, S. R., Renshaw, J. C., & Lead, J. R. (2009). Silver nanoparticle impact on bacterial growth: Effect of pH, CONCENTRATION, AND ORGANIC MATTER. *Environmental Science & Technology, 43*, 7285–7290

Farre, M., Sanchis, J., & Barcelo, D. (2011). Analysis and assessment of the occurrence, the fate and the behavior of nanomaterials in the environment. *Trac-Trends in Analytical Chemistry, 30*, 517–527

Feckler, A., Rosenfeldt, R. R., Seitz, F., Schulz, R., & Bundschuh, M. (2015). Photocatalytic properties of titanium dioxide nanoparticles affect habitat selection of and food quality for a key species in the leaf litter decomposition process. *Environmental Pollution, 196*, 276–283

Fouqueray, M., Dufils, B., Vollat, B., Chaurand, P., Botta, C., Abacci, K., Labille, J., Rose, J., & Garric, J. (2012). Effects of aged TiO_2 nanomaterial from sunscreen on Daphnia magna exposed by dietary route. *Environmental Pollution, 163*, 55–61

Frossard, A., Gerull, L., Mutz, M., & Gessner, M. O. (2012). Fungal importance extends beyond litter decomposition in experimental early-successional streams. *Environmental Microbiology, 14*, 2971–2983

Gavrilescu, M., Demnerova, K., Aamand, J., Agathoss, S., & Fava, F. (2015). Emerging pollutants in the environment: Present and future challenges in biomonitoring, ecological risks and bioremediation. *New Biotechnology, 32*, 147–156

Gessner, M. O., & Chauvet, E. (1994). Importance of stream microfungi in controlling breakdown rates of leaf-litter. *Ecology, 75*, 1807–1817

Gessner, M. O., Chauvet, E., & Dobson, M. (1999). A perspective on leaf litter breakdown in streams. *Oikos, 85*, 377–384

Gessner, M. O., Swan, C. M., Dang, C. K., McKie, B. G., Bardgett, R. D., Wall, D. H., & Hättenschwiler, S. (2010). Diversity meets decomposition. *Trends in Ecology & Evolution, 25*, 372–380

Gessner, M. O., & Tlili, A. (2016). Fostering integration of freshwater ecology with ecotoxicology. *Freshwater Biology, 61*, 1991–2001

Gil-Allue, C., Tlili, A., Schirmer, K., Gessner, M. O., & Behra, R. (2018). Long-term exposure to silver nanoparticles affects periphyton community structure and function. *Environmental Science-Nano, 5*, 1397–1407

Gionchetta, G., Romani, A. M., Oliva, F., & Artigas, J. (2019). Distinct responses from bacterial, archaeal and fungal streambed communities to severe hydrological disturbances. *Scientific Report, 9*, 13506

Gottschalk, F., Sonderer, T., Scholz, R. W., & Nowack, B. (2009). Modeled environmental concentrations of engineered nanomaterials (TiO_2, ZnO, Ag, CNT, fullerenes) for different regions. *Environmental Science & Technology, 43*, 9216–9222

Gottschalk, F., Sonderer, T., Scholz, R. W., & Nowack, B. (2010). Possibilities and limitations of modeling environmental exposure to engineered nanomaterials by probabilistic material flow analysis. *Environmental Toxicology and Chemistry, 29*, 1036–1048

Gottschalk, F., Sun, T., & Nowack, B. (2013). Environmental concentrations of engineered nanomaterials: Review of modeling and analytical studies. *Environmental Pollution, 181*, 287–300

Graça, M. A. S. (2001). The role of invertebrates on leaf litter decomposition in streams—A review. *International Review of Hydrobiology, 86*, 383–393

Hall, S., Bradley, T., Moore, J. T., Kuykindall, T., & Minella, L. (2009). Acute and chronic toxicity of nano-scale TiO_2 particles to freshwater fish, cladocerans, and green algae, and effects of organic and inorganic substrate on TiO_2 toxicity. *Nanotoxicology, 3*, 91–97

Handy, R. D., Owen, R., & Valsami-Jones, E. (2008). The ecotoxicology of nanoparticles and nanomaterials: Current status, knowledge gaps, challenges, and future needs. *Ecotoxicology, 17*, 315–325

Hartmann, N. B., Von der Kammer, F., Hofmann, T., Baalousha, M., Ottofuelling, S., & Baun, A. (2010). Algal testing of titanium dioxide nanoparticles—Testing considerations, inhibitory effects and modification of cadmium bioavailability. *Toxicology, 269*, 190–197

Hassellov, M., Readman, J. W., Ranville, J. F., & Tiede, K. (2008). Nanoparticle analysis and characterization methodologies in environmental risk assessment of engineered nanoparticles. *Ecotoxicology, 17*, 344–361

Holden, P. A., Gardea-Torresdey, J. L., Klaessig, F., Turco, R. F., Mortimer, M., Hund-Rinke, K., Cohen Hubal, E. A., Avery, D., Barcelo, D., Behra, R., Cohen, Y., Deydier-Stephan, L., Ferguson, P. L., Fernandes, T. F., Herr Harthorn, B., Henderson, W. M., Hoke, R. A., Hristozov, D., Johnston, J. M., … Nel, A. E. (2016). Considerations of environmentally relevant test conditions for improved evaluation of ecological hazards of engineered nanomaterials. *Environmental Science & Technology, 50*, 6124–6145

Holden, P. A., Klaessig, F., Turco, R. F., Priester, J. H., Rico, C. M., Avila-Arias, H., Mortimer, M., Pacpaco, K., & Gardea-Torresdey, J. L. (2014). Evaluation of exposure concentrations used in assessing manufactured nanomaterial environmental hazards: Are they relevant? *Environmental Science & Technology, 48*, 10541–10551

Jabiol, J., McKie, B. G., Bruder, A., Bernadet, C., Gessner, M. O., & Chauvet, E. (2013). Trophic complexity enhances ecosystem functioning in an aquatic detritus-based model system. *Journal of Animal Ecology, 82*, 1042–1051

Jain, A., Kumar, S., & Seena, S. (2019). Can low concentrations of metal oxide and Ag loaded metal oxide nanoparticles pose a risk to stream plant litter microbial decomposers? *Science of the Total Environment, 653*, 930–937

Kim, K. J., Sung, W. S., Moon, S. K., Choi, J. S., Kim, J. G., & Lee, D. G. (2008). Antifungal effect of silver nanoparticles on dermatophytes. *Journal of Microbiology and Biotechnology, 18*, 1482–1484

Lapresta-Fernández, A., Fernández, A., & Blasco, J. (2012). Nanoecotoxicity effects of engineered silver and gold nanoparticles in aquatic organisms. *TrAC Trends in Analytical Chemistry, 32*, 40–59

Laux, P., Riebeling, C., Booth, A. M., Brain, J. D., Brunner, J., Cerrillo, C., Creutzenberg, O., Estrela-Lopis, I., Gebel, T., Johanson, G., Jungnickel, H., Kock, H., Tentschert, J., Tlili, A., Schaffer, A., Sips, A. J. A. M., Yokel, R. A., & Luch, A. (2018). Challenges in characterizing the environmental fate and effects of carbon nanotubes and inorganic nanomaterials in aquatic systems. *Environmental Science-Nano, 5*, 48–63

Lee, W.-M., & An, Y.-J. (2013). Effects of zinc oxide and titanium dioxide nanoparticles on green algae under visible, UVA, and UVB irradiations: No evidence of enhanced algal toxicity under UV pre-irradiation. *Chemosphere, 91*, 536–544

Lin, L., & Webster, J. R. (2014). Detritus decomposition and nutrient dynamics in a forested headwater stream. *Ecological Modelling, 293*, 58–68

Liu, J., & Hurt, R. H. (2010). Ion release kinetics and particle persistence in aqueous nano-silver colloids. *Environmental Science & Technology, 44*, 2169–2175

Lorenzo, J. I., Nieto, O., & Beiras, R. (2002). Effect of humic acids on speciation and toxicity of copper to Paracentrotus lividus larvae in seawater. *Aquatic Toxicology, 58*, 27–41

Luderwald, S., Schell, T., Newton, K., Salau, R., Seitz, F., Rosenfeldt, R. R., Dackermann, V., Metreveli, G., Schulz, R., & Bundschuh, M. (2019). Exposure pathway dependent effects of titanium dioxide and silver nanoparticles on the benthic amphipod Gammarus fossarum. *Aquatic Toxicology, 212*, 47–53

Manier, N., Bado-Nilles, A., Delalain, P., Aguerre-Chariol, O., & Pandard, P. (2013). Ecotoxicity of non-aged and aged CeO_2 nanomaterials towards freshwater microalgae. *Environmental Pollution, 180*, 63–70

Mehennaoui, K., Cambier, S., Serchi, T., Ziebel, J., Lentzen, E., Valle, N., Guerold, F., Thomann, J. S., Giamberini, L., & Gutleb, A. C. (2018). Do the pristine physico-chemical properties of silver and gold nanoparticles influence uptake and molecular effects on Gammarus fossarum (Crustacea Amphipoda)? *Science of the Total Environment, 643*, 1200–1215

Mehennaoui, K., Georgantzopoulou, A., Felten, V., Andrei, J., Garaud, M., Cambier, S., Serchi, T., Pain-Devin, S., Guerold, F., Audinot, J. N., Giamberini, L., & Gutleb, A. C. (2016). Gammarus fossarum (Crustacea, Amphipoda) as a model organism to study the effects of silver nanoparticles. *Science of the Total Environment, 566–567*, 1649–1659

Navarro, E., Baun, A., Behra, R., Hartmann, N. B., Filser, J., Miao, A. J., Quigg, A., Santschi, P. H., & Sigg, L. (2008). Environmental behavior and ecotoxicity of engineered nanoparticles to algae, plants, and fungi. *Ecotoxicology, 17*, 372–386

Nowotny, J., Bak, T., Sheppard, L. R., & Nowotny, M. K. (2008). Reactivity of titanium dioxide with oxygen at room temperature and the related charge transfer. *Journal of the American Chemical Society, 130*, 9984–9993

Pascoal, C., & Cassio, F. (2004). Contribution of fungi and bacteria to leaf litter decomposition in a polluted river. *Applied and Environmental Microbiology, 70*, 5266–5273

Pascoal, C., Pinho, M., Cassio, F., & Gomes, P. (2003). Assessing structural and functional ecosystem condition using leaf breakdown: Studies on a polluted river. *Freshwater Biology, 48*, 2033–2044

Petersen, E. J., & Nelson, B. C. (2010). Mechanisms and measurements of nanomaterial-induced oxidative damage to DNA. *Analytical and Bioanalytical Chemistry, 398*, 613–650

Pradhan, A., Geraldes, P., Seena, S., Pascoal, C., & Cassio, F. (2015). Natural organic matter alters size-dependent effects of nanoCuO on the feeding behaviour of freshwater invertebrate shredders. *Science of the Total Environment, 535*, 94–101

Pradhan, A., Geraldes, P., Seena, S., Pascoal, C., & Cássio, F. (2016). Humic acid can mitigate the toxicity of small copper oxide nanoparticles to microbial decomposers and leaf decomposition in streams. *Freshwater Biology, 61*(12), 2197–2210.

Pradhan, A., Seena, S., Pascoal, C., & Cassio, F. (2011). Can metal nanoparticles be a threat to microbial decomposers of plant litter in streams? *Microbial Ecology, 62*, 58–68

Pradhan, A., Seena, S., Pascoal, C., & Cassio, F. (2012). Copper oxide nanoparticles can induce toxicity to the freshwater shredder Allogamus ligonifer. *Chemosphere, 89*, 1142–1150

Pradhan, A., Seena, S., Schlosser, D., Gerth, K., Helm, S., Dobritzsch, M., Krauss, G. J., Dobritzsch, D., Pascoal, C., & Cassio, F. (2015). Fungi from metal-polluted streams may have high ability to cope with the oxidative stress induced by copper oxide nanoparticles. *Environmental Toxicology and Chemistry, 34*, 923–930

Pradhan, A., Silva, C. O., Silva, C., Pascoal, C., & Cassio, F. (2016). Enzymatic biomarkers can portray nanoCuO-induced oxidative and neuronal stress in freshwater shredders. *Aquatic Toxicology, 180*, 227–235

Prado, R., Erdocia, X., & Labidi, J. (2013). Effect of the photocatalytic activity of TiO_2 on lignin depolymerization. *Chemosphere, 91*, 1355–1361

Riyami, S. A., Mahrouqi, D. A., Abed, R. M. M., Elshafie, A., Sathe, P., & Barry, M. J. (2019). Direct and indirect effects of zinc oxide and titanium dioxide nanoparticles on the decomposition of leaf litter in streams. *Ecotoxicology, 28*, 435–448

Rosi-Marshall, E. J., & Royer, T. V. (2012). Pharmaceutical compounds and ecosystem function: An emerging research challenge for aquatic ecologists. *Ecosystems, 15*, 867–880

Schäfer, R. B., Bundschuh, M., Rouch, D. A., Szöcs, E., von der Ohe, P. C., Pettigrove, V., Schulz, R., Nugegoda, D., & Kefford, B. J. (2012). Effects of pesticide toxicity, salinity and other environmental variables on selected ecosystem functions in streams and the relevance for ecosystem services. *Science of the Total Environment, 415*, 69–78

Schulz, R., Bundschuh, M., Gergs, R., Bruhl, C. A., Diehl, D., Entling, M. H., Fahse, L., Fror, O., Jungkunst, H. F., Lorke, A., Schafer, R. B., Schaumann, G. E., & Schwenk, K. (2015). Review on environmental alterations propagating from aquatic to terrestrial ecosystems. *Science of the Total Environment, 538*, 246–261

Scown, T. M., Santos, E. M., Johnston, B. D., Gaiser, B., Baalousha, M., Mitov, S., Lead, J. R., Stone, V., Fernandes, T. F., Jepson, M., van Aerle, R., & Tyler, C. R. (2010). Effects of aqueous exposure to silver nanoparticles of different sizes in rainbow trout. *Toxicological Sciences, 115*, 521–534

Seena, S., & Kumar, S. (2019). Short-term exposure to low concentrations of copper oxide nanoparticles can negatively impact the ecological performance of a cosmopolitan freshwater fungus. *Environmental Science: Processes & Impacts, 21*, 2001–2007

Seitz, F., Rosenfeldt, R. R., Schneider, S., Schulz, R., & Bundschuh, M. (2014). Size-, surface- and crystalline structure composition-related effects of titanium dioxide nanoparticles during their aquatic life cycle. *Science of the Total Environment, 493*, 891–897

Silva, T., Pokhrel, L. R., Dubey, B., Tolaymat, T. M., Maier, K. J., & Liu, X. F. (2014). Particle size, surface charge and concentration dependent ecotoxicity of three organo-coated silver nanoparticles: Comparison between general linear model-predicted and observed toxicity. *Science of the Total Environment, 468*, 968–976

Sondi, I., & Salopek-Sondi, B. (2004). Silver nanoparticles as antimicrobial agent: A case study on E-coli as a model for Gram-negative bacteria. *Journal of Colloid and Interface Science, 275*, 177–182

Steinberg, C. E. W., Kamara, S., Prokhotskaya, V. Y., Manusadzianas, L., Karasyova, T. A., Timofeyev, M. A., Jie, Z., Paul, A., Meinelt, T., Farjalla, V. F., Matsuo, A. Y. O., Burnison, B. K., & Menzel, R. (2006). Dissolved humic substances - ecological driving forces from the individual to the ecosystem level? *Freshwater Biology, 51*, 1189–1210

Stojak, A. R., Raftery, T., Klaine, S. J., & McNealy, T. L. (2011). Morphological responses of Legionella pneumophila biofilm to nanoparticle exposure. *Nanotoxicology, 5*, 730–742

Tangaa, S. R., Selck, H., Winther-Nielsen, M., & Khan, F. R. (2016). Trophic transfer of metal-based nanoparticles in aquatic environments: A review and recommendations for future research focus. *Environmental Science-Nano, 3*, 966–981

Tlili, A., Cornut, J., Behra, R., Gil-Allue, C., & Gessner, M. O. (2016). Harmful effects of silver nanoparticles on a complex detrital model system. *Nanotoxicology, 10*, 728–735

Tlili, A., Jabiol, J., Behra, R., Gil-Allué, C., & Gessner, M. O. (2017). Chronic exposure effects of silver nanoparticles on stream microbial decomposer communities and ecosystem functions. *Environmental Science & Technology, 51*, 2447–2455

Valko, M., Morris, H., & Cronin, M. T. (2005). Metals, toxicity and oxidative stress. *Current Medicinal Chemistry, 12*, 1161–1208

von der Kammer, F., Ferguson, P. L., Holden, P. A., Masion, A., Rogers, K. R., Klaine, S. J., Koelmans, A. A., Horne, N., & Unrine, J. M. (2012). Analysis of engineered nanomaterials in complex matrices (environment and biota): General considerations and conceptual case studies. *Environmental Toxicology and Chemistry, 31*, 32–49

Wagner, S., Gondikas, A., Neubauer, E., Hofmann, T., & von der Kammer, F. (2014). Spot the difference: Engineered and natural nanoparticles in the environment–release, behavior, and fate. *Angewandte Chemie (International Ed. in English), 53*, 12398–12419

Zhang, H., He, X., Zhang, Z., Zhang, P., Li, Y., Ma, Y., Kuang, Y., Zhao, Y., & Chai, Z. (2011). Nano-CeO_2 exhibits adverse effects at environmental relevant concentrations. *Environmental Science & Technology, 45*, 3725–3730

Zhang, Y., Yin, Y., Ma, H., Cao, X., Ma, B., Qv, M., Zhang, B., Akbar, S., & Du, J. (2019). Insight into chronic exposure effects of nanosized titanium dioxide on Typha angustifolia leaf litter decomposition. *Chemosphere, 224*, 680–688

Zhao, C. M., & Wang, W. X. (2012). Importance of surface coatings and soluble silver in silver nanoparticles toxicity to Daphnia magna. *Nanotoxicology, 6*, 361–370

Zubrod, J. P., Bundschuh, M., Feckler, A., Englert, D., & Schulz, R. (2011). Ecotoxicological impact of the fungicide tebuconazole on an aquatic decomposer-detritivore system. *Environmental Toxicology and Chemistry, 30*, 2718–2724

Part IV
Methodological Aspects and Applications of Measuring Plant Litter Decomposition

Chapter 19
The Construction of Plant Litter Decomposition Curves

Antoine Lecerf

Abstract Quantitative analysis of changes undergone by plant litter during decomposition is a main focus of theoretical and empirical studies of plant litter decomposition. Decomposition curves are most often described by systems of differential equations whose closed-form solutions enable simple estimation of fundamental parameters such as litter decay rate(s). Other potential applications of mathematical models of litter decomposition include analysis of the controls of plant traits *vs.* decomposers on decomposition and predictions of how litter decomposition responds to ecosystem changes. This chapter provides an overview of the main approaches to modelling plant litter decomposition and drawing decomposition curves produced under various assumptions. The Olson's negative exponential model has been widely used by freshwater ecologists to summarize and compare results of field and laboratory studies. Yet, it is still unclear where and when the assumption of time invariance of litter decay rate underlying the simple model is met. Process-based models incorporating litter heterogeneity and/or consumer-resource dynamics provide evidence that decomposition does not proceed at constant rate. Thus, relaxing the assumption of time invariance for litter decay rate is a necessary step towards a deeper mechanistic understanding of drivers and agents of plant litter decomposition in aquatic ecosystems.

19.1 Introduction

In plant litter decomposition research, much emphasis is placed on empirical and theoretical construction of curves depicting temporal pattern of detritus standing stock. Decomposition curves specifically refer to mass changes undergone by a pool of detrital organic matter since its appearance as a single-pulse input to the system, until its complete disappearance. The non-living nature of detritus makes detrital mass decreases through time. Empirical construction of decomposition curves is

A. Lecerf (✉)
Laboratoire Écologie Fonctionnelle et Environnement, CNRS, Université de Toulouse, Toulouse, France
e-mail: antoine.lecerf@univ-tlse3.fr

C. M. Swan et al. (eds.), *The Ecology of Plant Litter Decomposition in Stream Ecosystems*, https://doi.org/10.1007/978-3-030-72854-0_19

often achieved by the mean of litterbag experiments. This way of studying plant litter decomposition in the field enables to control for the amount and quality of detrital stock. Litterbags are also effective at collecting decomposers that are involved in decomposition, while preventing mixing of plant litter from the inside and outside of bags. Measurement of detrital mass remaining in litterbags sampled at different occasions is the basis of empirical construction of decomposition curves. Various mathematical models are then used for interpolation purpose and estimation of ecologically meaningful parameters underpinning the fate and effects of detritus in ecosystems.

Mathematical models of plant litter decomposition have been initially developed as part of broader models aiming to describe temporal dynamics of detrital organic matter dynamics in open systems fueled by natural plant litter inputs (e.g., Olson, 1963). Such a holistic approach focuses on organic matter budget and ecosystem stability and, therefore, it does not explicitly examine decomposition curves (Manzoni et al., 2009). Mathematical modelling here offers a realistic way to overcome limitations of empirical ecosystem research that is unable to cover all spatial and temporal scales relevant to analyze long-term detritus dynamics. There are few examples of applications of models of organic matter dynamics in freshwater ecosystem research. The comprehensive work by Boling et al. (1975) is worth noting here as their model of detritus standing stock in streams explicitly incorporated the heterogeneous nature of the detritus pool and qualitative changes undergone by detritus during decomposition. They used a discrete time approach whereby detrital mass change was expressed as a probability of transitioning from one stage to another in each time step. Quantification of organic matter dynamics in streams has also been achieved by the mean of continuous-time models (e.g., Richardson et al., 2009; Webster, 1983). The one developed by Webster (1983) was intended to assess the functional role of benthic invertebrates at the ecosystem level. Detritus stock was described based on physical models of transport and deposition of plant litter and fine particulate organic matter produced during decomposition, coupled with bioenergetic models for invertebrate detritivores (i.e., shredders and fine-particulate organic matter feeders). An interesting finding from this study was the ability of the model to reproduce observed pattern of plant litter mass change during decomposition in a closed system, such as litterbags (Webster, 1983).

Despite noteworthy efforts aimed at modelling patterns and processes of litter decomposition, the development of quantitative theories in this field has remained low on the research agenda for freshwater ecology. Common practices for analyzing empirical decomposition data currently largely rest on methods borrowed from terrestrial ecology, in particular from the seminal work published by Olson (1963). The leading role of terrestrial research here pertains to a numerical dominance of terrestrial ecologists studying plant litter decomposition as well as the importance of geochemical approaches to quantify organic matter stocks and flows at the ecosystem scale. In freshwater ecology, a naturalistic perspective is most often taken in studies on plant litter decomposition. However, in addition to empirical research, quantitative theories are required to achieve a robust understanding of patterns and processes underpinning the decomposition of submerged plant litter. This chapter is intended as a guide for freshwater ecologists to consolidate their knowledge on modelling plant litter decomposition and to deepen practices in quantitative analysis of decomposition data.

19.2 Overview of Kinetic Models for Plant Litter Decomposition

Many useful decomposition models are based on kinetic principles wherein detritus is represented by one or several compartments whose size(s) changes through time as a function of detrital mass itself (Manzoni et al., 2012). Compartments stand for homogeneous detritus pools and, therefore, the number of compartments reflects prior assumptions on the chemical diversity of organic matter. Rate of detrital mass loss from each compartment is expected to conform to first-order kinetics (Manzoni et al., 2009):

$$\dot{M}_{\mathrm{i}_t} = -k_i M_{\mathrm{i}_t} \tag{19.1}$$

where M_i is detrital mass in compartment i at time t (e.g., in days), $\dot{M}_{\mathrm{i}_t}$ is the absolute rate of detrital mass change through time (i.e., $\mathrm{d}M_{i_t}/\mathrm{d}t$), and k_i is a constant. The parameter k is broadly referred to as the litter decay rate, which is the fraction of litter mass remaining that is lost every instant of time. Unlike absolute mass change (i.e., $\dot{M}_{\mathrm{i}_t}$ in Eq. 19.1), k is dimensionless with respect to mass. As noted by Manzoni et al. (2009, 2012), the decay constant is analogous to a failure (or hazard) rate that is the likelihood that if something survives to one moment, it will not survive to the next. The probabilistic reasoning makes sense if detritus is viewed as a pool of molecules each facing a risk of being decomposed. The inverse of k (i.e., k^{-1}) quantifies the average time that molecules spend within the detrital pool.

The closed-form solution of Eq. 19.1 is an exponentially decreasing function wherein the pre-exponential factor M_{i_0} is defined a priori as the initial detrital mass in the compartment i:

$$M_{\mathrm{i}_t} = M_{\mathrm{i}_0} e^{-k_i t} \tag{19.2}$$

If a single homogeneous pool of detritus is assumed (see Table 19.1), Eq. 19.2 can be directly applied to model temporal pattern of detrital mass as proposed by Olson (1963). His model is also called "the negative exponential model" or "the single-pool model". The Olson's model has been broadly used in freshwater ecology to summarize results from litterbag experiments and to compare decomposition curves for different litter species or environmental conditions. It produces an upward concave curve, which is most often observed in studies of litter decomposition in freshwaters. An important asset of this model lies in its simplicity with respect to the number of parameters to be estimated (i.e., k-value) and the basic skills in statistics required for model estimation. It is worth noting here that the choice of the Olson's model is

Table 19.1 An assortment of kinetic-based models of plant litter decomposition. The selection is intended to illustrate how various assumptions on the time-dependence of litter decay rate lead to a diverse array of equations depicting litter decomposition curves

Model	M_t/M_0	$k_t = f(t)$	Types of decomposition curve
A	e^{-kt}	k	The negative exponential model from Olson (1963). k is a time-invariant decay rate
B	$1 - \kappa t$	$\frac{\kappa}{1-\kappa t}$	Model of linear decomposition trajectory. κ is the absolute mass loss per time unit (i.e., $\dot{M}_t$)
C	$\alpha e^{-k_1 t} + (1-\alpha)e^{-k_2 t}$	$\frac{\alpha k_1 e^{-k_1 t} + (1-\alpha)k_2 e^{-k_2 t}}{\alpha e^{-k_1 t} + (1-\alpha)e^{-k_2 t}}$	Bi-exponential (i.e., two-compartment) model where α is the fraction of labile organic matter and k_1 and k_2 are time-invariant decay rate for the labile and refractory detritus pools, respectively
D	$\frac{b^a}{(b+t)^a}$	$\frac{a}{b+t}$	Cf. model C1 in Manzoni et al. (2012). The detrital pool is assumed to be composed of detrital compounds decomposing independently at rates expected from a Gamma's two-parameters (a and b) function
E	$e^{-at-\frac{b}{m}(e^{-mt}-1)}$	$a + be^{-mt}$	This model proposed by Rovira and Rovira (2010) assumes that litter decay rate decreases exponentially from the initial value $a + b$ until stabilization near the minimal value a. m is a constant that gives the rate at which decay rate declines from early to late decomposition stage

rarely justified through rigorous comparison with concurrent models (e.g., Cornwell & Weedon, 2014).

When observed decomposition curves display no obvious curvature pattern, it can be tempting to draw a straight regression line through the data points (e.g., Baudoin et al., 2008). Linear decrease in detrital mass through time suggests that decomposition proceeds according to a zero-order kinetic model (see Model B in Table 19.1). In this case, the slope of the linear decomposition curve (κ) is equal to the time-derivative of detrital mass *vs* time and, therefore, it is not equivalent to the parameter k in the Olson's model (see Eq. 19.1). If we remember that $k = -\frac{\dot{M}_t}{M_t}$, then it is possible to find an expression that links k and κ (Model B in Table 19.1). At time $t = 0$, decay rate is equal to $\frac{\kappa}{M_0}$ and it increases with time asymptotically toward infinite value as detrital mass remaining gets close to zero. As a steady decrease of litter mass through time is a manifestation of time-varying decay rate, the linear model of plant litter decomposition should be considered as more complex than the Olson's model that assumes time-invariant decay rate.

An upward concave decomposition curve is expected to arise as a result of rapid initial loss of labile compounds (water-soluble molecules), and an opposite trend for increasing fraction of refractory compounds (fiber and lignin) in the detrital pool. This mechanism is not compatible with the key assumption underlying the Olson's model, that the detrital pool is made of homogeneous organic matter. Moreover, accumulation of refractory compounds through time should result in a decline in litter decay rate as decomposition proceeds, thus invalidating the assumption in the linear model (Model B). The heterogeneous nature of the detrital pool is conveniently represented by multi-compartment models of litter decomposition (Manzoni et al., 2009). The best-known example is the kinetic model with two compartments (also referred to as the bi-exponential model; see Table 19.1), one for labile compounds and one for refractory compounds, wherein decomposition proceeds independently following first-order kinetics (e.g., Chauvet, 1987). Other types of multi-compartment models include a larger number of compartments, connected or not to each other (see Manzoni et al., 2009 and 2012 for a comprehensive overview).

Another way to deal with detritus heterogeneity in a kinetic modelling framework is to substitute discrete compartments for a continuous distribution function of litter quality. In the general equation proposed by Carpenter (1981), litter decay rate is specified as a function of the quality of detrital compounds. A third function is to account for the transformation of detrital compounds into new compounds of lesser or greater quality during decomposition. The continuous modelling framework for plant litter decomposition thus offers a versatile toolbox for exploring the control of plant litter quality over decomposition. However, closed-form analytical solutions for detrital mass remaining (M_t) are only obtained under specific assumptions, which limits the relevance of such a model to analyze empirical data. For instance, two conditions must be satisfied to end up with equation D in Table 19.1: (1) the quality of detrital compounds does not change during decomposition and (2) the distribution of decay rates conforms to a Gamma probability density function (Manzoni et al., 2009).

19.3 A General Kinetic-Based Model of Litter Decomposition with Explicit Time-Varying Decay Rate

Time-varying decay rate is a common property of all litter decomposition models that produce non-exponential decreasing trajectories. The time-dependence of litter decay rate can be made explicit by rewriting model formulas in either of the following forms:

$$\dot{M}_t = -f(t)M_t \tag{19.3}$$

or

$$M_t = M_0 e^{-\int_0^t f(t)dt} \tag{19.4}$$

$f(t)$ is a mathematical expression for time-varying litter decay rate (i.e., the same as k_t). This function can be found a posteriori for all existing kinetic-based models of litter decomposition (Manzoni et al., 2012; see Table 19.1 for examples). If $f(t)$ is set to a constant, then Eq. 19.3 is equivalent to the first-order kinetic model (Eq. 19.1) and, therefore, Eq. 19.4 yields the Olson's model.

Rovira and Rovira (2010) suggested that f(t) could be specified in an intuitive manner so as to incorporate prior knowledge on expected pattern of litter decay rate over time and plausible values for upper and lower bounds. In Model E from Table 19.1, litter decay rate is assumed to decline from the initial value equal to $a + b$ to a final value b, following an asymptotic exponential function. The initial value can be taken equal to the decay rate for leachable compounds whereas the final value can be specified as the decay rate of the most refractory compound in the litter (i.e., lignin). The third parameter (m) indicates how fast litter decay rate changes through time, which may relate to the distribution of organic compounds along a degradability gradient as well as environmental constraints on plant litter decomposition.

Models with explicit time-varying decay rate are convenient to incorporate effects of environmental fluctuations on litter decomposition (Rovira & Rovira, 2010). Alp et al. (2016) applied this rationale to simulate temperature-dependent change in leaf litter standing stocks in lakes. A discrete-time modelling approach was used to reconstruct decomposition curves based on measured daily lake temperatures and empirical estimates of the temperature sensitivity of litter decomposition. In this study, the Arrhenius equation was used for this purpose (Alp et al., 2016). A continuous-time approach is conceivable if temperature variation is specified as a function of time. Then, Eq. 19.4 becomes:

$$M_t = M_0 e^{\int_0^t -f_0(t)e^{-Ea/[k_B\theta(t)]}dt} \tag{19.5}$$

where Ea is the activation energy of litter decomposition, k_B the Boltzmann constant, $f_0(t)$ is for temperature-independent variation in litter decay rate and $\theta(t)$ is a function that describes temperature variation.

If constant temperature-independent decay rate is also assumed in Eq. 19.5, $f_0(t)$ can be pulled out of the integral which can then be substituted by a sum of the values calculated by the formula $X_i = e^{-Ea/k_B\theta_i}$ for $i = 1$ to t. The resulting expression (Eq. 19.6) provides an alternative to the classical "degree-days correction" to estimate temperature-corrected decay rate out of empirical decomposition data (assuming that Ea is known).

$$M_t = M_0 e^{-k_0\left(\sum_{i=1}^{t} X_i\right)} \tag{19.6}$$

The general model of litter decomposition with explicit time-varying decay rate (Eq. 19.4) provides a basis for understanding the rationale of the "degree-days correction" commonly applied in aquatic decomposition studies. The equation describing this method is:

$$M_t = M_0 e^{-\lambda \int_0^t (t)dt} \tag{19.7}$$

where λ is a litter decay rate expressed in degree-day^{-1} and the integral is the temperature sum from day 1 to t^{th}. A key underlying assumption here is that litter decay rate increases proportionately with temperature [because $f(t) = \lambda\theta(t)$]. As litter decomposition is unlikely to cease at 0 °C, a more suitable expression for M_t is:

$$M_t = M_0 e^{-\int_0^t \lambda_0 + \lambda_1 \theta(t)dt} f_0(t) \tag{19.8}$$

Once rearranged and solved, Eq. 19.8 yields an expression that is equivalent to the model proposed by Hanson et al. (1984):

$$M_t = M_0 e^{-\lambda_0 t + \lambda_1 \int_0^t \theta(t)dt} \tag{19.9}$$

where $\lambda_0(\text{day}^{-1})$ and $\lambda_1(\text{degree.day}^{-1})$ are temperature-independent and dependent decay rates, respectively, and the integral is a temperature sum (i.e., degree-days). It is worth noting that Eq. 19.9 is potentially useful to take into account effects of fluctuation of other important environmental factors, such as nutrient concentration, on litter decay rate. In this case, θ(t) will be replaced by an appropriate function $x(t)$ and then $\int_0^t x(t)dt$ will be the sum of values taken by the focal factor (e.g., a sum of nutrient concentrations) during the timeframe of the study.

The three methods for removing temperature signal from litter decomposition curves presented here (Eqs. 19.6, 19.7, and 19.9) assume that temperature sensitivity of litter decay rate is constant over time. As the temperature sensitivity of decomposition is suspected to differ between labile and refractory compounds in litter (Fierer et al., 2005), variable temperature effects may arise as a result of change in their relative proportions during decomposition. Another underlying but untested assumption is the synchrony of decomposers' response to temperature fluctuations. One strength of the method based on the Arrhenius equation is that it takes into account the non-linear nature of temperature effect as expected from previous studies (Boyero et al., 2011; Follstad Shah et al., 2017). In contrast, the "degree-days correction" methods assume a linear temperature effect but do not require temperature-sensitivity of litter decomposition to be defined a priori (Hanson et al., 1984).

19.4 Process-Based Model of Plant Litter Decomposition

The view that litter decomposition is regulated by detritus quality is a cornerstone of the justification and formulation of kinetic-based models. In contrast, kinetic modelling lacks relevance to quantitatively analyze top-down control of plant litter decomposition mediated by abiotic and biotic agents of decomposition. Hieber and Gessner (2002) proposed a mechanistic framework for integrating decomposer effects on litter decomposition based on knowledge gained from freshwater research. In their model, biotically-driven detrital mass loss is expressed as a sum of instantaneous rates of organic matter flow from leaf litter to decomposers. Organic matter flows are assumed to depend on the biomass of three groups of decomposers: bacteria (*b*), filamentous fungi (*f*) and shredders (*s*). The corresponding mathematical expression is:

$$\dot{M}_t = -\left[g_b(B_{bt}) + g_f\left(B_{f_t}\right) + g_s(B_{st})\right]M_t \qquad (19.10)$$

In Eq. 19.10, $g_i(B_{it})$ are functions that estimate the fraction of M_t lost at each time step owing to the activity of decomposer group $i = \{b, f, s\}$. They are formulated as a product of decomposer biomass per unit of litter mass (equivalent to a consumer-to-resource biomass ratio) to biomass-specific (i.e., per capita) effect on detrital mass. It is interesting to note that the term in brackets in Eq. 19.10 is an expression of litter decay rate (i.e., $f(t)$ in Eq. 19.3) if decomposition were solely mediated by decomposers. Analysis of this model can thus give some hints about conditions under which time-invariance of litter decay rate could emerge. As decomposer biomass increases rapidly during the initial stage of litter decomposition (Gessner et al., 1999), the litter decay rate should increase rather than remain constant. A logistic function (i.e., S-shaped curve) have been used to describe colonization patterns of the main stream decomposers (e.g., Alemanno et al., 2007; Dang et al., 2009; Swan & Palmer, 2005). Accordingly, constant litter decay rate is most likely to arise when total decomposer biomass density gets stable near the upper limit (i.e., "carrying capacity") and *per capita* decomposer effects on decomposition do not change much through time. Both conditions are likely to be met an intermediate stage of litter decomposition. In a later stage, biomass density of fungi and shredders and per capita effect may decline as refractory compounds become a prominent fraction of the detritus pool.

Short-term feeding experiments provide quantified information about per capita effects of shredders on litter decomposition. Individual feeding activity is thought to be independent of detrital mass remaining, corresponding to type I functional response. However, shredders may cease feeding on detritus before complete disappearance due to low palatability of detrital compounds remaining a late decomposition stage (Hieber & Gessner, 2002). Litter palatability is also minimal an early stage of litter decomposition and increases as detritus gets colonized by microbial decomposers (Graça, 2001). Thus, Alemanno et al. (2007) introduced a Boolean function in their model to represent the possibility that shredders start feeding on detritus once microbial biomass has reached a certain threshold. Shredders feeding on conditioned leaf litter are documented to ingest between 10 and 80% of their body mass per day, though values beyond these bounds are sometimes reported. Feeding rate is likely to vary depending on individual phenotypes, intraguild interactions, and initial detritus chemistry (Danger et al., 2012; Lecerf et al., 2005; McKie et al., 2008; Rota et al., 2018; Santonja et al., 2017).

An energetic-based approach has been taken to estimate per capita effects of bacteria and fungi on litter decomposition. The rate of detrital mass loss is assumed to be proportional to carbon uptake rate by microbial decomposers, which is the sum of biomass production, exoenzyme synthesis and respiration (Geyer et al., 2016). The carbon uptake rate can be approximated by a ratio of gross growth rate to carbon use efficiency (Hieber & Gessner, 2002). Bacteria and fungi both allocate about one third of acquired carbon to the synthesis of new organic compounds; however, the former uptakes more carbon per unit of biomass than the latter (Hieber & Gessner, 2002). Alternative methods to estimate osmotroph activity include the use of enzymatic models as proposed by Sinsabaugh et al. (1993). Their method involves determining exoenzyme concentrations in detritus and finding statistical relationships between exoenzyme concentrations and the rate of detrital mass loss. As a result, the enzymatic-based approach to microbial decomposition takes into account both carbon transfer from detritus to decomposers and loss of partially-degraded compounds that escape decomposers (e.g., fragments of cell walls).

The study of Alemanno et al. (2007) constitutes the most comprehensive effort to incorporate patterns and processes underpinning decomposer effects on plant litter decomposition into a mechanistic model. This model was intended to reconstruct decomposition curves that were compared with empirical data of litter decomposition in a lake. The authors found that the best match between empirical and simulated decomposition curves arose when the model accounted for both effects of detritus heterogeneity, the activity of microbial and invertebrate decomposers and their interaction (Alemanno et al., 2007). Mechanistic models are also powerful tools to explore how decomposers mediate effects of environmental changes on process rates (e.g., Dang et al., 2009). However, there is only a handful of studies adopting such a mathematic approach to litter decomposition, conceivably due to the daunting complexity of existing models. For instance, the large number of parameters (>20) and equations makes the one developed by Alemanno et al. (2007) impracticable to analyze and run by many investigators.

19.5 A "Toy" Model of Plant Litter Decomposition

Here I present a "toy" model, a simple, idealized, analytical model based on a basic understanding of detritus and decomposer controls on plant litter decomposition. Detritus is assumed to be composed of two pools of organic matter (leachable and non-leachable) whose decomposition occurs in an independent fashion. Two equations describe the system:

$$\begin{cases} \dot{M}_{1_t} = -k_1 M_{1t} \\ \dot{M}_{2_t} = -g(M_{2t}) \end{cases} \tag{19.11}$$

where M_1 and M_2 are stocks of leachable and non-leachable organic matter, respectively. The sum of leachable (M_1) and non-leachable (M_2) organic matter is equal to detrital mass (M).

The dissolution process accounts for rapid loss of leachable organic matter during the initial stage of litter decomposition in aquatic ecosystems, which can be reasonably described by first-order kinetics (Alemanno et al., 2007; Dang et al., 2009). To obtain a closed-form expression for M_{1t}, the decay constant k_1 and the initial fraction of leachable organic matter in detritus $\left[\alpha = M_{10}/(M_{10} + M_{20})\right]$ must be known. Empirical determination of detrital mass loss in absence of biota can produce credible estimates for these parameters (e.g., Bärlocher, 1992; Gessner & Schwoerbel, 1989; Rodríguez-Pérez et al., 2018). Their values are likely to depend on the chemical quality but also the structural integrity of detritus. For instance, leachable compounds are released at faster rate from leaf litter that had previously exposed to drying or freezing conditions (Bärlocher, 1992).

In Eq. 19.11, $g(M_{2t})$ is a function used to describe how decomposer effect changes as litter decomposition proceeds. Following Hieber and Gessner (2002), it is expressed as the product of per capita decomposer effect (β), decomposer-to-detritus biomass ratio (R), and resource available (M_2). Substituting M_2 with M is required when R is determined based on total detrital mass, which is most often done in empirical studies. Due to rapid loss of leachable organic matter from submerged litter, the difference between M and M_2 is negligible most of the time detritus spends in the system and, thus, we have:

$$\dot{M}_{2_t} \approx -\beta_M R_M M_t \tag{19.12}$$

Here β and R are assumed to vary with detrital mass remaining and hence the product $\beta_M R_M$ is analogous to an implicit time-varying decay rate.

Decomposer colonization of detritus is driven by spatial movements (immigration and emigration) and local demographic processes (intrinsic growth rate). To avoid the complexity of deterministic models of decomposer biomass, an empirical expression for the product $R_M M_t$ (Eq. 19.12) is proposed here. A hump-shaped curve peaking at mid-decomposition stage describes how decomposer biomass changes with detrital mass. This pattern is expected because decomposer biomass is negligible at both

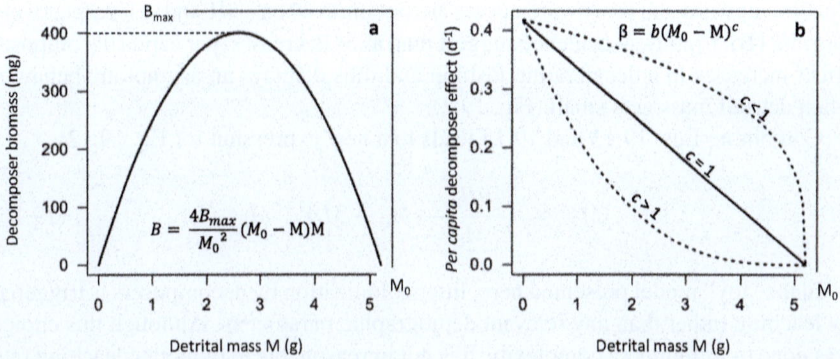

Fig. 19.1 Functions used to describe patterns of decomposer biomass (**a**) and per capita effect (**b**) along a detritus mass gradient (M)

ends of the range of detrital mass and the abundance and diversity of decomposers are thought to be the greatest when about half of the detrital mass had been lost. Though the exact nature of the relationship of decomposer biomass *vs* detrital mass is unknown (e.g., Swan & Palmer, 2005), it is convenient to use a quadratic function in this case (Fig. 19.1a):

$$R_M M_t = \frac{4B_{max}}{{M_0}^2}(M_0 - M_t)M_t \tag{19.13}$$

B_{max} is the maximal biomass reached over the lifespan of detritus patch of initial mass equal to M_0.

The per capita decomposer effect β is likely to depend on decomposer community structure owing to fundamental differences between bacteria, fungi and shredders. For instance, shredders or bacteria are likely to be much more efficient than fungi in degrading detritus (Hieber & Gessner, 2002). Thus, β can be expressed as a biomass weighted average of decomposer-specific per capita effect on litter decomposition:

$$\beta = \frac{\sum_{i=1}^{n} \beta_i B_i}{\sum_{i=1}^{n} B_i} \tag{19.14}$$

where β_i and B_i are per capita effect and biomass of decomposer group i, respectively.

The per capita decomposer effect β should increase as decomposition proceeds since early colonizers (fungi) are the least efficient decomposers. A power function is potentially relevant to describe the dependence of β upon M (Fig. 19.1b):

$$\beta_M = b(M_0 - M_t)^c \tag{19.15}$$

Here, b and c are positive constants. The scaling exponent c determines the shape of the relationship. Equation 19.15 produces a linear relationship when $c = 1$.

Concave upward and downward curves are obtained when $c > 1$ and < 1, respectively (Fig. 19.1B). Empirical evidence suggest that, as M decreases, per capita decomposer effect increases in a decelerating fashion and thus displays an asymptotic behavior when detrital mass gets small (Fig. 19.2).

Combining Eqs. 19.13 and 19.15 leads to a new expression for Eq. 19.12:

$$\dot{M}_{2_t} \approx -\frac{4bB_{max}}{{M_0}^2}(M_0 - M_t)^{1+c}M_t \tag{19.16}$$

In the "toy" model presented here, litter colonization by decomposers is triggered by leaching rather than any relevant demographic parameters. Although this choice was done to limit model complexity, it is not unreasonable to think that leaching rate could set the pace of microbial biomass accrual through a priming effect of soluble litter compounds on growth and activity of early colonizers (as suggested by results of Bärlocher, 1992). An explicit expression for this mechanism might be found but this will be at the expense of a greater complexity of the model.

Despite its inherent simplicity, the "toy" model described here is likely to produce realistic decomposition curves based on reasonable parameter values (Fig. 19.3). Furthermore, analysis of model equations can yield insights into patterns of litter decomposition and its determinants. For instance, Eq. 19.16 suggests that decay rate of non-leachable organic matter is likely to change during decomposition unless

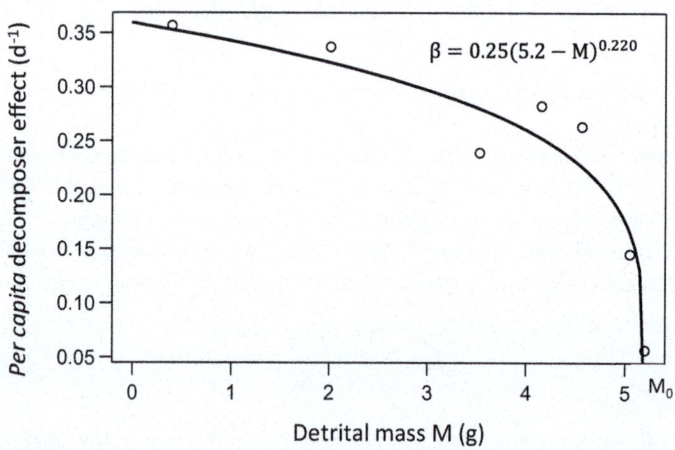

Fig. 19.2 Empirical relationship between per capita decomposer effect and detrital mass during decomposition of alder leaf litter. Data are extracted from results of the litterbag experiment published by Hieber and Gessner (2002). Points represent mean values of M and β for different sampling dates. β was recalculated based on decomposer biomass data (mean values) using Eq. 19.14. I assumed the per capita decomposer effect to be 0.057, 0.4 and 3.3 d^{-1} for fungi, shredders and bacteria, respectively. The dataset was then used to estimate parameters of Eq. 19.15 using nonlinear least-square regression

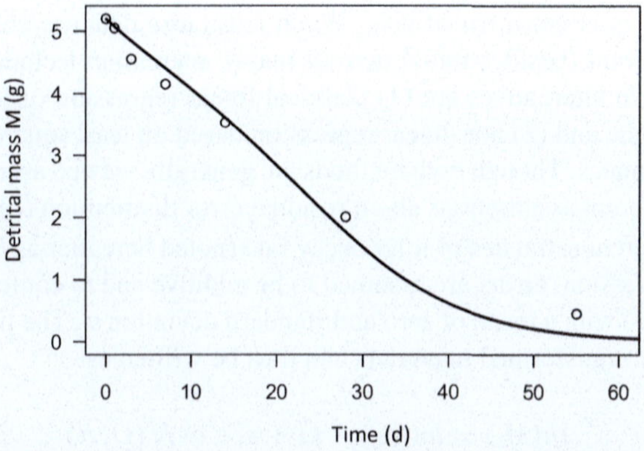

Fig. 19.3 Observed and reconstructed pattern of decomposition of alder (*Alnus glutinosa*) leaf litter. Points are mean values ($M_0 = 5.2$ g) of litter mass remaining in coarse mesh bags exposed in a stream. Data were extracted from the paper of Hieber and Gessner (2002). The curve was obtained by the mean of numerical simulation of the "toy" model presented in this chapter. The fraction of leachable organic matter (α) in alder leaf litter was set to 24%, a value used in a previous modelling study (Dang et al., 2009). In the latter work, a high leaching rate was assumed (ca. 0.6 d^{-1}) because alder leaf litter was previously air-dried. Empirical results displayed here were obtained with fresh alder leaves and therefore leaching rate was set to a substantially lower value ($k_l = 0.1$ d^{-1}). Model parameters for biotically-mediated decomposition were estimated based on decomposer biomass and per capita effect reported in Hieber and Gessner (2002): $B_{max} = 0.4$ g, $b = 0.25$ d^{-1}, $c = 0.22$ (see Fig. 19.2)

we have $c = -1$. The latter condition would be satisfied if the per capita decomposer effect decreased with detrital mass as predicted by the hyperbolic function $b/(M_0 - M_t)$ with $M_t < M_0$. This pattern can arise due to bottom-up control on litter decomposition; as decomposition proceeds, detritus quality decreases and so should the per capita decomposer effect. However, this is inconsistent with reports suggesting that least efficient decomposers (i.e., fungi) colonize detritus first whereas most efficient decomposers (shredders and bacteria) become prominent in intermediate and late decomposition stage. One may thus speculate that time-varying decay rate is a manifestation of predominantly top-down controlled litter decomposition whereas prevalence of bottom-up constraints (i.e., litter quality) across all decomposition stages results in constant litter decay rate.

19.6 What Everyone Should Know About Statistical Estimation of Litter Decay Rate

Virtually all empirical studies of litter decomposition aim to quantify how fast litter disappears from the system. Calculation of a single time-invariant decay rate (i.e., the Olson's *k*-value) has been by far the most common approach used by freshwater scientists. Choosing a calculation method among multiple possibilities offered by modern statistics is not a trivial task. When mass loss data are collected at more than one timepoint (besides initial detrital mass), regression technics can be used. Two best-known alternatives are (1) classical linear regression of log-transformed mass data *vs* time and (2) non-linear regression based on least squares or maximum likelihood estimates. Though both methods are generally viewed as interchangeable, they make different assumptions about residual error distribution (ε) and, therefore, may yield different estimates of litter decay rate (noted hereafter as $\widehat{k}$). In linear and non-linear regression, errors are assumed to be additive and to conform to a normal distribution ($\mathcal{N}$) with a mean of zero and standard deviation σ. The linear regression model with log-transformed mass data can thus be written as:

$$ln(M_t) = ln(M_0) - \widehat{k}t + \varepsilon, \varepsilon = \mathcal{N}(O, \sigma) \quad (19.17)$$

If we take the antilog of both sides of Eq. 19.17, ε becomes a multiplicative term:

$$M_t = M_0 e^{-\widehat{k}t} e^{\varepsilon} \quad (19.18)$$

Equation 19.18 can be rewritten in the form: $M_t = M_0 e^{-\widehat{k}t} + h(t)$ where $h(t) = (e^{\varepsilon} - 1)M_0 e^{-\widehat{k}t}$. This shows that residual error gets smaller as M decreases. In contrast, residual error is assumed to be constant over time when litter decay rate is estimated through non-linear regression analysis of untransformed mass data.

Log-transformation does not only influence the assumption of homoscedasticity but also the nature of residual distribution as e^{ε} (Eq. 19.18) has a log-normal distribution. It is unclear which assumption best matches the observed distribution and variance pattern of residual errors. As detrital mass values are bounded within a finite range (i.e., between 0 and M_0), the residual distribution may exhibit right skewness in the initial decomposition stage and left-skewness as detritus mass remaining approaches zero (Laliberté et al., 2012). In addition, errors may be smaller at both ends of the detrital mass range and higher on the mid-decomposition stage, when decomposers override more predictable physical factors (e.g., leaching) in mediating decomposition. Hence, if litter mass data are collected across all decomposition stages, it is plausible that neither linear regression of log-transformed mass data nor non-linear regression of untransformed data provides an optimal solution for estimating litter decay rate. Although advanced statistical technics make it possible the specification of complex error patterns (e.g., Laliberté et al., 2012), failure to correctly identify error distribution and variance patterns, as well as often low sample sizes, may lead to unreliable estimate of k-value. It has been argued that non-linear regression is more robust against departure from assumption of constant and normally distributed errors than linear regression (Adair et al., 2010; Laliberté et al., 2012). Alternatively, it is worth considering robust regression technics which might be even better to deal with complex error patterns and influential points (Maronna et al., 2019).

Selection of appropriate statistical methods for estimating k-value is only part of the problem since the assumption of constant decay rate might, in fact, be wrong. Investigators should be able to detect situations where the Olson's model fails to describe empirical decomposition data. Low precision (i.e., standard error and confidence intervals) of estimated k-value and low model R-square should be taken as evidence that litter decay rate is not constant. When replicate measurements are taken on each sampling time, a lack-of-fit test is useful to determine whether unexplained variance in litter mass data is due to significant deviation between observed and expected (here the Olson's model) decomposition curves or just due to pure error (i.e., time-independent variation among replicates; e.g., Hanson et al., 1984). Particular attention should also be paid to the pre-exponential constant in the negative exponential model (i.e., M_0), as a mismatch between observed and expected values may be a common manifestation of time-varying decay rate. Some authors have claimed that M_0 ought to be set to the expected value (Adair et al., 2010). This makes sense since estimated and expected values of M_0 should be identical when decomposition proceeds at a constant rate. In contrast, there is no reason to expect systematic convergence of estimated and expected parameters when litter decay rate varies within the timeframe of the study. As such M_0 should be freely estimated to ensure that model assumptions for regression analysis are met.

If a single decay rate is to be used to summarize temporal patterns of plant litter decomposition, it is important to ask whether $\hat{k}$ is representative of the true distribution of k_t. As the empirical and true distributions are generally unknown, it is unclear whether $\hat{k}$ represents a central tendency or any typical value of the distribution. This gap can be addressed through simulations and analysis of models of litter decomposition. An example is provided here (cf. Figure 19.4 and Table 19.2). Analysis of the simulated decomposition curve presented in Fig. 19.3 indicates that litter decay

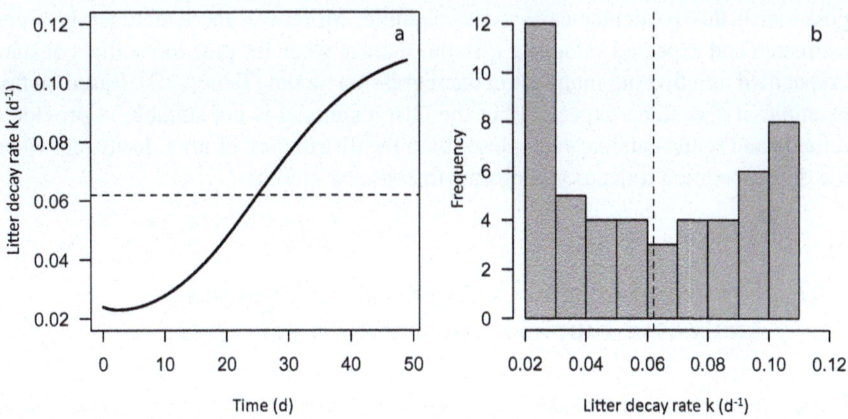

Fig. 19.4 Temporal pattern (**a**) and distribution (**b**) of litter decay rates calculated for the simulated decomposition curve presented in Fig. 19.3. The mean value is indicated by a dash line on each graph

Table 19.2 Effects of sampling scheme (number and location of timepoints) and curve fitting procedure (floating *vs* fixed intercept) on litter decay rate (k) estimated by the mean of the Olson's model fitted to simulated data. The decomposition curve shown in Fig. 19.3 was used to calculate expected litter mass remaining that was rescaled to $M_0 = 1$. Five replicate points were generated for each sampling date by adding random normal error (mean = 0, SD = 0.06) to predicted fraction of litter mass remaining. Maximum likelihood non-linear estimation was used to compute $\widehat{M_0}$ and $\hat{k}$. Unless specified (*ns*), all estimates are significant at $P < 0.05$

Sampling scheme		Floating intercept		Intercept forced to M_0
Number of time points	Sampling dates (days)	$\widehat{M_0}$	$\hat{k}$	$\hat{k}$
1	20			0.02931
1	50			0.05742
3	7, 20, 50	1.19	0.04418	0.03530
3	3, 7, 20	1.03^{ns}	0.03040	0.02795
5	7, 14, 28, 40, 50	1.25	0.04780	0.03743
5	3, 7, 14, 20, 28	1.06^{ns}	0.03447	0.03044
5	14, 20, 28, 40, 50	1.64	0.05712	0.03655

rate varies broadly (over ca. fivefold range; Fig. 19.4). Importantly, litter decay rate estimated using the Olson's model displays considerable variation depending on the sampling scheme (the number and distribution of timepoints at which litter mass remaining is determined) and whether or not the preexponential constant is fixed to initial litter mass (Table 19.2). Irrespective of the investigator's choices, the estimated k-value is always lower than the expected mean value (0.062 d^{-1}), conceivably due to a strong leverage effect of data points collected at early stage of litter decomposition in this particular case study example. Moreover, the difference between estimated and expected values for k is the smallest when the preexponential constant is specified as a floating intercept in the regression model (Table 19.2). Based on this example, it should be expected that the Olson's model is not suitable to provide a reliable and representative value describing the distribution of litter decay rate when the decomposition trajectory conforms to complex models.

19.7 Towards a Pragmatic Approach to Quantitative Analysis of Empirical Decomposition Curves

Fitting multiparameter models to empirical decomposition curves is the most widespread way of dealing with time-varying decay rate (e.g., Cornwell & Weedon, 2014). This approach requires that model complexity (and hence the number of parameters to estimate) is commensurated with amount of data available to describe decomposition curves. Investigators often rely on modest sample sizes, thus precluding the use of appropriate models with too many parameters. Besides, the use of most complex kinetic-based models should be refrained due to conceptual flaws owing to ecological rationale grounded in a bottom-view of plant litter decomposition. The two-compartment kinetic model (Model C in Table 19.1) ensures quantification of the dynamics of leachable and non-leachable organic matter assuming first-order kinetic principle and, therefore, can be viewed as a realistic extension of the Olson's model. Some parameters (i.e., fraction of leachable compounds and/or leaching rate) can be fixed a priori to limit the number of parameters to be estimated later. However, as discussed in Sect. 19.5 of this chapter, non-leachable organic matter may not disappear at a constant rate, notably if decomposers control litter decomposition.

Characterization of the distribution of litter decay rate provides an alternative strategy to analyze empirical decomposition curves. It is important to keep in mind that litter decay rate is defined in a way that ensure direct calculation without the requirement of parametric statistical estimation methods. As litter decay rate measures a proportional change in litter mass remaining through time, the following formula ensures its calculation at any time point or interval:

$$k_t = -\frac{d[\ln(M_t)]}{dt} \tag{19.19}$$

Equation 19.19 is implicitly used to calculate litter decay rate when decomposition curves are characterized solely by initial and final values of litter mass (e.g., Woodward et al., 2012). It is also relevant to analyze data of litter mass remaining determined at several time points (Fig. 19.5).

Plant litter decomposition research often aims at understanding the control of biotic and abiotic factors on the pace of litter decomposition. Many factors are likely to affect more strongly decomposer-mediated decomposition than leaching. It thus may be useful to obtain estimates of litter decay rate at different stages of litter decomposition, e.g., when leaching is thought to be the main driver of litter mass loss and, in later stages, when decomposer activity is a prevalent. As the magnitude of decomposer effect may change through time, several estimates of k may be required to properly describe later decomposition stages. Alternatively, the number of k-values to be extracted from data may be driven by the identification of breakpoints on graph of $\ln(M_t)$ *vs* time (Fig. 19.5). Piecewise regression analysis offers an integrative solution to estimate values for breakpoints and slope (i.e., $\hat{k}$) of each segment. It may also be used to test the hypothesis of constant decay rate, i.e., through a comparison of linear *vs* piecewise regression models. A reasonable advice is that the number of segments corresponding to stages of litter decomposition for which we want to estimate k should be proportionate to the number of sampling dates.

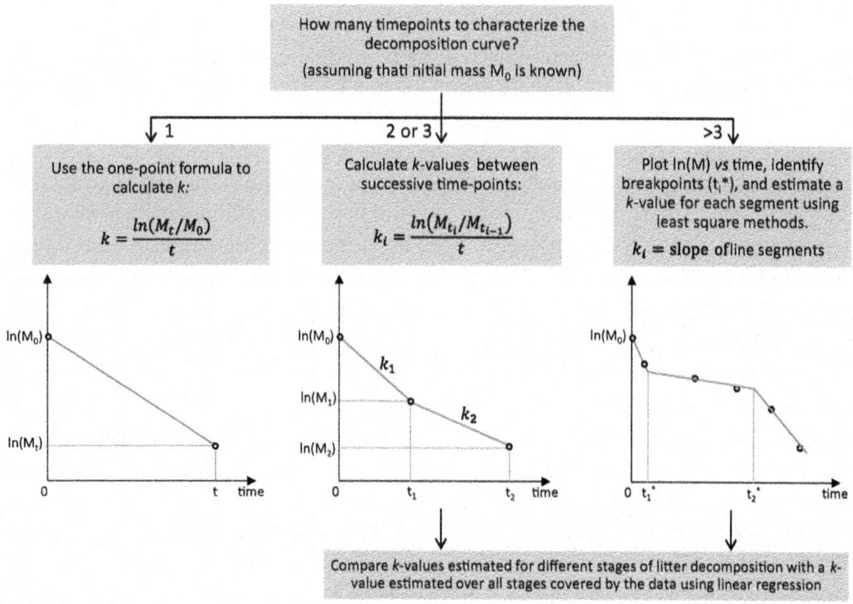

Fig. 19.5 Schematic description of a pragmatic approach for quantifying plant litter decomposition based on point measurements of litter mass during decomposition

19.8 Final Remarks

Despite major scientific advances in understanding spatial patterns and drivers of plant litter decomposition in freshwaters made over the past two decades, empirical research in this field still lacks solid grounding in quantitative theories. This pitfall manifests through a blind reliance on the constant-decay rate assumption when it comes to summarize results of empirical studies. Surely aquatic ecologists cannot be blamed for embracing the principle that "all models are wrong, but some (here the Olson's model) are useful". It is, however, important that everyone recognize that litter decomposition is unlikely to proceed at a constant rate and, therefore, complex models are required to achieve accurate description of decomposition curves. The assumption of time-varying decay rate is prevailing in terrestrial decomposition research wherein kinetic-based models are commonly used for quantitative assessment of stocks and flows of organic matter and major chemical elements (e.g., C) within (e.g., soil) and across ecosystem compartments. Such models largely developed by terrestrial ecologists are potentially useful to investigate how time-varying decay rate drives the polymorphism of decomposition curves, an issue that has received little attention in aquatic decomposition research.

The residence time of plant litter is presumably smaller in aquatic habitats than in soils and, therefore, litter decay rate is more likely to experience variations owing to natural environmental fluctuations (e.g., climatic factors) in the latter than former habitat. However, the frequent use of the "degree-days" correction in aquatic studies indicates that investigators are aware of the role of short-term environmental variability in mediating time-varying decay rate. Time-varying decay rate also pertains to detritus heterogeneity and litter colonization by decomposers. While decomposition models have been developed to incorporate the effect of detritus heterogeneity on decay rate (cf. compartment and continuous models), it is still unclear how decomposers mediate time-varying decay rate. This can be investigated by the mean of process-based models incorporating change in decomposer biomass, activity and diversity during decomposition. Efforts to develop such models are still required to take better account of all decomposer groups, including detritivores, and the way they interact with each other. Aquatic ecologists are expected to play a leading role in achieving this endeavor since top-down regulation of plant litter decomposition is a primary focus of research in aquatic ecosystems. As effects of decomposers and litter quality on decomposition are intricately linked, the development of quantitative theories of plant litter decomposition should ultimately aim at merging views of top-down and bottom-up regulation of decomposition.

The use of wrong models to analyze empirical data from studies on plant litter decomposition raises question about the actual meaning of model parameters presented in published articles. The Olson's model is likely to produce a range of estimates for litter decay rate depending on the sampling scheme and statistical estimation method. As litter quality and decomposer community may shape decomposition curves, the *k*-value yielded by the Olson's model may not ensure proper comparisons among litter species and/or across sites. These methodological flaws may plague global analyses of plant litter decomposition rate in freshwaters, wherein focal patterns of litter decay rate (e.g., effects of temperature or nutrient) are blurred in large unexplained variations (Follstad Shah et al., 2017; Woodward et al., 2012). Ways for improving the quality of decomposition data reported in primary studies include systematic evaluation of the time-invariance assumption, calculation of several estimates of litter decay rate instead of a single poorly defined constant, and identification of range(s) of decomposition stage covered by reported model parameters. Such information is key for furthering our understanding of broad-scale patterns of litter decomposition and linking biological rates to rates of biogeochemical cycling.

References

Adair, E. C., Hobbie, S. E., & Hobbie, R. K. (2010). Single-pool exponential decomposition models: Potential pitfalls in their use in ecological studies. *Ecology, 91*, 1225–1236

Alemanno, S., Mancinelli, G., & Basset, A. (2007). Effects of invertebrate patch use behaviour and detritus quality on reed leaf decomposition in aquatic systems: A modelling approach. *Ecological Modelling, 205*, 492–506

Alp, M., Cucherousset, J., Buoro, M., & Lecerf, A. (2016). Phenological response of a key ecosystem function to biological invasion. *Ecology Letters, 19*, 519–527

Bärlocher, F. (1992). Effects of drying and freezing autumn leaves on leaching and colonization by aquatic hyphomycetes. *Freshwater Biology, 28*, 1–7

Baudoin, J. M., Guérold, F., Felten, V., Chauvet, E., Wagner, P., & Rousselle, P. (2008). Elevated aluminium concentration in acidified headwater streams lowers aquatic hyphomycete diversity and impairs leaf litter breakdown. *Microbial Ecology, 56*, 260–269

Boling, R. H., Goodman, E. D., Van Sickle, J. A., Zimmer, J. O., Cummings, K. W., Petersen, R. C., & Reice, S. R. (1975). Toward a model of detritus processing in a woodland stream. *Ecology, 56*, 141–151

Boyero, L., Pearson, R. G., Gessner, M. O., Barmuta, L. A., Ferreira, V., Graça, M. A. S., Dudgeon, D., Boulton, A. J., Callisto, M., Chauvet, E., Helson, J. E., Bruder, A., Albariño, R. J., Yule, C. M., Arunachalam, M., Davies, J. N., Figueroa, R., Flecker, A. S., Ramírez, A., … West, D. C. (2011). A global experiment suggests climate warming will not accelerate litter decomposition in streams but might reduce carbon sequestration. *Ecology Letters, 14*, 289–294

Carpenter, S. R. (1981). Decay of heterogeneous detritus: A general model. *Journal of Theoretical Biology, 89*, 539–547

Chauvet, E. (1987). Changes in the chemical composition of alder, poplar and willow leaves during decomposition in a river. *Hydrobiologia, 148*, 34–44

Cornwell, W. K., & Weedon, J. T. (2014). Decomposition trajectories of diverse litter types: A model selection analysis. *Methods in Ecology and Evolution, 5*, 173–182

Dang, C. K., Schindler, M., Chauvet, E., & Gessner, M. O. (2009). Temperature oscillation coupled with fungal community shifts can modulate warming effects on litter decomposition. *Ecology, 90*, 122–131

Danger, M., Cornut, J., Elger, A., & Chauvet, E. (2012). Effects of burial on leaf litter quality, microbial conditioning and palatability to three shredder taxa. *Freshwater Biology, 57*, 1017–1030

Fierer, N., Craine, J. M., McLauchlan, K., & Schimel, J. P. (2005). Litter quality and the temperature sensitivity of decomposition. *Ecology, 86*, 320–326

Follstad Shah, J. J., Kominoski, J. S., Ardón, M., Dodds, W. K., Gessner, M. O., Griffiths, N. A., Hawkins, C. P., Johnson, S. L., Lecerf, A., LeRoy, C. J., Manning, D. W. P., Rosemond, A. D., Sinsabaugh, R. L., Swan, C. M., Webster, J. R., & Zeglin, L. H. (2017). Global synthesis of the temperature sensitivity of leaf litter breakdown in streams and rivers. *Global Change Biology, 23*, 3064–3075

Gessner, M. O., & Schwoerbel, J. (1989). Leaching kinetics of fresh leaf-litter with implications for the current concept of leaf-processing in streams. *Archiv Für Hydrobiologie, 15*, 81–90

Gessner, M. O., Chauvet, E., & Dobson, M. (1999). A perspective on leaf litter breakdown in streams. *Oikos, 85*, 377–384

Geyer, K. M., Kyker-Snowman, E., Grandy, A. S., & Frey, S. D. (2016). Microbial carbon use efficiency: Accounting for population, community, and ecosystem-scale controls over the fate of metabolized organic matter. *Biogeochemistry, 127*, 173–188

Graça, M. A. S. (2001). The role of invertebrates on leaf litter decomposition in streams—A review. *International Review of Hydrobiology, 86*, 383–393

Hanson, B. J., Cummins, K. W., Barnes, J. R., & Carter, M. W. (1984). Leaf litter processing in aquatic systems: A two variable model. *Hydrobiologia, 111*, 21–29

Hieber, M., & Gessner, M. O. (2002). Contribution of stream detritivores, fungi, and bacteria to leaf breakdown based on biomass estimates. *Ecology, 83*, 1026–1038

Laliberté, E., Adair, E. C., & Hobbie, S. E. (2012). Estimating litter decomposition rate in single-pool models using nonlinear beta regression. *PLoS ONE, 7*, e45140

Lecerf, A., Dobson, M., Dang, C. K., & Chauvet, E. (2005). Riparian plant species loss alters trophic dynamics in detritus-based stream ecosystems. *Oecologia, 146*, 432–442

Manzoni, S., Katul, G. G., & Porporato, A. (2009). Analysis of soil carbon transit times and age distribution using network theories. *Journal of Geophysical Research, 114*, G04025

Manzoni, S., Pineiro, G., Jackson, R. B., Jobbagy, E. G., Kim, J. H., & Porporato, A. (2012). Analytical models of soil and litter decomposition: Solution for mass loss and time-dependent decay rates. *Soil Biology and Biochemistry, 50*, 66–76

Maronna, R. A., Martin, D. R., & Yohai, V. J. (2019). *Robust statistics: Theory and methods* (with R) (2nd ed., 464 pp.). Wiley Series in Probability and Statistics. Wiley.

McKie, B. G., Woodward, G., Hladyz, S., Nistorescu, M., Preda, E., Popescu, C., Giller, P. S., & Malmqvist, B. (2008). Ecosystem functioning in stream assemblages from different regions: Contrasting responses to variation in detritivore richness, evenness and density. *Journal of Animal Ecology, 77*, 495–504

Olson, J. S. (1963). Energy storage and the balance of producers and decomposers in ecological systems. *Ecology, 44*, 322–331

Richardson, J. S., Hoover, T., & Lecerf, A. (2009). Coarse organic matter dynamics in small streams: Linking function to structural changes caused by catchment landuses [Special issue paper]. *Freshwater Biology, 54*, 2116–2126

Rodríguez-Pérez, H., Borrel, G., Leroy, C., Carrias, J. F., Corbara, B., Srivastava, D. S., & Céréghino, R. (2018). Simulated drought regimes reveal community resilience and hydrological thresholds for altered decomposition. *Oecologia, 187*, 267–279

Rota, T., Jabiol, J., Chauvet, E., & Lecerf, A. (2018). Phenotypic determinants of inter-individual variability of litter consumption rate in a detritivore population. *Oikos, 127*, 1670–1678

Rovira, P., & Rovira, R. (2010). Fitting litter decomposition datasets to mathematical curves: Towards a generalised exponential approach. *Geoderma, 150*, 329–343

Santonja, M., Pellan, L., & Piscart, C. (2017). Macroinvertebrate identity mediates the effects of litter quality and microbial conditioning on leaf litter recycling in temperate streams. *Ecology and Evolution, 8*, 2542–2553

Sinsabaugh, R. L., Osgood, M. P., & Findlay, S. (1993). Enzymatic models for estimating decomposition rates of particulate detritus. *Journal of the North American Benthological Society, 13*, 160–169

Swan, C. M., & Palmer, M. A. (2005). Leaf litter diversity leads to non-additivity in stream detritivore colonization dynamics. *Oceanological and Hydrobiological Studies, 34*, 19–38

Webster, J. R. (1983). The role of benthic macroinvertebrates in detritus dynamics of streams: A computer simulation. *Ecological Monographs, 53*, 383–404

Woodward, G., Gessner, M. O., Giller, P., Gulis, V., Hladyz, S., Lecerf, A., Malmqvist, B., McKie, B. G., Tiegs, S. D., Cariss, H., Dobson, M., Elosegi, A., Ferreira, V., Graça, M. A. S., Fleituch, T., Lacoursière, J., Nistorescu, M., Pozo, J., Risnoveanu, G., … Chauvet, E. (2012). Continental-scale effects of nutrient loading and eutrophication on stream ecosystem functioning. *Science, 336*, 1438–1440

Chapter 20
Design and Analysis of Laboratory Experiments on Aquatic Plant Litter Decomposition

Julia Reiss, R. A. Bailey, and Daniel M. Perkins

Abstract Microcosm studies are a useful tool when it comes to studying leaf litter decomposition but designing and analysing them can be a tricky path with many pitfalls. Because there is a plethora of drivers of leaf decomposition, it is important to be precise about the scientific questions that can be addressed with microcosm set-ups, and to use experimental designs that have minimal logistic implications but, at the same time, high statistical power. In this chapter, we first set the scene by introducing a hypothetical study that has the aim to estimate how leaf decomposition is driven by different decomposers and abiotic conditions. Following from this scenario, we give an overview of the main biotic and abiotic drivers of leaf decomposition that will play a role in laboratory settings (with special attention to consumer species identity, species richness, body size and metabolic capacity, and also temperature, time scales and stressors). We then explain how to design and analyse laboratory experiments on aquatic leaf litter decomposition including the mathematics for calculating the metabolic power of leaf decomposers and some statistical models. Further three case studies are given—highly controlled experiment that can be analysed by analysis of variance.

20.1 Introduction

Leaf litter falling into streams and rivers links energy flow from the terrestrial environments to the aquatic realm and is an important ecosystem process that mobilises large amounts of carbon and other nutrients (Abelho, 2001; Marks, 2019). The extent to which leaves are decomposed, and energy is recycled in the aquatic realm, depends

J. Reiss (✉) · D. M. Perkins
Department of Life Sciences, Whitelands College, Roehampton University, London SW15 4JD, UK
e-mail: julia.reiss@roehampton.ac.uk

R. A. Bailey
School of Mathematical Sciences, Queen Mary University of London, London E1 4NS, UK

School of Mathematics and Statistics, University of St Andrews, St Andrews, Fife KY16 9SS, UK

C. M. Swan et al. (eds.), *The Ecology of Plant Litter Decomposition in Stream Ecosystems*, https://doi.org/10.1007/978-3-030-72854-0_20

on a plethora of abiotic and biotic factors, and these might change and vary with time; and it is often these drivers that are the focus of a microcosm study in the laboratory.

Indeed, laboratory experiments, using aquatic microcosms, offer a way to disentangle which factors drive leaf decomposition (and if they act in synergy or antagonistically), given they are designed and analysed in a meaningful way. In general, tightly-controlled microcosm experiments offer a window into the complexity of nature (e.g., Bell et al., 2005), and many such studies have been carried out with leaf decomposers (e.g., Flores et al., 2016; Gonçalves et al., 2019; Jonsson & Malmqvist, 2000; Perkins et al., 2015; Risse-Buhl et al., 2015). A typical critique of microcosm studies is that they do not represent nature in a way field studies do; however the latter can often be criticised for not elucidating the mechanisms behind the observed patterns/phenomena. Highly controlled experiments can most relevantdeliver in this regard (Benton et al., 2007) because they can point to the mechanisms, i.e., they can produce the parameters that can be fitted in mathematical models (they inform modelling). Benton et al. (2007) point out how important it is to find these intrinsic mechanisms and how microcosm experiments can short-cut and 'speed up' insights from long-scale field work.

Many research groups have studied leaf litter decomposition in microcosm setups (Gonçalves et al., 2019; Pascoal et al., 2010; Perkins et al., 2015; Reiss et al., 2011) because not only is this an important ecosystem process but also it is possible to run experiments with little costs in terms of consumables. Leaf decomposition is further a process that can easily be monitored in microcosms and it is possible to replicate the decomposer communities. In fact, it is even difficult to exclude some leaf consumers such as bacteria and fungi from the laboratory as they 'travel' with the leaves from the field to the laboratory—in some cases even when the leaves are dried.

There is an unavoidable trade-off between replication and realism in all laboratory experiments and it is important to be precise about which questions can be answered in laboratory settings. Further, a central objective of these experiments is that they are planned and analysed in the best possible way and this chapter will give an overview of pitfalls and solutions to studying leaf litter decomposition in the laboratory.

Because this chapter will tackle only some aspects of designing and analysing leaf decomposition experiments, we have highlighted literature that has many relevant references cited within and that will help with further reading (see also Tables 20.1 and 20.2). It is helpful to read this chapter in conjunction with Chap. 21, where a number of approaches have been described for field experiments; and to consult the textbooks 'Methods to study litter decomposition: a practical guide' (Bärlocher et al., 2020) and 'Methods in stream ecology' (Hauer & Lamberti, 2007).

There are a number of exciting aspects of leaf decomposition we have not explored in detail in this chapter, such as other experimental designs that require other statistical analysis such as generalized additive mixed models (GAMM), explanations of consumer species in detail (Chaps. 9–11), or food web interactions and implications for decomposition (Chap. 7). We would like to refer the reader to other chapters and literature when it comes to the general theme of energy transfer from leaves to consumers which includes topics such as the role of nutritional value of leaves

Table 20.1 Drivers of leaf decomposition discussed in this chapter that should be considered in laboratory experiments with leaf decomposers

Drivers of leaf decomposition	Explanation	Examples
Species identity	Species vary in their contribution to leaf decomposition and in terms of their preferences (Bärlocher et al., 2020; Tachet et al., 2010)	Reiss et al. (2011), Swan and Palmer (2006) and Treton et al. (2004)
Biodiversity	If different species, different genotypes or different functional groups (and other entities of 'biodiversity') influence each other's performances, or contribute to different pathways in leaf decomposition, then biodiversity influences this ecosystem process (Reiss et al., 2009)	Andrade et al. (2016), Flores et al. (2016), Perkins et al. (2015) and Reiss et al. (2010, 2011)
Interactions between leaf decomposers	All individuals within an ecosystem are connected via vertical and horizontal linkages, for example within the food web. For instance, feeding interactions (trophic cascades) can alter leaf decomposition rates (Reiss et al., 2009)	Chambord et al. (2017), Duarte et al. (2006), Foucreau et al. (2013), Mille-Lindblom et al. (2006), Ribblett et al. (2005), Risse-Buhl et al. (2012, 2013) and Treton et al. (2004)
Body mass and metabolic rate	The performance of a leaf decomposer will depend on how large the organism is because large organisms have higher metabolic rates compared to smaller ones (Brown et al., 2004)	Flores et al. (2016) and Reiss et al. (2011)
Biomass	It is important to know whether total biomass is composed of small or large individuals, because the former will exhibit higher process rates for a given unit of mass (small organisms have a higher metabolic rate per unit of body mass) (Brown et al., 2004)	Reiss et al. (2011)

(continued)

Table 20.1 (continued)

Drivers of leaf decomposition	Explanation	Examples
Metabolic capacity	The metabolic capacity of an assemblage is its potential to contribute towards a certain process (e.g., to decompose leaves) (Brown et al., 2004; Gillooly et al., 2001; Peters, 1983).	Flores et al. (2016) and Perkins et al. (2010)
Temperature—metabolic capacity	The metabolic capacity of an assemblage increases with temperature until the organisms surpass their temperature optimum (Gillooly et al., 2001)	Martínez et al. (2014) and Perkins et al. (2010, 2015)
Temperature and time—reproductive potential of consumers	Generation time (and other related characteristics such as fungal sporulation rate) increases with temperature until the species' temperature optimum is reached (Gillooly et al., 2001; Reiss & Schmid-Araya, 2010)	Martínez et al. (2014)
Temperature—chemical processes	The higher the temperature, the quicker the chemical reactions, such as leaching of phenols from the leaves (Abelho, 2001; Bärlocher, 2005)	We are not aware of laboratory studies testing a range of temperatures in the absence of organisms
Effects of time and effects of chemical processes	Leaf decomposition changes over time (e.g., because consumer composition changes Abelho 2001). In a laboratory set-up it has been shown how leaching changes over time, at 20 °C (France et al., 1997) for example	France et al. (1997)
Stressors	Stressors such as water pollution (Ormerod et al., 2010) can affect leaf decomposers and hence decomposition. Further, they can have direct, chemical effects (e.g., on water pH)	Canhoto et al. (2017), Gonçalves et al. (2019) and Pascoal and Cássio (2004)

(continued)

Table 20.1 (continued)

Drivers of leaf decomposition	Explanation	Examples
Other	Other factors influence leaf decomposition. Examples are: habitat complexity (modulates interactions); nutritional value of leaves, water chemistry (e.g., oxygen concentration) or water current (sheer stress)	Flores et al. (2016), Larrañaga et al. (2014) and Risse-Buhl et al. (2013)

for consumers (Chap. 3; Jabiol et al., 2019; Larrañaga et al., 2014), for example by using stoichiometric approaches (Chapter 3; Farrell et al., 2018), a focus on leaf species diversity as such (Larrañaga et al., 2014; Martínez et al., 2013), the changes in resource quality over time (Chap. 19; Canhoto & Graça, 1996) and switch of consumer assemblage over time (e.g., because of seasonal changes). The focus of this chapter is very much on the consumers rather than the food resources (leaves).

In this chapter, we start with a hypothetical study that has the aim to estimate how biodiversity of consumers drivers leaf decomposition to 'set the scene'. Following from this scenario, we give a synthesis of possible drivers of leaf decomposition, followed by an overview of how to design and analyse leaf litter experiments in general. We illustrate the latter by explaining experimental set-ups that have been used in laboratory-based freshwater research and make suggestions for meaningful experimental designs with minimal logistic implications. Finally, we give three case studies to illustrate the importance of meaningful experimental design and analysis when it comes to leaf decomposition experiments. These case studies have in common that they are highly controlled and can be analysed by analysis of variance.

20.2 Planning Leaf Decomposition Experiments in the Laboratory—Where to Start?

Leaf decomposition is the sum of chemical and physical processes (Chap. 2), and organisms interacting with the leaves, while interacting with each other as well as with their abiotic environment. To approach this complex concept let us imagine an enthusiastic PhD student. She has identified that the freshwater shrimp *Gammarus* spp. and the freshwater hog louse *Asellus* spp. are abundant macroinvertebrates in the local streams, and are therefore likely to play an important role in leaf decomposition, but she is aware that other invertebrates such as insect larvae can graze on leaves and that tiny crustaceans such as *Cyclops* spp. can also graze on leaf biofilm. She wants to find out which species is the most efficient in terms of breaking leaves down, by feeding on them, and whether a combination of different species results in maximised leaf mass loss. The PhD student has a good understanding of the autecology of freshwater crustaceans, and she knows that within the species individuals vary. For

Table 20.2 Body mass as a driver of leaf decomposition and ways to measure the impact of this predictor

Calculation for…	Equation	Abbreviations	Worked example for an assemblage of 5 individuals of *Asellus aquaticus* (all 10 mm long and 2.5 mm wide), at 15 °C
Body mass (M) of one individual	The equation is species-specific. If the equation is unknown, the volume (V) of the individual can be calculated from a geometric shape (e.g., a spheroid) and V can be converted to mass by knowing the density of the organism (often 1.1). For example, the volume of a spheroid is $V = L*W^2 * \pi/6$. V is converted to WW by assuming a density of 1.1. DW can be assumed to be 25% of wet weight (WW). After (Reiss & Schmid-Araya, 2010)	Where V = volume in mm^3 (i.e., microlitres); L = length in mm; W = width in mm; WW = wet weight in µg (as 1 µg = 1 mm^3); DW = dry weight in µg—converted to mg by multiplying with 1000	*Asellus aquaticus* specific equation from Reiss et al. (2018) (mg, mm): $\log_{10}(M) = 2.652 * \log_{10}(L) - 1.841 = 6.4$ mg DW If calculated as a spheroid then DW = 9 mg
Assemblage biomass (AB)	$AB = \sum M * N$	Where M is individual body mass (e.g., dry weight, mg) and N is abundance (e.g., ind./m^2)	AB (mg) = 6.4 mg *5 = 32 mg DW
Metabolic capacity (MC) of a consumer assemblage, based upon general allometric-body size scaling relationships	$MC = \sum$ (per capita $M^{3/4}$)	Where M is individual body mass (e.g., dry weight, mg). The ¾ exponent used here describes a general relationship between basal metabolic rate and body size and has been applied to describe the allometric scaling of basal metabolic rate across wide range of organisms (see Brown et al., 2004; Peters, 1983)	MC (mg)= $\sum (6.4^{3/4})$ = 20 mg DW

(continued)

Table 20.2 (continued)

Calculation for…	Equation	Abbreviations	Worked example for an assemblage of 5 individuals of *Asellus aquaticus* (all 10 mm long and 2.5 mm wide), at 15 °C
Assemblage metabolism (I) is MC—but incorporates the effects of temperature. This equation yields predicted values for the expected rate at which resources should be consumed by each assemblage, based upon the sum of the metabolic capacities of all individual consumers in addition to the environmental temperature	$I = i_0 M^{3/4} e^{-Ei/kT}$	Where, i_0 is a normalisation constant that converts mass to energy and which is empirically derived, M = body mass, e = Euler's number, E_i is the activation energy of respiration (0.63 eV; after Gillooly et al., 2001), k = the Boltzmann constant in eV per Kelvin and T = temperature in Kelvin	This equation depends on results of a leaf decomposition experiment, i.e., depends on the empirical data and an empirically derived value for i_0; and hence we cannot give a worked example here—but see Perkins et al. (2010) and Fig. 20.2 for example data. i_0 can be derived through calculating the anti-log of the intercept for the relationship between metabolic capacity and temperature corrected decomposition on a log-log plot

example, the crustaceans will reproduce more than once a year in nature and different sized individuals will be present at different times of the year (i.e., there are small, still growing individuals, as well as males and females). The feeding activity of these individuals is strongly influenced by temperature as they are ectotherms, but other factors such as a suitable habitat play a role; and their growth and reproduction depend on many factors including the quality of their food and water temperature. For her experiments, it will be impossible to separate these crustaceans from bacteria and fungi in the water and these organisms also decompose leaves so she would like to include fungal species in her design.

Obviously, she is faced with a complex jigsaw puzzle if she tries to answer a seemingly simple question. Here we will make an attempt to assemble this puzzle, piece by piece (Fig. 20.1) with reference to pivotal literature (see also Tables 20.1 and 20.2), with the knowledge that it will make sense only from a particular point of view (for a selected pool of questions) and an appreciation for the fact that not all pieces of the puzzles are known (Fig. 20.1).

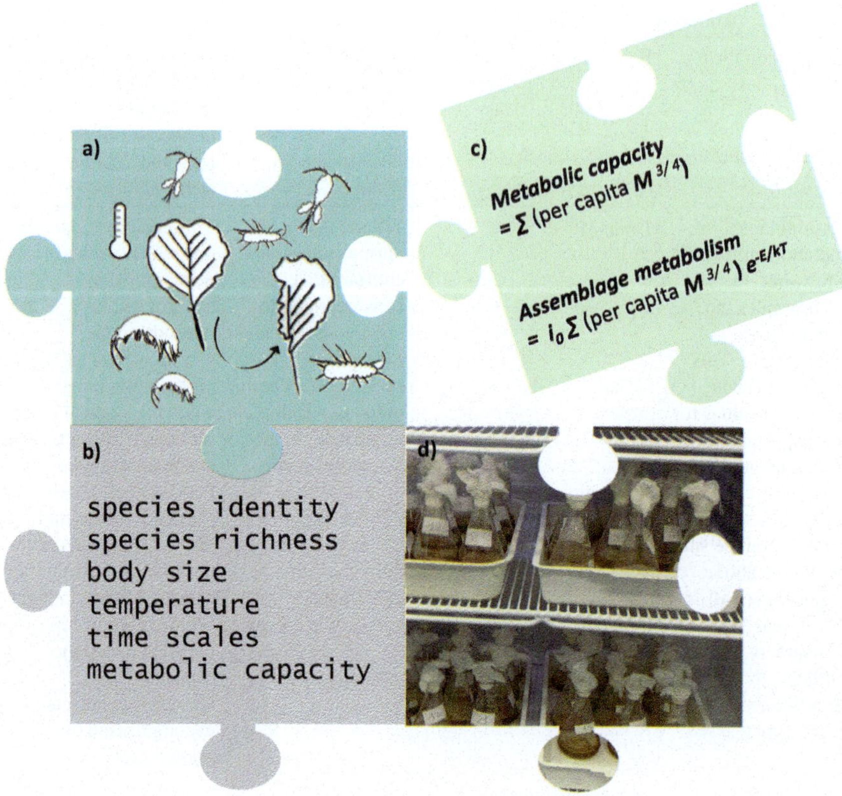

Fig. 20.1 **a** Leaf decomposition in aquatic microcosms is strongly driven by the presence of leaf decomposers and abiotic conditions; **b** the main drivers of leaf decomposition should be considered in leaf decomposition experiments; **c** ecological theory provides some tools for anticipating the scale of leaf decomposition because it considers organism traits such as body size and temperature; **d** laboratory experiments are designed to address particular questions, with particular leaf decomposers, and even complex set-ups can be run in logistically feasible ways

20.3 Biotic and Abiotic Factors to Consider in Leaf Decomposition Experiments

Many biotic and abiotic factors drive leaf decomposition and in the following sections some of them will be discussed in more detail (consumer species identity, species richness, body size and metabolic capacity, and also temperature and time scales; see Table 20.1).

20.3.1 Species Identity Drives Leaf Decomposition

A consortium of organisms decomposes leaves in streams and rivers; and bacteria, fungi and animals all play pivotal roles (Hieber & Gessner, 2002). Fungi are an important component of decomposer assemblages associated with plant litter in streams (Gessner et al., 2010) and macro-invertebrates feed on both fungi and leaf material (Canhoto & Graça, 2008). The fact that a group of macroinvertebrates has been dubbed 'shredders' indicates that there are species that are specialised in shredding leaf material and indeed we would expect such species to decompose leaves faster and more efficiently than other species. Clearly 'true' leaf litter feeding requires the existence of endogenous cellulases (fungi and shredders) or exocellulases (free living bacteria and endosymbiotic bacteria) (Zimmer, 2005). For example, snails might graze on leaf biofilm and indirectly aid the decomposition of leaves but they do not feed on as much leaf material as *Gammarus* which is a leaf shredding amphipod that has, just like the freshwater hog louse *Asellus*, endosymbiotic bacteria in its gut that can break down cellulose (Zimmer, 2006) (however—a word of caution: *Gammarus* and *Asellus* are strictly speaking omnivorous). Species identity, or at least a specific combination of traits, can therefore be an important driver of leaf decomposition—although surprisingly few studies have tested this. For example, the larvae of the cased caddisfly *Sericostoma* are very efficient in shredding leaves fast (e.g., González & Graça, 2003; Reiss et al., 2011), which is possibly due to its large size and high metabolic demands (see Sect. 3.2), but, of course, it is conceivable that they also possess very efficient cellulases, or gut symbionts.

20.3.2 Body Size, Biomass and Metabolic Rate Drive Ecosystem Processes: Calculating Metabolic Capacity

Earth's biota regulates numerous fluxes of energy and matter, including carbon uptake, nutrient cycling and oxygen production. When measured at local scales, these rates are referred to as 'ecosystem processes' and leaf decomposition is such a process. The activity of all organisms on Earth is constrained by the laws of physics and chemistry and this simple fact can be of immense help when studying ecosystem processes such as leaf decomposition because all organisms can operate only within the (mathematical!) limits of natural laws. This becomes apparent when studying metabolism and how this relates to the size of an organism, and, in turn, to how the organism can contribute towards ecosystem processes. Body size and metabolism are a hot topic in Ecology (Whitfield, 2004) because metabolism 'sets the pace of life' and drives processes across levels of biological organisation (from individuals to ecosystems; Brown et al., 2004). As a consequence, body mass is clearly a trait that needs to be considered when we study processes driven by organisms because body mass determines the basal metabolic rate, energy demands and ingestion rates of an individual (Brown et al., 2004; Perkins et al., 2012; Peters, 1983; Reiss et al., 2009;

Woodward et al., 2005). It follows that the performance of an assemblage is the sum of the metabolic power of its constituent individuals; or phrased in a different way; the body-mass vs biomass distribution in that assemblage. It is hence not surprising that there is a growing number of studies starting to consider the role of body size and biomass in experiments that address ecosystem process rates (Flores et al., 2016; McKie et al., 2008; Perkins et al., 2010; Reiss et al., 2010, 2011). Species identity is often confounded with body mass and using many differently sized individuals within a species or among species circumvents this problem and allows distinguishing between taxonomic and functional diversity and developing a more general mechanistic and predictive framework. Hence, we argue here that many future studies on leaf decomposition would profit from taking the body mass distribution of the leaf consumers into account and we encourage a more theoretical approach to leaf decomposition.

Because the activity of organisms can be understood through physical laws, such as laws of mass and energy balance, and thermodynamics, it is possible to use parameters such as body size and temperature (see Sect. 3.3) to predict how effective an assemblage of individuals will be when it comes to decomposition of leaf litter. A potentially promising approach to link the effects of body mass and temperature on ecosystem processes is through the application of the "Metabolic Theory of Ecology" (MTE) (sensu Brown et al., 2004). Building upon well-established body size allometries (Peters, 1983) and temperature scaling based upon first order physical principles (Gillooly et al., 2001), the MTE can be used to make quantitative predictions about a wide range of biological processes across levels of organisation (Brown et al., 2004). Leaf decomposition is the amalgamation of leaf feeding by different consumers. The feeding rate of each consumer species is, in turn, strongly connected to the body mass of individuals. If the body size distribution and abundance of leaf decomposers is known then it is possible to calculate metabolic rates (which correlate strongly with leaf decomposition) of individuals (Table 20.2) and those of entire assemblages (the so called 'metabolic capacity' [sensu Ruesink & Srivastava, 2001]; also called 'metabolic potential', see Table 20.2).

To summarise the above, given that allometric scaling relationship between individual body mass and metabolism is less than unity (Brown et al., 2004), the spectrum of individual body sizes characterizing a given assemblage is important in determining process rates. In the following sections we will shortly dive into more details about one important aspect of this, which is not necessarily intuitive. While an assemblage composed of larger individuals may have a higher total metabolic capacity, an assemblage composed of small individuals will exhibit higher mass-specific process rates (Peters, 1983). To illustrate the latter, let us imagine a hypothetical experiment with two crustaceans: *Asellus* and *Cyclops*. Individuals of *Asellus* are much larger than individuals of *Cyclops* but their metabolic rate per unit body mass is lower. This means that, if these two species are used in a laboratory setting, then *Cyclops* assemblages will consume more food over time compared to *Asellus*—given the biomass of the two assemblages is the same. Hence, small animals ingest more food in relation to their own body mass compared to larger organisms. Applying this knowledge to experimental microcosm set-up means it is extremely important to calculate the metabolic capacity (Table 20.2). Indeed, instead of adjusting for number or biomass (as done in most leaf decomposition experiments), the 'metabolic capacity' should be calculated (see Table 20.2). Metabolic capacity is a proxy for how much energy a given assemblage will use and how fast it will use it. In terms of adding individuals to microcosms, adjusting for metabolic capacity will always result in a lower number of small individuals (compared to larger individuals) than adjusting for the biomass, as the following example explains. Picture two microcosms, one for an *Asellus* only treatment and one for a *Cyclops* only treatment; and the aim is to control for body mass differences in these assemblages. Assuming we use very small *Asellus* individuals that are 6.6 mm long on average (~2.14 mg dry weight [DW]) means they weigh almost 71 times more than *Cyclops* (~1.6 mm long, 0.03 mg DW). Hence, when biomass is accounted for, 71 more *Cyclops* individuals are added for each *Asellus* individual. However, adjusting for metabolic capacity means that only 18 times more *Cyclops* are added for one *Asellus* individual (see Table 20.2 and for an example see Flores et al., 2016). Metabolic capacity can be a strong predictor of leaf decomposition (Fig. 20.2) and can also be extended to incorporate the effects of temperature (Fig. 20.2), as explained towards the end of Sect. 3.3.

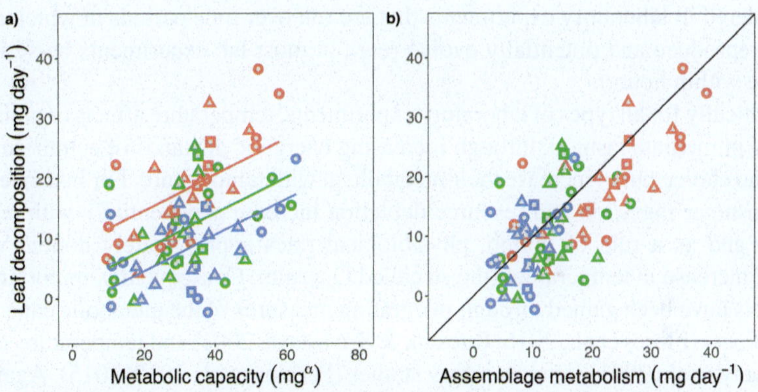

Fig. 20.2 Metabolic capacity of invertebrate shredders and environmental temperature drive leaf decomposition in microcosm experiments. Data are redrawn from Perkins et al. (2010). Symbols denote species richness levels (*circles* = mono-cultures, *triangles* = di-cultures and *squares* = tri-cultures) and symbol colours represent temperature treatment (*blue* = 5 °C, *green* = 10 °C and *red* = 15 °C). **a** Leaf decomposition increases linearly with metabolic capacity with greater absolute rates (higher intercept values) with increasing temperature. **b** When incorporating the effects of temperature through temperature scaling principles (see equation in Table 20.2), observed leaf decomposition increases significantly with predicted rates (i.e., assemblage metabolism), whereby the slope of this relationship does not differ from the fitted 1:1 line (*solid line*)

20.3.3 Temperature Affects Leaf Decomposition

Leaf mass loss in fresh water is strongly modulated by temperature, and a simple explanation is that the leaves 'leach' various substances when exposed to water, including phenols (e.g., tannins) (Bärlocher, 2005; Canhoto & Graça, 1996; Quinn et al., 2000). Leaching will follow chemical- and temperature rules and is unavoidable. Other abiotic factors such as light surely also play a role here and studies that manipulate light intensity in the laboratory must make sure that light intensity and temperature are not confounded.

Temperature also affects leaf decomposers, either over longer time scales (evolutionary response of organisms) or in a very direct, immediate way. Regarding the first point, numerous recent studies from fresh waters have suggested that environmental warming often favours smaller organisms, both within and among species, and that much of this can be ascribed to metabolic and physiological constraints that scale allometrically with body mass (see review by Perkins et al., 2010, and references therein). A high level of genetic variation that allows a species to adapt (Gamfeldt et al., 2005) is therefore an advantage when it comes to changes in the environment such as temperature regimes. Hence, evolutionary response to temperature should be considered in laboratory experiments that are run over time periods in which organisms reproduce and potentially evolve (e.g., in most lab experiments bacteria will evolve within hours).

Critically for all types of laboratory experiments, temperature affects organisms in a very immediate fashion, through increasing energetic demands of ectotherms that have no choice but to increase their metabolism with temperature. For instance, rates of consumer ingestion and resource depletion increase exponentially with temperature, and as a rule of thumb, physiological rates approximately double with a 10 °C increase in temperature (the so called Q_{10} rule). Consequently, important new insights have been gained through integrating measures of the metabolic capacity of consumers (Flores et al., 2016; Ruesink & Srivastava, 2001) and temperature scaling (Gillooly et al., 2001) into laboratory studies (Perkins et al., 2012, 2015). Again, the MTE offers the mathematical models behind temperature effects and much of this is based on the Boltzmann constant and Arrhenius equation (Gillooly et al., 2001; Table 20.2). For example, it is possible to predict a rate, such as leaf decomposition, for a hypothetical temperature, T_2, if the rate is known for a reference temperature, T_1 (Gillooly et al., 2001). Further, it is possible to incorporate the effects of temperature (in addition to body size) on metabolic capacity (Table 20.2, Fig. 20.2), and in doing so this approach can be extended to predict 'assemblage metabolism' (Table 20.2, Fig. 20.2). For instance, it is useful to calculate assemblage metabolism if leaf decomposition is run at different temperatures and the results from these temperatures are merged (see example in Fig. 20.2).

20.3.4 Biodiversity and Species Interactions Drive Leaf Decomposition

Metabolic power of decomposer assemblages and temperature are clearly strong predictors for plant litter decomposition but past litter decomposition experiments often had a strong focus on biodiversity effects (Perkins et al., 2010; Reiss et al., 2010, 2011). The general premise here is that assemblages that are very biodiverse will drive processes such as leaf litter decomposition better (e.g., faster) than assemblages that are less biodiverse. Biodiversity can be measured in different ways, with species richness being the most popular metric, and has indeed been shown to increase many ecosystem processes (see review by Reiss et al., 2009). When litter decomposition driven by a decomposer assemblage is measured a key question is: do species-rich assemblages exhibit faster rates than species-poor ones? There are three possible scenarios here: the different species within that assemblage can (1) perform in an additive fashion, i.e., as they would alone ('in monoculture'), (2) interact with each other and influence each other in a positive or negative way and (3) drive different processes that contribute to an overall effect (multifunctionality). In laboratory studies where species are offered a limited amount of food resource types and environmental conditions, they often perform in an additive fashion—meaning once metabolic capacity is accounted for, species decompose the same amount of leaf litter (Flores et al., 2016; Reiss et al., 2010, 2011). True biodiversity effects often only become apparent when a 'multifunctionality approach' is considered because species contribute to more than one process and respond differently to interactions with biotic and abiotic factors (Gamfeldt et al., 2008; Gamfeldt & Roger, 2017; Perkins et al., 2015). In the context of a leaf decomposition laboratory experiment this means that on top of measuring leaf decomposition, measuring other processes (e.g., production of faeces, respiration rates) will provide a more complete picture of the importance of biodiversity (Gamfeldt et al., 2008; Reiss et al., 2009).

So far we have focused on biodiversity within a single trophic level but of course leaf consumers are part of complex trophic interactions in fresh water ecosystems (Gessner et al., 2010; Reiss et al., 2009). While studying the macrofauna-fungi-leaf relationship has a longer tradition in laboratory experiments (e.g., Canhoto & Graça, 2008; Reiss et al., 2010), laboratory studies that involve microscopic consumers (such as protozoans and micro-metazoans) of leaf bacteria and fungi are on the rise (Chambord et al., 2017; Flores et al., 2016; Ribblett et al., 2005; Risse-Buhl et al., 2012); and they have shown that tropic interactions among these small organisms can have a substantial impact on leaf decomposition (e.g., Chambord et al., 2017; Risse-Buhl et al., 2015).

20.3.5 Other Abiotic Factors and Stressors

Species interactions and performance are influenced by abiotic factors other than temperature, of course, and an example is habitat complexity. For example, it is conceivable that different consumer species feeding on the same food resource interact less when habitat complexity is high or that complexity influences the overall performance of an assemblage because species can operate in their 'optimal' dimensional environment. However, to our knowledge, there are very few studies testing this hypothesis (Flores et al., 2016), but there is strong evidence that species are adapted (foraging and feeding) to the dimensionality of their environment (Pawar et al., 2012).

Most of the world's fresh waters are affected by global change and freshwater organisms are faced with a range of environmental changes such as pollution (including acidification), anoxia or light penetration (Ormerod et al., 2010). Leaf decomposition is clearly an ecosystem process that is heavily influenced by environmental change (see Chaps. 12–18) and laboratory experiments offer a way to target these issues (Canhoto et al., 2017; Gonçalves et al., 2019; Medeiros et al., 2009). To complicate matters, time scales are extremely important when it comes to estimating the effects of stressors because most species can endure non-optimal conditions for short periods, but not over longer (reproductive) time scales.

20.3.6 It Gets Complicated: A More Realistic Picture of What Drives Leaf Decomposition

We have expanded on only a few drivers of leaf decomposition here and, of course, are faced with the fact that they all play a role in this ecosystem process. For the purpose of a laboratory experiment, the questions must be very precise because it is rather obvious that it is extremely difficult to take all drivers into account. Drivers such as species identity, biomass or temperature can be confounded and influence each other. Moreover, leaf decomposition in nature is a very dynamic process where one driver might be important at one point in time but have negligible effects at some other. For example, leaf consumers will reproduce over time and generation time depends on body size and temperature. Further, reproduction will depend on the nutritional quality and the quantity of leaf material and the presence of other food resources. Clearly, in any case, leaf decomposition is strongly driven by the biomass and metabolic capacity of the consumers and possibly also driven by biodiversity and interactions between the consumers. Still, laboratory experiments can distil some of the mechanisms that operate in nature and we can ask meaningful questions that can be addressed with appropriate statistical analysis as described in the following section.

20.4 Statistical Approaches: Maximising Statistical Power While Reducing Logistics

In the following sections, some general approaches will be described that help to maximise statistical power and to minimise logistics when drawing up an experimental design for a laboratory study with leaf litter and decomposers. The research question will determine the experimental design for every experiment; hence we give examples for possible questions while explaining statistical approaches.

20.4.1 Analysis of Variance

In laboratory experiments as described here, the objective is to test which predictors describe a certain response best. In other words, we want to find out which factors contribute to leaf mass loss. Let us imagine a more complex experiment in which seven species are used (Fig. 20.3). The aim is to find out if species richness (e.g., 1, 2 and 3 species feeding on leaves together) or species identity (e.g., *Asellus*, *Gammarus*, etc.; see Fig. 20.3a) are responsible for leaf decomposition. In the example just given, there are 63 possible combinations or 'treatments' if all seven species are run as mono, di- and tri-cultures and, of course, they will have to be run in replication (e.g., let us assume 2 replicates, so 126 microcosms). These treatments represent 63 different 'assemblage identities'. Assemblage identity can be fitted as a predictor of leaf decomposition but really we are interested in whether species richness has an effect on the response or if species have particular effects. Assemblage identity and species richness represent mathematical 'models' that can be fitted in a statistical test. The overall aim is to compare means (of the replicates) for each level within a predictor (e.g., the predictor 'species richness' or 'assemblage identity') and to find out which predictor (or combination of predictors) describes the response best in terms of low variance around those means. For these aims and assumptions, a *t*-test would be the analysis of choice if we only compare two means but for more complex questions, as in our example, analysis of variance (ANOVA) can be an appropriate and popular tool when it comes to data analysis.

20.4.2 Running Designs That Are Not Fully Factorial—Statistical Power and Logistics

Ideally laboratory experiments are run in a fully factorial design because these deliver optimal statistical power (if combined with high replication). Clearly this is not always possible, especially when larger organisms are used that are laborious to sample in the field, and when laboratory space is limited. In this case, it is important to not randomly run certain treatments but to instead make sure that statistical power

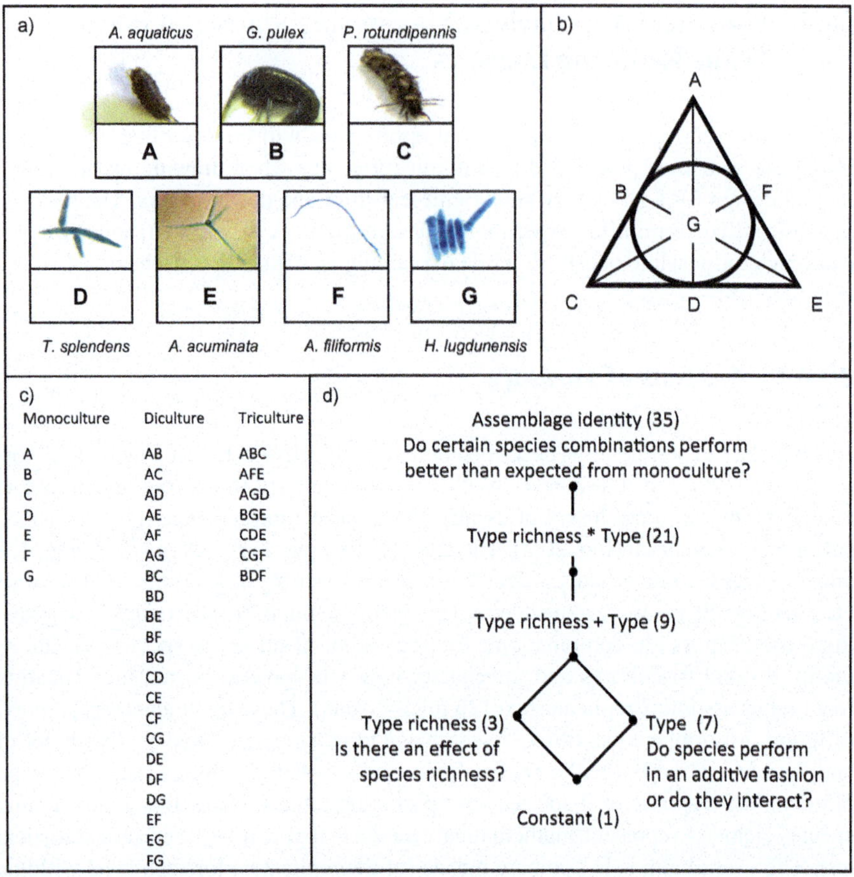

Fig. 20.3 Example for a statistical design of a leaf decomposition experiment (as used in Reiss et al., 2010). **a** The 7 species used in the experiment are called 'types' and assigned a letter. The aim is to assemble them in mono, di- and tri-cultures to assess both the effects of species identity (i.e., the effect of 'type') and species richness. **b** To reduce the number of microcosms, not all possible tri-cultures are run. Instead of 35 possible combinations, 7 are selected using the visualisation shown here. All types along a *straight line*, and along the circle are selected as tri-cultures and this means each pair of species occurs together exactly once. **c** The latter results in a non-fully factorial design with 7 tri-cultures instead of 35. **d** The questions will determine the statistical models that are fitted in the analysis of variance and the models can be visualised with a Hasse diagram. In these diagrams, there is one dot for each model and it is useful to show the dimension of each model—the number in the bracket, as well as its name. The diagram also contains edges linking some dots. The convention is that if model M1 contains model M2 then the dot for M1 is higher than the dot for M2 and there is a chain of generally downward edges linking the dot for M1 to the dot for M2. For example, 'Type richness' (3 parameters) and 'Type' (7 parameters) are not related models but they are nested in 'Assemblage identity' (35 parameters)

is maintained. For example, let us consider, again, the example of seven species (Fig. 20.3) in a leaf decomposition experiment with a biodiversity focus (see Reiss et al. [2010] for example). There are 35 possible tri-cultures from 7 species, making the set-up rather large with 63 treatments (mono, di- and tri-cultures). We want to keep the experiment manageable in size while obtaining clear information about the effect of each species on each other species. Rather than using a random collection of tri-cultures, statistical power is higher if each pair of species occurs together exactly once (Fig. 20.3) and this gives 7 tri-cultures instead of 35 (Fig. 20.3 and see Reiss et al., 2010). Those can be randomly assigned to replicated microcosms, or—even better—be run in blocks (see Sect. 4.5).

20.4.3 'Visualising ANOVA'—Hasse Diagrams

ANOVA cannot only be used when a predictor has more than two levels (e.g., the predictor 'species richness' has levels 1, 2 and 3) but more than one predictor (or 'model') can be fitted in a family of models (Bailey, 2008). For instance, we can fit both species richness and assemblage identity as predictors in the same analysis (Fig. 20.3). It should be noted that many ecologists will call a collection of models 'ANOVA model' and a single model 'independent variable' and we would like to point out that we are using terminology popular among statisticians (i.e., the term 'model' and 'predictor' instead of 'factor' or 'independent variable'). By fitting ANOVA, we can take into account that some predictors are related (they can be sub-sets of each other—they are 'nested', as in our example of assemblage identity and species richness) and that sometimes the response (leaf decomposition) is best explained by more than one model or even by an 'interaction' of models. The latter would tell us that a certain combination of predictors must be present to drive leaf mass loss best.

Bailey (2008, 2020) recommends showing the family of considered models in a Hasse diagram (see Fig. 20.3 and Bailey, 2020; Bailey & Reiss, 2014; Reiss et al., 2010 for examples). In these diagrams, there is one dot for each model and it is useful to show the dimension of each model as well as its name. The diagram also contains edges linking some dots. The convention is that if model M1 contains model M2 then the dot for M1 is higher than the dot for M2 and there is a chain of generally downward edges linking the dot for M1 to the dot for M2 (see example in Fig. 20.3d).

20.4.4 Fitting Statistical Models in ANOVA that Can Disentangle Additive Versus Facilitation or Antagonistic Effects

A typical question behind B-EF research is: do species-rich communities drive ecosystems better than species-poor ones? To test this in laboratory experiments, we must be able to find out what a species does in isolation, what it does in combination with another species and if any interactions between the species are antagonistic, additive or if the species even facilitate each other. This can be done by using a fully factorial experimental design (but see Sect. 4.2.) and the appropriate ANOVA models. One such model dubbed 'Type' (Bailey & Reiss, 2014; Reiss et al., 2010, 2011) is so called because it focusses on the effects of different 'types' rather than species per se. We could have called it 'species identity' but in some instances we want to use individuals of the same species that differ in terms of their traits (e.g., body mass). For example, within one species we could distinguish small and large individuals—two types. This 'type' model assumes that each type has a unique effect, which provokes a characteristic response irrespective of whether the type is combined with other types or not. For example, Reiss et al. (2011) used small and large individuals of the water hog louse *Asellus* in leaf decomposition experiments (in combination with other shredders in mono, di and tri-cultures). In this experiment, metabolic capacity was not accounted for, individuals were simply added in equal numbers when they were in polyculture (e.g., halved numbers in di-culture). The 'type' model that was fitted in the ANOVA therefore had a simple rationale (it was assumed that the response simply depends on additive effects of types) that can be illustrated with the following example. If 12 small *Asellus* feed on 0.6 g of leaf material over 28 days and 12 large *Asellus* feed on 0.7 g under the same conditions, then a di-culture of the two 'types' should feed on 0.65 g ([0.7 + 0.6]/2) if the abundances of each type are half in di-culture (that contains 6 small *Asellus* and 6 large *Asellus*). The response here is leaf mass loss but, of course, it can be replaced with other responses measured (such as algal grazing, amount of faeces produced etc.; see Perkins et al., 2015). Hence, the ANOVA essentially tests if the di-culture does indeed feed on 0.65 g.

In terms of the statistical model fitted, the response on monoculture A (the small *Asellus* monoculture) should be α_1; the response on monoculture B (the large *Asellus* monoculture) should be α_2; the response on di-culture AB should be $(\alpha_1 + \alpha_2)/2$. If we imagine a third type, such as small *Gammarus* (monoculture 'C'), and a tri-culture with 4 individuals in it of each type, the response on tri-culture ABC should be $(\alpha_1 + \alpha_2 + \alpha_3)/3$, and so forth. If this model, considered in the hierarchy of other models, explains the response best (and has the best AIC if calculated), we might conclude that there are no biodiversity effects (an no species interactions) on the particular response, under the lab conditions. However, a word of caution here: despite being counter-intuitive at first glance, although this model concludes that there are no species richness effects, this model can be significant in cases where biodiversity is indeed important for combined processes. This will be the case when multifunctionality (i.e., more than one process) is measured (Perkins et al., 2015).

20.4.5 Replication, Blocks, Randomisation and Pseudoreplication

When the experimental set up is so large that not all microcosms can fit into the same location at the same time (in a randomised arrangement) then high statistical power is maintained if sub-sets of the experiment are run in blocks where, generally speaking, it is the replicates that are run in blocks (e.g., one replicate per treatment in each block). These blocks can be rooms, shelves, or, most often, time. Block effects can then be accounted for in the ANOVA because 'block' can be fitted as a model in the analysis (see Reiss et al., 2010, 2011 for examples). If the experiment is not fully factorial, as for the tri-cultures in the example in Fig. 20.3, then statistical power is highest when (i) each species is present in the same number of tri-cultures and (ii) each pair of species is present in the same number of tri-cultures (Bailey, 2008, see Fig. 20.3). Simply using a random collection of tri-cultures will rarely achieve the same power.

Pseudoreplication (also called false replication) can occur when replicates are confounded with another variable that might have an influence on the response (Bailey & Greenwood, 2018; Hurlbert, 1984; Johnstone, 2013), such as a temperature cabinet or a location (e.g., a section of a river or a room). For example, Perkins et al. (2015) tested how leaf mass loss changed across three temperatures but temperature was not replicated as such because each temperature was confined to a particular temperature cabinet. In order to replicate temperature, more than one temperature cabinet should be run at the same time (e.g., two cabinets with the same temperature)—or a carefully planned block design should be used. For example, if there are three temperature cabinets that are set to three different temperatures, then it is possible to run the experiment with only some replicates at those three temperatures and to then repeat it at a later point, this time running the remaining replicates and switching cabinets (e.g., the cabinet that was used for 10 °C is used at 15 °C and so forth). In the analysis, time can be fitted as a block (block 1 is the first run, block 2 is the second run and so forth). Cabinets are fitted as sub-blocks with random effects. To test whether the predictor 'temperature' is needed in the model, the differences between temperatures are compared with the differences between cabinets within blocks.

20.5 Examples of Laboratory Experiments on Aquatic Leaf Litter Decomposition

In this chapter, we aim to convey that laboratory experiments on leaf decomposition are a popular tool for freshwater ecologists because they are logistically feasible, affordable ways to approach (rather theoretical) questions in ecology. In the following we will illustrate our earlier thoughts about drivers of leaf decomposition, experimental design and analysis by giving three examples for laboratory studies that measured leaf decomposition and which highlight the opportunities these approaches offer to ecologists.

20.5.1 Flores et al. (2016)—Effects of Biodiversity, Species Identity and Habitat Complexity on Leaf Decomposition

One such experiment is by Flores and colleagues (Flores et al., 2016) and it had two main questions: does habitat complexity have an effect on leaf decomposition and can a species-rich assemblage of macro-shredders drive leaf decomposition better than a single species? The authors knew that habitat complexity can provide refuge for prey and hence lower predation rates and they hypothesised that other species interactions and therefore ecosystem processes might be connected to habitat structure. They chose an experimental approach and manipulated habitat complexity using different configurations of plastic plants that they added to aquatic microcosms containing decomposing alder leaves. Three different detritivores species (*Asellus*, *Gammarus* and *Cyclops*) were subsequently introduced to these microcosms in a fashion that 3 levels of species richness were created (monocultures and all possible di- and tricultures). A focus here was to measure habitat complexity as fractal dimension but creating 4 levels of fractal dimension (including a treatment without plastic plants) meant that other predictors were confounded with habitat complexity and had to be accounted for. Rather than fitting 'habitat complexity' as a predictor, a more precise approach was taken by distinguishing: structure present *vs.* structure absent, amount (i.e., mass) of structures and fractal dimension of structures.

Another predictor was 'species richness' and to complicate matters, three response variables were measured: leaf mass loss, production of FPOM and pH in the microcosm. A seemingly simple experiment hence included complex assumptions about the traits of the species used and a rather intricate experimental design, that was not fully factorial, yet logistically feasible for the ecologists and optimised in terms of the analysis (admittedly with the help of a statistician, R. A. Bailey).

In previous experiments, they had found that two of these species, *Asellus* and *Gammarus*, did not interact when they were in combination with each other and hence a third crustacean, *Cyclops* was used. *Cyclops* is much smaller than the other two species and should feed in a different fashion (on leaf biofilm and faeces) and they also perceive structure differently. Importantly, in these experiments the number of crustaceans used was determined by calculating the metabolic capacity for mini-assemblages of these three species. Because small animals ingest more food in relation to their own body mass compared to larger organisms (see Sect. 3.1), adjusting for the metabolic capacity resulted in a smaller proportion of *Cyclops* versus *Asellus* individuals than adjusting for the biomass. In this case, this meant that in monocultures, 218 individuals were added to the '*Cyclops* only' treatments to give the same metabolic power as 15 individuals of *Gammarus* or 12 *Asellus* individuals in these respective monocultures. Half of these numbers were used when species were combined in di-cultures (e.g., 109 *Cyclops* individuals and 6 *Asellus* were present in the *Cyclops*/*Asellus* di-culture) and tri-cultures contained 1/3 of the monoculture individuals. This meant that all microcosms (potentially) had the same metabolic power.

With this set-up, it was possible to address two questions: (1) do processes associated with leaf decomposition increase with increasing complexity? and (2) is habitat complexity more important for processes than species interactions or identity per se? The rationale here was that more complex environments generally enhance foraging and feeding, and that polycultures would probably not perform in an unexpected way. Overall, the purpose was to show that habitat complexity influences plant litter decomposition and indeed the experiment provided some proof for this—two out of the three processes were linked to complexity. Microcosms with artificial plants in them had more FPOM and lower pH compared to microcosms without these artificial structures. The authors hypothesised that this could be caused by higher digestion and respiration when structures were present. Only taking the microcosms with artificial plants into account it became obvious that the amount of structure (i.e., amount of plastic plant added) was a stronger predictor of the response variables than the fractal dimension of the structures.

The experimental design and analysis of this experiment is potentially a good template for experiments that are designed to elucidate how stressors affect leaf decomposition. Although 'complexity' was not a stressor, it is an abiotic factor that can be replaced by another one (such as 'light' or 'micropollutant') in terms of experimental design.

20.5.2 Reiss et al. (2010)—Effects of Biodiversity on Leaf Decomposition

Reiss and colleagues (2010) designed and ran an experiment to address effects of biodiversity of fungi and shredders on leaf decomposition. The rationale here was that a large body of research has revealed (often) positive B-EF relationships in manipulative experiments. The vast majority of such studies have focused on either micro- or macro-organisms, but this was the first study to manipulate the diversity of both simultaneously under controlled laboratory conditions. Reiss et al. (2010) performed a microcosm experiment in which they manipulated species richness of aquatic fungi and invertebrates, two taxonomically distant sets of consumers that contribute to the same key ecosystem process in freshwaters, the decomposition of terrestrial leaf-litter. They used a statistical design to maximize parsimony and analytical power in an experiment with three levels of species richness (7 monoculture, 21 di-culture, and 7 tri-culture treatments). Litter decomposition was measured as both mass loss and the production of fine particulate organic matter (FPOM). They tested whether species richness affected these two processes or whether polycultures performed as predicted from their component monocultures. Further they calculated assemblage metabolism in each microcosm to test whether the processes were driven by the metabolic demands of fungi and invertebrates.

In general, across the 35 treatments, most species performed in an additive fashion and there was no effect of species richness on either process. There was evidence of assemblage identity effects (i.e., certain species combinations not performing as expected), with instances of significant differences for species combinations that contained both caddisfly larvae and fungi. These assemblages performed worse than expected, which might have been due to dual vertical and horizontal interactions, with the possibility that although both consumed litter directly the former may also have grazed on the latter. Apart from these particular species combinations, overall performance of a species in polyculture was effectively the same as in monoculture and reflected its metabolic demands. This suggests that even taxonomically distant consumers might exhibit a degree of functional redundancy for certain processes provided the remaining species can attain sufficient population biomass (and hence metabolic capacity) to compensate for the loss of other species, although whether such compensatory mechanisms operate in the field remains unknown. Further, species contribute to a multitude of ecosystem processes and progressively more species are needed to sustain the sum of them (Gamfeldt & Roger, 2017). This experiment highlighted how important it is to take metabolic demands into account and the dominance of additive effects of leaf decomposers (demonstrated by the good fit of the 'type' statistical model that was fitted in the ANOVA, see Fig. 20.4).

20.5.3 Perkins et al. (2015)—Species Contribute to More Than One Ecosystem Process (Multifunctionality)

The laboratory experiments by Perkins et al. (2015) were an extension of the B-EF focus by Reiss et al. (2010). Reiss et al. (2010) had found evidence for redundancy among leaf consumers, i.e., if a certain leaf decomposer was absent it could be compensated for by another. Species redundancy has been suggested as a widespread insurance mechanism against the effects of biodiversity loss on ecosystem functioning in the face of environmental change. Redundancy may be compromised when multiple ecosystem processes (termed multifunctionality) and environmental contexts are considered, yet very few studies have quantified this explicitly to date. Perkins et al. (2015) measured five key processes and their combined multifunctionality at three temperatures (5, 10 and 15 °C) in freshwater microcosms containing different animal assemblages (1–4 benthic macro-invertebrate species). For single processes, biodiversity effects were weak and were best predicted by additive-based models i.e., polyculture performances represented the sum of their monoculture

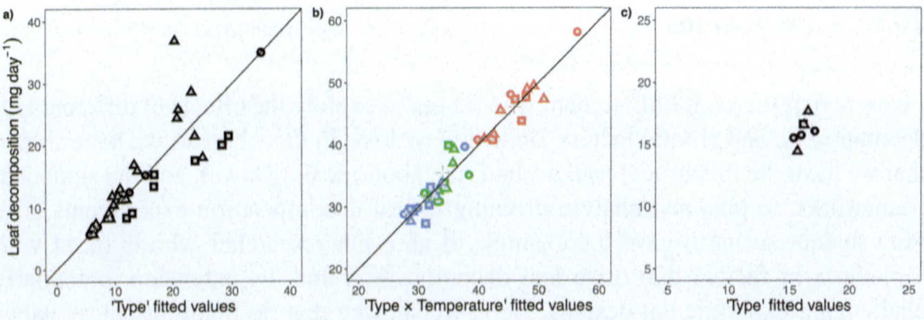

Fig. 20.4 The performance of statistical model 'Type' in leaf decomposition experiments. This model assumes that each type has a unique effect, which provokes a characteristic response irrespective of whether the type is combined with other types or not. **a** The type model describes the response extremely well in an experiment by Reiss et al. (2010). **b** Perkins et al. (2015) ran a similar experiment at 3 different temperatures and fitting all combinations of type and temperature describes the data best. **c** Data from Flores et al. (2016) showing that, when metabolic capacity is accounted for, the type model does not explain the data well because species effects that are based on body mass have been removed a priori. Each data point is a unique assemblage identity (averaged across replicates) where circles are mono-cultures, triangles are di-cultures, squares are tri-cultures and crosses are tetra-cultures. The fitted solid lines are 1:1 lines and are displayed in instances where the type model was significant in the original analyses

parts (see Figs. 20.2 and 20.4) and this echoed the results of Reiss et al. (2010) (Fig. 20.4). Indeed, if individuals are added to microcosms without adjusting for metabolic capacity, the 'type model' (see Sect. 3.4) will describe leaf decomposition best (Fig. 20.4) highlighting how important it is to fit this statistical model in the ANOVA and the additive effects of species. Yet, it makes little sense to fit the model if metabolic capacity is taken into account because effects that are due to species having different body sizes is removed (Fig. 20.4; Flores et al., 2016).

Perkins et al. (2015) did not account for metabolic capacity, however, and concluded that the performance of species was additive when they considered one process alone (e.g., leaf decomposition, Fig. 20.4). There were, however, significant effects of biodiversity on multifunctionality (all 5 processes are evaluated combined) at the low and high (but not intermediate) temperatures. Variation in the contribution of species to processes across temperatures meant that multifunctionality was promoted by multiple species: greater biodiversity was required to sustain multifunctionality across different temperatures than was the case for single processes or temperatures. This suggests that previous studies may have overestimated the scope for redundancy to buffer the consequences of biodiversity loss in a changing environment. Certainly, this experiment shows that temperature affects leaf decomposers in predictable ways (metabolic scaling laws) and illustrates how vital it is to record temperature in leaf decomposition studies.

20.6 Conclusions

Let us revisit the keen PhD student who set out to explore the effects of different leaf decomposers, and abiotic factors, on leaf mass loss. In this chapter, we have shown that we have the theoretical background (metabolic scaling laws), and the statistical frameworks, to plan and analyse meaningful leaf decomposition experiments. The PhD student scenario gave the example of an eager researcher who is faced with a plethora of factors that drive leaf decomposition and, by extension, potentially challenging experimental designs. Here, we showed that, in many instances, laboratory experiments can address precise questions when the metabolic demands of the decomposers are considered. For example, in our first case study, Flores and colleagues (2016) used metabolic theory to calculate metabolic capacity of the mini-assemblages they added to their aquatic microcosms (see Flores et al., 2016 and Sect. 5.1), which gave them the tools to disentangle species identity effects from body mass effects. Further, we showed that it is possible to streamline laboratory set-ups by making detailed plans about the statistical analysis. The overall aim here is to establish that the analysis can address the questions while minimising logistics.

For example, Flores et al. (2016) did not have to run a fully factorial design but the set-up still retained high statistical power. Hence it was possible for the experimenters to run fewer microcosms, and this allowed them to expand on questions and aims, as well as to include the effects of an abiotic driver (habitat complexity). In summary, in this chapter, we show that the experimental design and analysis will depend on which drivers of leaf decomposition are considered, the scientific questions and on whether the set-up in the laboratory can yield robust data. Because organisms can operate only within the restraints of natural laws, we can use theoretical frameworks to refine how we plan laboratory studies on leaf mass loss.

References

Abelho, M. (2001). From litterfall to breakdown in streams: A review. *The Scientific World Journal, 1*, 656–680. https://doi.org/10.1007/978-94-011-0729-7.

Andrade, R., Pascoal, C., & Cássio, F. (2016). Effects of inter and intraspecific diversity and genetic divergence of aquatic fungal communities on leaf litter decomposition—A microcosm experiment. *FEMS Microbiology Ecology, 92*(7).

Bailey, R. A. (2008). *Design of comparative experiments* (Vol. 25). Cambridge University Press.

Bailey, R. A. (2020). Hasse diagrams as a visual aid for linear models and analysis of variance. *Communications in Statistics—Theory and Methods*. https://doi.org/10.1080/03610926.2019.1676443.

Bailey, R. A., & Greenwood, J. J. D. (2018). Effects of neonicotinoids on bees: An invalid experiment. *Ecotoxicology, 27*(1), 1–7.

Bailey, R. A., & Reiss, J. (2014). Design and analysis of experiments testing for biodiversity effects in ecology. *Journal of Statistical Planning and Inference, 144*(1), 69–80. https://doi.org/10.1016/j.jspi.2012.09.008.

Bärlocher, F. (2005). Leaching. In M. A. Graça, F. Bärlocher, & M. O. Gessner (Eds.), *Methods to study litter decomposition: A practical guide* (1st ed., pp. 33–36). Springer.

Bärlocher, F., Gessner, M. O., & Graça, M. A. S. (2020). *Methods to study litter decomposition: A practical guide* (2nd ed.). Retrieved from https://www.springer.com/gp/book/9783030305147.

Bell, T., Newman, J. A., Silverman, B. W., Turner, S. L., & Lilley, A. K. (2005). The contribution of species richness and composition to bacterial services. *Nature, 36*(7054), 1157–1160. https://doi.org/10.1038/nature03891.

Benton, T. G., Solan, M., Travis, J. M. J., & Sait, S. M. (2007). Microcosm experiments can inform global ecological problems. *Trends in Ecology & Evolution, 22*(10), 516–521. https://doi.org/10.1016/j.tree.2007.08.003.

Brown, J. H., Gillooly, J. F., Allen, A. P., Savage, V. M., & West, G. B. (2004). Toward a metabolic theory of ecology. *Ecology, 85*(7), 1771–1789. https://doi.org/10.1890/03-9000.

Canhoto, C., & Graça, M. A. S. (1996). Decomposition of *Eucalyptus globulus* leaves and three native leaf species (*Alnus glutinosa,* Castanea sativa and *Quercus faginea*) in a Portuguese low order stream. *Hydrobiologia, 333*(2), 79–85. https://doi.org/10.1007/BF00017570.

Canhoto, C., & Graça, M. A. S. (2008). Interactions between fungi and stream invertebrates: Back to the future. *Novel Techniques and Ideas in Mycology: Fungal Diversity Research Series, 20,* 205–325.

Canhoto, C., Simões, S., Gonçalves, A. L., Guilhermino, L., & Bärlocher, F. (2017). Stream salinization and fungal-mediated leaf decomposition: a microcosm study. *Science of the Total Environment, 599,* 1638–1645.

Chambord, S., Tackx, M., Chauvet, E., Escolar, G., & Colas, F. (2017). Two microcrustaceans affect microbial and macroinvertebrate-driven litter breakdown. *Freshwater Biology, 62*(3), 530–543.

Duarte, S., Pascoal, C., Cássio, F., & Bärlocher, F. (2006). Aquatic hyphomycete diversity and identity affect leaf litter decomposition in microcosms. *Oecologia, 147*(4), 658–666. https://doi.org/10.1007/s00442-005-0300-4.

Farrell, K. J., Rosemond, A. D., Kominoski, J. S., Bonjour, S. M., Rüegg, J., Koenig, L. E., Baker, C. L., Trentman, M. T., Harms, T. K., & McDowell, W. H. (2018). Variation in detrital resource stoichiometry signals differential carbon to nutrient limitation for stream consumers across biomes. *Ecosystems, 21*(8), 1676–1691. https://doi.org/10.1007/s10021-018-0247-z.

Flores, L., Bailey, R. A., Elosegi, A., Larrañaga, A., & Reiss, J. (2016). Habitat complexity in aquatic microcosms affects processes driven by detritivores. *PLoS ONE, 11*(11), 1–15. https://doi.org/10.1371/journal.pone.0165065.

Foucreau, N., Puijalon, S., Hervant, F., & Piscart, C. (2013). Effect of leaf litter characteristics on leaf conditioning and on consumption by *Gammarus pulex*. *Freshwater Biology, 58*(8), 1672–1681.

France, R., Culbert, H., Freeborough, C., & Peters, R. (1997). Leaching and early mass loss of boreal leaves and wood in oligotrophic water. *Hydrobiologia, 345*(2–3), 209–214.

Gamfeldt, L., Hillebrand, H., & Jonsson, P. R. (2008). Multiple functions increase the importance of biodiversity for overall ecosystem functioning. *Ecology, 89*(5), 1223–1231. https://doi.org/10.1890/06-2091.1.

Gamfeldt, L., & Roger, F. (2017). Revisiting the biodiversity-ecosystem multifunctionality relationship. *Nature Ecology and Evolution, 1*(7), 0168. https://doi.org/10.1038/s41559-017-0168.

Gamfeldt, L., Wallén, J., Jonsson, P. R., Berntsson, K. M., & Havenhand, J. N. (2005). Increasing intraspecific diversity enhances settling success in a marine invertebrate. *Ecology, 86*(12), 3219–3224. https://doi.org/10.1890/05-0377.

Gessner, M. O., Swan, C. M., Dang, C. K., McKie, B. G., Bardgett, R. D., Wall, D. H., et al. (2010). Diversity meets decomposition. *Trends in Ecology & Evolution, 25*(6), 372–380. https://doi.org/10.1016/j.tree.2010.01.010.

Gillooly, J. F., Brown, J. H., West, G. B., Savage, V. M., & Charnov, E. L. (2001). Effects of size and temperature on metabolic rate. *Science, 293*(5538), 2248–2251. https://doi.org/10.1126/science.1061967.

Gonçalves, A. L., Simões, S., Bärlocher, F., & Canhoto, C. (2019). Leaf litter microbial decomposition in salinized streams under intermittency. *Science of the Total Environment, 653,* 1204–1212. https://doi.org/10.1016/j.scitotenv.2018.11.050.

González, J. M., & Graça, M. A. S. (2003). Conversion of leaf litter to secondary production by a shredding caddis-fly. *Freshwater Biology, 48*(9), 1578–1592. https://doi.org/10.1046/j.1365-2427.2003.01110.x.
Hauer, F. R., & Lamberti, G. A. (2007). *Methods in stream ecology* (2nd ed.). https://doi.org/10.1016/b978-0-12-332908-0.x5001-3.
Hieber, M., & Gessner, M. O. (2002). Contribution of stream detrivores, fungi, and bacteria to leaf breakdown based on biomass estimates. *Ecology, 83*(4), 1026–1038. https://doi.org/10.1890/0012-9658(2002)083%5b1026:cosdfa%5d2.0.co;2.
Hurlbert, S. H. (1984). Pseudoreplication and the design of ecological field experiments. *Ecological Monographs, 54*(2), 187–211. https://doi.org/10.2307/1942661.
Jabiol, J., Lecerf, A., Lamothe, S., Gessner, M. O., & Chauvet, E. (2019). Litter quality modulates effects of dissolved nitrogen on leaf decomposition by stream microbial communities. *Microbial Ecology, 77*(4), 959–966. https://doi.org/10.1007/s00248-019-01353-3.
Johnstone, P. (2013). *Planning and managing agricultural and ecological experiments*. Routledge.
Jonsson, M., & Malmqvist, B. (2000). Ecosystem process rate increases with animal species richness: Evidence from leaf-eating, aquatic insects. *Oikos, 89*(3), 519–523. https://doi.org/10.1034/j.1600-0706.2000.890311.x.
Larrañaga, A., Basaguren, A., & Pozo, J. (2014). Resource quality controls detritivore consumption, growth, survival and body condition recovery of reproducing females. *Marine & Freshwater Research, 65*(7), 910–917. https://doi.org/10.1071/MF13165.
Marks, J. C. (2019). Revisiting the fates of dead leaves that fall into streams. *Annual Review of Ecology Evolution and Systematics, 50,* 547–568. https://doi.org/10.1146/annurev-ecolsys-110218-024755.
Martínez, A., Larrañaga, A., Pérez, J., Basaguren, A., & Pozo, J. (2013). Leaf-litter quality effects on stream ecosystem functioning: A comparison among five species. *Fundamental and Applied Limnology, 183*(3), 239–248. https://doi.org/10.1127/1863-9135/2013/0514.
Martínez, A., Larrañaga, A., Pérez, J., Descals, E., & Pozo, J. (2014). Temperature affects leaf litter decomposition in low-order forest streams: field and microcosm approaches. *FEMS Microbiology Ecology, 87*(1), 257–267.
McKie, B. G., Woodward, G., Hladyz, S., Nistorescu, M., Preda, E., Popescu, C., Giller, P. S., & Malmqvist, B. (2008). Ecosystem functioning in stream assemblages from different regions: Contrasting responses to variation in detritivore richness, evenness and density. *Journal of Animal Ecology, 77*(3), 495–504. https://doi.org/10.1111/j.1365-2656.2008.01357.x.
Medeiros, A. O., Pascoal, C., & Graça, M. A. S. (2009). Diversity and activity of aquatic fungi under low oxygen conditions. *Freshwater Biology, 54*(1), 142–149.
Mille-Lindblom, C., Fischer, H., & Tranvik, J. L. (2006). Antagonism between bacteria and fungi: Substrate competition and a possible tradeoff between fungal growth and tolerance towards bacteria. *Oikos, 113*(2), 233–242.
Ormerod, S. J., Dobson, M., Hildrew, A. G., & Townsend, C. (2010). Multiple stressors in freshwater ecosystems. *Freshwater Biology, 55,* 1–4.
Pascoal, C., & Cássio, F. (2004). Contribution of fungi and bacteria to leaf litter decomposition in a polluted river. *Applied and Environmental Microbiology, 70*(9), 5266–5273. https://doi.org/10.1128/AEM.70.9.5266-5273.2004.
Pascoal, C., Cássio, F., Nikolcheva, L., & Bärlocher, F. (2010). Realized fungal diversity increases functional stability of leaf litter decomposition under zinc stress. *Microbial Ecology, 59*(1), 84–93. https://doi.org/10.1007/s00248-009-9567-z.
Pawar, S., Dell, A. I., & Savage, V. M. (2012). Dimensionality of consumer search space drives trophic interaction strengths. *Nature, 486*(7404), 485–489. https://doi.org/10.1038/nature11131.
Perkins, D. M., Bailey, R. A., Dossena, M., Gamfeldt, L., Reiss, J., Trimmer, M., et al. (2015). Higher biodiversity is required to sustain multiple ecosystem processes across temperature regimes. *Global Change Biology, 21*(1), 396–406. https://doi.org/10.1111/gcb.12688.

Perkins, D. M., McKie, B. G., Malmqvist, B., Gilmour, S. G., Reiss, J., & Woodward, G. (2010). Environmental warming and biodiversity-ecosystem functioning in freshwater microcosms. Partitioning the effects of species identity, richness and metabolism. *Advances in Ecological Research, 43*, 177–209. https://doi.org/10.1016/B978-0-12-385005-8.00005-8.

Perkins, D. M., Yvon-Durocher, G., Demars, B. O., Reiss, J., Pichler, D. E., Friberg, N., Trimmer, M., & Woodward, G. (2012). Consistent temperature dependence of respiration across ecosystems contrasting in thermal history. *Global Change Biology, 18*(4), 1300–1311. https://doi.org/10.1111/j.1365-2486.2011.02597.x.

Peters, R. H. (1983). *The ecological implications of body size*. Cambridge University Press.

Quinn, J. M., Burrell, G. P., & Parkyn, S. M. (2000). Influences of leaf toughness and nitrogen content on in-stream processing and nutrient uptake by litter in a Waikato, New Zealand, pasture stream and streamside channels. *New Zealand Journal of Marine and Freshwater Research, 34*(2), 253–271. https://doi.org/10.1080/00288330.2000.9516931.

Reiss, J., Bailey, R. A., Cássio, F., Woodward, G., & Pascoal, C. (2010). Assessing the contribution of micro-organisms and macrofauna to biodiversity-ecosystem functioning relationships in freshwater microcosms. *Advances in Ecological Research, 43*(C), 151–176. https://doi.org/10.1016/b978-0-12-385005-8.00004-6.

Reiss, J., Bailey, R. A., Perkins, D. M., Pluchinotta, A., & Woodward, G. (2011). Testing effects of consumer richness, evenness and body size on ecosystem functioning. *Journal of Animal Ecology, 80*(6), 1145–1154. https://doi.org/10.1111/j.1365-2656.2011.01857.x.

Reiss, J., Bridle, J. R., Montoya, J. M., & Woodward, G. (2009). Emerging horizons in biodiversity and ecosystem functioning research. *Trends in Ecology & Evolution, 24*(9), 505–514. https://doi.org/10.1016/j.tree.2009.03.018.

Reiss, J., & Schmid-Araya, J. M. (2010). Life history allometries and production of small fauna. *Ecology, 91*(2), 497–507. https://doi.org/10.1890/08-1248.1.

Ribblett, S. G., Palmer, M. A., & Wayne Coats, D. (2005). The importance of bacterivorous protists in the decomposition of stream leaf litter. *Freshwater Biology, 50*(3), 516–526.

Risse-Buhl, U., Karsubke, M., Schlief, J., Baschien, C., Weitere, M., & Mutz, M. (2012). Aquatic protists modulate the microbial activity associated with mineral surfaces and leaf litter. *Aquatic Microbial Ecology, 66*(2), 133–147.

Risse-Buhl, U., Herrmann, M., Lange, P., Akob, D. M., Pizani, N., Schönborn, W., Totsche, K. U., & Küsel, K. (2013). Phagotrophic protist diversity in the groundwater of a karstified aquifer–morphological and molecular analysis. *Journal of Eukaryotic Microbiology, 60*(5), 467–479.

Risse-Buhl, U., Schlief, J., & Mutz, M. (2015). Phagotrophic protists are a key component of microbial communities processing leaf litter under contrasting oxic conditions. *Freshwater Biology, 60*(11), 2310–2322.

Ruesink, J. L., & Srivastava, D. S. (2001). Numerical and per capita responses to species loss: Mechanisms maintaining ecosystem function in a community of stream insect detritivores. *Oikos, 93*(2), 221–234. https://doi.org/10.1034/j.1600-0706.2001.930206.x.

Swan, C. M., & Palmer, M. A. (2006). Preferential feeding by an aquatic consumer mediates non-additive decomposition of speciose leaf litter. *Oecologia, 149*(1), 107–114.

Tachet, H., Richoux, P., Bournaud, M., & Usseglio-Polatera, P. (2010). *Invertébrés d'eau douce: systématique, biologie, écologie*. CNRS editions.

Treton, C., Chauvet, E., & Charcosset, J.-Y. (2004). Competitive interaction between two aquatic hyphomycete species and increase in leaf litter breakdown. *Microbial Ecology, 48*(3), 439–446.

Whitfield, J. (2004). Ecology's big, hot idea. *PLoS Biology, 2*(12), https://doi.org/10.1371/journal.pbio.0020440.

Woodward, G., Ebenman, B., Emmerson, M., Montoya, J. M., Olesen, J. M., Valido, A., et al. (2005). Body size in ecological networks. *Trends in Ecology & Evolution, 20*(7), 402–409. https://doi.org/10.1016/j.tree.2005.04.005.

Zimmer, M. (2005). Cellulases. In F. Bärlocher, M. O. Gessner, & M. A. S. Graça (Eds.), *Methods to study litter decomposition* (pp. 249–254). Springer.

Zimmer, M. (2006). The role of animal-microbe interactions in isopod ecology and evolution. *Acta Biol Benrodis, 13,* 127–168.

Chapter 21
Plant Litter Decomposition as a Tool for Stream Ecosystem Assessment

André Frainer, Andreas Bruder, Fanny Colas, Verónica Ferreira, and Brendan G. McKie

Abstract The decomposition of plant litter in freshwaters is an integrative process involving multiple organism groups and connecting terrestrial and freshwater ecosystems. The quantification of leaf litter decomposition has been advocated as an effective indicator of ecosystem functional integrity in the bioassessment of freshwaters. Indeed, variation in litter decomposition rates has been used to detect the impacts of a wide range of anthropogenic disturbances on the functioning of detritus-based food webs in freshwater ecosystems, particularly in streams. However, these assessments have almost exclusively been undertaken as part of research projects, and the application of litter decomposition as a tool in routine biomonitoring remains limited. We evaluate the potential for litter decomposition as a tool for ecosystem assessment by environmental agencies and managers, drawing on insights and experiences from three lines of evidence: (i) a broad selection of published research projects, (ii) an

A. Frainer (✉)
Norwegian Institute for Nature Research (NINA), Tromsø, Norway
e-mail: andre.frainer@nina.no

Faculty of Biosciences, Fisheries and Economy, UiT The Arctic University of Norway, Tromsø, Norway

A. Bruder
Laboratory of Applied Microbiology, University of Applied Sciences and Arts of Southern Switzerland, Mendrisio, Switzerland
e-mail: andreas.bruder@supsi.ch

F. Colas
CNRS, ENTPE, UMR 5023 LEHNA, Univ Lyon, Université Claude Bernard Lyon 1, 69622 Villeurbanne, France
e-mail: fanny.colas@univ-lyon1.fr

V. Ferreira
Marine and Environmental Sciences Centre (MARE), Department of Life Sciences, University of Coimbra, Coimbra, Portugal
e-mail: veronica@ci.uc.pt

B. G. McKie
Department of Aquatic Sciences and Assessment, Swedish University of Agricultural Sciences (SLU), Uppsala, Sweden
e-mail: brendan.mckie@slu.se

C. M. Swan et al. (eds.), *The Ecology of Plant Litter Decomposition in Stream Ecosystems*, https://doi.org/10.1007/978-3-030-72854-0_21

existing national-scale monitoring program and (iii) a meta-analysis comparing litter decomposition rates between nutrient-enriched and reference sites. We use this as a basis for discussing *inter alia* common substrates used in decomposition assays, alternatives for field protocols and sampling designs, and the use of different indices and reference conditions when arriving at an assessment of functional status.

21.1 Background

21.1.1 The Promise of Litter Decomposition: Ecosystem Process Rates as a Tool for Stream Bioassessment and Management

Almost two decades ago, Gessner and Chauvet (2002) advanced the suggestion to use leaf litter decomposition to complement structural indicators for stream bioassessment, to address the lack of practical, cost-effective approaches available to management authorities for quantifying the functional aspects of ecosystems (see Sect. 21.1.2). Since then, numerous studies have tested the use of litter decomposition as a measure of freshwater functional integrity. Despite intense research, major challenges remain in the implementation of litter decomposition in bioassessment as a complementary tool to established structural measures of ecosystem integrity. We revisit the methods, experiences, and developments concerning the use of litter decomposition as a tool in stream bioassessment and discuss: when is litter decomposition most suitable for use in bioassessment and what are the remaining roadblocks and knowledge gaps precluding the widespread use of litter decomposition as a tool for management?

21.1.2 From Analyses of Structure to Functional Metrics

Ecologists and practitioners interested in assessing freshwater ecological condition mostly rely on structural measures of biological integrity. In benthic habitats of lotic and lentic systems, such measurements include counts of invertebrate species or families, the relative abundances of some key benthic invertebrate orders, e.g., Ephemeroptera (mayfly), Plecoptera (stonefly), and Trichoptera (caddisfly) (EPT index), or country/region specific indices that weight the relative abundance of sensitive versus tolerant taxa, including the British Biological Monitoring Working Party index (Armitage et al., 1983), the New Zealand Macroinvertebrate Community Index (Collier et al., 2014) and multimetric indices (Mondy et al., 2012). These structural measures capture variation in the composition and distribution of benthic invertebrate communities, and sometimes include information about their functional traits, and thus are widely used to assess the current ecological status of habitats and to track changes in status as management measures are applied (Burdon et al., 2020; Dahl & Johnson, 2004). Inferences of ecosystem level impacts (e.g., on ecosystem processes regulating fluxes of nutrients and energy) from these structural measurements are possible—for example a decline in abundance or diversity of invertebrates consuming algal biofilms might indicate a reduction in the importance of algae as an energy source in that system (McKie & Cranston, 2001). However, such inferences should always be heavily qualified because underlying assumptions simplify the ecological complexity of the system studied. For instance, tolerant species might adapt their functional role in a disturbed environment and partly compensate for the loss of more specialised and sensitive species, so that ecosystem process rates are maintained even as biodiversity declines (Ledger & Hildrew, 2005). Alternatively, a disturbance might have sublethal effects on the activities of functionally important species, so that functioning is affected without a marked loss of biodiversity (McKie & Malmqvist, 2009). Furthermore, structural measurements may be of little use in regions with naturally low abundance of certain groups or incomplete taxonomic information on some species. In all these cases, direct measures of ecological processes, which quantify fluxes in ecosystem functioning and are influenced by interactions of species with their environment (including resources), may serve as a complementary or even the primary approach for assessing ecosystem integrity.

Leaf litter decomposition is a pivotal ecological process, particularly in headwater streams and rivers, and also in ponds and littoral areas of lakes, and has been repeatedly proposed as an efficient bioassessment tool (Gessner & Chauvet, 2002; von Schiller et al., 2017; Young et al., 2008). Chauvet et al. (2016) provided an extended review and discussion of the use of litter decomposition for detecting impacts of different types of disturbances on ecosystem functioning. Nonetheless, leaf litter decomposition is not a one-size-fits-all tool. Boulton (1999) discussed the lack of a 'holy grail' tool for the assessment of ecosystem health (see Karr, 1999), and suggested that a combination of abiotic, structural, and functional measurements, including litter decomposition, should be used depending on the problem being addressed. Elosegi et al. (2017) further argued that ecologists should learn from millennia of development in medicine and use a combination of tools, including litter decomposition when suitable, to assess ecosystem health.

The importance of taking a multi-faceted approach to environmental assessment is explicitly recognized in the European Water Framework Directive (WFD), for example, which describes ecological status as "an expression of the quality of the structure and functioning of aquatic ecosystems associated with surface waters" (The European Parliament, 2015—Directive 2008/94/EC, pp. 423–426, Article 2, paragraph 21). Nevertheless, the biological elements mentioned in the WFD for the assessment of water quality are exclusively structural, i.e. composition and abundance of key organism groups. Environmental agencies continue to base ecological assessments on structural measurements alone (but see our Box 21.1 for an example of national-scale use of litter decomposition for bioassessment), although the desirability of developing appropriate functional metrics, including litter decomposition, as assessment tools is often acknowledged (e.g., USEPA, 2016).

The quantification of litter decomposition is fairly simple and practical, thus making it an efficient tool for assessments of ecosystem functional integrity. Furthermore, litter decomposition has already been tested across spatial (Chauvet et al., 2016; Tonin et al., 2017; Woodward et al., 2012) and temporal (Frainer & McKie, 2015; Frainer et al., 2014; Yeung et al., 2018) scales, and used to evaluate the effects of single and multiple stressors (Bruder et al., 2016; Castela et al., 2008; Colas et al., 2013; Pascoal et al., 2003), mostly in streams and rivers, but increasingly also in ponds and lakes (Quintão et al., 2013; Raposeiro et al., 2016; van Dokkum et al., 2002). Anthropogenic impacts studied include those associated primarily with changes in the abiotic environment, such as hydromorphological changes (Mendoza-Lera et al., 2012; Mollá et al., 2017; Sabater et al., 2018), nutrient loading (Ferreira et al., 2015; Woodward et al., 2012), acidification (Ferreira & Guérold, 2017), mining (Ferreira, Koricheva, Duarte et al., 2016), urbanization (Imberger et al., 2010), and pesticides (Rasmussen et al., 2012). Other studies have focused on effects of biotic changes on organic matter decomposition, such as the use of genetically modified crops (Rosi-Marshall et al., 2007), changes in forest composition driven by forest management (Ferreira, Koricheva, Pozo et al., 2016; Ferreira et al., 2019), species invasions (Alp et al., 2016; Hladyz et al., 2011), and changes in trophic interactions (Jabiol, McKie et al., 2013), including predation (Majdi et al., 2014) and parasitism (Hernández & Sukhdeo, 2008). Litter decomposition has also been used to assess the effectiveness of different types of ecological restoration and environmental mitigation measures (Entrekin et al., 2008; Flores et al., 2011; Frainer et al., 2018; Lepori et al., 2005). Below, we discuss the most prevalent methodological considerations when using litter decomposition as a bioassessment tool, including the use of different mesh sizes, litter types, temporal scales, habitats, and choice of reference conditions. We then revisit some proposed metrics for summarizing useful information from litter decomposition and suggest ways forward by addressing some of the main advantages and caveats of using litter decomposition as a tool for ecosystem assessment. We focus on the use of naturally abscised, dried leaf litter as the substrate in a litter decomposition assay because of its widespread use and the extended knowledge on this process gained over the past decades. However, we also address alternatives to the use of leaf litter to measure decomposition rates (Box 21.2).

Box 21.1: French case study

The National Office for Water and Aquatic Environments (Onema) is a French public institution created to support the implementation of public water policy in France as required by the European Water Framework Directive (WFD). Although the WFD defines the ecological status of surface water based on both *structure* and *functioning* of aquatic ecosystems, indicators developed under the WFD consider only ecosystem structure. In addition, WFD indicators have been mainly developed to focus on impacts of catchment landuse and chemical stressors. Addressing these shortcomings, Onema initiated and funded a research project (IDFun project, CNRS-Onema, 2012–2016) to develop an indicator reflecting responses of ecosystem functioning to hydromorphological changes of streams. Leaf decomposition was selected for this purpose because of its central role in river ecosystem functioning, the considerable scientific background on both the abiotic and biotic mechanisms involved and on the effects of various physical and chemical stressors, and the relative ease and low cost of the method.

A working group composed of scientists and agents from two Onema regional services was created to propose and test a protocol that fulfils the methodological requirements (i.e., being as standardized as possible, easy to use, efficient in time and costs). Alder (*Alnus glutinosa*) litter was used due to its ubiquity along French rivers and its fast decomposition rates, thus reducing the risks associated with extended field incubation time. Coarse-and fine-mesh litter bags were used to estimate total and microbial-mediated decomposition rates, respectively. Litter bags were placed in four areas near the stream bank in each site. These locations were selected because they are natural areas of detritus accumulation. Litter bags were retrieved after 7 and 21 days, and after 21 and 42 days for coarse- and fine-mesh bags, respectively, resulting in at least 50% of mass loss on the final sampling date. Leaves were then rinsed, dried, weighed, ashed, and reweighed to estimate ash-free dry mass (AFDM). Water temperature was recorded every 30 min during the incubation using data loggers and chemical and hydromorphological parameters were assessed using national standards. Software was developed to automatically estimate decomposition rates based on an exponential model fitting AFDM data to degree-days. The project team trained all regional services in the use of the protocol. Onema agents deployed the assay on 85 streams distributed all over France over three years.

The project provided a large database for scientists to propose thresholds on 'good ecosystem functioning' and to model relationships between hydromorphology and leaf decomposition (Colas et al., 2017). The protocol was well received by stakeholders overall, although there was a frequently expressed preference for a more easily standardised substrate than leaf litter, to reduce time for the collection of leaves. Twelve working-days are needed to apply the protocol, which is comparable in terms of efforts to other WFD protocols. Nonetheless, while the protocol is used for assessing restoration projects and by stakeholders who are less constrained by regulatory requirements and frameworks of national biomonitoring networks (e.g., water agency, natural parks, and water basins managers across the country), its implementation as routine bioassessment assay in the national WFD biomonitoring scheme is not planned yet. The main reasons are the need for further standardization of the protocol, the already high cost of the national biomonitoring scheme, therefore reducing possibilities for including new indicators, and the absence of leaf litter as a parameter to be assessed in the WFD. Thus, considering the high costs of biomonitoring, the French government is reluctant to integrate new indicators for ecosystem functioning into the national biomonitoring scheme if not specifically requested by the WFD.

21.2 Choosing the Appropriate Method

21.2.1 Litterbag: A Toolkit in Different Mesh Sizes

Different mesh sizes have been used to compare the contribution of differently sized organism groups to litter decomposition (Fig. 21.1). Large mesh sizes (often ranging between 5 and 10 mm mesh size) allow access to the litter by larger organisms (especially invertebrates that directly feed on the litter and/or litter associated biofilms). Here, the trade-off is often between a mesh size that allows enough invertebrate colonization, while retaining the leaf litter and most fragments. In some systems, larger crabs or crayfish might be important detritivores (Alp et al., 2016; Rincón & Covich, 2014), and thus larger mesh sizes (or even alternative approaches such as tying bundles of litter without enclosure in a mesh bag; Connolly & Pearson, 2013) might be needed. Small mesh sizes, often below 0.5 mm, block access by most detritivores, and thus focus on microbial contribution to decomposition, often resulting in slower decomposition rates than in large mesh. Mesh material varies. Often material such as garden trellis is used for coarse-mesh bags, whereas material with well-controlled mesh sizes (e.g., industrial filtration fabric) is used for fine-mesh bags. The role of mesh size in quantifying litter decomposition is crucial, as specific processes and organism groups contributing to litter decomposition can be studied individually (e.g., microbial processing; Bruder et al., 2014).

The effect of mesh size on decomposition rates can be illustrated by the cross-biome study conducted by Handa et al. (2014). They used three different mesh sizes, 0.25 mm, 1 mm, and 5 mm, in replicate field experiments in streams ranging from the subarctic to the tropics. The different mesh sizes allowed either (i) microorganisms alone, (ii) meiofauna (described as mesofauna by Handa et al., 2014) and microorganisms, or (iii) macrofauna, meiofauna, and microorganisms to access the enclosed leaf litter. They found that meiofauna contributed to 8.7% and macrofauna to 50.1% to carbon loss (a surrogate for leaf mass loss) in a temperate stream (Handa et al., 2014). The contribution of these size-groups to litter decomposition was substantially smaller in a subarctic stream, with 4.0% and 5.2% for the meiofauna and macrofauna, respectively. The differences in macrofauna-mediated decomposition rates between these two streams were interpreted as the consequence of differences in shredder densities in local communities. The temperate stream had a high density of *Gammarus pulex* (on average 273 individuals per m^2; Handa et al., 2014), a very efficient shredder that often reaches high densities in temperate streams (Woodward et al., 2012). In contrast, the subarctic stream supported lower densities of

Fig. 21.1 Illustration of litter-bags of two different mesh sizes, coarse and fine, placed on the bottom of a stream. Litter bags are often attached to a rope or chain that keeps them close to the substrate, where benthic invertebrates can colonize the leaf litter inside the coarse-mesh bags. The use of litter bags as a tool in bioassessment has shown promise to detect effects of anthropogenic stressors (right side of the illustration) relative to reference conditions (left side of the illustration)

stonefly shredders (on average 44 individuals per m^2), which predominately feed by scraping biofilms growing on litter surfaces, and hence are associated with lower leaf processing rates than taxa such as *G. pulex* and many Trichoptera that chew on the litter directly (McKie et al., 2008).

21.2.2 Leaf Litter Quality: From Recalcitrant to Labile and Nutrient Rich Leaves

The choice of the litter material has a decisive influence on decomposition rates, and thus potentially on the capacity of a decomposition assay to detect human impacts. This choice is thus an important methodological consideration, especially when comparing decomposition rates over substantial geographic or environmental gradients along which the dominant riparian vegetation and other environmental parameters (e.g., thermal regimes) may change. Litter quality, defined by the combination of physical and chemical characteristics, affects the colonization rates, biomass, and activity of microbial and invertebrate decomposers (Baldy et al., 1995; Ferreira et al., 2012; Frainer et al., 2015; Handa et al., 2014). Under similar environmental conditions, litter decomposition is faster for high-quality litter (i.e. high concentrations of nutrients, low concentrations of structural and secondary compounds) than for more recalcitrant litter. The most commonly considered parameters describing litter resource quality include %N, %Lignin, C:N, lignin:N, tannins, and litter toughness (Frainer et al., 2015; Lecerf & Chauvet, 2008; Ostrofsky, 1997; Schindler & Gessner, 2009). Other litter characteristics may be relevant for non-trophic effects on litter decomposition, and include those defining habitat structure in litter packs (Sanpera-Calbet et al., 2009) or the use for case-building by some groups of caddisflies (Moretti et al., 2009; Rincón & Martínez, 2006). Litter quality varies not only among litter species (Enriquez et al., 1993; Frainer et al., 2015; Ostrofsky, 1997), but also among litter from conspecific trees growing in different conditions (Graça & Poquet, 2014; Lecerf & Chauvet, 2008; Leroy et al., 2007).

For studies that aim at assessing the consequences of environmental conditions on decomposition rates, the heterogeneity in the litter material used should be minimized by standardizing the litter material across all sites sampled (Ferreira et al., 2019). Some extreme standardization procedures based on semi-natural substrates have also been applied (Box 21.2), which has the benefit of largely eliminating variability in decomposition rates due to uncontrolled (background) variation in litter quality, but at the cost of realism. Most commonly, researchers have used litter material from plant species that are present in the study area, but not necessarily collected locally, even when working across broad geographic scales (Handa et al., 2014; Irons et al., 1994; Woodward et al., 2012). This approach is justified by the very weak evidence for the so-called "home-field advantage" hypothesis, which postulates that local decomposer communities more effectively decompose litter from local species and sources, due to evolutionary adaptation. Empirical evidence for this effect is rare and might be limited to microbial adaptation to very recalcitrant litter (Yeung et al., 2019). Overall, litter quality seems to control the activity of decomposers more than litter source (Bruder et al., 2014; but see Kennedy & El-Sabaawi, 2017), although litter from contaminated sites should be avoided unless subject to specific research questions.

Differences in decomposition rates between litter species of contrasting quality has the additional advantage of yielding more information on the sensitivity of processes contributing to decomposition. For instance, high concentrations of tannins in leaf litter (e.g., in oak compared to alder leaves, Gessner & Chauvet, 1994) may reduce fungal activity due to complexation and inactivation of fungal exoenzymes (Gessner & Chauvet, 1994; McArthur et al., 1994) without affecting invertebrates directly. Differences in nutrient concentrations and specific nutrient ratios among litter species, as described by the ecological stoichiometry theory, may also be relevant for bioassessment, if, e.g., they favour or hinder feeding and growth of particular detritivore species (Abelho & Canhoto, 2020; Frainer et al., 2016; Halvorson et al., 2018).

21.2.3 Timing: Season and Duration

Litter decomposition rates may depend strongly on timing and duration of exposure. Broadly, the first days of litter decomposition in freshwaters are dominated by chemical leaching, followed by colonization by microbes. Microorganisms, fungi in particular, then initiate biological litter decomposition, thereby also increasing litter palatability for invertebrates. Although litter decomposition is often reported as a rate (thus standardized for time), the duration of exposure is relevant for the types of inferences that can be drawn. Decomposition of different litter constituents is not constant over the duration of the process, reflecting effects of litter quality (Grossman et al., 2020). Labile litter fractions (e.g., hemicellulose) are preferentially utilized before refractory fractions (e.g., lignin). The decomposer community also follows a successional pattern, with invertebrates often only gaining importance after microbial conditioning (Bruder et al., 2014; Jabiol, McKie et al., 2013). Decomposition rates differ substantially between these phases due to differences in lability of the resources but also biomass and activity of the decomposer groups. Studies have often accounted for leaching losses and calculated decomposition rates based on a measurement roughly half-way through the process (Handa et al., 2014) or based on several measurements at different stages (Bruder et al., 2014) to preclude unrepresentative estimates based on measurements during initial or terminal stages.

Seasonality and phenology are also important. Frainer et al. (2014) compared birch and alder litter decomposition rates between autumn and spring in streams of northern Sweden. They found that birch decomposed faster than alder in autumn, but in spring the difference depended on habitat characteristics, with birch decomposing faster in riffles, but alder decomposing faster in pools. These differences in decomposition rates were most likely due to differences in invertebrate community phenology, as detritivore composition differed between the two habitats. Other manifestations of phenology on litter decomposition rates may also ensue, e.g., between wet and dry seasons (Schlief & Mutz, 2011). Interannual differences in decomposition rates within the same litter species have also been reported, for instance due to differences in water current across years (Yeung et al., 2018). Overall, using litter material that is not synchronized in terms of quality and/or conditioning with background litter and decomposer community dynamics at the study site might yield unrealistic decomposition rates, e.g., due to preferential feeding of invertebrate decomposers on high-quality litter, or due to island effects, when a reduced amount of litter in the target habitat causes a disproportionately high concentration of invertebrates in the litter bags (Gjerløv & Richardson, 2004).

21.2.4 Habitat: From Lotic to Lentic Systems

Freshwaters cover a broad range of habitat types that are often characterized based on their water flow. Still water (*lentic)* habitats range from small water pools to larger ponds, lakes, and reservoirs, but also include still-water habitats found in caves and bromeliad tanks, for example, and in so-called "pool habitats" in streams and rivers, i.e. areas of deep, very slow flowing water typical of slow meanders and other "protected" channel sections. Running water (*lotic*) habitats range from small springs to the largest rivers, and include inlets and outlets of lakes and reservoirs. Litter input is particularly important in systems where autochthonous production is low, thus litter decomposition may be a useful measure of bioassessment in several of those freshwater habitats. To date, most work on litter decomposition has been done in streams, but water tanks in bromeliads (Benavides-Gordillo et al., 2019; Migliorini et al., 2018), caves (Galas et al., 1996; Silva et al., 2013), marshes (Flury & Gessner, 2011), small reservoirs (Colas et al., 2016), and shallow lakes (Alp et al., 2016; Carvalho et al., 2015; Pope et al., 1999) have also been tested for the effects of distinct biotic and abiotic stressors on litter decomposition. For example, Frainer et al. (2014) compared decomposition rates between pools and riffles within streams, and found large differences in decomposition rates between the two habitats, which were explained by differences in the benthic invertebrate community composition and phenology (see Sect. 21.2.3).

21.2.5 Selecting the Appropriate Reference Conditions

A great challenge in the application of bioassessment indicators lies in the evaluation of observed differences between impacted and non-impacted ecosystems (Elias et al., 2016; Feio et al., 2014). This challenge is equally relevant for litter decomposition assays. In many studies, impacts on litter decomposition and ecosystem functional integrity in general are inferred if a difference in decomposition rates is detected between impacted and reference sites (e.g., McKie & Malmqvist, 2009). Reference sites are chosen as genuinely pristine sites if available, or else as sites representing regionally "least disturbed" conditions, or as sites on a disturbance gradient, where the least disturbed sites are used as reference (Woodward et al., 2012). The effectiveness of this approach is seen in numerous publications detecting altered decomposition rates associated with increased nutrient levels, invasive species, mining pollution, hydro-morphological alterations, and pesticides, among others (see references in Sect. 21.1.2). The approach has also been used to assess the extent to which ecological mitigation and restoration have altered ecosystem functioning (see references in Sect. 21.1.2).

Within this framework, both faster and slower decomposition rates relative to a reference condition are considered as undesirable, and may indicate impaired functional integrity. However, in some cases, the lack of difference in decomposition rates between reference and impacted sites is also informative. A healthy microbial community should be able to respond to increased nutrients by increasing activity and biomass and in turn decomposition rates. Failure to do so might indicate an impairment of the microbial community due to some additional stressors (e.g., pesticides, Gardeström et al., 2016) or other limiting factors (Bruder et al., 2016).

Finally, different stressors may cancel each other, as when nutrient enrichment leads to oxygen depletion that negatively impacts detritivores and counteracts bottom-up stimulation of microbial activity. Such hidden stressor effects and stressor interactions need to be accounted for by measurement of additional biotic and abiotic parameters (Bruder et al., 2019).

Pitfalls are potentially associated with the definition of reference conditions. The reference condition is assumed to be representative of the natural condition of ecosystems in the absence of anthropogenic disturbances. The ecological status of an ecosystem is then defined according to the discrepancy between measurements in the impacted and reference situation. The choice of reference sites thus strongly affects interpretations of decomposition rates in bioassessment (Chauvet et al., 2016; Feio et al., 2010). Defining reference conditions can be challenging if streams and rivers that have similar characteristics but are free of any disturbance do not exist anymore (e.g., Feld et al., 2011). To address this issue, the concept of 'analytical reference condition' was developed (Downs et al., 2011) involving modelling the non-disturbed state of the ecosystem (i.e. 'hindcasting modelling') (Kilgour & Stanfield, 2006; Launois et al., 2011). The analytical reference condition is obtained by lowering the value of the human disturbance variables included in predictive models, thus providing an estimation of the value of the response variable with no or reduced disturbance (Soranno et al., 2011). Development of models of 'analytical reference conditions' for thresholds in litter decomposition are conceivable, but require spatially extensive, highly standardized data sets, of which only a few examples currently exist (Woodward et al., 2012; Tiegs et al., 2019; and Sect. 21.3). It might also be possible to model reference conditions from meta-analyses of smaller projects, but the challenges arising from variation in e.g., differences in litter substrates, decomposition periods, and disturbance intensities are substantial (see Sect. 21.3).

21.2.6 Ratios Between Coarse- and Fine-Mesh Bags

Decomposition rates (k) are in most cases extracted from first-order exponential decay models (Olson, 1963). Several studies have used the ratio in decomposition rates between coarse and fine-mesh bags (i.e. k_c:k_f) as a metric that gauges the relative contribution of shredders and microorganisms to litter decomposition (Lecerf, 2017; Pascoal et al., 2003). Based on literature values, Gessner and Chauvet (2002) proposed categories of ratios of decomposition rates to estimate ecosystem alteration in streams (for streams with high potential shredder abundance): k_c:k_f ratios ranging between 1.2 and 1.5 indicate no clear evidence of environmental alteration, ratios below 1.2 or ranging from 1.5 to 2 indicate moderate disturbance, whereas ratios >2 indicate severe disturbances. However, very high values of pollution may affect even tolerant shredder species, which may translate to low k_c and consequently low k_c:k_f. Importantly, whereas moderate levels of nutrient pollution might enhance litter nutritional quality and facilitate a greater invertebrate contribution to decomposition (i.e., increasing k_c relative to k_f), moderate levels of other disturbances, such as with hydromorphological degradation, riparian disturbances, and insecticides, may decrease invertebrate contribution to litter decomposition, thus decreasing k_c relative to k_f.

Applying Gessner and Chauvet (2002) categorization to the data from Handa et al. (2014) suggests severe alteration of the decomposition process in their temperate stream. A re-analysis of the data from Ferreira et al. (2015) suggests that a threshold ratio of 2 as an indication of severe alteration might be too low for many streams, including reference streams where shredder density and biomass may be naturally very high (see Sect. 21.3), especially when testing labile litter (Hieber & Gessner, 2002). Datasets with a broader gradient of environmental change than those used in Gessner and Chauvet (2002) suggest unimodal relationships between k_c:k_f ratio and nutrient pollution. For instance, a reanalysis of the data compiled by Ferreira et al. (2015) suggests a unimodal relationship with maximal ratios of k_c:k_f at 3.5 mg/L dissolved inorganic nitrogen (DIN) and 0.029 mg/L PO_4-P although with a high variability below the curve (see Sect. 21.3). Additionally, low k_c:k_f can also occur in stream naturally lacking efficient shredders, such as some tropical and insular streams (Bruder et al., 2014; Ferreira, Raposeiro et al., 2016).

Other approaches to the k_c:k_f ratio have been used. These include estimation of invertebrate-mediated decomposition rates ($k_{invertebrate}$) in isolation from decomposition mediated by microbes, and hence typically calculated based on the difference between percent litter mass remaining in coarse- and fine-mesh bags. A pan-European study, spanning stream sites along a very broad pollution gradient also suggests unimodal relationships of $k_{invertebrate}$ with nutrient concentrations (Woodward et al., 2012). These relationships had maximum values at approximately 3 mg/L DIN and 0.025 mg/L SRP. Unimodal relationships indicate that even tolerant shredder species become rare in highly polluted sites, e.g., due to exceedingly low levels of dissolved oxygen (Pascoal & Cássio, 2004), high concentration of ammonia (Lecerf et al., 2006), or other pollutants, resulting in low values of k_c:k_f and $k_{invertebrate}$.

21.3 Meta-Analysis Exemplifying Methodological Considerations In The Context Of Nutrient Enrichment: Reference Sites, Litter Quality and the Ratio Between Coarse and Fine-Mesh Bag Litter Decomposition Rates

21.3.1 Rationale

The effects of nutrient enrichment on litter decomposition in streams have been widely addressed, with empirical studies generally reporting a stimulation of litter decomposition with increases in dissolved nutrient concentration (Ferreira et al.,

2006; Gulis & Suberkropp, 2003; Rosemond et al., 2015). However, litter decomposition can be inhibited at high nutrient concentrations, which is generally attributed to toxicity associated with high concentrations of nitrite or ammonia, or to the concomitant change in other environmental factors (e.g., decrease in dissolved oxygen concentration or increase in fine sediment load and pesticide concentration) (Lecerf et al., 2006; Woodward et al., 2012). Even at lower to moderate nutrient concentrations, the effects of nutrient enrichment on litter decomposition have also been reported as non-significant when the stream is not nutrient limited or there are other limiting factors, such as low temperature, co-limitation by other nutrients, or low carbon quality of the litter (Baldy et al., 2007; Bruder et al., 2016; Chadwick & Huryn, 2003). Ferreira et al. (2015) summarized the effects of nutrient enrichment on litter decomposition in a meta-analysis of 99 studies that contributed 840 comparisons of litter decomposition rates in nutrient enriched and reference conditions and found an overall stimulation of litter decomposition by ~ 50% (95% CI: 41—58%). The effect was stronger when ambient nutrient concentration was lower and when the magnitude of the nutrient enrichment was higher (Ferreira et al., 2015). The magnitude of the effect also depended on litter identity (e.g., stronger stimulation of more recalcitrant oak than nitrogen-rich alder litter decomposition with nutrient enrichment), but not on climatic zone or type of decomposer community involved (microbes alone or microbes and invertebrates, assessed based on fine- or coarse-mesh bags, respectively) (Ferreira et al., 2015).

21.3.2 Methods

Here we revisit the database first analysed by Ferreira et al. (2015) to compare leaf litter decomposition rates in coarse- and fine-mesh bags and test the hypothesis that the magnitude of the effect of invertebrates on litter decomposition is (i) higher in nutrient enriched compared to reference streams as invertebrates take advantage of increased microbial biomass and activity in nutrient enriched conditions (Gulis et al., 2006), (ii) higher for more labile and nutrient-rich than for recalcitrant litter species as invertebrates play a greater role on the decomposition of the former (Hieber & Gessner, 2002), and (iii) higher for boreal than temperate regions as many shredders (e.g., from the orders Trichoptera and Plecoptera) have evolved as cold water species and microbial activity is limited at lower water temperature (Boyero, Pearson, Dudgeon et al., 2011; Irons et al., 1994; Taylor & Chauvet, 2014; Tiegs et al., 2019). The database was modified to contain only field correlative studies (not field manipulative studies) that used leaves (not woody substrates) and both coarse and fine-mesh bags (not studies that used only one mesh size, or no mesh bags). This resulted in 14 studies being retained. The matrix was then reorganized to contrast litter decomposition rates in coarse × fine-mesh bags, which resulted in 218 comparisons.

The effect of invertebrate activity on leaf litter decomposition rates (k, /d) was estimated per stream site as a response ratio (k_c:k_f; Hedges et al., 1999) and combined using the random effects model of meta-analysis (with the restricted maximum likelihood method for estimation of between-study variance) in R (R Core Team, 2015), using the *metafor* package (Viechtbauer, 2010). The effects of invertebrate activity on the decomposition of alder (*A. glutinosa*) and oak (*Quercus robur*) litter incubated in reference and nutrient enriched streams in boreal and temperate regions were further compared by estimating effect sizes for each contrast.

21.3.3 Results and Discussion

Leaf litter decomposition was higher in coarse- than fine-mesh bags by a factor of 2.42 (95% CI: 2.12—2.66). Thirty-three k_c:*kf* values were detected missing to the right of the global k_c:k_f by the Duval and Tweedie's trim and fill method, which quantifies publication bias in meta-analyses (Duval & Tweedie, 2000). When these missing values are inputted into the analysis the new estimate is even higher (k_c:k_f: 2.82; 95% CI: 2.57—3.09), suggesting that the results based on the matrix are conservative. As hypothesized, the higher litter decomposition rates in coarse over fine-mesh bags was stronger for alder than for oak ($p = 0.0001$), and stronger in temperate than in boreal regions ($p < 0.001$) (Fig. 21.2).

Decomposition rates were higher in coarse- than fine-mesh bags across all sites, and this difference was observed both in nutrient enriched and reference streams ($p = 0.105$) (Fig. 21.2). This may, however, reflect our reliance on classifications of

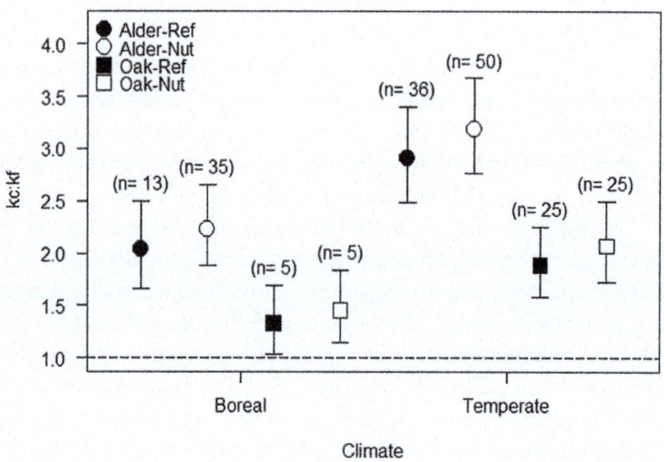

Fig. 21.2 Effects of invertebrate activity (k_c:k_f, ± 95% CI) on alder and oak leaf litter decomposition incubated in reference (Ref) and nutrient enriched (Nut) streams in boreal and temperate regions ($n = 194$). The dashed line (k_c:k_f = 1) indicates no significant effect of invertebrates on litter decomposition, while R > 1 indicates stimulation of litter decomposition in the presence of invertebrates. Effects of invertebrates' activities on litter decomposition are significant when the 95%CI does not include 1. Treatments significantly differ when their 95%CI do not overlap

reference and nutrient enriched streams provided by the authors of individual studies, since it is possible that reference streams in some regions have dissolved nutrient concentrations similar to those of nutrient enriched streams in other regions and vice versa (see previous section and Table S1 in Woodward et al., 2012).

To overcome this potential artifact, k_c:k_f was regressed against DIN and PO_4-P concentration (μg/L; ln-transformed) for the entire database, and separately for boreal and temperate regions using meta-regression. The ratio of litter decomposition in coarse- over fine-mesh bags increased with increasing DIN concentration, both in analyses of the entire database (slope = 0.116, $p < 0.0001$, $r^2 = 0.09$) and for boreal regions (slope = 0.131, $p = 0.005$, $r^2 = 0.07$), with a similar response also observed for temperate regions (slope = 0.09, $p = 0.058$, $r^2 = 0.02$) (Fig. 21.3). We found a clear positive relationship between k_c:k_f and PO_4-P concentration in boreal regions (slope = 0.371, $p < 0.0001$, $r^2 = 0.49$), but not in temperate regions (slope = 0.06, $p = 0.11$, $r^2 = 0.01$) or when considering the entire database (slope < 0.01, $p = 0.99$, $r^2 = 0$) (Fig. 21.3).

The hump-shaped distribution of k_c:k_f values along the nutrient gradient, particularly when values are computed for the entire dataset or only for the temperate region, is similar to that found by Woodward et al. (2012). Hence, although our meta-analysis indicates that invertebrate-mediated litter decomposition may respond positively to

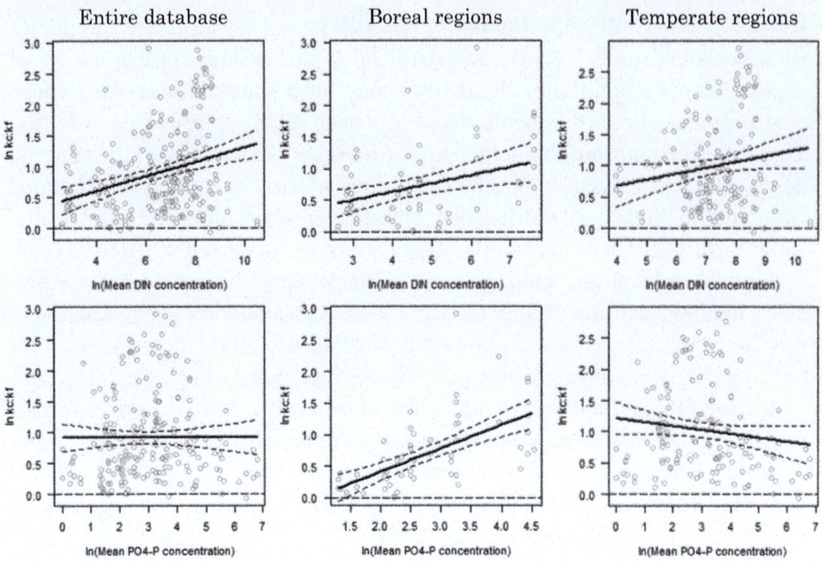

Fig. 21.3 Relationship between the effects of invertebrate activity on litter decomposition (ln k_c:k_f) and DIN and PO_4-P concentration in stream water (ln-transformed). The horizontal dashed lines (ln k_c:k_f = 0) indicate no significant effect of invertebrates on litter decomposition, while ln k_c:k_f > 0 indicates stimulation of litter decomposition in the presence of invertebrates. The relationships (meta-regression) are shown by the solid lines and associated 95% CI by the dashed lines

nutrient enrichment, it also shows large variability in the response of litter decomposition at medium levels of nutrient pollution. This may be due to the larger sample pool found around those medium values, which may reflect larger variability in overall background conditions, resulting in larger ecological variation as well. At the extreme levels of nutrient availability, where fewer sites are normally found, either nutrient limitation or excess nutrients seem to reduce the feeding capacity of invertebrates on leaf litter. The boreal region, which in comparison to the temperate region is often less nutrient polluted but also had fewer sites included in our analysis, did not show a hump-shaped distribution of k_c:k_f along the nutrient gradient. Instead, it had a clearer positive effect of nutrient enrichment on invertebrate-mediated decomposition rates.

Overall, the results from our meta-analysis indicate a potential positive effect of nutrient availability on litter decomposition up to a level where invertebrate-mediated decomposition starts slowing down, possibly due to deleterious effects arising from high levels of nutrient or other pollution. Our results also show a strong regional component (boreal vs temperate), with distinct responses of leaf litter and invertebrates to nutrient availability. Finally, our results illustrate the shortcomings in comparing reference *vs.* treatment streams, particularly when extrapolating the results across studies or regions that may differ in their criteria for classifying reference conditions or the availability thereof.

Box 21.2: Alternatives to the use of leaf litter

Bioassessment tools require standardization to ensure comparability of measurements among sites. Leaf litter may have shortcomings that reduce comparability of measurements, especially over larger spatio-temporal scales. These shortcomings might include low cross-scale replicability of the material being used, since even within the same litter species, spatial and interannual variability in litter C and nutrient concentration or other sources of intraspecific variation might undermine comparability of assays conducted in different years or at different locations. One approach for addressing this would be to undertake a detailed analysis of litter characteristics, to account for such variation in post-hoc analyses. Alternatively, highly standardized organic substrates have been proposed as substitutes for natural leaf litter, and include the following:

Cotton fabrics have received most attention as alternative material to quantify organic matter decomposition and have recently been used for large-scale comparisons of decomposition rates in streams (Tiegs et al., 2019). Cotton fibres are mainly composed of cellulose, and lack significant amounts of nutrients, lignin, tannins, etc. (Colas et al., 2019), which are important litter characteristics that control decomposition rates. Moreover, the physical characteristics of cotton fibres largely preclude invertebrate feeding (van Gestel et al., 2003) and their consequences for growth of fungal decomposers are also unknown. Nonetheless, cotton fabrics have been shown sensitive to dissolved concentrations of nitrogen and phosphorus, as well as to pH and water temperature (Boulton & Quinn, 2000; Hildrew et al., 1984; Jenkins et al., 2013) and to overall effects of agriculture and urbanization (Clapcott et al., 2012).

Wood veneers and sticks (e.g., commercially available ice cream sticks) have been used to quantify decomposition rates of plant material of low resource quality. Wood has lower nutrient but higher lignin concentration than leaves of most tree species (Arroita et al., 2012). Not surprisingly, the low resource quality is reflected in extremely low decomposition rates and low decomposer biomass compared to leaf litter (Arroita et al., 2012), and decomposition might be dominated by microbes (McTammany et al., 2008). Decomposition of wood sticks seems to respond to anthropogenic effects (Abril et al., 2015; Arroita et al., 2012; McTammany et al., 2008), and may show a hump-shaped relationship with gradients of anthropogenic impacts (Abril et al., 2015; McTammany et al., 2008), being reduced at elevated nutrients similar to leaf litter (McTammany et al., 2008). Wood veneers might provide a promising tool for studies that aim at estimating the process over longer time scales and/or in situations where physical abrasion by flow or suspended sediment is substantial.

Agar tablets, known as DECOTABs, are decomposed by microorganisms whose activity depends on the concentrations of cellulose, nutrients and minerals mixed into DECOTABs (Hunting et al., 2016; Kampfraath et al., 2012), but DECOTABs are also readily colonized and consumed by shredders and collector/gatherers (Kampfraath et al., 2012). DECOTABs can be purposely produced to reflect different organic matter resource quality (Hunting et al., 2016) and can include contaminants to the mix (Zhai et al., 2018).

Polymer sticks: decomposition in streams is governed by physical and biological processes (driven by microbial extracellular enzymes) and can be estimated by changes in the relative abundance of individual polymer ions estimated by mass spectrometry (Rivas et al., 2016).

Tea bags have been used due to the ease of implementation and high level of standardisation. Tea leaves contained within tea-bags can be considered an extreme example of leaf litter exposed in very small fine-mesh bags. Often tea bags of two contrasting types are used, and the difference in mass loss between the two may reflect differences in nutrient limitation in the system (Seelen et al., 2019). Currently, we are not aware of any data of microbial biomass and community composition measured from tea bags exposed in freshwaters. It remains to be seen if the particular litter quality often used as tea results in representative decomposition rates and decomposer activities.

21.4 Final Considerations

Litter decomposition has been tested and used as a measure of functional integrity in freshwater ecosystems for more than two decades in various ecological contexts. Most of these studies have shown negative effects of anthropogenic stressors on microbial and invertebrate-mediated decomposition rates. Nonetheless, questions regarding its suitability as a tool for stream ecosystem assessment remain. An important aspect is the lack of consensus on the consequences of altered (increased or decreased) decomposition rates for ecosystem functioning, for the integrity of local food webs, and for fluxes of energy into adjacent ecosystems.

Examples where a disturbance affects key decomposer groups and reduces decomposition rates seem straight-forward. Such cases are often interpreted as indicating impairments in ecosystem functioning, likely associated with reduced fluxes of carbon and nutrients from the litter into secondary production (Frainer et al., 2016; Halvorson et al., 2018; Kominoski et al., 2018; Rosemond et al., 2015). This can lead to accumulation of organic matter which might then either be broken down anaerobically (often associated with a greater production of CO_2 and CH_4) or else washed downstream and lost from the local food web (Lepori et al., 2005). The consequences of reduced decomposition rates on secondary production and entire food webs are still largely unknown, as compensatory mechanisms, including the use of alternative resources by consumers, could mask negative effects brought about by reduction in leaf litter availability, palatability, or nutritional quality. On the other hand, increased decomposition rates in response to stressors is not always a sign of good ecosystem integrity, and might therefore in itself be an indication of anthropogenically altered processes (McKie & Malmqvist, 2009). For example, production of fine particulate organic matter as litter fragments or of shredder fecal particles that exceeds the uptake capacity of local consumers might result in greater downstream export of carbon and nutrients, and longer, more leaky, nutrient spirals (Bundschuh & McKie, 2016).

The importance of these different scenarios depends on the goals of ecological assessment. If the goal is to assess whether ecosystem functioning has changed following a disturbance, the approach of comparing decomposition rates between impacted and reference sites, and before and after the disturbance when possible, will often be sufficient. Beyond this, development of a more specific assessment framework will require calibration of the litter decomposition assay. Such calibration requires not only accounting for how different disturbances typically affect decomposition rates (increase or decrease), but also understanding what response would be expected from a well-functioning biota compared to an impacted biota, and ultimately the consequences of changed functioning in interlinked ecosystems.

A more advanced framework for litter decomposition assays should also include guidelines on and harmonization of methodological choices, including of e.g., mesh sizes and litter species, and duration of the assay. For example, if addressing the effects of nutrient enrichment, low-nutrient litter with high quality carbon is better suited than high-nutrient litter where microbes are not nutrient limited. If physical fragmentation might be relevant at the site, the use of fine-mesh bags may be advised; if invertebrates are an important organism group, coarse-mesh bags are required. Temporal dynamics of the decomposition process are key, since the impacts of different types of human disturbances might be best assessed at different points in the decomposition process. For example, an impact primarily on microbes might need to be assessed over a different duration than impacts on detritivores; however, this issue has not received much attention in research. Finally, regional variation is also important. Larger scale studies using litter decomposition (e.g., Boyero, Pearson, Gessner et al., 2011; Woodward et al., 2012) indicate high inter-regional variability of decomposition rates, even when identical litter material is used and hydromorphological conditions among sites are similar. This suggests that region-specific guidelines, for example for litter species or the season and length of the decomposition assay, might be required. Standardization by degree-days is one approach for increasing comparability among regions where the duration of the assay might need to vary.

The shortcomings described here are not exclusive to the use of litter decomposition as a bioassessment tool. Many countries and environmental agencies rely solely on structural measures for stream bioassessment, although they oftentimes still lack proper national or regional classification criteria for their target organisms. Also, structural measures are most commonly taken as snapshots in time, without accounting for temporal variability in community composition. These shortcomings do not prevent the use of structural measures in bioassessment and should also not prevent the use of litter decomposition assays when this measure suits the goals of bioassessment. National- or regional-level monitoring programs are likely to be the best suited for using litter decomposition in bioassessment due to the more practical evaluation and definition of reference conditions and easier standardization of the litter type (see Box 21.1). Helping to reduce costs, knowledge from these cases may even allow litter decomposition to be used as alternative to structural measures when suitable, instead of as a complementary tool.

There is increasing consensus that litter decomposition is enhanced in streams with low to moderate nutrient enrichment. Woodward et al. (2012) pointed out that this is in fact the range of nutrient enrichment levels for which structural measures (e.g., those based on EPT-taxa) may be less effective, highlighting the potential for an environmental assay based on litter decomposition to address a gap in the sensitivity of current ecological assessment approaches. At higher nutrient levels, decomposition is inhibited by toxic effects of ammonia or nitrite and other stressors, resulting in similar k_c:k_f ratios as those of reference conditions. Other stressors, e.g., habitat loss, vegetation change, hydro-morphological modifications, siltation, salinization, temperature alterations, parasites, species invasion, and other consequences of land use modification have not been assessed to the same detail as those of nutrient enrichment, but ongoing research is rapidly filling these knowledge gaps. Despite these challenges, the use of litter decomposition as a complementary tool, and in some cases, as substitute for structural measures, may provide much detail and mechanistic understanding of effects of anthropogenic stressors in freshwater ecosystems. Litter decomposition assays are fairly easy to implement and the abundant scientific literature and interest all highlight their value as an important measure of ecosystem functioning in ecosystem assessments.

References

Abelho, M., & Canhoto, C. (2020). The role of carbon, nitrogen, and phosphorus in leaf decomposition mediated by aquatic fungi. *Limnetica, 39*, 275–282

Abril, M., Muñoz, I., Casas-Ruiz, J. P., Gómez-Gener, L., Barceló, M., Oliva, F., & Menéndez, M. (2015). Effects of water flow regulation on ecosystem functioning in a Mediterranean river network assessed by wood decomposition. *Science of the Total Environment, 517*, 57–65

Alp, M., Cucherousset, J., Buoro, M., & Lecerf, A. (2016). Phenological response of a key ecosystem function to biological invasion. *Ecology Letters, 19*, 519–527

Armitage, P., Moss, D., & Wright, J. (1983). The performance of a new biological water quality score system based on macroinvertebrates over a wide range of unpolluted running-water sites. *Water Research, 17*, 333–347

Arroita, M., Aristi, I., Flores, L., Larrañaga, A., Díez, J., Mora, J., Romaní, A. M., & Elosegi, A. (2012). The use of wooden sticks to assess stream ecosystem functioning: Comparison with leaf breakdown rates. *Science of the Total Environment, 440*, 115–122

Baldy, V., Gessner, M. O., & Chauvet, E. (1995). Bacteria, fungi and the breakdown of leaf litter in a large river. *Oikos, 74*, 93–102

Baldy, V., Gobert, V., Guerold, F., Chauvet, E., Lambrigot, D., & Charcosset, J. Y. (2007). Leaf litter breakdown budgets in streams of various trophic status: Effects of dissolved inorganic nutrients on microorganisms and invertebrates. *Freshwater Biology, 52*, 1322–1335

Benavides-Gordillo, S., Farjalla, V. F., González, A. L., & Romero, G. Q. (2019). Changes in rainfall level and litter stoichiometry affect aquatic community and ecosystem processes in bromeliad phytotelmata. *Freshwater Biology, 64*, 1357–1368

Boulton, A. J. (1999). An overview of river health assessment: Philosophies, practice, problems and prognosis. *Freshwater Biology, 41*, 469–479

Boulton, A. J., & Quinn, J. M. (2000). A simple and versatile technique for assessing cellulose decomposition potential in floodplain and riverine sediments. *Archiv für Hydrobiologie, 150*(1), 133–151.

Boyero, L., Pearson, R. G., Dudgeon, D., Graça, M. A. S., Gessner, M. O., Albariño, R. J., Ferreira, V., Yule, C. M., Boulton, A. J., Arunachalam, M., Callisto, M., Chauvet, E., Ramirez, A., Chara, J., Moretti, M. S., Gonçalves, J. F., Jr., Helson, J. E., Chará-Serna, A. M., Encalada, A. C., … Pringle, C. M. (2011). Global distribution of a key trophic guild contrasts with common latitudinal diversity patterns. *Ecology, 92*, 1839–1848

Boyero, L., Pearson, R. G., Gessner, M. O., Barmuta, L. A., Ferreira, V., Graça, M. A. S., Dudgeon, D., Boulton, A. J., Callisto, M., Chauvet, E., Helson, J. E., Bruder, A., Albariño, R. J., Yule, C. M., Arunachalam, M., Davies, J. N., Figueroa, R., Flecker, A. S., Ramirez, A., … West, D. C. (2011). A global experiment suggests climate warming will not accelerate litter decomposition in streams but might reduce carbon sequestration. *Ecology Letters, 14*, 289–294

Bruder, A., Frainer, A., Rota, T., & Primicerio, R. (2019). The Importance of ecological networks in multiple-stressor research and management. *Frontiers in Environmental Science, 7*, 59

Bruder, A., Salis, R. K., McHugh, N. J., & Matthaei, C. D. (2016). Multiple-stressor effects on leaf litter decomposition and fungal decomposers in agricultural streams contrast between litter species. *Functional Ecology, 30*(7), 1257–1266.

Bruder, A., Schindler, M. H., Moretti, M. S., & Gessner, M. O. (2014). Litter decomposition in a temperate and a tropical stream: The effects of species mixing, litter quality and shredders. *Freshwater Biology, 59*, 438–449

Bundschuh, M., & McKie, B. G. (2016). An ecological and ecotoxicological perspective on fine particulate organic matter in streams. *Freshwater Biology, 61*, 2063–2074

Burdon, F. J., Ramberg, E., Sargac, J., Forio, M. A. E., de Saeyer, N., Mutinova, P. T., Moe, T. F., Pavelescu, M. O., Dinu, V., Cazacu, C., Witing, F., Kupilas, B., Grandin, U., Volk, M., Rîşnoveanu, G., Goethals, P., Friberg, N., Johnson, R. K., & McKie, B. G. (2020). Assessing the benefits of forested riparian zones: A qualitative index of riparian integrity is positively associated with ecological status in European streams. *Water, 12*, 1178

Carvalho, C., Hepp, L. U., Palma-Silva, C., & Albertoni, E. F. (2015). Decomposition of macrophytes in a shallow subtropical lake. *Limnologica, 53*, 1–9

Castela, J., Ferreira, V., & Graça, M. A. S. (2008). Evaluation of stream ecological integrity using litter decomposition and benthic invertebrates. *Environmental Pollution, 153*, 440–449

Chadwick, M., & Huryn, A. (2003). Effect of a whole-catchment N addition on stream detritus processing. *Journal of the North American Benthological Society, 22*, 194–206

Chauvet, E., Ferreira, V., Giller, P. S., McKie, B. G., Tiegs, S. D., Woodward, G., Elosegi, A., Dobson, M., Fleituch, T., Graça, M. A. S., Gulis, V., Hladyz, S., Lacoursière, J. O., Lecerf, A., Pozo, J., Preda, E., Riipinen, M., Rîşnoveanu, G., Vadineanu, A., … Gessner, M. O. (2016). Litter decomposition as an indicator of stream ecosystem functioning at local-to-continental scales: Insights from the European RivFunction project. In A. J. Dumbrell, R. L. Kordas, & G. Woodward (Eds.), *Advances in Ecological Research* (Chap. 3, pp. 99–182). Academic Press.

Clapcott, J. E., Collier, K. J., Death, R. G., Goodwin, E. O., Harding, J. S., Kelly, D., Leathwick, J. R., & Young, R. G. (2012). Quantifying relationships between land-use gradients and structural and functional indicators of stream ecological integrity. *Freshwater Biology, 57*, 74–90

Colas, F., Baudoin, J.-M., Chauvet, E., Clivot, H., Danger, M., Guérold, F., & Devin, S. (2016). Dam-associated multiple-stressor impacts on fungal biomass and richness reveal the initial signs of ecosystem functioning impairment. *Ecological Indicators, 60*, 1077–1090

Colas, F., Baudoin, J.-M., Danger, M., Usseglio-Polatera, P., Wagner, P., & Devin, S. (2013). Synergistic impacts of sediment contamination and dam presence on river functioning. *Freshwater Biology, 58*, 320–336

Colas, F., Baudoin, J.-M., Gob, F., Tamisier, V., Valette, L., Kreutzenberger, K., Lambrigot, D., & Chauvet, E. (2017). Scale dependency in the hydromorphological control of a stream ecosystem functioning. *Water Research, 115*, 60–73

Colas, F., Woodward, G., Burdon, F. J., Guérold, F., Chauvet, E., Cornut, J., Cébron, A., Clivot, H., Danger, M., Danner, M. C., Pagnout, C., & Tiegs, S. D. (2019). Towards a simple global-standard bioassay for a key ecosystem process: Organic-matter decomposition using cotton strips. *Ecological Indicators, 106*, 105466

Collier, K., Clapcott, J., & Neale, M. (2014). *A macroinvertebrate attribute to assess ecosystem health for New Zealand waterways for the national objectives framework—Issues and options*. University of Waikato.

Connolly, N. M., & Pearson, R. G. (2013). Nutrient enrichment of a heterotrophic stream alters leaf litter nutritional quality and shredder physiological condition via the microbial pathway. *Hydrobiologia, 718*, 85–92

Dahl, J., & Johnson, R. K. (2004). A multimetric macroinvertebrate index for detecting organic pollution of streams in southern Sweden. *Archiv Für Hydrobiologie, 160*, 487–513

Downs, P., Ms, S., Bk, O., Ze, D., Tc, C., & Jc, S. (2011). Restoring ecological integrity in highly regulated rivers: The role of baseline data and analytical references. *Environmental Management, 48*, 847–864

Duval, S., & Tweedie, R. (2000). Trim and fill: A simple funnel-plot-based method of testing and adjusting for publication bias in meta-analysis. *Biometrics, 56*, 455–463

Elias, C. L., Calapez, A. R., Almeida, S. F. P., Chessman, B., Simões, N., & Feio, M. J. (2016). Predicting reference conditions for river bioassessment by incorporating boosted trees in the environmental filters method. *Ecological Indicators, 69*, 239–251

Elosegi, A., Gessner, M. O., & Young, R. G. (2017). River doctors: Learning from medicine to improve ecosystem management. *Science of the Total Environment, 595*, 294–302

Enriquez, S., Duarte, C., & Sand-Jensen, K. (1993). Patterns in decomposition rates among photosynthetic organisms: The importance of detritus C:N: P content. *Oecologia, 94*, 457–471

Entrekin, S. A., Tank, J. L., Rosi-Marshall, E. J., Hoellein, T. J., & Lamberti, G. A. (2008). Responses in organic matter accumulation and processing to an experimental wood addition in three headwater streams. *Freshwater Biology, 53*, 1642–1657

Feio, M. J., Aguiar, F. C., Almeida, S. F. P., Ferreira, J., Ferreira, M. T., Elias, C., Serra, S. R. Q., Buffagni, A., Cambra, J., Chauvin, C., Delmas, F., Dörflinger, G., Erba, S., Flor, N., Ferréol, M.,

Germ, M., Mancini, L., Manolaki, P., Marcheggiani, S., ... Vieira, C. (2014). Least disturbed condition for European Mediterranean rivers. *Science of the Total Environment, 476–477*, 745–756

Feio, M. J., Alves, T., Boavida, M., Medeiros, A., & Graça, M. A. S. (2010). Functional indicators of stream health: A river-basin approach. *Freshwater Biology, 55*, 1050–1065

Feld, C. K., Birk, S., Bradley, D. C., Hering, D., Kail, J., Marzin, A., Melcher, A., Nemitz, D., Pedersen, M. L., Pletterbauer, F., Pont, D., Verdonschot, P. F. M., & Friberg, N. (2011). From natural to degraded rivers and back again: A test of restoration ecology theory and practice. In G. Woodward (ed.), *Advances in Ecological Research* (Chap. 3, pp. 119–209). Academic Press.

Ferreira, V., Boyero, L., Calvo, C., Correa, F., Figueroa, R., Gonçalves, J. F., Goyenola, G., Graça, M. A. S., Hepp, L. U., Kariuki, S., López-Rodríguez, A., Mazzeo, N., M'Erimba, C., Monroy, S., Peil, A., Pozo, J., Rezende, R., & Teixeira-de-Mello, F. (2019). A global assessment of the effects of eucalyptus plantations on stream ecosystem functioning. *Ecosystems, 22*, 629–642

Ferreira, V., Castagneyrol, B., Koricheva, J., Gulis, V., Chauvet, E., & Graça, M. A. S. (2015). A meta-analysis of the effects of nutrient enrichment on litter decomposition in streams. *Biological Reviews, 90*, 669–688

Ferreira, V., Encalada, A. C., & Graça, M. A. S. (2012). Effects of litter diversity on decomposition and biological colonization of submerged litter in temperate and tropical streams. *Freshwater Science, 31*, 945–962

Ferreira, V., & Guérold, F. (2017). Leaf litter decomposition as a bioassessment tool of acidification effects in streams: Evidence from a field study and meta-analysis. *Ecological Indicators, 79*, 382–390

Ferreira, V., Gulis, V., & Graça, M. A. S. (2006). Whole-stream nitrate addition affects litter decomposition and associated fungi but not invertebrates. *Oecologia, 149*, 718–729

Ferreira, V., Koricheva, J., Duarte, S., Niyogi, D. K., & Guérold, F. (2016). Effects of anthropogenic heavy metal contamination on litter decomposition in streams—A meta-analysis. *Environmental Pollution, 210*, 261–270

Ferreira, V., Koricheva, J., Pozo, J., & Graça, M. A. S. (2016). A meta-analysis on the effects of changes in the composition of native forests on litter decomposition in streams. *Forest Ecology and Management, 364*, 27–38

Ferreira, V., Raposeiro, P. M., Pereira, A., Cruz, A. M., Costa, A. C., Graça, M. A. S., & Gonçalves, V. (2016). Leaf litter decomposition in remote oceanic island streams is driven by microbes and depends on litter quality and environmental conditions. *Freshwater Biology, 61*, 783–799

Flores, L., Larrañaga, A., Díez, J., & Elosegi, A. (2011). Experimental wood addition in streams: Effects on organic matter storage and breakdown. *Freshwater Biology, 56*, 2156–2167

Flury, S., & Gessner, M. O. (2011). Experimentally simulated global warming and nitrogen enrichment effects on microbial litter decomposers in a marsh. *Applied and Environmental Microbiology, 77*, 803–809

Frainer, A., Jabiol, J., Gessner, M. O., Bruder, A., Chauvet, E., & McKie, B. G. (2016). Stoichiometric imbalances between detritus and detritivores are related to shifts in ecosystem functioning. *Oikos, 125*, 861–871

Frainer, A., & McKie, B. G. (2015). Shifts in the diversity and composition of consumer traits constrain the effects of land use on stream ecosystem functioning. *Advances in Ecological Research, 52*, 169–200

Frainer, A., McKie, B. G., & Malmqvist, B. (2014). When does diversity matter? Species functional diversity and ecosystem functioning across habitats and seasons in a field experiment. *Journal of Animal Ecology, 83*, 460–469

Frainer, A., Moretti, M. S., Xu, W., & Gessner, M. O. (2015). No evidence for leaf trait dissimilarity effects on litter decomposition, fungal decomposers, and nutrient dynamics. *Ecology, 96*, 550–561

Frainer, A., Polvi, L. E., Jansson, R., & McKie, B. G. (2018). Enhanced ecosystem functioning following stream restoration: The roles of habitat heterogeneity and invertebrate species traits. *Journal of Applied Ecology, 55*, 377–385

Galas, J., Bednarz, T., Dummicka, E., Starzecka, A., & Wojtan, K. (1996). Litter decomposition in a mountain cave water. *Archiv Für Hydrobiologie, 138*, 199–211

Gardeström, J., Ermold, M., Goedkoop, W., & McKie B. G. (2016). Disturbance history influences stressor impacts: effects of a fungicide and nutrients on microbial diversity and litter decomposition. *Freshwater Biology, 61*, 2171–2184

Gessner, M. O., & Chauvet, E. (1994). Importance of stream microfungi in controlling breakdown rates of leaf-litter. *Ecology, 75*, 1807–1817

Gessner, M. O., & Chauvet, E. (2002). A case for using litter breakdown to assess functional stream integrity. *Ecological Applications, 12*, 498–510

Gjerløv, C., & Richardson, J. S. (2004). Patchy resources in a heterogeneous environment: Effects of leaf litter and forest cover on colonisation patterns of invertebrates in a British Columbian stream. *Archiv Für Hydrobiologie, 161*, 307–327

Graça, M. A. S., & Poquet, J. M. (2014). Do climate and soil influence phenotypic variability in leaf litter, microbial decomposition and shredder consumption? *Oecologia, 174*, 1021–1032

Grossman, J. J., Cavender-Bares, J., & Hobbie, S. E. (2020). Functional diversity of leaf litter mixtures slows decomposition of labile but not recalcitrant carbon over two years. *Ecological Monographs, 90*, e01407.

Gulis, V., Ferreira, V., & Graça, M. A. S. (2006). Stimulation of leaf litter decomposition and associated fungi and invertebrates by moderate eutrophication: Implications for stream assessment. *Freshwater Biology, 51*, 1655–1669

Gulis, V., & Suberkropp, K. (2003). Leaf litter decomposition and microbial activity in nutrient-enriched and unaltered reaches of a headwater stream. *Freshwater Biology, 48*, 123–134

Halvorson, H. M., Fuller, C. L., Entrekin, S. A., Scott, J. T., & Evans-White, M. A. (2018). Detrital nutrient content and leaf species differentially affect growth and nutritional regulation of detritivores. *Oikos, 127*, 1471–1481

Handa, I. T., Aerts, R., Berendse, F., Berg, M. P., Bruder, A., Butenschoen, O., Chauvet, E., Gessner, M. O., Jabiol, J., Makkonen, M., McKie, B. G., Malmqvist, B., Peeters, E. T. H. M., Scheu, S., Schmid, B., van Ruijven, J., Vos, V. C. A., & Hättenschwiler, S. (2014). Consequences of Biodiversity Loss for Litter Decomposition Across Biomes. *Nature, 509*, 218–221

Hedges, L. V., Gurevitch, J., & Curtis, P. S. (1999). The meta-analysis of response ratios in experimental ecology. *Ecology, 80*, 1150–1156

Hernández, A. D., & Sukhdeo, M. V. K. (2008). Parasite effects on isopod feeding rates can alter the host's functional role in a natural stream ecosystem. *International Journal for Parasitology, 38*, 683–690

Hieber, M., & Gessner, M. O. (2002). Contribution of stream detrivores, fungi, and bacteria to leaf breakdown based on biomass estimates. *Ecology, 83*, 1026–1038

Hildrew, A. G., Townsend, C. R., Francis, J., & Finch, K. (1984). Cellulolytic decomposition in streams of contrasting pH and its relationship with invertebrate community structure. *Freshwater Biology, 14*, 323–328

Hladyz, S., Åbjörnsson, K., Giller, P. S., & Woodward, G. (2011). Impacts of an aggressive riparian invader on community structure and ecosystem functioning in stream food webs. *Journal of Applied Ecology, 48*, 443–452

Hunting, E. R., Vonk, J. A., Musters, C. J. M., Kraak, M. H. S., & Vijver, M. G. (2016). Effects of agricultural practices on organic matter degradation in ditches. *Scientific Reports, 6*, 1–9

Imberger, S. J., Thompson, R. M., & Grace, M. R. (2010). Searching for effective indicators of ecosystem function in urban streams: Assessing cellulose decomposition potential. *Freshwater Biology, 55*, 2089–2106

Irons, J. G., Oswood, M. W., Stout, R. J., & Pringle, C. M. (1994). Latitudinal patterns in leaf litter breakdown: Is temperature really important? *Freshwater Biology, 32*, 401–411

Jabiol, J., McKie, B. G., Bruder, A., Bernadet, C., Gessner, M. O., & Chauvet, E. (2013). Trophic complexity enhances ecosystem functioning in an aquatic detritus-based model system. *Journal of Animal Ecology, 82*, 1042–1051

Jenkins, G. B., Woodward, G., & Hildrew, A. G. (2013). Long-term amelioration of acidity accelerates decomposition in headwater streams. *Global Change Biology, 19*, 1100–1106

Kampfraath, A. A., Hunting, E. R., Mulder, C., Breure, A. M., Gessner, M. O., Kraak, M. H. S., & Admiraal, W. (2012). DECOTAB: A multipurpose standard substrate to assess effects of litter quality on microbial decomposition and invertebrate consumption. *Freshwater Science, 31*, 1156–1162

Karr, J. (1999). Defining and measuring river health. *Freshwater Biology, 41*, 221–234

Kennedy, K. T. M., & El-Sabaawi, R. W. (2017). A global meta-analysis of exotic versus native leaf decay in stream ecosystems. *Freshwater Biology, 62*, 977–989

Kilgour, B. W., & Stanfield, L. W. (2006). Hindcasting reference conditions in streams. *American Fisheries Society Symposium, 48*, 623–639

Kominoski, J. S., Rosemond, A. D., Benstead, J. P., Gulis, V., & Manning, D. W. P. (2018). Experimental nitrogen and phosphorus additions increase rates of stream ecosystem respiration and carbon loss. *Limnology and Oceanography, 63*, 22–36

Launois, L., Veslot, J., Irz, P., & Argillier, C. (2011). Development of a fish-based index (FBI) of biotic integrity for French lakes using the hindcasting approach. *Ecological Indicators, 11*, 1572–1583

Lecerf, A. (2017). Methods for estimating the effect of litterbag mesh size on decomposition. *Ecological Modelling, 362*, 65–68

Lecerf, A., & Chauvet, E. (2008). Intraspecific variability in leaf traits strongly affects alder leaf decomposition in a stream. *Basic and Applied Ecology, 9*, 598–605

Lecerf, A., Usseglio-Polatera, P., Charcosset, J.-Y., Lambrigot, D., Bracht, B., & Chauvet, E. (2006). Assessment of functional integrity of eutrophic streams using litter breakdown and benthic macroinvertebrates. *Archiv Für Hydrobiologie, 165*, 105–126

Ledger, M., & Hildrew, A. (2005). The ecology of acidification and recovery: Changes in herbivore-algal food web linkages across a stream pH gradient. *Environmental Pollution, 137*, 103–118

Lepori, F., Palm, D., & Malmqvist, B. (2005). Effects of stream restoration on ecosystem functioning: Detritus retentiveness and decomposition. *Journal of Applied Ecology, 42*, 228–238

Leroy, C. J., Whitham, T. G., Wooley, S. C., & Marks, J. C. (2007). Within-species variation in foliar chemistry influences leaf-litter decomposition in a Utah river. *Journal of the North American Benthological Society, 26*, 426–438

Majdi, N., Boiché, A., Traunspurger, W., & Lecerf, A. (2014). Predator effects on a detritus-based food web are primarily mediated by non-trophic interactions. *Journal of Animal Ecology, 83*, 953–962

McArthur, J., Aho, J., Rader, R., & Mills, G. L. (1994). Interspecific leaf interactions during decomposition in aquatic and floodplain ecosystems. *Journal of the North American Benthological Society, 13*, 57–67

McKie, B., & Cranston, P. (2001). Colonisation of experimentally immersed wood in south eastern Australia: Responses of feeding groups to changes in riparian vegetation. *Hydrobiologia, 452*, 1–14

McKie, B. G., & Malmqvist, B. (2009). Assessing ecosystem functioning in streams affected by forest management: Increased leaf decomposition occurs without changes to the composition of benthic assemblages. *Freshwater Biology, 54*, 2086–2100

McKie, B. G., Woodward, G., Hladyz, S., Nistorescu, M., Preda, E., Popescu, C., Giller, P. S., & Malmqvist, B. (2008). Ecosystem functioning in stream assemblages from different regions: Contrasting responses to variation in detritivore richness, evenness and density. *Journal of Animal Ecology, 77*, 495–504

McTammany, M. E., Benfield, E. F., & Webster, J. R. (2008). Effects of agriculture on wood breakdown and microbial biofilm respiration in southern Appalachian streams. *Freshwater Biology, 53*, 842–854

Mendoza-Lera, C., Larrañaga, A., Pérez, J., Descals, E., Martínez, A., Moya, O., Arostegui, I., & Pozo, J. (2012). Headwater reservoirs weaken terrestrial-aquatic linkage by slowing leaf-litter processing in downstream regulated reaches. *River Research and Applications, 28*, 13–22

Migliorini, G. H., Srivastava, D. S., & Romero, G. Q. (2018). Leaf litter traits drive community structure and functioning in a natural aquatic microcosm. *Freshwater Biology, 63*, 341–352

Mollá, S., Casas, J. J., Menéndez, M., Basaguren, A., Casado, C., Descals, E., González, J. M., Larrañaga, A., Lusi, M., Martínez, A., Mendoza-Lera, C., Moya, O., Pérez, J., Riera, T., Roblas, N., & Pozo, J. (2017). Leaf-litter breakdown as an indicator of the impacts by flow regulation in headwater streams: Responses across climatic regions. *Ecological Indicators, 73*, 11–22

Mondy, C. P., Villeneuve, B., Archaimbault, V., & Usseglio-Polatera, P. (2012). A new macroinvertebrate-based multimetric index (I2M2) to evaluate ecological quality of French wadeable streams fulfilling the WFD demands: A taxonomical and trait approach. *Ecological Indicators, 18*, 452–467

Moretti, M. S., Loyola, R. D., Becker, B., & Callisto, M. (2009). Leaf abundance and phenolic concentrations codetermine the selection of case-building materials by Phylloicus sp. (Trichoptera, Calamoceratidae). *Hydrobiologia, 630*, 199–206

Olson, J. S. (1963). Energy storage and the balance of producers and decomposers in ecological systems. *Ecology, 44*, 322–331

Ostrofsky, M. (1997). Relationship between chemical characteristics of autumn-shed leaves and aquatic processing rates. *Journal of the North American Benthological Society, 16*, 750–759

Pascoal, C., & Cássio, F. (2004). Contribution of fungi and bacteria to leaf litter decomposition in a polluted river. *Applied and Environmental Microbiology, 70*, 5266–5273

Pascoal, C., Pinho, M., Cássio, F., & Gomes, P. (2003). Assessing structural and functional ecosystem condition using leaf breakdown: Studies on a polluted river. *Freshwater Biology, 48*, 2033–2044

Pope, R. J., Gordon, A. M., & Kaushik, N. K. (1999). Leaf litter colonization by invertebrates in the littoral zone of a small oligotrophic lake. *Hydrobiologia, 392*, 99–112

Quintão, J. M. B., Rezende, R. S., & Júnior, J. F. G. (2013). Microbial effects in leaf breakdown in tropical reservoirs of different trophic status. *Freshwater Science, 32*, 933–950

R Core Team. (2015). *R: A language and environment for statistical computing*. https://www.R-project.org/.

Raposeiro, P. M., Ferreira, V., Guri, R., Gonçalves, V., & Martins, G. M. (2016). Leaf litter decomposition on insular lentic systems: Effects of macroinvertebrate presence, leaf species, and environmental conditions. *Hydrobiologia*, 1–15.

Rasmussen, J. J., Wiberg-Larsen, P., Baattrup-Pedersen, A., Monberg, R. J., & Kronvang, B. (2012). Impacts of pesticides and natural stressors on leaf litter decomposition in agricultural streams. *Science of the Total Environment, 416*, 148–155

Rincón, J., & Covich, A. (2014). Effects of insect and decapod exclusion and leaf litter species identity on breakdown rates in a tropical headwater stream. *Revista De Biología Tropical, 62*, 143–154

Rincón, J., & Martínez, I. (2006). Food quality and feeding preferences of Phylloicus sp. (Trichoptera:Calamoceratidae). *Journal of the North American Benthological Society, 25*, 209–215

Rivas, D., Ginebreda, A., Elosegi, A., Pozo, J., Pérez, S., Quero, C., & Barceló, D. (2016). Using a polymer probe characterized by MALDI-TOF/MS to assess river ecosystem functioning: From polymer selection to field tests. *Science of the Total Environment, 573*, 532–540

Rosemond, A. D., Benstead, J. P., Bumpers, P. M., Gulis, V., Kominoski, J. S., Manning, D. W. P., Suberkropp, K., & Wallace, J. B. (2015). Experimental nutrient additions accelerate terrestrial carbon loss from stream ecosystems. *Science, 347*, 1142–1145

Rosi-Marshall, E. J., Tank, J. L., Royer, T. V., Whiles, M. R., Evans-White, M., Chambers, C., Griffiths, N. A., Pokelsek, J., & Stephen, M. L. (2007). Toxins in transgenic crop byproducts may affect headwater stream ecosystems. *Proceedings of the National Academy of Sciences of the USA, 104*, 16204–16208

Sabater, S., Bregoli, F., Acuña, V., Barceló, D., Elosegi, A., Ginebreda, A., Marcé, R., Muñoz, I., Sabater-Liesa, L., & Ferreira, V. (2018). Effects of human-driven water stress on river ecosystems: A meta-analysis. *Scientific Reports, 8*, 11462

Sanpera-Calbet, I., Lecerf, A., & Chauvet, E. (2009). Leaf diversity influences in-stream litter decomposition through effects on shredders. *Freshwater Biology, 54*, 1671–1682

von Schiller, D., Acuña, V., Aristi, I., Arroita, M., Basaguren, A., Bellin, A., Boyero, L., Butturini, A., Ginebreda, A., Kalogianni, E., Larrañaga, A., Majone, B., Martínez, A., Monroy, S., Muñoz, I., Paunović, M., Pereda, O., Petrovic, M., Pozo, J., … Elosegi, A. (2017). River ecosystem processes: A synthesis of approaches, criteria of use and sensitivity to environmental stressors. *Science of the Total Environment, 596–597*, 465–480

Schindler, M. H., & Gessner, M. O. (2009). Functional leaf traits and biodiversity effects on litter decomposition in a stream. *Ecology, 90*, 1641–1649

Schlief, J., & Mutz, M. (2011). Leaf decay processes during and after a supra-seasonal hydrological drought in a temperate lowland stream. *International Review of Hydrobiology, 96*, 633–655

Seelen, L. M. S., Flaim, G., Keuskamp, J., Teurlincx, S., Arias Font, R., Tolunay, D., Fránková, M., Šumberová, K., Temponeras, M., Lenhardt, M., Jennings, E., & de Senerpont Domis, L. N. (2019). An affordable and reliable assessment of aquatic decomposition: Tailoring the Tea Bag Index to surface waters. *Water Research, 151*, 31–43

Silva, M., Rezende, R., & Ferreira, R. L. (2013). Detritus processing in lentic cave habitats in the neotropics. *Subterranean Biology, 11*, 3–14

Soranno, P. A., Wagner, T., Martin, S. L., McLean, C., Novitski, L. N., Provence, C. D., & Rober, A. R. (2011). Quantifying regional reference conditions for freshwater ecosystem management: A comparison of approaches and future research needs. *Lake and Reservoir Management, 27*, 138–148

Taylor, B. R., & Chauvet, E. E. (2014). Relative influence of shredders and fungi on leaf litter decomposition along a river altitudinal gradient. *Hydrobiologia, 721*, 239–250

The European Parliament. (2015). Directive 2008/94/EC of the European Parliament and of the Council of 22 October 2008. In *Core EU Legislation* (pp. 423–426). Macmillan Education.

Tiegs, S. D., Costello, D. M., Isken, M. W., Woodward, G., McIntyre, P. B., Gessner, M. O., Chauvet, E., Griffiths, N. A., Flecker, A. S., Acuña, V., Albariño, R., Allen, D. C., Alonso, C., Andino, P., Arango, C., Aroviita, J., Barbosa, M. V. M., Barmuta, L. A., Baxter, C. V., … and Zwart, J. A. (2019). Global patterns and drivers of ecosystem functioning in rivers and riparian zones. *Science Advances, 5,* eaav0486.

Tonin, A. M., Gonçalves, J. F., Bambi, P., Couceiro, S. R. M., Feitoza, L. A. M., Fontana, L. E., Hamada, N., Hepp, L. U., Lezan-Kowalczuk, V. G., Leite, G. F. M., Lemes-Silva, A. L., Lisboa, L. K., Loureiro, R. C., Martins, R. T., Medeiros, A. O., Morais, P. B., Moretto, Y., Oliveria, P. C. A., Pereira, E. B., … Boyero, L. (2017). Plant litter dynamics in the forest-stream interface: Precipitation is a major control across tropical biomes. *Scientific Reports, 7*, 1–14

USEPA. (2016). A practitioner's guide to the biological condition gradient: A framework to describe incremental change in aquatic ecosystems. In U.S. environmental protection agency (p. 250).

van Gestel, C. A. M., Kruidenier, M., & Berg, M. P. (2003). Suitability of wheat straw decomposition, cotton strip degradation and bait-lamina feeding tests to determine soil invertebrate activity. *Biology and Fertility of Soils, 37*, 115–123

van Dokkum, H. P., Slijkerman, D. M. E., Rossi, L., & Costantini, M. L. (2002). Variation in the decomposition of Phragmites australis litter in a monomictic lake: The role of gammarids. *Hydrobiologia, 482*, 69–77

Viechtbauer, W. (2010). Conducting meta-analyses in R with the metafor package. *Journal of Statistical Software, 36*, 1–48

Woodward, G., Gessner, M. O., Giller, P. S., Gulis, V., Hladyz, S., Lecerf, A., Malmqvist, B., McKie, B. G., Tiegs, S. D., Cariss, H., Dobson, M., Elosegi, A., Ferreira, V., Graça, M. A. S., Fleituch, T., Lacoursière, J., Nistorescu, M., Pozo, J., Risnoveanu, G., … Chauvet, E. (2012). Continental-scale effects of nutrient pollution on stream ecosystem functioning. *Science, 336*, 1438–1440

Yeung, A. C. Y., Kreutzweiser, D. P., & Richardson, J. S. (2019). Stronger effects of litter origin on the processing of conifer than broadleaf leaves: A test of home-field advantage of stream litter breakdown. *Freshwater Biology, 64*, 1755–1768

Yeung, A. C. Y., Musetta-Lambert, J. L., Kreutzweiser, D. P., Sibley, P. K., & Richardson, J. S. (2018). Relations of interannual differences in stream litter breakdown with discharge: Bioassessment implications. *Ecosphere, 9*, e02423

Young, R., Matthaei, C., & Townsend, C. R. (2008). Organic matter breakdown and ecosystem metabolism: Functional indicators for assessing river ecosystem health. *Journal of the North American Benthological Society, 27*, 605–625

Zhai, Y., Brun, N. R., Bundschuh, M., Schrama, M., Hin, E., Vijver, M. G., & Hunting, E. R. (2018). Microbially-mediated indirect effects of silver nanoparticles on aquatic invertebrates. *Aquatic Sciences, 80*, 44

Chapter 22
Leaf Litter Decomposition as a Contributor to Ecosystem Service Provision

John S. Richardson and Dalal E.L. Hanna

Abstract Leaf litter decomposition supports the provision of numerous ecosystem services—that is, the benefits nature provides to humans—through its contribution to the reduction and storage of organic materials that could negatively impact water supplies, and its support of the productivity of freshwater ecosystems. The magnitude of these services varies spatially, and with land use. Leaf litter decomposition contributes a large fraction of energy supporting production in small streams, and to downstream production through supplies of finer particles produced by consumers. Decomposition results in storage of carbon as biomass in biological communities, and reduces the amounts of organic materials that could impact water quality downstream. Changes in water quality that result from decomposition can also influence the aesthetic and recreational value of waterways. Land use can influence decomposition, for instance elevated nutrient concentrations can result in higher decomposition rates, but high temperatures and sediment loads may diminish those rates. Changes in the rates of such ecosystem functions may be proportional to changes in the associated benefits provided to humans. While decomposition is a key ecosystem function that can contribute to the provision of numerous services, there are few estimates of the human value of this process in freshwaters.

22.1 What Are Ecosystem Services, and How Can Decomposition of Litter in Freshwater Contribute to Service Provision?

Ecological processes and phenomena contribute to the well-being of humans in many ways. These have been considered as ecosystem services to highlight the

J. S. Richardson (✉)
University of British Columbia, Vancouver, BC V6T 1Z4, Canada
e-mail: john.richardson@ubc.ca

D. E.L. Hanna
McGill University, MacDonald Campus, Sainte-Anne-de-Bellevue, QC H9X 3V9, Canada
e-mail: dalal.hanna@mail.mcgill.ca

C. M. Swan et al. (eds.), *The Ecology of Plant Litter Decomposition in Stream Ecosystems*, https://doi.org/10.1007/978-3-030-72854-0_22

diverse benefits nature provides to people (Díaz et al., 2018; Millennium Ecosystem Assessment, 2005). The value of ecosystem services can be determined using a suite of indicators, ranging from biophysical, to qualitative, to monetary. Replacing the contributions nature makes to people using human technology can be expensive (Costanza et al., 1997), and is not always possible (Raudsepp-Hearne et al., 2010). Leaf litter decomposition in freshwaters is an ecosystem function that supports the provision of several ecosystem services, as we will discuss in this chapter.

Ecosystem services, also referred to as nature's contributions to people (Díaz et al., 2018), are generally defined as the benefits humans accrue from nature, such as production of fiber and timber, pollination of crops, provision of clean water, as well as access to areas within which to recreate or to feel a connection with nature. The provision of ecosystem services is dependent on different forms of capital, which can range from natural, to social, to financial (Natural Capital Coalition, 2019). The history of the use of the term ecosystem services covers a half century, but has been focused in popular discourse by inclusion as a central aspect of the Millennium Ecosystem Assessment (2005). There are several different ecosystem services typologies (e.g., CICES, MA, TEEB). One of the most commonly used ones was outlined in the Millennium Ecosystem Assessment (MA), and defines four main categories of types of ecosystem services, including: cultural ecosystem services (nonmaterial benefits such as the aesthetic beauty of streams, or their emotional and spiritual significance to certain people, and the access they provide to an area to do recreational activities); provisioning ecosystem services (material products obtained from ecosystems such as freshwater for drinking and fish for consumption); regulating ecosystem services (benefits derived from the regulation of ecosystem processes such as the purification of water or prevention of erosion); and supporting ecosystem services (necessary processes for the production of other services such as habitat provision and nutrient cycling).

Numerous resources now reclassify supporting services as ecological processes rather than services per se. This distinction was made because these processes, which may also be considered supporting services, are often essential first steps for the successful provision of other services, making it difficult to disentangle their provision and monetary values (Carpenter et al., 2009; Hein et al., 2006; Mancinelli & Mulder, 2015). Distinguishing between the two can help to identify which services (or processes) should be considered when assessing the monetary value of the contributions nature makes to people because it helps avoid issues associated with double-counting—an erroneous practice whereby the monetary value of services that are not distinct from each other are summed (Fu et al., 2011). Considering the relationship between services, and which among these are intermediate (i.e. required processes for the provision of a benefit humans derive from nature), as opposed to final (i.e. services that directly lead to the provision of a benefit), is a useful approach to determine which services can be considered in a monetary valuation without double-counting (Fisher et al., 2009).

Ecosystem services can be quantified using a range of indicators spanning from biophysical, to qualitative, to monetary, each assessing value in unique ways (Seppelt et al., 2011). The indicators used to assess an ecosystem service may also reflect different aspects of that service—the capacity of an ecosystem to provide services (i.e. the type, amount, and quality of services available) may be distinct to the demand for services (i.e. the type, amount, and quality of services used or wanted by people) (Villamagna et al., 2013). The most appropriate indicators to quantify services are therefore context dependent, and should be determined based on a range of considerations including what information those quantifying services are interested in, how they intend to use their quantification, who is involved in the quantification or likely to be affected by it, and the resources at the disposal of those undertaking the quantification (e.g., Felipe-Lucia et al., 2015; Hanna et al., 2018).

Decomposition in freshwaters contributes to ecosystem service provisioning in several ways (Table 22.1). First, decomposition contributes to regulating water quality by reducing the amounts of organic materials in water, which provides cleaner water to downstream users, requiring less processing to prepare for human consumption. Next, the role of decomposers in clarifying water also helps maintain the aesthetic value of freshwater systems, which people may consider to be unclean and "muddy" (e.g., Macadam & Stocken, 2015) without decomposers reducing amounts of organic materials found in water. The aesthetic appearance of waterways also influences the ways in which they are used by people for recreation. Third, decomposition contributes to secondary production of freshwater ecosystems, primarily valued as fish production, a provisioning service. Fourth, decomposition may be considered a mechanism for carbon sequestration, largely in the biomass of organisms, thereby contributing to the regulation of climate. Various approaches can be used to quantify how decomposition contributes to the provision of these services. When working to assign them a monetary value it is especially important to be explicit about how they do or do not provide independent values, and to avoid double counting of the

Table 22.1 Summary of ecosystem services supported by leaf litter decomposition

Ecosystem services supported by decomposition[a]	Brief explanation of how decomposition supports service provision
Water quality regulation	Decomposition removes leaf detritus from aquatic ecosystems, reducing the particulate carbon load in the water and therefore regulating its quality by improving clarity and reducing some costs of removal
Aesthetic and recreational value of waterways	Leaf litter decomposition removes leaves from aquatic ecosystems giving them the clear appearance that is often valued by recreational users
Provisioning of food	Through the consumption of carbon and conversion of leaf litter, decomposition contributes to the production of tissue which in turn becomes available as food for higher trophic levels, including humans
Carbon sequestration	Decomposition degrades detritus and stores it as biomass which can sequester carbon over a range of time scales

[a]This list is an example of ecosystem services supported by decomposition and is not necessarily exhaustive

different outcomes (Fu et al., 2011). In the next section we discuss how decomposition contributes to the provision of each of these ecosystem services in turn.

Ecosystem functions contribute to the provision of services where humans access the benefits derived from function or gain value from it. With decomposition, these benefits can be transported from where they occur to hydrologically connected areas where people may also benefit from them. For example, stream decomposition can result in downstream improvements in water quality (Emelko et al., 2011), and the contribution of decomposition to secondary production may benefit fish in different spatial patches relative to where the initial process takes place (Losey & Vaughan, 2006; Wipfli & Gregovich, 2002). The spatial disconnect between the locations where ecosystem services supported by decomposition are produced and the locations where they produce tangible benefits to humans emphasizes the importance of considering hydrological connectivity when working to manage hydrological ecosystem services, such as those supported by decomposition (Brauman et al., 2007; Sutherland et al., 2018).

22.2 Evidence for the Relative Importance of Decomposition for Ecosystem Service Provision: Mineralisation, Production, and Storage

1. Water quality regulation

Leaf detritus is decomposed and some of that stored energy is respired as CO_2 (mineralised), or CH_4 under anaerobic conditions, removing it from the system and reducing the particulate carbon load that might be carried downstream. This process removes leaf litter from the water and reduces water processing for other uses downstream, therefore contributing to the regulation of water quality. Decomposition in tributary streams can reduce the biological oxygen demand in reservoirs if there were large inputs of leaf litter, which further compromises water quality. In particular, aquatic insects provide a large role in decomposition of leaf litter in freshwaters, at least in temperate zones (Macadam & Stockan, 2015). Perhaps 50% or more of decomposition is due to invertebrate feeding (Cuffney et al., 1990; Hieber & Gessner, 2002). However, microbes (bacteria and fungi) are considered to be more important to these processes in tropical regions (Boyero et al., 2011). Detritivorous invertebrates may consume a large portion of leaf litter annually. One estimate based on litterbag studies suggests 17–45% of leaf litter is consumed by detritivores (Eggert & Wallace, 2003). Wallace et al. (1995) estimate that of leaf litter entering small streams, only about 2–5% is transported downstream as coarse particles, and the rest is biologically processed within the local stream system. Of course, some fraction of leaf litter is converted to fine particles (e.g., faeces), which form important food resources for other organisms (microbes and invertebrates) and are sequestered in consumer biomass (see points 3 and 4 below) or flushed downstream where they might need to be filtered from drinking water supplies.

2. Aesthetic and recreational value of waterways

Clean water and clear, cobble-bottom stream beds have an aesthetic value (cultural ecosystem service) that is supported by decomposition. This value can in turn influence the provision of other recreational services—perceptions of collections of organic materials along stream and lake shorelines might affect how people use waterways. Indeed, research shows that in the United States, people are willing to travel almost an hour further and spend more money to access clearer waterways to recreate (Keeler et al., 2015). In Finland, people prefer to swim and fish in clearer waterways (Vesterinen et al., 2010), and in New Zealand water clarity affects the overall suitability people assign to waterways for their use (Smith et al., 1995). Although no studies to date have assessed the specific monetary value of decomposition as a contributor to the aesthetic and recreational value of waterway shorelines, it is clear that decomposition is essential for transforming the leaves and organic matter found in waterways. Through this process, leaf litter decomposition helps secure the clear appearance of water that is valued by numerous people. Because waterways are hydrologically connected, this process not only contributes to the aesthetic and recreational value of waterways in the regions it takes place in, but also in connected downstream regions.

3. Provisioning of food

Carbon consumed contributes to production of new tissue, which is available as food for higher trophic levels, including humans. One attempt to quantify this as a provisioning service considered aquatic insects as food for recreationally and commercially valuable fish (Losey & Vaughn, 2006). The proportion of this flow of detritivores to fish production around the world has not been evaluated, although detritivores of leaf litter, or fine particles produced from leaf litter, can make up a substantial portion of the biomass in freshwaters. The adult stages of aquatic shredding insects also contribute to production in the terrestrial environment, but whether these contribute to service provision depends on the terrestrial consumers, and whether they contribute to food chains that humans are a part of.

The process of decomposition by detritivores may also convert litter into other kinds of particles, including dissolved organic matter (DOM) consumed by microbes. These small particles can contribute to productivity downstream, and there is evidence that production of many animals depends on finer particles produced by the actions of larger invertebrate species, particularly by consumers of leaf litter (Heard & Richardson, 1995). This flow of energy to organisms feeding on finer particles (FPOM—fine particulate organic matter) may be a greater contributor to the production of fish and other animals (e.g., waterfowl) than the direct contribution of leaf litter detritivores.

4. Carbon sequestration

Detritus that is processed and stored as biomass contributes to retention and storage of carbon in riverine ecosystems (Sutfin et al., 2016; Wohl et al., 2012). This latter function is also related to production (above). It has been estimated that in excess of 75% of total production in small streams is based on decomposition of terrestrial leaf litter and incorporation into stream food webs (e.g., Fisher & Likens, 1973). Removal of litter inputs reduced overall stream production to <20% of when leaves contribute to support of the food web (Wallace et al., 1999). Wallace et al.'s (1995) removal of detritivores resulted in a ~25% decrease in supply of FPOM to downstream when insects were killed with an insecticide.

22.3 Variation in the Provisioning of Services Across Geographies and Seasons

Rates (and values) of the decomposition function differ with ecosystem type and size, climate, landscape type, etc. A large fraction of decomposition of leaf litter occurs in small streams, especially those with complex channels, but the landscape-level proportion of this has not been calculated. Decomposition rates are also highly dependent on temperatures (season), life cycle timing of consumers, and input timing (e.g., Tank et al., 2010).

Organic matter inputs may be exported from source streams, although having been transformed into prey organisms that can be consumed by predators downstream (Richardson, 2020). Export of invertebrates will have less impact on water quality downstream. For instance, Wipfli and Gregovich (2002) demonstrated that export of invertebrates from small streams, and based on consumption of leaf litter, could support half the production of fishes in receiving reaches. The rate of decomposition depends on types of consumers available, and there is evidence that the composition of the consumer assemblage varies with latitude and hence the service can vary latitudinally (e.g., Boyero et al., 2011).

Hill et al. (2014) estimated the value of some ecosystem services in headwater streams, primarily nutrient retention and mineralization. They found enormous variation in the values of N sequestration and N mineralisation in headwaters across the USA, from 22.2 to 64.9 kg/ha/y and 8.08 to 31.8 kg/ha/y, respectively. These differences across landscapes were related to land cover and land use, as well as climatic variation. Whether there is such variation in provision of services from decomposition across landscapes is currently unexplored. Moreover, even if a function contributes to a change, it may not support an ecosystem service unless there are humans to benefit from it, so in many regions there would be no service generated from decomposition, although there may be capacity to provide services (Sutherland et al., 2018).

There are few, if any, studies of the importance of decomposition as a contributor to ecosystem service provision in lakes. Several studies mention carbon sequestration in lakes and wetlands as supporting an ecosystem service (e.g., Villa & Bernal, 2018), but there are no studies of decomposition as an ecosystem function itself. Villa & Bernal (2018) estimate that freshwater wetlands may contribute to storage of ~700 Mg C/ha, and low decomposition rates attributed to recalcitrance of organic matter, limiting nutrient supplies, and anaerobic conditions in wetlands. In this case, the apparently minor role of decomposition may contribute to an alternative ecosystem function, i.e., storage, in these freshwaters. However, these accumulations can also reduce the value of the service it supports by causing anoxia and potentially enhancing methanogenesis, with methane being a more potent greenhouse gas than CO_2.

22.4 Influence of Protection and Land Use on the Contribution of Leaf Litter Decomposition to Provisioning Rates

The protection status, land use, and cover of the landscape surrounding streams affects the ecosystem services and biodiversity supported by streams (Hanna et al., 2020). Stream decomposition rates and the services they support are also likely to be affected by land use as described below and elsewhere in this book. Often land use that removes canopy from stream riparian zones has negative impacts by reducing habitat quality and altering rates of several ecosystem functions, including decomposition. Alterations of temperature regimes (often warmer in summer), changes to bank stability and wood inputs, different hydrological patterns and reduction in litter inputs are all consequences of streamside modification, which have known effects on decomposition rates. Hydrological alterations to streams from land use can exacerbate peak flows, and result in greater export rates of undecomposed litter to downstream where it could impair water quality (e.g., Hoover et al., 2006; Wallace et al., 1995). Similarly, reductions of flow also affects leaf litter decomposition rates (Huang et al., 2018). Changes in the rates of these processes above can in turn feedback to influence the provision of related services.

Landscape-scale land-use intensity can also influence decomposition and related services: in a study in Germany, land-use intensity across the landscape (agriculture and viticulture) reduced microbial rates of decomposition of leaf litter by about 30%, but there was no significant effect when macroinvertebrates were allowed access to leaves by using coarse-mesh nets (Voss et al., 2015). Other studies have demonstrated effects of watershed-scale canopy cover on macroinvertebrate communities—a study in Canada found that streams that benefit from riparian zone protection but have varying levels of development (i.e. canopy cover) in their watersheds support different communities of macroinvertebrates (Hanna et al., 2020). As further discussed below, shifts in macroinvertebrate communities can affect decomposition rates (Srivastava et al., 2009).

Stream-channel complexity or heterogeneity is important to providing diverse habitats, and is one objective of stream restoration activities. Adding to the physical structure of degraded streams has been shown to increase decomposition rates and storage of fine particulate materials in a study in northern Sweden (Frainer et al., 2018). They found decomposition rates of leaf litter were ~60% higher in restored and reference streams compared to channelized streams, mostly due to higher abundance and diversity of macroinvertebrates (Frainer et al., 2018). They also found that retention of FPOM nearly doubled across their restored treatment streams.

Loss of species diversity of detritus consumers can lower decomposition rates. Srivastava et al. (2009) found using meta-analysis that higher taxonomic diversity of consumers resulted in an average 57% higher decomposition rate of leaf litter. Hence, loss of biodiversity may reduce rates of decomposition in freshwaters, thereby affecting the services this process supports.

Changes in diversity of detritus consumers can result from a variety of factors, ranging from land-use change to the establishment of invasive species. Indeed, research shows that stream watershed land-use change also results in macroinvertebrate species assemblages shifts (Hanna et al., 2020). Research also points to examples of invasive species establishment decreasing the provision of ecosystem services (Walsh et al., 2016). However, invasion may also result in an increase in service provision. In Europe, an invasive crayfish and crab were able to increase rates of decomposition (by ~75%), and increased production of FPOM relative to the native crayfish (Doherty-Bone et al., 2018).

Streamside (or lakes or wetlands) clearing may increase temperatures by reducing canopy cover, increasing rates of decomposition. At the same time, decreasing canopy cover reduces inputs, which are essential for sustaining decomposer communities (e.g., Richardson, 1991). Vegetation change through management or succession through natural disturbances alters litter quality, and the rates of decomposition (Bastias et al., 2018; Kominoski et al., 2011), altering potential for service provision. The leaf litter of many plantation tree species tends to be of low quality and decomposes very slowly, such as eucalyptus, pine, Douglas-fir and others (e.g., García et al., 2012; Seena et al., 2017), reducing the potential of decomposition prior to arriving at water intake facilities. Mild nutrient enrichment can accelerate decomposition (e.g., Rosemond et al., 2015). However, pollution in many forms reduces decomposition rates and can alter the capacity of the process to contribute to service provision. Experimental additions to mesocosms of chlorothalonil, the most commonly applied fungicide used in agriculture in the USA, resulted in ~28% decrease in decomposition rates (McMahon et al., 2012). A common insecticide, chlorpyrifos, reduced decomposition rates by ~21% in a mesocosm experiment (Chará-Serna & Richardson, 2018). Water abstraction can also result in lower decomposition rates, as rates were 36–60% lower in streams with water withdrawals, although this was only detected in winter and not spring (Arroita et al., 2015). Together, these examples demonstrate that land-use impacts can dramatically alter the rates of this ecosystem function, and the potential for the provision of related ecosystem services.

22.5 Trade-off and Synergies of Ecosystem Service Provision Associated to Decomposition

The ecosystem services supported by decomposition can interact with each other, resulting in trade-offs or synergies (Rodríguez et al., 2006). For example, an increase in water quality regulation via the process of leaf litter decomposition is likely to result in positive outcomes for perceived aesthetic value, which can in turn increase recreational benefits people derive from a waterway. The relationship between water quality regulation and the aesthetic and recreational value of waterways is an example of a synergy between ecosystem services, whereby the increased provision of one service enables the increased provision of other services. High decomposition may also lead to lower rates of other processes that support the provision of ecosystem services. For instance, reconstructed wetlands in Illinois showed an inverse relationship between the magnitude of carbon storage as leaf litter and decomposition rates. Using a cotton strip assay, this study demonstrates that bigger wetlands with more organic matter lead to more anoxic conditions and lower decomposition rates (Jessop et al., 2015), as anaerobic decomposition is slower than aerobic processing. This is an example of how the provision of a service like carbon storage may trade-off with the provision of others supported by decomposition, such as aesthetics. Considering how different ecosystem services interact is important when making management decisions, as efforts to increase the provision of single services can lead to unexpected outcomes for the provision of others (Bennett et al., 2009). In the context of leaf litter decomposition, considering how waterway management or restoration strategies will affect decomposition and related services helps understand how actions are likely to affect ecosystem service provision.

Increasing degrees of stream intermittence by water use and diversion, and by climate change, may reduce the capacity of stream decomposers to contribute to ecosystem service provision, and in the end transfer more of the organic material downstream where it may perturb water quality (e.g., Datry et al., 2018). Service value will depend on how intermittency progresses, whether in the headwaters or truncated where larger streams lose surface flow (interrupted).

22.6 Conclusion

"In particular, the decomposers as a group must be food-limited, since by definition they comprise the trophic level which degrades organic debris" (Hairston et al., 1960). Hairston et al. (1960) go on to note that if not for decomposers of organic matter, detrital material "…would accumulate rapidly". It has been shown in several ways that decomposers are in fact limited by their organic matter resources and generally increase (Richardson, 1991) or decrease (Rowe & Richardson, 2001; Wallace et al., 1999) proportionally to the abundance of their resources. Fortunately, decomposers provide a function that contributes to human well-being.

Decomposition of organic matter in freshwater supports the provision of important ecosystem services. Provision of cleaner water, the aesthetics of cleaner shorelines, supporting secondary production, and increased carbon storage are all outcomes of this ecosystem function that are valued by humans.

References

Arroita, M., Aristi, I., Diez, J., Martinez, M., Oyarzun, G., & Elosegi, A. (2015). Impact of water abstraction on storage and breakdown of coarse organic matter in mountain streams. *Science of the Total Environment, 503,* 233–240.

Bastias, E., Ribot, M., Romaní, A. M., Sabater, F., López, P., & Martí, E. (2018). Responses of microbially driven leaf litter decomposition to stream nutrients depend on litter quality. *Hydrobiologia, 806,* 333–346.

Bennett, E. M., Peterson, G. D., & Gordon, L. J. (2009). Understanding relationships among multiple ecosystem services. *Ecology Letters, 12,* 1394–1404.

Boyero, L., Pearson, R. G., Gessner, M. O., Barmuta, L. A., Ferreira, V., Graca, M. A. S., Dudgeon, D., Boulton, A. J., Callisto, M., Chauvet, E., Helson, J. E., Bruder, A., Albariño, R. J., Yule, C. M., Arunachalam, M., Davies, J. N., Figueroa, R., Flecker, A. S., Rarnirez, A., … West, D. C. (2011). A global experiment suggests climate warming will not accelerate litter decomposition in streams but might reduce carbon sequestration. *Ecology Letters, 14,* 289–294.

Brauman, K. A., Daily, G. C., Duarte, T. K., & Mooney, H. A. (2007). The nature and value of ecosystem services: An overview highlighting hydrologic services. *Annual Review of Environment and Resources, 32,* 67–98.

Carpenter, S. R., Mooney, H. A., Agard, J., Capistrano, D., DeFries, R. S., Díaz, S., Dietz, T., Duraiappah, A. K., Oteng-Yeboah, A., Pereira, H. M., Perrings, C., Reid, W. V., Sarukhan, J., Scholes, R. J., & Whyte, A. (2009). Science for managing ecosystem services: Beyond the millennium ecosystem assessment. *Proceedings of the National Academy of Sciences of the United States of America, 106,* 1305–1312.

Chará-Serna, A. M., & Richardson, J. S. (2018). Chlorpyrifos interacts with other agricultural stressors to alter stream invertebrate community in laboratory microcosms. *Ecological Applications, 28,* 162–176.

Costanza, R., d'Arge, R., de Groot, R., Farberk, S., Grasso, M., Hannon, B., Limburg, K., Naeem, S., O'Neill, R. V., Paruelo, J., Raskin, R. G., Sutton, P., & van den Belt, M. (1997). The value of the world's ecosystem services and natural capital. *Nature, 387,* 253–260.

Cuffney, T. F., Wallace, J. B., & Lugthart, G. H. (1990). Experimental evidence quantifying the role of benthic invertebrates in organic matter dynamics of headwater streams. *Freshwater Biology, 23,* 281–299.

Datry, T., Boulton, A. J., Bonada, N., Fritz, K., Leigh, C., Sauquet, E., Tockner, K., Hugueny, B., & Dahm, C. N. (2018). Flow intermittence and ecosystem services in rivers of the Anthropocene. *Journal of Applied Ecology, 55,* 353–364.

Díaz, S., Pascual, U., Stenseke, M., Martin-Lopez, B., Watson, R. T., Molnar, Z., Hill, R., Chan, K. M. A., Baste, I. A., Brauman, K. A., Polasky, S., Church, A., Lonsdale, M., Larigauderie, A., Leadley, P. W., van Oudenhoven, A. P. E., van der Plaat, F., Schroter, M., Lavorel, S., … Shirayama, Y. (2018). Assessing nature's contributions to people. *Science, 359,* 270–272.

Doherty-Bone, T. M., Dunn, A. M., Liddell, C., & Brown, L. E. (2018). Transformation of detritus by a European native and two invasive alien freshwater decapods. *Biological Invasions, 20,* 1799–1808.

Eggert, S. L., & Wallace, J. B. (2003). Litter breakdown and invertebrate detritivores in a resource-depleted Appalachian stream. *Archiv für Hydrobiologie, 156,* 315–338.

Emelko, M. B., Silins, U., Bladon, K. D., & Stone, M. (2011). Implications of land disturbance on drinking water treatability in a changing climate: Demonstrating the need for "source water supply and protection" strategies. *Water Research, 45,* 461–472. https://doi.org/10.1016/j.watres.2010.08.051.

Felipe-Lucia, M. R., Martín-López, B., Lavorel, S., Berraquero-Díaz, L., Escalera-Reyes, J., & Comín, F. A. (2015). Ecosystem services flows: Why stakeholders' power relationships matter. *PloS ONE, 10*(7), e0132232.

Fisher, B., Turner, R. K., & Morling, P. (2009). Defining and classifying ecosystem services for decision making. *Ecological Economics, 68,* 643–653.

Fisher, S. G., & Likens, G. E. (1973). Energy flow in bear brook, New Hampshire—integrative approach to stream ecosystem metabolism. *Ecological Monographs, 43,* 421–439.

Frainer, A., Polvi, L. E., Jansson, R., & McKie, B. G. (2018). Enhanced ecosystem functioning following stream restoration: The roles of habitat heterogeneity and invertebrate species traits. *Journal of Applied Ecology, 55,* 377–385.

Fu, B. J., Su, C. H., Wei, Y. P., Willett, I. R., Lü, Y. H., & Liu, G. H. (2011). Double counting in ecosystem services valuation: Causes and countermeasures. *Ecological Research, 26,* 1–14.

García, L., Richardson, J. S., & Pardo, I. (2012). Leaf quality influences invertebrate colonization and drift in a temperate rainforest stream. *Canadian Journal of Fisheries and Aquatic Sciences, 69,* 1663–1673.

Hairston, N. G., Smith, F. E., & Slobodkin, L. B. (1960). Community structure, population control, and competition. *The American Naturalist, 94,* 421–425.

Hanna, D. E. L., Raudsepp-Hearne, C., & Bennett, E. M. (2020). Effects of land use, cover, and protection on stream and riparian ecosystem services and biodiversity. *Conservation Biology, 34,* 244–255. https://doi.org/10.1111/cobi.13348.

Hanna, D. E. L., Tomscha, S. A., Dallaire C. O., & Bennett, E. M. (2018). A review of riverine ecosystem service quantification: Research gaps and recommendations. *Journal of Applied Ecology, 55,* 1299–1311.

Heard, S. B., & Richardson, J. S. (1995). Shredder-collector facilitation in stream detrital food webs: Is there enough evidence? *Oikos, 72,* 359–366.

Hein, L., Koppen, K. V., DeGroot, R. S., & Van Ierland, E. C. (2006). Spatial scales, stakeholders and the valuation of ecosystem services. *Ecological Economics, 57,* 209–228.

Hieber, M., & Gessner, M. O. (2002). Contribution of stream detritivores, fungi, and bacteria to leaf breakdown based on biomass estimates. *Ecology, 83,* 1026–1038.

Hill, B. H., Kolka, R. K., McCormick, F. H., & Starry, M. A. (2014). A synoptic survey of ecosystem services from headwater catchments in the United States. *Ecosystem Services, 7,* 106–115.

Hoover, T., Richardson, J. S., & Yonemitsu, N. (2006). Flow-substrate interactions create and mediate leaf litter resource patches in streams. *Freshwater Biology, 51,* 435–447.

Huang, W., Liu, X., Peng, W., Ma, W., & Zhang, J. (2018). Quantitative response of leaf-litter decomposition rate to water abstraction in a gradient: Implications for environmental flow management. *Ecohydrology, 11,* article UNSP e1919.

Jessop, J., Spyreas, G., Pociask, G. E., Benson, T. J., Ward, M. P., Kent, A. D., & Matthews, J. W. (2015). Tradeoffs among ecosystem services in restored wetlands. *Biological Conservation, 191,* 341–348.

Keeler, B. L., Wood, S. A., Polasky, S., Kling, C., Filstrup, C. T., & Downing, J. A. (2015). Recreational demand for clean water: Evidence from geotagged photographs by visitors to lakes. *Frontiers in Ecology and the Environment, 13,* 76–81.

Kominoski, J. S., Marczak, L. B., & Richardson, J. S. (2011). Riparian forest composition affects stream litter decomposition despite similar microbial and invertebrate communities. *Ecology, 92,* 151–159.

Losey, J. E., & Vaughan, M. (2006). The economic value of ecological services provided by insects. *BioScience, 56,* 311–323.

Macadam, C. R., & Stockan, J. A. (2015). More than just fish food: Ecosystem services provided by freshwater insects. *Ecological Entomology, 40,* 113–123.

Mancinelli, G., & Mulder, C. (2015). Detrital dynamics and cascading effects on supporting ecosystem services. *Advances in Ecological Research, 53,* 97–160.

McMahon, T. A., Halstead, N. T., Johnson, S., Raffel, T. R., Romansic, J. M., Crumrine, P. W., & Rohr, J. R. (2012). Fungicide-induced declines of freshwater biodiversity modify ecosystem functions and services. *Ecology Letters, 15,* 714–722.

Millennium Ecosystem Assessment. (2005). *Ecosystems and human well-being: Synthesis.* Washington, DC: Island Press.

Natural Capital Coalition. (2019, October 28). *What is natural capital?* Available via: https://naturalcapitalcoalition.org/natural-capital-2/.

Raudsepp-Hearne, C., Peterson, G. D., Tengö, M., Bennett, E. M., Holland, T., Benessaiah, K., Donal, G. K. M., & Pfeifer, L. (2010). Untangling the environmentalist's paradox: Why is human well-being increasing as ecosystem services degrade? *BioScience, 60,* 576–589.

Richardson, J. S. (2020). Headwater streams. In M. Goldstein & D. DellaSala (Eds.), *Encyclopedia of the world's biomes.* Reference module in earth systems and environmental sciences. Elsevier. https://doi.org/10.1016/B978-0-12-409548-9.11957-8.

Richardson, J. S. (1991). Seasonal food limitation of detritivores in a montane stream: An experimental test. *Ecology, 72,* 873–887.

Rodríguez, J. P., Beard Jr., T. D., Bennett, E. M., Cumming, G. S., Cork, S. J., Agard, J., Dobson, A. P., & Peterson, G. D. (2006). Trade-offs across space, time, and ecosystem services. *Ecology and Society, 11,* 28. http://www.ecologyandsociety.org/vol11/iss1/art28/.

Rosemond, A. D., Benstead, J. P., Bumpers, P. M., Gulis, V., Kominoski, J. S., Manning, D. W. P., Suberkropp, K., & Wallace, J. B. (2015). Experimental nutrient additions accelerate terrestrial carbon loss from stream ecosystems. *Science, 347,* 1142–1145.

Rowe, L., & Richardson, J. S. (2001). Community responses to experimental food depletion: Resource tracking by stream invertebrates. *Oecologia, 129,* 473–480.

Seena, S., Carvalho, F., Cássio, F., & Pascoal, C. (2017). Does the developmental stage and composition of riparian forest stand affect ecosystem functioning in streams? *Science of the Total Environment, 609,* 1500–1511.

Seppelt, R., Dormann, C. F., Eppink, F. V., Lautenbach, S., & Schmidt, S. (2011). A quantitative review of ecosystem service studies: Approaches, shortcomings and the road ahead. *Journal of Applied Ecology, 48,* 630–636.

Smith, D. G., Croker, G. F., & Mcfarlane, K. (1995). Human perception of water appearance. *New Zealand Journal of Marine and Freshwater Research, 29,* 29–43.

Srivastava, D. S., Cardinale, B. J., Downing, A. L., Duffy, J. E., Jouseau, C., Sankaran, M., & Wright, J. P. (2009). Diversity has stronger top-down than bottom-up effects on decomposition. *Ecology, 90,* 1073–1083.

Sutfin, N. A., Wohl, E. E., & Dwire, K. A. (2016). Banking carbon: A review of organic carbon storage and physical factors influencing retention in floodplains and riparian ecosystems. *Earth Surface Processes and Landforms, 41,* 38–60.

Sutherland, I. J., Villamagna, A. M., Dallaire, C. O., Bennett, E. M., Chin, A. T. M., Yeung, A. C. Y., Lamothe, K. A., Tomscha, S. A., & Cormier, R. (2018). Undervalued and under pressure: A plea for greater attention toward regulating ecosystem services. *Ecological Indicators, 94,* 23–32.

Tank, J. L., Rosi-Marshall, E. J., Griffiths, N. A., Entrekin, S. A., & Stephen, M. L. (2010). A review of allochthonous organic matter dynamics and metabolism in streams. *Freshwater Science, 29,* 118–146.

Vesterinen, J., Pouta, E., Huhtala, A., & Neuvonen, M. (2010). Impacts of changes in water quality on recreation behavior and benefits in Finland. *Journal of Environmental Management, 91,* 984–994.

Villa, J. A., & Bernal, B. (2018). Carbon sequestration in wetlands, from science to practice: An overview of the biogeochemical process, measurement methods, and policy framework. *Ecological Engineering, 114,* 115–128.

Villamagna, A. M., Angermeier, P. L., & Bennett, E. M. (2013). Capacity, pressure, demand, and flow: A conceptual framework for analyzing ecosystem service provision and delivery. *Ecological Complexity, 15,* 114–121.

Voss, K., Fernandez, D., & Schäfer, B. (2015). Organic matter breakdown in streams in a region of contrasting anthropogenic land use. *Science of the Total Environment, 527,* 179–184.

Wallace, J. B., Eggert, S. L., Meyer, J. L., & Webster, J. R. (1999). Effects of resource limitation on a detrital-based ecosystem. *Ecological Monographs, 69,* 409–442.

Wallace, J. B., Whiles, M. R., Eggert, S., Cuffney, T. F., Lugthart, G. H., & Chung, K. (1995). Long-term dynamics of coarse particulate organic matter in 3 Appalachian Mountain streams. *Journal of the North American Benthological Society, 14,* 217–232.

Walsh, J. R., Carpenter, S. R., & Vander, M. J. (2016). Invasive species triggers a massive loss of ecosystem services through a trophic cascade. *Proceedings of the National Academy of Sciences of the United States, 113,* 4081–4085.

Wipfli, M. S., & Gregovich, D. P. (2002). Export of invertebrates and detritus from fishless headwater streams in southeastern Alaska: Implications for downstream salmonid production. *Freshwater Biology, 47,* 957–969.

Wohl, E., Dwire, K., Sutfin, N., Polvi, L., & Bazan, R. (2012). Mechanisms of carbon storage in mountainous headwater rivers. *Nature Communications, 3,* 126.

Correction to: Salt Modulates Plant Litter Decomposition in Stream Ecosystems

Cristina Canhoto, Felix Bärlocher, Miguel Cañedo-Argüelles, Rosa Gómez, and Ana Lúcia Gonçalves

Correction to:
Chapter 15 in: C. M. Swan et al. (eds.),
***The Ecology of Plant Litter Decomposition in Stream Ecosystems*, https://doi.org/10.1007/978-3-030-72854-0_15**

The original version of this chapter was inadvertently published with incorrect author name Ros Gómez in Chapter 15, which has now been corrected to "Rosa Gómez". The erratum chapter and the book have been updated with the changes.

The updated version of this chapter can be found at
https://doi.org/10.1007/978-3-030-72854-0_15

C. M. Swan et al. (eds.), *The Ecology of Plant Litter Decomposition in Stream Ecosystems*, https://doi.org/10.1007/978-3-030-72854-0_23

The manufacturer's authorised representative in the EU is Springer Nature Customer Service Centre GmbH, Europaplatz 3, 69115 Heidelberg, Germany. If you have any concerns regarding our products, please contact ProductSafety@springernature.com

Printed and bound by CPI Group (UK) Ltd, Croydon, CR0 4YY
30/09/2026
02225491-0018